Shock Formation in Small-Data Solutions to 3D Quasilinear Wave Equations

Mathematical
Surveys
and
Monographs

Volume 214

Shock Formation in Small-Data Solutions to 3D Quasilinear Wave Equations

Jared Speck

American Mathematical Society
Providence, Rhode Island

2010 *Mathematics Subject Classification*. Primary 35L67;
Secondary 35L05, 35L10, 35L72, 35Q31, 35L15.

For additional information and updates on this book, visit
www.ams.org/bookpages/surv-214

Library of Congress Cataloging-in-Publication Data

Names: Speck, Jared, 1980–
Title: Shock formation in small-data solutions to 3D quasilinear wave equations / Jared Speck.
Description: Providence, Rhode Island : American Mathematical Society, [2016] | Series: Mathematical surveys and monographs ; volume 214 | Includes bibliographical references and index.
Identifiers: LCCN 2016022109 | ISBN 9781470428570 (alk. paper)
Subjects: LCSH: Wave equation–Numerical solutions. | Shock waves–Mathematics. | Differential equations, Nonlinear–Numerical solutions. | Quasilinearization. | AMS: Partial differential equations – Hyperbolic equations and systems – Shocks and singularities. msc | Partial differential equations – Hyperbolic equations and systems – Wave equation. msc | Partial differential equations – Hyperbolic equations and systems – Second-order hyperbolic equations. msc | Partial differential equations – Hyperbolic equations and systems – Quasilinear second-order hyperbolic equations. msc | Partial differential equations – Equations of mathematical physics and other areas of application – Euler equations. msc | Partial differential equations – Hyperbolic equations and systems – Initial value problems for second-order hyperbolic equations. msc
Classification: LCC QC174.26.W28 S72 2016 | DDC 515/.3535—dc23
LC record available at https://lccn.loc.gov/2016022109

∞ The paper used in this book is acid-free and falls within the guidelines
established to ensure permanence and durability.
Visit the AMS home page at `http://www.ams.org/`

10 9 8 7 6 5 4 3 2 1   21 20 19 18 17 16

To Genevieve, for sleeping (just) long enough
to allow me to complete this project,
and to Teddy, the comeback kid with a heart of steel.

Contents

Preface

A central issue surrounding the study of quasilinear hyperbolic PDEs is that classical solutions, generated by smooth initial conditions, can develop singularities in finite time. In principle, many different kind of singularities are possible. The subject of this monograph is perhaps the most well-known type: shocks. Our work here primarily concerns scalar quasilinear wave equations, the main reasons being **1)** they arise in many important mathematical, physical, and geometric contexts; **2)** the last few decades have led to the development of advanced machinery tailored to such equations; and **3)** their characteristic hypersurfaces[1] are relatively simple[2] and can be analyzed using tools and concepts from the well-developed theory of Lorentzian geometry.

Before the groundbreaking work of S. Alinhac [**2**, **4**–**6**], very little was known about shock formation except in problems that are effectively one (spatial) dimensional. One-dimensional shock formation results are of course classical. We describe some of the most important such results in Chapter 1. As we describe in detail in Chapter 2, Alinhac proved finite-time shock formation for a class of wave equations in two and three spatial dimensions. Roughly, his proof applied to wave equations of the form[3]

$$(h^{-1})^{\alpha\beta}(\partial\Phi)\partial_\alpha\partial_\beta\Phi = 0$$

whenever the nonlinear terms *fail* to satisfy the null condition, which was first formulated by S. Klainerman [**45**] in three spatial dimensions and later in two spatial dimensions by Alinhac [**1**, **2**]. In the above wave equation, $h = h(\partial\Phi)$ is a Lorentzian metric such that $h(\partial\Phi = 0) = m$, where m is the standard Minkowski metric on $\mathbb{R}^{1+n}$, $n \in \{2, 3\}$. Alinhac's main results concern solutions generated by small initial data that belong to a suitable Sobolev space and that verify a non-degeneracy condition. The foundation of his proof was a new system of *geometric coordinates* tied to an *eikonal function* u, which by definition is a solution to the *eikonal equation*, that is, the hyperbolic PDE

$$(h^{-1})^{\alpha\beta}(\partial\Phi)\partial_\alpha u\partial_\beta u = 0,$$

supplemented by appropriate initial conditions. The level sets of u are true characteristics (as opposed to approximate ones) corresponding to the nonlinear wave equation.

[1]In the study of wave equations, characteristic hypersurfaces are often referred to as "null hypersurfaces" in view of their connection to the Lorentzian notion of a null vectorfield. More generally, they are often referred to as simply "the characteristics."

[2]Roughly, in our study of wave equations, the characteristics are a family of "true curved cones" corresponding to the dynamic Lorentzian metric of the wave equation.

[3]The equation is to be interpreted as an equation given relative to standard rectangular coordinates, which we describe at the beginning of Chapter 2. Moreover, here and throughout, we use Einstein's summation convention.

Eikonal functions are perhaps best known for the central role they played in Christodoulou-Klainerman's celebrated proof [**19**] of the stability of Minkowski spacetime.[4] That work was the first instance in which eikonal functions were used to prove a global nonlinear result for a hyperbolic PDE. In the study of shock formation, the behavior of u, its properties, and their connection to the behavior of the solution variable[5] lie at the heart of the analysis. This was the case in Alinhac's work and in Christodoulou's work (described two paragraphs below), and it remains true in the present monograph as well. As we will see, in the problem of shock formation, the eikonal function plays an even more important role than it does in the proof of stability of Minkowski spacetime; there is an alternate proof [**59**] of the stability of Minkowski spacetime, due to Lindblad-Rodnianski, that relies on an approximate eikonal function corresponding to the Minkowski metric rather than a true eikonal function. This alternate approach leads to remarkable simplifications in the analysis because the characteristics associated to the Minkowski metric are much simpler.[6] In contrast, in the problem of shock formation, the formation of the shock and the corresponding blow-up of the solution are *exactly tied to the blow-up of the first rectangular coordinate partial derivatives of a true eikonal function.* Thus, there is little hope of finding an alternate proof that avoids the use of a true eikonal function (and, as we will see, the weighty baggage that accompanies it).

It is well-known that energy estimates are an unavoidable aspect of the study of quasilinear wave equations in more than one spatial dimension. To close the energy estimates in his proof of shock formation, Alinhac relied on a Nash-Moser iteration scheme featuring a free boundary. The presence of the free boundary is connected to the blow-up time of the iterates, which can vary, albeit slightly. A fundamental aspect of his proof, which is also present in Christodoulou's work and the present monograph, is that *the solution remains regular*[7] *relative to the geometric coordinates* mentioned two paragraphs above. In the equations studied by Alinhac, the singularity occurs in the second rectangular coordinate partial derivatives of Φ and is tied to the degeneration of the geometric coordinate system relative to the rectangular one. At the same time, the degeneracy is tied to the intersection of the characteristics. In discussing Alinhac's results and related ones, we often refer to the intersection of the characteristics as "the formation of a shock." Our intention in using this terminology is to highlight the following fundamentally important aspect of Alinhac's work: he proved that Φ and its first rectangular coordinate partial derivatives $\partial_\alpha \Phi$ remain bounded all the way up to the singularity. In particular, the singularity occurs at the level of the first rectangular coordinate partial derivatives of the Lorentzian metric components $h_{\alpha\beta}(\partial\Phi)$ and not in the $h_{\alpha\beta}(\partial\Phi)$ themselves; this feature is of fundamental importance for closing the proof.

Although Alinhac's approach is compellingly short, it has some limitations, which we describe in detail in Sect. 2.11.1. In particular, his framework allows one

---

[4]Roughly, [**19**] is a small-data global existence result for the Einstein-vacuum equations.

[5]In the present work, the solution variable is the solution to the wave equation.

[6]In this context, the characteristics associated to the Minkowski metric are the usual flat Minkowski light cones.

[7]Actually, the solution does not necessarily remain regular at the high derivative levels: a fundamental aspect of the proof is that the high-order energies are allowed to blow-up as the shock forms; see three paragraphs below.

to follow the solution only to the first singularity[8] *and not further*. In his 2007 monograph [**17**], D. Christodoulou proved, for a sub-class of Alinhac's wave equations,[9] a breakthrough result that significantly sharpened Alinhac's results and eliminated the drawbacks of his approach. Specifically, Christodoulou's work applies to the wave equations that arise in irrotational relativistic[10] (compressible) fluid mechanics in three spatial dimensions. In this context, the equations are known as the irrotational relativistic Euler equations. Christodoulou assumed that the data have small H^N norm, where N is a sufficiently large (nonexplicit) integer. To deduce the shock formation, he also assumed that the data verify a signed integral inequality, distinct from the nondegeneracy condition of Alinhac mentioned above. Christodoulou's framework allows one to do much more than follow the solution to the first singularity: it provides a complete picture of a portion of the maximal development[11] of the data including a description of the behavior of the solution along the boundary;[12] this is the main advantage of Christodoulou's framework. One of the key reasons that Christodoulou was able to sharpen Alinhac's results is that he was able to close the energy estimates without invoking a Nash-Moser iteration scheme. Moreover, his estimates do not involve a free boundary. Instead, Christodoulou developed a forwards approach relative to a set of geometric coordinates that, like Alinhac's, are tied to a true eikonal function. By forwards approach, we mean that Christodoulou derives traditional global-existence-type estimates for a Cauchy problem in the geometric coordinates, relative to which the solution remains rather smooth (see, however, footnote 7). As in Alinhac's results and those of the present monograph, the blow-up occurs in certain rectangular coordinate partial derivatives of the solution and is tied to the degeneracy of the change of variables map from geometric to rectangular coordinates.

In Christodoulou's framework, the degeneracy mentioned at the end of the previous paragraph and the corresponding blow-up of the solution are mediated by the vanishing of a quantity known as the *inverse foliation density*, which we denote by μ. Roughly, μ is the reciprocal of a derivative of the eikonal function u. Geometrically, $1/\mu$ is a measure of the density of the level sets of u. As we will see starting in Chapter 2, the vanishing of μ is equivalent to the intersection of the characteristics and the blow-up of the eikonal function's first rectangular coordinate partial derivatives, as we mentioned above. Moreover, *these degeneracies are exactly tied to the formation of a singularity in the solution to the wave equation*, much like in the classic example of Burgers' equation (see Sect. 1.1.2). The study of μ and the prospect of its vanishing are the main themes of [**17**] and the present work. For reasons that we describe two paragraphs below, the most compelling advantage of Christodoulou's framework is that it allows one to construct the portion of the set

[8]Roughly speaking, Alinhac's proof works when there is such a unique first singularity; his nondegeneracy conditions on the data ensure that this is the case.

[9]One has to take into account some simple differences in normalization, described in Sect. 2.11.2, in order to see that Christodoulou's equations fall under the scope of Alinhac's work.

[10]In [**21**], Christodoulou-Miao extended the result to the nonrelativistic case.

[11]Roughly speaking, the maximal development is the largest classical solution that is uniquely determined by the data.

[12]The boundary can be very complicated and, in particular, it is not contained in a constant-time hypersurface.

$\{\mu = 0\}$ that corresponds to the part of the boundary[13] along which the solution blows up.[14] Generally, the set $\{\mu = 0\}$ "evolves" into a spacetime region lying to the future of the constant-time hypersurface of first blow-up and thus it lies to the future of the region that Alinhac was able to probe.

It turns out that Christodoulou's sharp description is accompanied by severe technical difficulties: the high-order energies are allowed to blow-up as $\mu \to 0$. This difficulty is fundamentally tied to the regularity theory of the eikonal function. As was first shown in [**19**], to control the top derivatives of u, one must invoke a nontrivial procedure based on elliptic estimates and modified quantities, the latter being special combinations of terms that satisfy evolution equations with a good structure. In the problem of shock formation, this procedure introduces, at the top order, a difficult factor of $1/\mu$ into the energy identities; *this factor is the reason that the high-order energies are allowed to blow up.* A related but distinct difficulty is that *one needs to show that the low-order energies remain bounded all the way up to* $\{\mu = 0\}$. This latter step is essential for establishing, via Sobolev embedding, the basic uniform L^∞ estimates that allow one to control error terms and to treat the problem as a traditional one in which one derives global-existence-type estimates (relative to the geometric coordinates). These are the main technical difficulties that one encounters in the problem of shock formation à la Christodoulou and they are a primary reason that the work is technical and lengthy. In Chapter 2, we provide an extended overview of these issues, especially in view of the fact that they do not arise in any other context in the study of nonlinear wave equations.

Christodoulou's framework is compelling for the geometric insight it provides into the formation of shocks and for the sharp description that it yields. In addition, his approach is fundamentally important for a related problem: it turns out that his sharp description of the solution near the boundary of the maximal development is an essential ingredient in setting up the *shock development problem*, which is the problem of **1)** continuing the solution to Euler's equations (in a weak sense, subject to appropriate jump and entropy conditions) beyond the first singularity and **2)** at the same time, constructing the shock hypersurface[15] across which the solution is discontinuous.[16] The shock development problem in relativistic fluid mechanics was recently solved in spherical symmetry [**20**]. Away from symmetry, the problem remains open and is expected to be exceptionally difficult.

We mention here an important problem related to the shock development problem, which A. Majda solved [**62–64**] in the early 1980s. Specifically, for hyperbolic systems of conservation laws with suitable structure in more than one spatial dimension, Majda solved the *shock front problem.* That is, in a suitable Sobolev

[13]Roughly, a subset of $\{\mu = 0\}$ corresponds to the singular portion of the maximal development of the data.

[14]The boundary of the maximal development also contains another portion, along which the solution does not blow up; see Sect. 2.11.2 and Theorem 2.19 in particular.

[15]Part of the set $\{\mu = 0\}$ turns out to be a hypersurface portion along which, in the classical formulation of the irrotational Euler equations, certain solution derivatives blow up. However, $\{\mu = 0\}$ does not correspond to the physically correct hypersurface of discontinuity. The physically correct hypersurface of discontinuity propagates at supersonic speed starting from the first spacetime point where μ vanishes and develops "before" the set $\{\mu = 0\}$ has a chance to form. The physically correct hypersurface can be derived only by imposing the weak formulation of the full compressible Euler equations (without the assumption of irrotationality) starting from the time of first blow-up and by assuming suitable jump and entropy conditions.

[16]One expects the solution to be smooth on either side of the shock hypersurface.

framework, he proved a local existence result starting from an initial discontinuity given across a smooth hypersurface[17] subset of the Cauchy hypersurface. We stress that the initial hypersurface of discontinuity is prescribed. In contrast, in the shock development problem mentioned in the previous paragraph, the spacetime hypersurface of discontinuity is fully dynamic, emerging from singularity-free initial data. In the shock front problem, the data must verify suitable jump conditions, entropy conditions, and higher-order compatibility conditions. As in the shock development problem, the shock front problem features a free boundary: the shock hypersurface,[18] which is one of the unknowns. Majda's work also required an additional assumption[19] on the data that seems to be necessary for the stability of the corresponding linearized problem.

We now describe the origin and motivation behind the present monograph. In his work [**17**], Christodoulou exploited various special structures enjoyed by the wave equations of irrotational relativistic fluid mechanics, structures which Alinhac did not use in his proof of shock formation. In particular, the wave equations in [**17**] derive from a Lagrangian[20] and are invariant under the Poincaré group; just below equation (2.11), we describe some ways in which Christodoulou used these structures in his proof. In studying Christodoulou's work [**17**], the author discovered that it is possible to use his framework to close the proof of shock formation for a larger class of equations and without relying on these special structures. In particular, his framework can be extended to treat all of the wave equations studied by Alinhac. A somewhat surprising fact, which plays a fundamental role in our analysis, is that they all have a special null structure, *even though they fail to satisfy Klainerman's null condition.* This special structure is not visible relative to the standard formulation of the wave equation, *but becomes visible upon reformulating it as a system of geometric wave equations*; see Lemmas A.9 and A.16 for the main results in this direction. It is this realization that led to the present work. Our work here also generalizes and unifies earlier work on singularity formation initiated by F. John in the 1970s and continued by L. Hörmander and many others.

More precisely, in the present monograph, we extend Christodoulou's framework and use it to prove that shock singularities often develop in initially small, regular solutions to two important classes[21] of quasilinear wave equations in three spatial dimensions. Specifically, we study **i)** covariant scalar wave equations of the form $\square_{g(\Psi)}\Psi = 0$ and **ii)** Alinhac's noncovariant scalar wave equations, that is, wave equations of the form $(h^{-1})^{\alpha\beta}(\partial\Phi)\partial_\alpha\partial_\beta\Phi = 0$. Our main result shows that whenever the nonlinear terms fail Klainerman's (classic) null condition,[22] shocks develop in solutions arising from an open set of small data. Hence, within the classes **i)** and **ii)**, our work can be viewed as a sharp converse to a fundamental result, due separately to Christodoulou [**15**] and Klainerman [**47**], which showed that

[17] This hypersurface is co-dimension two when viewed as a subset of spacetime.

[18] The shock hypersurface is a co-dimension one subset of spacetime.

[19] The assumption is automatically verified for the nonrelativistic Euler equations under the adiabatic equations of state $p = A\rho^\gamma$, where $A > 0$ and $\gamma > 1$ are constants.

[20] That is, the equations in [**17**] are Euler-Lagrange equations.

[21] It turns out that the two classes of equations are more closely related than one might expect; see the discussion below equation (2.9) and in Appendix A.

[22] Readers should take care not to confuse Klainerman's null condition with the future strong null condition and past strong null condition introduced in Appendix A and mentioned in Remark 2.3.

when the null condition is verified in three spatial dimensions, small-data global existence holds. Roughly, we give the same sharp description of the solution that Christodoulou gave in [**17**]. However, to avoid lengthening the monograph, we did not give a full description of the boundary of the maximal development nor the behavior of the solution along it. For readers interested in those details, we remark that the estimates proved in our main Theorem 22.1 are sufficient for invoking the arguments of [**17**, Chapter 15] in which Christodoulou reveals properties of the maximal development. That is, with modest additional effort, our results could be extended to give the same sharp description of the maximal development that Christodoulou gave in [**17**, Chapter 15].

In proving our main results, we have taken substantial steps that go beyond replicating the proofs given by Christodoulou in [**17**]. This is partly out of necessity, as the general class of equations that we treat leads to new kinds of error terms that are not present in [**17**]. However, we have also developed alternate strategies that greatly simplify certain aspects of the proof. One big simplification is that we are more selective in our use of geometry. That is, we use sharp, fully geometric decompositions only for treating the most delicate terms. Another simplification is that our bootstrap argument is very straightforward in view of the fact that we have organized the monograph in a linear fashion (see two paragraphs below). We have also developed alternate approaches to deriving some of the difficult top-order estimates by reducing them to other top-order estimates. This spares one a great deal of effort; see, for example, the discussion at the beginning of Sect. 15.1 in which we describe a simplified approach for obtaining estimates for the top-order derivatives of $\upmu$.

We now give an overview of the content and organization of the monograph. Chapter 1 sets the stage for the rest of monograph but is independent of the remaining chapters. It contains historical background, a discussion of shock formation in solutions to Burgers' equation, a discussion of singularity formation in 2×2 strictly hyperbolic genuinely nonlinear systems, an overview of wave dispersion in higher dimensions and its connection to global and almost global existence results, an overview of the vectorfield method (including the multiplier and commutator methods) for deriving generalized energy estimates, and a discussion of the null condition. In Chapter 2, we describe the main results of the monograph, place them in context, and provide an extended overview of the most important aspects of the proofs. In the remaining chapters, we develop the machinery and estimates needed to prove the two theorems of the monograph, which are located in Chapters 22 and 23. Roughly, in the first theorem (the main one, which is difficult to prove), we show that the solution must persist unless $\upmu$ vanishes, and we derive sharp a priori estimates that hold as long as $\upmu$ remains positive. In the second theorem, which is a relatively easy consequence of the first one, we exhibit an open set of data such that $\upmu$ does in fact vanish in finite time, thus yielding a shock singularity. Our analysis in Chapters 3–23 applies to covariant wave equations of the form $\square_{g(\Psi)}\Psi = 0$, while in Appendix A, we outline how to extend the results to the class of noncovariant wave equations studied by Alinhac. In Appendix B, we summarize the notation and conventions used in Chapters 2–23.

Chapters 3–23 are interdependent and are designed to be read consecutively. That is, this part of the monograph constitutes one long bootstrap-type proof, presented in chronological order. In Sect. 2.13, we give an overview of the contents of each chapter and provide suggestions on how to read the monograph, both for expert and novice readers.

This monograph is mostly self-contained but relies extensively on basic concepts from Lorentzian and Riemannian geometry such as Levi-Civita connections, fundamental forms, curvature, pullbacks, etc. We anticipate that there are readers with knowledge in fluid mechanics, conservation laws, and/or PDEs but who are unfamiliar with those geometric concepts. Such readers can find introductory geometric material, suitable for reading almost all[23] of the present monograph, in select portions of the books [**66**, **67**, **78**]. Novice geometers should bear in mind that at the end of the day, one aims to derive estimates, and that the geometry merely provides a framework for organizing calculations and revealing analytic structural features that would otherwise be difficult to detect.

We close by highlighting the following wide-open question:

> In more than one spatial dimension, to what extent can the results of this monograph be generalized to quasilinear systems featuring multiple speeds of propagation, such as the equations of elasticity, the equations of magnetohydrodynamics, the equations of crystal optics, the Euler–Einstein equations of cosmology, or even coupled systems of wave equations featuring two or more metrics with strictly separated[24] speeds of propagation?

Jared Speck

[23] Our short proof of geometric Sobolev embedding, presented in Chapter 18, relies on a handful of more advanced results from geometry.

[24] This roughly corresponds to the presence of two or more distinct families of characteristic hypersurfaces.

Acknowledgments

I am grateful for the support offered by NSF grant #DMS-1162211 and by a Solomon Buchsbaum grant administered by the Massachusetts Institute of Technology. I would like to thank the American Institute of Mathematics for funding three SQuaREs workshops on the formation of shocks, which greatly aided the development of many of the ideas in this monograph. I would like to thank Gustav Holzegel, Sergiu Klainerman, Jonathan Luk, Willie Wong, and Shiwu Yang for participating in the workshops and for their helpful contributions. I offer special thanks to Willie Wong for creating Figure 4. I am grateful for the support of my PhD advisors, Michael Kiessling and Shadi Tahvildar-Zadeh, who encouraged me to read Christodoulou's monograph on shock formation. I would also like to thank Hans Lindblad for sharing his insight on Alinhac's work.

CHAPTER 1

Introduction

As we described in the Preface, the purpose of this chapter is to provide some historical background and introductory material. The remaining chapters in the monograph are independent of the present one. Our main goal here is to set the stage for our study of the difficult problem of shock formation in more than one spatial dimension. Readers may also consult the companion survey article [**28**] for additional introductory material.

1.1. Shock formation in one spatial dimension

In Sect. 1.1, we provide a brief overview of shock formation in one spatial dimension. We give some historical background and provide two proofs that shocks often form in solutions to Burgers' equation. The second proof is sharper and has important philosophical parallels with our later study of shock formation in three spatial dimensions. For illustration, we also provide a proof of blow-up for a model quasilinear wave equation under the assumption of plane symmetry.[1] To this end, we embed the equation into the theory of 2×2 strictly hyperbolic genuinely nonlinear systems and provide a standard proof of blow-up for such systems.

1.1.1. A very brief history of shock formation in one spatial dimension. The development of the modern theory of shock waves in solutions to nonlinear hyperbolic PDEs has a rich history filled with false starts and tantalizing turns. For fascinating descriptions of the events leading to the modern theory, we refer the reader to the introduction of [**22**] as well as the survey article [**72**]. Here we only describe the historical results that are most directly connected to the results of the present monograph.

The earliest known observation of shock formation[2] was made by the British physicist/astronomer James Challis [**11**] in his study of the evolution of an ideal gas in one spatial dimension under the isothermal equation of state $p = c_s^2 \rho$, where p is the pressure, ρ is the density, and the constant $c_s > 0$ is the speed of sound. More precisely, Challis considered the compressible Euler equations for ρ and the velocity v:

$$\partial_t \rho + \partial_x (\rho v) = 0, \tag{1.1a}$$

$$\partial_t (\rho v) + \partial_x (\rho v^2) + c_s^2 \partial_x \rho = 0 \tag{1.1b}$$

[1] By plane symmetry, we mean that the solution depends only on a time variable t and a single real spatial variable x.

[2] In PDE literature, shock formation is also known as *wave breaking*.

in conjunction with Bernoulli's equation[3] for the fluid potential Φ, which verifies $v = \partial_x \Phi$:

$$\partial_t \Phi + \frac{1}{2}(\partial_x \Phi)^2 + c_s^2 \ln \rho = 0. \tag{1.2}$$

Forty years earlier, Poisson [**69**] had shown that as a consequence of (1.1a)-(1.2), Φ verifies the following[4] closed equation:

$$\partial_t^2 \Phi + 2(\partial_x \Phi)\partial_x \partial_t \Phi + (\partial_x \Phi)^2 \partial_x^2 \Phi - c_s^2 \partial_x^2 \Phi = 0. \tag{1.3}$$

Poisson also showed that any function Φ verifying an identity of the form

$$\partial_x \Phi = f\left(x + (c_s - \partial_x \Phi)t\right), \tag{1.4}$$

where f is a smooth function, is a solution[5] to (1.3). Note that (1.4) implies that f is the initial condition of $\partial_x \Phi$.

Challis's key observation was that in the case $f(x) = -\sin\left(\frac{\pi}{2}x\right)$, the solution to (1.4) verifies $\partial_x \Phi = 0$ along the line $x = -c_s t$ and $\partial_x \Phi = 1$ along the line $x = -1 - (c_s - 1)t$. Since these lines intersect at the point $(t, x) = (1, -c_s)$, one concludes that *the classical solution must break down there.*

1.1.2. Shock formation in solutions to Burgers' equation and the method of characteristics. Results in the spirit of Challis' blow-up result of Sect. 1.1.1 have been derived for many nonlinear hyperbolic PDEs and initial conditions in one spatial dimension.[6] Here is a far-from-exhaustive collection of examples: Riemann's foundational work [**71**] on the compressible Euler equations in which he invented the method of Riemann invariants, Lax's work [**55**] on scalar conservation laws, his use of Riemann invariants in his study [**54**] of 2×2 strictly hyperbolic genuinely nonlinear systems (see Sect. 1.1.3), Jeffrey's work [**31**] on magnetoacoustics, Jeffrey-Korobeinikov's work [**33**] on nonlinear electromagnetism, Jeffrey-Teymur's work [**32**] on hyperelastic solids, John's extension [**35**] of Lax's work to a larger class of systems in which the method of Riemann invariants cannot be used, Liu's further refinement [**60**] of John's work, John's work [**37**] on spherically symmetric

---

[3]From straightforward computations based on the assumption $v = \partial_x \Phi$ and equations (1.1a)-(1.1b), we find that $\partial_x \left\{\partial_t \Phi + \frac{1}{2}(\partial_x \Phi)^2 + c_s^2 \ln \rho\right\} = 0$. Thus, $\partial_t \Phi + \frac{1}{2}(\partial_x \Phi)^2 + c_s^2 \ln \rho$ is a function of t, which we have set equal to 0 in (1.2).

[4]Equation (1.3) is the Euler-Lagrange equation for Φ. More precisely, in one spatial dimension, and more generally for irrotational solutions in higher dimensions, the compressible Euler equations are equivalent to a quasilinear wave equation for Φ that can written in Euler-Lagrange form. The Lagrangian is $\mathcal{L} = p$, where the pressure p is viewed as a function of the spacetime gradient of Φ; see, for example, [**17**] and [**21**] for further details. For the equation of state $p = c_s^2 \rho$ in one spatial dimension, equation (1.2) implies that $p = c_s^2 \exp\left(-c_s^{-2}\left\{\partial_t \Phi + \frac{1}{2}(\partial_x \Phi)^2\right\}\right)$. The Euler-Lagrange equation is $\partial_t \left(\frac{\partial \mathcal{L}}{\partial(\partial_t \Phi)}\right) + \partial_x \left(\frac{\partial \mathcal{L}}{\partial(\partial_x \Phi)}\right) = 0$, and when $p = c_s^2 \rho$, it can be written in the form (1.3).

[5]To derive the family of solutions (1.4), one can first compute that the function $v + c_s \ln \rho$ is a Riemann invariant (see Sect. 1.1.3) for the system (1.1a)-(1.1b). A special class of solutions is such that this Riemann invariant is constant, which implies that $\partial_x(v + c_s \ln \rho) = 0$. Using this identity and equations (1.1a)-(1.1b), we deduce that these special solutions verify the equation $\partial_t v + (v - c_s)\partial_x v = 0$. This equation, when supplemented with the initial condition $v|_{t=0} = f$, has the solution $v = f\left(x + (c_s - v)t\right)$. Recalling that $v = \partial_x \Phi$, we arrive at Poisson's solutions (1.4).

[6]Under the umbrella of "one spatial dimension," we include problems that are effectively one dimensional, such as the study of spherically symmetric solutions in higher dimensions.

solutions to the equations of elasticity, Klainerman-Majda's work [**49**] on nonlinear vibrating string equations, Bloom's work [**10**] on nonlinear electrodynamics, and Cheng-Young-Zhang's work [**13**] on magnetohydrodynamics and related systems.

In the present section, we use the model case of Burgers' equation to illustrate the essential features of modern proofs of blow-up. We provide two proofs. The first is standard and is effectively a proof by contradiction showing that nontrivial compactly supported (in the spatial variable) smooth solutions cannot be continued indefinitely. The second is much more refined, shows that the singularity is a shock, and is more closely aligned with our later proof of shock formation for solutions to wave equations in three spatial dimensions.

In the following discussion. (t, x) are standard rectangular coordinates on $\mathbb{R}^2$ and ∂_t, ∂_x are the corresponding coordinate partial derivatives. The Cauchy problem for Burgers' equation in one spatial dimension with unknown $\Psi(t, x)$ and initial data $\mathring{\Psi}$ is:

$$\partial_t \Psi + \Psi \partial_x \Psi = 0, \tag{1.5a}$$

$$\Psi(0, x) = \mathring{\Psi}(x). \tag{1.5b}$$

We first present the standard crude analysis, which shows that the solution blows up in finite time but provides limited information about the true nature of the singularity. The argument is based on the method of characteristics, which by definition are the solutions $\gamma(t; x) := (\gamma^0(t; x), \gamma^1(t; x))$ of the following ODE initial value problem:

$$\frac{d}{dt}\gamma^0(t; x) = 1, \qquad \frac{d}{dt}\gamma^1(t; x) = \Psi \circ \gamma(t; x), \tag{1.6a}$$

$$\gamma^0(0; x) = 0, \qquad \gamma^1(0; x) = x. \tag{1.6b}$$

Note that $\{\gamma(\cdot; x)\}_{x \in \mathbb{R}}$ is a set of curves in the plane parametrized by x. To exhibit the blow-up of smooth solutions, we first differentiate (1.5a) with ∂_x and use equation (1.6a) to deduce that along the characteristics, we have

$$\frac{d}{dt}\{(\partial_x \Psi) \circ \gamma(t; x)\} = -\{(\partial_x \Psi) \circ \gamma(t; x)\}^2 . \tag{1.7}$$

Note that for each fixed x, (1.7) is a Riccati-type[7] ODE for $(\partial_x \Psi) \circ \gamma(\cdot; x)$. In fact, (1.7) is equivalent to $\frac{d}{dt}\left\{\frac{1}{(\partial_x \Psi) \circ \gamma(t; x)}\right\} = 1$. In view of the initial conditions (1.6b), it follows that if $(\partial_x \mathring{\Psi})(x_0) < 0$, then $(\partial_x \Psi) \circ \gamma(t; x_0)$ must blow up by the time $t = -1/(\partial_x \mathring{\Psi})(x_0)$. Note that all nontrivial compactly supported data must lead to blow-up because such data always have at least one point x_0 with $\partial_x \mathring{\Psi}(x_0) < 0$.

One criticism of the above argument is that in its present form, it is actually a proof by contradiction showing that smooth solutions do not persist for all time. That is, additional arguments must be given in order to guarantee that $\partial_x \Psi$ is the true quantity that blows up first and that no other kind of singularity occurs before that. Of course, in the case of Burgers' equation, the additional arguments are simple; we leave the details to the reader. A major difficulty that we must overcome in the present monograph is that the crude proof of blow-up by contradiction presented above does not seem to generalize to quasilinear wave equations in more than

[7] The standard Riccati ODE is $\dot{y} = y^2$.

one spatial dimension. The main reason is that in two or more spatial dimensions, one needs to supplement the method of characteristics with energy estimates (see Sect. 1.2.1). In order to close the energy estimates in the shock formation problem, we must obtain a detailed description of the dynamics near the shock going far beyond a crude description; see Sect. 2.10.4 for an outline of the argument, which is quite involved, even at the heuristic level.

In view of the limitations of the above crude proof of blow-up for solutions to Burgers' equation, we now give a sharper proof that serves as a slightly more realistic caricature of the way in which we prove the main shock formation results of the monograph. Our argument here is based on a version of Christodoulou's geometric framework [**17**], which applies in three spatial dimensions and is the framework that we use throughout most of the monograph. More precisely, because our analysis here involves only one spatial dimension, we use only a small portion of Christodoulou's framework in the present section, a portion that is essentially equivalent to Majda's geometric approach [**61**] to proving shock formation in one spatial dimension. We remark that Majda's work has its roots in the geometric approach of Keller-Ting [**43**]. Our argument here is also similar to the one given by Alinhac in [**3**].

We begin by constructing the key ingredient, which is a new dynamic coordinate $u = u(t, x)$ obtained by solving the following Cauchy problem for a transport equation:

$$\frac{d}{dt}\{u \circ \gamma(t; x)\} = \partial_t u + \Psi \partial_x u = 0, \tag{1.8a}$$

$$u(0, x) = u \circ \gamma(0; x) = x. \tag{1.8b}$$

In (1.8a)-(1.8b), the characteristics γ are the solutions to (1.6a)-(1.6b). The variable u, which we call an *eikonal function*, is an analog of the well-known Lagrangian coordinates that are often used in fluid mechanics. The main idea of the analysis is to study Burgers' equation relative to the coordinate system (t, u), which will reveal important structural features that are not visible relative to the coordinates (t, x). Note that[8]

$$\frac{\partial}{\partial t}|_u = \partial_t + \Psi \partial_x \tag{1.9}$$

and that $\frac{\partial}{\partial t}|_u$ corresponds to differentiation along the characteristics. The most important property of the coordinates (t, u) is that relative to them, Burgers' equation (1.5a) becomes a *linear* PDE:

$$\frac{\partial}{\partial t}|_u \Psi = 0. \tag{1.10}$$

It follows from (1.10) that when the data are smooth, Ψ *and its derivatives of all orders* with respect to $\frac{\partial}{\partial t}|_u$ and $\frac{\partial}{\partial u}|_t$ remain finite for all time. In fact, all of these quantities are constant in t at fixed u. Thus, a singularity can develop in Ψ relative to the (t, x) coordinates *only if the change of variables map from* (t, u) *coordinates to* (t, x) *coordinates becomes degenerate.*

---

[8]Throughout this section, $\frac{\partial}{\partial t}|_u$ denotes the partial derivative with respect to t at fixed u, and $\frac{\partial}{\partial u}|_t$ has the analogous meaning.

To expand upon the ideas from the previous paragraph, we introduce the *inverse foliation density* μ, which we define to be

$$\mu := \frac{1}{\partial_x u}. \tag{1.11}$$

The function $1/\mu$ measures the density of the level sets of u relative to the lines $\{x = \text{const}\}$. By (1.8b), μ is initially 1, and shock formation (that is, the intersection of the characteristics) exactly corresponds to infinite density of the level sets, or equivalently, $\mu \to 0$. It is not immediately clear that there should be a connection between $\mu \to 0$ and the blow-up of $\partial_x \Psi$. There is in fact a strong connection, and to illustrate it, we need only to use the chain rule, (1.10), and (1.11) to deduce that

$$\partial_x \Psi = \frac{1}{\mu} \frac{\partial}{\partial u}|_t \Psi. \tag{1.12}$$

Thus, as long as $\frac{\partial}{\partial u}|_t \Psi \neq 0$, we see that $\partial_x \Psi$ blows up $\iff$ $\mu \to 0$. Moreover, as we show in the next paragraph, the vanishing of μ is exactly tied to having a solution with $\frac{\partial}{\partial u}|_t \Psi < 0$.

To show that shocks indeed form, we first use (1.8a), (1.9), (1.11), and (1.12) to compute that μ verifies the following evolution equation, where $[\partial_x, \frac{\partial}{\partial t}|_u]$ denotes the commutator of ∂_x and $\frac{\partial}{\partial t}|_u$:

$$\frac{\partial}{\partial t}|_u \mu = -\mu^2 \frac{\partial}{\partial t}|_u \partial_x u = \mu^2 [\partial_x, \frac{\partial}{\partial t}|_u] u = \mu \partial_x \Psi = \frac{\partial}{\partial u}|_t \Psi. \tag{1.13}$$

To obtain information about the source term $\frac{\partial}{\partial u}|_t \Psi$ on the right-hand side of (1.13), we simply recall that $\frac{\partial}{\partial u}|_t \Psi$ is constant in t at fixed u. Thus, if $\frac{\partial}{\partial u}|_t \Psi$ is initially negative, we conclude from (1.13) that μ will necessarily vanish in finite time along the corresponding characteristic, and by (1.12), $\partial_x \Psi$ will blow-up like $1/\mu$. In summary, our second proof of blow-up is sharper than the first because *we have identified that the vanishing of the geometric quantity* μ *is the precise condition causing blow-up and because the solution remains regular in* (t, u) *coordinates*.

REMARK 1.1 (μ **plays a key role in the monograph**). To show that shocks form in the wave equations of interest in the present monograph, the main object of study is a quantity that is analogous to the one defined in (1.11) (see (2.27)). In fact, this monograph is primarily about the extension of the above argument to two classes of quasilinear wave equations in three spatial dimensions and the many additional complications that arise in the presence of angular derivatives and dispersion.

We end this section by highlighting another way to think about the dynamics of solutions Ψ to Burgers' equation: Ψ remains regular along the directions tangent to the characteristics but become singular in the transversal directions. As we will see, these basic features are also partially present in the shock forming wave equation solutions that we study later in the monograph. In fact, these features play a critically important role in the analysis.

1.1.3. Two by two strictly hyperbolic genuinely nonlinear systems and their connection to plane symmetric quasilinear wave equations. As we described in the Preface, the main result of this monograph is that shock formation often occurs in small-data solutions to two classes of quasilinear wave equations in three spatial dimensions. A simple example of an equation on $\mathbb{R}^{1+3}$ to which our main result applies is

$$-\partial_t^2\Phi + c^2\Delta\Phi = 0, \tag{1.14}$$

where the factor $c = c(\partial_t\Phi) > 0$ is a smooth function of $\partial_t\Phi$ verifying the following structural assumptions:

$$c(0) = 1, \qquad c'(0) \neq 0. \tag{1.15}$$

In (1.14), Δ denotes the standard Laplacian on $\mathbb{R}^3$, and in (1.15), $c'(p) = \frac{d}{dp}c(p)$. The proof of shock formation for solutions to (1.14) is incredibly more complicated than the proofs of blow-up for Burgers' equation solutions given in Sect. 1.1.2. Not surprisingly, under the assumption of plane symmetry (that is, that the solution depends only on $(t, x) \in \mathbb{R}^{1+1}$), the proof drastically simplifies because we do not need to derive energy estimates and because plane symmetric solutions do not exhibit dispersion. However, the proof still has some new features not found in either of the proofs of blow-up that we gave for Burgers' equation. For this reason, we provide a proof of plane-symmetric blow-up in this section. Our proof is a standard adaption of the first (crude) proof (by contradiction) of blow-up for Burgers' equation given in Sect. 1.1.2. We note that is possible to give a sharper proof of plane-symmetric shock formation in the spirit of our second proof of blow-up for Burgers' equation; one could proceed by making straightforward modifications to the sharp proof of shock formation for spherically symmetric solutions to quasilinear wave equations in three spatial dimensions given in the companion survey article [**28**].

A simple way to prove blow-up for plane symmetric solutions to (1.14) is to show that under the assumptions (1.15), if $\partial_t\Phi$ is sufficiently small, then (1.14) is equivalent to a member of an important class of PDEs known as 2×2 *strictly hyperbolic genuinely nonlinear systems*; we will verify the equivalence below. In [**54**], Lax proved the first general blow-up results for such systems by extending, with the help of Riemann invariants, the method of characteristics that we used in Sect. 1.1.2 in our crude proof of blow-up for solutions to Burgers' equation. John later developed an approach that allowed him to extend [**35**] Lax's blow-up results to a larger class of systems in one spatial dimension in which the method of Riemann invariants cannot be used. Moreover, in the case of the nonlinear vibrating string equations in one spatial dimension, Klainerman and Majda showed [**49**] that the genuine nonlinearity condition is not necessary.

We now provide a proof of Lax's blow-up results [**54**] and show that it implies blow-up for plane symmetric solutions to (1.14). In the remainder of this section, lowercase Latin indices take on the values 1 and 2, lowercase Greek indices take on the values 0 and 1, and we use Einstein's summation convention. Moreover, we use the notation $(x^0, x^1) = (t, x)$, $\partial_0 = \partial_t$, and $\partial_1 = \partial_x$.

We will study the following Cauchy problem for a 2×2 nonlinear hyperbolic system:

$$\partial_t U_j + M_j^a(U)\partial_x U_a = 0, \qquad (j = 1, 2), \tag{1.16a}$$

$$U_j|_{t=0} = \mathring{U}_j. \tag{1.16b}$$

We say that (1.16a) is a *strictly hyperbolic system* on the domain $\mathcal{U} \subset \mathbb{R}^2$ if for every $U \in \mathcal{U}$, the 2×2 matrix $M_j^i(U)$ has two distinct eigenvalues $\lambda_1(U) < \lambda_2(U)$.

Under the assumption of strict hyperbolicity, for $i = 1, 2$, we let $r^{(i)}$ and $l_{(i)}$ respectively denote the Euclidean-unit length right eigencovector and the Euclidean-unit length left eigenvector (which are unique up to an overall sign) corresponding to λ_i. In particular, these quantities verify the following systems of equations (with no summation over i):

$$M_b^a r_a^{(i)} = \lambda_i r_b^{(i)}, \qquad (b, i = 1, 2), \tag{1.17a}$$

$$M_a^b l_{(i)}^a = \lambda_i l_{(i)}^b, \qquad (b, i = 1, 2). \tag{1.17b}$$

We use the following critically important identity in our subsequent analysis:

$$r_a^{(i)} l_{(j)}^a = 0, \qquad (i \neq j). \tag{1.18}$$

To derive (1.18), we contract equation (1.17a) against $l_{(j)}^b$ and use equation (1.17b) to deduce the identity $\lambda_j r_a^{(i)} l_{(j)}^a = \lambda_i r_a^{(i)} l_{(j)}^a$ (with no summation over i or j). The desired result (1.18) follows from this identity and the strict hyperbolicity assumption.

We say that the strictly hyperbolic system (1.16a) is a *genuinely nonlinear system* on the domain $\mathcal{U} \subset \mathbb{R}^2$ if for every $U \in \mathcal{U}$ and $i = 1, 2$, we have (with no summation over i):

$$r^{(i)} \cdot D_U \lambda_i \neq 0. \tag{1.19}$$

In (1.19) and in the remainder of this section, $D_U f = \left(\frac{\partial f}{\partial U_1}, \frac{\partial f}{\partial U_2}\right)$ denotes the gradient of f viewed as a function of U and $\cdot$ denotes the Euclidean dot product. Thus, in component form, equation (1.19) reads $r_a^{(i)} \frac{\partial \lambda_i}{U_a} \neq 0$ (with summation over a but *not* over i).

We now show that under the assumption of plane symmetry and under the assumptions (1.15), when $\partial_t \Phi$ is sufficiently small, equation (1.14) is equivalent to a 2×2 genuinely nonlinear strictly hyperbolic system. In plane symmetry, equation (1.14) reduces to

$$-\partial_t^2 \Phi + c^2(\partial_t \Phi)\partial_x^2 \Phi = 0. \tag{1.20}$$

To derive the equivalent 2×2 genuinely nonlinear strictly hyperbolic system, we introduce the variables

$$U_1 := \partial_t \Phi, \qquad U_2 := \partial_x \Phi. \tag{1.21}$$

It is straightforward to see that for C^2 solutions, equation (1.20) is equivalent to equation (1.16a), where the matrix $M = M(U)$ has the components

$$\begin{pmatrix} M_1^1 & M_1^2 \\ M_2^1 & M_2^2 \end{pmatrix} = \begin{pmatrix} 0 & -c^2 \\ -1 & 0 \end{pmatrix}. \tag{1.22}$$

To verify the genuine nonlinearity and strict hyperbolicity conditions for the matrix (1.22) when $\partial_t\Phi$ is sufficiently small, we simply recall that $c = c(\partial_t\Phi) = c(U_1)$ and compute that

$$\lambda_1 = -c, \qquad \lambda_2 = c, \tag{1.23a}$$

$$r^{(1)} = \frac{1}{\sqrt{1+c^{-2}}}(1, c^{-1}), \qquad r^{(2)} = \frac{1}{\sqrt{1+c^{-2}}}(1, -c^{-1}), \tag{1.23b}$$

$$l_{(1)} = \frac{1}{\sqrt{1+c^2}}(1, c), \qquad l_{(2)} = \frac{1}{\sqrt{1+c^2}}(1, -c), \tag{1.23c}$$

$$r_a^{(1)}\frac{\partial\lambda_1}{U_a} = -\frac{1}{\sqrt{1+c^{-2}}}c', \qquad r_a^{(2)}\frac{\partial\lambda_2}{U_a} = \frac{1}{\sqrt{1+c^{-2}}}c'. \tag{1.23d}$$

As is well known, to analyze genuinely nonlinear strictly hyperbolic systems, it is convenient to use *Riemann invariants*. Given such a system, for $i = 1, 2$, we define a Riemann invariant $w_i = w_i(U_1, U_2)$ to be any function that is constant along the integral curves[9] of $r^{(i)}$. Equivalently, w_i verifies the following equation (with no summation over i):

$$r^{(i)} \cdot D_U w_i = 0. \tag{1.24}$$

Under the assumption that $D_U w_i \neq 0$ for $i = 1, 2$, w_1 and w_2 form a system of *state space coordinates* that can be used in place of U_1 and U_2. The main advantage is that relative to a coordinate system of Riemann invariants, the Cauchy problem (1.16a)-(1.16b) takes the following simple form:[10]

$$L_{(2)}w_1 = 0, \qquad L_{(1)}w_2 = 0, \tag{1.25a}$$

$$w_1|_{t=0} = \mathring{w}_1, \qquad w_2|_{t=0} = \mathring{w}_2, \tag{1.25b}$$

where the two *characteristic vectorfields* $L_{(i)}$ are defined as follows:

$$L_{(i)} := \partial_t + \lambda_i\partial_x. \tag{1.26}$$

We now verify the equivalence of the systems (1.16a) and (1.25a) when w_1 and w_2 form a coordinate system of Riemann invariants. To this end, we first claim that the genuine nonlinearity condition (1.19) is equivalent to

$$\frac{\partial\lambda_i}{\partial w_j} = \frac{\partial\lambda_i}{\partial U_a}\frac{\partial U_a}{\partial w_j} \neq 0, \qquad (i \neq j). \tag{1.27}$$

The equality in (1.27) follows from the chain rule. To verify the $\neq 0$ aspect of the claim, we note that (1.19) and (1.24) imply that $D_U\lambda_i$ and $D_U w_i$ are not parallel or anti-parallel. Moreover, by the chain rule, when $i \neq j$, we have $\frac{\partial w_i}{\partial U_a}\frac{\partial U_a}{\partial w_j} = 0$. Thus, when $i \neq j$, the one-form with components $\left(\frac{\partial U_1}{\partial w_j}, \frac{\partial U_2}{\partial w_j}\right)$ is perpendicular to $D_U w_i$ and hence, by the previous observation, not perpendicular to $D_U\lambda_i$. Clearly this is equivalent to the desired $\neq 0$ statement in (1.27).

---

[9]More precisely, w_i is constant along the integral curves of the dual of the covector $r^{(i)}$ with respect to the standard Euclidean metric on the state space $\mathbb{R}^2$.

[10]Throughout the monograph, if X is a vectorfield and f is a scalar-valued function, then $Xf := X^\alpha\partial_\alpha f$ denotes the derivative of f in the direction X. The X^α are the components of X relative to the spacetime coordinate partial derivative vectorfield frame $\{\partial_\alpha\}$. In particular, the first equation in (1.25a) is $L_{(2)}^0\partial_0 w_1 + L_{(2)}^1\partial_1 w_1 = \partial_t w_1 + \lambda_i\partial_x w_1 = 0$.

We now give the proof that the first equation in (1.25a) is a consequence of the system (1.16a). The proofs that the second equation in (1.25a) also follows from (1.16a) and that the reverse implication holds are similar, and we omit those details. To proceed, we first use the chain rule and equation (1.16a) to deduce that

$$\begin{aligned}(1.28)\qquad \partial_t w_1 + \lambda_2 \partial_x w_1 &= \frac{\partial w_1}{\partial U_a}\partial_t U_a + \lambda_2 \frac{\partial w_1}{\partial U_a}\partial_x U_a \\ &= -\frac{\partial w_1}{\partial U_a} M_a^b \partial_x U_b + \lambda_2 \frac{\partial w_1}{\partial U_a}\partial_x U_a \\ &= \frac{\partial w_1}{\partial U_a}\left\{-M_a^b + \lambda_2 \delta_a^b\right\}\partial_x U_b,\end{aligned}$$

where δ_a^b is the standard Kronecker delta. We now claim that $\frac{\partial w_1}{\partial U_a}\left\{-M_a^b + \lambda_2 \delta_a^b\right\} = 0$, which implies the desired first equation in (1.25a). This claim follows from the orthogonality condition (1.18) and the definition (1.24) of a Riemann invariant, which together guarantee that $D_U w_1$ is proportional to $l_{(2)}$, and from equation (1.17b).

In the remainder of this section, we assume that the solution remains inside of a compact subset of the region of state space in which the system is strictly hyperbolic and genuinely nonlinear. Under this assumption, we will show that when the data $(\mathring{w}_1, \mathring{w}_2)$ (given in (1.25b)) are compactly supported and nontrivial, the solution to (1.25a) blows up in finite time. This shows in particular that the plane symmetric wave equation (1.20) exhibits finite-time blow-up. For definiteness, in view of the $\neq 0$ statement in (1.27), we assume that there is a constant $C > 0$ and a point x_0 such that

$$(1.29)\qquad \frac{\partial \lambda_2}{\partial w_1} > C,$$

$$(1.30)\qquad \partial_x \mathring{w}_1(x_0) < 0.$$

We will show that (1.29)-(1.30) imply finite-time blow-up for w_1. Moreover, by making straightforward modifications to our proof, we could deduce that related blow-up results hold for w_1 if the signs in (1.29)-(1.30) are altered in a compatible fashion, and that w_2 blows up if (1.29) is replaced with a signed condition on $\frac{\partial \lambda_1}{\partial w_2}$ and (1.30) is replaced with a compatible signed condition on $\partial_x \mathring{w}_2$. It is straightforward to check that for nontrivial compactly supported data, at least one of these conditions leading to blow-up must occur.

Our proof of the blow-up of w_1 is an adaption of our first (crude) proof of blow-up for Burgers' equation given in Sect. 1.1.2. In the remaining discussion in this section, we view the characteristic speeds λ_i as functions of the Riemann invariants (w_1, w_2) but we often suppress this functional dependence. Moreover, we view w_1 and w_2 as functions of the standard coordinates (t, x). As in our first proof of blow-up for Burgers' equation, the main idea is to exploit the fact that $\partial_x w_1$ verifies a Riccati-type equation. The new complication is that the equation is coupled to w_2. Specifically, we differentiate the first equation in (1.25a) with ∂_x and use the chain rule, definition (1.26), and the identity $\partial_x w_2 = \frac{1}{\lambda_2 - \lambda_1} L_{(2)} w_2$

(valid when w_2 verifies the second equation in (1.25a)) to deduce that

$$L_{(2)}\partial_x w_1 + \frac{\partial\lambda_2}{\partial w_1}(\partial_x w_1)^2 + \left\{\frac{1}{\lambda_2-\lambda_1}\frac{\partial\lambda_2}{\partial w_2}L_{(2)}w_2\right\}\partial_x w_1 = 0. \tag{1.31}$$

To derive the desired blow-up result for $\partial_x w_1$, we control the term in braces on the left-hand side of (1.31) and show that it remains bounded. To this end, we define, much as in (1.8a)-(1.8b), the family of characteristic curves $\gamma_{(2)}(t;x) = (\gamma_{(2)}^0(t;x), \gamma_{(2)}^1(t;x))$ corresponding to the vectorfield $L_{(2)}$ to be the solutions to the following ODE initial value problems:

$$\frac{d}{dt}\gamma_{(2)}^0(t;x) = 1, \qquad \frac{d}{dt}\gamma_{(2)}^1(t;x) = \lambda_2\circ\gamma_{(2)}(t;x), \tag{1.32a}$$

$$\gamma_{(2)}^0(0;x) = 0, \qquad \gamma_{(2)}^1(0;x) = x. \tag{1.32b}$$

As a first quantitative step in our proof of blow-up, we now derive simple uniform bounds from above and below for w_1 and w_2. To this end, we note that by the chain rule and (1.32a), the first equation in (1.25a) can be expressed as

$$\frac{d}{dt}\left\{w_1\circ\gamma_{(2)}^0(t;x)\right\} = 0. \tag{1.33}$$

Integrating (1.33) starting from $t = 0$, we find that at any time t of classical existence, we have

$$\min_{x\in\mathbb{R}} w_1\circ\gamma_{(2)}^0(t;x) = \min_{x\in\mathbb{R}} \mathring{w}_1(x), \qquad \max_{x\in\mathbb{R}} w_1\circ\gamma_{(2)}^0(t;x) = \max_{x\in\mathbb{R}} \mathring{w}_1. \tag{1.34}$$

Similarly, by integrating along characteristic curves corresponding to the vectorfield $L_{(1)}$, we find that (1.34) also holds with w_1 everywhere replaced with w_2. Thus, we deduce that there exists a constant $C > 0$ depending on the data such that in any region of classical existence, we have

$$-C \le w_i \le C. \tag{1.35}$$

Our proof of blow-up relies on using an integrating factor $\iota(t;x)$ to simplify the structure of equation (1.31). Specifically, we define

$$\iota(t;x) := \exp\left(\int_{w=\mathring{w}_2(x)}^{w=w_2\circ\gamma_{(2)}(t;x)}\left\{\frac{1}{\lambda_2-\lambda_1}\frac{\partial\lambda_2}{\partial w_2}\right\}(\mathring{w}_1(x), w)\,dw\right). \tag{1.36}$$

Note that by (1.32b), we have $\iota(0;x) = 1$. Note also that (1.35) and the strict hyperbolicity assumption $\lambda_1 < \lambda_2$ imply that there exists a constant $c > 1$ (depending on the data) such that in any region of classical existence, we have

$$c^{-1} < \iota(t;x) < c. \tag{1.37}$$

The main step in the proof of blow-up for solutions $\partial_x w_1$ to equation (1.31) is to derive the following equation:

(1.38)

$$\begin{aligned}(\partial_x w_1)\circ\gamma_{(2)}(t;x) &= (\partial_x\mathring{w}_1)(x)\iota^{-1}(t;x)\\ &\quad\times\left\{1 + (\partial_x\mathring{w}_1)(x)\int_{s=0}^t \iota^{-1}(s;x)\frac{\partial\lambda_2}{\partial w_1}\circ\gamma_{(2)}(s;x)\,ds\right\}^{-1}.\end{aligned}$$

Once we have obtained (1.38), the desired finite-time blow-up of $\partial_x w_1$ easily follows from (1.29)-(1.30) and (1.37).

It remains for us to derive equation (1.38). To this end we use equations (1.33) and (1.32b) to deduce that

$$w_1 \circ \gamma_{(2)}(t,x) = \mathring{w}_1(x). \tag{1.39}$$

Hence, from definition (1.26), equations (1.32a) and (1.39), and the chain rule, we deduce the following identity involving the term in braces on the left-hand side of (1.31):

$$\begin{aligned} &\frac{d}{dt} \int_{w=\mathring{w}_2(x)}^{w=w_2 \circ \gamma_{(2)}(t;x)} \left\{ \frac{1}{\lambda_2 - \lambda_1} \frac{\partial \lambda_2}{\partial w_2} \right\} (\mathring{w}_1(x), w) \, dw \\ &= \left\{ \frac{1}{\lambda_2 - \lambda_1} \frac{\partial \lambda_2}{\partial w_2} L_{(2)} w_2 \right\} \circ \gamma_{(2)}(t;x). \end{aligned} \tag{1.40}$$

Using (1.32a), (1.36) (1.40), and the chain rule, we can rewrite the Riccati-type equation (1.31) as

$$\frac{d}{dt} \left(\frac{1}{\iota(t;x)(\partial_x w_1) \circ \gamma_{(2)}(t;x)} \right) = \frac{1}{\iota(t;x)} \frac{\partial \lambda_2}{\partial w_1} \circ \gamma_{(2)}(t;x). \tag{1.41}$$

Finally, integrating equation (1.41) with respect to time from 0 to t, using the initial conditions (1.32b), and carrying out straightforward computations, we arrive at the desired equation (1.38). This concludes our proof of blow-up of solutions to (1.25a)-(1.25b).

1.2. New aspects in more than one spatial dimension

In Sect. 1.2, we discuss some fundamental issues that distinguish the study of quasilinear hyperbolic PDEs in two or more spatial dimensions from the case of one spatial dimension. In particular, we discuss the need for energy estimates and the possible presence of dispersion. Both of these features play a key role in our derivation of the main shock formation results of this monograph. Next, for nonlinear wave equations, we discuss the role that dispersion plays in modern proofs of small-data global existence. Moreover, in the case of three spatial dimensions, we discuss Klainerman's (classic) null condition and outline why nonlinearities verifying it allow for small-data global existence. The wave equations that we study in Chapters 2-23 have nonlinearities that fail the null condition, which ultimately leads to small-data shock formation.

1.2.1. The energy method and local well-posedness. For quasilinear hyperbolic equations, a fundamental difference between one and more than one spatial dimension is that in the latter case, one cannot rely exclusively on the method of characteristics. Even basic local well-posedness[11] results are based on the availability of a priori L^2-based energy estimates; there are no known approaches that avoid them. The use of the energy method to prove local well-posedness has a long history with a huge number of contributors. In Sect. 1.2.1, we review basic local well-posedness theory for the well-known class of symmetric hyperbolic systems, with an emphasis on the role of the energy method. As we describe in Sect. 1.2.2, wave equations can be viewed as a special case of symmetric hyperbolic systems for which a *much larger* class of energy estimates is available. Although in the present section we discuss only symmetric hyperbolic systems in detail, we briefly mention

[11]By "well-posedness", we mean existence, uniqueness, and continuous dependence on the initial conditions.

several notable classes of equations (with some overlap) to which the energy method has been applied.

- In the case of quasilinear wave equations, relevant for the present monograph, Schauder [**73**] applied energy methods to linearized versions of the equations and employed a fixed point argument to prove local well-posedness. This is the earliest known example of this kind of argument, which is now standard.
- For *strictly hyperbolic* first-order systems, which by definition have characteristic speeds that are strictly separated, Petrovskii [**68**] discovered a coercive energy identity based on the Fourier transform and used it to prove local well-posedness. Gårding [**24, 25**], using ideas from Leray's work [**56**] (see the next item), extended the well-posedness result to higher-order strictly hyperbolic scalar equations.
- Leray developed the theory of *Leray hyperbolicity* [**56**], which allows for operators having principal parts of different orders within the same system.
- As we have mentioned, there is a well-posedness theory for the well-known *symmetric hyperbolic systems*, which are not generally strictly hyperbolic. The general theory for such systems was developed by Friedrichs [**23**]. We review this theory later in this section.
- Christodoulou introduced [**16**] *regularly hyperbolic* PDEs, which are a class of well-posed Euler-Lagrange equations for maps between a domain manifold $\mathcal{M}$ and a target manifold $\mathcal{N}$. He developed a geometric energy method framework for such PDEs that extends the multiplier method described in Sect. 1.2.2 and leads to a large family of energy estimates. The full theory takes into account the structure of both $\mathcal{M}$ and $\mathcal{N}$.

We now review the basic theory of local well-posedness for symmetric hyperbolic systems. We start by providing their definition.

DEFINITION 1.2 (**Symmetric hyperbolic systems**). A PDE system[12]

$$A_I^{J\alpha}(U)\partial_\alpha U_J = F_I(U), \qquad (I = 1, 2, \cdots, m) \tag{1.42}$$

on the domain[13] $\mathbb{R}^{1+n}$ in the unknowns $U = (U_1, \cdots, U_m)$ is said to be *symmetric hyperbolic* in an open subset $\mathcal{U} \subset \mathbb{R}^m$ if for each fixed $U \in \mathcal{U}$, we have the symmetry property

$$A_I^{J\alpha}(U) = A_J^{I\alpha}(U), \qquad (I, J = 1, 2, \cdots, m), \ (\alpha = 0, 1, \cdots, n), \tag{1.43}$$

and furthermore, there exists a one-form $\xi = (\xi_0, \xi_1, \cdots, \xi_n)$ on $\mathbb{R}^{1+n}$ (generally varying from point to point) such that

$$\text{the } m \times m \text{ matrices } \xi_\alpha A_I^{J\alpha}(U), (I, J = 1, \cdots, m), \text{ are positive definite.} \tag{1.44}$$

REMARK 1.3 (**Symmetrizable systems**). The theory of symmetric hyperbolic systems can also be extended to apply to symmetrizable systems, which are

[12]Throughout the remainder of the monograph, we use Einstein's summation convention. Lowercase Greek "spacetime" indices vary over $0, 1, \cdots, n$ and lowercase Latin "spatial" indices vary over $1, 2, \cdots, n$. Moreover, starting in Chapter 2, we have $n = 3$. In Sect. 1.2.1, capital Latin indices correspond to the target and vary over $1, 2, \cdots, m$.

[13]Throughout Sect. 1.2.1, $(x^0 = t, x^1, \cdots, x^n)$ are standard rectangular coordinates on $\mathbb{R}^{1+n}$ and $\partial_\alpha = \dfrac{\partial}{\partial x^\alpha}$ are the corresponding coordinate partial derivatives.

systems that can be transformed into symmetric hyperbolic form; see, for example, [**22**]. In many cases, the transformation procedure involves a nonlinear state-space change of variables of the form $U \to f(U)$, where $f : \mathbb{R}^m \to \mathbb{R}^m$.

The main step in proving well-posedness for the system (1.42) is to obtain Sobolev estimates for solutions to the following linearized version of (1.42):

$$A_I^{J\alpha}(\widetilde{U})\partial_\alpha U_J = F_I(\widetilde{U}), \qquad (I = 1, 2, \cdots, m). \tag{1.45}$$

In (1.45), $\widetilde{U} : \mathbb{R}^{1+n} \to \mathcal{U} \subset \mathbb{R}^m$ is a background function[14] and U is the unknown. The Sobolev estimates follow from an energy identity for solutions to (1.45), which we now explain. In the ensuing discussion, we abbreviate

$$|U|_E := \sqrt{(E^{-1})^{IK} U_I U_K},$$

$$\langle A^\alpha(\widetilde{U})U, U\rangle_E := (E^{-1})^{IK} A_I^{J\alpha}(\widetilde{U}) U_J U_K,$$

and

$$\left\langle \left\{\partial_\alpha \left(A^\alpha(\widetilde{U})\right)\right\} U, U\right\rangle_E := (E^{-1})^{IK} \left\{\partial_\alpha \left(A_I^{J\alpha}(\widetilde{U})\right)\right\} U_J U_K,$$

where $E_{IJ} = \mathrm{diag}(1, 1, \cdots, 1)$ is the standard Euclidean metric on $\mathbb{R}^m$. To derive the energy identities, we use a framework that is related to a robust, more powerful geometric framework for deriving generalized energy estimates for wave equations. The geometric framework plays a critical role in this monograph, and we provide an overview of it in Sect. 1.2.2. The main idea behind deriving the energy identities is to apply the divergence theorem in a suitable spacetime region with the help of the *compatible energy current* vectorfield $J^\alpha_{(Energy)}[U]$ on $\mathbb{R}^{1+n}$ defined by

$$J^\alpha_{(Energy)}[U] := \langle A^\alpha(\widetilde{U})U, U\rangle_E, \qquad (\alpha = 0, 1, \cdots, n). \tag{1.46}$$

To proceed, we use (1.43) and (1.45) to compute that

$$\partial_\alpha J^\alpha_{(Energy)}[U] = \left\langle \left\{\partial_\alpha A^\alpha(\widetilde{U})\right\} U, U\right\rangle_E + 2\langle U, F(\widetilde{U})\rangle_E. \tag{1.47}$$

We can obtain coercive energy identities in spacetime regions whose boundary is the union of "spacelike" hypersurfaces, which are hypersurfaces with co-normals ξ that verify the analog of (1.44) for the linear system (1.45). We now explain a particular case, sufficient for deducing local well-posedness on a small spacetime patch. We consider spacetime subsets foliated by the level sets of a scalar-valued *time function* τ. That is, we assume that there is a C^1 scalar-valued function τ defined on a subset of $\mathbb{R}^{1+n}$ such that the level sets

$$H_\lambda := \{p \in \mathbb{R}^{1+n} \mid \tau(p) = \lambda\} \tag{1.48}$$

have the property that $\xi_\alpha A_I^{J\alpha}(\widetilde{U})$ is positive definite at every point in H_λ, where ξ is the one-form that is (Euclidean) metric-dual[15] to the future-directed[16] Euclidean-unit normal vectorfield along H_λ. Here, by Euclidean metric, we mean the standard one on $\mathbb{R}^{1+n}$ defined by $e_{\alpha\beta} = \mathrm{diag}(1, 1, \cdots, 1)$. For convenience, we assume here

---

[14] In a proof of well-posedness via an iteration scheme, the role of the background function is played by a known iterate, and the role of U is played by the next iterate, which is to be solved for; see just below Prop. 1.5.

[15] That is, $\xi_\alpha = e_{\alpha\beta} N^\beta$, where N is the future-directed Euclidean unit normal vectorfield to H_λ.

[16] Throughout the monograph, future-directed vectors V are such that $V^0 > 0$, where, relative to the rectangular coordinates, $V = V^\alpha \partial_\alpha$.

that the H_λ are compact and have a common C^1 boundary, denoted by ∂H (see Figure 1). We now derive energy estimates on regions $\mathcal{M}_\lambda \subset \mathbb{R}^{1+n}$ defined by (see Figure 1)

$$\mathcal{M}_\lambda := \cup_{\lambda' \in [0,\lambda]} H_{\lambda'}. \tag{1.49}$$

The most geometric way of formulating the divergence theorem in the present context involves the well-known *co-area formula* (see, for example, [**53**]), which states that for functions f, we have the following integral identity on the region (1.49):

$$\int_{\mathcal{M}_\lambda} f \, d\varpi_e = \int_{\lambda'=0}^{\lambda} \int_{H_{\lambda'}} f |d\tau|_e^{-1} \, d\sigma_{\lambda'} \, d\lambda'. \tag{1.50}$$

In (1.50), $d\varpi_e$ is the spacetime volume form[17] corresponding to the Euclidean metric $e_{\alpha\beta}$ and $d\sigma_\lambda$ is the volume form corresponding to the Riemannian metric ${}^{(\lambda)}h_{\alpha\beta}$ induced on[18] H_λ by $e_{\alpha\beta}$. Moreover, $d\tau = (\partial_t \tau, \partial_1 \tau, \cdots, \partial_n \tau)$ is the spacetime gradient one-form of τ and $|d\tau|_e = \sqrt{(e^{-1})^{\alpha\beta}\partial_\alpha \tau \partial_\beta \tau}$ is its Euclidean norm.

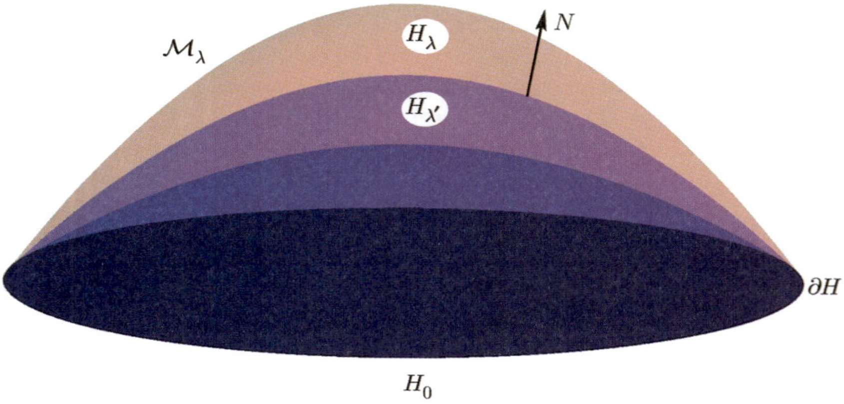

FIGURE 1. Foliation of the spacetime region $\mathcal{M}_\lambda$ by $H_{\lambda'}$, $\lambda' \in [0,\lambda]$. N denotes the Euclidean-unit normal to $H_{\lambda'}$, which is dual to ξ via the Euclidean metric.

The fundamental energy corresponding to the linear system (1.45) and the hypersurface H_λ is

$$\mathbb{E}(\lambda) := \int_{H_\lambda} \langle \xi_\alpha A^\alpha(\widetilde{U})U, U\rangle_E \, d\sigma_\lambda = \int_{H_\lambda} \xi_\alpha J^\alpha_{(Energy)}[U] \, d\sigma_\lambda, \tag{1.51}$$

where $J^\alpha_{(Energy)}[U]$ is defined in (1.46). In view of our above assumptions on H_λ, we see that there exists a constant $C > 1$ depending on $\widetilde{U}$ such that

$$C^{-1} \int_{H_\lambda} |U|_E^2 |d\tau|_e^{-1} \, d\sigma_\lambda \leq \mathbb{E}(\lambda) \leq C \int_{H_\lambda} |U|_E^2 |d\tau|_e^{-1} \, d\sigma_\lambda. \tag{1.52}$$

---

[17] Relative to the standard rectangular coordinates on $\mathbb{R}^{1+n}$, we have $d\varpi_e = dx^0 \cdots dx^n$.

[18] We have ${}^{(\lambda)}h_{\alpha\beta} = e_{\alpha\beta} - \xi_\alpha \xi_\beta$. Furthermore, relative to rectangular coordinates on $\mathbb{R}^{1+n}$, we have $\xi_\alpha = \dfrac{\partial_\alpha \tau}{|d\tau|_e}$.

Applying the divergence theorem with the current (1.46) over the region $\mathcal{M}_\lambda$ and using the identity (1.47), we obtain the following *energy identity*:

(1.53)

$$\mathbb{E}(\lambda) = \mathbb{E}(0) + \int_{\lambda'=0}^{\lambda} \int_{H_{\lambda'}} \left\{ \left\langle \left(\partial_\alpha A^\alpha(\widetilde{U})\right) U, U \right\rangle_E + 2\langle U, F(\widetilde{U})\rangle_E \right\} |d\tau|_e^{-1} \, d\sigma_{\lambda'} \, d\lambda'.$$

We remark that in our proof of shock formation for solutions to wave equations, we rely on intricate geometric energy identities in the spirit of (1.53); see Prop. 10.13.

Using (1.52), (1.53), and the elementary inequality $ab \leq \frac{1}{2}(a^2+b^2)$, we see that there exists a constant $C > 1$ depending on $\widetilde{U}$ and its C^1 norm such that

$$\mathbb{E}(\lambda) \leq \mathbb{E}(0) + C \int_{\lambda'=0}^{\lambda} \mathbb{E}(\lambda') \, d\lambda' + C \int_{\lambda'=0}^{\lambda} \int_{H_{\lambda'}} F^2(\widetilde{U}) |d\tau|_e^{-1} \, d\sigma_{\lambda'} \, d\lambda'. \tag{1.54}$$

From (1.54) and Gronwall's inequality, we deduce the fundamentally important a priori estimate

$$\mathbb{E}(\lambda) \leq \left\{ \mathbb{E}(0) + C \int_{\lambda'=0}^{\lambda} \int_{H_{\lambda'}} F^2(\widetilde{U}) |d\tau|_e^{-1} \, d\sigma_{\lambda'} \, d\lambda' \right\} \exp(C\lambda). \tag{1.55}$$

As we describe below, the estimate (1.55) is the main ingredient in the proof of local well-posedness for symmetric hyperbolic systems. In the next proposition, we provide a basic local well-posedness result. For convenience, we restrict our attention to the case in which the one-form ξ from Def. 1.2 has the simple form $\xi = (1, \mathbf{0}_{1\times n})$ relative to rectangular coordinates and hence $H_\lambda = \Sigma_\lambda$, where in the remaining discussion,

$$\Sigma_t := \left\{ (s, x^1, x^2, \cdots, x^n) \in \mathbb{R}^{1+n} \mid s = t \right\} \simeq \mathbb{R}^n \tag{1.56}$$

denotes the flat hypersurface in $\mathbb{R}^{1+n}$ of constant time t.

REMARK 1.4 (**The spaces $H_e^s(\Sigma_t)$ and $L_e^2(\Sigma_t)$**). Throughout the monograph, we use the notation $H_e^s(\Sigma_t)$ to denote the standard Sobolev space of order s corresponding to rectangular spatial coordinate partial derivatives along Σ_t, where the volume form inherent in the corresponding norm $\|\cdot\|_{H_e^s(\Sigma_t)}$ is the one induced by the standard Euclidean metric e on Σ_t; see Sect. B.17. A similar remark applies to the Lebesgue space $L_e^2(\Sigma_t)$ and the corresponding norm $\|\cdot\|_{L_e^2(\Sigma_t)}$. This distinction is important because later in the monograph (see Sect. 3.19), we define another norm, denoted by $\|\cdot\|_{L^2(\Sigma_t)}$, where the volume form that we use in defining it is, near a shock singularity, drastically different than the one induced by e.

PROPOSITION 1.5 (**Local well-posedness and continuation criteria for symmetric hyperbolic systems**). *Assume that the system (1.42) is symmetric hyperbolic on the open set $\mathcal{U} \subset \mathbb{R}^m$ and has vanishing inhomogeneous term*[19] *$F_I(U) \equiv 0$. Let $s > n/2+1$ and let $\mathring{U} \in H_e^s(\Sigma_0)$ be initial data. Assume that $\mathring{U}(\Sigma_0)$ is contained in a compact subset $\mathfrak{K}$ of interior($\mathcal{U}$) such that for $U \in \mathfrak{K}$, the $m \times m$ matrix $A_J^{I0}(U)$, $(I, J = 1, \cdots, m)$, is positive definite. Then there exists a $T > 0$,*

[19]The proposition can easily be adapted so as to apply when $F_I(U) \not\equiv 0$, under suitable assumptions on $F_I(U)$.

depending only on $\|\mathring{U}\|_{H^s_e}(\Sigma_0)$ *and* $\mathfrak{K}$*, such that these data launch a unique classical solution* U *existing on the slab* $[0,T)\times\mathbb{R}^n$ *which has the regularity property*

$$U \in C([0,T), H^s_e(\Sigma_t)). \tag{1.57}$$

The solution depends continuously on the data. Moreover, there is a compact set $\mathfrak{K}' \subset$ *interior*$(\mathcal{U})$ *with* $\mathfrak{K}\subset$ *interior*$(\mathfrak{K}')$ *such that* $U([0,T)\times\mathbb{R}^n)\subset\mathfrak{K}'$*. Furthermore,* U *verifies the following estimate for* $t\in[0,T)$*:*

$$\begin{aligned}\|U\|_{H^s_e(\Sigma_t)} \leq f\left(\left\|\mathring{U}\right\|_{H^s_e(\Sigma_0)};\mathfrak{K}';s\right)\\ \times \exp\left(C_{\mathfrak{K}';s}\int_{t'=0}^{t}\sum_{\alpha=0}^{n}\|\partial_\alpha U\|_{L^\infty(\Sigma_{t'})}\,dt'\right),\end{aligned} \tag{1.58}$$

where f *is a function that is continuous and increasing in its first argument and that vanishes when* $\mathring{U}=0$*. Moreover, there exists a maximal slab* $[0,T_{(Lifespan)})\times\mathbb{R}^n$ *on which the above properties, including the regularity property* (1.57)*, hold. Finally, if* $T_{(Lifespan)}<\infty$*, then either* $\sum_{\alpha=0}^{n}\int_{t'=0}^{T_{(Lifespan)}}\|\partial_\alpha U\|_{L^\infty(\Sigma_{t'})}\,dt'=\infty$ *or* $U([0,T)\times\mathbb{R}^n)$ *escapes every compact subset of interior*$(\mathcal{U})$ *as* $T\uparrow T_{(Lifespan)}$*.*

The proof of the proposition is standard and most of the details can be found, for example, in [**30**, Chapter VI]. The main idea is to define iterates[20] $\{U^{(k)}\}_{k=0}^{\infty}$ where $U^{(0)} := \mathring{U}$ and $U^{(k+1)}$ is determined in terms of $U^{(k)}$ by solving the linearized system $A_I^{J\alpha}(U^{(k)})\partial_\alpha(U_J^{(k+1)}-\mathring{U}) = -A_I^{Ja}(U_J^{(k)})\partial_a\mathring{U}$ with initial data $U_J^{(k+1)}|_{\Sigma_0}-\mathring{U}=0$. One then derives energy estimates along the lines of (1.55) to deduce, with the help of the standard Sobolev calculus, uniform estimates for[21] $\{\sup_{t\in[0,T)}\{\|U^{(k)}-\mathring{U}\|_{H^s_e(\Sigma_t)}\}_{k=0}^{\infty}$ whenever T is sufficiently small. In particular, standard Sobolev embedding implies that $\sum_{\alpha=0}^{n}\|\partial_\alpha U\|_{L^\infty(\Sigma_{t'})}\leq C\|U\|_{H^s_e(\Sigma_{t'})}$, which provides control over the argument of exp on the right-hand side of (1.58). Next, one can derive similar uniform estimates based on the previous estimates and use them to show that the sum $\sum_{k=0}^{\infty}\sup_{t\in[0,T)}\left\|U^{(k+1)}-U^{(k)}\right\|_{L^2_e(\Sigma_t)}$ is finite, which implies the convergence of the iterates in L^2. One can then show that the limit U is a classical solution and obtain additional information, including estimates on its derivatives up to order s. The proofs are based on similar arguments and standard results from functional analysis.

1.2.2. Square integral-type dispersion for wave equations based on generalized energy estimates. A major aspect permeating the study of quasilinear hyperbolic PDEs in higher dimensions is that dispersion is sometimes present, and it can cause the solution to decay (at least initially) and thus delay or prevent the formation of singularities. In the opposite direction, depending on the structure of the equations, it is possible that other kinds of singularities besides shocks might develop. For example, a major open problem is whether or not vorticity blow-up occurs in the compressible Euler equations. Although Sideris gave, in the case of three spatial dimensions, a proof by contradiction that an open set of initial data generate solutions that blow-up [**74**], we have reason to suspect that his result is

[20] Alternatively, one could formulate a similar proof based on the contraction mapping principle.

[21] In a rigorous proof, one has to smooth the data during this step to compensate for the loss of one derivative of the data in the term on the right-hand side of the linearized system.

detecting shocks rather than vorticity blow-up; below we elaborate on this suspicion. Specifically, Sideris showed that under suitable convexity assumptions on the fluid equation of state, small-data blow-up occurs; see also [**26**] for related large-data blow-up results for the relativistic Euler equations. Sideris' proof was based on using virial identity arguments to indirectly rule out the possibility of global smooth solutions. That is, he constructed spatially averaged quantities that, on the one hand are smooth when the solution is smooth, but on the other hand must blow up in finite time as a consequence of some differential inequalities verified by solutions. In particular, Sideris' proof did not reveal the blow-up mechanism. It is important to note that Sideris' condition for blow-up involves an integral condition on the data that resembles the one given by Christodoulou in [**17**] for the full vorticity-containing relativistic Euler equations in the small-data regime. The key point is that Christodoulou showed [**17**, Chapter 14] that his condition leads to the development of a large irrotational region in which a shock singularity forms; see also [**21**] for a similar result in the case of the irrotational nonrelativistic compressible Euler equations. For these reasons, we speculate that Siderian blow-up is caused by the same mechanism.

Dispersive estimates play an especially important role in the study of solutions to nonlinear wave equations in[22] two or more spatial dimensions. In particular, the issue of global existence versus finite-time blow-up for solutions is often decided by a competition between the dispersive nature of the corresponding linear problem and the strength of the nonlinearities.[23] In the shock forming wave equation solutions (in three spatial dimensions) that we study later in this monograph, the dispersion just barely fails to suppress the singularity formation. However, an important and perhaps surprising fact is that dispersion, both L^2-type and pointwise, plays a key role in our proof of the shock formation. In fact, relative to a dynamic system of coordinates related to the eikonal function u (see (1.3a)) used in our study of Burgers' equation, the shock forming solutions exhibit dispersive behavior at the lower derivative levels, all the way up to and including the shock; see Sect. 2.6 for an overview. Because the dispersive behavior is so important, we dedicate Sects. 1.2.2 and 1.2.3 to providing some general background information on wave dispersion. In this section, we focus on L^2-type dispersion.

We begin by motivating some of the quasilinear wave equations that we study in detail starting in Chapter 2. To avoid the difficult problem, mentioned above, of analyzing vorticity in solutions to the compressible Euler equations, one can study irrotational solutions. As we mentioned in Footnote 4 on pg. 2, for such solutions, the compressible Euler equations reduce to single quasilinear wave equation of Euler-Lagrange type for the scalar-valued fluid potential. In Sect. 2.11.2, we provide additional details on the structure of the wave equation in the case of the relativistic Euler equations. In the present section, rather than studying the equations of irrotational fluid mechanics, we instead study a closely related (as described in Appendix A) but seemingly more geometric family of covariant quasilinear wave equations of the form

$$\square_{g(\Psi)}\Psi = \mathcal{N}(\Psi, \partial\Psi) \tag{1.59}$$

---

[22] In one spatial dimension, the linear wave equation $-\partial_t^2\Phi + \partial_x^2\Phi = 0$ is effectively a transport equation whose solutions do not disperse.

[23] That is, often the nonlinear terms do not enhance the decay rates corresponding to the linear problem.

on the domain $\mathbb{R}^{1+n}$. In equation (1.59), $\square_g := (g^{-1})^{\alpha\beta}\mathscr{D}^2_{\alpha\beta}$ is the covariant wave operator[24] corresponding to g, $\mathscr{D}$ is the Levi-Civita connection corresponding to g, and $\mathcal{N}(\Psi, \partial\Psi)$ is a nonlinearity consisting of quadratic and higher-order terms. The notation $g = g(\Psi)$ means that relative to standard rectangular coordinates $(x^0 = t, x^1, x^2, \cdots, x^n)$ on $\mathbb{R}^{1+n}$, the components $g_{\mu\nu}(\Psi)$ are given smooth functions of Ψ. We assume that there is an open set $\mathcal{H}$ (of "hyperbolicity") such that for $\Psi \in \mathcal{H}$, the metric $g(\Psi)$ is Lorentzian[25] and, for convenience, that relative to the rectangular coordinates, that[26] $g_{00}(\Psi) < 0$.

Note that $\square_{g(\Psi)}\Psi$, when expanded relative to the rectangular coordinates, involves both quasilinear and semilinear terms. We remark that under certain assumptions that we describe starting in Chapter 2, when $n = 3$, the main shock formation results of this monograph apply to solutions of (1.59).

Before discussing L^2-type dispersion, we first discuss local well-posedness for equation (1.59). One can prove a basic well-posedness result for equation (1.59) by introducing the "new" independent variables $\partial_\nu\Psi := U_\nu$, $(\nu = 0, 1, \cdots, n)$, and then reformulating the equation as a (first-order) symmetric hyperbolic system in Ψ and U; one can then apply Prop. 1.5, which leads to the following proposition; see also Prop. 21.1, which provides a detailed version of local well-posedness that is relevant for the main results of the monograph.

PROPOSITION 1.6 (**Local well-posedness and continuation criteria for quasilinear wave equations of type** (1.59)). *Let $s > n/2 + 1$ and let $\mathcal{H} \subset \mathbb{R}$ be the set mentioned just below equation* (1.59). *Let $(\mathring{\Psi}, \mathring{\Psi}_0) = (\Psi|_{\Sigma_0}, \partial_t\Psi|_{\Sigma_0}) \in H_e^{s+1}(\Sigma_0) \times H_e^s(\Sigma_0)$ be initial data for the wave equation* (1.59) *and assume that there is a compact subset $\mathfrak{K} \subset$ interior$(\mathcal{H})$ such that $\Psi(\Sigma_0) \subset \mathfrak{K}$. Then there exists a $T > 0$, depending only on $\sum_{a=1}^n \|\partial_a\mathring{\Psi}_0\|_{H_e^s(\Sigma_0)} + \|\mathring{\Psi}_0\|_{H_e^s(\Sigma_0)}$ and $\mathfrak{K}$, such that these data launch a unique classical solution existing on the slab $[0,T) \times \mathbb{R}^n$ which has the regularity property*

$$\Psi \in C([0,T), H_e^{s+1}(\Sigma_t)), \qquad \partial_t\Psi \in C([0,T), H_e^s(\Sigma_t)). \tag{1.60}$$

The solution depends continuously on the data. Moreover, there is a compact set $\mathfrak{K}' \subset$ interior$(\mathcal{H})$ with $\mathfrak{K} \subset$ interior$(\mathfrak{K}')$ such that $\Psi([0,T)\times\mathbb{R}^n) \subset \mathfrak{K}'$. Furthermore, Ψ verifies the following estimate for[27] *$t \in [0,T)$:*

$$\sum_{\alpha=0}^{n} \|\partial_\alpha\Psi\|_{H_e^s(\Sigma_t)} \leq f\left(\sum_{a=1}^{n} \left\|\partial_a\mathring{\Psi}\right\|_{H_e^s(\Sigma_0)} + \left\|\mathring{\Psi}_0\right\|_{H_e^s(\Sigma_0)}; \mathfrak{K}'; s\right) \times \exp\left(C_{\mathfrak{K}';s}\int_{t'=0}^{t}\sum_{\alpha=0}^{n}\|\partial_\alpha\Psi\|_{L^\infty(\Sigma_{t'})}\,dt'\right), \tag{1.61}$$

where f is a function that is continuous and increasing in its first argument and that vanishes when the data are trivial. Moreover, there exists a maximal slab

---

[24] Relative to an arbitrary coordinate system, $\square_g\Psi = \frac{1}{\sqrt{|\det g|}}\partial_\alpha(\sqrt{|\det g|}(g^{-1})^{\alpha\beta}\partial_\beta\Psi)$.

[25] By Lorentzian, we mean that relative to arbitrary coordinates, $g_{\alpha\beta}$ is an $(1+n)\times(1+n)$ matrix of signature $(-,+,+,\cdots,+)$.

[26] The inequality $g_{00}(\Psi) < 0$ is equivalent to the timelike character of the vectorfield $\partial_t = \frac{\partial}{\partial x^0}$.

[27] An estimate for $\|\Psi\|_{L_e^2(\Sigma_t)}$ can be obtained by integrating the estimate (1.61) for $\|\partial_t\Psi\|_{L_e^2(\Sigma_t)}$ with respect to time.

$[0, T_{(Lifespan)}) \times \mathbb{R}^n$ *on which the above properties, including the regularity property* (1.60), *hold. Finally, if* $T_{(Lifespan)} < \infty$, *then either*

$$\sum_{\alpha=0}^{n} \int_{t'=0}^{T_{(Lifespan)}} \|\partial_\alpha \Psi\|_{L^\infty(\Sigma_{t'})} \, dt' = \infty$$

or $\Psi([0,T) \times \mathbb{R}^n)$ *escapes every compact subset of interior*$(\mathcal{H})$ *as* $T \uparrow T_{(Lifespan)}$.

As we described below Prop. 1.5, the standard proof of Prop. 1.6 relies on using Sobolev embedding to estimate the norm in the argument of exp in equation (1.61) as follows:

$$\sum_{\alpha=0}^{n} \|\partial_\alpha \Psi\|_{L^\infty(\Sigma_{t'})} \leq C \sum_{\alpha=0}^{n} \|\partial_\alpha \Psi\|_{H_e^s(\Sigma_{t'})}.$$

It is important to note that in some cases, the Sobolev exponent s from Prop. 1.6 has been markedly improved by using refined tools such as Strichartz and bilinear estimates to replace the above crude Sobolev estimate. The most advanced result along these lines is the recent proof of the bounded L^2 curvature conjecture [**52**], which for the Einstein-vacuum equations of general relativity with $n = 3$ essentially leads to local well-posedness for $(\mathring{\Psi}, \mathring{\Psi}_0) \in H_e^2(\Sigma_0) \times H_e^1(\Sigma_0)$. This remarkable work is a major extension of earlier work [**51**] of Klainerman-Rodnianski, which proved local well-posedness for the Einstein-vacuum equations with $n = 3$ relative to wave coordinates[28] for data verifying, for any $\epsilon > 0$, $(\mathring{\Psi}, \mathring{\Psi}_0) \in H_e^{2+\epsilon}(\Sigma_0) \times H_e^{1+\epsilon}(\Sigma_0)$. Their work was extended by Smith-Tataru [**75**], who showed that when $n \in \{3, 4, 5\}$, quasilinear wave equations of the form

$$(g^{-1})^{\alpha\beta}(\Psi)\partial_\alpha\partial_\beta\Psi = \mathcal{N}^{\alpha\beta}(\Psi)\partial_\alpha\Psi\partial_\beta\Psi$$

are locally well-posed for data verifying, for any $\epsilon > 0$, $(\mathring{\Psi}, \mathring{\Psi}_0) \in H_e^{(n+1)/2+\epsilon}(\Sigma_0) \times H_e^{(n-1)/2+\epsilon}(\Sigma_0)$; see also [**79**] for an alternate proof in the case $n = 3$. It is also important to note that Lindblad showed [**57**] that when $n = 3$, it is not possible to further lower the Sobolev exponent without additional structure on the nonlinearities (such as the structure present in the case of the Einstein-vacuum equations).

The symmetric hyperbolic framework, which we outlined in Sect. 1.2.1, is generally not well-suited for proving results going beyond local well-posedness. For wave equations, one can derive significantly more sophisticated L^2-type estimates based on the *vectorfield multiplier* and *vectorfield commutator methods*, which are collectively known as the *vectorfield method*. The family of *generalized energy estimates* afforded by these methods is much larger than the family of energy estimates afforded by the symmetric hyperbolic framework in the sense that one has great freedom in the choice of multipliers, differential operator commutators, and surfaces of integration. Both methods are important for deriving the dispersive properties of waves and play a key role in our proofs of the main results of this monograph. In the present section, we focus on the multiplier method. We first review the basic framework, which is applicable to general nonlinear wave equations. For illustration, we then show how the multiplier method can be used to derive L^2-type dispersive estimates for solutions to the linear wave equation. In Sect. 1.2.3, we focus on the commutator method which, when combined with the multiplier method,

[28]In wave coordinates, the Einstein-vacuum equations are equivalent to a system of wave equations of the form $(g^{-1})^{\alpha\beta}(\Psi)\partial_\alpha\partial_\beta\Psi^I = \mathcal{N}^{\alpha\beta}(\Psi)\partial_\alpha\Psi\partial_\beta\Psi^I$, where $\Psi = \{\Psi^I\}$ is an array consisting of the components of the metric.

can be used to derive refined pointwise decay estimates exhibiting the directionally dependent dispersive properties of linear waves.

We now outline the main ideas behind the multiplier method. We consider the nonlinear wave equation (1.59) because of its relevance for the main results of this monograph; see Chapter 10 for a detailed version of the method for this equation in the context of the problem of shock formation. At the heart of the approach lies the *energy-momentum tensorfield* $Q[\Psi]$, a type $\binom{0}{2}$ tensorfield on $\mathbb{R}^{1+n}$ defined by (see Footnote 12 on pg. 12 regarding our index conventions)

$$Q_{\mu\nu}[\Psi] := \mathscr{D}_\mu \Psi \mathscr{D}_\nu \Psi - \frac{1}{2} g_{\mu\nu} (g^{-1})^{\alpha\beta} \mathscr{D}_\alpha \Psi \mathscr{D}_\beta \Psi. \tag{1.62}$$

The quantity $Q[\Psi]$ has the following well-known key properties (the first is easy to prove while for the second, readers may consult [**18**] for the main ideas behind the proof):

(1) For solutions to (1.59), we have[29]

$$\mathscr{D}_\alpha Q^\alpha{}_\nu[\Psi] = \mathscr{N}(\Psi, \partial\Psi) \mathscr{D}_\nu \Psi. \tag{1.63}$$

(2) For future-directed (see Footnote 16 on pg. 13), causal[30] vectors V and W, we have the *dominant energy condition*

$$Q_{\alpha\beta}[\Psi] V^\alpha W^\beta \geq 0. \tag{1.64}$$

As we will now explain, the properties (1.63) and (1.64), when combined with the divergence theorem, allow one to derive a large family of coercive L^2-type energy estimates. This is the multiplier method in action.

For bookkeeping purposes in the divergence theorem, it is convenient to introduce the following *compatible current* vectorfield ${}^{(V)}J^\nu[\Psi]$, which depends on an auxiliary *multiplier vectorfield* V:

$${}^{(V)}J^\nu[\Psi] := Q^\nu{}_\alpha[\Psi] V^\alpha. \tag{1.65}$$

Using (1.63) and the symmetry of $Q[\Psi]$, we compute that for *solutions* to (1.59), we have the following identity:

$$\mathscr{D}_\alpha {}^{(V)}J^\alpha[\Psi] = \frac{1}{2} Q^{\alpha\beta}[\Psi] {}^{(V)}\pi_{\alpha\beta} + \mathscr{N}(\Psi, \partial\Psi) V\Psi. \tag{1.66}$$

In (1.66), $V\Psi = V^\alpha \partial_\alpha \Psi$ denotes the V directional derivative of Ψ (see Footnote 10 on pg. 8),

$${}^{(V)}\pi_{\mu\nu} := \mathcal{L}_V g_{\mu\nu} = \mathscr{D}_\mu V_\nu + \mathscr{D}_\nu V_\mu \tag{1.67}$$

denotes the deformation tensor[31] of V, and $\mathcal{L}_V$ denotes Lie differentiation with respect to V (see Def. 3.55).

---

[29] Here and throughout the remainder of the monograph, we lower and raise indices with g and g^{-1}.

[30] By definition, causal vectors V are such that $g(V, V) := g_{\alpha\beta} V^\alpha V^\beta \leq 0$.

[31] The second equality in (1.67) is a consequence of the torsion-free property of the connection $\mathscr{D}$.

By integrating the identity (1.66) over a suitable spacetime region $\mathcal{M} \subset \mathbb{R}^{1+n}$ bounded by spacelike[32] and/or null[33] hypersurfaces and applying the divergence theorem, we obtain an energy identity, in analogy with (1.53). By (1.65), the corresponding integrals along the bounding hypersurfaces appearing in the divergence theorem feature the integrands $Q_{\alpha\beta}[\Psi]V^\alpha N^\beta$, where N is the unit normal[34] vectorfield N along the bounding hypersurface. Hence, by (1.64) and the fact that $Q[\Psi]$ is quadratic in the derivatives of Ψ, we infer that *the integrals along the bounding hypersurfaces are coercive in the derivatives of* Ψ whenever V and N are both future-directed and causal. It is because the integrals are coercive that we refer such identities as *generalized energy identities* and the resulting estimates as *generalized energy estimates.* In contrast, *the theory of symmetric hyperbolic systems provides (up to scalar function multiples) only one compatible current that leads to a coercive energy identity,* namely the compatible energy current (1.46).

REMARK 1.7 (**Connection between** $^{(V)}\pi$ **and Noether's theorem**). Note that by the identity (1.66) and the divergence theorem argument described above, the tensorfield $^{(V)}\pi_{\mu\nu}$ is connected to the availability (or not) of *conservation laws* for solutions to the wave equation. In particular, if the right-hand side of (1.59) vanishes and if V is a Killing field[35] of g, then (1.66) implies that $\mathscr{D}_\alpha {}^{(V)}J^\alpha[\Psi] = 0$. In particular, there is a conserved quantity associated to the current (1.65); see below for some simple but important examples. This phenomenon may be viewed as a geometric version of *Noether's theorem*, tailored to wave equations. We note that most metrics g do not admit any Killing fields. Thus, generally speaking, in order to prove a global result for a nonlinear problem, one must construct vectorfields V that not only yield coercive energies (upon integrating the corresponding compatible current over a suitable domain), but that also are such that one can control the error integral generated by the first term[36] on the right-hand side of (1.66).

For the purpose of illustration, we now derive some coercive energy identities for solutions to the standard linear wave equation[37]

$$\square_m \Psi = 0, \tag{1.68}$$

where m is the Minkowski metric on $\mathbb{R}^{1+n}$. The simplest nontrivial example occurs when $V = \partial_t$ (relative to standard rectangular coordinates) and $\mathcal{M} = \cup_{t'=0}^{t} \Sigma_{t'}$ (see (1.56) for the definition of Σ_t). It is straightforward to compute that $^{(\partial_t)}\pi = 0$, $N = \partial_t$, and $^{(\partial_t)}J^\alpha[\Psi]N_\alpha = \frac{1}{2}\sum_{\alpha=0}^{n}(\partial_\alpha \Psi)^2$. The resulting energy identity yields the

[32] A spacelike hypersurface Σ is such that at each point, the future-directed unit normal N is timelike (that is, it has a negative length as measured by g). In particular, we have $g(N, N) = -1$.

[33] A *null hypersurface* has normal vectors that are tangent to the hypersurface and that have length 0 as measured by g. A particular kind of null hypersurface, namely null cones, forms the lateral boundaries of the integration region that we use in our proof of shock formation for wave equations in three spatial dimensions; see Prop. 10.13.

[34] The phrase "unit normal" is not accurate if the bounding hypersurface is null, since in this case any normal is null (that is, it has length 0 as measured by g).

[35] By definition, Killing fields of g are vectorfields such that $^{(V)}\pi = 0$.

[36] In some cases, this first term contains a piece with a favorable sign that can be used to control other error terms. This is the case in the shock formation problem that we study later in the monograph; see the discussion surrounding equation (2.53).

[37] Relative to standard rectangular coordinates $\{x^\alpha\}_{\alpha=0,1,\cdots,n}$ on $\mathbb{R}^{1+n}$, $\square_m = (m^{-1})^{\alpha\beta}\partial_\alpha\partial_\beta$, where $(m^{-1})^{\alpha\beta} = \mathrm{diag}(-1, 1, 1\cdots, 1)$ is the inverse Minkowski metric.

well-known conservation of the L^2 norm of the gradient:

$$\sum_{\alpha=0}^{n} \int_{\Sigma_t} (\partial_\alpha \Psi)^2 \, d^n x = \sum_{\alpha=0}^{n} \int_{\Sigma_0} (\partial_\alpha \Psi)^2 \, d^n x, \tag{1.69}$$

where $d^n x := dx^1 \cdots dx^n$.

Although it is of fundamental importance, the energy identity (1.69) does not contain any information about the dispersive behavior of linear waves. A much more refined L^2-type estimate, exhibiting some aspects of wave dispersion, can be obtained by replacing the multiplier ∂_t with the Minkowskian *Morawetz multiplier* K, defined by

$$K := (t^2 + r^2)\partial_t + 2tr\partial_r = \frac{1}{2}(t+r)^2 L_{(Flat)} + \frac{1}{2}(t-r)^2 \underline{L}_{(Flat)}. \tag{1.70}$$

In (1.70),

$$L_{(Flat)} := \partial_t + \partial_r, \qquad \underline{L}_{(Flat)} := \partial_t - \partial_r \tag{1.71}$$

are the standard (future) outgoing/ingoing radial Minkowski-null[38] vectorfields. Above and throughout,

$$r = \sqrt{\sum_{a=1}^{n} (x^a)^2} \tag{1.72}$$

is the standard Euclidean radial coordinate on $\mathbb{R}^n$ and

$$\partial_r = \frac{x^a}{r} \partial_a \tag{1.73}$$

is the standard Euclidean radial vectorfield. The vectorfield K is a conformal Killing field of m, which means that its deformation tensor is a scalar function multiple of m. More precisely, simple calculations yield (see (1.67)) that

$${}^{(K)}\pi_{\mu\nu} = 4t m_{\mu\nu}. \tag{1.74}$$

As a consequence of (1.66) and (1.74), we have that[39] $\mathscr{D}_\alpha {}^{(K)}J^\alpha[\Psi] \neq 0$, which, in light of Remark 1.7, means that ${}^{(K)}J$ does not directly lead to a conservation law. However, as is described in [**48**], we can modify ${}^{(K)}J[\Psi]$ by adding correction terms to create a divergence-free vectorfield. Specifically, we define (see Footnote 29 on pg. 20 regarding the notation)

$${}^{(K+Correction)}J^\nu[\Psi] := {}^{(K)}J^\nu[\Psi] + (n-1)t\Psi\partial^\nu\Psi - \frac{n-1}{2}\Psi^2 \partial^\nu t, \tag{1.75}$$

and straightforward calculations based on (1.66) and (1.74) yield that for solutions to (1.68), we have

$$\mathscr{D}_\alpha {}^{(K+Correction)}J^\alpha[\Psi] = 0. \tag{1.76}$$

[38]By Minkowski-null, we mean that $m(L_{(Flat)}, L_{(Flat)}) = m(\underline{L}_{(Flat)}, \underline{L}_{(Flat)}) = 0$, where $m(V,W) := m_{\alpha\beta}V^\alpha W^\beta$.

[39]Note that when the metric is equal to the Minkowski metric m, the operator $\mathscr{D}_\alpha$ agrees, relative to rectangular coordinates, with the standard partial derivative operator ∂_α.

As in (1.69), we can integrate the identity (1.76) over the spacetime domain $[0,t] \times \mathbb{R}^n$ and invoke the divergence theorem to obtain the following conservation law (again with $N = \partial_t$):

$$\int_{\Sigma_t} {}^{(K+Correction)}J^{\alpha}[\Psi] N_{\alpha}\, d^n x = \int_{\Sigma_0} {}^{(K+Correction)}J^{\alpha}[\Psi] N_{\alpha}\, d^n x. \tag{1.77}$$

The identity (1.77) is useful only if the integrals $\int_{\Sigma_t} \cdots$ are coercive. As we now describe, they are in fact coercive in a rather strong sense when $n \geq 3$ due to the weights inherent in the definition (1.70) of K. The main estimate of interest to us is that when $n \geq 3$, there is a constant $C > 0$ depending on n such that

$$\begin{aligned}
&\int_{\Sigma_t} {}^{(K+Correction)}J^{\alpha}[\Psi] N_{\alpha}\, d^n x \\
&\geq \frac{1}{C}\int_{\Sigma_t} (t+r)^2 (L_{(Flat)}\Psi)^2\, d^n x + \frac{1}{C}\int_{\Sigma_t} (t-r)^2 (\underline{L}_{(Flat)}\Psi)^2\, d^n x \\
&\quad + \frac{1}{2}\int_{\Sigma_t} (t^2+r^2)|\nabla\!\!\!/_{\not{h}}\Psi|^2\, d^n x + \frac{1}{C}\int_{\Sigma_t} \frac{t^2+r^2}{r^2}\Psi^2\, d^n x.
\end{aligned} \tag{1.78}$$

In (1.78), $\nabla\!\!\!/_{\not{h}}$ is the Levi-Civita connection of $\not{h}$, the Riemannian metric on the Euclidean spheres of constant t and r induced by the Minkowski metric m. Thus, $|\nabla\!\!\!/_{\not{h}}\Psi|$ is the size[40] of the angular derivatives of Ψ. The proof of (1.78) is more involved than the proof of (1.69) and requires some new ingredients. Below we will provide a proof using arguments given in [**48**]. Equation (1.77) and inequality (1.78) together exhibit some important L^2-type dispersive properties of linear waves showing that, for example, the gradient of Ψ cannot remain concentrated (in the $L^2(\mathbb{R}^n)$ sense) in any compact set of $\mathbb{R}^n$ as $t \to \infty$. It also shows that $L_{(Flat)}\Psi$ and $\nabla\!\!\!/_{\not{h}}\Psi$ enjoy stronger decay properties than $\underline{L}_{(Flat)}\Psi$. Estimates in this vein play an important role in many nonlinear problems, including our analysis of shock forming waves.[41]

To prove (1.78), we first use (1.62), (1.65), (1.75), and the fact that $N = \partial_t$ along Σ_t to compute that (see the proof of Lemma 10.7 for inspiration on how to carry out the computations)

$$\begin{aligned}
&{}^{(K+Correction)}J^{\alpha}[\Psi] N_{\alpha} \\
&= \frac{1}{4}(t+r)^2(L_{(Flat)}\Psi)^2 + \frac{1}{4}(t-r)^2(\underline{L}_{(Flat)}\Psi)^2 + \frac{1}{2}(t^2+r^2)|\nabla\!\!\!/_{\not{h}}\Psi|^2 \\
&\quad + (n-1)t\Psi\partial_t\Psi - \frac{n-1}{2}\Psi^2.
\end{aligned} \tag{1.79}$$

[40] The implicit metric in the norm $|\nabla\!\!\!/_{\not{h}}\Psi|$ is $\not{h}$.

[41] Our analysis in the shock formation problem for wave equations takes place in a region trapped in between two outgoing null cones. Because the width of the region is not too large, it turns out that in order to close our estimates, we can rely on a simplified Morawetz multiplier that does not involve an analog of the term $\frac{1}{2}(t-r)^2\underline{L}_{(Flat)}$ from the definition (1.70); see (10.11b). However, our analog of the identity (1.77) also involves square integrals along the outgoing cones that provide new coercive information going beyond that provided by the Σ_t integrals; see Prop. 10.13 and Lemma 14.1. These additional "cone fluxes" play a critical role in closing the energy estimates in the shock formation problem.

It is not immediately apparent from (1.79) that $\int_{\Sigma_t} {}^{(K+Correction)}J^{\alpha}[\Psi]N_{\alpha}\, d^n x$ is coercive in the sense of (1.78). Our proof of (1.78) relies on the following two identities for the product $t\Psi\partial_t\Psi$, which are easy to verify (see Footnote 12 on pg. 12 regarding our summation convention):

(1.80a) $$t\Psi\partial_t\Psi = \Psi(t\partial_t\Psi + r\partial_r\Psi) + \frac{n}{2}\Psi^2 - \frac{1}{2}\partial_a\left\{x^a\Psi^2\right\},$$

(1.80b) $$t\Psi\partial_t\Psi = \frac{t}{r}\Psi(r\partial_t\Psi + t\partial_r\Psi) + \frac{n-2}{2}\frac{t^2}{r^2}\Psi^2 - \frac{1}{2}\partial_a\left\{t^2\frac{x^a}{r^2}\Psi^2\right\},$$

as well as the identity

(1.81) $$\begin{aligned}&\frac{1}{4}(t+r)^2(L_{(Flat)}\Psi)^2 + \frac{1}{4}(t-r)^2(\underline{L}_{(Flat)}\Psi)^2\\ &= \frac{1}{2}(t\partial_t\Psi + r\partial_r\Psi)^2 + \frac{1}{2}(r\partial_t\Psi + t\partial_r\Psi)^2.\end{aligned}$$

To proceed, we decompose the factor $n-1$ from (1.79) as

$$n - 1 = \alpha + \beta,$$

where we will choose the positive constants α and β below. We then rewrite the product $(n-1)t\Psi\partial_t\Psi$ from (1.79) as $\alpha t\Psi\partial_t\Psi + \beta t\Psi\partial_t\Psi$, and we express $\alpha t\Psi\partial_t\Psi$ as α times the right-hand side of (1.80a) and $\beta t\Psi\partial_t\Psi$ as β times the right-hand side of (1.80a). Next, integrating by parts over Σ_t on the terms $\alpha t\Psi\partial_t\Psi$ and $\beta t\Psi\partial_t\Psi$ and using (1.80a), (1.80b), and (1.81), we derive the following identity:

(1.82) $$\begin{aligned}&\int_{\Sigma_t} {}^{(K+Correction)}J^{\alpha}[\Psi]N_{\alpha}\, d^n x\\ &= \frac{1}{2}\int_{\Sigma_t}(t\partial_t\Psi + r\partial_r\Psi)^2 + 2\alpha\Psi(t\partial_t\Psi + r\partial_r\Psi) + \{\alpha n - (n-1)\}\Psi^2\, d^n x\\ &+ \frac{1}{2}\int_{\Sigma_t}(r\partial_t\Psi + t\partial_r\Psi)^2 + 2\beta\frac{t}{r}\Psi(r\partial_t\Psi + t\partial_r\Psi) + \beta(n-2)\frac{t^2}{r^2}\Psi^2\, d^n x\\ &+ \frac{1}{2}\int_{\Sigma_t}(t^2+r^2)|\nabla\!\!\!/_{m\!\!\!/}\Psi|^2\, d^n x.\end{aligned}$$

We now choose

(1.83) $$\alpha := \frac{n}{2}, \qquad \beta := \frac{n-2}{2}$$

in (1.82). It is straightforward to check that under this choice, when $n \geq 3$, the cross terms $2\alpha\cdots$ and $2\beta\cdots$ in the first two integrands on the right-hand side of (1.82) are dominated by the positive definite terms $(t\partial_t\Psi + r\partial_r\Psi)^2$, $(r\partial_t\Psi + t\partial_r\Psi)^2$, and Ψ^2. In view of the identity (1.81), it follows that when $n \geq 3$, there exists a $C > 0$ such that the desired estimate (1.78) holds.

1.2.3. Pointwise dispersion in solutions to wave equations. In this section, we complement the L^2-type dispersive estimates derived in Sect. 1.2.2, which were based on the multiplier method, with pointwise dispersive (decay) estimates that rely on the multiplier and commutator methods. As is well known [**48**], solutions to the linear wave equation $\square_m\Psi = 0$ in $\mathbb{R}^{1+n}$ with initial data having

Fourier support[42] contained in the dyadic shell $\{\xi \in \mathbb{R}^n \mid \frac{1}{2} \leq |\xi| \leq 2\}$ verify the *basic dispersive estimate*[43]

$$(1.84) \qquad |\Psi(t,x)| \leq C(1+t)^{\frac{1-n}{2}} \sum_{\alpha=0}^{n} \|\partial_\alpha \Psi\|_{W_e^{\frac{n-1}{2},1}(\Sigma_t)}, \qquad (t,x) \in [0,\infty) \times \mathbb{R}^n.$$

The standard proof of (1.84) is based on the method of stationary phase applied to the Fourier representation of the solution. Although the estimate (1.84) is useful in linear theory, there are two ways in which it is inadequate for studying the kinds of quasilinear equations of interest to us in this monograph. The first is that L^1-type norms are not suitable, for in quasilinear problems, the only kinds of regularity propagated at top-order are L^2-based. The second is that (1.84) fails to capture more refined, directionally-dependent dispersive behavior, which often plays a critical role in the study of solutions to nonlinear equations. In particular, as we describe in Sect. 2.6, the refined behavior plays an important role in our proof of shock formation.

Fortunately, Klainerman developed [**46**] the *commutator method*, a robust L^2-type vectorfield approach to deriving refined pointwise dispersive estimates. To explain his approach, which is connected to the one that we use throughout this monograph, we first introduce a *Minkowskian null (vectorfield) frame*

$$(1.85) \qquad \{L_{(Flat)} := \partial_t + \partial_r,\ \underline{L}_{(Flat)} := \partial_t - \partial_r,\ X_{1;(Flat)}, \cdots, X_{n-1;(Flat)}\},$$

where $\cup_{a=1}^{n-1}\{X_{a;(Flat)}\}$ is a locally defined vectorfield frame spanning the tangent space of the $n-1$-dimensional Euclidean spheres $\{t = \text{const}\} \cap \{r = \text{const}\}$ whenever $r > 0$. In its most basic form, the commutator method is based on commuting the wave equation with a subset of Killing and conformal Killing fields[44] of the Minkowski metric. A convenient set of commutators is the $\mathscr{Z}_{(Flat)}$, whose elements can be expressed relative to rectangular coordinates as follows:

$$(1.86)$$
$$\mathscr{Z}_{(Flat)} := \cup_{\alpha=0}^{n}\{\partial_\alpha\} \bigcup \cup_{1 \leq i < j \leq n}\{x^i \partial_j - x^j \partial_i\} \bigcup \cup_{i=1}^{n}\{x^i \partial_t + t\partial_i\} \cup \{x^\alpha \partial_\alpha\}.$$

In (1.86), the ∂_α are the translations, the $x^i\partial_j - x^j\partial_i$ are the Euclidean rotations, the $x^i\partial_t + t\partial_i$ are the Lorentz boosts, and $x^\alpha\partial_\alpha$ is the scaling vectorfield. All vectorfields in $\mathscr{Z}_{(Flat)}$ are Killing except for the scaling vectorfield, which is conformal Killing and verifies $\mathcal{L}_{x^\alpha \partial_\alpha} m = {}^{(x^\alpha \partial_\alpha)}\pi = 2m$.

---

[42]The restriction on the Fourier support of the data can be removed by replacing the Euclidean Sobolev norm $\|\cdot\|_{W_e^{\frac{n-1}{2},1}(\Sigma_t)}$ on the right-hand side of (1.84) with an appropriate $L^1(\mathbb{R}^n)$-type Besov norm of the data.

[43]We note that $\|f\|_{W_e^{\frac{n-1}{2},1}(\Sigma_t)} := \sum_{|\vec{I}| \leq \frac{n-1}{2}} \left\|\partial^{\vec{I}} f\right\|_{L_e^1(\Sigma_t)}$, where $\vec{I}$ denotes a multi-index corresponding to repeated differentiation with respect to the rectangular spatial coordinate vectorfields ∂_i.

[44]We recall that Killing fields are vectorfields whose deformation tensors (1.67) vanish while conformal Killing fields have deformation tensors that are a scalar-valued function times the Minkowski metric.

A simple calculation reveals that a general vectorfield V has the following commutation property with the Minkowskian wave operator $\square_m$:

$$(1.87)\qquad [\square_m, V]\Psi = (\mathscr{D}_\alpha {}^{(V)}\pi^{\alpha\beta})\mathscr{D}_\beta\Psi - \frac{1}{2}(\mathscr{D}^\alpha \mathrm{tr}_m {}^{(V)}\pi)\mathscr{D}_\alpha\Psi + {}^{(V)}\pi^{\alpha\beta}\mathscr{D}^2_{\alpha\beta}\Psi,$$

where $[P, Q]$ denotes the commutator of P and Q and $\mathrm{tr}_m {}^{(V)}\pi = (m^{-1})^{\alpha\beta(V)}\pi_{\alpha\beta}$ is the trace of ${}^{(V)}\pi$ (see (1.67) for the definition of ${}^{(V)}\pi$). An important consequence of (1.87) is that the vectorfields in $\mathscr{Z}_{(Flat)}$ enjoy the following good commutation properties with $\square_m$:

$$(1.88a)\qquad [\square_m, Z_{(Flat)}] = 0, \qquad \text{if } Z_{(Flat)} \in \mathscr{Z}_{(Flat)}\backslash\{x^\alpha\partial_\alpha\},$$

$$(1.88b)\qquad [\square_m, x^\alpha\partial_\alpha] = 2\square_m.$$

In particular, if

$$(1.89)\qquad \square_m\Psi = 0,$$

then $\square_m \mathscr{Z}^{\vec{I}}_{(Flat)}\Psi = 0$ for any multi-index $\mathscr{Z}^{\vec{I}}_{(Flat)}$ of vectorfield operators constructed out of the elements of $\mathscr{Z}_{(Flat)}$. Thus, for any integer $N \geq 0$, we deduce from (1.69) that solutions to (1.89) verify (see Remark 1.4 regarding the notation for the norms)

$$(1.90)\qquad \sum_{|\vec{I}|\leq N}\sum_{\alpha=0}^{n}\left\|\partial_\alpha \mathscr{Z}^{\vec{I}}_{(Flat)}\Psi\right\|_{L^2_e(\Sigma_t)} = \sum_{|\vec{I}|\leq N}\sum_{\alpha=0}^{n}\left\|\partial_\alpha \mathscr{Z}^{\vec{I}}_{(Flat)}\Psi\right\|_{L^2_e(\Sigma_0)},$$

where the right-hand side of (1.90) is determined by the data $(\Psi|_{\Sigma_0}, \partial_t\Psi|_{\Sigma_0})$.

Some important pointwise dispersive properties of linear waves now follow immediately from (1.90) and the well-known Klainerman-Sobolev inequality,[45] the first version of which was proved in [**46**], while the version (1.91) is found in [**76**]:

(1.91)

$$(1 + |t + r|)^{\frac{n-1}{2}}(1 + |t - r|)^{1/2}\sum_{\alpha=0}^{n}|\partial_\alpha\Psi(t, x)| \leq C\sum_{|\vec{I}|\leq\frac{n+2}{2}}\sum_{\alpha=0}^{n}\left\|\partial_\alpha \mathscr{Z}^{\vec{I}}_{(Flat)}\Psi\right\|_{L^2_e(\Sigma_t)}.$$

Moreover, the estimate (1.91) also holds with $\partial_\alpha\Psi$ replaced by Ψ on the left and $\partial_\alpha \mathscr{Z}^{\vec{I}}_{(Flat)}\Psi$ replaced by $\mathscr{Z}^{\vec{I}}_{(Flat)}\Psi$ on the right. The proof of (1.91) is based on modifying the standard proof of Sobolev embedding to take advantage of the weights inherent in the definitions of the Euclidean rotations, Lorentz boosts, and the scaling vectorfield. Moreover, as a corollary, in the region $\{t \geq 0\}$, we have the following decay estimates for the derivatives of $\partial_\alpha\Psi$ with respect to the null frame vectorfields

[45]Inequality (1.91) is also known as the global Sobolev inequality.

(see [**76**] for a proof):

$$(1.92a)\qquad (1+|t+r|)^{\frac{n+1}{2}}(1+|t-r|)^{1/2}\sum_{\alpha=0}^{n}\left|L_{(Flat)}\partial_\alpha\Psi(t,x)\right|$$
$$\leq C\sum_{|\vec{I}|\leq\frac{n+4}{2}}\sum_{\alpha=0}^{n}\left\|\partial_\alpha\mathscr{Z}_{(Flat)}^{\vec{I}}\Psi\right\|_{L_e^2(\Sigma_t)},$$

$$(1.92b)\qquad (1+|t+r|)^{\frac{n+1}{2}}(1+|t-r|)^{1/2}\sum_{\alpha=0}^{n}\sum_{a=1}^{n-1}\left|X_{a;(Flat)}\partial_\alpha\Psi(t,x)\right|$$
$$\leq C\sum_{|\vec{I}|\leq\frac{n+4}{2}}\sum_{\alpha=0}^{n}\left\|\partial_\alpha\mathscr{Z}_{(Flat)}^{\vec{I}}\Psi\right\|_{L_e^2(\Sigma_t)},$$

$$(1.92c)\qquad (1+|t+r|)^{\frac{n-1}{2}}(1+|t-r|)^{3/2}\sum_{\alpha=0}^{n}\left|\underline{L}_{(Flat)}\partial_\alpha\Psi(t,x)\right|$$
$$\leq C\sum_{|\vec{I}|\leq\frac{n+4}{2}}\sum_{\alpha=0}^{n}\left\|\partial_\alpha\mathscr{Z}_{(Flat)}^{\vec{I}}\Psi\right\|_{L_e^2(\Sigma_t)}.$$

Note that (1.92a) and (1.92b) imply that the derivatives of $\partial_\alpha\Psi$ with respect to the vectorfields $\{L_{(Flat)}, X_{1;(Flat)}, \cdots, X_{n-1;(Flat)},\}$, which span the tangent space of the outgoing Minkowski null cones $\{t-r=\text{const}\}$ when $r>0$, enjoy the extra decay rate factor of $(1+t+r)^{-1}$ relative to the Klainerman-Sobolev inequality (1.91). In contrast, $\underline{L}_{(Flat)}$ is transversal to the outgoing Minkowski null cones, and $\underline{L}_{(Flat)}\partial_\alpha\Psi$ gains only the decay rate factor $(1+|t-r|)^{-1}$ relative to the Klainerman-Sobolev inequality. The proofs of (1.92a)-(1.92c) are based on the Klainerman-Sobolev inequality (1.91) together with algebraic identities involving weighted combinations of the vectorfields in $\mathscr{Z}_{(Flat)}$. For example, the derivations of (1.92a) and (1.92c) rely in part on the identities

$$(1.93a)\qquad (t+r)L_{(Flat)}=x^\alpha\partial_\alpha+\frac{x^a}{r}\{x^a\partial_t+t\partial_a\},$$

$$(1.93b)\qquad (t-r)\underline{L}_{(Flat)}=x^\alpha\partial_\alpha-\frac{x^a}{r}\{x^a\partial_t+t\partial_a\}.$$

1.2.4. Almost global existence and global existence via the classic null condition. In Sects. 1.2.2 and 1.2.3, we studied some of the dispersive properties of linear waves. In the present section, we discuss the effect that dispersion has on small-data solutions to nonlinear wave equations, with an emphasis on the case of three spatial dimensions. For convenience, we restrict our attention to Cauchy problems of the form

$$(1.94a)\qquad (g^{-1})^{\alpha\beta}(\Phi,\partial\Phi)\partial_\alpha\partial_\beta\Phi=\mathscr{N}(\Phi,\partial\Phi),$$

$$(1.94b)\qquad (\Phi|_{t=0},\partial_t\Phi|_{t=0})=(\mathring{\Phi},\mathring{\Phi}_0).$$

We assume that relative to rectangular coordinates $(x^0=t,x^1,\cdots,x^n)$ on $\mathbb{R}^{1+n}$,

$$(1.95)\qquad g_{\alpha\beta}(0,0)=m_{\alpha\beta},$$

$$(1.96)\qquad \mathscr{N}(0,0)=\frac{\partial\mathscr{N}}{\partial\Phi}(0,0)=\frac{\partial\mathscr{N}}{\partial(\partial\Phi)}(0,0)=0,$$

where $m_{\alpha\beta} = \mathrm{diag}(-1, 1, 1, \cdots, 1)$ is the Minkowski metric. Note that (1.95)-(1.96) imply that equation (1.94a) is a quadratic perturbation of the linear wave equation. We also note that the local well-posedness proposition (Prop. 1.6) can be extended to apply to equation (1.94a) with one notable change: the Sobolev exponent must be increased to $s > n/2 + 2$ because of the dependence of g on the first derivatives of Φ in (1.94a).

A primary feature of the solution to the nonlinear problem (1.94a)-(1.94b) is that when the data are small, the linear dispersive effects described in Sects. 1.2.2 and 1.2.3 initially dominate the dynamics. Consequently, the solution decays for a long time, even if a singularity eventually forms. The first results capturing the effect of dispersion on the solution's lifespan in three spatial dimensions were proved by John and Klainerman [**34, 44**]. Under some mild assumptions on the structure of the nonlinearities, they used dispersive estimates to show that if the data and a certain number their derivatives are of small size $\mathring{\epsilon}$ relative to a suitable weighted Sobolev norm,[46] then the classical lifespan of the solution is at least $\mathcal{O}(\exp(c\mathring{\epsilon}^{-1}))$. This kind of long-time existence result is often referred to as *almost global existence*. The shock forming solutions that we study in this monograph in fact form their first singularity around the time $\mathcal{O}(\exp(c\mathring{\epsilon}^{-1}))$. Klainerman later simplified and extended these results [**46**] using the aforementioned vectorfield commutator and multiplier methods. His proof of almost global existence is similar to the proof of Prop. 1.6, but with some important new features. It is based on commuting the wave equation (1.94a) with the vectorfields in $\mathscr{Z}_{(Flat)}$ defined in (1.86), exploiting the commutation properties (1.88a)-(1.88b), deriving energy estimates using the multiplier ∂_t (much as in (1.69)) in order to control the quantity on the left-hand side of (1.90), and using the Klainerman-Sobolev inequality (1.91) to bound the factor

$$\exp\left(\int_{t'=0}^{t} \sum_{\alpha=0}^{n} \|\partial_\alpha \Psi\|_{L^\infty(\Sigma_{t'})}\, dt'\right) \tag{1.97}$$

from (1.61), which also appears in the energy estimates of [**46**].

In four spatial dimensions, the pointwise dispersive decay provided by the Klainerman-Sobolev inequality (1.91) suggests that the factor (1.97) is uniformly bounded, which in turn suggest small-data global existence. In fact, (1.91) is strong enough to yield small-data global existence for all quadratic nonlinearities except those of the form[47] Φ^2. In five or more spatial dimensions, the inequality (1.91) is strong enough to yield small-data global existence for all quadratic nonlinearities.

In three spatial dimensions, the question of which quadratic nonlinearities allow for small-data global existence is much more delicate. Klainerman identified an important class of nonlinearities with a now-famous structural property allowing for small-data global existence. He called this property the (classic) *null condition* [**45**] (see Footnote 22 on pg. xix regarding the terminology). Distinct proofs of

---

[46]Roughly, the norms are equivalent to the right-hand side of (1.91) at $t = 0$.

[47]The main obstacle in handling nonlinearities of the form Φ^2 is that energy estimates in the spirit of (1.69) for the nonlinear equation do not directly provide bounds for the quantities $\|\mathscr{Z}_{(Flat)}^{\vec{I}}\Phi\|_{L^2_e(\Sigma_t)}$, which are needed to control solutions to the $\mathscr{Z}_{(Flat)}^{\vec{I}}$-commuted equation. One can derive estimates for these terms by integrating estimates for $\|\partial_t \mathscr{Z}_{(Flat)}^{\vec{I}}\Phi\|_{L^2_e(\Sigma_t)}$ in time, but in four spatial dimensions, the time integration results in a t-dependent factor that is large enough to spoil the proof of global existence.

small-data global existence for nonlinearities verifying the null condition were given by Klainerman [47] and Christodoulou [15]. There are several equivalent ways to describe it. Our description here is different than Klainerman's original one and is instead closely tied to the decay rates for linear waves outlined in Sect. 1.2.2. We remark that although the null condition can be formulated for systems of equations (see [47]), we restrict our attention here to the scalar wave equation (1.94a).

Specifically, the null condition is verified by the nonlinearities in equation (1.94a) if one Taylor expands them around **0** to quadratic order, decomposes all derivatives relative to the null frame (1.85), and finds that the following terms are *absent*: **i)** $(\underline{L}_{(Flat)}\Phi)^2$, $(L_{(Flat)}\Phi)^2$; **ii)** $\Phi\underline{L}_{(Flat)}\underline{L}_{(Flat)}\Phi$, $\Phi L_{(Flat)}L_{(Flat)}\Phi$. The canonical example of a quadratic term verifying Klainerman's null condition is $(m^{-1})^{\alpha\beta}\partial_\alpha\Phi\partial_\beta\Phi$, where m is the Minkowski metric; see Sect 2.4 for a proof of this fact. The main point is that when these terms are absent, the only kinds of quadratic terms that are present have at least one factor that, thanks to the estimates (1.92a)-(1.92c) and their higher-order analogs, are expected to decay faster than $(1+t)^{-1}$. For clarity, we note that the products $(\underline{L}_{(Flat)}\Phi)^2$ and $\Phi\underline{L}_{(Flat)}\underline{L}_{(Flat)}\Phi$ are possible obstructions to global existence in the region $\{t \geq 0\}$, while the products $(L_{(Flat)}\Phi)^2$ and $\Phi L_{(Flat)}L_{(Flat)}\Phi$ are possible obstructions to global existence in the region $\{t \leq 0\}$.

In three spatial dimensions, a glaring question is: what is the global behavior of small-data solutions to equation (1.94a) when the null condition fails? Note that the Klainerman-Sobolev inequality (1.91) suggests that when $n = 3$ and the data are small, the factor

$$\sum_{\alpha=0}^{n} \int_{t'=0}^{t} \|\partial_\alpha \Psi\|_{L^\infty(\Sigma_{t'})}\, dt'$$

from (1.61) is logarithmically divergent as $t \to \infty$. One suspects that in the absence of special structure such as the null condition, this divergence can lead to the formation of a singularity. This suspicion is in fact borne out in many cases: for many scalar wave equations failing the null condition, including the ones that we study in this monograph, singularities eventually form. At the beginning of Chapter 2, we survey the blow-up results that are known when the classic null condition fails and place the sharp small-data shock formation results of the present monograph in context.

CHAPTER 2

Overview of the Two Main Theorems

In [**17**], Christodoulou proved a breakthrough result giving a detailed description of the formation of shocks in small-data solutions to the relativistic Euler equations in three spatial dimensions. As we describe in Sect. 2.11.2, his main results concerned irrotational regions, where the relativistic Euler equations are equivalent to a scalar quasilinear wave equation that fails the classic null condition; see Sects. 1.2.4 and 2.4 for an overview of this material. In this monograph, we develop an extended version of Christodoulou's framework that yields sharp information about the global behavior of solutions to two important classes of quasilinear wave equations (described just below) in three spatial dimensions. In particular, we recover the most important aspects of Christodoulou's results as a special case. In this chapter, we overview our results and provide sketches of the most important parts of the proofs; the detailed proofs occupy the rest of the monograph. Our results apply to data belonging to H_e^N (we describe this function space in Remark 1.4) for a sufficiently large integer[1] N and they show that for equations belonging to the two classes, initially small, regular solutions often[2] develop shock singularities[3] in finite time.

We assume that the equations are quadratic perturbations of the standard linear wave equation $\square_m \Psi = 0$ on Minkowski spacetime $(\mathbb{R}^{1+3}, m)$, where m is the Minkowski metric and the linear wave operator $\square_m$ is given by $\square_m = -\partial_t^2 + \Delta = -\partial_t^2 + \sum_{a=1}^3 \partial_a^2$ relative to standard Minkowski rectangular coordinates $(x^0 := t, x^1, x^2, x^3)$ on $\mathbb{R}^{1+3}$. The two classes of wave equations that we treat are: **i)** covariant wave equations of the form

$$\square_{g(\Psi)} \Psi = 0,$$

where $\square_{g(\Psi)}$ is the covariant wave operator of the Lorentzian metric $g(\Psi)$ (see Footnote 24 on pg. 18 for an explicit expression valid relative to arbitrary coordinates), and **ii)** noncovariant wave equations (see Footnote 12 on pg. 12 regarding our summation convention) of the form

$$(h^{-1})^{\alpha\beta}(\partial\Phi)\partial_\alpha\partial_\beta\Phi = 0,$$

where $h(\partial\Phi)$ is a Lorentzian metric and $\partial_\alpha := \dfrac{\partial}{\partial x^\alpha}$ denotes a rectangular coordinate partial derivative. We show below that these two classes of equations are much more intimately related than first appearances suggest. We restrict our attention to initial data that are compactly supported in the Euclidean unit ball centered at the origin in $\Sigma_0 := \{(t, x^1, x^2, x^3) \in \mathbb{R}^4 \mid t = 0\} \simeq \mathbb{R}^3$, and we study the solution only in the region determined by the portion of the data lying in the

[1] As is shown by (2.7), $N = 25$ suffices for the class of covariant wave equations that we treat.

[2] In Remark 2.4, we flesh out the meaning of "often."

[3] We must make suitable assumptions on the nonlinear terms to ensure that shocks form.

exterior of the Euclidean sphere S_{0,U_0} of radius $1 - U_0$ centered at the origin in Σ_0 (see Figure 1 on pg. 37), where $0 < U_0 < 1$ is a parameter. We expect that our work could be extended to apply to noncompactly supported data and to a larger solution region.[4] An extensive account of related work as well as an overview of our proofs are provided in the companion survey article [**28**]. Hence, in the present introduction, we provide a relatively sparse account of the relevant literature and a briefer overview our proofs.

F. John gave a nonconstructive proof [**36**] showing the existence of a large class of equations that are quadratic perturbations of the standard linear wave equation on $\mathbb{R}^{1+3}$ and that exhibit finite-time blow-up for a large set[5] of smooth data. Interestingly, the data were not constrained by a size assumption. His proof applied to some semilinear equations including (relative to rectangular coordinates)

$$\square_m \Phi = -(\partial_t \Phi)^2,$$

to some quasilinear equations including

$$\square_m \Phi = -(\partial_t \Phi)\partial_t^2 \Phi,$$

and to related equations with nonlinearities verifying a signed technical condition. Later work by Alinhac[6] [**2**] and Christodoulou [**17**] (see also the follow-up work [**21**] by Christodoulou-Miao) revealed that for some related classes of quasilinear wave equations having a nontrivial intersection with John's class, for solutions launched by a set of small data, the singularity is caused by the intersection of characteristic hypersurfaces. That is, the singularity is a shock. In Sect. 2.11, we discuss the precise classes of equations and data covered by Alinhac and Christodoulou and the differences in their approaches. Although John's proof [**36**] relied on assumptions regarding the positive definiteness of the nonlinearity, no such assumptions were needed in the proofs of Alinhac or Christodoulou, nor are they needed in the present work. It is interesting to note that as of the present, the nature of the breakdown of solutions to the semilinear equations covered by John's work [**36**] remains poorly understood. Our main objectives in the monograph are the following.

(I) We identify criteria for the structure of the equations and *stable* criteria on small initial data that guarantee that the solution develops a shock singularity in finite time. In particular, our results recover, as special cases, the most important aspects of the shock formation results of Christodoulou and Alinhac. Specifically, for the equations we study, the global behavior of small-data solutions is determined by the presence or absence of certain tensorial components in the wave equations. If the tensorial components are absent, then the nonlinearities verify Klainerman's (classic) null condition [**45**], and the methods of [**47**] and [**15**] yield small-data global existence. If they are present, then the null condition fails, and our work yields shock formation for solutions launched by an open set of small data (see Remark 2.4). Moreover, we show that for general small data,

[4] In particular, we expect that our results could be extended to yield sharp information about the behavior of the solution in the interior of the cone $\mathcal{C}_{U_0}$ depicted in Figure 1.

[5] For some nonlinearities, John's results yield blow-up for *all* nontrivial, smooth, compactly supported data.

[6] Despite the title, Alinhac's article [**2**] addresses both the cases of two and three spatial dimensions.

shocks are the only kind of singularities that can develop in the region of interest.

(II) For the shock forming solutions, we provide a detailed description of the dynamics from $t = 0$ all the way up to and including the constant-time hypersurface subset $\Sigma_{T_{(Lifespan)};U_0}^{U_0}$ (see definition (3.6b) and (22.1)) where the first shock singularity point lies. To provide this description, we extend the groundbreaking framework developed by Christodoulou [**17**], which he used to prove sharp shock formation results for irrotational regions of relativistic fluids. We have also significantly simplified some aspects of his approach, in part by being more selective in our use of geometry. That is, we perform fully geometric decompositions only when deriving the most delicate estimates. In deriving many of the less delicate estimates, we use a simpler, less geometric approach, which has spared us a great deal of analysis.

As does [**17**], our results provide, in particular, a sharp description of the quantities that remain regular up to and including the time of first shock formation. An important virtue of our estimates is that as in [**17**, Chapter 15], they could be extended to give a detailed description of the maximal future development of the portion of the data in the exterior of $S_{0,U_0} \subset \Sigma_0$; see Figure 4 on pg. 72. This is one key advantage of Christodoulou's approach over Alinhac's, which is tailored only to see the first singularity point.[7] One would need a precise description of the solution up to the boundary of the maximal development, not just information near the first singularity point, if one wanted to try to weakly continue[8] the solution under suitable extension criteria.

2.1. First description of the two theorems

Our work is divided into two main theorems.

(1) **Main Theorem of the Monograph: A Sharp Classical Lifespan Theorem.** In Theorem 22.1, we show that for small-data solutions to covariant wave equations of the form $\square_{g(\Psi)}\Psi = 0$, there is a scalar function μ that dictates the global behavior of the solution in the region of interest. The function μ, which we discuss in great detail below, has a simple geometric interpretation: $1/\mu$ measures the foliation density of a family of outgoing null cones (which are characteristic hypersurfaces of the dynamic metric). It is an analog of the quantity (1.11) that we used in our sharp analysis of shock formation for solutions to Burgers' equation (see Sect. 1.1.2). In short, μ starts out near 1 and either it goes to 0 in finite time and a shock forms due to the intersection of characteristic hypersurfaces (that is, due to the blow-up of the foliation density), or it remains positive and, because of dispersive effects, no singularity forms

[7]Hence, Alinhac's results only apply to data for which there is a unique first singularity point.

[8]An important aspect of the weak continuation problem (which, in the Preface, we referred to as the *shock development problem*) is that one must *dynamically construct* the hypersurface across which the solution is discontinuous (the solution is expected be smooth on either side of the hypersurface). That is, the hypersurface of (jump-type) discontinuity is an unknown in the problem.

in the region. Furthermore, the theorem provides many quantitative estimates that are verified up to and including the constant-time hypersurface subset $\Sigma_{T_{(Lifespan)};U_0}^{U_0}$ (see definition (3.6b) and (22.1)) where the first shock singularity point lies. The proof of Theorem 22.1 is based on a long bootstrap argument that occupies the majority of the monograph. Our proof is based on Christodoulou's framework [**17**], but we develop some alternate strategies that significantly shorten and simplify some aspects of the analysis and allow us to treat a larger class of equations. In Appendix (A), we outline how to extend Theorem 22.1 so that it applies to noncovariant wave equations of the form $(h^{-1})^{\alpha\beta}(\partial\Phi)\partial_\alpha\partial_\beta\Phi = 0$.

REMARK 2.1 (**μ^{-1} degeneracy is responsible for the length**). The main reasons that the present work and the work [**17**] are so long are **i)** the high-order L^2 estimates of Theorem 22.1 are rather degenerate with respect to powers of $1/\mu$ (see Sect. 2.10.3 for an overview) and **ii)** an intimately related difficulty: the top-order L^2 estimates are very difficult to derive without "losing derivatives" (see Sect. 2.10.2 for an overview). The degeneracy of the high-order L^2 estimates with respect to powers of $1/\mu$ **is the main new feature** that distinguishes the problem of small-data shock formation from other global results for nonlinear wave equations that are found in the literature.

REMARK 2.2 (**Key innovations of Christodoulou's framework**). The most important innovations of the framework of [**17**] are that it provides a means to establish the degenerate high-order L^2 estimates as well as a means to show that the degeneracy does not propagate down to the lower derivative levels.

REMARK 2.3 (**Strong null condition**). Our work can easily be extended to allow for the presence of additional "harmless" quadratic semilinear terms in the wave equations. For example, a sufficient condition in the region $\{t \geq 0\}$ is that the terms verify the future strong null condition of Def. A.11. Such terms remain small throughout the future evolution and do not interfere with the shock formation processes. In contrast, the results of our theorems are not stable under the addition of general cubic semilinear terms. Roughly, the reason is that our framework is tailored to precisely control the dangerous quadratic interactions, whereas for solutions that blow-up, general cubic terms would dominate the dynamics near the singularity.

(2) **Second Theorem: Shock Formation for Nearly Spherically Symmetric Small Data.** We state this theorem as Theorem 23.9. It is relatively easy to prove thanks to the difficult estimates of Theorem 22.1. Theorem 23.9 shows that for equations of the form $\square_{g(\Psi)}\Psi = 0$ with nonlinearities that fail the classic null condition, there exists an open set of small data that launch solutions such that μ vanishes in finite time, thus yielding the onset of a shock. Specifically, the theorem shows that shocks form in solutions corresponding to small, nontrivial, compactly supported, nearly spherically symmetric data. For technical reasons, the theorem applies to data specified at time $-1/2$ and supported in a Euclidean ball

of radius[9] 1/2. Moreover, with some additional effort, we could extend Theorem 23.9 to show shock formation in solutions corresponding to a significantly larger class of data; see Remark 2.4.

REMARK 2.4 (**Shock formation for small compactly supported data**). In Sect. 2.11.3, we outline how to extend Theorem 23.9 to show finite-time shock formation in solutions corresponding to all sufficiently small compactly supported nontrivial data. More precisely, the discussion in Sect. 2.11.3 leaves open the possibility that the small size of the data that is sufficient for proving the sharp classical lifespan theorem (Theorem 22.1) might not be sufficient for proving shock formation. That is, our proof of shock formation depends on the profile of the data, which we might need to multiply by a small rescaling factor in order to ensure the blow-up.

2.2. The basic structure of the equations

As we mentioned at the beginning, the first class of problems that we study is Cauchy problems for covariant scalar wave equations of the form

$$\square_g \Psi = 0, \tag{2.1a}$$

$$(\Psi|_{t=0}, \partial_t \Psi|_{t=0}) = (\mathring{\Psi}, \mathring{\Psi}_0). \tag{2.1b}$$

Above, $g = g(\Psi)$ is a Lorentzian metric,

$$\square_g := (g^{-1})^{\alpha\beta} \mathscr{D}^2_{\alpha\beta} \tag{2.2}$$

is the covariant wave operator corresponding to g (see Footnote 24 on pg. 18), and $\mathscr{D}$ is the Levi-Civita connection corresponding to g. In order to make more precise statements, we now formally introduce one of the two coordinate systems that we use in our analysis. Specifically, throughout the remainder of the monograph, $\{x^\mu\}_{\mu=0,1,2,3}$ denotes a fixed rectangular coordinate system on Minkowski spacetime $(\mathbb{R}^{1+3}, m)$ relative to which $m_{\mu\nu} = \text{diag}(-1,1,1,1)$. x^0 is the time coordinate and (x^1, x^2, x^3) are the spatial coordinates. Greek "spacetime" indices vary from 0 to 3 and lowercase Latin "spatial" indices vary from 1 to 3. Repeated indices are summed over their respective ranges. The symbol ∂_ν denotes the rectangular coordinate vectorfield $\frac{\partial}{\partial x^\nu}$. We often use the alternate notation $t := x^0$, $\partial_t := \frac{\partial}{\partial x^0}$. Upon rescaling the metric, we may assume[10] that

$$(g^{-1})^{00}(\Psi) = -1, \tag{2.3}$$

which simplifies some of our calculations.

We assume that relative to standard Minkowski-rectangular coordinates, g is a linearly small perturbation of the Minkowski metric m, $(\mu, \nu = 0,1,2,3)$:

$$g_{\mu\nu} = g_{\mu\nu}(\Psi) := m_{\mu\nu} + g^{(Small)}_{\mu\nu}(\Psi), \tag{2.4}$$

---

[9]By making straightforward modifications throughout the monograph, this result can be extended to allow for an arbitrary finite radius of support.

[10]Actually, rescaling the metric leads to the presence of an additional semilinear term on the right-hand side of equation (2.1a). However, this additional term has a very good null structure, in the sense of Lemma A.16. In particular, this term satisfies the "future strong null condition" (see Def. A.11), which is relevant for the region $\{t \geq 0\}$ treated in detail in this monograph. As will become clear, such a good term makes only a tiny contribution to the future dynamics of small-data solutions (even those solutions that form shocks) and hence we ignore it.

where

$$m_{\mu\nu} := \mathrm{diag}(-1, 1, 1, 1), \tag{2.5}$$

and the $g_{\mu\nu}^{(Small)}$ are given smooth functions of Ψ (at least when Ψ is sufficiently small) verifying

$$g_{\mu\nu}^{(Small)}(0) = 0. \tag{2.6}$$

We prove our main theorems under the assumption that the size $\mathring{\epsilon}$ of the data is sufficiently small, where

$$\mathring{\epsilon} = \mathring{\epsilon}[(\mathring{\Psi}, \mathring{\Psi}_0)] := \|\mathring{\Psi}\|_{H_e^{25}(\Sigma_0^1)} + \|\mathring{\Psi}_0\|_{H_e^{24}(\Sigma_0^1)}. \tag{2.7}$$

In (2.7), $\|\cdot\|_{H_e^M(\Sigma_0^1)}$ denotes the standard Euclidean Sobolev norm corresponding to order $\leq M$ rectangular spatial coordinate partial derivatives on the Euclidean unit ball $\Sigma_0^1 \subset \mathbb{R}^3$. The number of derivatives in (2.7), 25, could be modestly reduced with some additional effort. We were motivated to explicitly keep track of the number of derivatives that we use in proving our results because, as we will see, the number is connected to certain crucially important structural features of our equations and our estimates.

REMARK 2.5 ($N_{(Christodoulou)}$). In [**17**], Christodoulou did not provide an explicit estimate for the number of derivatives $N_{(Christodoulou)}$ on the data needed for his main results. In his bootstrap argument, $N_{(Christodoulou)}^{-1}$ was a parameter that he chose to be sufficiently small to counter other large, nonexplicit constants.

For simplicity, we do not study the solution along complete constant-time hypersurfaces. Rather, we fix the parameter $U_0 \in (0, 1)$ and study the future-behavior of the solution in the region evolutionarily determined by the portion of the data lying in the exterior of the sphere S_{0,U_0} of Euclidean radius $1 - U_0$ centered at the origin in $\Sigma_0 := \{(t, x^1, x^2, x^3) \in \mathbb{R}^4 \mid t = 0\} \simeq \mathbb{R}^3$. In the exterior of S_{0,U_0}, the data are nontrivial in the following annular region: $\Sigma_0^{U_0} := \{(0, x^1, x^2, x^3) \in \mathbb{R}^4 \mid 1 - U_0 \leq r \leq 1\}$, where

$$r = \sqrt{\sum_{a=1}^{3} (x^a)^2} \tag{2.8}$$

is the standard Euclidean radial coordinate. In particular, we avoid the origin, where we would encounter mild technical difficulties associated to the vanishing of r. Standard domain of dependence considerations imply that the solution Ψ completely vanishes in the exterior of the flat cone $\mathcal{C}_0 := \{(t,x^1,x^2,x^3) | 1 - r + t = 0\}$. We are interested in the nontrivial portion of the solution, which is trapped between the two outgoing null cones $\mathcal{C}_{U_0}$ and $\mathcal{C}_0$ (see Figure 1); we explain this region in much greater detail below.

In brief, our main goal in this monograph is to show that if the nonlinearities in (2.1a) fail Klainerman's (classic) null condition [**45**] (which we discuss in Sect. 2.4), then nontrivial data of arbitrarily small size verifying an *open* condition launch a solution that forms a shock in finite time. Furthermore, we give a detailed description of the various phases of the dynamics of all small-data solutions, up to and including the time of first shock formation (for those that form shocks). We stress that the covariant wave equation (2.1a) features both quasilinear terms and semilinear terms when it is written relative to rectangular coordinates; see equation

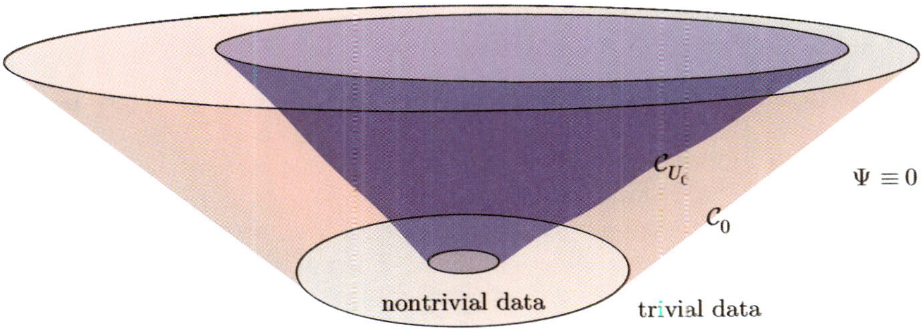

FIGURE 1. The region of interest

(2.17) However, for equation (2.1a), failure of the classic null condition for both the quasilinear terms and the semilinear terms is *determined by a single factor*, a function ${}^{(+)}\aleph$ that we define below in (2.22).

We now briefly discuss the second class of equations that we study, that is, noncovariant equations of the form

$$(h^{-1})^{\alpha\beta}(\partial\Phi)\partial_\alpha\partial_\beta\Phi = 0. \tag{2.9}$$

In analogy with the case of equation (2.1a), we assume that

$$h_{\mu\nu} := m_{\mu\nu} + h_{\mu\nu}^{(Small)}(\partial\Phi), \tag{2.10}$$

where the $h_{\mu\nu}^{(Small)}$ are given smooth functions of[11] $\partial\Phi$ (at least when $\partial\Phi$ is sufficiently small) verifying

$$h_{\mu\nu}^{(Small)}(0) = 0. \tag{2.11}$$

It turns out that the study of solutions to (2.10) can be effectively reduced to the study of equation (2.1a). We provide an outline of the reduction in Appendix A. In short, by differentiating (2.9) with rectangular coordinate derivatives ∂_ν, the question of the long-time behavior of solutions to (2.9) can be transformed into an equivalent question of the long-time behavior of solutions to a coupled system of equations in the unknowns $\Psi_\nu := \partial_\nu\Phi$. The main point is that the system comprises scalar equations that are closely related to equation (2.1a). Furthermore, in the small-data regime, the system turns out to be rather weakly coupled except in a few key aspects. This structure was first observed by Christodoulou in [**17**] for a class of wave equations that derive from a Lagrangian. In this monograph, we show that the structure survives for general wave equations of the form (2.9)-(2.10). This fact is based on the availability of some good (null) structure present in certain semilinear terms (see Lemma A.16), which completely vanished[12] in Christodoulou's work [**17**]. In total, except for a handful of key aspects, the difference between the coupled system and the scalar equation (2.1a) is minimal. Hence, in this monograph, we provide detailed proofs only in the case of the scalar equation (2.1a). We remark that one needs an additional important ingredient to understand the long-time

---

[11]Throughout, $\partial\Phi$ denotes the gradient of Φ with respect to the rectangular spacetime coordinates.

[12]More precisely, in [**17**], Christodoulou rescaled the physical metric by a scalar function multiple in order to eliminate the semilinear terms.

behavior of solutions to (2.9). We flesh out this remark in Sect. A.2 and provide the additional ingredient in Prop. A.20.

REMARK 2.6 (**The number of derivatives needed to treat equation (2.9)**). In view of the discussion below equation (2.7), one might believe that to treat equation (2.9), it suffices to consider small data with one more derivative compared to equation (2.1a), that is, data such that $(\Phi|_{t=0}, \partial_t \Phi|_{t=0}) \in H_e^{26}(\Sigma_0^1) \times H_e^{25}(\Sigma_0^1)$. However the argument that we sketch in Appendix A might be slightly nonoptimal in that it might require some additional differentiability on the data due to increased degeneracy of the top-order energy estimates; see Remark A.17.

2.3. The structure of the equation relative to rectangular coordinates

Relative to the rectangular coordinates introduced in Sect. 2.2, equation (2.1a) can be expressed as

$$(g^{-1})^{\alpha\beta}\partial_\alpha\partial_\beta\Psi - (g^{-1})^{\alpha\beta}(g^{-1})^{\kappa\lambda}\Gamma_{\alpha\kappa\beta}\partial_\lambda\Psi = 0, \tag{2.12}$$

where (recall the decomposition (2.4))

$$\Gamma_{\alpha\kappa\beta} = \Gamma_{\alpha\kappa\beta}(\Psi, \partial\Psi) := \frac{1}{2}\left\{\partial_\alpha g_{\kappa\beta}^{(Small)} + \partial_\beta g_{\alpha\kappa}^{(Small)} - \partial_\kappa g_{\alpha\beta}^{(Small)}\right\} \tag{2.13}$$

are the lowered Christoffel symbols of g relative to the rectangular coordinates. In order to fully clarify the structure of the nonlinearities, we introduce the following smooth functions of Ψ, which play a fundamental role in our analysis.

DEFINITION 2.7 (**Derivatives of the metric component functions with respect to Ψ**). Relative to the rectangular coordinates, we define the smooth functions $G_{\mu\nu}(\cdot)$ as follows:

$$G_{\mu\nu} = G_{\mu\nu}(\Psi) := \frac{d}{d\Psi} g_{\mu\nu}^{(Small)}(\Psi). \tag{2.14}$$

We also define the following smooth functions $G'_{\mu\nu}(\cdot)$, which play a supporting role in our analysis:

$$G'_{\mu\nu} = G'_{\mu\nu}(\Psi) := \frac{d}{d\Psi} G_{\mu\nu}(\Psi). \tag{2.15}$$

Using (2.14), we can express

$$2\Gamma_{\alpha\kappa\beta} = G_{\kappa\beta}\partial_\alpha\Psi + G_{\alpha\kappa}\partial_\beta\Psi - G_{\alpha\beta}\partial_\kappa\Psi \tag{2.16}$$

and hence equation (2.12) can be rewritten as

$$(g^{-1})^{\alpha\beta}\partial_\alpha\partial_\beta\Psi - \frac{1}{2}(g^{-1})^{\alpha\beta}(g^{-1})^{\kappa\lambda}\left\{2G_{\alpha\kappa}\partial_\beta\Psi - G_{\alpha\beta}\partial_\kappa\Psi\right\}\partial_\lambda\Psi = 0. \tag{2.17}$$

2.4. The (classic) null condition

We now explain the connection between the wave equations under study and Klainerman's (classic) null condition,[13] which we first introduced in Sect. 1.2.4. As we described, he used the vectorfield method [**47**] to show that (2.17) has global small-data solutions if the nonlinearities verify the null condition [**45**]; see also [**15**] for Christodoulou's alternate proof, which is based on an approach known as

[13]Klainerman's condition is most often referred to as simply the "null condition." We sometimes refer to it as the "classic null condition" in order to avoid confusing it with the future strong null condition and past strong null condition introduced in Appendix A.

the conformal method. Klainerman's definition [45] of the null condition involves Taylor expanding the nonlinearities around $(\Psi, \partial\Psi, \partial^2\Psi) = 0$ and keeping only the quadratic part. In the case of equation (2.17), we compute that the quadratic terms are, up to constant factors, as follows:

$$G_{\kappa\lambda}(\Psi = 0)(m^{-1})^{\alpha\kappa}(m^{-1})^{\beta\lambda}\Psi\partial_\alpha\partial_\beta\Psi, \tag{2.18a}$$

$$G_{\kappa\lambda}(\Psi = 0)(m^{-1})^{\kappa\lambda}(m^{-1})^{\alpha\beta}\partial_\alpha\Psi\partial_\beta\Psi, \tag{2.18b}$$

$$G_{\kappa\lambda}(\Psi = 0)(m^{-1})^{\alpha\kappa}(m^{-1})^{\beta\lambda}\partial_\alpha\Psi\partial_\beta\Psi. \tag{2.18c}$$

By definition, the quasilinear terms (2.18a) are said to verify the (classic) null condition if and only if for every Minkowski-null covector[14] ξ, we have $G_{\kappa\lambda}(\Psi = 0)(m^{-1})^{\alpha\kappa}(m^{-1})^{\beta\lambda}\xi_\alpha\xi_\beta = 0$. For the semilinear terms (2.18b), the defining condition is $G_{\kappa\lambda}(\Psi = 0)(m^{-1})^{\kappa\lambda}(m^{-1})^{\alpha\beta}\xi_\alpha\xi_\beta = 0$, while for (2.18c) the defining condition is $G_{\kappa\lambda}(\Psi = 0)(m^{-1})^{\alpha\kappa}(m^{-1})^{\beta\lambda}\xi_\alpha\xi_\beta = 0$. It follows that (2.18b) always verifies the null condition, while (2.18a) and (2.18c) verify it if and only if $G_{\alpha\beta}(\Psi = 0)\ell^\alpha\ell^\beta = 0$ for every Minkowski-null vector ℓ, that is, for every vector ℓ verifying $m_{\alpha\beta}\ell^\alpha\ell^\beta = 0$.

We now examine the (classic) null condition from a different perspective, one that is related to the perspective of Sect. 1.2.4 and that is closely connected to our analysis of shock forming solutions in the region $\{t \geq 0\}$. To proceed, we consider, for the purpose of illustration, the vectorfield frame

$$\{L_{(Flat)} := \partial_t + \partial_r, R_{(Flat)} := -\partial_r, X_{(Flat);1}, X_{(Flat);2}\}, \tag{2.19}$$

where $X_{(Flat);1}$ and $X_{(Flat);2}$ are a local frame on the Euclidean spheres of constant t and r and

$$\partial_r := \frac{x^a}{r}\partial_a \tag{2.20}$$

is the standard Euclidean radial vectorfield. Note that relative to the frame (2.19), we can decompose the inverse Minkowski metric as follows:

$$\begin{aligned}(m^{-1})^{\alpha\beta} = &-L^\alpha_{(Flat)}L^\beta_{(Flat)} - (L^\alpha_{(Flat)}R^\beta_{(Flat)} + R^\alpha_{(Flat)}L^\beta_{(Flat)}) \\ &+ (\not{m}^{-1})^{AB}X^\alpha_{(Flat);A}X^\beta_{(Flat);B},\end{aligned} \tag{2.21}$$

where the 2×2 matrix $(\not{m}^{-1})^{AB}$ is the inverse of the 2×2 matrix $\not{m}_{AB} := m_{\alpha\beta}X^\alpha_{(Flat);A}X^\beta_{(Flat);B}$. With the help of (2.21) and (2.18a)-(2.18c), it is straightforward to see that for equation (2.17), the null condition is verified if and only the quadratic part of the nonlinearities, when expressed in terms of the frame derivatives, does not contain any quasilinear terms proportional to[15] $\Psi R_{(Flat)}R_{(Flat)}\Psi$ or semilinear terms proportional to $(R_{(Flat)}\Psi)^2$. Simple calculations yield that up to constant factors, both of these terms have the same coefficient, namely the *future null condition failure factor* ${}^{(+)}\aleph$ defined by

$$^{(+)}\aleph := \underbrace{G_{\alpha\beta}(\Psi = 0)}_{\text{constants}} L^\alpha_{(Flat)}L^\beta_{(Flat)}. \tag{2.22}$$

Note that ${}^{(+)}\aleph$ can be viewed as a function depending only on $\theta = (\theta^1, \theta^2)$, where θ^1 and θ^2 are local angular coordinates corresponding to spherical coordinates (t, r, θ)

[14] By definition, ξ is Minkowski-null if and only if $(m^{-1})^{\alpha\beta}\xi_\alpha\xi_\beta = 0$.

[15] See Footnote 10 on pg. 8 regarding the notation.

on Minkowski spacetime. The following important result follows from the methods of [**47**] and [**15**]: if the terms $\Psi R_{(Flat)} R_{(Flat)} \Psi$ and $(R_{(Flat)}\Psi)^2$ are absent from equation (2.17) when it is expanded relative to the frame (2.19), then small-data *future-global* existence holds. Hence, up to constant factors, ${}^{(+)}\aleph$ is the coefficient of the terms in equation (2.17) that are possible obstructions to future-global existence. The relevant point, which essentially follows from the techniques outlined in Sect. 1.2.3, is that in the region $\{t \geq 0\}$, $\Psi R_{(Flat)} R_{(Flat)} \Psi$ and $(R_{(Flat)}\Psi)^2$ decay more slowly than the other quadratic nonlinear terms and hence have the potential to eventually cause singularities to form (see however, Remark 2.9 concerning the product $\Psi R_{(Flat)} R_{(Flat)} \Psi$). All other quadratic terms decay sufficiently fast as $t \to \infty$ to allow for small-data future-global solutions. Moreover, from the discussion in the previous paragraph, we conclude that the null condition is verified by the nonlinear terms in equation (2.17) if and only if[16] ${}^{(+)}\aleph \equiv 0$. Our main shock formation theorem, Theorem 23.9, provides a converse to the results of [**47**] and [**15**] for the equations under consideration: if ${}^{(+)}\aleph \not\equiv 0$, then equation (2.17) exhibits small-data future shock formation caused by the presence of *both* of the quadratic terms $\Psi R_{(Flat)} R_{(Flat)} \Psi$ and $(R_{(Flat)}\Psi)^2$, *which appear in the correct proportions* due to the geometric structure of the equation.

REMARK 2.8 (**Most equations of type** (2.17) **have** ${}^{(+)}\aleph \not\equiv 0$). It follows from the above discussion that for equation (2.17), ${}^{(+)}\aleph$ completely vanishes if and only if, relative to the rectangular coordinates, $g_{\alpha\beta} = (1 + f(\Psi))m_{\alpha\beta} + \mathcal{O}(\Psi^2)$ for some smooth function $f(\Psi)$ verifying $f(0) = 0$. Hence, for equations of type (2.17), the null condition is restrictive and holds only in very special cases.

REMARK 2.9 (**The quasilinear terms by themselves do not cause small--data singularities**). The results of Alinhac [**7**] and Lindblad [**58**] show that if we modify equation (2.12) by deleting the semilinear terms that fail the (classic) null condition, then the modified equation admits small-data global solutions.[17] Their result holds even though the modified equation can contain a quasilinear term proportional to $\Psi R_{(Flat)} R_{(Flat)} \Psi$, which fails the null condition. In the presence of such a term, the global solutions can have distorted asymptotics. In particular, the true characteristics can significantly deviate from the Minkowskian characteristics as $t \to \infty$. Roughly, the analog of μ in [**7**] and [**58**] can become arbitrarily small as $t \to \infty$, but it never vanishes in finite time. On the other hand, if we delete the quasilinear terms from equation (2.12) but retain the semilinear terms, then the results of John [**36**] imply that under certain structural assumptions on the semilinear terms, all nontrivial solutions must break down in finite time. However, the singularity formation mechanism is not well-understood.

We stress that *the results of Alinhac* [**7**] *and Lindblad* [**58**] *do not extend to equation* (2.9). More precisely, even though there are no semilinear terms present in this equation, it exhibits small-data shock formation when its nonlinearities fail the null condition; see Remark A.1.

REMARK 2.10 (**Shock formation for** $\{t \leq 0\}$). We could also study past shock formation. In this case, the function ${}^{(+)}\aleph$ from (2.22) needs to be modified

[16] The key point is that all possible future-directed Minkowski-null vectors ℓ from the previous paragraph are achieved by the vectorfield $L_{(Flat)}$ in (2.22) as it varies over the region $\{t \geq 0\}$.

[17] More precisely, Lindblad's work [**58**] treats the general case, while Alinhac's work [**7**] addresses the specific equation $-\partial_t^2\Psi + (1 + \Psi)^2\Delta\Psi = 0$.

to account for the fact that $L_{(Flat)}$ is *not* tangent to the outgoing Minkowski null cones as we head towards the past; see Remark 3.35.

2.5. Basic geometric constructions

In deriving our main results, we adopt the framework of [**17**] and perform the vast majority of our analysis relative to a new system of *geometric coordinates*:

$$(t, u, \vartheta^1, \vartheta^2). \tag{2.23}$$

The coordinate t is the Minkowski time coordinate from Sect. 2.2, while the "dynamic coordinate" u is an *outgoing eikonal function*, an analog of the coordinate (1.8a) that we used in analyzing Burgers' equation. Specifically, u a solution to the eikonal equation

$$(g^{-1})^{\alpha\beta}(\Psi)\partial_\alpha u \partial_\beta u = 0 \tag{2.24}$$

with level sets that are outgoing to the future (see Figure 2). We define the initial value of u by

$$u|_{t=0} = 1 - r, \tag{2.25}$$

where $r = \sqrt{\sum_{a=1}^{3}(x^a)^2}$ is the standard Euclidean radial coordinate. The solution to (2.24) is a perturbation of the flat eikonal function $u_{(Flat)} = 1 + t - r$. The level sets of u, which we denote by $\mathcal{C}_u$, are null (characteristic) hypersurfaces of the Lorentzian metric g. The $\mathcal{C}_u$ intersect the constant Minkowski-time hypersurfaces Σ_t in spheres $S_{t,u}$. We denote the Riemannian metric that g induces on $S_{t,u}$ by $\not{g}$. We denote the annular region in Σ_t that is trapped between the inner sphere $S_{t,u}$ and the outer sphere $S_{t,0}$ by Σ_t^u. We denote the portion of $\mathcal{C}_u$ in between Σ_0 and Σ_t by $\mathcal{C}_u^t$. We denote the solid spacetime region in between Σ_0, Σ_t, $\mathcal{C}_u^t$, and $\mathcal{C}_0^t$ by $\mathcal{M}_{t,u}$ (see Figure 2). More precisely, we define $\mathcal{M}_{t,u}$ to be "open at the top." That is, Σ_t^u is not part of $\mathcal{M}_{t,u}$. In contrast, we define $\mathcal{C}_u^t$ to be "closed at the top." The functions ϑ^1, ϑ^2 from (2.23) are local coordinates on the $S_{t,u}$ that are easy to construct; see below.

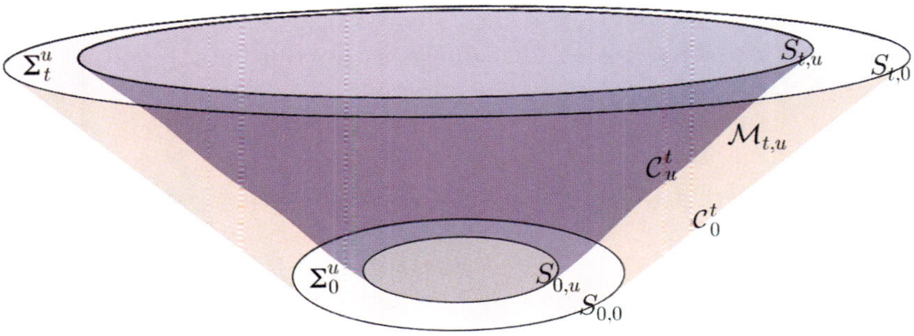

FIGURE 2. Surfaces and regions

We stress the following important point: the eikonal equation (2.24) is a hyperbolic evolution equation for u that must be solved *in conjunction* with the wave equation (2.1a). That is, we are studying the wave equation (2.1a) coupled to (2.24). Clearly, in rectangular coordinates, the wave equation completely decouples. However, as we describe below, it is extremely useful to study the wave equation relative to a dynamic vectorfield frame constructed with the help of u. Relative to this frame, (2.1a) and (2.24) and their higher-order versions become highly coupled.

Intuitively, one might expect that the intersection of the level sets of u should correspond to a shock singularity, as is the case for solutions to Burgers' equation (see Sect. 1.1.2). Recall that in our study of Burgers' equation, the analysis was based entirely on transport equations, that is, on deriving estimates along the characteristics. In contrast, in more than one spatial dimension, wave equations do not reduce to transport equations along the characteristics. Nonetheless, it is well-established that for certain nonlinear wave-like equations in more than one spatial dimension, there are solution regimes such that eikonal functions can be used to derive sharp estimates. For example, for small-data solutions to wave equations in three spatial dimensions with nonlinearities that verify the (classic) null condition, the solution's lower-order derivatives verify a transport equation with sources that decay at an integrable-in-time rate along the null generators[18] of the characteristic hypersurfaces, and this structure allows one to derive sharp pointwise decay estimates at the lower derivative levels. The first instance of the use of an eikonal function to solve a global nonlinear problem is found in the celebrated proof of Christodoulou-Klainerman [**19**] of the global stability of Minkowski spacetime as a solution to Einstein's equations. Eikonal functions also played a central role in the Klainerman-Rodnianski [**50**] proof of low regularity local well-posedness for a class of quasilinear wave equations and the Klainerman-Rodnianski-Szeftel proof of the bounded L^2 curvature conjecture [**52**]. They also played a fundamental role in both Alinhac's and Christodoulou's proofs of small-data shock formation. Moreover, in the shock formation proofs, the eikonal function played a more essential role than it did in the proof of the stability of Minkowski spacetime. Specifically, Lindblad-Rodnianski gave a second proof [**59**] of the stability of Minkowski spacetime, relative to wave coordinates, that did not rely on a true eikonal function. Instead, they closed their small-data global existence proof by deriving estimates relative to the background Minkowskian geometry[19] with the help of a Minkowski eikonal function $u_{(Flat)} = t - r$. In contrast, as we will see, for the solutions studied in this monograph, small-data shock formation *exactly corresponds* to the intersection of the level sets of a true eikonal function verifying (2.24), as in the case of Burgers' equation; see Sects. 2.8 and 2.9 for an outline of the argument. Recall that in the case of Burgers' equation, we were able to provide a crude proof of blow-up without introducing an eikonal function. In contrast, it is difficult to imagine an analogous crude proof of small-data blow-up for quasilinear wave equations in more than one spatial dimension; as we describe below in great detail, in order to close the energy estimates, it seems that one needs at least a very close approximation of a true eikonal function and that the background rectangular coordinates and Minkowskian geometry are not precise enough to close the proof.

Intimately connected to the eikonal function is the following gradient vectorfield, which can be expressed relative to rectangular coordinates as follows:

$$L_{(Geo)}^{\nu} := -(g^{-1})^{\nu\alpha}\partial_{\alpha}u. \tag{2.26}$$

[18]Null generators are geometric analogs of the integral curves of the vectorfield $L_{(Flat)}$ from (2.19), tailored to the true characteristics of the equation.

[19]More precisely, some of the transport equation estimates in [**59**] were derived along first-order corrections to the outgoing Minkowskian characteristics.

It is easy to verify that $L_{(Geo)}$ is null and geodesic, that is, that $g(L_{(Geo)}, L_{(Geo)}) = 0$ and $\mathscr{D}_{L_{(Geo)}} L_{(Geo)} = 0$ (recall that $\mathscr{D}$ is the Levi-Civita connection of g and thus $\mathscr{D}g = 0$). Furthermore, a related quantity that we mentioned above, the *inverse foliation density* $\upmu$, is the most important object in this monograph from the point of view of small-data shock formation:

$$\upmu := -\frac{1}{(g^{-1})^{\alpha\beta}\partial_\alpha u \partial_\beta t} = \frac{1}{L^0_{(Geo)}}. \tag{2.27}$$

The function $1/\upmu$ is a measure of the density of the stacking of the level sets of u relative to Σ_t. It is an analog[20] of the quantity (1.11) that we used in our proof of shock formation for solutions to Burgers' equation. In the case of the background solution $\Psi \equiv 0$, we have $\upmu \equiv 1$. For perturbed solutions, the stacking density becomes infinite (that is, the level sets of u intersect and a shock forms) when $\upmu = 0$. In Figure 3 on pg. 45, we exhibit a solution in which $\upmu$ has become very small in a certain region because the density of the level sets of u has become large (that is, a shock has nearly formed in this figure). One of the primary results of this monograph is our proof that for a class of small-data solutions, $\upmu$ becomes 0 in finite time before any other kind of singularity occurs. In addition, in Theorem 22.1, we prove a deeper result, which an analog of the main result of [**17**]: we show that for small-data solutions, **in the region of interest, the only possible way a singularity can form is for $\upmu$ to become 0 in finite time.**

We now introduce the following rescaled version of $L_{(Geo)}$, which plays a fundamental role in our analysis:

$$L := \upmu L_{(Geo)}. \tag{2.28}$$

Note that $Lt = L^0 = 1$ and thus L is future-directed (see Footnote 16 on pg. 13). Our interest in L lies in the fact that our proof shows that when a shock forms, the rectangular components $L^\nu_{(Geo)}$ blow-up, while the rectangular components L^ν remain near those of $L_{(Flat)} := \partial_t + \partial_r$. For this reason and many others that will become apparent, L is useful for analyzing solutions.

To complete our geometric coordinate system, we complement t, u with local coordinates $(\vartheta^1, \vartheta^2)$ on the initial Euclidean sphere $S_{0,0}$ and propagate them inward to the $S_{0,u}$ by solving $-\partial_r \vartheta^A = 0$ and then to the future by solving $L\vartheta^A = 0$, $(A = 1, 2)$. We denote the coordinate partial derivative vectorfield corresponding to ϑ^1 by $X_1 := \frac{\partial}{\partial \vartheta^1}|_{t,u,\vartheta^2}$ and similarly for X_2. Since $X_A t = X_A u = 0$, $(A = 1, 2)$, it follows that the X_A are tangent to the spheres $S_{t,u}$.

REMARK 2.11 (**$\upmu$ is connected to the Jacobian determinant of the change of variables map**). For the solutions under consideration, the Jacobian determinant of the change of variables map Υ from geometric to rectangular coordinates vanishes precisely when $\upmu$ vanishes (see Lemma 3.67). Hence, small-data shock formation can alternatively be viewed as a breakdown in the map Υ^{-1}.

---

[20] More precisely, the quantity (1.11) measures the density of the characteristics relative to the lines of constant x rather than lines of constant t; this discrepancy is not important.

2.6. The rescaled frame and dispersive sup-norm estimates

The following basic principle, first exhibited by Christodoulou [**17**], is a key ingredient in our analysis: the lower-order derivatives of the solutions in the directions L, X_1, X_2, which are tangent to the outgoing null cones $\mathcal{C}_u$, exhibit dispersive behavior, even as a shock forms! Furthermore, when expressed in terms of the Minkowskian time coordinate t, the decay rates of these derivatives are the same[21] as those of solutions to the linear wave equation (see Sect. 1.2.3). This is somewhat analogous to the regular behavior that we observed in solutions to Burgers' equation along the characteristics (see Sect. 1.1.2). Furthermore, solutions' derivatives in directions transversal to the $\mathcal{C}_u$ also behave in a linearly dispersive fashion,[22] but *only after one rescales the transversal direction by* $\upmu$. The main reason that the solution's rescaled transversal derivative exhibits dispersive behavior is that *the rescaling by* $\upmu$ *eliminates the semilinear term in the wave equation that is most dangerous in the region* $\{t \geq 0\}$, *due to its slow decay*; see Remark 2.12.

Hence, as will become apparent, a good choice for a rescaled transversal vectorfield is the *radially inwardly pointing* vectorfield $\breve{R}$, which verifies

$$\breve{R}u = 1, \qquad \breve{R}t = 0, \qquad g(\breve{R}, X_1) = g(\breve{R}, X_2) = 0. \tag{2.29}$$

It is not difficult to verify using (2.3) that (see Lemma 3.18)

$$g(\breve{R}, \breve{R}) = \upmu^2, \qquad g(L, \breve{R}) = -\upmu. \tag{2.30}$$

An important aspect of our proof is showing that rectangular components of the vectorfield

$$R := \upmu^{-1}\breve{R} \tag{2.31}$$

remain near those of $R_{(Flat)} := -\partial_r$ throughout the evolution. Thus, when $\upmu$ vanishes, the vectorfield $\breve{R}$, when viewed as a vectorfield on $\mathbb{R}^{1+3}$ expressed relative to rectangular coordinates, also vanishes.

In total, the above vectorfields form a useful *rescaled frame* with span equal to $\text{span}\{\partial_\alpha\}_{\alpha=0,1,2,3}$ at each point where $\upmu > 0$ (see Figure 3 on pg. 45):

$$\{L, \breve{R}, X_1, X_2\}. \tag{2.32}$$

We carry out most of our analysis relative to the frame (2.32). Occasionally, when performing calculations, we find it convenient to replace $\breve{R}$ with the following vectorfield:

$$\breve{\underline{L}} := \upmu L + 2\breve{R}, \tag{2.33}$$

which is future-directed (see Footnote 16 on pg. 13), g-orthogonal to the $S_{t,u}$, ingoing, g-null, and normalized by $g(L, \breve{\underline{L}}) = -2\upmu$.

In the proof of our sharp classical lifespan theorem (Theorem 22.1), to capture the solution's dispersive behavior relative to the rescaled frame (2.32), we make the

---

[21] In contrast to Sect. 1.2.3, we are now working in a region of bounded u-width. Hence, we are concerned only with decay in t, and not with decay in u.

[22] The analogous statement in the case of Burgers' equation is that $\frac{\partial}{\partial u}|_t \Psi$ is constant in t at fixed u.

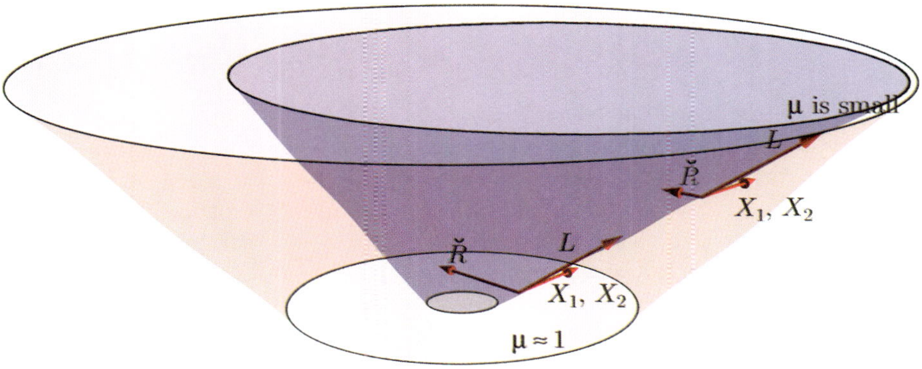

FIGURE 3. The rescaled frame at two distinct points, embedded in $(m, \mathbb{R}^{1+3})$

following bootstrap assumptions (see Chapter 12 for a more precise statement) on a spacetime region of the form $\mathcal{M}_{T_{(Bootstrap)},U_0}$ (see Figure 2 on pg. 41):

$$\|L\Psi\|_{C^0(\Sigma_t^{U_0})}, \qquad \|\not{d}\Psi\|_{C^0(\Sigma_t^{U_0})} \leq \frac{\varepsilon}{(1+t)^2}, \tag{2.34a}$$

$$\|\Psi\|_{C^0(\Sigma_t^{U_0})}, \qquad \left\|\breve{R}\Psi\right\|_{C^0(\Sigma_t^{U_0})} \leq \frac{\varepsilon}{1+t}. \tag{2.34b}$$

In (2.34a) and throughout, $\not{d}f$ denotes the angular differential of the scalar-valued function f, viewed as a function of the geometric angular coordinates $(\vartheta^1, \vartheta^2)$. In addition, $\|\cdot\|_{C^0(\Sigma_t^{U_0})}$ is the standard sup-norm (see Def. 3.70). Furthermore, ε is a small number whose smallness is shown, near the end of the proof of the sharp classical lifespan theorem, to be controlled by the size of the data (see (2.7)). We show that the assumptions (2.34a)-(2.34b), which reflect the dispersive decay rates of solutions to the *linear wave equation*, hold all the way up to and including the time of first shock formation. We make similar assumptions for some of the higher derivatives of Ψ, but not near the top order since, as we already mentioned in Remark 2.1, our high-order estimates are allowed to degenerate when μ becomes small. More precisely, our analysis shows that the following analog of the linear behavior described in Sect. 1.2.3 holds: each additional L or angular differentiation results in a decay rate gain of $(1+t)^{-1}$. These pointwise decay estimates help us control the inessential error terms that arise in our analysis. At the end of the proof of Theorem 22.1, we derive improvements of the above bootstrap assumptions from our L^2 estimates with the help of Sobolev embedding.

It is important to understand how the covariant wave operator looks relative to the rescaled frame. In Prop. 5.4, we show that

$$\mu\square_{g(\Psi)}\Psi = -L(\mu L\Psi + 2\breve{R}\Psi) + \mu\not{\Delta}\Psi - \mathrm{tr}_{\not{g}}\chi\breve{R}\Psi + \mathrm{Error}, \tag{2.35}$$

where $\not{\Delta}$ is the covariant Laplacian corresponding to the metric $\not{g}$ on the $S_{t,u}$ induced by g, and Error denotes quadratically small terms that decay at an integrable-in-time rate. The error terms depend on the up-to-first-order derivatives Ψ and the up-to-second-order derivatives of u. In (2.35), χ is the null second fundamental form, a crucially important symmetric type $\binom{0}{2}$ tensorfield on the spheres that verifies

$$\chi_{AB} := \chi(X_A, X_B) = g(\mathscr{D}_A L, X_B), \tag{2.36}$$

$\mathscr{D}_A L$ denotes the covariant derivative of L in the direction X_A, and $\mathrm{tr}_{\not g}\chi = (\not g^{-1})^{AB}\chi_{AB}$. Roughly, χ can be viewed as a second derivative of u that involves an angular derivative. In proving our sharp classical lifespan theorem, we encounter the following serious difficulty: it takes a huge amount of effort to derive top-order L^2 estimates for χ. One issue is that a naive approach results in a loss of derivatives. Another issue is that even though we are able to find way to avoid derivative loss, our estimates for the top derivatives are very degenerate and can blow-up like a power of $1/\upmu$; see Sects. 2.10.2 and 2.10.3 for additional details. In the case $\Psi \equiv 0$ we have $\mathrm{tr}_{\not g}\chi_{(Flat)} = 2r^{-1}$, whereas for perturbed solutions, we prove that $\mathrm{tr}_{\not g}\chi = 2\varrho^{-1} + \text{Error}$, where Error is of size $\varepsilon \ln(e+t)(1+t)^{-2}$ and

$$\varrho(t,u) := 1 - u + t \tag{2.37}$$

is a geometric radial variable, tailored to the eikonal function. On the bootstrap region $\mathcal{M}_{T_{(Bootstrap)},U_0}$, we have the following estimate, which we often use: $\varrho \approx 1+t$. In view of the fact that $L\varrho = 1$ and $\breve{R}\varrho = -1$, we deduce from (2.35) that

$$\upmu \square_{g(\Psi)}\Psi = -L\left\{\upmu L(\varrho\Psi) + 2\breve{R}(\varrho\Psi)\right\} + \varrho\upmu \not\Delta\Psi + \text{Error}, \tag{2.38}$$

where Error denotes quadratically small terms that decay at an integrable-in-time rate. Equation (2.38) plays an important role in our proof of small-data shock formation; see Sect. 2.8 and Lemma 23.1.

REMARK 2.12 (**"Eliminating" the dangerous semilinear term by rescaling**). A crucially important structural feature of equations (2.35) and (2.38) is that they do not contain any quadratic term proportional to $(\breve{R}\Psi)^2$. This is quite remarkable since the wave equation (2.17), when expressed relative to the Minkowskian frame (2.19), can contain a term proportional to $(R_{(Flat)}\Psi)^2$. The point is that by using the rescaled frame (2.32) and bringing the $\upmu$ factor "inside" the outer L differentiation in (2.35) and (2.38), we have generated cancellation that "eliminates" the term $(\breve{R}\Psi)^2$. That is, we have been able to eliminate the quadratic semilinear term with the slowest decay rate as $t \to \infty$; see the discussion in Sects. 1.2.4 and 2.4. The price we pay is that **i)** our rescaled frame degenerates relative to the rectangular coordinate frame $\{\partial_\alpha\}_{\alpha=0,1,2,3}$ when $\upmu$ becomes small and **ii)** the terms Error on the right-hand side of (2.38) have a somewhat complicated structure because they depend not only on Ψ and its first derivatives, but also on the up-to-second-order derivatives of u; see Prop. 5.4. Hence, in order to verify that the terms Error are in fact error terms, we must obtain a good understanding of the asymptotic behavior of u and its derivatives.

2.7. The basic structure of the coupled system and sup-norm estimates for the eikonal function quantities

As we stressed above, we are studying the wave equation (2.1a) coupled to the eikonal equation (2.24). In practice, rather than directly studying the eikonal equation, we study evolution equations verified by its first derivatives, which are represented by the rectangular components $L^\nu_{(Geo)}$ (see (2.26)). More precisely, as we will see, it is convenient to instead study the equations verified by the quantities

μ and L^i, $(i = 1, 2, 3)$, defined by (2.27) and (2.28). We can derive evolution equations for these quantities by writing down the geodesic equation $\mathscr{D}_{L_{(Geo)}} L_{(Geo)} = 0$ relative to the rectangular coordinates and then translating it into equations for μ and L^i. We carry this out in full detail in Chapter 4. We find that these "eikonal function quantities" verify the following transport equations:

$$L\mu = \frac{1}{2} G_{LL} \breve{R}\Psi + \mu \mathrm{Error}(L\Psi, L^1, L^2, L^3), \tag{2.39a}$$

$$LL^i = \mathrm{Error}^i(L\Psi, \not{d}\Psi, L^1, L^2, L^3), \tag{2.39b}$$

where $G_{LL} := G_{\alpha\beta}(\Psi) L^\alpha L^\beta$, $G_{\alpha\beta}$ is defined in (2.14), and Error denotes small terms that decay at an integrable-in-time rate.

In total, we can use bootstrap assumptions in the spirit of (2.34), the evolution equations (2.39), and a small-data assumption to derive C^0 estimates for μ, L^i, and their lower-order derivatives. Like the bootstrap assumptions (2.34), these estimates complement our L^2 estimates (which we discuss below) and allow us to control various error terms. An important aspect of these C^0 estimates is that $\mu - 1$ and its lower-order derivatives can shrink or grow like $\mathcal{O}(\varepsilon \ln(e + t))$. The possible logarithmic-in-time shrinking behavior of μ is in fact the source of small-data shock formation; we illustrate this in more detail in Sect. 2.9. The same statement holds for the difference between ϱL^i and its Minkowskian value x^i, that is, for $\varrho L^i - x^i$. Similar statements hold for many other scalar and tensorial quantities that play a role in this monograph. We derive most of these C^0 estimates in Chapter 12.

2.8. Lower bounds for the rescaled radial derivative of the solution in the case of shock formation

We now heuristically explain the behavior of Ψ for the shock forming solutions of interest. The behavior that we describe in this section is one of the key ingredients in the proof of our small-data shock formation theorem, Theorem 23.9. The main idea is that there exists an open set of data[23] whose corresponding solutions verify the following *signed lower bound*, which is valid for sufficiently large times and for some fixed choices of u and ϑ (that is, along some integral curves of L):

$$\breve{R}\Psi(t, u, \vartheta) \gtrsim \frac{\varepsilon}{1 + t}. \tag{2.40}$$

Furthermore, for solutions corresponding to a different open set of data, the opposite signed bound holds: $\breve{R}\Psi(t, u, \vartheta) \lesssim -\varepsilon(1 + t)^{-1}$. Since $\breve{R} = \mu R$, (2.40) is equivalent to

$$R\Psi(t, u, \vartheta) \gtrsim \frac{1}{\mu} \frac{\varepsilon}{(1 + t)}. \tag{2.41}$$

Clearly, the bound (2.41) implies that the near-Euclidean-unit-length derivative $R\Psi$ blows up when $\mu \to 0$. Moreover, the bound (2.41) (with the correct sign, depending on the nonlinearities) is in fact the main ingredient needed[24] to show

[23] More precisely, in Chapter 23, to show that a shock forms, we assume that the data are prescribed at time $t = -1/2$. We make this assumption only for technical convenience; this detail is not important for understanding the heuristic description that we are providing here.

[24] In the case of Burgers' equation, the analogous statement is that the vanishing of μ is exactly tied to having a solution with $\frac{\partial}{\partial u}|_t \Psi < 0$; see the discussion surrounding equation (1.13).

that $\upmu$ vanishes in finite time. We elaborate upon this in Sect. 2.9. All of our shock forming solutions are such that $R\Psi$ blows up for exactly these reasons.

To explain how we derive (2.41), we impose the wave equation by setting the right-hand side of (2.38) equal to 0, which yields

$$L\left\{\upmu L(\varrho\Psi) + 2\breve{R}(\varrho\Psi)\right\} = \varrho\upmu \not\Delta\Psi + \text{Error}, \tag{2.42}$$

where the terms Error are quadratically small and decay at an integrable-in-time rate. For convenience, in this monograph, we prove the bound (2.41) in complete detail only for an open set of nearly spherically symmetric small data (see the proof of Theorem 23.9). However, in Sect. 2.11.3, we provide an outline of how to enlarge the class of small data to which our shock formation arguments apply. By "nearly spherically symmetric," we mean that we assume that the lower-order angular derivatives of the data are even smaller than the small $\breve{R}$-directional derivative, and we prove that the relative smallness is propagated in the solution by the flow of the equations (see Cor. 22.4). Furthermore, from a higher-order analog of (2.34), we find that even though $\varrho\upmu\not\Delta\Psi$ is only linearly small, it decays at an integrable-in-time rate. The net effect is that for such solutions, the right-hand side of (2.42) has a tiny amplitude and is integrable in time, and (2.42) can effectively be treated as a transport equation for the quantity $\upmu L(\varrho\Psi) + 2\breve{R}(\varrho\Psi)$ along the integral curves of $L = \frac{\partial}{\partial t}$. Hence, integrating (2.42) with respect to time at fixed u and ϑ and ignoring the small right-hand side, we obtain, relative to the geometric coordinates:

$$\left\{\upmu L(\varrho\Psi) + 2\breve{R}(\varrho\Psi)\right\}(t,u,\vartheta) \approx f[\mathring{\Psi}, \mathring{\Psi}_0](u,\vartheta), \tag{2.43}$$

where $f[\mathring{\Psi}, \mathring{\Psi}_0](u,\vartheta)$ is equal to the term in braces on the left-hand side of (2.42) evaluated at $t = 0$, and $(\mathring{\Psi}, \mathring{\Psi}_0)$ are the data (2.1b). Furthermore, using (2.34a) and (2.43) and the identities $L\varrho = 1$ and $\breve{R}\varrho = -1$, we find that for sufficiently large times, we have

$$\breve{R}\Psi(t,u,\vartheta) \approx \frac{1}{2}\frac{1}{\varrho(t,u)} f[\mathring{\Psi}, \mathring{\Psi}_0](u,\vartheta). \tag{2.44}$$

Recalling that $\varrho(t,u) \approx 1 + t$, we see that for suitable data, (2.44) implies the desired bound (2.41).

2.9. The main ideas behind the vanishing of the inverse foliation density

We now explain why the estimates of Sect. 2.8 imply that $\upmu$ vanishes in finite time. We first address the important factor G_{LL} in equation (2.39a). In Lemma 12.49, we show that $G_{LL}(t,u,\vartheta)$ is well-approximated by ${}^{(+)}\aleph(t = 0, u = 0, \vartheta)$, where ${}^{(+)}\aleph$, the future null condition failure factor from (2.22), is actually independent[25] of u at $t = 0$. We remark that one significant difference between our work and that of Christodoulou's **[17]** is that in **[17]**, the analog of ${}^{(+)}\aleph$ is a (nonzero) constant; the presence of an angle-dependent ${}^{(+)}\aleph$ in our work here creates additional complications in the analysis. In view of the above discussion and the

[25]When $t = 0$, the geometric angular coordinates ϑ^A coincide with Euclidean spherical coordinates θ^A and hence, as we explained just below equation (2.22), ${}^{(+)}\aleph(0,u,\vartheta)$ can be viewed as a function of ϑ alone.

transport equation (2.39a), we see that for angles ϑ with ${}^{(+)}\aleph(t = 0, u = 0, \vartheta) < 0$ and for the class of data that lead to the bound (2.40) (with the same angle ϑ), we have

$$L\upmu(t, u, \vartheta) \lesssim -\varepsilon \left| {}^{(+)}\aleph(t = 0, u = 0, \vartheta) \right| \frac{1}{1+t} + \text{Error}, \tag{2.45}$$

where the error terms are small in magnitude compared to the first term. Integrating (2.45) along the integral curves of L, ignoring the error terms, and using the initial condition $\upmu = 1 + \mathcal{O}(\varepsilon)$, we obtain that

$$\upmu(t, u, \vartheta) \lesssim 1 - \varepsilon \left| {}^{(+)}\aleph(t = 0, u = 0, \vartheta) \right| \ln(e + t). \tag{2.46}$$

Hence, for such data, we conclude from inequality (2.46) that $\upmu$ vanishes at a time $T_{(Lifespan)} \sim \exp(c\varepsilon^{-1})$. Furthermore, for angles ϑ with ${}^{(+)}\aleph(t = 0, u = 0, \vartheta) > 0$, we can also derive the bound (2.46) and the vanishing of $\upmu$ in finite time by using the remark made in the first sentence after equation (2.40).

2.10. The role of Theorem 22.1 in justifying the heuristics

In Sects. 2.8 and 2.9, we sketched a proof of our small-data shock formation theorem, Theorem 23.9. However, in order to justify the heuristic arguments given above, we must **i)** show that coupled system (2.1a), (2.39a), (2.39b) for Ψ, $\upmu$, L^i is well-posed in the sense that we can control suitable numbers of derivatives of all quantities without losing derivatives and **ii)** show that all of the "error terms" mentioned above contribute only small corrections to the heuristic analysis. In particular, we have to make sure that the blow-up behavior (2.41) does not couple back into the error terms and cause them to become large near the shock, that is, in regions where $\upmu$ is small. The task **ii)** is highly coupled to **i)** and occupies the bulk of our work. As we have mentioned, our main results in this direction are provided by Theorem 22.1. Roughly, the theorem states that the solution persists unless $\upmu$ vanishes, at which point a shock has formed. Furthermore, the theorem provides a slew of quantitative estimates that must hold until $\upmu$ vanishes. In fact, many quantities can be extended relative to the geometric coordinates as C^k functions for some $k > 0$, all the way to the constant-time hypersurface subset $\Sigma_{T_{(Lifespan)};U_0}^{U_0}$ where the first shock singularity point lies.

2.10.1. Square integral estimates for the wave variable. The only known methods for controlling solutions to quasilinear wave equations are heavily based on L^2 estimates. To derive such estimates for Ψ, we use both the vectorfield multiplier and commutator methods, which we outlined in Sects. 1.2.2 and 1.2.3. The multiplier method roughly corresponds to applying a well-chosen vectorfield differential operator to Ψ, multiplying both sides of the wave equation by this quantity, and integrating by parts over the spacetime region $\mathcal{M}_{t,u}$ with the help of the divergence theorem. We present the details in Chapter 10. In this monograph, we use two distinct multipliers:

$$T := (1 + 2\upmu)L + 2\breve{R}, \tag{2.47a}$$

$$\widetilde{K} := \varrho^2 L. \tag{2.47b}$$

The timelike multiplier (2.47a) was used in [**17**] while the Morawetz-type multiplier (2.47b) is a slightly modified version of a multiplier used in [**17**].

We first discuss the estimates generated by (2.47a). In this case, an energy $\mathbb{E}[\Psi](t,u)$ and a cone flux $\mathbb{F}[\Psi](t,u)$ naturally arise from applying the divergence theorem to solutions of $\square_{g(\Psi)}\Psi = 0$ on the spacetime region $\mathcal{M}_{t,u}$ bounded by Σ_0^u, Σ_t^u, $\mathcal{C}_u^t$, and $\mathcal{C}_0^t$ (see Figure 2 on pg. 41). These quantities have the following L^2 coerciveness properties[26] (see Lemma 14.1):

$$\text{(2.48a)}\quad \mathbb{E}[\Psi](t,u) \geq C^{-1}\left\|\Psi\right\|_{L^2(S_{t,u})}^2 + C^{-1}\left\|\Psi\right\|_{L^2(\Sigma_t^u)}^2 + C^{-1}\left\|\sqrt{\mu}L\Psi\right\|_{L^2(\Sigma_t^u)}^2 + C^{-1}\left\|\mu L\Psi\right\|_{L^2(\Sigma_t^u)}^2 + \left\|\breve{R}\Psi\right\|_{L^2(\Sigma_t^u)}^2 + C^{-1}\left\|\sqrt{\mu}{d\!\!/}\Psi\right\|_{L^2(\Sigma_t^u)}^2 + C^{-1}\left\|\mu{d\!\!/}\Psi\right\|_{L^2(\Sigma_t^u)}^2,$$

$$\text{(2.48b)}\quad \mathbb{F}[\Psi](t,u) \geq C^{-1}\left\|L\Psi\right\|_{L^2(\mathcal{C}_u^t)}^2 + C^{-1}\left\|\sqrt{\mu}L\Psi\right\|_{L^2(\mathcal{C}_u^t)}^2 + C^{-1}\left\|\sqrt{\mu}{d\!\!/}\Psi\right\|_{L^2(\mathcal{C}_u^t)}^2.$$

In most of the monograph, the relevant area and volume forms on $S_{t,u}$, Σ_t^u, $\mathcal{C}_u^t$, and $\mathcal{M}_{t,u}$ (which, as we rigorously define in Sect. 3.19, are inherent in the definitions of the L^2 norms in (2.48a)-(2.48b)), are respectively[27]

(2.49)

$$d\upsilon_{g\!\!/} := \sqrt{\det g\!\!/}\, d\vartheta^1 d\vartheta^2, \qquad d\underline{\varpi} := d\upsilon_{g\!\!/}\, du', \qquad d\overline{\varpi} := d\upsilon_{g\!\!/}\, dt', \qquad d\varpi := d\upsilon_{g\!\!/} du'\, dt'.$$

The volume forms $d\underline{\varpi}$ and $d\varpi$ and "rescaled" in the sense that the canonical forms induced by g on Σ_t^u and $\mathcal{M}_{t,u}$ are respectively $\mu\, d\underline{\varpi}$ and $\mu\, d\varpi$. The advantage of the rescaled forms is that they remain regular (that is, finite and nondegenerate) up to and including the shock singularity. Hence, any degeneration of the energies and cone fluxes at the shock is caused by an explicitly displayed μ-dependent factor going to 0.

In the case of the multiplier T, the divergence theorem yields (see Prop. 10.13) the following energy-cone flux identity for functions Ψ vanishing along the flat outer cone[28] $\mathcal{C}_0$:

$$\text{(2.50)}\qquad \mathbb{E}[\Psi](t,u) + \mathbb{F}[\Psi](t,u) = \mathbb{E}[\Psi](0,u) - \int_{\mathcal{M}_{t,u}} (T\Psi)(\mu\square_{g(\Psi)}\Psi)\, d\varpi - \frac{1}{2}\int_{\mathcal{M}_{t,u}} \mu Q^{\alpha\beta}[\Psi]\,{}^{(T)}\pi_{\alpha\beta}\, d\varpi.$$

In (2.50), $Q[\Psi]$ denotes the energy-momentum tensorfield corresponding to Ψ (see (1.62)) and ${}^{(T)}\pi$ denotes the deformation tensor of T (see (1.67)). The two $\mathcal{M}_{t,u}$ integrals on the right-hand side of (2.50) can be decomposed, relative to the rescaled frame $\{L, \breve{R}, X_1, X_2\}$, into a collection of error terms that can be bounded in terms of integrals of $\mathbb{E}[\Psi]$ and $\mathbb{F}[\Psi]$ (note that the first integral in fact vanishes for solutions to the wave equation). Then from Gronwall's inequality, we can derive, for solutions

---

[26] It is important that the coefficient of $\left\|\breve{R}\Psi\right\|_{L^2(\Sigma_t^u)}^2$ in (2.48a) is equal to 1. This constant affects the number of derivatives that we need to close our estimates; see the sentence just above inequality (2.64).

[27] For clarity, we note, for example, that in the expression $\|\Psi\|_{L^2(S_{t,u})}$, the integrand Ψ is considered to be a function of $(t,u,\vartheta^1,\vartheta^2)$ and the integration variables are ϑ^1 and ϑ^2. Similarly, in the expression $\|L\Psi\|_{L^2(\mathcal{C}_u^t)}$, the integrand $L\Psi$ is considered to be a function of $(t',u,\vartheta^1,\vartheta^2)$ and the integration variables are t', ϑ^1, and ϑ^2.

[28] Recall that this vanishing is a simple consequence of our assumption that the data are supported in Σ_0^1.

to $\square_{g(\Psi)}\Psi = 0$, an a priori estimate for $\mathbb{E}[\Psi](t,u) + \mathbb{F}[\Psi](t,u)$. This strategy, which at this level of generality is standard, is the main idea of the proof of Theorem 22.1.

To close our estimates, we also use the aforementioned vectorfield commutator method. More precisely, we commute the wave equation (2.1a) with a family of vectorfields and derive the identity (2.50) for the commuted quantities. Our commutation vectorfields are the set

$$\mathscr{Z} := \{\varrho L, \breve{R}, O_{(1)}, O_{(2)}, O_{(3)}\}, \tag{2.51}$$

where the $O_{(l)}$ are geometric rotation vectorfields tangent to the spheres $S_{t,u}$. The set (2.51) has span equal to $\text{span}\{\partial_\alpha\}_{\alpha=0,1,2,3}$ at each point where $\mu > 0$. Christodoulou used a similar commutation set in [**17**], the difference being that we use ϱL in place of Christodoulou's commutation vectorfield $(1+t)L$ because ϱL exhibits simpler commutation properties with various operators. Specifically, to prove our sharp classical lifespan theorem, we commute the wave equation up to 24 times with all possible combinations of vectorfields in $\mathscr{Z}$. In stark contrast to our earlier study of the linear wave equation, where commuting with the elements of $\mathscr{Z}_{(Flat)}$ produced no error terms (see (1.88a)-(1.88b)), commuting the elements of $\mathscr{Z}$ through the operator $\square_g$ generates a large number of error terms that we must bound. The set $\mathscr{Z}$ is carefully constructed so that the error terms on the right-hand side of the identity (2.50) for the differentiated quantities $\mathscr{Z}^M\Psi$, $M \leq 24$, are "controllable." However, as we describe below, some of the error terms are extremely difficult to bound, which is the main reason that this monograph is lengthy. It is not practical for us, in this introduction, to fully describe the properties that controllable error terms should have; see Remarks 7.6 and 9.9 for an overview of the most important properties. We construct the geometric rotation vectorfields $O_{(l)}$ (see Chapter 6) by projecting the Euclidean rotations[29] $O_{(Flat;l)} := \epsilon_{lab}x^a\partial_b$ onto the spheres $S_{t,u}$, that is, by writing them as linear combinations of the non-rescaled frame vectors $\{R, X_1, X_2\}$ and then discarding the R component. Because the $O_{(Flat;l)}$ generally contain a (small) nonzero component of R, they are unsuitable for studying solutions near points where μ vanishes; by (2.41), $O_{(Flat;l)}\Psi$ can blow-up at such points. It is important to note that all commutation vectorfields $Z \in \mathscr{Z}$ depend on the first derivatives of the eikonal function u. Hence, the Z are *dynamically constructed* and depend on the solution itself, through u and its first derivatives.

When we use the energy-cone flux identity (2.50) to derive a priori L^2 estimates, we encounter the following four main difficulties.

(**1**) The energy-cone flux quantities (2.48a)-(2.48b) are not sufficient for controlling some of the error integrals appearing on the right-hand side of (2.50). One reason is that some of the error terms lack good t-weights. The same remarks apply to the error integrals corresponding to the higher-order quantities $\mathscr{Z}^M\Psi$.

(**2**) The energy defined by (2.48a) **controls only μ-weighted versions of $L\Psi$ and $\not{d}\Psi$** and hence becomes very weak if μ decays to 0 (and similarly for the higher-order analogs of (2.48a)). However, **some of the error terms that we encounter upon commuting the wave equation lack μ weights** and thus seem to be uncontrollable when μ is near 0. In view of the first term on the right-hand side of (2.48b), we see that the

[29]Here and throughout, ϵ_{ijk} is the fully antisymmetric tensorfield normalized by $\epsilon_{123} = 1$.

cone flux $\mathbb{F}[\Psi]$ overcomes this difficulty for the L derivatives. However, the difficulty persists for the $\not{d}$ derivatives because the $\not{d}\Psi$-controlling term on the right-hand side of (2.48b) features a $\upmu$ weight.

(3) To control error term products using the Sobolev calculus, we need good C^0 estimates for the lower-order derivatives of Ψ and good C^0 estimates for the lower-order derivatives of quantities constructed out of the eikonal function, such as $\upmu$, L^i, and χ.

(4) We need to derive suitable L^2 estimates to control the higher-order derivatives of quantities constructed out of the eikonal function.

The first difficulty above is standard and often appears in the global analysis of solutions to nonlinear wave equations. However, in the problem of small-data shock formation, the last three difficulties are quite challenging, especially (**4**). In his work [**17**], Christodoulou developed new strategies to overcome them. We have extended them so that they apply to the equations of interest in this monograph.

To handle difficulty (**1**), the lack of good t-weights, we also derive an analog of the energy-cone flux identity (2.50) in the case of the Morawetz-type multiplier (2.47b). Actually, the correct analog of the divergence theorem identity (2.50) is somewhat more involved in this case because we need some correction terms to obtain useful estimates; see Prop. 10.13. Note that we encountered Morawetz multipliers and the need for correction terms in Sect. 1.2.2, when we reviewed generalized energy estimates in the context of the linear wave equation. The corresponding energies $\widetilde{\mathbb{E}}$ and cone fluxes $\widetilde{\mathbb{F}}$ for the solution to equation (2.1a) are coercive in the following sense (see Lemma 14.1):

(2.52a)
$$\widetilde{\mathbb{E}}[\Psi](t,u) \geq C^{-1}(1+t)^2 \left\| \sqrt{\upmu}\left(L\Psi + \frac{1}{2}\mathrm{tr}_{\not{g}}\chi\Psi \right) \right\|^2_{L^2(\Sigma_t^u)} + \|\varrho\sqrt{\upmu}\not{d}\Psi\|^2_{L^2(\Sigma_t^u)},$$

(2.52b)
$$\widetilde{\mathbb{F}}[\Psi](t,u) \geq C^{-1} \left\| (1+t')\left(L\Psi + \frac{1}{2}\mathrm{tr}_{\not{g}}\chi\Psi \right) \right\|^2_{L^2(\mathcal{C}_u^t)}.$$

The good t-weights in (2.52a)-(2.52b) are suitable for handling most error terms.

To handle difficulty (**2**), the unfortunate presence of $\upmu$-weights in front of $|\not{d}\Psi|^2$ in the energies and cone fluxes, we use the remarkable observation made in [**17**]: in the case of the Morawetz multiplier (2.47b), the divergence theorem yields an *additional positive spacetime integral* that is activated when $\upmu$ becomes small and that provides control over the angular derivatives, *without the problematic* $\upmu$ *weights*. Specifically, the spacetime integral $\widetilde{\mathbb{K}}[\Psi]$ defined by[30]

$$\widetilde{\mathbb{K}}[\Psi](t,u) := \frac{1}{2}\int_{\mathcal{M}_{t,u}} \varrho^2[L\upmu]_-|\not{d}\Psi|^2\,d\varpi, \tag{2.53}$$

appears on the right-hand side of the analog of the identity (2.50) with a negative sign. Bringing it over to the left-hand side, we obtain a spacetime Morawetz integral that is coercive in the following sense (see Lemma 14.5):

$$\widetilde{\mathbb{K}}[\Psi](t,u) \gtrsim \int_{\mathcal{M}_{t,u}} \mathbf{1}_{\{\upmu\leq 1/4\}}\frac{1+t'}{\ln(e+t')}|\not{d}\Psi|^2\,d\varpi, \tag{2.54}$$

[30] In (2.53), $[L\upmu]_- := |L\upmu|$ when $L\upmu < 0$ and $[L\upmu]_- := 0$ when $L\upmu \geq 0$.

where $\mathbf{1}_{\{\mu \leq 1/4\}}$ denote the characteristic function of the spacetime set $\{(t,u,\vartheta) \mid \mu(t,u,\vartheta) \leq 1/4\}$. Not only does the estimate (2.54) solve the problem of μ weights for $|\not{d}\Psi|^2$, it also has favorable t-weights. We stress that we must use the integrals $\widetilde{\mathbb{K}}[\mathscr{Z}^M\Psi]$ to close all M^{th} order a priori L^2 estimates. That is, we use the Morawetz integrals at all orders, not just the top one. The coerciveness estimate (2.54) follows directly from the following critically important[31] estimate for $L\mu$ (see (13.28)):

$$\mu(t,u,\vartheta) \leq \frac{1}{4} \implies L\mu(t,u,\vartheta) \leq -c\frac{1}{(1+t)\ln(e+t)}, \tag{2.55}$$

where $c > 0$ is a constant.

REMARK 2.13 (**The "point of no return" implication of** (2.55)). Inequality (2.55) has another interesting consequence. Since the right-hand side of (2.55) is not integrable in t, if $\mu(t,u,\vartheta) \leq 1/4$, then μ must continue to shrink (along the integral curve of L corresponding to fixed u and ϑ) as t increases until it eventually becomes 0 and a shock forms.

To handle difficulty (**3**), that of obtaining good C^0 estimates for lower-order derivatives, we use different strategies for Ψ and the eikonal function quantities. For Ψ, instead of using an analog of the Klainerman-Sobolev inequality (1.91), we use a less refined Sobolev embedding estimate (see Cor. 18.11) to improve the C^0 bootstrap assumptions (2.34) after we derive suitable L^2 estimates. As we mentioned in Sect. 2.7, to derive C^0 estimates for the eikonal function quantities, we integrate the transport equations (2.39a)-(2.39b) for μ and L^i and use a small-data assumption together with the nondegenerate C^0 bootstrap assumptions for Ψ. Actually, to control the error terms, we have to derive C^0 estimates for a huge number of quantities related to the eikonal function, and the previous sentence, though correct in spirit, is only a caricature of the estimates needed. We derive many of these C^0 estimates in Chapter 12. Although they are tedious to derive, they are not difficult.

The final difficulty (**4**) is by far the most challenging one. The main challenge, which was first identified in [**19**] and [**50**] and which appears in [**17**] and the present monograph in a much more severe form,[32] is that one has to work very hard to control the top-order derivatives of u in L^2; a naive approach would result in a loss of one derivative of u relative to Ψ and would seem to suggest that the estimates will not close. To further explain this difficulty, we commute the wave equation (2.1a) with a commutation vectorfield $Z \in \mathscr{Z}$ (see (2.51)) to obtain the following schematic expression (see (9.1) for a precise expression):

$$\mu\square_{g(\Psi)}(Z\Psi) = \mu\mathscr{D}^{(Z)}\pi \cdot \mathscr{D}\Psi + \mu^{(Z)}\pi \cdot \mathscr{D}^2\Psi, \tag{2.56}$$

where ${}^{(Z)}\pi$ is the deformation tensor of Z (see (1.67)). It is crucially important to note that for $Z \in \mathscr{Z}$, the term $\mu\mathscr{D}^{(Z)}\pi$ from (2.56) depends on the *third derivatives* of the eikonal function u. For example, in the case that Z is a geometric rotation

[31]The value 1/4 in inequality (2.55) is convenient but not optimal; it can be enlarged to 1 minus a function that depends on the size of the data.

[32]As we explain in Sect. 2.10.4, the new difficulty in the small-data shock formation problem is that the price one pays for avoiding top-order derivative loss is the introduction of a factor of $1/\mu$ into the top-order energy identities, which leads to degenerate top-order L^2 estimates.

vectorfield O, we obtain (see Prop. 7.7, Lemma 9.4, and Prop. 9.6)

$$\mu\square_{g(\Psi)}(O\Psi) = (\breve{R}\Psi)O\mathrm{tr}_{\not{g}}\chi + \cdots \sim (\breve{R}\Psi)\partial^3 u + \cdots, \tag{2.57}$$

where χ is as in (2.36). Furthermore, one can show (see Cor. 11.4 and the proof of Cor. 11.7) that $O\mathrm{tr}_{\not{g}}\chi$ verifies the transport equation

$$LO\mathrm{tr}_{\not{g}}\chi = \frac{1}{2}G_{LL}\not{\Delta}O\Psi + \cdots. \tag{2.58}$$

By applying the divergence theorem identity (2.50) with $O\Psi$ in the role of Ψ and using equation (2.57), we can obtain control over some second-order derivatives of Ψ in L^2, assuming that we can control the right-hand side of (2.57) in L^2. However, to obtain L^2 estimates for $O\mathrm{tr}_{\not{g}}\chi$, we must integrate equation (2.58) along the integral curves of L, which seems to require that we have control over *three derivatives* of Ψ in L^2. This derivative-loss problem cannot be overcome by simply further commuting the wave equation. Hence, our estimates will not close without an additional ingredient, which we describe in Sect. 2.10.2.

REMARK 2.14 (**The importance of error terms that never appear**). The following property of our commutation vectorfield set $\mathscr{Z}$ is crucially important: after commuting the operator $\mu\square_{g(\Psi)}$ one time with $Z \in \mathscr{Z}$ (see (2.56)), we never produce the terms $\not{d}\breve{R}\mu$ or $\breve{R}\breve{R}\mu$; this fact follows from the proof of Prop. 16.4. Like the term $O\mathrm{tr}_{\not{g}}\chi$ in equation (2.57), $\not{d}\breve{R}\mu$ and $\breve{R}\breve{R}\mu$ are third-order derivatives of the eikonal function u and thus could cause derivative loss for the reasons explained just above. However, there is a serious discrepancy between $O\mathrm{tr}_{\not{g}}\chi$ and the other two terms. As we describe in Sect. 2.10.2, there is a way to avoid the derivative loss corresponding to the term $O\mathrm{tr}_{\not{g}}\chi$. In fact, the argument we provide can be extended to show that for all $Z \in \mathscr{Z}$, we can avoid the derivative loss corresponding to $Z\mathrm{tr}_{\not{g}}\chi$. In contrast, we are not aware of any method that would allow us to avoid the derivative loss that would occur in the presence of $\not{d}\breve{R}\mu$ or $\breve{R}\breve{R}\mu$.

2.10.2. Top-order square integral estimates for the eikonal function. As an example, we now explain how we overcome the derivative loss for the eikonal function quantity $O\mathrm{tr}_{\not{g}}\chi$ in (2.57), which depends on third derivatives of u. For illustration, we imagine that (2.57) is an equation for a top-order quantity. The main idea, which was first revealed in [**50**], is that by virtue of the wave equation $\square_{g(\Psi)}\Psi = 0$, $L\mathrm{tr}_{\not{g}}\chi$ has a remarkable structure. Specifically, up to lower-order derivative terms, $L\mathrm{tr}_{\not{g}}\chi$ is equal to the curvature component $-(\not{g}^{-1})^{AB}\mathscr{R}_{LALB}$, where $\mathscr{R}$ is the Riemann curvature tensor of g (see Def. 11.1). The special structure revealed in [**50**] is: by using the wave equation for Ψ, one can show that $-(\not{g}^{-1})^{AB}\mathscr{R}_{LALB}$ can be written as perfect L derivative of the first derivatives of Ψ plus lower-order terms; see Cor. 11.4. Putting these L derivatives back on the left, we obtain a transport equation along the integral curves of L for a "modified" version of $\mathrm{tr}_{\not{g}}\chi$ that does not lose derivatives relative to Ψ (see Cor. 11.7). Furthermore, this special structure essentially survives under commutations, and in total, we obtain an equation of the following schematic form (see Prop. 11.10 for the precise version):

$$L\left\{\mu O\mathrm{tr}_{\not{g}}\chi - O(G_{LL}\breve{R}\Psi) + \cdots\right\} \sim \mu\partial^2\Psi + \cdots. \tag{2.59}$$

Since the right-hand side of (2.59) depends on only two derivatives of Ψ, it thus appears that we have solved the problem of losing derivatives. However, on the right-hand side of (2.59), present in the $\cdots$, lies another difficult term, roughly of the form $\mu \hat{\not{\mathcal{L}}}_O \hat{\chi} \cdot \hat{\chi}$, where $\hat{\chi}$ is the trace-free part of χ and $\hat{\not{\mathcal{L}}}$ is the projection of the trace-free Lie derivative operator onto the $S_{t,u}$ (see Def. 8.2). The difficulty is that like $O\mathrm{tr}_{\not{g}}\chi$, $\hat{\not{\mathcal{L}}}_O \hat{\chi}$ is (schematically) of the form $\partial^3 u$ and hence, as we described above, has the potential to cause derivative loss. The saving grace is that, as was first shown in [**50**] based in part on the ideas of [**19**], one can derive an elliptic PDE on the spheres $S_{t,u}$ of the form $\mu \mathrm{d}\not{\mathrm{i}}\mathrm{v} \hat{\chi} = \cdots$, where $\cdots$ denotes controllable source terms that do not create derivative loss. The elliptic estimates imply that[33] $\mu \not{\nabla} \hat{\chi}$ can be suitably bounded in L^2 in terms of $\mu \mathrm{d}\not{\mathrm{i}}\mathrm{v} \hat{\chi}$ (see Lemma 18.9), and from that estimate, one can easily recover the desired L^2 estimates for $\hat{\not{\mathcal{L}}}_O \hat{\chi}$.

REMARK 2.15 (**More than one kind of modified quantity**). As we describe in more detail at the beginning of Chapter 11, in order to close our top-order energy estimates, we rely on more than one kind of modified quantity: fully modified quantities, and partially modified quantities. The fully modified quantities are the ones we described above and are analogs of the terms in braces in (2.59). We use the partially modified for a different reason: to avoid the presence of top-order error integrals with damaging time growth.

2.10.3. A hierarchy of square integral estimates. Now that we have sketched how we avoid losing derivatives in both Ψ and u we now provide a quantitative overview of the L^2 estimates that we derive in our sharp classical lifespan theorem (see Theorem 22.1 for the complete details). Our estimates involve the quantity

$$\mu_\star(t, u) := \min\{1, \min_{\Sigma_t^u} \mu\}, \tag{2.60}$$

which captures the "worst-case" behavior of μ on the constant-time hypersurface subset Σ_t^u. Much like in [**17**], we derive the following hierarchy of L^2 estimates for Ψ, where $\mathbb{E}[\cdot]$ and $\mathbb{F}[\cdot]$ are L^2-controlling in the sense of (2.48a) and (2.48b), $\mathring{\epsilon}$ is the size of the data (which we assume to be small) as defined in (2.7), $u \in [0, U_0]$, and *the estimates hold up to and including the first time that* $\mu_\star(\cdot, u)$ *vanishes*,[34] which corresponds to the time of first shock formation in a strip of eikonal function width u:

$$\mathbb{E}^{1/2}[\mathscr{Z}^{24}\Psi](t,u) + \mathbb{F}^{1/2}[\mathscr{Z}^{24}\Psi](t,u) \lesssim \mathring{\epsilon} \ln^{A_*}(e+t)\mu_\star^{-8.75}(t,u), \tag{2.61a}$$

$$\mathbb{E}^{1/2}[\mathscr{Z}^{23}\Psi](t,u) + \mathbb{F}^{1/2}[\mathscr{Z}^{23}\Psi](t,u) \lesssim \mathring{\epsilon}\mu_\star^{-7.75}(t,u), \tag{2.61b}$$

$$\cdots,$$

$$\mathbb{E}^{1/2}[\mathscr{Z}^{16}\Psi](t,u) + \mathbb{F}^{1/2}[\mathscr{Z}^{16}\Psi](t,u) \lesssim \mathring{\epsilon}\mu_\star^{-.75}(t,u), \tag{2.61c}$$

$$\mathbb{E}^{1/2}[\mathscr{Z}^{15}\Psi](t,u) + \mathbb{F}^{1/2}[\mathscr{Z}^{15}\Psi](t,u) \lesssim \mathring{\epsilon}, \tag{2.61d}$$

$$\cdots,$$

$$\mathbb{E}^{1/2}[\Psi](t,u) + \mathbb{F}^{1/2}[\Psi](t,u) \lesssim \mathring{\epsilon}. \tag{2.61e}$$

In the above estimates, $A_* > 0$ is a large constant, and $\mathscr{Z}^M$ denotes an arbitrary M^{th} order operator corresponding to differentiating M times with respect to vectorfields Z belonging to the commutation set $\mathscr{Z}$ from (2.51). The exponent 8.75

[33]Here and throughout, $\not{\nabla}$ denotes the Levi-Civita connection corresponding to the metric $\not{g}$ on $S_{t,u}$.

[34]Hence, the estimates hold for all time if no shock forms.

in (2.61a) is connected to certain structural constants in the equations that we describe below, and its size is intimately tied to the number of derivatives we need to close our estimates. We also derive a similar hierarchy for the Morawetz multiplier quantities $\widetilde{\mathbb{E}}$, $\widetilde{\mathbb{F}}$, and $\widetilde{\mathbb{K}}$ (see (2.52a)-(2.52b) and (2.54)) and for the norm $\|\cdot\|_{L^2(\Sigma_t^u)}$ of quantities such as $\upmu$, L^i, and χ constructed out of the eikonal function u.

REMARK 2.16 (**The number of derivatives needed**). We have used 25 derivatives to close our estimates partially out of convenience. We need many derivatives because of the degenerate top-order estimate (2.61a) and because our analysis allows us to gain only one power of $\upmu_\star$ as we descend in our L^2 estimates. As we explain below, the gain of one power at each step seems to be a fundamental, unavoidable feature, as does the need to descend to a non$\upmu_\star^{-1}$-degenerate level (such as (2.61d)). However, we expect that the total number of derivatives could be reduced with additional effort.

The proof of (2.61a)-(2.61e) *differs in some crucially important ways from the familiar bootstrap-type argument for deriving L^2 estimates.* One new feature, inherited from the framework of [**17**], is that even though $\upmu_\star$ is a "0th order" quantity, we cannot control all features of its behavior with a standard bootstrap argument because we cannot generally say how it will behave at late times. Hence, we have to find a way to derive the estimates (2.61a)-(2.61e) independent of whether or not $\upmu_\star$ goes to 0. As we describe in Sect. 2.10.5. some of this analysis is based on a posteriori estimates.

A second feature of crucial importance is that the lower-order estimates (2.61d)-(2.61e) do not degenerate at all, even if $\upmu_\star$ goes to 0. That is, from the vantage point of the rescaled frame, the lower-order derivatives of Ψ remain regular *all the way to the shock.* Equivalently, relative to the geometric coordinates $(t, u, \vartheta^1, \vartheta^2)$, Ψ remains regular. The nondegenerate estimates (2.61d)-(2.61e) are especially important because they can be used to derive, through a Sobolev embedding result (see Cor. 18.11), improvements of the C^0 bootstrap assumptions (2.34a)-(2.34b). As we have noted, in practice, we use the C^0 bootstrap assumptions to show that the terms we have deemed "error terms" are in fact small.

2.10.4. Top-order square integral estimates. We now explain how we derive the top-order estimate (2.61a) and in particular why the top-order quantities can blow up like a power of $1/\upmu_\star$. We consider the commuted wave equation (2.57) and, for the sake of illustration, we imagine that $\upmu\Box_{g(\Psi)}O\Psi = (\breve{R}\Psi)O\mathrm{tr}_{\not g}\chi + \cdots$ is the top-order wave equation. We now sketch a proof of why the difficult term $(\breve{R}\Psi)O\mathrm{tr}_{\not g}\chi$ leads to the degenerate energy estimate (2.61a). To proceed, we apply the divergence theorem identity (2.50) with $O\Psi$ in place of Ψ and retain only the difficult error integral generated by the term $(\breve{R}\Psi)O\mathrm{tr}_{\not g}\chi$ (which comes from the first integral on the right-hand side of (2.50)). Furthermore, we split $TO\Psi = 2\breve{R}O\Psi + (1+2\upmu)LO\Psi$ (see (2.47a)) and retain only the most difficult piece $2\breve{R}O\Psi$ involving transversal derivatives of Ψ. In total, we arrive at the following integral inequality (recall that $\mathring{\epsilon}$ is the size of the data; see (2.7)):

$$\mathbb{E}[O\Psi](t,u) + \mathbb{F}[O\Psi](t,u) \leq C\mathring{\epsilon}^2 - \int_{\mathcal{M}_{t,u}} (2\breve{R}\Psi)(O\mathrm{tr}_{\not g}\chi)\breve{R}O\Psi\, d\varpi + \cdots . \tag{2.62}$$

As we explained in Sect. 2.10.2, in order to avoid losing derivatives in (2.62), we have to replace the $O\mathrm{tr}_{\not g}\chi$ term with a modified quantity, that is, the term in braces on the left-hand side of (2.59). This replacement results in the appearance of a factor of $1/\mu$ into the $\mathcal{M}_{t,u}$ integral; as we describe below, this factor is of fundamental importance. We then have to estimate the error integrals corresponding to both the modified quantity and the difference between it and $O\mathrm{tr}_{\not g}\chi$. The analysis of the error integral generated by the modified quantity is extremely technical (see the proof of Lemma 20.25); it is arguably the most difficult analysis in the monograph. Thus, in order to avoid hijacking the discussion, we focus only on the discrepancy error integral, which is also difficult and is sufficient to illustrate the main ideas behind our derivation of (2.61a). The discrepancy is equal to $1/\mu$ multiplied by the Ψ-dependent quantities in braces on the left-hand side of (2.59). The factor $1/\mu$ is of supreme importance and is at the heart of many of the technical difficulties that we encounter in this monograph. In (2.59), the only explicitly written discrepancy term is $O(G_{LL}\breve{R}\Psi)$ because the remaining terms in $\cdots$ involve $\mathcal{C}_u$-tangential derivatives of Ψ and are much easier handle. Furthermore, only the top-order part $G_{LL}\breve{R}O\Psi$ is difficult to analyze, so we only discuss this part. Upon making these substitutions in (2.62), we obtain

$$\text{(2.63)}\quad \mathbb{E}[O\Psi](t,u) + \mathbb{F}[O\Psi](t,u) \leq C\mathring{\epsilon}^2 + 2\int_{\mathcal{M}_{t,u}} \frac{1}{\mu} G_{LL}(\breve{R}\Psi)(\breve{R}O\Psi)^2\, d\varpi + \cdots .$$

If we were to simply substitute the dispersive bound (2.34b) for $|\breve{R}\Psi|$ into (2.63), then the resulting inequality would lead, by a difficult Gronwall argument (based on an inequality in the spirit of (2.75) below), to the a priori estimate $\mathbb{E}^{1/2}[O\Psi](t,u) + \mathbb{F}^{1/2}[O\Psi](t,u) \lesssim \mathring{\epsilon}\mu_\star^{-c\mathring{\epsilon}\ln(e+t)}$, which is inadequate for closing our estimates due to the growing exponent. In particular, this line of reasoning would not allow us derive the a priori estimate (2.61a). To circumvent this difficulty, we adopt Christodoulou's strategy and use equation (2.39a) to replace $G_{LL}\breve{R}\Psi$ with $2L\mu$ plus some additional error integrals that are relatively easy to control because their integrands are decaying sufficiently fast in time and because they do not involve a singular factor of $1/\mu$ (thanks to the factor of μ multiplying the second term on the right-hand side of (2.39a)). Then, using the coerciveness estimate (2.48a), we see that (2.63) implies

$$\begin{aligned}\text{(2.64)}\quad \mathbb{E}[O\Psi](t,u) + \mathbb{F}[O\Psi](t,u) &\leq C\mathring{\epsilon}^2 - 4\int_{\mathcal{M}_{t,u}} \frac{L\mu}{\mu}(\breve{R}O\Psi)^2\, d\varpi + \cdots \\ &\leq C\mathring{\epsilon}^2 + 4\int_{t'=0}^{t} \sup_{\Sigma_{t'}^u}\left|\frac{L\mu}{\mu}\right| \mathbb{E}[O\Psi](t',u)\, dt' + \cdots .\end{aligned}$$

We now highlight two crucially important features of the estimate (2.64): **i)** the factor 4 is a universal "structural constant" that does not depend on how many times we commute the equations and **ii)** the "Gronwall factor" $\sup_{\Sigma_{t'}^u}\left|\frac{L\mu}{\mu}\right|$ on the right-hand side of (2.64) has a special structure that allows us to close our estimates. To see the first hint of the structure, we imagine that μ depends only on t and that μ is small and decreasing to 0 along the integral curves of $L = \frac{\partial}{\partial t}$, that is, that $L\mu < 0$. Then $\frac{1}{\mu}|L\mu| = \frac{\partial}{\partial t}\ln\frac{1}{\mu}$. Hence, applying Gronwall's inequality to

(2.64) and ignoring the terms $\cdots$, we would deduce that

$$\mathbb{E}[O\Psi](t,u) + \mathbb{F}[O\Psi](t,u) \lesssim \mathring{\epsilon}^2 \upmu^{-4}(t). \tag{2.65}$$

Roughly, this drastically simplified argument captures the main reason that the top-order estimate (2.61a) is degenerate. The reason that the power of $1/\upmu_\star$ in (2.61a) is larger than the power suggested by inequality (2.65) is that we have ignored the presence of some additional difficult error integrals that arise in the analysis of (2.64) and also in the analysis of the Morawetz energy-cone flux quantities $\widetilde{\mathbb{E}}$ and $\widetilde{\mathbb{F}}$. We also see why it is important that the structural constant 4 does not change as we repeatedly commute the equations; if the constant rapidly grew, then the power of $1/\upmu_\star$ on the right-hand side of the L^2 estimates (2.61a)-(2.61b) etc. would also grow, and we would have no reason to expect that we could eventually descend to a nondegenerate level (2.61d) and close the whole process.

2.10.5. The behavior of the critical Gronwall factor. It is not easy to rigorously derive an estimate in the spirit of (2.65) from inequality (2.64). Because the estimate is at the heart of our sharp classical lifespan theorem (Theorem 22.1), we now provide some additional details. As we will see, the most serious complication is that $\upmu(t,u,\vartheta)$ is highly (u,ϑ)-dependent, even though it is approximately monotonic at fixed (u,ϑ). In particular, $\upmu$ can either logarithmically grow or shrink in t depending on (u,ϑ). Consequently, as we explain below, in order to derive suitable estimates for $\upmu$ that account for its full spectrum of possible behaviors, we must complement our a priori estimates with *a posteriori estimates.*

The key ingredient driving the approximate monotonicity of $\upmu$ is its *smaller-than-expected weighted acceleration* along the integral curves of L. More precisely, we have the following better-than-expected estimate:

$$|L(\varrho L\upmu)| \lesssim \mathring{\epsilon} \frac{\ln(e+t)}{(1+t)^2}, \tag{2.66}$$

where the right-hand side is integrable in t. By "better-than-expected," we mean that a naive approach based only on the dispersive-type bootstrap assumptions (2.34) would lead to the nonintegrable rate $(1+t)^{-1}$ on the right-hand side of (2.66), which is not sufficient for deriving any of our main results. The reason that we can derive the estimate (2.66) is that the only potentially slowly decaying term on the right-hand side of the equation $L(\varrho L\upmu) = \cdots$ (see equation (2.39a)) is $G_{LL} L(\varrho \breve{R}\Psi)$, and furthermore, for solutions to $\square_{g(\Psi)}\Psi = 0$, $L(\varrho\breve{R}\Psi)$ decays at an integrable-in-time rate (see Lemma 12.58). This fact is familiar from the case of the linear wave equation in three spatial dimensions, whose solutions Ψ verify $\left|L_{(Flat)}(rR_{(Flat)}\Psi)\right| \lesssim (1+t)^{-2}$ (see (2.19) for definitions of $L_{(Flat)}$ and $R_{(Flat)}$). By twice integrating (2.66) along the integral curves of L from the initial data hypersurface to a late time t and using[35] $\upmu(0,\cdot) \sim 1$, we see that for times $0 \le s \le t$, $L\upmu$ and $\upmu$ can be heuristically modeled by

$$L\upmu(s,u,\vartheta) \sim -\delta_{t,u,\vartheta} \frac{1}{\varrho(s,u)}, \tag{2.67a}$$

$$\upmu(s,u,\vartheta) \sim 1 - \delta_{t,u,\vartheta} \ln\left(\frac{\varrho(s,u)}{\varrho(0,u)}\right), \tag{2.67b}$$

[35]Throughout Chapter 2, we often write "$A \sim B$" to imprecisely indicate that A is well-approximated by B.

where $\delta_{t,u,\vartheta} := -[\varrho L\mu](t,u,\vartheta)$. We stress that the estimates (2.67a)-(2.67b) are not purely a priori in nature; they involve the quantity $\delta_{t,u,\vartheta}$, which is a posteriori in nature. For the sake of illustration, we imagine that $\delta_{t,u,\vartheta} \geq 0$ for all t, u, ϑ and set $\delta_t := \sup_{(u',\vartheta)\in[0,u]\times\mathbb{S}^2} \delta_{t,u',\vartheta}$. This corresponds to a scenario in which μ is shrinking along all integral curves of L at a curve-dependent rate. We then find that the difficult factor on the right-hand side of (2.64) can be bounded as follows:

$$\int_{t'=0}^{t} \sup_{\Sigma_{t'}^{u}} \left| \frac{L\mu}{\mu} \right| dt' \leq \int_{s=0}^{t} \frac{\delta_t}{\varrho(s,u)\left\{1 - \delta_t \ln\left(\frac{\varrho(s,u)}{\varrho(0,u)}\right)\right\}} ds \tag{2.68}$$
$$= \ln\left\{1 - \delta_t \ln\left(\frac{\varrho(t,u)}{\varrho(0,u)}\right)\right\} \sim \ln \mu_\star^{-1}(t,u).$$

We stress that the important feature that allows us to arrive at the right-hand side of (2.68) is that the *same constant* δ_t appears in the both the numerator and denominator in the ds integral. Clearly, the bound (2.68), when combined with inequality (2.64) and Gronwall's inequality, leads to the desired estimate $\mathbb{E}[O\Psi](t,u) + \mathbb{F}[O\Psi](t,u) \lesssim \mathring{\epsilon}\mu_\star^{-4}(t,u)$. In summary, the only feature of (2.61a) not captured by our heuristic analysis is the growth factor $\ln^{A_*}(e+t)$. The factor $\ln^{A_*}(e+t)$, which is not very important and in any case is present only in the top-order estimates, appears because along some integral curves of L, μ can grow and the ratio $\frac{L\mu}{\mu}$ can decay at the slow rate $\{(1+t)\ln(e+t)\}^{-1}$ (see inequality (13.26)).

2.10.6. The descent scheme for the below-top-order square integral estimates. If we used the strategy of Sect. 2.10.4 to estimate $\mathbb{E}[\cdot]$ and $\mathbb{F}[\cdot]$ at all derivative levels, then all of our L^2 estimates would experience the same degeneracy with respect to powers of $1/\mu_\star$ as the top-order estimate (2.61a). If this were the best we could do, then there would be no hope of improving the bootstrap assumptions (2.34) through Sobolev embedding, and our proof would fall apart. Hence, it is extremely important that the below-top-order L^2 estimates (2.61b)-(2.61e) become less degenerate in powers of $1/\mu_\star$ as we descend. The main idea behind deriving the improved estimates is that at the lower levels, we can avoid using modified quantities and simply allow the loss of one derivative. We then need to make sure that this procedure somehow results in a gain of a power of $\mu_\star$. The reasons that we can in fact gain are **i)** we have sharp information about how $\mu_\star$ behaves (in the difficult case where μ is shrinking, we roughly have the caricature estimate (2.67b)) and **ii)** by avoiding the use of modified quantities, we see a large gain in the powers of t available in the dangerous error integrand on the right-hand side of (2.62).

To illustrate the descent scheme, let us imagine that the third-order derivatives of Ψ are top-order and that we are trying to deduce improved L^2 estimates for the second-order derivatives of Ψ, that is, estimates that are less degenerate with respect to powers of $\mu_\star^{-1}(t,u)$. For simplicity, we imagine that we have proved the following top-order L^2 estimates for the third-order derivatives, which are consistent with (2.61a) up to the unimportant factor $\ln^{A_*}(e+t)$:

$$\mathbb{E}^{1/2}[\mathscr{Z}^2\Psi](t,u) - \mathbb{F}^{1/2}[\mathscr{Z}^2\Psi](t,u) \lesssim \mathring{\epsilon}\mu_\star^{-B}(t,u), \tag{2.69}$$

$$\widetilde{\mathbb{E}}^{1/2}[\mathscr{Z}^2\Psi](t,u) - \widetilde{\mathbb{F}}^{1/2}[\mathscr{Z}^2\Psi](t,u) \lesssim \mathring{\epsilon}\mu_\star^{-B}(t,u), \tag{2.70}$$

where $B > 1$. We now sketch a proof of how to use (2.69)-(2.70) to derive estimates for the just-below-top-order quantities $\mathbb{E}^{1/2}[O\Psi](t,u) + \mathbb{F}^{1/2}[O\Psi](t,u)$ that are similar to (2.69) *but with the exponent B reduced by one*, as in (2.61b).

To derive the desired estimate, we estimate the error integral

$$-\int_{\mathcal{M}_{t,u}} (2\breve{R}\Psi)(O\mathrm{tr}_{\not g}\chi)\breve{R}O\Psi\, d\varpi$$

on the right-hand side of (2.62) in a different way, one which involves a loss of one derivative (which is permissible below top order). To proceed, we commute (2.58) with ϱ^2 to derive[36] the equation[37]

$$L(\varrho^2 O\mathrm{tr}_{\not g}\chi) = \frac{1}{2}\varrho^2 G_{LL}\not\Delta O\Psi + \cdots, \tag{2.71}$$

use the following property (see (12.46c)) of our rotation vectorfields (familiar from the case of Minkowski spacetime):

$$\varrho^2 |\not\Delta\Psi| \lesssim (1+t)\sum_{l=1}^{3} |\not d O_{(l)}\Psi|, \tag{2.72}$$

and use the coerciveness property (2.52a) to deduce the schematic inequality

(2.73)
$$\begin{aligned}\left\|L(\varrho^2 O\mathrm{tr}_{\not g}\chi)\right\|_{L^2(\Sigma_t^u)} \lesssim (1+t)\left\|\not d O\Psi\right\|_{L^2(\Sigma_t^u)} + \cdots &\lesssim \upmu_\star^{-1/2}(t,u)\widetilde{\mathbb{E}}^{1/2}[O\Psi](t,u) + \cdots \\ &\lesssim \mathring{\epsilon}\upmu_\star^{-B-1/2}(t,u) + \cdots .\end{aligned}$$

Note that we have generated an additional factor $\upmu_\star^{-1/2}$ on the right-hand side of (2.73) because of the factor $\sqrt{\upmu}$ in the second term on the right-hand side of (2.52a). It is not too difficult (see Lemma 12.57) to integrate (2.73) in time (recall that $L = \frac{\partial}{\partial t}$) and to account for the size of the $S_{t,u}$ area form $d\upsilon_{\not g} \sim \varrho^2\, d\vartheta$ to deduce that

$$\left\|\varrho^2 O\mathrm{tr}_{\not g}\chi\right\|_{L^2(\Sigma_t^u)} \lesssim \mathring{\epsilon}(1+t)\int_{t'=0}^{t} \frac{1}{(1+t')\upmu_\star^{B+1/2}(t',u)}\, dt' + \cdots . \tag{2.74}$$

The important point concerning the time integral in (2.74) is that we can prove the following estimate, which allows us to gain a power of $\upmu_\star$ by integrating in time, at the expense of introducing a factor $\ln(e+t)$, which is harmless below top order because of the good weight ϱ^2 available on the left-hand side of (2.74):

$$\int_{t'=0}^{t} \frac{1}{(1+t')\upmu_\star^{B+1/2}(t',u)}\, dt' \lesssim \ln(e+t)\upmu_\star^{1/2-B}(t,u). \tag{2.75}$$

The estimate (2.75) can be proved by using arguments similar to but much simpler than the ones we used in sketching the proof of (2.68); see inequality (13.147) and its proof. Inserting (2.75) into (2.74) and using $\varrho \approx 1+t$, we find that

$$\left\|O\mathrm{tr}_{\not g}\chi\right\|_{L^2(\Sigma_t^u)} \lesssim \mathring{\epsilon}\frac{\ln(e+t)}{1+t}\upmu_\star^{1/2-B}(t,u) + \cdots . \tag{2.76}$$

---

[36] Recall that $\varrho \approx 1+t$ in the region of interest.

[37] The point of including the factor ϱ^2 is that $\varrho^2 O\mathrm{tr}_{\not g}\chi$ verifies a good equation because ϱ^2 leads to the cancellation of a linear term; see the proof of Lemma 12.45.

We now insert the estimate (2.76) into the error integral on the right-hand side of (2.62) and use Cauchy-Schwarz, the bootstrap assumption (2.34b), and the coerciveness estimate (2.48a) to deduce

(2.77)
$$\begin{aligned}\left|\int_{\mathcal{M}_{t,u}} 2(\breve{R}\Psi)(O\mathrm{tr}_{\not g}\chi)\breve{R}O\Psi\, d\varpi\right| &\lesssim \mathring{\epsilon}\int_{t'=0}^{t}\frac{1}{1+t'}\left\|O\mathrm{tr}_{\not g}\chi\right\|_{L^2(\Sigma_{t'}^u)}\left\|\breve{R}O\Psi\right\|_{L^2(\Sigma_{t'}^u)}\,dt' \\ &\lesssim \mathring{\epsilon}^2\int_{t'=0}^{t}\frac{\ln(e+t')}{(1+t')^2\mu_\star^{B-1/2}(t',u)}\mathbb{E}^{1/2}[O\Psi](t',u)\,dt' \\ &\quad +\cdots.\end{aligned}$$

Hence, from (2.62) and (2.77), we have

(2.78)
$$\begin{aligned}&\sup_{s\in[0,t]}\{\mathbb{E}[O\Psi](s,u)-\mathbb{F}[O\Psi](s,u)\} \\ &\leq C\mathring{\epsilon}^2+\sup_{s\in[0,t]}\mathbb{E}^{1/2}[O\Psi](s,u)\mathring{\epsilon}^2\int_{t'=0}^{t}\frac{\ln(e+t')}{(1+t')^2\mu_\star^{B-1/2}(t',u)}\,dt' \\ &\quad+\cdots.\end{aligned}$$

We now use inequality (2.78) and an argument similar to the one used to deduce (2.75) in order to gain a power of $\mu_\star$ through time integration, but this time taking advantage of the additional decay available in the integrand of (2.78); the decay allows us to avoid the $\ln(e+t)$ factor in (2.75) (see (13.146) for the precise estimate). We thus conclude the desired reduction in the power of $1/\mu_\star$:

(2.79)
$$\mathbb{E}^{1/2}[O\Psi](t,u)+\mathbb{F}^{1/2}[O\Psi](t,u)\lesssim\mathring{\epsilon}\mu_\star^{3/2-B}(t,u)+\cdots.$$

We remark that the estimate (2.79) is slightly misleading in the sense that it suggests that we can gain $\mu_\star^{3/2}$ at each stage in the descent. In reality, there are additional error integrals on the right-hand side of (2.62) that only allow us to gain a single power of $\mu_\star$, and hence we have the hierarchy (2.61a)-(2.61e).

2.11. Comparison with related work

We now compare our work with that of Alinhac and Christodoulou.

2.11.1. Alinhac's shock formation results. Alinhac obtained some important results that significantly advanced our understanding of singularity formation in solutions to wave equations. Specifically, he was the first to understand the nature of the first singularity in small-data solutions to a class of quasilinear wave equations [**2**, **4**–**6**]. As we describe below, he proved finite-time shock formation when the data are small, compactly supported, and verify some nondegeneracy conditions that generically hold.[38] He studied equations in both two and three spatial dimensions, but we discuss here only the case of three spatial dimensions, which, despite the title, is addressed in [**2**]. We now briefly summarize Alinhac's

[38] For equations that are invariant under Euclidean rotations, some small data containing a spherically symmetric angular sector do not verify Alinhac's nondegeneracy assumptions.

results. His work applied to equation (2.9) under the assumptions (2.10)-(2.11). That is, he studied the Cauchy problem[39]

$$(h^{-1})^{\alpha\beta}(\partial\Phi)\partial_\alpha\partial_\beta\Phi = 0, \tag{2.80a}$$

$$(\Phi|_{t=0}, \partial_t\Phi|_{t=0}) = (\mathring{\Phi}, \mathring{\Phi}_0). \tag{2.80b}$$

More precisely, he studied solutions corresponding to a one-parameter family of compactly supported data of the form $(\lambda\mathring{\Phi}, \lambda\mathring{\Phi}_0)$, where $\lambda \in (0,1)$ was chosen to be sufficiently small (and the amount of smallness needed depends on $(\mathring{\Phi}, \mathring{\Phi}_0)$). The correct analog of the future null condition failure factor (2.22) for equation (2.80a) is

$$^{(+)}\aleph := m_{\kappa\lambda}H^{\kappa}_{\mu\nu}(\partial\Phi = 0)L^{\lambda}_{(Flat)}L^{\mu}_{(Flat)}L^{\nu}_{(Flat)}, \tag{2.81}$$

where $m_{\mu\nu}$ is the Minkowski metric and relative to rectangular coordinates, we have

$$H^{\lambda}_{\mu\nu}(\partial\Phi) := \frac{\partial}{\partial(\partial_\lambda\Phi)}h^{(Small)}_{\mu\nu}(\partial\Phi). \tag{2.82}$$

In particular, when $^{(+)}\aleph \equiv 0$, Klainerman's (classic) null condition [**45**] is verified and his work [**47**] (and Christodoulou's work [**15**]) yields small-data global existence. As in the case of (2.22), the quantity $^{(+)}\aleph$ defined in (2.81) can be viewed as a function depending only on $\theta = (\theta^1, \theta^2)$, where θ^1 and θ^2 are local angular coordinates corresponding to spherical coordinates on Minkowski spacetime.

Alinhac's results partially confirmed a conjecture of John, where by "partially," we mean that his proof used, in an essential fashion, his aforementioned nondegeneracy assumptions on the data. To further explain these assertions, we first state the definition of the Radon transform of a function f on $\mathbb{R}^3$. Given any such f and points $q \in \mathbb{R}$, $\theta \in \mathbb{S}^2 \subset \mathbb{R}^3$, we define

$$\mathcal{R}[f](q,\theta) := \int_{P_{q,\theta}} f(y)\,d\sigma_{q,\theta}(y), \tag{2.83}$$

where $P_{q,\theta} := \{y \in \mathbb{R}^3 \mid e(\theta, y) = q\}$, $d\sigma_{q,\theta}(y)$ denotes the area form induced on the two-dimensional plane $P_{q,\theta}$ by the Euclidean metric e on $\mathbb{R}^3$, and $e(\theta, y)$ is the Euclidean inner product of θ and y (where both are viewed as vectors in $\mathbb{R}^3$, the former being unit length). Alinhac's condition for shock formation involves the following function of (q,θ), which also depends on the data $(\mathring{\Phi}, \mathring{\Phi}_0)$:

$$\digamma[(\mathring{\Phi}, \mathring{\Phi}_0)](q,\theta) := -\frac{1}{4\pi}\frac{\partial}{\partial q}|_\theta \mathcal{R}[\mathring{\Phi}](q,\theta) + \frac{1}{4\pi}\mathcal{R}[\mathring{\Phi}_0](q,\theta). \tag{2.84}$$

The function $\digamma[(\mathring{\Phi}, \mathring{\Phi}_0)]$ is Friedlander's radiation field for the solution to the *linear* wave equation $\square_m\Phi = 0$ corresponding to the data $(\mathring{\Phi}, \mathring{\Phi}_0)$. That is, the r-weighted linear solution $r\Phi_{(Linear)}$, relative to spherical coordinates (t, r, θ) on Minkowski spacetime, is asymptotic to $\digamma[(\mathring{\Phi}, \mathring{\Phi}_0)](q = r - t, \theta)$ as $t \to \infty$ in regions with $|r - t|$ uniformly bounded. The relevance of $\digamma$ for the problem of shock formation lies in the fact that $\digamma[(\mathring{\Phi}, \mathring{\Phi}_0)](q = r - t, \theta)$ provides a good approximation to the *nonlinear* solution at time $\mathring{\epsilon}^{-1}$ (where $\mathring{\epsilon}$ is the small size of the data), long before any singularity has formed. Roughly speaking, at time $\mathring{\epsilon}^{-1}$, for a class of small-data solutions that form shocks, the dangerous term that eventually causes the shock

[39] Following the conventions of [**17**], we denote the dynamic metric by $h = h(\partial\Phi)$ in this section and the next one.

formation has acquired the "shock driving" sign, at least along some integral curves of the outgoing null vectorfield L. It takes a certain amount of time to see the sign of the dangerous term because one must wait for certain $\mathcal{C}_u$-tangent derivatives to sufficiently decay before its sign becomes visible; see Sect. 2.11.3 for additional discussion. Hence, for a class of small-data solutions, whether or not shock formation occurs can be detected from the state of the dangerous term relatively early in the evolution, at time $\mathring{\epsilon}^{-1}$. Furthermore, the state of the dangerous term at time $\mathring{\epsilon}^{-1}$ can in turn be determined from the data with the help of $F[(\mathring{\Phi}, \mathring{\Phi}_0)](q = r - t, \theta)$.

A connection between F and the lifespan of the solution to the nonlinear equation (2.80a) was first observed by John [**39**] and Hörmander [**29**]. In these works, they proved that for initial data of the form $(\lambda\mathring{\Phi}, \lambda\mathring{\Phi}_0)$ with $\lambda > 0$, the classical lifespan $T_{(Lifespan);\lambda}$ of the solution to equation (2.80a) verifies

$$\liminf_{\lambda \downarrow 0} \lambda \ln T_{(Lifespan);\lambda} \geq \frac{1}{\sup_{(q,\theta)\in\mathbb{R}\times\mathbb{S}^2} \frac{1}{2} {}^{(+)}\aleph(\theta) \frac{\partial^2}{\partial q^2} |_\theta F[(\mathring{\Phi}, \mathring{\Phi}_0)](q, \theta)}, \tag{2.85}$$

where ${}^{(+)}\aleph$ is defined in (2.81). John conjectured that inequality (2.85) is sharp and that in the limit $\lambda \downarrow 0$, the lower bound also serves as an upper bound. He made significant progress towards proving his conjecture [**40, 41**] by showing that near the expected blow-up time, the second rectangular derivatives of the solution start to grow.

We now motivate John's conjecture and the result (2.85). We denote the solution to (2.80a) corresponding to the rescaled data $(\lambda\mathring{\Phi}, \lambda\mathring{\Phi}_0)$ by Φ_λ. We first note that an important aspect of the works [**39**] and [**29**] is that they show that the dispersive decay rates corresponding to the linear wave equation (see Sect. 1.2.3) are verified by small-data solutions to the nonlinear equation (2.80a) until very close to the conjectured singularity time. Their proof was based on Klainerman's Minkowskian vectorfield method, the Minkowskian eikonal function $r - t$, and F, whose role we further explain below. Hence, one can obtain a good approximation to equation (2.80a), valid until near the conjectured first singularity time, by expanding it relative to the Minkowskian frame (2.19) and keeping only the quadratic term that fails the (classic) null condition (see the discussion in Sects. 1.2.4 and 2.4) and the related linear term that drives its evolution along the integral curves of $L_{(Flat)} = \partial_t + \partial_r$. Also using the approximation $\frac{r}{t} \sim 1$ we find that a suitable approximate equation is as follows (where we recall that $R_{(Flat)} := -\partial_r$):

$$L_{(Flat)}(rR_{(Flat)}\Phi_\lambda) - \frac{1}{2}\frac{1}{t}{}^{(+)}\aleph(rR_{(Flat)}\Phi_\lambda)R_{(Flat)}(rR_{(Flat)}\Phi_\lambda) = 0. \tag{2.86}$$

The approximation $\frac{r}{t} \sim 1$ used in motivating (2.86) is accurate at large times in the Minkowskian wave zone $|r - t| \approx 1$, which, in the case of compactly supported data, is the region where one expects the first singularity to form. Note also that in obtaining (2.86), we have discarded Euclidean angular derivatives (which, as we described in Sect. 1.2.3, decay quickly, at least until near the shock). Equation (2.86) is a Burgers-type equation in the unknown $rR_{(Flat)}\Phi_\lambda$ whose *first inward radial derivative* $R_{(Flat)}(rR_{(Flat)}\Phi_\lambda)$ can blow up along the integral curves of $L_{(Flat)}$.

From the point of view of understanding the blow-up of solutions to the nonlinear wave equation (2.80a), the most relevant data for the approximating equation (2.86) is not $rR_{(Flat)}\Phi_\lambda|_{t=0}$. The reason is that equation (2.86) does not take into account the influence of some of the linear terms in equation (2.80a), such as the Euclidean angular derivatives, which can be influential in the early phase of the dynamics. As we explain in more in more detail in Sect. 2.11.3, this early phase lasts roughly until time λ^{-1} (note that λ can roughly be viewed as the size of the data). Moreover, the "data" induced at time λ^{-1} by the approximately linear evolution is effectively determined, up to small errors, by $\digamma$ as follows relative to spherical coordinates (t, r, θ) on Minkowski spacetime:

$$rR_{(Flat)}\Phi_{\lambda}(\lambda^{-1}, r, \theta) \sim -\lambda^{(+)}\aleph(\theta)\frac{\partial}{\partial q}|_{\theta}\digamma[(\mathring{\Phi}, \mathring{\Phi}_0)](r - \lambda^{-1}, \theta). \tag{2.87a}$$

Similarly,

$$R_{(Flat)}(rR_{(Flat)}\Phi_{\lambda})(\lambda^{-1}, r, \theta) \sim \lambda^{(+)}\aleph(\theta)\frac{\partial^2}{\partial q^2}|_{\theta}\digamma[(\mathring{\Phi}, \mathring{\Phi}_0)](r - \lambda^{-1}, \theta). \tag{2.87b}$$

By modifying our first (crude) proof of blow-up for solutions to Burgers' equation (given just below equation (1.5a)), we compute that the solution to the model equation (2.86) corresponding to the data (2.87a)-(2.87b) is such that $R_{(Flat)}(rR_{(Flat)}\Phi_{\lambda})$ blows up along the integral curve of $L_{(Flat)}$ emanating from the point with Minkowskian spherical coordinates $(\lambda^{-1}, r, \theta)$ near the time $\sim \exp\left\{\frac{2}{\lambda^{(+)}\aleph(\theta)\frac{\partial^2}{\partial q^2}|_{\theta}\digamma[(\mathring{\Phi}, \mathring{\Phi}_0)](r - \lambda^{-1}, \theta)}\right\}$, whenever the argument of exp is positive. Hence, the lifespan of the solution to equation (2.86) corresponding to the data (2.87a)-(2.87b) is well-approximated by the time corresponding to solving for $T_{(Lifespan);\lambda}$ in inequality (2.85).

It is easy to see that for compactly supported data, the denominator on the right-hand of (2.85) is ≥ 0. It is natural to wonder whether or not there exist any nontrivial compactly supported data such that the denominator on the right-hand side of (2.85) is equal to 0. The results of John [**39**] and Hörmander [**29**] suggest that such data would lead to a global solution (since in this case, the right-hand side of (2.85) would be infinite). However, John showed [**39**, pg. 98] that when ${}^{(+)}\aleph \not\equiv 0$, there are no such data. We state his important observation as a proposition and sketch its simple proof.

PROPOSITION 2.17 (**Only trivial data cause the right-hand side of** (2.85) **to vanish**). *Let $\mathring{\Phi}, \mathring{\Phi}_0 \in C_c^{\infty}(\mathbb{R}^3)$. Assume that ${}^{(+)}\aleph \not\equiv 0$ and that*

$$\sup_{(q,\theta)\in\mathbb{R}\times\mathbb{S}^2} {}^{(+)}\aleph(\theta)\frac{\partial^2}{\partial q^2}|_{\theta}\digamma[(\mathring{\Phi}, \mathring{\Phi}_0)](q, \theta) = 0. \tag{2.88}$$

Then $(\mathring{\Phi}, \mathring{\Phi}_0) = (0, 0)$.

SKETCH OF PROOF. We will deduce from the assumptions of the proposition that $\digamma[(\mathring{\Phi}, \mathring{\Phi}_0)] \equiv 0$. Then using the fact that $\mathcal{R}[\cdot](q, \theta) = \mathcal{R}[\cdot](-q, -\theta)$ and the two identities

$$\digamma[(\mathring{\Phi}, \mathring{\Phi}_0)](q, \theta) \pm \digamma[(\mathring{\Phi}, \mathring{\Phi}_0)](-q, -\theta) = 0,$$

where $-\theta$ denotes the polar opposite of θ on $\mathbb{S}^2$, it is straightforward to see from (2.84) that $\mathring{\Phi}$ and $\mathring{\Phi}_0$ have vanishing Radon transforms. The proposition then follows from the fact that a compactly supported function with a vanishing Radon transform must completely vanish (see, for example, [**27**, Corollary 2.8]).

To show that $\mathcal{F}[(\mathring{\Phi}, \mathring{\Phi}_0)] \equiv 0$, we partition $\mathbb{S}^2$ into an open set $\mathcal{O}$ and a closed set $\mathcal{C}$ as follows:

$$\mathbb{S}^2 = \mathcal{O} \cup \mathcal{C}, \tag{2.89}$$

$$\mathcal{O} := \{\theta \in \mathbb{S}^2 \mid {}^{(+)}\aleph(\theta) \neq 0\}, \qquad \mathcal{C} := \{\theta \in \mathbb{S}^2 \mid {}^{(-)}\aleph(\theta) = 0\}. \tag{2.90}$$

For any fixed $\theta \in \mathcal{O}$, we deduce from (2.88) that $\frac{\partial^2}{\partial q^2}|_\theta \mathcal{F}[(\mathring{\Phi}, \mathring{\Phi}_0)](q, \theta)$ is either nonpositive or nonnegative for $q \in \mathbb{R}$. Since $\mathcal{F}[(\mathring{\Phi}, \mathring{\Phi}_0)](q, \theta)$ vanishes for large $|q|$ (this fact is essentially the sharp version of Huygens' principle), it is straightforward to see that $\mathcal{F}[(\mathring{\Phi}, \mathring{\Phi}_0)](q, \theta) = 0$ for $q \in \mathbb{R}$. We next note that since ${}^{(+)}\aleph$ is a nontrivial analytic function on $\mathbb{S}^2$, $\mathcal{C}$ must have an empty interior and thus $\mathcal{C} = \partial\mathcal{C}$. Hence, given any $\theta \in \mathcal{C}$, we conclude from continuity and the property $\mathcal{F}[(\mathring{\Phi}, \mathring{\Phi}_0)]|_{\mathbb{R}\times\mathcal{O}} \equiv 0$ that $\mathcal{F}[(\mathring{\Phi}, \mathring{\Phi}_0)](q, \theta) = 0$ for $q \in \mathbb{R}$. We have thus shown that $\mathcal{F}[(\mathring{\Phi}, \mathring{\Phi}_0)] \equiv 0$ as desired. □

We now summarize the main features of Alinhac's results as Theorem 2.18. The results stated in the theorem are a partial summary of Theorems 2 and 3 of [**2**] in the case of three spatial dimensions. We stress that *his main contribution was confirming that John's conjecture holds for a large set of data verifying uniqueness and nondegeneracy conditions*; see just below equation (2.91).

THEOREM 2.18 (**Alinhac**). *Let* $(\lambda\mathring{\Phi}, \lambda\mathring{\Phi}_0) \in C^\infty(\mathbb{R}^3) \times C^\infty(\mathbb{R}^3)$ *be a one-parameter family of compactly supported data for the quasilinear wave equation* (2.80a) *under the assumptions* (2.10)-(2.11). *Let* Φ_λ *denote the corresponding solution. Assume that the (data-dependent) function*

$$\frac{1}{2}{}^{(+)}\aleph(\theta)\frac{\partial^2}{\partial q^2}|_\theta \mathcal{F}[(\mathring{\Phi}, \mathring{\Phi}_0)](q, \theta) \tag{2.91}$$

has a unique, strictly positive, nondegenerate maximum at (q_*, θ_*), *where the future null condition failure factor* ${}^{(+)}\aleph$ *is defined in* (2.81). *If* λ *is sufficiently small and positive (where the amount of smallness needed depends on* $(\mathring{\Phi}, \mathring{\Phi}_0)$*), then we have the following conclusions.*

Sharp classical lifespan result. *The classical lifespan* $T_{(Lifespan);\lambda}$ *of* Φ_λ *is finite and verifies*

$$\lim_{\lambda\downarrow 0} \lambda \ln T_{(Lifespan);\lambda} = \frac{1}{\frac{1}{2}{}^{(+)}\aleph(\theta_*)\frac{\partial^2}{\partial q^2}|_\theta \mathcal{F}[(\mathring{\Phi}, \mathring{\Phi}_0)](q_*, \theta_*)}. \tag{2.92}$$

Sharp description near the unique first blow-up point. *Along the constant-time hypersurface* $\Sigma_{T_{(Lifespan);\lambda}}$ *of first shock formation, there is a unique point* $p_{(Blow-up);\lambda}$ *where the solution blows up. In addition, with* $C = C[(\mathring{\Phi}, \mathring{\Phi}_0)]$, Φ_λ *and its first rectangular derivatives verify the following pointwise bound for* $\{0 \leq t \leq T_{(Lifespan);\lambda}\}$*:*

$$|\Phi_\lambda| + \sum_{\alpha=0}^{3} |\partial_\alpha \Phi_\lambda| \leq C\lambda\frac{1}{1+t}. \tag{2.93}$$

In particular, Φ_λ *and its first rectangular derivatives remain finite along* $\Sigma_{T_{(Lifespan);\lambda}}$. *Moreover, in the* **complement** *of a small neighborhood of* $p_{(Blow-up);\lambda}$ *intersected with* $\{0 \leq t \leq T_{(Lifespan);\lambda}\}$, *the second-order rectangular derivatives of* Φ_λ *are* *bounded in magnitude by* $\leq$ *the right-hand side of* (2.93). *In contrast, the following blow-up behavior occurs*[40] *for* t *close to and* $\leq T_{(Lifespan);\lambda}$:

$$
\begin{aligned}
(2.94) \qquad & C^{-1}\left\{t \ln\left(\frac{T_{(Lifespan);\lambda}}{t}\right)\right\}^{-1} \\
& \leq \sum_{\alpha,\beta=0}^{3} \|\partial_\alpha \partial_\beta \Phi_\lambda\|_{C^0(\Sigma_t)} \\
& \leq C\left\{t \ln\left(\frac{T_{(Lifespan);\lambda}}{t}\right)\right\}^{-1}.
\end{aligned}
$$

Furthermore, in the intersection of $\{t \leq T_{(Lifespan);\lambda}\}$ *with some neighborhood of* $p_{(Blow-up);\lambda}$, *there exist an eikonal function* u *and a related system of geometric coordinates, one of which is* u, *such that relative to the geometric coordinates,* Φ_λ *and its higher derivatives extend smoothly to* $\Sigma_{T_{(Lifespan);\lambda}}$, *even at the point* $p_{(Blow-up);\lambda}$. *Within this region, the change of variables map from geometric to rectangular coordinates is smooth and invertible except at the point* $p_{(Blow-up);\lambda}$, *where its Jacobian determinant vanishes.*

We now highlight two merits of Alinhac's results. First, he was the first to show that failure of the (classic) null condition in equation (2.80a) often leads, in the small-data regime, to the formation of a singularity caused by the intersection of the level sets of a true eikonal function. Second, for reasons explained below, his proof is relatively short.

The main limitation of Alinhac's framework is that it is fundamentally tailored to the first point of intersection of the characteristics. Thus, his results cannot easily be extended to provide information about the maximal development of the data. We further discuss the origin of this limitation two paragraphs below. In contrast, as we describe in Sects. 2.11.2 and 2.11.3, Christodoulou's approach and the approach of the present monograph allow one to obtain a detailed description of the maximal future development of the data in the exterior of the sphere $S_{0,U_0} \subset \Sigma_0^{U_0}$. As we explained in the Preface, this sharp information is essential for attacking the problem of extending the solution, in a generalized sense, beyond the shock. Furthermore, Alinhac's results are fundamentally based on his nondegeneracy assumptions on the data, which are stated just below (2.91) and which "force" the solution to form a shock. Thus, his results do not provide a true analog of the sharp classical lifespan theorem proved in [**17**] and in Theorem 22.1 of the present monograph, which imply that singularities in small-data solutions can only be caused by the vanishing of μ. Moreover, his assumptions are not verified by some data containing a spherically symmetric sector when the wave equation is invariant under Euclidean rotations.

We now sketch some of the main ideas behind Alinhac's approach, which is quite different than the approach of [**17**] and the present monograph. As we will describe in the next paragraph, the most important part of his analysis is based on a type of "backwards approach," as opposed to the "forwards approach" associated to solving a pure Cauchy problem, which was adopted in [**17**] and the present monograph. To proceed with our sketch, we first recall John's conjecture that the lifespan lower bound (2.85) proved in [**39**] and [**29**] is also an upper bound in the small-data

[40]The norm $\|\cdot\|_{C^0(\Sigma_t)}$ is the standard sup-norm (see Def. 3.70).

limit. Alinhac's proof is grounded in the belief that one should expand the actual lifespan $T_{(Lifespan);\lambda}$ of the solution to (2.80a)-(2.80b) as the lifespan predicted by John's conjecture plus a small error to be solved for. Thus, he splits the nonlinear evolution into two stages. In the first stage, he simply quotes the results that we described above when motivating John's conjecture via the approximating equation (2.86). That is, Alinhac uses the results of [**39**] and [**29**], which were based on Klainerman's Minkowskian vectorfield method, the Minkowskian eikonal function $r - t$, and Friedlander's radiation field for the linear wave equation, to follow the solution almost all the way to the lifespan predicted by John's conjecture.

In the second stage, near the very end of the solution's classical lifespan, Alinhac constructs, via an iteration scheme based on Nash-Moser-type estimates, a true eikonal function that enables him to follow the solution all the way to the first singularity. The "initial" data for this stage are of course inherited from the state of the solution at the end of the first stage. An important technical difficulty arising in Alinhac's approach is that the blow-up time $T_{(Lifespan);\lambda}$ of the nonlinear solution is not precisely known. Consequently, his proof involves a free boundary corresponding to the unknown constant-time hypersurface of first blow-up. A related difficulty is that the lifespan can slightly vary from iterate to iterate and thus there is an extra step involved to fix the spacetime domain. The reason that Alinhac's proof is so short is that he avoids using many of the intricate geometric structures present in Christodoulou's framework. For example, his iteration scheme produces a true eikonal function for the nonlinear wave equation only in the limit; the iterates themselves involve approximate eikonal functions whose level sets are null hypersurfaces for the metric evaluated along the *previous* iterate. Furthermore, his commutation vectorfields also depend on the previous iterate. The net effect is that his linearized equations and commutation vectorfields do not have the same good structure enjoyed by the nonlinear equations studied in [**17**] and the present monograph, and thus he does not recover (or need) the same sharp top-order L^2 estimates for the eikonal function required by those works. Consequently, Alinhac's energy estimates for the iterates lose derivatives relative to the previous iterate. Nonetheless, he is able to close his proof by deriving tame L^2-type estimates for his linearized equations and using the Nash-Moser framework. Alinhac's iteration scheme is fundamentally based on his condition "(H);" see [**4**, pg. 15]. Roughly speaking, condition (H) requires that each iterate has characteristics that intersect at exactly one point belonging to its constant-time hypersurface of first blow-up. In particular, for spherically symmetric data in the case of equations invariant under the Euclidean rotations, condition (H) already fails for the zeroth iterate and the next iterate cannot be constructed. For similar reasons, his framework does not allow one to study portions of constant-time hypersurfaces Σ_t containing a submanifold along which μ is 0, as can happen when Σ_t lies to the future of the constant-time hypersurface of first blow-up; see the "singular part" of the boundary of the maximal development described in Theorem 2.19. In total, *the restrictive nature of condition (H) is the reason that Alinhac's framework applies only to data that verify his uniqueness and nondegeneracy assumptions and that it does not reveal the complete structure of the maximal development of the data.*

2.11.2. Christodoulou's shock formation results. In [**17**] Christodoulou proved several landmark results in relativistic fluid mechanics; see also the work

[**21**] of Christodoulou-Miao for the same results proved in the case of nonrelativistic compressible fluid mechanics. Specifically, Christodoulou proved analogs of our Theorem 22.1 and Theorem 23.9 for all of the physically relevant scalar quasilinear wave equations that arise in irrotational relativistic fluid mechanics in Minkowski spacetime. In addition, his work went somewhat beyond these two theorems in that he also gave a detailed description of the maximal future development of the data given in the exterior of the sphere $S_{0,U_0} \subset \Sigma_0^{U_0}$ whenever $0 \leq U_0 < 1/2$. Moreover, he extended his result to cover a class of small fluid data that have nonvanishing vorticity. However, the main aspects of his work addressed only a region in which the fluid is irrotational. In the irrotational region, the relativistic Euler equations (that is, the fluid equations) reduce to a scalar quasilinear wave equation for a potential function Φ with a past-directed spacetime gradient $\partial\Phi$. We stress that the difficult part of Christodoulou's argument was his proof of an analog of Theorem 22.1, and that the remaining aspects his work are easier to derive. We also stress that although Alinhac had already proved his small-data shock formation results (summarized in Theorem 2.18) for a larger class of equations,[41] there was great novelty in Christodoulou's thoroughness of his description of the dynamics and in particular, in his description of the solution up to the boundary of the maximal development of the data. A particularly attractive feature of Christodoulou's detailed description is that it is suitable as a starting point for trying to extend the solution, in a generalized sense, beyond the shock; see the Preface for additional discussion.

We now provide some details. Christodoulou's wave equations were the Euler-Lagrange equations for Lagrangians $\mathcal{L}$ of the form $\mathcal{L} = \mathcal{L}(\sigma)$, where[42]

$$\sigma := -(m^{-1})^{\alpha\beta}\partial_\alpha\Phi\partial_\beta\Phi \tag{2.95}$$

and $(m^{-1})^{\alpha\beta} = \mathrm{diag}(-1,1,1,1)$ is the standard reciprocal Minkowski metric. As is explained in [**17**, Chapter 1], in order to obtain a relativistic fluid interpretation from $\mathcal{L}(\sigma)$, it suffices to make the following five positivity assumptions:

$$\sigma,\ \mathscr{L}(\sigma),\ \frac{d\mathscr{L}}{d\sigma},\ \frac{d}{d\sigma}\left(\mathscr{L}/\sqrt{\sigma}\right),\ \frac{d^2\mathscr{L}}{d\sigma^2} > 0. \tag{2.96}$$

The assumptions (2.96) imply that Φ can be interpreted as a potential function for an irrotational relativistic fluid with physically reasonable properties such as having a speed of sound in between 0 and 1, having a positive proper energy density, etc. The Euler-Lagrange equation corresponding to $\mathcal{L}(\sigma)$ is

$$\partial_\alpha\left(\frac{\partial\mathcal{L}}{\partial(\partial_\alpha\Phi)}\right) = -2\partial_\alpha\left(\frac{\partial\mathcal{L}}{\partial\sigma}(m^{-1})^{\alpha\beta}\partial_\beta\Phi\right) = 0, \tag{2.97}$$

and the background solutions to (2.97) of interest are $\Phi = kt$, where k is a nonzero constant; these are the solutions that correspond to nonvacuum constant fluid states.[43] Relative to rectangular coordinates, equation (2.97) can be expressed

---

[41] One has to take into account the simple differences in normalization, noted below, in order to see that Christodoulou's equations fall under the scope of Alinhac's work.

[42] $\sqrt{\sigma}$ is a physical quantity known as the *relativistic enthalpy per particle*.

[43] Perturbations of the nonvacuum constant states are much easier to study than compact perturbations of the vacuum state. The reason is that the Euler equations become degenerate along vacuum boundaries.

as

$$(h^{-1})^{\alpha\beta}(\partial\Phi)\partial_\alpha\partial_\beta\Phi = 0, \tag{2.98}$$

where the *reciprocal acoustical metric* h^{-1} is defined by

$$(h^{-1})^{\alpha\beta} = (h^{-1})^{\alpha\beta}(\partial\Phi) := (m^{-1})^{\alpha\beta} - F(m^{-1})^{\alpha\kappa}(m^{-1})^{\beta\lambda}\partial_\kappa\Phi\partial_\lambda\Phi, \tag{2.99}$$

$$F = F(\sigma) := \frac{2}{G}\frac{dG}{d\sigma}, \tag{2.100}$$

$$G = G(\sigma) := 2\frac{d\mathcal{L}}{d\sigma}. \tag{2.101}$$

Because Christodoulou's background solutions are nonzero, there are some superficial differences between the way that his solutions look and the way that ours and Alinhac's look. The differences essentially correspond to different normalization choices and are not of fundamental importance. For example, the propagation speed corresponding to Christodoulou's background solution is not 1 as in our work and that of Alinhac, but is instead

$$\eta_0 := \eta(\sigma = k^2), \tag{2.102}$$

where the *speed of sound* $\eta > 0$ is the function of σ defined by

$$\eta^2 = \eta^2(\sigma) := 1 - \sigma H, \tag{2.103}$$

$$H = H(\sigma) := \frac{F}{1 + \sigma F}. \tag{2.104}$$

Using (2.96), it is straightforward to show that η verifies $0 < \eta < 1$. Christodoulou did not assume that $(h^{-1})^{00}$ is equal to -1 as we did in (2.3). Instead, his work features a dynamic lapse function $\alpha > 0$ defined by

$$\alpha^{-2} = \alpha^{-2}(\partial\Phi) := -(h^{-1})^{00}(\partial\Phi). \tag{2.105}$$

In the study of small perturbations of the background solutions, the physically and mathematically relevant inverse background metric is not the standard inverse Minkowski metric in the form $(m^{-1})^{\alpha\beta} = \text{diag}(-1, 1, 1, 1)$, but is instead the flat inverse metric $(h^{-1})^{\alpha\beta}(\partial_t\Phi = k, \partial_1\Phi = \partial_2\Phi = \partial_3\Phi = 0)$. More precisely, relative to rectangular coordinates such that $\Phi = kt$, the flat metric h takes the form

$$h(\partial_t\Phi = k, \partial_1\Phi = \partial_2\Phi = \partial_3\Phi = 0) = -\eta_0^2 dt^2 + \sum_{a=1}^{3}(dx^a)^2. \tag{2.106}$$

The eikonal function corresponding to the background solution is

$$u_{(Flat)} = 1 - r + \eta_0 t, \tag{2.107}$$

where $r = \sqrt{\sum_{a=1}^{3}(x^a)^2}$. The inverse foliation density corresponding to the background solution is

$$\mu_{(Flat)} = \eta_0. \tag{2.108}$$

The outgoing and ingoing null vectorfields corresponding to the background solution are

$$L_{(Flat)} = \partial_t - \eta_0\partial_r, \qquad \underline{\check{L}}_{(Flat)} = \eta_0^{-1}\partial_t - \partial_r. \tag{2.109}$$

The analog of the future null condition failure factor (2.81) is

$$\frac{dH}{d\sigma}(\sigma = k^2). \tag{2.110}$$

Unlike in the general case of (2.81), the quantity in (2.110) is a constant. Christodoulou showed that $\frac{dH}{d\sigma} \equiv 0$ if and only if, up to normalization constants,

$$\mathcal{L}(\sigma) = 1 - \sqrt{1-\sigma}. \tag{2.111}$$

The Lagrangian (2.111) is therefore exceptional in that it is the only member of the above family of Lagrangians such that the quadratic nonlinearities that arise in expanding its Euler-Lagrange equation (2.97) around the background solutions $\Phi = kt$ verify Klainerman's null condition.

We now summarize Christodoulou's results. The results stated below as Theorem 2.19 are a conglomeration of [**17**, Theorem 13.1 on pg. 888, Theorem 14.1 on pg. 903, Proposition 15.3 on pg. 974, and the Epilogue on pg. 977]. The quantities such as L, $\breve{\underline{L}}$, etc. that appear in the theorem are essentially the same as the quantities that we use throughout our monograph, up to the differences in normalization outlined in the previous paragraph.

THEOREM 2.19 (**Christodoulou**). *Let σ be as defined in (2.95). Assume that the Lagrangian $\mathcal{L}(\sigma)$ verifies the positivity conditions (2.96) in a neighborhood of $\sigma = k^2$, where k is a nonzero constant, but that $\mathcal{L}(\sigma)$ is not the exceptional Lagrangian (2.111). Consider the following Cauchy problem for the quasilinear (Euler-Lagrange) wave equation corresponding to $\mathcal{L}$, expressed relative to rectangular coordinates:*

$$\partial_\alpha \left(\frac{\partial \mathcal{L}}{\partial(\partial_\alpha \Phi)} \right) = 0, \tag{2.112a}$$

$$(\Phi|_{t=0}, \partial_t \Phi|_{t=0}) = (\mathring{\Phi}, \mathring{\Phi}_0). \tag{2.112b}$$

Assume that the data are small perturbations of the data corresponding to the nonzero constant-state solution $\Phi = kt$ and that the perturbations are compactly supported in the Euclidean unit ball. Let $U_0 \in (0, 1/2)$ and let

$$\mathring{\epsilon} = \mathring{\epsilon}[(\mathring{\Phi}, \mathring{\Phi}_0)] := \|\mathring{\Phi}_0 - k\|_{H^N(\Sigma_0^{U_0})} + \sum_{i=1}^{3} \|\partial_i \mathring{\Phi}\|_{H^N(\Sigma_0^{U_0})} \tag{2.113}$$

denote the size of the data, where N is a sufficiently large integer.[44]

Sharp classical lifespan result. *If $\mathring{\epsilon}$ is sufficiently small, then a sharp classical lifespan theorem in analogy with Theorem 22.1 holds.*

Small-data shock formation. *We define the following data-dependent functions of $u|_{\Sigma_0} = 1 - r$:*

(2.114)

$$\mathcal{E}[(\mathring{\Phi}, \mathring{\Phi}_0)](u) := \sum_{\Psi \in \{\partial_t \Phi - k, \partial_1 \Phi, \partial_2 \Phi, \partial_3 \Phi\}} \int_{\Sigma_0^u} \left\{ \alpha^{-2} \mu (\eta_0^{-1} + \alpha^{-2} \mu)(L\Psi)^2 + (\breve{\underline{L}}\Psi)^2 + (\eta_0^{-1} + 2\alpha^{-2}\mu)\mu |{\not{d}}\Psi|^2 \right\} d\underline{\varpi},$$

[44] A numerical value of N was not provided in [**17**]; see Remark 2.5.

(2.115)
$$\mathcal{S}[(\mathring{\Phi},\mathring{\Phi}_0)](u) := \int_{S_{0,u}} r\left\{(\mathring{\Phi}_0 - k) - \eta_0\partial_r\mathring{\Phi}\right\} d\upsilon_{\not g} + \int_{\Sigma_0^u}\left\{2(\mathring{\Phi}_0 - k) - \eta_0\partial_r\mathring{\Phi}\right\} d^3x,$$

where $d\underline{\varpi}$ is defined in (2.49), $d\upsilon_{\not g}$ denotes the Euclidean area form on the sphere $S_{0,u}$ of Euclidean radius $r = 1 - u$, and d^3x denotes the standard flat volume form on $\mathbb{R}^3$. Assume that

$$\ell := \frac{dH}{d\sigma}(\sigma = k^2) > 0. \tag{2.116}$$

There exist constants $C > 0$ and $C' > 0$, independent of $U \in [0, U_0]$, such that if $\mathring{\epsilon}$ is sufficiently small and if for some $U \in (0, U_0]$ we have

$$\mathcal{S}[(\mathring{\Phi},\mathring{\Phi}_0)](U) \leq -C\mathring{\epsilon}\mathcal{E}^{1/2}[(\mathring{\Phi},\mathring{\Phi}_0)](U) < 0, \tag{2.117}$$

then a shock forms in the solution[45] *Φ and the first shock in the maximal development of the portion of the data in the exterior of $S_{0,U} \subset \Sigma_0^U$ originates in the hypersurface region $\Sigma_{T_{(Lifespan)};U}^U$ (see Def. 3.7), where*

$$T_{(Lifespan);U} < \exp\left(C'\frac{U}{\left|k^3\ell\mathcal{S}[(\mathring{\Phi},\mathring{\Phi}_0)](U)\right|}\right). \tag{2.118}$$

A similar result holds if $\ell < 0$; in this case, in (2.117), we delete the "$-$" sign and change "$\leq$" and "$<$" respectively to "$\geq$" and "$>$."

Description of the boundary of the maximal development. *For a large set of small-data shock forming solutions verifying some technical nondegeneracy conditions, the boundary $\mathcal{B}$ of the maximal development of the data in the exterior of $S_{0,U} \subset \Sigma_0^U$ is a union $\mathcal{B} = (\partial_-\mathcal{H} \cup \mathcal{H}) \cup \underline{\mathcal{C}}$, where $\partial_-\mathcal{H} \cup \mathcal{H}$ is the singular part and $\underline{\mathcal{C}}$ is the regular part; see Figure 4. μ vanishes along the singular part and is positive on the regular part, and the solution and its rectangular coordinate partial derivatives up to a certain order extend continuously (in the rectangular coordinates) to the regular part. Each component of $\partial_-\mathcal{H}$ is a smooth 2-dimensional embedded submanifold of Minkowski spacetime that is spacelike with respect to the dynamic metric h. The corresponding component of $\mathcal{H}$ is a smooth, embedded, 3-dimensional submanifold in Minkowski spacetime ruled by curves of vanishing arc length as measured by h with past endpoints on $\partial_-\mathcal{H}$. The corresponding component $\underline{\mathcal{C}}$ is the incoming null hypersurface corresponding to $\partial_-\mathcal{H}$, and it is ruled by incoming h-null geodesics with past endpoints on $\partial_-\mathcal{H}$.*

Again, we stress that a key advantage of Christodoulou's approach is that the estimates of his sharp classical lifespan theorem allow him to understand the maximal development of the data in the exterior of the sphere $S_{0,U_0} \subset \Sigma_0^{U_0}$, including the behavior of the solution up to the boundary. We also mention again that based on the estimates of Theorem 22.1, our analysis of shock forming solutions to equations (2.1a) and (2.9) could be extended, without too much additional effort, to provide an analogous description of the maximal development.

[45] That is, Φ and its first rectangular derivatives remain bounded, while some second-order rectangular derivative of Φ blows up due to the vanishing of μ.

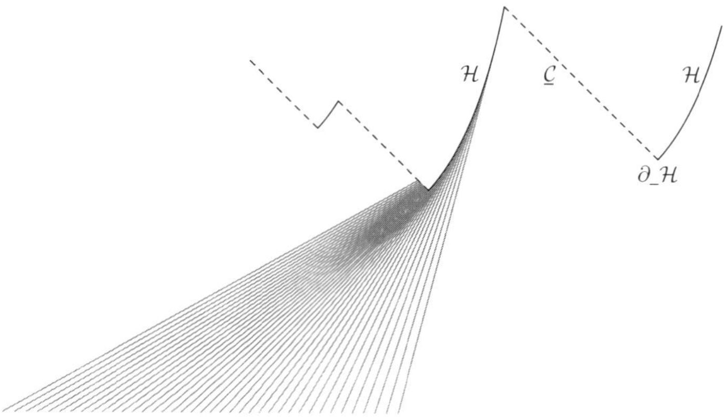

FIGURE 4. A cross section of the maximal development. The gray lines are level sets of the eikonal function u near the first blow-up point. The dotted lines are the regular boundary $\underline{\mathcal{C}}$. The black curves are the boundary $\mathcal{H}$, whose past endpoints are $\partial_{-}\mathcal{H}$.

2.11.3. Further discussion on the shock forming data. We now compare our conditions on the data, sufficient to imply shock formation, with those of Alinhac and Christodoulou. We first note that there is no obvious connection between any of the three classes of shock forming data identified in Theorems 2.18, 2.19, and 23.9. However, the previous discussion essentially shows that among the three theorems, Alinhac's treats the largest set of data: recall that John conjectured (see the discussion just below (2.85)), based on the results of his Prop. 2.17, that finite-time blow-up should occur in all solutions to equation (2.80a) corresponding to nontrivial small compactly supported data, and that Alinhac was able to partially confirm the conjecture by proving it under some nondegeneracy assumptions on the data (see just below (2.91)). Below we outline how to extend the results of the present monograph to complete the proof of John's conjecture by eliminating Alinhac's nondegeneracy assumptions. Using the same ideas, we could also prove an analogous result for Christodoulou's wave equations (2.112a) and our equations $\square_{g(\Psi)}\Psi = 0$. However, there is a sense in which our results are not complete; as was the case for Alinhac, in order for our proof of shock formation to go through, we may need to shrink the amplitude of the data to an extent that depends on their profile. Hence, our analysis leaves open the possibility that there exist nontrivial data small enough such that the sharp classical lifespan (Theorem 22.1) applies, but such that a shock does not form in the solution; such a solution, if it exists, would be future-global.

We can also derive a "localized very-long-time existence result" for data such that the quantity (2.91) (or its analogs for the other equations) is nonpositive in a suitable annular region. For solutions launched by such data, there is a region of spacetime in which the term that typically dominates the behavior of $L\mu$ has a nonnegative sign and thus cannot cause μ to decay. Later in this section, we outline how to use this fact to prove that in a certain region of spacetime, these solutions exist, in an annular region of thin u-width, beyond the lifespan lower-bound $> \exp\left(\frac{1}{C\mathring{\epsilon}}\right)$, which always holds in the small-data regime (see inequality (22.4)).

Before addressing the issues noted above, we first note that under some structural assumptions on the nonlinearities, one can derive an analog of Christodoulou's shock formation criterion (2.117) for equations (2.1a) and (2.9). Specifically, a sufficient assumption on the nonlinearities would be that the future null condition failure factor ${}^{(+)}\aleph = {}^{(+)}\aleph(\theta)$ from (2.22)/(2.81) takes on a strictly positive or negative sign for all $\theta \in \mathbb{S}^2$. Then, thanks to the sharp estimates of Theorem 22.1, an analog of inequality (2.117) should be sufficient to allow Christodoulou's arguments leading to his shock formation theorem [**17**, Theorem 14.1 on pg. 903] to go through nearly verbatim. His arguments are based on using the estimates of his analog of Theorem 22.1 to derive estimates for quantities that are averaged over the spheres $S_{t,u}$. More precisely, using the estimates of Theorem 22.1 and applying Christodoulou's arguments to solutions of (2.1a) under the assumption of an analog of inequality (2.117), we could deduce the existence of some *unknown* integral curve of L along which we have a large-time lower bound of the form $\pm \breve{R}\Psi \gtrsim \mathring{\epsilon}(1+t)^{-1}$ (we can achieve either sign for $\breve{R}\Psi$, depending on our choice of data). Inserting this estimate into equation (2.39a) and using the fact that $G_{LL}(t,u,\vartheta)$ is well-approximated by ${}^{(+)}\aleph(t=0,u=0,\vartheta)$ (see Lemma 12.49), we could deduce the following large-time estimate:

$$\pm L\mu(t,u_*,\vartheta_*) \geq c\,{}^{(+)}\aleph(t=0,u=0,\vartheta_*)\mathring{\epsilon}\frac{1}{1+t} + \text{Error}, \tag{2.119}$$

where (u_*,ϑ_*) are the coordinates corresponding to the unknown integral curve, and for $\mathring{\epsilon}$ sufficiently small, the term Error is dominated by the first term on the right-hand side of (2.119). The sign $\pm$ in (2.119) depends on the sign of the analog of the left-hand side of (2.117). Hence, when ${}^{(+)}\aleph(t=0,u=0,\vartheta)$ takes on a definite sign[46] for all $\vartheta \in \mathbb{S}^2$, the data are sufficiently small, and the data are chosen to generate the sign "$-$" on the left-hand side of (2.119), we can integrate inequality (2.119) in time to deduce that μ must vanish in finite time.

We now return to the issue noted in the first paragraph of Sect. 2.11.3: sketching a proof that we can use the analog of Theorem 22.1 for equation (2.80a) to eliminate Alinhac's uniqueness and nondegeneracy assumptions on the data (see just below (2.91)), thus completing the proof of John's conjecture. Specifically, we outline how to prove small-data shock formation for equation (2.80a) by assuming only John's criterion on the initial data:

$$\text{the function from (2.91) is positive at one point } (q_*,\theta_*), \tag{2.120}$$

and by perhaps shrinking the amplitude of the data by a small constant factor if necessary, as Alinhac did in his proof Theorem 2.18.[47] Furthermore, by Prop. 2.17, the condition (2.120) is *always satisfied when the data are nontrivial and compactly supported.* John's criterion (2.120) can also be modified in a straightforward fashion to apply to Christodoulou's equations (2.112a) and our equations $\square_{g(\Psi)}\Psi = 0$.

[46] Recall that at $t=0$, the geometric angular coordinates $(\vartheta^1,\vartheta^2)$ coincide with Euclidean angular coordinates (θ^1,θ^2).

[47] As will become clear, shrinking the amplitude of the data can help the shock driving term dominate error terms.

Moreover, John's criterion is sharp in a way that we make precise in the next paragraph. We now sketch proofs of these claims. For definiteness, we focus only on the equation $\square_{g(\Psi)}\Psi = 0$, and we consider data $(\mathring{\Psi}, \mathring{\Psi}_0)$ that are compactly supported in the Euclidean unit ball Σ_0^1. Throughout this discussion, $\mathring{\epsilon}$ denotes the (small) size of the data $(\mathring{\Psi}, \mathring{\Psi}_0)$. The main idea of the proof is that under the appropriate version of John's criterion for the equation $\square_{g(\Psi)}\Psi = 0$, we can find a point p belonging to a set of the form $\Sigma_{\mathring{\epsilon}^{-1}}^{U_0}$ such that *along the integral curve* of L passing through p, we have the following estimate for $t \geq \mathring{\epsilon}^{-1}$:

$$L\upmu(t, u, \vartheta) = -\frac{1}{2}\varrho^{-1}(t, u)\left\{{}^{(+)}\aleph\frac{\partial}{\partial q}|_{\theta}\mathcal{F}[(\mathring{\Psi}, \mathring{\Psi}_0)]\right\}|_p + \text{Error}, \tag{2.121}$$

where $\mathcal{F}[(\mathring{\Psi}, \mathring{\Psi}_0)]$ denotes Friedlander's radiation field (see (2.84)), the constant-valued term $\left\{{}^{(+)}\aleph\frac{\partial}{\partial q}|_{\theta}\mathcal{F}[(\mathring{\Psi}, \mathring{\Psi}_0)]\right\}|_p$ in (2.121) is *positive*, and[48] Error $= o(\mathring{\epsilon})(1+t)^{-1}$. Hence, integrating (2.121) with respect to t starting from time $\mathring{\epsilon}^{-1}$, we infer that $\upmu$ will vanish in finite time, thus yielding shock formation.[49] To find a viable point p, we first note that, as will become clear, the quantity

$$\frac{1}{2}{}^{(+)}\aleph(\theta)\frac{\partial}{\partial q}|_{\theta}\mathcal{F}[(\mathring{\Psi}, \mathring{\Psi}_0)](q = r - \mathring{\epsilon}^{-1}, \theta)$$

is the appropriate analog of the quantity (2.91) in the present context. Hence, if the data verify John's criterion, that is, they are such that ${}^{(+)}\aleph(\theta)\frac{\partial}{\partial q}|_{\theta}\mathcal{F}[(\mathring{\Psi}, \mathring{\Psi}_0)](q, \theta)$ is positive at some point (q_*, θ_*), then there exists a point p belonging to a set of the form[50] $\Sigma_{\mathring{\epsilon}^{-1}}^{U_0}$ such that $-\frac{1}{2}\left\{{}^{(+)}\aleph\frac{\partial}{\partial q}|_{\theta}\mathcal{F}[(\mathring{\Psi}, \mathring{\Psi}_0)]\right\}|_p$ is negative; this point p will suffice for our purposes. In view of the evolution equation (2.39a) for $\upmu$, to conclude (2.121), the main task is to show that for times beyond $\mathring{\epsilon}^{-1}$, the important term $\frac{1}{2}[G_{LL}\varrho\breve{R}\Psi](t, r, \theta)$ is well-approximated by

$$-\frac{1}{2}\varrho^{-1}(t, u)\left\{{}^{(+)}\aleph\frac{\partial}{\partial q}|_{\theta}\mathcal{F}[(\mathring{\Psi}, \mathring{\Psi}_0)]\right\}|_p$$

along the integral curve of L passing through p. There are two main steps in the proof of the approximation. The first step is to show that on the time interval $[0, \mathring{\epsilon}^{-1}]$, the linear dynamics dominate the behavior of the nonlinear solution and

---

[48]Here, $o(\mathring{\epsilon})$ denotes a term such that $\lim_{\mathring{\epsilon}\downarrow 0}\frac{o(\mathring{\epsilon})}{\mathring{\epsilon}} = 0$.

[49]This is the point in the argument where we might need to rescale the amplitude of the data in order to ensure that the first term on the right-hand side of (2.121) dominates the second; the first term shrinks linearly in the scaling factor, while the second shrinks at a super-linear rate.

[50]One minor difficulty that would have to be addressed in a complete proof is that $\Sigma_{\mathring{\epsilon}^{-1}}^{U_0}$ might correspond to a value $U_0 > 1$, which is a set that we have not even defined. However, our results can be extended to apply to such sets. The main task would be to modify the construction of the eikonal function u so that on $\Sigma_{\mathring{\epsilon}^{-1}}$, it is allowed to take on some values larger than 1. Alternatively, we could avoid extending the construction of the eikonal function by instead starting with data that are given along $\Sigma_{-1/2}$ and compactly supported in the Euclidean ball of radius 1/2 centered at the origin; see also Footnote 23 on pg. 47. We adopt this latter approach in Theorem 23.9.

as a consequence, the product $\frac{1}{2}[G_{LL}\varrho \breve{R}\Psi](t = \mathring{\epsilon}^{-1}, r, \theta)$ is well-approximated by

$$-\frac{1}{2}{}^{(+)}\aleph(\theta)\frac{\partial}{\partial q}|_{\theta}\mathcal{F}[(\mathring{\Psi}, \mathring{\Psi}_0)](q = r - \mathring{\epsilon}^{-1}, \theta).$$

Here, (t, r, θ) are spherical coordinates on Minkowski spacetime. We describe this step in more detail two paragraphs below. To explain the second step, we let $(\mathring{\epsilon}^{-1}, u, \vartheta)$ be the geometric coordinates of p. We now integrate equation (2.42) along the integral curve of L emanating from p and use (2.34a) and its higher-order analogs as well as the estimates of Lemma 12.49 to derive refined version of (2.44), showing that for $t \geq \mathring{\epsilon}^{-1}$, the product $\frac{1}{2}G_{LL}(t, u, \vartheta)\breve{R}\Psi(t, u, \vartheta)$ is well-approximated by

$$-\frac{1}{2}\varrho^{-1}(t, u)G_{LL}(\mathring{\epsilon}^{-1}, u, \vartheta)\breve{R}\Psi(\mathring{\epsilon}^{-1}, u, \vartheta).$$

Note that by (2.34a) and its higher-order analogs, for times beyond $\mathring{\epsilon}^{-1}$, the term $\varrho\upmu\Delta\!\!\!/\Psi$ from equation (2.42) has had time to sufficiently decay and thus it only makes a small contribution when we integrate starting from p. Combining these two steps, we deduce the desired approximation, which completes our sketch of a proof of (2.121)

We are now in a position to sketch the proof of the "localized very-long-time existence result" mentioned above, which holds in regions where ${}^{(+)}\aleph\frac{\partial}{\partial q}|_{\theta}\mathcal{F}[(\mathring{\Psi}, \mathring{\Psi}_0)]$ is nonpositive. To illustrate what we mean, we assume that the data are such that there exists a $q_1 \in (0, 1)$ and an annular region $\mathcal{A} := \{(q, \theta) \mid q \in [q_1, 1] \times \mathbb{S}^2\}$ with $\min_{\mathcal{A}} -{}^{(+)}\aleph\frac{\partial}{\partial q}|_{\theta}\mathcal{F}[(\mathring{\Psi}, \mathring{\Psi}_0)] \geq 0$. We consider the family of integral curves of L that start at time $\mathring{\epsilon}^{-1}$ and emanate from the subset of points in $\Sigma_{\mathring{\epsilon}^{-1}}$ that intersect the spacetime region corresponding to[51] $\mathcal{A}$. Fixing any point p belonging to the subset and integrating inequality (2.121) (which also holds for this point p) along the corresponding integral curve (starting from time $\mathring{\epsilon}^{-1}$), we deduce that any shrinking of $\upmu$ could only be caused by the term $\mathrm{Error} = o(\mathring{\epsilon})(1+t)^{-1}$. In particular, the amplitude of Error goes to 0 strictly faster than $\mathring{\epsilon}$. Hence, it takes at least $\exp\left\{\frac{1}{o(\mathring{\epsilon})}\right\}$ amount of time for $\upmu$ to vanish in this region, and by Theorem 22.1, the sharp classical lifespan theorem, no singularity of any kind will form while $\upmu$ is positive. As we mentioned above, this existence time is longer than the time $\exp\left(\frac{1}{C\mathring{\epsilon}}\right)$ that holds for general small data (see inequality (22.4)).

We now give a brief overview explaining the so-called "first step" stated two paragraphs above, namely showing that relative to spherical coordinates (t, r, θ) on Minkowski spacetime, $[G_{LL}\varrho\breve{R}\Psi](t, r, \theta)$ is well-approximated by

$$-{}^{(+)}\aleph\frac{\partial}{\partial q}|_{\theta}\mathcal{F}[(\mathring{\Psi}, \mathring{\Psi}_0)](q = r - t, \theta)$$

[51]Here, q should be thought of as the Minkowskian null coordinate $q = r - t$. Thus, the family of integral curves of L under study emanate precisely from the subset of points in $\Sigma_{\mathring{\epsilon}^{-1}}$ that intersect the "Minkowskian region" trapped between the flat inner cone $\{r - t = q_1\}$ and the flat outer cone $\mathcal{C}_0 = \{r - t = 1\}$.

at time $t = \mathring{\epsilon}^{-1}$. Since Lemma 12.49 shows that G_{LL} is well-approximated by ${}^{(+)}\aleph$, it remains only for us to explain the following estimate, valid for small data:

$$\left\| \varrho \breve{R}\Psi + \frac{\partial}{\partial q}|_{\theta} F[(\mathring{\Psi}, \mathring{\Psi}_0)] \right\|_{C^0(\Sigma^{U_0}_{\mathring{\epsilon}^{-1}})} \leq C\mathring{\epsilon}^2 \ln\left(\frac{1}{\mathring{\epsilon}}\right). \tag{2.122}$$

In (2.122), $\|\cdot\|_{C^0(\Sigma^{U_0}_{\mathring{\epsilon}^{-1}})}$ denotes the standard sup norm (see Def. 3.70). To derive (2.122), one can rely on two standard kinds of estimates. First, one needs an estimate showing that at time $\mathring{\epsilon}^{-1}$, the solution $\Psi_{(Linear)}$ to the *linear* wave equation $\square_m \Psi = 0$ with initial data $(\mathring{\Psi}, \mathring{\Psi}_0)$ is well-approximated by $r^{-1}F[(\mathring{\Psi}, \mathring{\Psi}_0)]$ and that similar results hold for the higher-order $(t, r, \theta^1, \theta^2)$ coordinate partial derivatives of $\Psi_{(Linear)}$. Such estimates can be derived with the help of the fundamental solution to the linear wave equation; see, for example, [**30**]. The second kind are estimates showing that at time $\mathring{\epsilon}^{-1}$, *long before any singularity can form*, the nonlinear solution Ψ is well-approximated by $\Psi_{(Linear)}$, that ϱ is well-approximated by r, that μ is well-approximated by 1, that L is well-approximated by $\partial_t + \partial_r$, that R is well-approximated by $-\partial_r$, etc. The estimates for Ψ can be derived with the help of Klainerman's Minkowskian vectorfield method approach, as described in [**47**]. The estimates for μ and L^i can then be derived with the help of the transport equations (2.39a)-(2.39b) and the dispersive C^0 estimates for Ψ (see Sect. 2.6). The remaining estimates then follow without much difficulty; all of these estimates can be derived by using arguments that are explained in detail in Chapter 12. Combining such estimates, one can deduce (2.122) without much difficulty.

We conclude this section by mentioning an open problem: deciding whether or not *all* (nontrivial) compactly supported data verifying the smallness assumptions of Theorem 22.1 must necessarily lead to shock formation; the limitation of the argument outlined in this section is that we might need to shrink the amplitude of the data in order to know that a shock will form (see just below (2.120) and Footnote 49).

2.12. Outline of the monograph

In Appendix A, we outline how to extend our results for covariant equations $\square_{g(\Psi)}\Psi = 0$ to noncovariant equations $(h^{-1})^{\alpha\beta}(\partial\Phi)\partial_\alpha\partial_\beta\Phi = 0$. In Appendix B, we collect some of the important notation and conventions that we use throughout the monograph.

As we have emphasized, our main goal in this monograph is to prove Theorem 22.1, the sharp classical lifespan theorem. Our proof of shock formation for nearly spherically symmetric small data, Theorem 23.9, then follows without much additional effort. We prove Theorem 23.9 in the final chapter of the monograph, Chapter 23. The monograph and indeed, the main bootstrap argument culminating in the proof of Theorem 22.1, are organized in an essentially linear fashion. We now outline the main steps in its proof.

Step 1: Deriving equations. Chapters 3-11 are dedicated to geometry and algebra. In short, in these chapters, we define all of the quantities[52] that play a role in our analysis and we derive the equations that they satisfy. In particular, in Prop. 10.13, we derive energy-cone flux identities, which are the starting point for our derivation of a priori L^2-type estimates for Ψ and its derivatives. The

[52] A few of the definitions are located in Chapter 2.

most difficult of this analysis occurs in Chapter 11, where we take special care to construct suitable modified quantities and to derive evolution equations for them. As we described in Sect. 2.10.2, we must use the modified quantities in order to avoid derivative loss and other difficulties when controlling the top-order derivatives of the eikonal function quantities.

Step 2: Bootstrap assumptions and easy C^0 and pointwise estimates. In Chapter 12, we state C^0 bootstrap assumptions for Ψ and its lower order derivatives on a "bootstrap region" of the form $\mathcal{M}_{T_{(Bootstrap)},U_0}$ (see Figure 2 on pg. 41). We also state similar assumptions for $\upmu$, L^i, and χ (more precisely, for re-centered versions of these variables in which we subtract off their background values), as well as the positivity assumption $\upmu > 0$. We then use these bootstrap assumptions, the equations from Step 1, and a small-data assumption to derive C^0 and pointwise estimates for most of the quantities defined in Step 1. In particular, we derive C^0 estimates for the re-centered versions of $\upmu$, L^i, χ, and their lower-order derivatives that are improvements over the bootstrap assumptions, thus closing this portion of the bootstrap argument. These pointwise estimates are tedious but relatively easy to derive. We save the most difficult pointwise estimates for Steps 3 and 5.

Step 3: Sharp estimates for $\upmu$. In Chapter 13, we derive sharp estimates, far more detailed than those of Step 2, for the inverse foliation density $\upmu$ (see (2.68) for a model example of the kinds of estimates that we derive). These estimates play a key role in our Gronwall argument for the energy-cone flux quantities. The most difficult aspects of this analysis involve a posteriori estimates for $\upmu$, as we described in Sect. 2.10.5.

Step 4: Coercive L^2-controlling quantities. In Chapter 14, we exhibit the coerciveness of the L^2-type energy-cone flux quantities defined in Step 1. We then use the energy-cone flux quantities as building blocks to construct our fundamental L^2-controlling quantities, which we use to control Ψ and its derivatives in L^2.

Step 5: Pointwise estimates for the error terms. In Step 5, we derive pointwise estimates for the error integrands appearing in the energy-cone flux identities of Prop. 10.13. More precisely, we derive these pointwise estimates in the case of Ψ and also in the cases of the higher-order derivatives $\mathscr{Z}^N\Psi$, where $\mathscr{Z}^N$ is an Nth order string of commutation vectorfields, which we constructed in Step 1. Many of the error integrands that we must bound arise because when we commute the wave equation $\upmu\square_{g(\Psi)}\Psi = 0$ with $\mathscr{Z}^N$, we generate a huge number of error terms. These pointwise estimates are a precursor to Step 8, in which we derive estimates for the corresponding error integrals. We divide the pointwise estimates into Chapter 16 (easy estimates) and Chapter 17 (difficult estimates). In Chapter 16, we identify those integrand factors that are "harmless." The harmless terms have a negligible effect on the dynamics and are easy to treat in Step 8. The remaining terms, which are the ones that would cause derivative loss and other problems if handled improperly, require special care. In Chapter 17, we derive pointwise estimates for these remaining difficult error integrand terms. To derive these difficult estimates, we use the modified quantities constructed in Step 1.

Step 6: Elliptic estimates and Sobolev embedding. Chapter 18 is an interlude. We derive elliptic estimates for various quantities on the Riemannian manifolds $(S_{t,u}, \gslash)$. We use the elliptic estimates in Step 8 to control some of the top-order error terms in L^2. We also derive a Sobolev embedding result for functions on $(S_{t,u}, \gslash)$. This is the tool that we eventually use, after we have derived suitable L^2 estimates, to improve the C^0 bootstrap assumptions for Ψ made in Step 2.

Step 7: Below-top-order L^2 estimates for the eikonal function quantities. In Chapter 19, we use the equations from Step 1 and the C^0 and pointwise estimates from Step 2 to derive a priori L^2 estimates for the below-top-order derivatives of quantities constructed out of the eikonal function, such as the re-centered versions of μ, L^i, etc; such terms appear in Step 5 when we commute the wave equation. The right-hand sides of these estimates involve the fundamental L^2-controlling quantities constructed in Step 4. These estimates are relatively easy to derive without using the modified quantities constructed in Step 1 because we allow them to lose one derivative relative to Ψ.

Step 8: A priori L^2 estimates for Ψ and the eikonal function quantities up to top-order. This step, which we carry out in Chapter 20, is the hardest one. We estimate, in L^2, the remaining error terms appearing in the commuted wave equation. In particular, we estimate the difficult top-order eikonal function quantities that we ignored in Step 7. As in Step 7, the right-hand sides of these estimates involve the fundamental L^2-controlling quantities constructed in Step 4. We then use the estimates for the error terms to derive, by a long Gronwall argument based on integral identities such as (2.50), a priori estimates for the fundamental L^2-controlling quantities on the bootstrap region $\mathcal{M}_{T_{(Bootstrap)},U_0}$. This chapter contains the main estimates we need to prove Theorem 22.1, our sharp classical lifespan theorem.

Step 9: Local well-posedness. Chapter 21 is another interlude. We sketch a proof of local well-posedness for the equations of interest and we establish continuation criteria for avoiding the blow-up of solutions. In particular, the results of Chapter 21 imply that the bootstrap assumptions from Step 2 are satisfied for at least a short time.

Step 10: The sharp classical lifespan theorem. In Chapter 22, we state and prove Theorem 22.1, which is the main theorem of the monograph. The difficult part of the argument was carried out in Step 8. Roughly, the theorem shows that for small data, the *only* way a singularity could form is for μ to go to 0 in finite time, which would necessarily signify the onset of shock formation. The theorem also provides a collection of estimates that are verified by the solution, whether or not it forms a shock. We also prove Cor. 22.4, which shows that if the initial data have "very small" angular derivatives, then this condition is propagated by the solution; we use the corollary in Step 11.

Step 11: Shock formation. In Chapter 23, we prove Theorem 23.9, which is our small-data shock formation theorem. Roughly, the theorem states that sufficiently small, nearly spherically symmetric data lead to finite-time shock formation. The theorem is not difficult to prove, thanks to the estimates of Theorem 22.1 and Cor. 22.4. With a bit of additional effort, we could have extended Theorem 23.9 to show shock formation in solutions corresponding to a significantly larger class of data; see Sect. 2.11.3 for an outline of a proof.

2.13. Suggestions on how to read the monograph

We extensively use standard concepts from differential geometry in this monograph. For an introduction to the basic concepts in Lorentzian and Riemannian geometry that play a role in our analysis, readers may consult [**66**, **67**, **78**].

To become acquainted with the new difficulties that set this work apart from more standard global results for nonlinear wave equations, we suggest starting with Prop. 7.7, Lemma 9.4, and Prop. 9.6. From these, one can see the kind of error terms that arise when we commute the wave equation with our commutation vectorfields. The error terms lead to the presence of error integrals in the L^2-type energy-cone flux identities of Prop. 10.13. We identify the most difficult error terms in Lemma 16.1, Prop. 16.4, and Cor. 16.6. We bound the corresponding difficult error integrals in Lemma 20.25 and Lemma 20.26. Because of the degenerate nature of the Gronwall-type Lemma 20.10, the bounds from these two lemmas lead to a priori estimates for the high-order L^2 quantities that are allowed to blow-up as $\mu \to 0$. **These degenerate a priori estimates are the primary nonstandard aspects of our work and the work [17].**

CHAPTER 3

Initial Data, Basic Geometric Constructions, and the Future Null Condition Failure Factor

In this chapter, we discuss the initial data of interest for the covariant wave equation[1] (2.1a), that is, for the equation $\square_{g(\Psi)}\Psi = 0$. We then define the eikonal function u, some intimately related subsets of spacetime, a related set of geometric coordinates, several vectorfield frames, the change of variables map between the geometric and rectangular coordinates, and many related geometric constructions, all of which play an important role in our analysis. In particular, we define the inverse foliation density $\upmu$, which in a certain sense is the main object of study in this monograph. As we discussed in Sect. 2.8, for the solutions of interest, the intersection of the characteristics and the blow-up of a geometric radial derivative of Ψ are exactly tied to the vanishing of $\upmu$. We also define the *future null condition failure factor* ${}^{(+)}\aleph$, which, roughly speaking, is the coefficient of the terms that drive the shock formation in the region $\{t \geq 0\}$. Finally, we introduce some schematic notation that we often use to capture the essential features of our equations and estimates.

REMARK 3.1 (**There is some redundancy**). Some of the discussion and definitions in Chapter 3 also appear in Chapter 2. For pedagogical reasons, we have chosen to repeat some material here.

REMARK 3.2 (**Notation and conventions**). We refer the reader to Appendix B for a summary of the notation and conventions that we use throughout the remainder of the monograph.

We recall that $\{x^\mu\}_{\mu=0,1,2,3}$ (with $t = x^0$) denotes the fixed rectangular coordinate system from Sect. 2.2 and that in Sects. 2.2-2.3, we described our assumptions on the metric g and the nonlinear terms in the wave equation $\square_{g(\Psi)}\Psi = 0$.

REMARK 3.3 (**Implicit assumption on the smallness of Ψ**). Throughout the monograph, when deriving equations and identities, we silently assume that $|\Psi|$ is small enough so that the nonlinearities are well-defined and the metric $g(\Psi)$ is Lorentzian.

3.1. Initial data

The Cauchy hypersurface of interest is $\Sigma_0 := \{(t, x^1, x^2, x^3) \in \mathbb{R}^4 \mid t = 0\}$ and the data for the wave equation (2.1a) are

$$\mathring{\Psi} := \Psi|_{\Sigma_0}, \qquad \mathring{\Psi}_0 := \partial_t\Psi|_{\Sigma_0}. \tag{3.1}$$

[1] We formally define the covariant wave operator $\square_{g(\Psi)}$ in Def. 3.12.

We assume that $\mathring{\Psi}$, $\mathring{\Psi}_0$ are compactly supported in the Euclidean unit ball centered at the origin in Σ_0. Let U_0 be any real number verifying

$$0 < U_0 < 1. \tag{3.2}$$

We view U_0 as a parameter that is fixed until Sect. 23.2. Our use of the notation "U_0" is connected to the eikonal function u, which we define in Sect. 3.2.

Let $r = \sqrt{\sum_{a=1}^3 (x^a)^2}$ denote the standard Euclidean radial coordinate. We study the future-behavior of the solution in the region that is evolutionarily determined by the portion of the data lying in the exterior of the sphere S_{0,U_0} of Euclidean radius $1 - U_0$ centered at the origin. In the exterior of S_{0,U_0}, the data are nontrivial only in the annular region $\Sigma_0^{U_0}$ centered at the origin with inner Euclidean radius $1 - U_0$ and outer Euclidean radius 1 (and thus the thickness of the region is U_0):

$$\Sigma_0^{U_0} := \{(0, x^1, x^2, x^3) \in \mathbb{R}^4 \mid 1 - U_0 \leq r \leq 1\}. \tag{3.3}$$

Standard domain of dependence considerations imply that the solution Ψ completely vanishes in the exterior of the flat cone $\mathcal{C}_0 := \{(t, x^1, x^2, x^3) \mid 1 - r + t = 0\}$. We are interested in the nontrivial portion of the solution, which is trapped between the two outgoing null cones $\mathcal{C}_{U_0}$ and $\mathcal{C}_0$ (see Figure 1 on pg. 37).

REMARK 3.4 (**The role of U_0**). In Sect. 23.2, for given initial data, we choose the data-dependent parameter U_0 so that the shock singularity forms in the region trapped between the two outgoing null cones $\mathcal{C}_{U_0}$ and $\mathcal{C}_0$.

3.2. The eikonal function and the geometric radial variable

As we explained in Sect. 2.5, the eikonal function plays a fundamental role in our analysis of solutions.

DEFINITION 3.5 (**The eikonal function**). The eikonal function u increases towards the future (that is, $\partial_t u > 0$) and verifies

$$(g^{-1})^{\alpha\beta} \partial_\alpha u \partial_\beta u = 0. \tag{3.4a}$$

We define the initial condition of u as follows, where (x^1, x^2, x^3) are the rectangular coordinates on Σ_0:

$$u|_{\Sigma_0}(x^1, x^2, x^3) = 1 - r(x^1, x^2, x^3), \qquad r(x^1, x^2, x^3) := \sqrt{\sum_{a=1}^{3} (x^a)^2}. \tag{3.4b}$$

We often use the following geometric radial variable in our analysis.

DEFINITION 3.6 (**The geometric radial variable**). We define the geometric radial variable ϱ by

$$\varrho := 1 - u + t. \tag{3.5}$$

The following subsets of spacetime play an important role in our analysis; see Figure 2 on pg. 41.

DEFINITION 3.7 (**Subsets of spacetime**). We define the following spacetime subsets:

$$\Sigma_{t'} := \{(t, x^1, x^2, x^3) \in \mathbb{R}^4 \mid t = t'\}, \tag{3.6a}$$

$$\Sigma_{t'}^{u'} := \{(t, x^1, x^2, x^3) \in \mathbb{R}^4 \mid t = t',\ 0 \leq u(t, x^1, x^2, x^3) \leq u'\}, \tag{3.6b}$$

$$\mathcal{C}_{u'}^{t'} := \{(t, x^1, x^2, x^3) \in \mathbb{R}^4 \mid 0 \leq t \leq t',\ u(t, x^1, x^2, x^3) = u'\}, \tag{3.6c}$$

$$S_{t',u'} := \mathcal{C}_{u'}^{t'} \cap \Sigma_{t'}^{u'} = \{(t, x^1, x^2, x^3) \in \mathbb{R}^4 \mid t = t',\ u(t, x^1, x^2, x^3) = u'\}, \tag{3.6d}$$

$$\mathcal{M}_{t',u'} := \cup_{u \in [0,u']} \mathcal{C}_u^{t'} \cap \{(t, x^1, x^2, x^3) \in \mathbb{R}^4 \mid t < t'\}. \tag{3.6e}$$

We refer to the Σ_t and Σ_t^u as "constant time slices," the $\mathcal{C}_u^t$ as "outgoing null cones," and the $S_{t,u}$ as "spheres." We sometimes use the notation $\mathcal{C}_u$ in place of $\mathcal{C}_u^t$ when we are not concerned with the truncation time t.

REMARK 3.8. We make the following remarks.

- It follows from standard domain of dependence considerations that for any $t \geq 0$, the solution corresponding to the data from Sect. 3.1 agrees with the trivial solution $\Psi \equiv 0$ in the exterior of $\mathcal{C}_0^t$. We therefore do not comment further on this region.
- The $S_{t,u}$ are two-dimensional topological spheres. They are not generally round Euclidean spheres, even though the $S_{0,u}$ are.
- Note that $\mathcal{M}_{t,u}$ is "open at the top."

3.3. First fundamental forms and Levi-Civita connections

We now define some basic geometric objects related to the subsets from Def. 3.7.

DEFINITION 3.9 (**First fundamental forms**). We define the following first fundamental forms.

- $\underline{g}$ denotes the Riemannian metric on Σ_t induced by g. That is, $\underline{g}(X, Y) = g(X, Y)$ for all Σ_t-tangent vectors X and Y. Equivalently, we have $\underline{g}_{ij} = g_{ij}$, $(i, j = 1, 2, 3)$. $\underline{g}^{-1}$ denotes the corresponding inverse metric.
- $g\!\!\!/$ denotes the Riemannian metric on $S_{t,u}$ induced by g. That is, $g\!\!\!/(X, Y) = g(X, Y)$ for all $S_{t,u}$-tangent vectors X and Y. $g\!\!\!/^{-1}$ denotes the corresponding inverse metric.

REMARK 3.10 (**Meaning of** $g(X, Y)$). Throughout, we use the notation $g(X, Y) = g_{\alpha\beta} X^\alpha Y^\beta$ to denote the inner product of the vectors X and Y with respect to the metric g. Occasionally, we write $\langle X, Y \rangle_g$ instead of $g(X, Y)$. We use similar notation for inner products with respect to the metrics $g\!\!\!/$ or $\underline{g}$.

DEFINITION 3.11 (**Levi-Civita connections**). We use the following notation for the Levi-Civita connections associated to g, m, and $g\!\!\!/$.

- $\mathscr{D}$ denotes the Levi-Civita connection of the spacetime metric g.
- ∇ denotes the Levi-Civita connection of the Minkowski metric m.
- $\nabla\!\!\!\!/$ denotes the Levi-Civita connection of $g\!\!\!/$.

DEFINITION 3.12 (**Covariant wave operators and Laplacians**). We use the following standard notation.

- $\square_g := (g^{-1})^{\alpha\beta} \mathscr{D}_{\alpha\beta}^2$ denotes the covariant wave operator corresponding to the spacetime metric g.

- $\not{\Delta} := (\not{g}^{-1})^{AB} \not{\nabla}_{AB}^2$ denotes the covariant "angular" Laplacian corresponding to $\not{g}$, where the indices A, B vary over $1, 2$ and correspond to components relative to local coordinates on $S_{t,u}$; we construct local coordinates in Sect. 3.5.

REMARK 3.13 (**Meaning of $\mathscr{D}_{XY}^2 W$**). Throughout, we use the convention that if X, Y, and W are spacetime vectorfields, then $\mathscr{D}_{XY}^2 W := X^\alpha Y^\beta \mathscr{D}_\alpha \mathscr{D}_\beta W$. That is, the contractions against X and Y are taken *after* the two covariant differentiations act on W. Similar remarks apply to the connection $\not{\nabla}$.

3.4. Frame vectorfields and the inverse foliation density

We now define the gradient vectorfield $L_{(Geo)}$ associated to the eikonal function.

DEFINITION 3.14 (**The outgoing null geodesic vectorfield**). Let u be the eikonal function (3.4a). We define the vectorfield $L_{(Geo)}$ by

$$L_{(Geo)}^\nu := -(g^{-1})^{\nu\alpha} \partial_\alpha u = -\mathscr{D}^\nu u. \tag{3.7}$$

Applying the operator $\mathscr{D}^\nu := (g^{-1})^{\nu\alpha} \mathscr{D}_\alpha$ to the eikonal equation (3.4a) and using the Leibniz rule, the identity $\mathscr{D}^\nu (g^{-1})^{\alpha\beta} = 0$, the symmetry property $\mathscr{D}^\nu \mathscr{D}_\alpha u = \mathscr{D}_\alpha \mathscr{D}^\nu u$, and definition (3.7), we find that $0 = \mathscr{D}^\alpha u \mathscr{D}_\alpha \mathscr{D}^\nu u = -L_{(Geo)}^\alpha \mathscr{D}_\alpha \mathscr{D}^\nu u = \mathscr{D}_{L_{(Geo)}} L_{(Geo)}^\nu$. That is, $L_{(Geo)}$ obeys the *geodesic equation*

$$\mathscr{D}_{L_{(Geo)}} L_{(Geo)} = 0. \tag{3.8}$$

Also note that by (3.4a), $L_{(Geo)}$ is g-null, that is, $g(L_{(Geo)}, L_{(Geo)}) = 0$. Moreover, since $L_{(Geo)}$ is proportional to the metric dual of the one-form du, which is conormal to the level sets $\mathcal{C}_u$ of the eikonal function, it follows that $L_{(Geo)}$ is g-orthogonal to $\mathcal{C}_u$. Hence, the $\mathcal{C}_u$ have null normals. Such hypersurfaces are known as *null hypersurfaces*.

We now introduce the most important quantity in our analysis: the inverse foliation density of the level sets of u relative to the hypersurfaces Σ_t.

DEFINITION 3.15 (**The inverse foliation density of the outgoing cones**). Let t be the Minkowskian time coordinate and let u be the eikonal function. We define the inverse foliation density (of the level sets of u relative to the hypersurfaces Σ_t) μ by

$$\mu := -\frac{1}{(g^{-1})^{\alpha\beta} \partial_\alpha t \partial_\beta u} = -\frac{1}{(g^{-1})^{0\alpha} \partial_\alpha u} = \frac{1}{L_{(Geo)}^0}, \tag{3.9}$$

where the last two equalities hold in the **rectangular** coordinate system.

Later in the monograph, we will see that when μ vanishes, many quantities, including the rectangular components $L_{(Geo)}^\nu$, blow-up like μ^{-1}. The outgoing null vectorfield L, defined just below, is a rescaled version of $L_{(Geo)}$ that "removes the singular factor μ^{-1}." We will show that L remains regular and near

$$L_{(Flat)} := \partial_t + \partial_r \tag{3.10}$$

throughout the evolution. Similarly, the vectorfield $\underline{L}$ defined below is an ingoing null vectorfield that remains regular and near

$$\underline{L}_{(Flat)} := \partial_t - \partial_r \tag{3.11}$$

throughout the evolution. In contrast, the rectangular components of the ingoing null vectorfield $\breve{\underline{L}} = \mu \underline{L}$ vanishes precisely when the shock forms.

DEFINITION 3.16 (L, $\breve{\underline{L}}$, **and** $\underline{L}$). We define the *rescaled outgoing null vectorfield* L as follows:

$$L := \upmu L_{(Geo)}. \tag{3.12}$$

We then define $\breve{\underline{L}}$ to be the (unique) null vectorfield that is g-orthogonal to $S_{t,u}$ and normalized by

$$g(\breve{\underline{L}}, L) = -2\upmu. \tag{3.13}$$

Finally, we define

$$\underline{L} := \upmu^{-1}\breve{\underline{L}}. \tag{3.14}$$

REMARK 3.17 (L **is** g**-orthogonal to** $\mathcal{C}_u$). Note that like $L_{(Geo)}$, L is g-orthogonal to $\mathcal{C}_u$.

In the next lemma, we reveal some basic properties of L and $\breve{\underline{L}}$.

LEMMA 3.18 (**Basic properties of the null pair** L $\breve{\underline{L}}$). *The following identities hold, where $t = x^0$ is the rectangular time coordinate function and u is the eikonal function:*

$$Lu = 0, \qquad Lt = L^0 = 1, \tag{3.15a}$$

$$\breve{\underline{L}}u = 2, \qquad \breve{\underline{L}}t = \breve{\underline{L}}^0 = \upmu. \tag{3.15b}$$

PROOF. From (3.4a) and (3.7), we see that $L_{(Geo)}u = 0$. It thus follows from (3.12) that $Lu = 0$ as desired. The identity $Lt = 1$ follows from (3.7), (3.9), and (3.12). We have thus shown (3.15a). We next note that $\breve{\underline{L}}u = -g(\breve{\underline{L}}, L_{(Geo)}) = -\upmu^{-1}g(\breve{\underline{L}}, L)$. We conclude from (3.13) that $\breve{\underline{L}}u = 2$ as desired. We have thus shown (3.15b).

To show that $\breve{\underline{L}}t = \upmu$, we define (relative to rectangular coordinates) $V^\nu := -(g^{-1})^{\nu\alpha}\partial_\alpha t$. We note that V is future-directed and g-orthogonal to Σ_t and hence to $S_{t,u}$. Since L and $\breve{\underline{L}}$ span the g-orthogonal complement of $S_{t,u}$, there exist scalars a and b such that

$$V^\nu = aL^\nu + b\breve{\underline{L}}^\nu. \tag{3.16}$$

Contracting (3.16) against L_ν, using the fact that L is null, and using (3.13) and (3.15a), we find that $1 = Lt = -L_\nu V^\nu = 2\upmu b$. Similarly, we find that $\breve{\underline{L}}t = 2\upmu a$. Also using the fact that $\breve{\underline{L}}$ is null, we deduce that

$$g_{\alpha\beta}V^\alpha V^\beta = -4\upmu ab = -2a. \tag{3.17}$$

On the other hand, by (2.3), we have that

$$g_{\alpha\beta}V^\alpha V^\beta = (g^{-1})^{\alpha\beta}\partial_\alpha t\partial_\beta t = (g^{-1})^{00} = -1. \tag{3.18}$$

From (3.17) and (3.18) it follows that $a = 1/2$ and hence $\breve{\underline{L}}t = \upmu$ as desired. □

We now define an *inward-pointing* Σ_t-tangent vectorfield $\breve{R}$ and an *upward pointing* Σ_t-normal vectorfield N.

DEFINITION 3.19 (**The vectorfields $\breve{R}$, R, and N**). We define the "radial" vectorfields $\breve{R}$, R, and the vectorfield N (which we show in Lemma 3.20 to be the future-directed timelike unit-normal to Σ_t) as follows:

$$\breve{R} := \frac{1}{2}\{-\upmu L + \breve{\underline{L}}\}, \tag{3.19a}$$

$$R := \upmu^{-1}\breve{R} = \frac{1}{2}\{-L + \underline{L}\}, \tag{3.19b}$$

$$N := \frac{1}{2}\{L + \underline{L}\} = L + R. \tag{3.19c}$$

The Minkowskian analogs of (3.19b) and (3.19c) are $-\partial_r$ and ∂_t respectively. In the next lemma, we reveal some basic properties of L, $\breve{R}$, R, and N.

LEMMA 3.20 (**Basic properties of L, $\breve{R}$, R, and N**). *$\breve{R}$ is g-orthogonal to the $S_{t,u}$ and tangent to Σ_t. N is timelike, future-directed, g-orthogonal to $S_{t,u}$, and g-orthogonal to Σ_t.*

In addition, the following identities hold:

$$g(L, L) = 0, \tag{3.20a}$$

$$g(\breve{R}, \breve{R}) = \upmu^2, \tag{3.20b}$$

$$g(R, R) = 1, \tag{3.20c}$$

$$g(L, \breve{R}) = -\upmu, \tag{3.20d}$$

$$g(L, R) = -1. \tag{3.20e}$$

Furthermore, we have

$$g(N, N) = -1. \tag{3.21}$$

In addition, the rectangular components of N are given by

$$N^\nu = -(g^{-1})^{0\nu}. \tag{3.22}$$

Furthermore, we have

$$\breve{R}t = 0, \qquad \breve{R}u = 1, \tag{3.23a}$$

$$Nt = 1, \qquad Nu = 1. \tag{3.23b}$$

Finally, the commutator vectorfield $[L, \breve{R}]$ is $S_{t,u}$-tangent.

PROOF. (3.20a) is a restatement of the fact that L is null. (3.20b) follows from (3.19a), the fact that L and $\breve{\underline{L}}$ are null, and (3.13). (3.20c) follows from (3.19b) and (3.20b). (3.20d) follows from (3.19a), (3.20a), and (3.13). (3.20e) follows from (3.19b) and (3.20d).

(3.21) follows from (3.19c), (3.20a), (3.20c), and (3.20e) .

(3.23a) follows from (3.19a), (3.15a), and (3.15b). (3.23b) follows from (3.19c), (3.15a), and (3.15b).

The fact that $\breve{R}$ is Σ_t-tangent follows from the identity $\breve{R}t = 0$. The fact that $\breve{R}$ is g-orthogonal to $S_{t,u}$ follows from (3.19a) and the fact that L and $\breve{\underline{L}}$ are g-orthogonal to $S_{t,u}$.

To obtain the desired properties of N including the identity (3.22), we note that the proof of Lemma 3.18 reveals that N is equal to the vectorfield $V^\nu := -(g^{-1})^{\nu\alpha}\partial_\alpha t$, which is timelike, future-directed, g-orthogonal to the $S_{t,u}$, and g-orthogonal to Σ_t.

To obtain the fact that $[L, \breve{R}]$ is $S_{t,u}$-tangent, we use (3.15a) and (3.23a) to conclude that $0 = [L, \breve{R}]t = [L, \breve{R}]u$. □

In the next lemma, we derive some important identities involving the rectangular spatial derivatives of the eikonal function.

LEMMA 3.21 (**Identities involving the rectangular spatial derivatives of** u). *The eikonal function* u *verifies the following equation:*

$$(\underline{g}^{-1})^{ab}\partial_a u \partial_b u = \mu^{-2}, \tag{3.24}$$

where $\underline{g}^{-1}$ *is defined in Def. 3.9.*

Furthermore, the rectangular spatial derivatives of u *verify the following equation,* $(i = 1, 2, 3)$*:*

$$\mu\partial_i u = R_i. \tag{3.25}$$

PROOF. To prove (3.24), we use definition (3.9) and the assumption $(g^{-1})^{00} = -1$ (that is, (2.3)) to deduce that $\mu^{-2} = (\partial_t u)^2 - 2(g^{-1})^{0a}\partial_t u\partial_a u + ((g^{-1})^{0a}\partial_a u)^2$. From this identity and the eikonal equation (3.4a), we deduce the identity

$$(g^{-1})^{ab}\partial_a u\partial_b u + ((g^{-1})^{0a}\partial_a u)^2 = \mu^{-2}. \tag{3.26}$$

Furthermore, it is straightforward to verify that under the assumption $(g^{-1})^{00} = -1$, we have the following identity for 3×3 matrices, $(i, j = 1, 2, 3)$:

$$(\underline{g}^{-1})^{ij} = (g^{-1})^{ij} + (g^{-1})^{0i}(g^{-1})^{0j}. \tag{3.27}$$

From (3.26) and (3.27), we deduce the desired identity (3.24).

Next, we consider the pair of one-forms on Σ_t with rectangular spatial components $(\partial_1 u, \partial_2 u, \partial_3 u)$ and (R_1, R_2, R_3). By construction, both one-forms are inward pointing[2] and $\underline{g}$-orthogonal to the $S_{t,u}$. Hence, we must have $\partial_i u = zR_i$ with $z > 0$. Note that since $R^0 = 0$, we have $R_i = \underline{g}_{ia}R^a$. Hence, we have $(\underline{g}^{-1})^{ab}R_a R_b = \underline{g}(R, R) = g(R, R) = 1$, which implies that $z = \sqrt{(\underline{g}^{-1})^{ab}\partial_a u\partial_b u}$. The desired identity (3.25) now follows from this identity and (3.24). □

3.5. Geometric coordinates

In this section, we construct the geometric coordinate system $(t, u, \vartheta^1, \vartheta^2)$ that plays a fundamental role in our analysis.

As a first step, we construct a smooth oriented atlas $\{(\mathbb{D}_i, \vartheta_i^1, \vartheta_i^2)\}_{i=1,2,3,4}$ on the *Euclidean* unit sphere $S_{0,0} \simeq \mathbb{S}^2$. The precise structure of the atlas is not very important for our analysis, but on occasion we will derive estimates for the coordinate functions themselves. Hence, for concreteness, we let $0 < \widetilde{\vartheta}_1^1 < \pi$ and $0 < \widetilde{\vartheta}_1^2 < 2\pi$ be standard spherical coordinate functions defined on all of $\mathbb{S}^2$ (viewed as a subset of $\mathbb{R}^3$) except for a single closed great half-circle joining the north and south poles. We let $\widetilde{\mathbb{D}}_1$ be the corresponding open subset of $\mathbb{S}^2$. Let $\mathbb{D}_1 \subset \widetilde{\mathbb{D}}_1$ be the subset corresponding to restricting the range of $\widetilde{\vartheta}_1^1$ to $(\pi/8, 7\pi/8)$ and the range of $\widetilde{\vartheta}_1^2$ to $(0, 15\pi/8)$, and for $j = 1, 2$, let ϑ_1^j be the restriction of $\widetilde{\vartheta}_1^j$ to $\mathbb{D}_1$. We have therefore constructed the chart $(\mathbb{D}_1, \vartheta_1^1, \vartheta_1^2)$. Next, it is easy to see that by moving the chart $(\mathbb{D}_1, \vartheta_1^1, \vartheta_1^2)$ with Euclidean rotations, we can obtain three similar charts $\{(\mathbb{D}_i, \vartheta_i^1, \vartheta_i^2)\}_{i=2,3,4}$ such that $\mathbb{S}^2 = \cup_{i=1}^4 \mathbb{D}_i$. We have therefore constructed

[2] Note that at $t = 0$, $\partial_i u = -x^i/r$.

our atlas. Note that for $i = 1, 2, 3, 4$, the $(\vartheta_i^1, \vartheta_i^2)$ are coordinate functions on $\mathbb{D}_i$ that could be smoothly extended to a larger open region containing the closure of $\mathbb{D}_i$ in its interior.

DEFINITION 3.22 (**Standard atlas on $\mathbb{S}^2$**). We refer to the above atlas as the standard atlas on $\mathbb{S}^2$.

REMARK 3.23 (**Suppression of the precise coordinate chart**). Throughout the monograph, we typically suppress the set $\mathbb{D}_i$ and the index i and simply refer to the above coordinates as $(\vartheta^1, \vartheta^2)$.

We now (inwardly) extend the local coordinate functions $(\vartheta^1, \vartheta^2)$ from $S_{0,0} \simeq \mathbb{S}^2$ to the region $\Sigma_0^{U_0} = \cup_{u\in[0,U_0]} S_{0,u} \subset \Sigma_0^{U_0}$ by using the *Euclidean* radial vectorfield $-\partial_r = -\frac{x^a}{r}\partial_a$ to transport them, $(A = 1, 2)$:

$$-\partial_r \vartheta^A|_{\Sigma_0^{U_0}} = 0. \tag{3.28}$$

REMARK 3.24 (**How to think about the ϑ^A along Σ_0**). Equation (3.28) implies that the ϑ^A are local coordinates on the Euclidean spheres of constant r in Σ_0. That is, along[3] Σ_0, we may identify $(\vartheta^1, \vartheta^2)$ with Euclidean angular coordinates (θ^1, θ^2), and we may think of $(r, \vartheta^1, \vartheta^2) = (r, \theta^1, \theta^2)$ as a standard Euclidean spherical coordinate system.

We now extend $(\vartheta^1, \vartheta^2)$ to $S_{t,u}$ for $t > 0$ and $u \in [0, U_0]$ by using the g-null vectorfield L to transport them. That is, for $A = 1, 2$, we impose

$$L\vartheta^A = 0. \tag{3.29}$$

DEFINITION 3.25 (**Geometric coordinates**). We refer to the functions $(t, u, \vartheta^1, \vartheta^2)$ induced on spacetime regions of the form $\mathcal{M}_{T,U_0}$ as the *geometric coordinates*.

We denote the geometric coordinate partial derivative vectorfields by

$$\frac{\partial}{\partial t}, \frac{\partial}{\partial u}, \frac{\partial}{\vartheta^1}, \frac{\partial}{\vartheta^2}. \tag{3.30}$$

REMARK 3.26 (**Different ways to think about ϑ**). Note that we are slightly abusing notation by using the symbols ϑ^1, ϑ^2 to denote local coordinate functions on $\mathbb{S}^2$ and also the corresponding coordinate functions induced on $\mathcal{M}_{T,U_0}$, obtained by solving (3.29) subject to the initial conditions on $\Sigma_0^{U_0}$ described just above (3.29). Equivalently, we are identifying the coordinate function pair $(\vartheta^1, \vartheta^2)$ on $\mathcal{M}_{T,U_0}$ with the corresponding point ϑ belonging to $S_{0,0} \simeq \mathbb{S}^2$. We often make these identifications throughout the monograph; the precise meaning of the symbol "ϑ" will always be clear from context.

Since $Lt = 1$ and $Lu = L\vartheta^A = 0$, $(A = 1, 2)$, it follows that relative to the geometric coordinates, we have

$$L = \frac{\partial}{\partial t} = \frac{\partial}{\partial t}|_{u,\vartheta^1,\vartheta^2}. \tag{3.31}$$

We often use the identity (3.31) in our analysis.

---

[3]Just below, we will construct an extension of $(\vartheta^1, \vartheta^2)$ to Σ_t for $t > 0$. Our construction is such that the $(\vartheta^1, \vartheta^2)$ do not generally coincide with Euclidean angular coordinates on Σ_t.

DEFINITION 3.27 (X_1 **and** X_2). X_1 and X_2 are the locally defined $S_{t,u}$-tangent vectorfields

$$X_1 = \frac{\partial}{\partial \vartheta^1} := \frac{\partial}{\partial \vartheta^1}|_{t,u,\vartheta^2}, \qquad X_2 = \frac{\partial}{\partial \vartheta^2} := \frac{\partial}{\partial \vartheta^2}|_{t,u,\vartheta^1}. \tag{3.32}$$

REMARK 3.28 (**The meaning of** ξ_A). We often denote the contraction of a one-form ξ against the $S_{t,u}$ frame vectors $\{X_1, X_2\}$ by using the abbreviated notation $\xi_A := \xi_{X_A} = \xi_a X_A^a$. We use similar abbreviations when contracting other types of tensors against vectors; see Sect. 3.8.

REMARK 3.29 (**C^k-equivalent differential structures until shock formation**). We often silently identify spacetime regions of the form $\mathcal{M}_{t,U_0}$ (see definition (3.6e)) with the region $[0,t) \times [0,U_0] \times \mathbb{S}^2$ corresponding to the geometric coordinates. This identification is justified by the fact that during the classical lifespan of the solution, the differential structure on $\mathcal{M}_{t,U_0}$ corresponding to the geometric coordinates is C^k-equivalent, for some large positive integer k, to the differential structure on $\mathcal{M}_{t,U_0}$ corresponding to the rectangular coordinates. The equivalence is captured by the fact that the change of variables map Υ (see Sect. 3.18) from geometric to rectangular coordinates is differentiable with a differentiable inverse, until a shock forms; see Theorem 22.1. However, at points where μ vanishes and the rectangular derivatives of Ψ blow-up (see inequality (22.11)), the inverse map Υ^{-1} becomes singular and the equivalence of the differential structures breaks down as well.

3.6. Frames

In this section, we define the vectorfield frames that we use in our analysis.

DEFINITION 3.30 (**Vectorfield frames**). We define the following vectorfield frames, where the vectorfield elements are defined in Sects. 3.4 and 3.5.

- $\{L, \breve{R}, X_1, X_2\}$ denotes the rescaled frame.
- $\{L, R, X_1, X_2\}$ denotes the nonrescaled frame.
- $\{L, \breve{\underline{L}}, X_1, X_2\}$ denotes the rescaled null frame.
- $\{L, \underline{L}, X_1, X_2\}$ denotes the nonrescaled null frame.

REMARK 3.31 (**The span of the frame vectorfields**). The analysis of Sect. 3.18 can be used to show that for the small-data solutions that we study, $\{L, \breve{R}, X_1, X_2\}$ and $\{L, \breve{\underline{L}}, X_1, X_2\}$ have span equal to[4] $\text{span}\{\partial_\alpha\}_{\alpha=0,1,2,3}$ at each point where $\mu > 0$. However, these two frames degenerate relative to $\{\partial_\alpha\}_{\alpha=0,1,2,3}$ when μ vanishes in the sense that the Euclidean length of $\breve{R}$ and $\breve{\underline{L}}$, viewed as vectorfields on $\mathbb{R}^4$, goes to 0 as $\mu \to 0$; see also Remark 2.11. In contrast, the estimates of Theorem 22.1 can be used to show that for small-data solutions, the frames $\{L, R, X_1, X_2\}$ and $\{L, \underline{L}, X_1, X_2\}$ do not degenerate[5] relative to $\{\partial_\alpha\}_{\alpha=0,1,2,3}$ as $\mu \to 0$.

[4]This fact is a simple consequence of equation (3.83) and our proof (see Theorem 22.1) that in the small-data regime, the Jacobian determinant of the change of variables map from geometric to rectangular coordinates is nonvanishing whenever $\mu > 0$.

[5]However, as is the case for many other quantities, the rectangular components of the non-rescaled frame vectorfields can become multi-valued functions of the rectangular coordinates at points where $\mu = 0$.

3.7. The future null condition failure factor

We now define the function ${}^{(+)}\aleph$, which depends on the structure of the nonlinearities and is of fundamental importance in determining whether or not shocks can form to the future.

DEFINITION 3.32 (**The future null condition failure factor**). We define the scalar-valued *future null condition failure factor* ${}^{(+)}\aleph$ by

$$(3.33)\qquad {}^{(+)}\aleph := \underbrace{G_{\alpha\beta}(\Psi = 0)}_{\text{constants}} L^{\alpha}_{(Flat)} L^{\beta}_{(Flat)},$$

where $G_{\alpha\beta}$ is defined in (2.14), $L_{(Flat)} = \partial_t + \frac{x^a}{r}\partial_a$ is the standard outgoing Minkowski-null, Minkowski-geodesic vectorfield and $r = \sqrt{\sum_{a=1}^{3}(x^a)^2}$ is the standard Euclidean radial coordinate.

We also define ${}^{(+)}\mathring{\aleph}$, relative to the geometric coordinates, by (see Remark 3.33)

$$(3.34)\qquad {}^{(+)}\mathring{\aleph}(t,u,\vartheta) = {}^{(+)}\mathring{\aleph}(\vartheta) := {}^{(+)}\aleph(t=0,u=0,\vartheta) = {}^{(+)}\aleph(t=0,u,\vartheta).$$

As we explained in Sect. 2.3, ${}^{(+)}\aleph$ is the coefficient of the dangerous slow decaying quadratic terms in the wave equation $\square_{g(\Psi)}\Psi = 0$ in the region $\{t \geq 0\}$. In particular, when ${}^{(+)}\aleph \equiv 0$, Klainerman's work [**47**] and Christodoulou's work [**15**] both yield small-data global existence.

REMARK 3.33 (**The dependence of ${}^{(+)}\aleph$ on various coordinates**). Note that ${}^{(+)}\aleph$ can be viewed as a function depending only on $\theta = (\theta^1, \theta^2)$, where θ^1 and θ^2 are local angular coordinates corresponding to spherical coordinates on Minkowski spacetime. Furthermore, at $t = 0$, our geometric coordinates $(\vartheta^1, \vartheta^2)$ coincide with (θ^1, θ^2), and $L^{\alpha}_{(Flat)}|_{t=0}$ does not depend on u. It follows that relative to the geometric coordinates (t, u, ϑ), the right-hand side of (3.34) is a function of ϑ alone.

REMARK 3.34 (**The role of ${}^{(+)}\mathring{\aleph}$**). The point of introducing ${}^{(+)}\mathring{\aleph}$ is that it is a good approximation to ${}^{(+)}\aleph$ (see Lemma 12.49) that has the added advantage of being constant along the integral curves of L (along which u and ϑ are fixed). This property sometimes makes ${}^{(+)}\mathring{\aleph}$ slightly easier to work with compared to ${}^{(+)}\aleph$.

REMARK 3.35 (**Past null condition failure factor**). We could also study shock formation in the region $\{t \leq 0\}$. In this case, the relevant analog of ${}^{(+)}\aleph$ is the *past null condition failure factor* ${}^{(-)}\aleph$, which is defined by replacing $L_{(Flat)}$ with $-\partial_t + \partial_r$ in equation (3.33). Note that $-\partial_t + \partial_r$ is an outgoing Minkowski-null vectorfield in the region $\{t \leq 0\}$. It is straightforward to see that ${}^{(+)}\aleph$ is nontrivial if and only if ${}^{(-)}\aleph$ is nontrivial. In fact, because the quantities $G_{\alpha\beta}(\Psi = 0)$ in (3.33) are constants, ${}^{(+)}\aleph$ and $-{}^{(-)}\aleph$ have the same range.

3.8. Contraction and component notation

In this section, we define some contraction and component notation that we use throughout the monograph.

DEFINITION 3.36 (**Contraction and component notation**). If ξ is a type $\binom{0}{2}$ spacetime tensor and V, W are vectors, then we define the contraction

$$\xi_{VW} := \xi_{\alpha\beta}V^{\alpha}W^{\beta}. \tag{3.35}$$

Similarly, if ξ is a type $\binom{2}{0}$ spacetime tensor, then we define the contraction

$$\xi_{VW} := \xi^{\alpha\beta}V_{\alpha}W_{\beta}. \tag{3.36}$$

If V is a spacetime vector, then we write

$$V = V^L L + V^{\breve{R}}\breve{R} + V^A X_A \tag{3.37}$$

to denote the decomposition of V relative to the rescaled frame $\{L, \breve{R}, X_1, X_2\}$ of Def. 3.30. In particular, V^L, $V^{\breve{R}}$, and V^A, $(A = 1, 2)$, are scalars.

We use similar contraction and component notation for tensors ξ of any type.

REMARK 3.37 (**$S_{t,u}$ contraction abbreviations**). As we noted in Remark 3.28, we often use abbreviations such as $\xi_A := \xi_{X_A}$ when contracting against the $S_{t,u}$ frame vectors X_1 and X_2.

REMARK 3.38 (**Expansion of a vectorfield relative to the rescaled frame**). It is straightforward to compute, with the help of Lemma 3.20, that

$$V^L = -V_L - \upmu^{-1}V_{\breve{R}}, \qquad V^{\breve{R}} = -\upmu^{-1}V_L, \qquad V^A = (\not{g}^{-1})^{AB}V_B. \tag{3.38}$$

3.9. Projection operators and tensors along submanifolds

In this section, we define two important projection operators. We then provide the closely related definitions of $\Sigma_t^{U_0}$ tensors and $S_{t,u}$ tensors.

DEFINITION 3.39 (**Projection operators**). Let $N = L + R$ be the future-directed unit-normal to $\Sigma_t^{U_0}$ (see Lemma 3.20) and let $\delta_{\nu}^{\ \mu}$ be the standard Kronecker delta. We define the $\Sigma_t^{U_0}$ projection operator $\underline{\Pi}$ and the $S_{t,u}$ projection operator $\not{\Pi}$ to be the following type $\binom{1}{1}$ spacetime tensorfields $(\mu, \nu = 0, 1, 2, 3)$:

$$\underline{\Pi}_{\nu}^{\ \mu} := \delta_{\nu}^{\ \mu} - N_{\nu}N^{\mu} = \delta_{\nu}^{\ \mu} + \delta_{\nu}^{\ 0}N^{\mu}, \tag{3.39a}$$

$$\not{\Pi}_{\nu}^{\ \mu} := \delta_{\nu}^{\ \mu} + R_{\nu}L^{\mu} + L_{\nu}(L^{\mu} + R^{\mu}) = \delta_{\nu}^{\ \mu} - \delta_{\nu}^{\ 0}L^{\mu} + L_{\nu}R^{\mu}. \tag{3.39b}$$

We note that the second equality in (3.39b) holds by virtue of (2.3) (see Lemma 4.3 for additional details).

It is straightforward to verify that $\underline{\Pi}_{\nu}^{\ \mu}N^{\nu} = \underline{\Pi}_{\nu}^{\ \mu}N_{\mu} = 0$ and that $\underline{\Pi}_{\nu}^{\ \mu}V^{\nu} = V^{\mu}$ for $\Sigma_t^{U_0}$-tangent vectors V. Similarly, it is straightforward to verify that $\not{\Pi}_{\nu}^{\ \mu}L^{\nu} = \not{\Pi}_{\nu}^{\ \mu}L_{\mu} = \not{\Pi}_{\nu}^{\ \mu}R^{\nu} = \not{\Pi}_{\nu}^{\ \mu}R_{\mu} = 0$ and that $\not{\Pi}_{\nu}^{\ \mu}Y^{\nu} = Y^{\mu}$ for $S_{t,u}$-tangent vectors Y.

Furthermore, it is straightforward to verify (see Lemma 4.3 for additional details) that the following alternate expressions hold relative to the rectangular coordinates $(i, j = 1, 2, 3$ and $\mu, \nu = 0, 1, 2, 3)$:

$$\not{\Pi}_{j}^{\ i} = \delta_{j}^{\ i} - R_j R^i, \qquad \not{\Pi}_{\nu}^{\ 0} = 0, \qquad \not{\Pi}_{0}^{\ \mu} = 0, \tag{3.40}$$

where $\delta_{j}^{\ i}$ is the standard Kronecker delta. In particular, all "0-containing components" of $\not{\Pi}$ completely vanish, a fact that we often use in our analysis.

DEFINITION 3.40 (**Projections of tensors onto $\Sigma_t^{U_0}$ and $S_{t,u}$**). If $\xi^{\mu_1\cdots\mu_m}_{\nu_1\cdots\nu_n}$ is a type $\binom{m}{n}$ spacetime tensor, then the projections of ξ onto $\Sigma_t^{U_0}$ and $S_{t,u}$ are respectively the type $\binom{m}{n}$ tensors $\underline{\Pi}\xi$ and $\not{\Pi}\xi$ with the following components:

$$(\underline{\Pi}\xi)^{\mu_1\cdots\mu_m}_{\nu_1\cdots\nu_n} := \underline{\Pi}_{\widetilde{\mu}_1}^{\ \mu_1}\cdots\underline{\Pi}_{\widetilde{\mu}_m}^{\ \mu_m}\underline{\Pi}_{\nu_1}^{\ \widetilde{\nu}_1}\cdots\underline{\Pi}_{\nu_n}^{\ \widetilde{\nu}_n}\xi^{\widetilde{\mu}_1\cdots\widetilde{\mu}_m}_{\widetilde{\nu}_1\cdots\widetilde{\nu}_n}, \tag{3.41a}$$

$$(\not{\Pi}\xi)^{\mu_1\cdots\mu_m}_{\nu_1\cdots\nu_n} := \not{\Pi}_{\widetilde{\mu}_1}^{\ \mu_1}\cdots\not{\Pi}_{\widetilde{\mu}_m}^{\ \mu_m}\not{\Pi}_{\nu_1}^{\ \widetilde{\nu}_1}\cdots\not{\Pi}_{\nu_n}^{\ \widetilde{\nu}_n}\xi^{\widetilde{\mu}_1\cdots\widetilde{\mu}_m}_{\widetilde{\nu}_1\cdots\widetilde{\nu}_n}. \tag{3.41b}$$

DEFINITION 3.41 (**$S_{t,u}$ projection notation**). If ξ is a spacetime tensor, then we define

$$\not{\xi} := \not{\Pi}\xi. \tag{3.42}$$

If ξ is a symmetric type $\binom{0}{2}$ spacetime tensor and V is a spacetime vectorfield, then we define

$$\not{\xi}_V := \not{\Pi}(\xi_V), \tag{3.43}$$

where ξ_V is the spacetime one-form with components $\xi_{\alpha\nu}V^\alpha$, $(\nu = 0, 1, 2, 3)$.

In our analysis, we estimate two kinds of quantities: scalar functions and $S_{t,u}$ tensorfields.

DEFINITION 3.42 (**$\Sigma_t^{U_0}$ and $S_{t,u}$ tensors**). Let ξ be a spacetime tensor. We say that ξ is a $\Sigma_t^{U_0}$ tensor if

$$\underline{\Pi}\xi = \xi. \tag{3.44}$$

Similarly, we say that ξ is an $S_{t,u}$ tensor if

$$\not{\Pi}\xi = \xi. \tag{3.45}$$

It is easy to show that ξ is a $\Sigma_t^{U_0}$ tensor if and only if any contraction of any downstairs index of ξ against N necessarily results in 0, and any contraction of any upstairs index of ξ against the g-dual of N necessarily results in 0. Similarly, it is easy to show that ξ is an $S_{t,u}$ tensor if and only if any contraction of any downstairs index of ξ against either L or R necessarily results in 0, and any contraction of any upstairs index of ξ against either the g-dual of L or the g-dual of R necessarily results in 0.

REMARK 3.43 (**Inherent $S_{t,u}$ tensors vs. $S_{t,u}$ tensors embedded in spacetime**). Throughout this monograph, we alternate back and forth between viewing $S_{t,u}$ tensors as tensors that are inherent to the two-dimensional spheres $S_{t,u}$ and viewing $S_{t,u}$ tensors as spacetime tensors with the property[6] (3.45). That is, roughly speaking, we identify $\xi^{A_1\cdots A_m}_{B_1\cdots B_n} \simeq \xi^{\mu_1\cdots\mu_m}_{\nu_1\cdots\nu_n}$. When it comes to performing calculations, both points of view can be helpful, depending on the situation at hand. Similar remarks apply to $\Sigma_t^{U_0}$ tensors. For example, in this way, we may identify the tensorfields $\underline{g}$ and $\not{g}$ from Def. 3.9 respectively with $\underline{\Pi}g$ and $\not{\Pi}g$.

[6] Note that given any $S_{t,u}$-inherent tensor ξ defined at a point $p \in S_{t,u}$, there is a unique spacetime tensor $\widetilde{\xi}$ defined at p such that $\widetilde{\xi} = \not{\Pi}\widetilde{\xi}$ and such that, under the conventions of Def. 3.36 and Remark 3.37, we have the identity $\xi^{A_1\cdots A_m}_{B_1\cdots B_n} = \widetilde{\xi}^{A_1\cdots A_m}_{B_1\cdots B_n}$ for $A_1, \cdots, A_m, B_1, \cdots B_n \in \{1, 2\}$. When we alternate back and forth between the two points of view, it is always with this unique extension in mind. Similar remarks apply to $\Sigma_t^{U_0}$ tensors.

3.10. Expressions for the metrics and volume form factors

In this section, we provide some expressions for the metrics and volume form factors that we use in our analysis. We start with a lemma that clarifies that the vectorfield $\breve{R}$ is related to, but not generally equal to the geometric coordinate partial derivative $\frac{\partial}{\partial u}$.

LEMMA 3.44 (**The vectorfield Ξ**). *There exists an $S_{t,u}$-tangent vectorfield Ξ such that*

$$\breve{R} = \frac{\partial}{\partial u} - \Xi, \tag{3.46}$$

where the $-$ sign in front of Ξ is a convention.

PROOF. The lemma is a simple consequence of (3.23a). $\square$

We now derive an evolution equation for Ξ.

LEMMA 3.45 (**Transport equation verified by Ξ**). *The $S_{t,u}$-tangent vectorfield Ξ verifies the following transport equation:*

$$L\Xi^A = [L, \Xi]^A = -[L, \breve{R}]^A, \tag{3.47}$$

where $[\cdot,\cdot]$ denotes the Lie bracket of two vectorfields.

PROOF. Lemma 3.45 follows easily from equation (3.46) and the identities $[\frac{\partial}{\partial u}, L] = [L, X_A] = [\frac{\partial}{\partial u}, X_A] = 0$. $\square$

We now provide the components of g and $\underline{g}$ relative to the geometric coordinates.

LEMMA 3.46 (**Components of g and $\underline{g}$ relative to the geometric coordinates**). *We can express the spacetime metric g relative to the geometric coordinates $(t, u, \vartheta^1, \vartheta^2)$ as*

$$g = -\mu dt \otimes du - \mu du \otimes dt + \mu^2 du \otimes du + \not{g}_{AB}(d\vartheta^A + \Xi^A du) \otimes (d\vartheta^B + \Xi^B du), \tag{3.48}$$

where Ξ is the $S_{t,u}$-tangent vectorfield from Lemma 3.44.

Furthermore, we can express the first fundamental form[7] *$\underline{g}$ of $\Sigma_t^{U_0}$ as follows relative to the geometric coordinates $(u, \vartheta^1, \vartheta^2)$ induced on $\Sigma_t^{U_0}$:*

$$\underline{g} = \mu^2 du \otimes du + \not{g}_{AB}(d\vartheta^A + \Xi^A du) \otimes (d\vartheta^B + \Xi^B du). \tag{3.49}$$

PROOF. We prove only (3.48) since (3.49) can be proved in a similar fashion. Since $L = \frac{\partial}{\partial t}$ is null, it follows from (3.31) that the coefficient of $dt \otimes dt$ in (3.48) is 0. Next, since $g(L, X_A) = 0$, it follows from (3.31) that the coefficients of $dt \otimes d\vartheta^A$ and $d\vartheta^A \otimes dt$ in (3.48) are 0. Next, using (3.31), (3.20d), and (3.46), we compute that $g(\frac{\partial}{\partial t}, \frac{\partial}{\partial u}) = g(L, \breve{R}) - \Xi^A g(L, X_A) = g(L, \breve{R}) = -\mu$. This yields

[7] We stress that in equation (3.49), we are viewing $\underline{g}$ as a tensorfield on $\Sigma_t^{U_0}$ and ignoring that it can be extended to $\Sigma_t^{U_0}$-transversal vectorfields by imposing that it annihilates the $\Sigma_t^{U_0}$-normal vectorfield N; see Remark 3.43. Similar remarks apply to $\underline{g}$ and $\not{g}$ in Cor. 3.47 and Lemma 3.67 below.

the terms $-\mu dt \otimes du - \mu du \otimes dt$. Similarly, it follows from (3.20b) and (3.46) that $g(\frac{\partial}{\partial u}, \frac{\partial}{\partial u}) = \mu^2 + \Xi^A \Xi^B g(X_A, X_B) = \mu^2 + \mathscr{g}_{AB}\Xi^A\Xi^B$, which yields the term $(\mu^2 + \mathscr{g}_{AB}\Xi^A\Xi^B)du \otimes du$. Next, since $g(X_A, X_B) = \mathscr{g}(X_A, X_B) = \mathscr{g}_{AB}$, it follows that the coefficient of $d\vartheta^A \otimes d\vartheta^B$ in (3.48) is $\mathscr{g}_{AB}$. Finally, since (3.46) implies that $0 = g(\breve{R}, X_A) = g(\frac{\partial}{\partial u}, X_A) - \Xi^B g(X_B, X_A) = g(\frac{\partial}{\partial u}, X_A) - \mathscr{g}_{AB}\Xi^B$, we conclude that the coefficients of $du \otimes d\vartheta^A$ and $d\vartheta^A \otimes du$ in (3.48) are equal to $\mathscr{g}_{AB}\Xi^B$. □

We now provide expressions for the geometric volume form factors of g and $\underline{g}$; see Def. 3.68 for the definitions of the volume forms.

COROLLARY 3.47 (**The geometric volume form factors of g and $\underline{g}$**). *Let $\underline{g}$ and $\mathscr{g}$ be the first fundamental forms from Def. 3.9 (see Footnote 7). Then we have the following identities:*

$$|\det g| = \mu^2 \det \mathscr{g}, \tag{3.50a}$$

$$\det \underline{g} = \mu^2 \det \mathscr{g}. \tag{3.50b}$$

In (3.50a), the determinant on the left-hand side is taken relative to the geometric coordinates $(t, u, \vartheta^1, \vartheta^2)$ and the determinant on the right-hand side is taken relative to the geometric coordinates $(\vartheta^1, \vartheta^2)$ induced on $S_{t,u}$. Similarly, in (3.50b), the determinant on the left-hand side is taken relative to the geometric coordinates $(u, \vartheta^1, \vartheta^2)$ induced on $\Sigma_t^{U_0}$, while the determinant on the right-hand side is taken relative to the geometric coordinates $(\vartheta^1, \vartheta^2)$ induced on $S_{t,u}$.

PROOF. To prove (3.50a), we first note that since both sides of (3.50a) have the same transformation properties with respect to changes of angular coordinates of the form $\widetilde{\vartheta}^1 = f^1(u, \vartheta^1, \vartheta^2)$, $\widetilde{\vartheta}^2 = f^2(u, \vartheta^1, \vartheta^2)$ (where the f^A are the coordinate transformation functions), it suffices to show that (3.50a) holds for a well-chosen version $(\widetilde{\vartheta}^1, \widetilde{\vartheta}^2)$ of such angular coordinates. To this end, we fix t and endow $\Sigma_t^{U^0}$ with local coordinates $(u, \widetilde{\vartheta}^1, \widetilde{\vartheta}^2)$ (where u is the usual eikonal function) in such a way that $\breve{R} = \frac{\partial}{\partial u}|_{\widetilde{\vartheta}^1, \widetilde{\vartheta}^2}$. To achieve this construction, we set $\widetilde{\vartheta}^A := \vartheta^A$ along $S_{t,0}$ and then (inwardly) transport $\widetilde{\vartheta}^A$ with $\breve{R}$, that is, $\breve{R}\widetilde{\vartheta}^A = 0$. We then extend the coordinates $\widetilde{\vartheta}^A$ off of $\Sigma_t^{U_0}$ by solving the transport equation $L\widetilde{\vartheta}^A = 0$. Relative to these new local coordinates $(t, u, \widetilde{\vartheta}^1, \widetilde{\vartheta}^2)$, the identity (3.48) holds along $\Sigma_t^{U^0}$, but with $\mathscr{g}_{AB}$ replaced by $g(\frac{\partial}{\partial \widetilde{\vartheta}^A}, \frac{\partial}{\partial \widetilde{\vartheta}^B})$, $d\vartheta$ replaced by $d\widetilde{\vartheta}$, and Ξ replaced by 0. Hence, relative to these new coordinates, the identity (3.50a) follows from a simple computation based on the block form of the metric (along $\Sigma_t^{U_0}$).

The identity (3.50b) can be proved in a similar fashion with the help of the identity (3.49), and we omit the details. □

LEMMA 3.48 (**The metrics in terms of the frame vectorfields**). *Let L and $\breve{\underline{L}}$ be the vectorfields from Def. 3.16 and let $\mathscr{g}$ be the first fundamental form of $S_{t,u}$ from Def. 3.9. Then the following identities hold (we note that the meaning of*

$\not{g}_{\mu\nu}$ and $(\not{g}^{-1})^{\mu\nu}$ is clarified in Remark 3.43):

$$g_{\mu\nu} = -\frac{1}{2}\upmu^{-1}\left(L_\mu \breve{\underline{L}}_\nu + \breve{\underline{L}}_\mu L_\nu\right) + \not{g}_{\mu\nu}, \tag{3.51a}$$

$$(g^{-1})^{\mu\nu} = -\frac{1}{2}\upmu^{-1}\left(L^\mu \breve{\underline{L}}^\nu + \breve{\underline{L}}^\mu L^\nu\right) + (\not{g}^{-1})^{\mu\nu}. \tag{3.51b}$$

In addition, let $\breve{R}$ and R be the vectorfields from Def. 3.19. Then the following identities hold:

$$g_{\mu\nu} = -L_\mu L_\nu - (L_\mu R_\nu + R_\mu L_\nu) + \not{g}_{\mu\nu} \tag{3.52a}$$

$$= -L_\mu L_\nu - \upmu^{-1}(L_\mu \breve{R}_\nu + \breve{R}_\mu L_\nu) + \not{g}_{\mu\nu}, \tag{3.52b}$$

$$(g^{-1})^{\mu\nu} = -L^\mu L^\nu - (L^\mu R^\nu + R^\mu L^\nu) + (\not{g}^{-1})^{\mu\nu} \tag{3.52c}$$

$$= -L^\mu L^\nu - \upmu^{-1}(L^\mu \breve{R}^\nu + \breve{R}^\mu L^\nu) + (\not{g}^{-1})^{\mu\nu}. \tag{3.52d}$$

Moreover, let $\underline{g}$ and $\underline{g}^{-1}$ be as in Def. 3.9. Then we have the following identities:

$$\underline{g}_{ij} = R_i R_j + \not{g}_{ij} \tag{3.53a}$$

$$= \upmu^{-2}\breve{R}_i \breve{R}_j + \not{g}_{ij}, \tag{3.53b}$$

$$(\underline{g}^{-1})^{ij} = R^i R^j + (\not{g}^{-1})^{ij} \tag{3.53c}$$

$$= \upmu^{-2}\breve{R}^i \breve{R}^j + (\not{g}^{-1})^{ij}. \tag{3.53d}$$

Finally, we have

$$(\not{g}^{-1})^{\mu\nu} = (\not{g}^{-1})^{AB} X_A^\mu X_B^\nu. \tag{3.54}$$

PROOF. The identity (3.51a) can easily be verified by contracting both sides against all possible pairs of vectors belonging to the rescaled null frame $\{L, \breve{\underline{L}}, X_1, X_2\}$ and computing, with the help of Lemma 3.20, that both sides agree. The identity (3.54) follows in the same way. The identity (3.51b) then follows from raising the indices of both sides of (3.51a) with g^{-1} and from the identity $(g^{-1})^{\mu\alpha}(g^{-1})^{\nu\beta}\not{g}_{\alpha\beta} = (\not{g}^{-1})^{\mu\nu}$.

Similar arguments yield (3.53a)-(3.53d), except that in deriving them, we contract against all possible pairs of vectors belonging to $\{R, X_1, X_2\}$.

The identities (3.52a)-(3.52d) then follow from using (3.19a)-(3.19b) to substitute for $\breve{\underline{L}}$ in (3.51a)-(3.51b). □

3.11. The trace and trace-free parts of tensors

We now provide some standard definitions connected to the trace and trace-free part of tensors.

DEFINITION 3.49 (**Trace and trace-free parts of tensors**). If ξ is a type $\binom{0}{2}$ spacetime tensor, then

$$\mathrm{tr}_g \xi := (g^{-1})^{\alpha\beta}\xi_{\alpha\beta} \tag{3.55}$$

denotes its g-trace.

If ξ is a type $\binom{0}{2}$ $S_{t,u}$ tensor, then

$$\mathrm{tr}_{\not{g}} \xi := (\not{g}^{-1})^{AB}\xi_{AB} \tag{3.56a}$$

denotes its $\not{g}$-trace and

$$\hat{\xi}_{AB} := \xi_{AB} - \frac{1}{2}\mathrm{tr}_{\not{g}}\xi \not{g}_{AB} \tag{3.56b}$$

denotes its trace-free part.

If ξ and ω are $S_{t,u}$ covectors, then

$$(\xi \hat{\otimes} \omega)_{AB} := \frac{1}{2}(\xi_A \omega_B + \xi_B \omega_A) - \frac{1}{2}\mathrm{tr}_{\not{g}}(\xi \otimes \omega)\not{g}_{AB} \tag{3.57}$$

denotes the symmetrized trace-free part of the type $\binom{0}{2}$ $S_{t,u}$ tensor $\xi \otimes \omega$, where $(\xi \otimes \omega)_{AB} := \xi_A \omega_B$.

3.12. Angular differential

We now define the angular differential of a scalar-valued function.

DEFINITION 3.50 (**Angular differential**). If f is a function, we define $\not{d}f$ to be the $S_{t,u}$ one-form

$$\not{d}f := \not{\Pi} df, \tag{3.58}$$

where df is the standard spacetime differential of f and we are using the notation of Def. 3.41.

Note that $\not{d}_A f = X_A f$ and hence $\not{d}f$ can be viewed as the standard angular differential of f viewed as a function of the geometric angular coordinates $(\vartheta^1, \vartheta^2)$.

DEFINITION 3.51 (**The meaning of $\not{d}_i$**). If f is a function, then for $i = 1, 2, 3$,

$$\not{d}_i f := (\not{d}f) \cdot \partial_i = \not{\Pi}_i^{\ a} \partial_a f \tag{3.59}$$

denotes the ith component of the $S_{t,u}$ one-form $\not{d}f$ relative to the rectangular coordinate frame.

In the next lemma, we compute the angular differential of the rectangular coordinate functions.

LEMMA 3.52 (**Angular differential of the x^i**). *Let $\not{\Pi}$ be the $S_{t,u}$ projection from Def. 3.39 and let x^i be the rectangular spatial coordinate function. Then the following identities hold relative to the rectangular coordinates ($i, j = 1, 2, 3$):*

$$\not{d}_j x^i = \not{\Pi}_j^{\ i}. \tag{3.60}$$

Furthermore, we have (for $A = 1, 2$ and $i = 1, 2, 3$)

$$\not{d}_A x^i = X_A^i. \tag{3.61}$$

PROOF. To prove (3.60), we use the definition of $\not{d}x^j$ and the identity $\partial_\alpha x^i = \delta_\alpha^i$ to compute that $\not{d}_j x^i = \not{\Pi}_j^{\ \alpha} \partial_\alpha x^i = \not{\Pi}_j^{\ i}$ as desired. To prove (3.61), we contract X_A^j against both sides of (3.60). The left-hand side clearly results in $\not{d}_A x^i$, while since X_A is $S_{t,u}$-tangent, the right-hand side results in X_A^i as desired. □

3.13. Musical notation

In this section, we define our use of the symbols "#" and "♭."

DEFINITION 3.53 (**Sharp and flat notation**). If ξ is an $S_{t,u}$ covector, then we define $\xi^{\#}$ to be the $\not{g}$-dual of ξ, which is an $S_{t,u}$-tangent vector. That is, $(\xi^{\#})^A := (\not{g}^{-1})^{AB}\xi_B$. We often abbreviate $\xi^A := (\xi^{\#})^A$.

If Y is an $S_{t,u}$-tangent vector, then we define $Y_\flat$ to be the $\not{g}$-dual of Y, which is an $S_{t,u}$ covector. That is, $(Y_\flat)_A := (\not{g}^{-1})_{AB}Y^B$. We often abbreviate $Y_A := (Y_\flat)_A$.

Similarly, if ξ is a symmetric type $\binom{0}{2}$ $S_{t,u}$ tensor, then we define its $\not{g}$-dual $\xi^{\#}$ to be the type $\binom{1}{1}$ $S_{t,u}$ tensor with A, B component equal to $(\not{g}^{-1})^{AC}\xi_{CB}$, and its $\not{g}$-double dual $\xi^{\#\#}$ to be the symmetric type $\binom{2}{0}$ $S_{t,u}$ tensor with A, B component equal to $(\not{g}^{-1})^{AC}(\not{g}^{-1})^{BD}\xi_{CD}$.

We use similar notation to denote the $\not{g}$-duals of general type $\binom{m}{0}$ and type $\binom{0}{n}$ $S_{t,u}$ tensors, and we use abbreviations similar to the ones mentioned above for vectors and covectors.

3.14. Pointwise norms

Unless we explicitly state otherwise, we measure the pointwise norms of $S_{t,u}$ tensors relative to the metric $\not{g}$. In the following definition, we make this precise.

DEFINITION 3.54 (**Norm of $S_{t,u}$ tensors relative to $\not{g}$**). Let ξ be a type $\binom{m}{n}$ $S_{t,u}$ tensor with rectangular components $\xi_{\nu_1\cdots\nu_n}^{\mu_1\cdots\mu_m}$ and $S_{t,u}$ components $\xi_{B_1\cdots B_n}^{A_1\cdots A_m}$. We define[8] $|\xi|^2$, the square of the norm of ξ, as follows:

$$
\begin{aligned}
|\xi|^2 &:= \not{g}_{\mu_1\widetilde{\mu}_1}\cdots\not{g}_{\mu_m\widetilde{\mu}_m}(\not{g}^{-1})^{\nu_1\widetilde{\nu}_1}\cdots(\not{g}^{-1})^{\nu_n\widetilde{\nu}_n}\xi_{\nu_1\cdots\nu_n}^{\mu_1\cdots\mu_m}\xi_{\widetilde{\nu}_1\cdots\widetilde{\nu}_n}^{\widetilde{\mu}_1\cdots\widetilde{\mu}_m} \\
&= \not{g}_{A_1\widetilde{A}_1}\cdots\not{g}_{A_m\widetilde{A}_m}(\not{g}^{-1})^{B_1\widetilde{B}_1}\cdots(\not{g}^{-1})^{B_n\widetilde{B}_n}\xi_{B_1\cdots B_n}^{A_1\cdots A_m}\xi_{\widetilde{B}_1\cdots\widetilde{B}_n}^{\widetilde{A}_1\cdots\widetilde{A}_m}.
\end{aligned}
\tag{3.62}
$$

3.15. Lie derivatives and projected Lie derivatives

In this section, we provide the standard definition of the Lie derivative of a tensorfield. We then define several related Lie derivative operators involving projections.

DEFINITION 3.55 (**Lie derivatives**). If V^μ is a spacetime vectorfield and $\xi_{\nu_1\cdots\nu_n}^{\mu_1\cdots\mu_m}$ is a type $\binom{m}{n}$ spacetime tensorfield, then relative to the rectangular coordinates, the Lie derivative of ξ with respect to V is the type $\binom{m}{n}$ spacetime tensorfield $\mathcal{L}_V\xi$ with the following components:

$$
\begin{aligned}
\mathcal{L}_V\xi_{\nu_1\cdots\nu_n}^{\mu_1\cdots\mu_m} &:= V^\alpha\partial_\alpha\xi_{\nu_1\cdots\nu_n}^{\mu_1\cdots\mu_m} \\
&\quad -\sum_{a=1}^{m}\xi_{\nu_1\cdots\nu_n}^{\mu_1\cdots\mu_{a-1}\alpha\mu_{a+1}\cdots\mu_m}\partial_\alpha V^{\mu_a}+\sum_{b=1}^{n}\xi_{\nu_1\cdots\nu_{b-1}\alpha\nu_{b+1}\cdots\nu_n}^{\mu_1\cdots\mu_m}\partial_{\nu_b}V^\alpha.
\end{aligned}
\tag{3.63}
$$

In addition, when V and W are both vectorfields, we often use the standard Lie bracket notation $[V, W] := \mathcal{L}_V W$.

---

[8]Throughout the monograph, we treat indices decorated with a tilde, such as $\widetilde{A}$, in the same way as their nondecorated counterparts.

REMARK 3.56 (**Pullback formulation of Lie differentiation**). On occasion, we will use the following equivalent formula for Lie differentiation with respect to a spacetime vectorfield V (see, for example, [**78**, Appendix C] and note that a different notation for the pullback operator is used there):

$$\mathcal{L}_V \xi_{\nu_1 \cdots \nu_n}^{\mu_1 \cdots \mu_m} = \frac{d}{d\lambda}|_{\lambda=0} (\varphi_{(\lambda)}^* \xi)_{\nu_1 \cdots \nu_n}^{\mu_1 \cdots \mu_m}. \tag{3.64}$$

In (3.64), $\varphi_{(\lambda)} : \mathcal{M}_{t,u} \to \mathcal{M}_{t,u}$ is the flow map of V and $\varphi_{(\lambda)}^* \xi$ is the pullback of ξ by $\varphi_{(\lambda)}$, which takes the following form relative to an arbitrary coordinate system:

(3.65)

$$\begin{aligned} &(\varphi_{(\lambda)}^* \xi)_{\nu_1 \cdots \nu_n}^{\mu_1 \cdots \mu_m} \\ &= (\partial_{\alpha_1} \varphi_{(-\lambda)}^{\mu_1}) \circ \varphi_{(\lambda)} \cdots (\partial_{\alpha_m} \varphi_{(-\lambda)}^{\mu_m}) \circ \varphi_{(\lambda)} (\partial_{\nu_1} \varphi_{(\lambda)}^{\beta_1}) \cdots (\partial_{\nu_n} \varphi_{(\lambda)}^{\beta_n}) \left\{ \xi_{\beta_1 \cdots \beta_n}^{\alpha_1 \cdots \alpha_m} \circ \varphi_{(\lambda)} \right\}. \end{aligned}$$

By "the flow map of V," we mean that for each real number λ belonging to a small neighborhood of 0 and each spacetime point $p \in \mathcal{M}_{t,u}$, the four components of $\varphi_{(\lambda)}(p)$ verify the ODE system $\frac{d}{d\lambda} \varphi_{(\lambda)}^{\nu}(p) = V^{\nu} \circ \varphi_{(\lambda)}(p)$ and $\varphi_{(0)} = I$, where I is the identity map on $\mathcal{M}_{t,u}$. A fundamental result from ODE theory (see, for example, [**8**, Chapter 2]) is that $\varphi_{(-\lambda)} = \varphi_{(\lambda)}^{-1}$.

It is a standard fact that Lie differentiation is coordinate invariant[9] and obeys the Leibniz rule as well as the Jacobi-type identity

$$\mathcal{L}_V \mathcal{L}_W \xi - \mathcal{L}_W \mathcal{L}_V \xi = \mathcal{L}_{[V,W]} \xi = \mathcal{L}_{\mathcal{L}_V W} \xi. \tag{3.66}$$

In our analysis, we often Lie differentiate contractions of tensorial products of $S_{t,u}$ tensors and apply the Leibniz rule. Due in part to the special properties (such as those proved below in Lemma 4.10) of the vectorfields that we use to differentiate, the non-$S_{t,u}$ components of the differentiated factor in the products typically cancel. Therefore, projecting each factor onto the $S_{t,u}$ after one Lie differentiation does not change the contracted output but has the advantage of forcing all differentiated factors in the product to remain $S_{t,u}$-tangent and thus "ready" to be Lie differentiated again. This motivates the following definition.

DEFINITION 3.57 (**$\Sigma_t^{U_0}$-projected and $S_{t,u}$-projected Lie derivatives**). Let ξ be a spacetime tensorfield. We define $\underline{\mathcal{L}}_V \xi$ and $\not{\mathcal{L}}_V \xi$ to respectively be the following $\Sigma_t^{U_0}$ and $S_{t,u}$ tensorfields:

$$\underline{\mathcal{L}}_V \xi := \underline{\Pi} \mathcal{L}_V \xi, \tag{3.67a}$$

$$\not{\mathcal{L}}_V \xi := \not{\Pi} \mathcal{L}_V \xi, \tag{3.67b}$$

where we are using the notation of Def. 3.40. We refer to $\underline{\mathcal{L}}_V$ as a $\Sigma_t^{U_0}$-projected Lie derivative operator and $\not{\mathcal{L}}_V$ as an $S_{t,u}$-projected Lie derivative operator.

LEMMA 3.58 (**Alternate expression for Lie derivatives of $S_{t,u}$ tensorfields**). *If $\xi_{\nu_1 \cdots \nu_n}^{\mu_1 \cdots \mu_m}$ is a type $\binom{m}{n}$ $S_{t,u}$ tensorfield and X is an $S_{t,u}$-tangent vectorfield,*

[9] That is, the formula (3.63) holds in an arbitrary coordinate system.

then relative to the rectangular coordinates, we have

(3.68)

$$\begin{aligned}\not{\mathcal{L}}_X \xi^{\mu_1\cdots\mu_m}_{\nu_1\cdots\nu_n} &= X^{\alpha} \nabla\!\!\!/_{\alpha} \xi^{\mu_1\cdots\mu_m}_{\nu_1\cdots\nu_n} \\ &- \sum_{a=1}^{m} \xi^{\mu_1\cdots\mu_{a-1}\alpha\mu_{a+1}\cdots\mu_m}_{\nu_1\cdots\nu_n} \nabla\!\!\!/_{\alpha} X^{\mu_a} + \sum_{b=1}^{n} \xi^{\mu_1\cdots\mu_m}_{\nu_1\cdots\nu_{b-1}\alpha\nu_{b+1}\cdots\nu_n} \nabla\!\!\!/_{\nu_b} X^{\alpha}.\end{aligned}$$

PROOF. By the torsion-free property $\mathscr{D}_W V - \mathscr{D}_V W = \mathcal{L}_W V$ (valid for all spacetime vectorfields W and V), equation (3.63) holds for $V := X$ and with ∂ replaced by $\mathscr{D}$ on the right-hand side. We then project both sides of the identity onto $S_{t,u}$ using $\Pi\!\!\!/$. For $S_{t,u}$ tensorfields ξ, $\mathscr{D}\xi$ and $\nabla\!\!\!/\xi$ differ only by terms that are g-orthogonal to $S_{t,u}$. Hence, the $S_{t,u}$ projection allows us to replace $\mathscr{D}$ with $\nabla\!\!\!/$. We have thus proved (3.68). □

3.16. Second fundamental forms

In this section, we provide the standard definition of the second fundamental forms of $\Sigma_t^{U_0}$ and $S_{t,u}$ relative to the metric g. We then provide some useful expressions for these tensorfields.

DEFINITION 3.59 (**Second fundamental form of $\Sigma_t^{U_0}$**). Let $N = L + R$ be the future-directed unit normal to Σ_t (see Lemma 3.20). We define the second fundamental form k of $\Sigma_t^{U_0}$ relative to g to be the following type $\binom{0}{2}$ $\Sigma_t^{U_0}$ tensorfield:

$$k := \frac{1}{2}\mathcal{L}_N g. \tag{3.69}$$

DEFINITION 3.60 (**Null second fundamental form of $S_{t,u}$**). We define the null second fundamental form χ of $S_{t,u}$ relative to g to be the following type $\binom{0}{2}$ $S_{t,u}$ tensorfield:

$$\chi := \frac{1}{2}\not{\mathcal{L}}_L g. \tag{3.70}$$

It follows in a straightforward fashion from definitions (3.69)-(3.70), Def. 3.9, and Remark 3.43 that the following alternate expressions hold:

$$k = \frac{1}{2}\mathcal{L}_N \underline{g}, \tag{3.71}$$

$$\chi = \frac{1}{2}\not{\mathcal{L}}_L g\!\!\!/. \tag{3.72}$$

In the next lemma, we provide additional expressions for χ, $k\!\!\!/$, and $k\!\!\!/_R$, where $k\!\!\!/$ and $k\!\!\!/_R$ are defined in terms of k by Def. 3.41.

LEMMA 3.61 (**Alternate expressions for χ, $k\!\!\!/$, and $k\!\!\!/_R$**). *χ and $k\!\!\!/$ are symmetric type $\binom{0}{2}$ $S_{t,u}$ tensorfields that verify the following identities:*

$$\chi_{AB} = g(\mathscr{D}_A L, X_B), \tag{3.73}$$

$$k\!\!\!/_{AB} = g(\mathscr{D}_A N, X_B). \tag{3.74}$$

Furthermore, $k\!\!\!/_R$ is an $S_{t,u}$ one-form that verifies the following identity:

$$k\!\!\!/_{RA} = g(\mathscr{D}_A L, R). \tag{3.75}$$

PROOF. The symmetry properties follow trivially from the definitions. We now prove (3.74). Using definition (3.69), the Leibniz rule, the torsion-free property $\mathscr{D}_V W - \mathscr{D}_W V = [V, W]$, the identity $[X_A, X_B] = 0$, and the identity $g(N, X_A) = 0$, we deduce (3.74) as follows:

(3.76)
$$\begin{aligned}
2\not{k}_{AB} &= (\mathcal{L}_N g)(X_A, X_B) \\
&= N[g(X_A, X_B)] - g([N, X_A], X_B) - g(X_A, [N, X_B]) \\
&= g(\mathscr{D}_N X_A, X_B) + g(X_A, \mathscr{D}_N X_B) - g([N, X_A], X_B) - g(X_A, [N, X_B]) \\
&= g(\mathscr{D}_A N, X_B) + g(\mathscr{D}_B N, X_A) = g(\mathscr{D}_A N, X_B) - g(N, \mathscr{D}_B X_A) \\
&= g(\mathscr{D}_A N, X_B) - g(N, \mathscr{D}_A X_B) = 2g(\mathscr{D}_A N, X_B).
\end{aligned}$$

The proofs of (3.73) and (3.75) are similar (the proof of the former is based on definition (3.70)), and we omit the details. □

3.17. Frame components, relative to the nonrescaled frame, of the derivatives of the metric with respect to the solution

In this section, we define the components of symmetric type $\binom{0}{2}$ spacetime tensorfields relative to the nonrescaled frame $\{L, R, X_1, X_2\}$.

DEFINITION 3.62 (**Components of H relative to the nonrescaled frame $\{L, R, X_1, X_2\}$**). Given any symmetric type $\binom{0}{2}$ spacetime tensorfield H with rectangular components $H_{\mu\nu}$, we define the frame components of $H_{\mu\nu}$ (relative to the frame $\{L, R, X_1, X_2\}$) to be the scalar-valued functions $H_{LL} := H_{\alpha\beta} L^\alpha L^\beta$, $H_{LR} := H_{\alpha b} L^\alpha R^b$, and $H_{RR} := H_{ab} R^a R^b$, the $S_{t,u}$ one-forms with frame components $\not{H}_{LA} := H_{\alpha b} L^\alpha X_A^b$, $\not{H}_{RA} := H_{\alpha b} R^\alpha X_A^b$, and the symmetric type $\binom{0}{2}$ $S_{t,u}$ tensorfield with frame components $\not{H}_{AB} := H_{ab} X_A^a X_B^b$. We often respectively use the following abbreviations for the latter three tensorfields:

$$\not{H}_L, \not{H}_R, \not{H}. \tag{3.77}$$

We rarely need to distinguish between the various frame components of H. Hence, to simplify the presentation, it is convenient to place all of these components into an array and to view them as a single entity. Specifically, we define $H_{(Frame)}$ and $H_{(Frame)}^{\#}$ to be the following arrays:

$$H_{(Frame)} := (H_{LL}, H_{LR}, H_{RR}, \not{H}_L, \not{H}_R, \not{H}), \tag{3.78a}$$

$$H_{(Frame)}^{\#} := \left(H_{LL}, H_{LR}, H_{RR}, \not{H}_L^{\#}, \not{H}_R^{\#}, \not{H}^{\#}\right). \tag{3.78b}$$

REMARK 3.63 (**H is always G or G'**). In this monograph, the tensorfield $H_{\mu\nu}$ from Def. 3.62 will always be equal to either $G_{\mu\nu}$ or $G'_{\mu\nu}$, which are defined in (2.14) and (2.15).

DEFINITION 3.64 (**Norm of $H_{(Frame)}$**). We define

$$|H_{(Frame)}| := |H_{LL}| + |H_{LR}| + |H_{RR}| + |\not{H}_L| + |\not{H}_R| + |\not{H}|, \tag{3.79}$$

and similarly for $\left|H_{(Frame)}^{\#}\right|$.

DEFINITION 3.65 (**Derivatives of $H_{(Frame)}$**). If V is a vectorfield, then we define

$$\mathcal{L}_V H_{(Frame)} := (V(H_{LL}), V(H_{LR}), V(H_{RR}), \not{\mathcal{L}}_V \not{H}_L, \not{\mathcal{L}}_V \not{H}_R, \not{\mathcal{L}}_V \not{H}), \tag{3.80}$$

and similarly for $\mathcal{L}_V H^{\#}_{(Frame)}$. We note that in (3.80) and throughout, $\not{\mathcal{L}}_V \not{H}_L := \not{\mathcal{L}}_V(\not{H}_L)$. We use similar notation if the Lie derivative operator is replaced with the $S_{t,u}$ covariant derivative operator $\not{\nabla}$.

3.18. The change of variables map

In this section, we introduce the change of variables map from geometric to rectangular coordinates and reveal its basic properties.

DEFINITION 3.66 (**The change of variables map**). We define $\Upsilon : [0, T) \times [0, U_0] \times \mathbb{S}^2 \to \mathcal{M}_{T,U_0}$, $\Upsilon(t, u, \vartheta^1, \vartheta^2) = (x^0, x^1, x^2, x^3)$, to be the change of variables map from geometric to rectangular coordinates.

In the next lemma, we compute the Jacobian determinant of Υ. Later, we use the lemma to show that for the solutions of interest, Υ is a diffeomorphism onto its image as long as $\mu > 0$.

LEMMA 3.67 (**The Jacobian determinant of the map Υ**). *Let Υ be the change of variables map from Def. 3.66 and let $\underline{g}$ and $\not{g}$ be as in Def. 3.9 (see Footnote 7 on pg. 93). Then the Jacobian determinant of Υ can be expressed as*

$$\det \frac{\partial(x^0, x^1, x^2, x^3)}{\partial(t, u, \vartheta^1, \vartheta^2)} = \mu (\det \underline{g})^{-1/2} \sqrt{\det \not{g}}, \tag{3.81}$$

where $(\det \underline{g})^{-1/2}$ is a smooth function of Ψ in a neighborhood of 0 that verifies $(\det \underline{g})^{-1/2}(\Psi = 0) = 1$. In (3.81), $\det \underline{g}$ is taken relative to the rectangular spatial coordinates and $\det \not{g}$ is taken relative to the geometric angular coordinates $(\vartheta^1, \vartheta^2)$ induced on $S_{t,u}$.

PROOF. We claim that for $A = 1, 2$ and $i = 1, 2, 3$, we have

$$\begin{aligned} &\frac{\partial x^0}{\partial t} = 1, \quad \frac{\partial x^0}{\partial u} = 0, \quad \frac{\partial x^0}{\partial \vartheta^A} = 0, \\ &\frac{\partial x^i}{\partial t} = L^i, \quad \frac{\partial x^i}{\partial u} = \breve{R}^i + \Xi^i, \quad \frac{\partial x^i}{\partial \vartheta^A} = \not{d}_A x^i = X_A^\alpha \partial_\alpha x^i = X_A^i, \end{aligned} \tag{3.82}$$

where Ξ is the $S_{t,u}$-tangent vectorfield from (3.46). The identities in (3.82) are a straightforward consequence of Def. 3.27, (3.31), and (3.46).

Using (3.82), we compute that

$$\det \frac{\partial(x^0, x^1, x^2, x^3)}{\partial(t, u, \vartheta^1, \vartheta^2)} = \det \begin{pmatrix} 1 & 0 & 0 & 0 \\ L^1 & \breve{R}^1 + \Xi^1 & X_1^1 & X_2^1 \\ L^2 & \breve{R}^2 + \Xi^2 & X_1^2 & X_2^2 \\ L^3 & \breve{R}^3 + \Xi^3 & X_1^3 & X_2^3 \end{pmatrix} \tag{3.83}$$

$$= \det \begin{pmatrix} \breve{R}^1 & X_1^1 & X_2^1 \\ \breve{R}^2 & X_1^2 & X_2^2 \\ \breve{R}^3 & X_1^3 & X_2^3 \end{pmatrix} + \det \begin{pmatrix} \Xi^1 & X_1^1 & X_2^1 \\ \Xi^2 & X_1^2 & X_2^2 \\ \Xi^3 & X_1^3 & X_2^3 \end{pmatrix}$$

$$= \mu \det \begin{pmatrix} R^1 & X_1^1 & X_2^1 \\ R^2 & X_1^2 & X_2^2 \\ R^3 & X_1^3 & X_2^3 \end{pmatrix},$$

where to deduce the last equality, we used the fact that $\Xi \in \text{span}\{X_1, X_2\}$. The last determinant on the right-hand side of (3.83) can be interpreted as $\upsilon_e(R, X_1, X_2)$, where υ_e is the volume form of the Euclidean metric e on Σ_t (where $e_{ij} = \delta_{ij}$ relative to the rectangular spatial coordinates). We next recall the following standard fact: $\upsilon_{\underline{g}} = \sqrt{\det \underline{g}}\, \upsilon_e$, where $\upsilon_{\underline{g}}$ is the volume form of $\underline{g}$ and the determinant of $\underline{g}$ is taken relative to the rectangular spatial coordinates. Since $\underline{g}_{ij} = \delta_{ij} + g_{ij}^{(Small)}(\Psi)$ where $g_{ij}^{(Small)}(0) = 0$ (see (2.4) and (2.6)), it follows that $(\det \underline{g})^{-1/2} = 1 + f(\Psi)$, where f is smooth in a neighborhood of 0 and vanishes at $\Psi = 0$. We furthermore note that since R is the $\underline{g}$-unit normal to $S_{t,u}$, it follows that $\upsilon_{\underline{g}}(R, X_1, X_2) = \upsilon_{\not{g}}(X_1, X_2) = \sqrt{\det \not{g}}$. Here, $\upsilon_{\not{g}}$ is the area form of $\not{g}$ and $\det \not{g}$ is taken relative to the geometric angular coordinates $(\vartheta^1, \vartheta^2)$. Combining these identities, we conclude the desired identity (3.81). □

3.19. Area forms, volume forms, and norms

In this section, we define the area forms, volume forms, and norms that we use during our L^2 analysis of solutions. We begin by noting that the results of Cor. 3.47 imply that relative to the geometric coordinates, the geometric area and volume forms induced on $S_{t,u}$, $\Sigma_t^{U_0}$, and $\mathcal{M}_{T_{(Bootstrap)}, U_0}$ are respectively $d\upsilon_{\not{g}(t',u',\vartheta)} := \sqrt{\det \not{g}(t', u', \vartheta)} d\vartheta$, $\mu d\upsilon_{\not{g}(t',u',\vartheta)}\, du'$, and $\mu d\upsilon_{\not{g}(t',u',\vartheta)}\, du'\, dt'$. However, because the factor μ in the latter two forms is extremely important, we now define rescaled volume forms in which the factor is missing. Throughout the monograph, all integrals are defined relative to the rescaled forms. In particular, **we always explicitly indicate the factor in integrands whenever it is present**. We stress that we do *not* rescale the area form on $S_{t,u}$.

DEFINITION 3.68 (**Area forms and <u>rescaled</u> volume forms on $S_{t,u}$, Σ_t^u, $\mathcal{C}_u^t$, and $\mathcal{M}_{t,u}$**). We define the area form $d\upsilon_{\not{g}}$ on $S_{t,u}$, the rescaled volume form $d\underline{\varpi}$ on Σ_t^u, the volume form $d\overline{\varpi}$ on $\mathcal{C}_u^t$, and the rescaled volume form $d\varpi$ on $\mathcal{M}_{t,u}$ as follows (relative to the geometric coordinates):

$$d\upsilon_{\not{g}} = d\upsilon_{\not{g}(t,u,\vartheta)} := \sqrt{\det \not{g}(t, u, \vartheta)}\, d\vartheta = \sqrt{\det \not{g}(t, u, \vartheta^1, \vartheta^2)}\, d\vartheta^1 d\vartheta^2, \tag{3.84a}$$

$$d\underline{\varpi} = d\underline{\varpi}(t, u', \vartheta) := d\upsilon_{\not{g}(t,u',\vartheta)}\, du', \tag{3.84b}$$

$$d\overline{\varpi} = d\overline{\varpi}(t', u, \vartheta) := d\upsilon_{\not{g}(t',u,\vartheta)}\, dt', \tag{3.84c}$$

$$d\varpi = d\varpi(t', u', \vartheta) := d\upsilon_{\not{g}(t',u',\vartheta)}\, du'\, dt', \tag{3.84d}$$

where in (3.84a), the determinant is taken relative to the geometric angular coordinates $(\vartheta^1, \vartheta^2)$ induced on $S_{t,u}$.

We now define our Lebesgue norms. The important point is that all norms are defined relative to the area and volume forms of Def. 3.68.

DEFINITION 3.69 (**Lebesgue norms relative to the area and rescaled volume forms**). We define the norms $\|\cdot\|_{L^2(S_{t,u})}$, $\|\cdot\|_{L^2(\Sigma_t^u)}$, $\|\cdot\|_{L^2(\mathcal{C}_u^t)}$, and $\|\cdot\|_{L^2(\mathcal{M}_{t,u})}$ of $S_{t,u}$ tensorfields ξ as follows:

$$\|\xi\|^2_{L^2(S_{t,u})} := \int_{\mathbb{S}^2} |\xi|^2(t,u,\vartheta) d\upsilon_{\not g} = \int_{S_{t,u}} |\xi|^2 d\upsilon_{\not g}, \tag{3.85a}$$

$$\|\xi\|^2_{L^2(\Sigma_t^u)} := \int_{u'=0}^{u} \int_{\mathbb{S}^2} |\xi|^2(t,u',\vartheta) d\upsilon_{\not g}\, du' = \int_{\Sigma_t^u} |\xi|^2\, d\underline{\varpi}, \tag{3.85b}$$

$$\|\xi\|^2_{L^2(\mathcal{C}_u^t)} := \int_{t'=0}^{t} \int_{\mathbb{S}^2} |\xi|^2(t',u,\vartheta) d\upsilon_{\not g}\, dt' = \int_{\mathcal{C}_u^t} |\xi|^2\, d\overline{\varpi}, \tag{3.85c}$$

$$\|\xi\|^2_{L^2(\mathcal{M}_{t,u})} := \int_{t'=0}^{t} \int_{u'=0}^{u} \int_{\mathbb{S}^2} |\xi|^2(t',u',\vartheta) d\upsilon_{\not g}\, du'\, dt' = \int_{\mathcal{M}_{t,u}} |\xi|^2\, d\varpi, \tag{3.85d}$$

where $|\xi|$ defined in Def. 3.54.

We similarly define the L^2 norms of ξ over subsets Ω of the above sets. For example, if $\Omega \subset \Sigma_t^u$, then

$$\|\xi\|^2_{L^2(\Omega)} := \int_{\Omega} |\xi|^2\, d\underline{\varpi}. \tag{3.86}$$

The relevant area/volume form corresponding to the norm $\|\cdot\|_{L^2(\Omega)}$ will always be clear from context.

We now define our C^k spaces and the usual sup-norm on the space C^0. Note that we do not define a norm on C^k for $k > 1$. The reason is that our estimates for the higher derivatives of various $S_{t,u}$ tensorfields ξ are more naturally expressed in terms of the C^0 norm of the $S_{t,u}$-projected Lie derivatives of ξ.

DEFINITION 3.70 (**C^k spaces and norms**). Given any subset Ω of the spacetime region $\{t \geq 0\}$ and any integer $k \geq 0$, $C^k(\Omega)$ denotes the set of $S_{t,u}$ tensorfields that, relative to the coordinates induced on Ω by the geometric coordinates $(t, u, \vartheta) \in [0,\infty) \times [0, U_0] \times \mathbb{S}^2$, are k-times continuously differentiable.

We define the following norm $\|\cdot\|_{C^0(\Omega)}$ for elements of $C^0(\Omega)$:

$$\|\xi\|_{C^0(\Omega)} := \sup_{p \in \Omega} |\xi(p)|. \tag{3.87}$$

REMARK 3.71 (**We typically do not explicitly state that $f \in C^0(\Omega)$**). Whenever we state an estimate that implies that $\|f\|_{C^0(\Omega)} < \infty$, we adopt the convention that the estimate also carries with it the implication $f \in C^0(\Omega)$. That is, by our conventions, we write estimates of the form $\|f\|_{C^0(\Omega)} < \infty$ only when $f \in C^0(\Omega)$.

3.20. Schematic notation

We often use schematic notation in our effort to highlight only the important features of our equations and inequalities. In particular, we often encounter products of terms such that the numerical coefficients and the precise tensorial structure are irrelevant from the point of view of proving our main sharp classical lifespan theorem. We indicate the basic structure of such terms in a very crude fashion by using "schematic array notation" *without explicit reference to the precise frame components of the terms and without indicating contractions.* For the terms whose precise structure is important, *we always either explicitly write their frame components or explicitly mention that the term is "exact."* We now give an example. Below, we derive equation (5.5b), which precisely reads

(3.88)

$$\slashed{k}_{AB}^{(Tan-\Psi)} = \frac{1}{2}\slashed{G}_{AB}\, L\Psi - \frac{1}{2}\slashed{G}_{LB}\, \slashed{d}_A\Psi - \frac{1}{2}\slashed{G}_{LA}\, \slashed{d}_B\Psi - \frac{1}{2}\slashed{G}_{RB}\, \slashed{d}_A\Psi - \frac{1}{2}\slashed{G}_{RA}\, \slashed{d}_B\Psi.$$

In view of Def. 3.62, in order to schematically indicate (3.88), we write

$$\slashed{k} = G_{(Frame)}\begin{pmatrix} L\Psi \\ \slashed{d}\Psi \end{pmatrix}. \tag{3.89}$$

Note that we have suppressed all numerical factors in (3.89). Whenever we use such schematic notation, the reader should interpret this as meaning that the numerical factors have no substantial effect on the estimates that we derive for the schematic terms. Also note that certain combinations suggested by (3.89) do not actually occur on the right-hand side of the precise formula (3.88). For example, there is no such term of the form $G_{LR}L\Psi$, and in fact, such a term does not even make sense since the left-hand side of (3.88) is a type $\binom{0}{2}$ $S_{t,u}$ tensorfield. The reader should interpret this as meaning that even if there were such a term, it would not affect our inequalities except for possibly altering them by unimportant constant factors.

We now give a few more examples. If we want to indicate the precise structure of the left-hand side of (3.88) and also the first term on the right, we write

$$\slashed{k}_{AB} = \frac{1}{2}\slashed{G}_{AB}\, L\Psi + G_{(Frame)}\slashed{d}\Psi. \tag{3.90}$$

To schematically indicate the type $\binom{1}{0}$ $S_{t,u}$ tensorfield $\epsilon_{lca} g_{ab}^{(Small)} x^c \slashed{d}^{\#} x^b$, where $\epsilon_{...}$ is the fully antisymmetric symbol normalized by $\epsilon_{123} = 1$ and $g_{ab}^{(Small)} = g_{ab}^{(Small)}(\Psi)$ is defined in (2.4), we write

$$f(\Psi)\Psi x \slashed{d}^{\#} x, \tag{3.91}$$

where f is a smooth function of Ψ (recall that $g_{ab}^{(Small)}(0) = 0$).

To schematically indicate the type $\binom{1}{1}$ $S_{t,u}$ tensorfield $\frac{1}{2}\slashed{G}_B^{\ A}\, L\Psi - \frac{1}{2}\slashed{G}_L^{\ A}\, \slashed{d}_B\Psi + \frac{1}{2}\slashed{G}_L'^{\ A}\, \slashed{d}_B\Psi$, we write

$$\begin{pmatrix} G_{(Frame)}^{\#} \\ G_{(Frame)}'^{\#} \end{pmatrix}\begin{pmatrix} L\Psi \\ \slashed{d}\Psi \end{pmatrix}. \tag{3.92}$$

Again, we stress that even though the notation (3.92) suggests the presence of some terms that are not actually in the precise expression, the reader should interpret this as meaning that the presence or absence of such terms will not affect any of the estimates that we derive except for possibly altering them by unimportant constant factors.

The justification of our schematic treatment of many terms is based on the fact that our estimates are, with only a few exceptions that we clearly point out, *not sensitive* to the precise tensorial structures present; this will become clear when we carry out the analysis.

CHAPTER 4

Transport Equations for the Eikonal Function Quantities

In this chapter, we first introduce the "re-centered" variables $L^i_{(Small)}$, $R^i_{(Small)}$, and $\chi^{(Small)}$, which are closely connected to L^i, R^i, and χ but vanish for the background solution $\Psi \equiv 0$ (because to form the re-centered variables, we subtract off the background values from L^i, R^i, and χ). We then calculate the Christoffel symbols of g relative to the rectangular coordinates and, based on these calculations, derive various transport equations for and angular differentials of μ and the rectangular components L^i, R^i of the frame vectorfields L and R. These quantities are first derivatives of the eikonal function, so in effect, we are deriving commuted versions of the eikonal equation. We also derive transport equations for and angular differentials of some of the re-centered variables. These transport equations are the evolution equations that we use to estimate the eikonal function quantities, except at the top order. We then provide an expression for $\chi^{(Small)}$ in terms of Ψ and $L^i_{(Small)}$. In the remainder of the monograph, we use the expression for analyzing all derivatives of $\chi_{(Small)}$ except those of top order. Finally, we provide some Lie derivative identities that we use later in the monograph.

4.1. Re-centered variables and the eikonal function quantities

We start by defining re-centered versions $L^i_{(Small)}$, $R^i_{(Small)}$, and $\chi^{(Small)}$ of L, R, and χ. The re-centered quantities vanish when $\Psi \equiv 0$ because to form them, we subtract off the background values from L^i, R^i, and χ.

DEFINITION 4.1 ($L^i_{(Small)}$, $R^i_{(Small)}$, **and** $\chi^{(Small)}$). We define the scalar-valued functions $L^i_{(Small)}$ and $R^i_{(Small)}$, $(i = 1, 2, 3)$, and the symmetric type $\binom{0}{2}$ $S_{t,u}$ tensorfield $\chi^{(Small)}$ as follows:

$$L^i_{(Small)} := L^i - \frac{x^i}{\varrho}, \tag{4.1a}$$

$$R^i_{(Small)} := R^i + \frac{x^i}{\varrho}, \tag{4.1b}$$

$$\chi^{(Small)} := \chi - \frac{\not{g}}{\varrho}. \tag{4.1c}$$

In (4.1a)-(4.1c) the vectorfields L and R and the $S_{t,u}$ tensorfield χ are defined in (3.12), (3.19b), and (3.70).

REMARK 4.2 (**Eikonal function quantities**). We informally refer to any of μ L^i, $L^i_{(Small)}$, $\breve{R}^i$ R^i, $R^i_{(Small)}$, χ, or $\chi^{(Small)}$ as an "eikonal function quantity," since they are constructed out of the derivatives of the eikonal function.

We often use the following lemma to reduce the analysis of the rectangular components of R to those of L.

LEMMA 4.3 (**Algebraic relationships between the rectangular components of L and R**). *The following identities hold relative to the rectangular coordinates, ($\mu = 0, 1, 2, 3$):*

$$R_\mu = -L_\mu - \delta_\mu^0, \tag{4.2a}$$

$$R^\mu = -L^\mu - (g^{-1})^{0\mu}, \tag{4.2b}$$

where δ_μ^0 is the standard Kronecker delta.

Moreover, we have the following identity, ($i = 1, 2, 3$):

$$R^i_{(Small)} = -L^i_{(Small)} - (g^{-1})^{0i}. \tag{4.3}$$

PROOF. To prove (4.2a), we use (3.19c) and (3.21) to deduce that the one-form with rectangular components $R_\mu + L_\mu$ is future-directed, of length -1, and g-orthogonal to Σ_t. We now note that the one-form ξ with rectangular components $\xi_\mu = -\delta_\mu^0$ also is future-directed, g-normal to Σ_t, and has length $(g^{-1})^{00} = -1$ (see (2.3)). Hence, the identity (4.2a) holds.

To prove (4.2b), we simply raise the indices of (4.2a) with g^{-1}.

(4.3) then follows as a simple consequence of (4.1a), (4.1b), and (4.2b). □

4.2. Covariant derivatives and Christoffel symbols relative to the rectangular coordinates

We often use the following lemma to compute the covariant derivatives of various vectorfields relative to the rectangular coordinates.

LEMMA 4.4 (**Covariant derivatives and Christoffel symbols relative to the rectangular coordinates**). *Let V be a spacetime vectorfield. Then relative to the rectangular coordinates, we have*

$$\mathscr{D}_\mu V^\nu = \partial_\mu V^\nu + (g^{-1})^{\nu\kappa}\Gamma_{\mu\kappa\alpha}V^\alpha, \tag{4.4a}$$

where the fully lowered Christoffel symbols $\Gamma_{\mu\kappa\alpha}$ verify

(4.4b)

$$\Gamma_{\alpha\kappa\beta} = \frac{1}{2}\left\{\partial_\alpha g_{\kappa\beta} + \partial_\beta g_{\alpha\kappa} - \partial_\kappa g_{\alpha\beta}\right\} = \frac{1}{2}\left\{G_{\kappa\beta}\partial_\alpha\Psi + G_{\alpha\kappa}\partial_\beta\Psi - G_{\alpha\beta}\partial_\kappa\Psi\right\}.$$

PROOF. The identity (4.4a) and the first equality in (4.4b) are standard identities from differential geometry. The second equality in (4.4b) follows from the first equality, the chain rule, (2.4), and (2.14). □

REMARK 4.5 (**Index conventions for the Christoffel symbols**). In many other works, the authors adopt a different convention for the placement of the indices on the Christoffel symbols.

4.3. Transport equation for the inverse foliation density

In this section, we derive the transport equation verified by $\upmu$.

LEMMA 4.6 (**The transport equation verified by $\upmu$**). *The inverse foliation density $\upmu$ defined in (3.9) verifies the following transport equation:*

$$L\upmu = \omega^{(Trans-\Psi)} + \upmu\omega^{(Tan-\Psi)} := \omega, \tag{4.5}$$

where

$$\omega^{(Trans-\Psi)} := \frac{1}{2}G_{LL}\breve{R}\Psi, \tag{4.6a}$$

$$\omega^{(Tan-\Psi)} := -\frac{1}{2}G_{LL}L\Psi - G_{LR}L\Psi. \tag{4.6b}$$

PROOF. Relative to rectangular coordinates, the 0 component of equation (3.8) is

$$L_{(Geo)}L^0_{(Geo)} = -(g^{-1})^{0\gamma}\Gamma_{\alpha\gamma\beta}L^\alpha_{(Geo)}L^\beta_{(Geo)}. \tag{4.7}$$

Multiplying (4.7) by μ^3, referring to definition (3.12), and using the identity (4.2b), we deduce that

$$\begin{aligned}\mu^2 LL^0_{(Geo)} &= \frac{1}{2}\left\{\mu L^\gamma + \breve{R}^\gamma\right\}\{G_{\gamma\beta}\partial_\alpha\Psi + G_{\alpha\gamma}\partial_\beta\Psi - G_{\alpha\beta}\partial_\gamma\Psi\}L^\alpha L^\beta \\ &= \frac{1}{2}\mu G_{LL}L\Psi + \mu G_{LR}L\Psi - \frac{1}{2}G_{LL}\breve{R}\Psi.\end{aligned} \tag{4.8}$$

The desired equation (4.5) now follows from (4.8) and the identities $\mu^2 LL^0_{(Geo)} = \mu^2 L\left(\frac{1}{\mu}\right) = -L\mu$. □

4.4. Transport equations for the rectangular components of the frame vectorfields

In this section, we derive transport equations for and angular differentials of the rectangular components of L, R, and their re-centered versions.

PROPOSITION 4.7 (**Transport equations for and angular differentials of the rectangular frame components**). *Let L^i, R^i, $L^i_{(Small)}$, and $R^i_{(Small)}$ be the (scalar-valued) rectangular spatial components from Defs. 3.16, 3.19, and 4.1. Then the following transport equations are verified by L^i and $L^i_{(Small)}$, $(i = 1, 2, 3)$:*

$$\begin{aligned}LL^i &= \frac{1}{2}G_{LL}(L\Psi)R^i - G\!\!\!/_{LA}(g\!\!\!/^{-1})^{AB}(d\!\!\!/_B x^i)L\Psi \\ &\quad + \frac{1}{2}G_{LL}(g\!\!\!/^{-1})^{AB}(d\!\!\!/_A x^i)d\!\!\!/_B\Psi,\end{aligned} \tag{4.9a}$$

$$\begin{aligned}L(\varrho L^i_{(Small)}) &= -\frac{1}{2}\varrho G_{LL}(L\Psi)L^i_{(Small)} \\ &\quad - \frac{1}{2}G_{LL}(L\Psi)x^i - \frac{1}{2}\varrho G_{LL}(L\Psi)(g^{-1})^{0i} \\ &\quad - \frac{1}{2}\varrho G_{LL}(g\!\!\!/^{-1})^{AB}(d\!\!\!/_A\Psi)d\!\!\!/_B x^i - \varrho(g\!\!\!/^{-1})^{AB}G\!\!\!/_{LA}(d\!\!\!/_B x^i)L\Psi.\end{aligned} \tag{4.9b}$$

Furthermore, there exist $S_{t,u}$ one-forms $\lambda^{(Tan-\Psi)}$ and $\theta^{(Tan-\Psi)}$ and type $\binom{0}{2}$ $S_{t,u}$ tensorfields $\Lambda^{(Tan-\Psi)}$ and $\Theta^{(Tan-\Psi)}$ such that the following expressions hold for *the angular differential (see Def. 3.50) of* L^i, R^i, $L^i_{(Small)}$, *and* $R^i_{(Small)}$, $(i = 1, 2, 3)$:

$$\not{d}_A L^i = (\not{g}^{-1})^{BC}\chi_{AB}\not{d}_C x^i + (\not{g}^{-1})^{BC}\Lambda_{AB}^{(Tan-\Psi)}\not{d}_C x^i + \lambda_A^{(Tan-\Psi)} R^i, \tag{4.10a}$$

$$\not{d}_A R^i = -(\not{g}^{-1})^{BC}\chi_{AB}\not{d}_C x^i + (\not{g}^{-1})^{BC}\Theta_{AB}^{(Tan-\Psi)}\not{d}_C x^i + \theta_A^{(Tan-\Psi)} R^i, \tag{4.10b}$$

(4.10c)

$$\begin{aligned}\not{d}_A L^i_{(Small)} &= (\not{g}^{-1})^{BC}\chi_{AB}^{(Small)}\not{d}_C x^i + \lambda_A^{(Tan-\Psi)} R^i_{(Small)} \\ &\quad - \frac{x^i}{\varrho}\lambda_A^{(Tan-\Psi)} + (\not{g}^{-1})^{BC}\Lambda_{AB}^{(Tan-\Psi)}\not{d}_C x^i,\end{aligned}$$

(4.10d)

$$\begin{aligned}\not{d}_A R^i_{(Small)} &= -(\not{g}^{-1})^{BC}\chi_{AB}^{(Small)}\not{d}_C x^i + \theta_A^{(Tan-\Psi)} R^i_{(Small)} \\ &\quad - \frac{x^i}{\varrho}\theta_A^{(Tan-\Psi)} + (\not{g}^{-1})^{BC}\Theta_{AB}^{(Tan-\Psi)}\not{d}_C x^i,\end{aligned}$$

where

$$\lambda_A^{(Tan-\Psi)} := -G_{LR}\not{d}_A\Psi - \frac{1}{2}G_{RR}\not{d}_A\Psi, \tag{4.11a}$$

$$\theta_A^{(Tan-\Psi)} := -\frac{1}{2}G_{RR}\not{d}_A\Psi, \tag{4.11b}$$

$$\Lambda_{AB}^{(Tan-\Psi)} := -\frac{1}{2}\not{G}_{AB}\, L\Psi + \frac{1}{2}\not{G}_{LA}\,\not{d}_B\Psi - \frac{1}{2}\not{G}_{LB}\,\not{d}_A\Psi, \tag{4.11c}$$

$$\Theta_{AB}^{(Tan-\Psi)} := \frac{1}{2}\not{G}_{AB}\, L\Psi - \frac{1}{2}\not{G}_{LA}\,\not{d}_B\Psi - \frac{1}{2}\not{G}_{LB}\,\not{d}_A\Psi - \not{G}_{RB}\,\not{d}_A\Psi. \tag{4.11d}$$

In the above formulas, χ *is the* $S_{t,u}$ *tensorfield defined by* (3.70) *and* $\chi^{(Small)}$ *is the* $S_{t,u}$ *tensorfield defined by* (4.1c).

Proof. We first prove (4.9a). With ∇ denoting the Levi-Civita connection of the Minkowski metric (so that $\nabla_\nu = \partial_\nu$ relative to rectangular coordinates), we use Lemma 4.4 to compute that for $\nu = 0, 1, 2, 3$, we have

(4.12)

$$\begin{aligned}LL^\nu = \nabla_L L^\nu &= \mathscr{D}_L L^\nu - (g^{-1})^{\nu\kappa}L^\alpha L^\beta \Gamma_{\alpha\kappa\beta} \\ &= \mathscr{D}_L L^\nu - \frac{1}{2}(g^{-1})^{\nu\kappa}L^\alpha L^\beta \left\{G_{\kappa\beta}\partial_\alpha\Psi + G_{\alpha\kappa}\partial_\beta\Psi - G_{\alpha\beta}\partial_\kappa\Psi\right\}.\end{aligned}$$

Since $L^0 = 1$, it follows that $\nabla_L L^\nu$ is Σ_t-tangent. Hence, for $i = 1, 2, 3$, we can expand

$$\nabla_L L^i = zR^i + Y^i, \tag{4.13}$$

where z is a scalar and Y is an $S_{t,u}$-tangent vectorfield. Next, using (3.8), (3.12), and (4.5), we deduce that

$$\mathscr{D}_L L^\nu = \mu^{-1}\omega^{(Trans-\Psi)}L^\nu + \omega^{(Tan-\Psi)}L^\nu. \tag{4.14}$$

Contracting (4.13) against $\breve{R}_i$ and using (4.6a), (4.6b), (4.12), (4.14), and the identities $\breve{R}_i R^i = \upmu$ and $\breve{R}_i L^i = -\upmu$ (see Lemma 3.20), we deduce

(4.15)
$$\begin{aligned}\upmu z = g(\nabla_L L, \breve{R}) &= g(\mathscr{D}_L L, \breve{R}) - \frac{1}{2}\breve{R}^\kappa L^\alpha L^\beta \{G_{\kappa\beta}\partial_\alpha\Psi + G_{\alpha\kappa}\partial_\beta\Psi - G_{\alpha\beta}\partial_\kappa\Psi\} \\ &= -\omega^{(Trans-\Psi)} - \upmu\omega^{(Tan-\Psi)} - \upmu G_{LR}L\Psi + \frac{1}{2}G_{LL}\breve{R}\Psi = \frac{1}{2}\upmu G_{LL}L\Psi.\end{aligned}$$

Similarly, we compute that $Y^i = (\not{g}^{-1})^{AB}g(Y, X_A)X_B^i$, where

$$\begin{aligned}(4.16)\quad g(Y, X_A) &= g(\nabla_L L, X_A) \\ &= g(\mathscr{D}_L L, X_A) - \frac{1}{2}X_A^\kappa L^\alpha L^\beta \{G_{\kappa\beta}\partial_\alpha\Psi + G_{\alpha\kappa}\partial_\beta\Psi - G_{\alpha\beta}\partial_\kappa\Psi\} \\ &= -\not{G}_{LA}\, L\Psi + \frac{1}{2}G_{LL}\not{d}_A\Psi.\end{aligned}$$

From (4.13), (4.15), (4.16), and the identity $\not{d}_A x^i = X_A^i$ (see Lemma 3.52), we conclude that

$$(4.17)\qquad LL^i = \nabla_L L^i = \frac{1}{2}G_{LL}(L\Psi)R^i - \not{G}_L^A\,(\not{d}_A x^i)L\Psi + \frac{1}{2}G_{LL}\not{d}^A\Psi\not{d}_A x^i.$$

We have thus proved (4.9a).

The desired identity (4.9b) then follows from (4.9a), Def. 4.1, the identities $Lx^i = L^i$, $L\varrho = 1$, and (4.3), and straightforward computations.

The proofs of (4.10a) and (4.10c) are similar and rely on equations (3.71), (3.73), (3.75), and the identity

$$(4.18)\qquad (\underline{\mathcal{L}}_N \underline{g})_{RA} = \not{G}_{RA}\, L\Psi - \not{G}_{LA}\, R\Psi - G_{LR}\not{d}_A\Psi - G_{RR}\not{d}_A\Psi$$

in place of (4.14). The identity (4.18) follows from (2.3), (3.19c), (3.22), (3.52c), (3.63), (3.67a), and the chain rule identity $-V(g^{-1})^{0i} = (g^{-1})^{0\alpha}(g^{-1})^{i\beta}G_{\alpha\beta}V\Psi$ (valid for any spacetime vectorfield V).

The identities (4.10b) and (4.10d) then follow from (4.10a), (4.10c), Lemma 4.3, the chain rule identity from the previous paragraph, and straightforward computations. □

4.5. An expression for the re-centered null second fundamental form in terms of other quantities

In the next lemma, we show that the $S_{t,u}$ tensorfield $\chi_{(Small)}$ from Def. 4.1 is an auxiliary variable in the sense that it can be expressed in terms of the up-to-second-order derivatives of Ψ and the up-to-first-order derivatives of $L^i_{(Small)}$, $(i = 1, 2, 3)$.

LEMMA 4.8 (**Expression for $\chi_{(Small)}$ in terms of other quantities**). *The quantities $\chi^{(Small)}$ and $\mathrm{tr}_{\not{g}}\chi^{(Small)}$ can be expressed as follows, where the operator $\not{d}$ is as in Def. 3.51:*

$$(4.19a)\qquad \chi_{AB}^{(Small)} = g_{cb}(\not{d}_A x^a)\not{d}_B L^b_{(Small)} - \Lambda_{AB}^{(Tan-\Psi)},$$

$$(4.19b)\qquad \mathrm{tr}_{\not{g}}\chi^{(Small)} = \not{d}_a L^a_{(Small)} - (\not{g}^{-1})^{AB}\Lambda_{AB}^{(Tan-\Psi)}.$$

Above, $\Lambda_{AB}^{(Tan-\Psi)}$ is the $S_{t,u}$ tensorfield defined in (4.11c).

PROOF. To derive (4.19a), we contract (4.10c) against $g_{ij} \not{d}_D x^j$ and use the identities $g_{ij}(\not{d}_C x^i)\not{d}_D x^j = g(X_C, X_D) = \not{g}_{CD}$ (see (3.61)) and

$$g_{ij}(R^i_{(Small)} - \frac{x^i}{\varrho})\not{d}_D x^j = g(R, X_D) = 0$$

(see also (4.1b)).

To derive (4.19b), we contract (4.19a) against $(\not{g}^{-1})^{AB}$ and use the identity $(\not{g}^{-1})^{AB} g_{ab}(\not{d}_A x^a)\not{d}_B = \not{d}_b$. □

4.6. Identities involving deformation tensors and Lie derivatives

In this section, we provide some deformation tensor and Lie derivative identities that play a role in our analysis. We begin by providing the standard definition of a deformation tensor of a vectorfield.

DEFINITION 4.9 (**Deformation tensor**). We associate the following type $\binom{0}{2}$ tensorfield ${}^{(V)}\pi$ to a vectorfield V:

$$ {}^{(V)}\pi_{\mu\nu} := \mathcal{L}_V g_{\mu\nu} = \mathscr{D}_\mu V_\nu + \mathscr{D}_\nu V_\mu, \tag{4.20}$$

where the second equality in (4.20) is a simple consequence of (3.63) and the torsion-free property of the connection $\mathscr{D}$.

LEMMA 4.10 (**Vectorfield commutator properties**). *Let Y be an $S_{t,u}$-tangent vectorfield. Then $[L, \breve{R}]$, $[L, Y]$, and $[\breve{R}, Y]$ are also $S_{t,u}$-tangent and the following identities hold:*

$$[L, \breve{R}] = \mathcal{L}_L \breve{R} = \not{\mathcal{L}}_L \breve{R} = {}^{(\breve{R})}\not{\pi}_L^{\#} = -{}^{(L)}\not{\pi}_{\breve{R}}^{\#}, \tag{4.21a}$$

$$[L, Y] = \mathcal{L}_L Y = \not{\mathcal{L}}_L Y = {}^{(Y)}\not{\pi}_L^{\#}, \tag{4.21b}$$

$$[\breve{R}, Y] = \mathcal{L}_{\breve{R}} Y = \not{\mathcal{L}}_{\breve{R}} Y = {}^{(Y)}\not{\pi}_{\breve{R}}^{\#}. \tag{4.21c}$$

Above, the vectorfields ${}^{(V)}\not{\pi}_W^{\#}$ are the $\not{g}$-duals of the one-forms ${}^{(V)}\not{\pi}_W$ defined by (3.43) and (4.20).

PROOF. We first prove the equalities in (4.21c). To prove the first two, it suffices to show that $[\breve{R}, Y]t = [\breve{R}, Y]u = 0$. These identities follow easily from the identities $Yt = Yu = \breve{R}t = 0$ and $\breve{R}u = 1$ (see (3.23a)). To deduce the final equality in (4.21c), it suffices to show that ${}^{(Y)}\pi_{\breve{R}A} = g([\breve{R}, Y], X_A)$ for $A = 1, 2$. To this end, we use the torsion-free property $[\breve{R}, Y] = \mathscr{D}_{\breve{R}} Y - \mathscr{D}_Y \breve{R}$, the identity $g(\breve{R}, X_A) = 0$, and the fact that $[X_A, Y]$ is $S_{t,u}$-tangent (since it is the Lie bracket of two $S_{t,u}$-tangent vectorfields) to compute that

$$\begin{aligned} {}^{(Y)}\pi_{\breve{R}A} := g(\mathscr{D}_{\breve{R}} Y, X_A) + g(\mathscr{D}_A Y, \breve{R}) &= g([\breve{R}, Y], X_A) + g(\mathscr{D}_Y \breve{R}, X_A) + g(\mathscr{D}_A Y, \breve{R}) \\ &= g([\breve{R}, Y], X_A) - g(\mathscr{D}_Y X_A, \breve{R}) + g(\mathscr{D}_A Y, \breve{R}) \\ &= g([\breve{R}, Y], X_A) + g([X_A, Y], \breve{R}) \\ &= g([\breve{R}, Y], X_A) \end{aligned} \tag{4.22}$$

as desired. The identities in (4.21b) follow similarly with the help of the identities $Lt = 1$ and $Lu = 0$ (see (3.15a)). The identities in (4.21a) then follow similarly with the help of the fact that by (4.21b) and (4.21c), $[L, X_A]$ and $[\breve{R}, X_A]$ are $S_{t,u}$-tangent. □

LEMMA 4.11 (**Basic properties of the $S_{t,u}$ projection operator**). *Let $\Pi\!\!\!/_{\nu}{}^{\mu}$ be the $S_{t,u}$ projection operator from Def. 3.39. If Y is an $S_{t,u}$-tangent vectorfield, and if $V \in \{L, \varrho L, \breve{R}\}$ or V is also an $S_{t,u}$-tangent vectorfield, then we have, ($\mu = 0, 1, 2, 3$):*

$$Y^{\alpha} \mathcal{L}_V \Pi\!\!\!/_{\alpha}{}^{\mu} = 0. \tag{4.23}$$

Furthermore, we have

$$(\not{d}_A x^a) \mathcal{L}_V \Pi\!\!\!/_{a}{}^{\mu} = 0. \tag{4.24}$$

Finally, we have, ($\mu, \nu = 0, 1, 2, 3$):

$$\Pi\!\!\!/_{\nu}{}^{\alpha} \mathcal{L}_V \Pi\!\!\!/_{\alpha}{}^{\mu} = 0. \tag{4.25}$$

PROOF. Since $Y^{\mu} = Y^{\alpha} \Pi\!\!\!/_{\alpha}{}^{\mu}$, the Leibniz rule implies that $\mathcal{L}_V Y^{\mu} = \mathcal{L}_V (Y^{\alpha} \Pi\!\!\!/_{\alpha}{}^{\mu}) = (\mathcal{L}_V Y^{\alpha}) \Pi\!\!\!/_{\alpha}{}^{\mu} + Y^{\alpha} \mathcal{L}_V \Pi\!\!\!/_{\alpha}{}^{\mu}$. On the other hand, Lemma 4.10 implies that $\mathcal{L}_V Y^{\mu} = (\mathcal{L}_V Y^{\alpha}) \Pi\!\!\!/_{\alpha}{}^{\mu}$. The desired identity (4.23) now follows from equating these two identities for $\mathcal{L}_V Y^{\mu}$.

To prove (4.24), we note that $\not{d}_A x^a = X_A^a$ by (3.61). Since for each fixed A the vectorfield with rectangular spatial components X_A^a is $S_{t,u}$-tangent, (4.24) follows from (4.23).

To prove (4.25), we simply note that for each fixed ν, $\Pi\!\!\!/_{\nu}{}^{\alpha}$ can be viewed as an $S_{t,u}$-tangent vectorfield. Hence, (4.25) follows from (4.23). ☐

COROLLARY 4.12 (**Sometimes $S_{t,u}$ projection is redundant**). *If ξ is a type $\binom{0}{n}$ spacetime tensorfield and if $V \in \{L, \varrho L, \breve{R}\}$ or V is an $S_{t,u}$-tangent vectorfield, then*

$$\not{\mathcal{L}}_V \xi = \not{\mathcal{L}}_V \not{\xi}. \tag{4.26}$$

PROOF. The identity (4.26) is an easy consequence of definitions (3.42) and (3.67b) and the identity (4.25). ☐

COROLLARY 4.13 (**Basic commutation formula for $S_{t,u}$-projected Lie derivatives**). *If ξ is a type $\binom{0}{n}$ $S_{t,u}$ tensorfield, if $V \in \{L, \varrho L, \breve{R}\}$ or V is an $S_{t,u}$-tangent vectorfield, and if $W \in \{L, \varrho L, \breve{R}\}$ or W is an $S_{t,u}$-tangent vectorfield, then*

$$[\not{\mathcal{L}}_V, \not{\mathcal{L}}_W] \xi = \not{\mathcal{L}}_{[V,W]} \xi. \tag{4.27}$$

PROOF. We give the proof in the case that ξ is an $S_{t,u}$ one-form; the general case can be handled in the same way. We first recall the standard commutation property (3.66) for Lie derivatives: $[\mathcal{L}_V, \mathcal{L}_W]\xi = \mathcal{L}_{[V,W]}\xi$. Thus, a simple calculation yields that

$$[\not{\mathcal{L}}_V, \not{\mathcal{L}}_W]\xi_{\mu} = \not{\mathcal{L}}_{[V,W]}\xi_{\mu} + \Pi\!\!\!/_{\mu}{}^{\alpha} (\mathcal{L}_V \Pi\!\!\!/_{\alpha}{}^{\nu}) \mathcal{L}_W \xi_{\nu} - \Pi\!\!\!/_{\mu}{}^{\alpha} (\mathcal{L}_W \Pi\!\!\!/_{\alpha}{}^{\nu}) \mathcal{L}_V \xi_{\nu}, \tag{4.28}$$

where $\Pi\!\!\!/_{\mu}{}^{\alpha}$ is the $S_{t,u}$ projection tensorfield from Def. 3.39. Using (4.25), we see that the last two terms on the right-hand side of (4.28) vanish. We have thus proved (4.27). ☐

LEMMA 4.14 (**Expressions for $S_{t,u}$-projected Lie derivatives**). *Let ξ be a spacetime one-form with rectangular components $\xi_{\alpha} = \xi_{\alpha}(\Psi)$ that are functions of Ψ, and let ξ' be the one-form with rectangular components $\xi'_{\alpha} = \xi'_{\alpha}(\Psi) := \frac{d}{d\Psi}\xi_{\alpha}(\Psi)$.*

Let $\xi\!\!/$, $\xi\!\!/'$ denote the projections of ξ, ξ' onto $S_{t,u}$ (see Def. 3.41). Let $Z \in \{L, \varrho L, \breve{R}\}$ or let Z be $S_{t,u}$-tangent. Then

$$\mathcal{L}\!\!\!/_Z \xi\!\!/_A := (\mathcal{L}\!\!\!/_Z \xi\!\!/)_A = \xi\!\!/'_A Z\Psi + \xi_a \mathrm{d}\!\!\!/_A Z^a. \tag{4.29}$$

Furthermore, if $V \in \left\{L, R, \frac{\partial}{\partial x^j}\right\}_{j=1,2,3}$, then with $\xi_V := \xi_\alpha V^\alpha$ and $\xi'_V := \xi'_\alpha V^\alpha$, we have

$$Z\xi_V := Z(\xi_V) = \xi'_V Z\Psi + \xi_a Z V^a. \tag{4.30}$$

In addition, if $V \in \left\{L, R, \frac{\partial}{\partial x^j}\right\}_{j=1,2,3}$, then with $G_{\mu\nu}(\Psi)$ and $G'_{\mu\nu}(\Psi)$ as defined in Def. 2.7, we have

$$ZG_{VW} := Z(G_{VW}) = G'_{VW} Z\Psi + G_{aW} ZV^a + G_{Va} ZW^a, \tag{4.31a}$$

$$\mathcal{L}\!\!\!/_Z G\!\!\!/_{VA} := (\mathcal{L}\!\!\!/_Z(G\!\!\!/_V))_A = G\!\!\!/'_{VA} Z\Psi + G\!\!\!/_{Aa} ZV^a + G_{Va} \mathrm{d}\!\!\!/_A Z^a, \tag{4.31b}$$

$$\mathcal{L}\!\!\!/_Z G\!\!\!/_{AB} := (\mathcal{L}\!\!\!/_Z G\!\!\!/)_{AB} = G\!\!\!/'_{AB} Z\Psi + G\!\!\!/_{Aa} \mathrm{d}\!\!\!/_B Z^a + G\!\!\!/_{aB} \mathrm{d}\!\!\!/_A Z^a. \tag{4.31c}$$

Above, $G\!\!\!/_V$ is the $S_{t,u}$ one-form formed by projecting the one-form with rectangular components $G_{V\nu} = G_{\alpha\nu} V^\alpha$ onto $S_{t,u}$, $G\!\!\!/$ is the symmetric type $\binom{0}{2}$ $S_{t,u}$ tensorfield formed by projecting the symmetric type $\binom{0}{2}$ tensor with rectangular components $G_{\mu\nu}$ onto $S_{t,u}$, $G\!\!\!/'$ is the symmetric type $\binom{0}{2}$ $S_{t,u}$ tensorfield formed by projecting the symmetric type $\binom{0}{2}$ tensor with rectangular components $G'_{\mu\nu}$ onto $S_{t,u}$, and similarly for the other quantities (see Def. 3.41).

PROOF. We prove only (4.31b); the proofs of the remaining identities are essentially the same. We first note that the ith rectangular component of the $S_{t,u}$ one-form $G\!\!\!/_V$ is $G_{\alpha\beta} V^\alpha \Pi\!\!\!/_i^{\ \beta}$, where $\Pi\!\!\!/$ is the $S_{t,u}$ projection from Def. 3.39. Hence, using the Leibniz and chain rules, the fact that the 0-containing components of $\Pi\!\!\!/$ are trivial, and the fact that the 0 component of V is constant, we have

(4.32)

$$\begin{aligned}
\mathcal{L}_Z(G_{\alpha\beta} V^\alpha \Pi\!\!\!/_i^{\ \beta}) &= V^\alpha \Pi\!\!\!/_i^{\ \beta} \mathcal{L}_Z G_{\alpha\beta} + \Pi\!\!\!/_i^{\ \beta} G_{\alpha\beta} \mathcal{L}_Z V^\alpha + G_{\alpha\beta} V^\alpha \mathcal{L}_Z \Pi\!\!\!/_i^{\ \beta} \\
&= V^\alpha \Pi\!\!\!/_i^{\ \beta} \underbrace{ZG_{\alpha\beta}}_{G'_{\alpha\beta} Z\Psi} + \Pi\!\!\!/_i^{\ \beta} V^\alpha G_{\alpha\gamma} \partial_\beta Z^\gamma + \Pi\!\!\!/_i^{\ \beta} G_{\beta\gamma} V Z^\gamma + \Pi\!\!\!/_i^{\ \beta} G_{\alpha\beta} Z V^\alpha \\
&\quad - \Pi\!\!\!/_i^{\ \beta} G_{\alpha\beta} V Z^\alpha + G_{\alpha\beta} V^\alpha \mathcal{L}_Z \Pi\!\!\!/_i^{\ \beta}.
\end{aligned}$$

Contracting the left-hand side of (4.32) against $\mathrm{d}\!\!\!/_A x^i$ and using (3.61) yields $(\mathcal{L}\!\!\!/_Z(G\!\!\!/_V))_A$. On the other hand, contracting the right-hand side of (4.32) against $\mathrm{d}\!\!\!/_A x^i$ and using (3.61), the identities $ZV^0 = \mathrm{d}\!\!\!/_A Z^0 = 0$, and $(\mathrm{d}\!\!\!/_A x^i)\mathcal{L}_Z \Pi\!\!\!/_i^{\ \beta} = 0$ (that is, (4.24)), we obtain the sum $G\!\!\!/'_{VA} Z\Psi + G_{Va} \mathrm{d}\!\!\!/_A Z^a + G\!\!\!/_{Aa} VZ^a + G\!\!\!/_{aA} ZV^a - G\!\!\!/_{aA} VZ^a$. Noting the cancellation of the third and fifth terms, we obtain the desired identity (4.31b). □

COROLLARY 4.15 (**The L derivatives of some components of $G_{(Frame)}$**). *The following identities hold:*

$$LG_{LL} = \begin{pmatrix} G_{(Frame)}, G^{\#}_{(Frame)} \\ G'_{(Frame)} \end{pmatrix} \begin{pmatrix} L\Psi \\ \not{d}\Psi \end{pmatrix}, \tag{4.33a}$$

$$\not{\mathcal{L}}_L \not{G}_{LA} = \not{G}_L^{\ B} \chi_{AB} + \begin{pmatrix} G_{(Frame)} G^{\#}_{(Frame)} \\ G'_{(Frame)} \end{pmatrix} \begin{pmatrix} L\Psi \\ \not{d}\Psi \end{pmatrix}, \tag{4.33b}$$

$$\not{\mathcal{L}}_L \not{G}_{AB} = \not{G}_{AC} \chi_B^{\ C} + \not{G}_{BC} \chi_A^{\ C} + \begin{pmatrix} G_{(Frame)} G^{\#}_{(Frame)} \\ G'_{(Frame)} \end{pmatrix} \begin{pmatrix} L\Psi \\ \not{d}\Psi \end{pmatrix}, \tag{4.33c}$$

where the first term on the right-hand side of (4.33b) *and the first two terms on the right-hand side of* (4.33c) *are exact and the remaining ones are schematic, and* $G_{(Frame)}$, $G^{\#}_{(Frame)}$ *and* $G'_{(Frame)}$ *are as in Defs.* 2.7 *and* 3.62.

PROOF. We prove only (4.33b) since the proofs of the other identities are similar. We first use (4.31b) with $V = Z := L$ to deduce that

$$\not{\mathcal{L}}_L \not{G}_{LA} = \not{G}'_{LA} L\Psi + \not{G}_{Aa} LL^a + G_{La} \not{d}_A L^a. \tag{4.34}$$

Inserting (4.9a) and (4.10a) into (4.34), we deduce that

$$\begin{aligned} \not{\mathcal{L}}_L \not{G}_{LA} &= \not{G}'_{LA} L\Psi + \frac{1}{2} G_{LL} \not{G}_{AR} L\Psi - \not{G}_A^{\ B} \not{G}_{LB} L\Psi + \frac{1}{2} G_{LL} \not{G}_A^{\ B} \not{d}_B \Psi \\ &\quad + \not{G}_L^{\ B} \chi_{AB} + \not{G}_L^{\ B} \Lambda_{AB}^{(Tan-\Psi)} + G_{LR} \lambda_A^{(Tan-\Psi)}. \end{aligned} \tag{4.35}$$

The desired identity (4.33b) now follows from (4.35), (4.11a), and (4.11c). ∎

CHAPTER 5

Connection Coefficients of the Rescaled Frames and Geometric Decompositions of the Wave Operator

In this chapter, we compute the connection coefficients of the two rescaled frames $\{L, \breve{R}, X_1, X_2\}$ and $\{L, \breve{\underline{L}}, X_1, X_2\}$. We refer to these frames as "rescaled" because the vectorfields $\breve{R}$ and $\breve{\underline{L}}$ are adapted to the behavior of μ and hence can significantly deviate from their Minkowskian counterparts, which are $-\partial_r$ and $\partial_t - \partial_r$. We then provide several decompositions of the μ-weighted covariant wave operator $\mu\square_{g(\Psi)}$ relative to these frames.

5.1. Connection coefficients of the rescaled frame

In the next lemma, we compute the connection coefficients of the rescaled frame $\{L, \breve{R}, X_1, X_2\}$.

LEMMA 5.1 (**Connection coefficients of the rescaled frame $\{L, \breve{R}, X_1, X_2\}$ and their decomposition into μ^{-1}-singular and μ^{-1}-regular pieces**). *Let ζ be the $S_{t,u}$ one-form defined by (see the identity (3.75))*

(5.1)
$$\zeta_A := k\!\!\!/_{RA} = g(\mathscr{D}_A L, R) = \mu^{-1} g(\mathscr{D}_A L, \breve{R}).$$

Then the covariant derivatives of the frame vectorfields can be expressed as follows, where the tensorfields k, χ, and ω are defined in (3.69), (3.70), and (4.5):

(5.2a)
$$\mathscr{D}_L L = \mu^{-1}\omega L,$$

(5.2b)
$$\mathscr{D}_{\breve{R}} L = -\omega L + \mu\zeta^A X_A + (d\!\!\!/^A \mu) X_A,$$

(5.2c)
$$\mathscr{D}_A L = -\zeta_A L + \chi_A^{\ B} X_B,$$

(5.2d)
$$\mathscr{D}_L \breve{R} = -\omega L - \mu\zeta^A X_A,$$

(5.2e)
$$\mathscr{D}_{\breve{R}} \breve{R} = \mu\omega L + \left\{\mu^{-1}\breve{R}\mu + \omega\right\}\breve{R} - \mu(d\!\!\!/^A \mu) X_A,$$

(5.2f)
$$\mathscr{D}_A \breve{R} = \mu\zeta_A L + \zeta_A \breve{R} + \mu^{-1}(d\!\!\!/_A \mu)\breve{R} + \mu k\!\!\!/_A^{\ B} X_B - \mu\chi_A^{\ B} X_B,$$

(5.2g)
$$\mathscr{D}_L X_A = \mathscr{D}_A L,$$

(5.2h)
$$\mathscr{D}_A X_B = \nabla\!\!\!\!/_A X_B + k\!\!\!/_{AB} L + \mu^{-1}\chi_{AB}\breve{R}.$$

Furthermore, we can decompose the frame components of the tensorfields ζ and $k\!\!\!/$ into μ^{-1}-singular and μ^{-1}-regular pieces as follows:

(5.3a)
$$\zeta_A = \mu^{-1}\zeta_A^{(Trans-\Psi)} + \zeta_A^{(Tan-\Psi)},$$

(5.3b)
$$k\!\!\!/_{AB} = \mu^{-1} k\!\!\!/_{AB}^{(Trans-\Psi)} + k\!\!\!/_{AB}^{(Tan-\Psi)},$$

where

(5.4a) $$\zeta_A^{(Trans-\Psi)} := -\frac{1}{2}\mathcal{G}_{LA}\,\breve{R}\Psi,$$

(5.4b) $$\not{k}_{AB}^{(Trans-\Psi)} := \frac{1}{2}\mathcal{G}_{AB}\,\breve{R}\Psi,$$

and

(5.5a)
$$\zeta_A^{(Tan-\Psi)} := \frac{1}{2}\mathcal{G}_{RA}\,L\Psi - \frac{1}{2}G_{LR}\not{d}_A\Psi - \frac{1}{2}G_{RR}\not{d}_A\Psi,$$

(5.5b)
$$\not{k}_{AB}^{(Tan-\Psi)} := \frac{1}{2}\mathcal{G}_{AB}\,L\Psi - \frac{1}{2}\mathcal{G}_{LB}\,\not{d}_A\Psi - \frac{1}{2}\mathcal{G}_{LA}\,\not{d}_B\Psi - \frac{1}{2}\mathcal{G}_{RB}\,\not{d}_A\Psi - \frac{1}{2}\mathcal{G}_{RA}\,\not{d}_B\Psi.$$

PROOF. We prove two representative identities; the remaining identities in the lemma can be proved using similar arguments with the help of Lemma 4.10 and the torsion-free property $\mathscr{D}_V W - \mathscr{D}_W V = [V, W]$, valid for arbitrary vectorfields V and W. We first prove (5.2f). To proceed, we expand $\mathscr{D}_A \breve{R}$ relative to the frame as follows:

(5.6) $$\mathscr{D}_A \breve{R} = a^L L + a^{\breve{R}}\breve{R} + a^B X_B,$$

where a^L, $a^{\breve{R}}$, and a^B are scalars to be determined. To compute $a^{\breve{R}}$, we first take the inner product of the left-hand side of (5.6) with L and compute, with the help of Lemma 3.20, that $g(\mathscr{D}_A\breve{R}, L) = g(\mathscr{D}_A(\upmu R), L) = (\not{d}_A\upmu)g(R, L) - \upmu g(R, \mathscr{D}_A L) = -\not{d}_A\upmu - \upmu\zeta_A$. On the other hand, the inner product of the right-hand side of (5.6) with L is $-\upmu a^{\breve{R}}$. Equating these two quantities, we deduce that $a^{\breve{R}} = \upmu^{-1}\not{d}_A\upmu + \zeta_A$ as desired.

To compute a^L, we first take the inner product of the left-hand side of (5.6) with $\breve{R}$ and compute, with the help of Lemma 3.20, that $g(\mathscr{D}_A\breve{R}, \breve{R}) = \frac{1}{2}\not{d}_A g(\breve{R}, \breve{R}) = \frac{1}{2}\not{d}_A(\upmu^2) = \upmu\not{d}_A\upmu$. On the other hand, the inner product of the right-hand side of (5.6) with $\breve{R}$ is $-\upmu a^L + \upmu^2 a^{\breve{R}}$. Equating these two quantities and using the above expression for $a^{\breve{R}}$, we deduce that $a^L = \upmu\zeta_A$ as desired.

To compute a^B, we take the inner product of (5.6) with X_B and use (3.19c), (3.73), and (3.74) to compute that $a^C \not{g}_{BC} = g(\mathscr{D}_A\breve{R}, X_B) = g(\mathscr{D}_A(\upmu R), X_B) = \upmu g(\mathscr{D}_A R, X_B) = \upmu g(\mathscr{D}_A N, X_B) - \upmu g(\mathscr{D}_A L, X_B) = \upmu\not{k}_{AB} - \upmu\chi_{AB}$ as desired. We have thus proved (5.2f).

As our second example, we prove (5.3b), (5.4b), and (5.5b). To proceed, we use definition (3.69) and the chain rule identity $N g_{ij} = G_{ij} N\Psi$ to deduce that relative to rectangular coordinates, we have

(5.7) $$2k_{ij} = G_{ij} N\Psi + g_{i\alpha}\partial_j N^\alpha + g_{j\alpha}\partial_i N^\alpha.$$

Contracting (5.7) against $X_A^i X_B^j$, recalling that $N = L + R = L + \upmu^{-1}\breve{R}$, and using the decomposition (3.52a), we deduce that

(5.8) $$2\not{k}_{AB} = \mathcal{G}_{AB}\left\{\upmu^{-1}\breve{R}\Psi + L\Psi\right\} + \not{g}_{Aa}\not{d}_B\{L^a + R^a\} + \not{g}_{Ba}\not{d}_A\{L^a + R^a\}.$$

To conclude the desired identities (5.3b), (5.4b), and (5.5b), we use equations (4.10a) and (4.10b) to substitute for $\not{d}_A L^a$, $\not{d}_B L^a$, $\not{d}_A R^a$, and $\not{d}_B R^a$ in (5.8), use the identities $(\not{g}^{-1})^{BC}\not{g}_{Aa}\not{d}_C x^a = \delta_A^B$ (see (3.61)) and $\not{g}_{Aa}R^a = 0$, and perform straightforward calculations. This concludes our proof of the lemma. □

COROLLARY 5.2 (**An expression for $[L, \breve{R}]$**). *The following identity holds:*

$$\mathcal{L}_L \breve{R}^A = [L, \breve{R}]^A = -(\not{g}^{-1})^{AB}\not{d}_B \upmu - 2\upmu(\not{g}^{-1})^{AB}\zeta_B. \tag{5.9}$$

PROOF. The corollary follows from the torsion-free property $[L, \breve{R}] = \mathscr{D}_L \breve{R} - \mathscr{D}_{\breve{R}} L$, (5.2b), and (5.2d). □

5.2. Connection coefficients of the rescaled null frame

For convenience, we perform some of our computations relative to the rescaled null frame $\{L, \breve{\underline{L}}, X_1, X_2\}$. In the following lemma, we provide the connection coefficients of this frame.

LEMMA 5.3 (**Connection coefficients of the rescaled null frame $\{L, \breve{\underline{L}}, X_1, X_2\}$**). *We define the $S_{t,u}$ tensorfields*

$$\eta := \zeta + \upmu^{-1}\not{d}\upmu, \tag{5.10a}$$

$$\underline{\chi} := 2\upmu \not{k} - \upmu\chi, \tag{5.10b}$$

where the tensorfields k, χ, and ζ are defined by (3.69), (3.70), and (5.1). Then the covariant derivatives of the rescaled null frame vectorfields can be expressed as follow:

$$\mathscr{D}_L L = \upmu^{-1}(L\upmu)L, \tag{5.11a}$$

$$\mathscr{D}_{\breve{\underline{L}}} L = -(L\upmu)L + 2\upmu\eta^A X_A, \tag{5.11b}$$

$$\mathscr{D}_A L = -\zeta_A L + \chi_A{}^B X_B, \tag{5.11c}$$

$$\mathscr{D}_L \breve{\underline{L}} = -2\upmu\zeta^A X_A, \tag{5.11d}$$

$$\mathscr{D}_{\breve{\underline{L}}} \breve{\underline{L}} = \{\upmu^{-1}\breve{\underline{L}}\upmu + L\upmu\}\breve{\underline{L}} - 2\upmu(\not{d}^A\upmu)X_A, \tag{5.11e}$$

$$\mathscr{D}_A \breve{\underline{L}} = \eta_A \breve{\underline{L}} + \underline{\chi}_A{}^B X_B, \tag{5.11f}$$

$$\mathscr{D}_L X_A = \mathscr{D}_A L, \tag{5.11g}$$

$$\mathscr{D}_A X_B = \nabla\!\!\!\!/_A X_B + \frac{1}{2}\upmu^{-1}\underline{\chi}_{AB} L + \frac{1}{2}\upmu^{-1}\chi_{AB}\breve{\underline{L}}. \tag{5.11h}$$

PROOF. Lemma 5.3 follows from the identity $\breve{\underline{L}} = \upmu L + 2\breve{R}$, Lemma 5.1, and straightforward computations. □

5.3. Frame decomposition of the inverse-foliation-density-weighted wave operator

In the next proposition, we decompose the $\upmu$-weighted wave operator $\upmu\square_{g(\Psi)}$ relative to the rescaled frame $\{L, \breve{R}, X_1, X_2\}$.

PROPOSITION 5.4 (**Frame decomposition of $\upmu\square_{g(\Psi)}f$**). *Let f be a scalar-valued function. Then relative to the frame $\{L, \breve{R}, X_1, X_2\}$, $\upmu\square_{g(\Psi)}f$ can be expressed in either of the following two forms, where $\varrho = 1 - u + t$ is the geometric* *radial variable from definition* (3.5)*:*

$$\mu\square_{g(\Psi)}f = -L(\mu Lf + 2\breve{R}f) + \mu\Delta\!\!\!/ f \underbrace{-\mathrm{tr}_{g\!\!\!/}\chi\breve{R}f}_{\text{slowest decay}} - \mathrm{tr}_{g\!\!\!/}k\!\!\!/^{(Trans-\Psi)}Lf - \mu\mathrm{tr}_{g\!\!\!/}k\!\!\!/^{(Tan-\Psi)}Lf - 2\zeta^{(Trans-\Psi)\#}\cdot d\!\!\!/ f - 2\mu\zeta^{(Tan-\Psi)\#}\cdot d\!\!\!/ f, \tag{5.12a}$$

$$\mu\square_{g(\Psi)}f = -(\mu L + 2\breve{R})(Lf) + \mu\Delta\!\!\!/ f \underbrace{-\mathrm{tr}_{g\!\!\!/}\chi\breve{R}f}_{\text{slowest decay}} - \omega^{(Trans-\Psi)}Lf - \mu\omega^{(Tan-\Psi)}Lf - \mathrm{tr}_{g\!\!\!/}k\!\!\!/^{(Trans-\Psi)}Lf - \mu\mathrm{tr}_{g\!\!\!/}k\!\!\!/^{(Tan-\Psi)}Lf + 2\zeta^{(Trans-\Psi)\#}\cdot d\!\!\!/ f + 2\mu\zeta^{(Tan-\Psi)\#}\cdot d\!\!\!/ f + 2(d\!\!\!/^{\#}\mu)\cdot d\!\!\!/ f. \tag{5.12b}$$

Furthermore, with $\breve{\underline{L}} = \mu L + 2\breve{R}$*, we also have the following alternate decompositions:*

$$L\left\{\breve{\underline{L}}(\varrho f)\right\} = -\varrho\mu\square_{g(\Psi)}f + \varrho\Delta\!\!\!/ f + \varrho(\mu-1)\Delta\!\!\!/ f + 2(\mu-1)Lf + \omega^{(Trans-\Psi)}f + \mu\omega^{(Tan-\Psi)}f - \varrho\mathrm{tr}_{g\!\!\!/}k\!\!\!/^{(Trans-\Psi)}Lf - \varrho\mu\mathrm{tr}_{g\!\!\!/}k\!\!\!/^{(Tan-\Psi)}Lf - \varrho\mathrm{tr}_{g\!\!\!/}\chi^{(Small)}\breve{R}f - 2\varrho\zeta^{(Trans-\Psi)\#}\cdot d\!\!\!/ f - 2\varrho\mu\zeta^{(Tan-\Psi)\#}\cdot d\!\!\!/ f, \tag{5.13a}$$

$$\breve{R}\left\{\varrho\left(Lf + \frac{1}{2}\mathrm{tr}_{g\!\!\!/}\chi f\right)\right\} = -\frac{1}{2}\mu L(\varrho Lf) + \frac{1}{2}\varrho\mu\Delta\!\!\!/ f - \frac{1}{2}\varrho\mu\square_{g(\Psi)}f - \frac{1}{2}\varrho\omega^{(Trans-\Psi)}Lf - \frac{1}{2}\varrho\mu\omega^{(Tan-\Psi)}Lf - \frac{1}{2}\varrho\mathrm{tr}_{g\!\!\!/}k\!\!\!/^{(Trans-\Psi)}Lf - \frac{1}{2}\varrho\mu\mathrm{tr}_{g\!\!\!/}k\!\!\!/^{(Tan-\Psi)}Lf + \varrho\zeta^{(Trans-\Psi)\#}\cdot d\!\!\!/ f + \varrho\mu\zeta^{(Tan-\Psi)\#}\cdot d\!\!\!/ f + \varrho(d\!\!\!/^{\#}\mu)\cdot d\!\!\!/ f + \frac{1}{2}\mu Lf - Lf + \frac{1}{2}\varrho(\breve{R}\mathrm{tr}_{g\!\!\!/}\chi^{(Small)})f - \frac{1}{2}\mathrm{tr}_{g\!\!\!/}\chi^{(Small)}f. \tag{5.13b}$$

In the above expressions, the $S_{t,u}$ *tensorfields* χ, $\chi^{(Small)}$, $\omega^{(Trans-\Psi)}$, $\omega^{(Tan-\Psi)}$, $\zeta^{(Trans-\Psi)}$, $k\!\!\!/^{(Trans-\Psi)}$, $\zeta^{(Tan-\Psi)}$, *and* $k\!\!\!/^{(Tan-\Psi)}$ *are defined by* (3.70), (4.1c), (4.6a), (4.6b), (5.4a), (5.4b), (5.5a), *and* (5.5b).

PROOF. Using (3.51b), Lemmas 5.1 and 5.3, and the Leibniz rule identities $X_A(X_B f) = (\mathscr{D}_A X_B)f + \mathscr{D}^2_{AB} f = (\nabla\!\!\!/_A X_B)f + \nabla\!\!\!/^2_{AB} f$, we compute that

(5.14)

$$\begin{aligned}
\mu\Box_{g(\Psi)} f &= \mu(g^{-1})^{\alpha\beta}\mathscr{D}^2_{\alpha\beta} f = -\mathscr{D}^2_{L\breve{\underline{L}}} f + \mu(g\!\!\!/^{-1})^{AB}\mathscr{D}^2_{AB} f \\
&= -L(\breve{\underline{L}} f) + (\mathscr{D}_L \breve{\underline{L}})f + \mu(g\!\!\!/^{-1})^{AB}\nabla\!\!\!/^2_{AB} f \\
&\quad - \mu(g\!\!\!/^{-1})^{AB} k\!\!\!/_{AB} Lf - (g\!\!\!/^{-1})^{AB}\chi_{AB}\breve{R} f \\
&= -L(\breve{\underline{L}} f) - 2\mu\zeta^{\#}\cdot d\!\!\!/ f + \mu\Delta\!\!\!\!/ f - \mu \mathrm{tr}_{g\!\!\!/} k\!\!\!/ Lf - \mathrm{tr}_{g\!\!\!/}\chi \breve{R} f \\
&= -L(\breve{\underline{L}} f) + \mu\Delta\!\!\!\!/ f - \mathrm{tr}_{g\!\!\!/}\chi\breve{R} f - \mathrm{tr}_{g\!\!\!/} k\!\!\!/^{(Trans-\Psi)} Lf - \mu\mathrm{tr}_{g\!\!\!/} k\!\!\!/^{(Tan-\Psi)} Lf \\
&\quad - 2\zeta^{(Trans-\Psi)\#}\cdot d\!\!\!/ f - 2\mu\zeta^{(Tan-\Psi)\#}\cdot d\!\!\!/ f.
\end{aligned}$$

From (5.14) and the identity $\breve{\underline{L}} = \mu L + 2\breve{R}$, we deduce (5.12a). The proof of (5.12b) is similar; we simply interchange the order of L and $\breve{\underline{L}}$ in (5.14).

To prove (5.13a), we multiply both sides of (5.12a) by ϱ, use the identities $L\varrho = 1$ and $\breve{R}\varrho = -1$ together with the decomposition $\mathrm{tr}_{g\!\!\!/}\chi = 2\varrho^{-1} + \mathrm{tr}_{g\!\!\!/}\chi^{(Small)}$, use Lemma 4.6 to substitute for $L\mu$, and carry out straightforward computations. To prove (5.13b), we multiply both sides of (5.12b) by ϱ and use a similar argument. □

CHAPTER 6

Construction of the Rotation Vectorfields and Their Basic Properties

In this chapter, we construct the $S_{t,u}$-tangent rotation vectorfields O and derive some of their basic properties. Our construction is not difficult: we simply project away the dangerous R component present in the Euclidean rotations. If we did not remove this dangerous component, then the rotational derivatives of Ψ could blow-up like $\upmu^{-1}$ at the shock formation points. Hence, they would be useless for detecting dispersive behavior that persists up until the shock.

6.1. Construction of the rotation vectorfields

We first note that the type $\binom{1}{1}$ $S_{t,u}$ tensorfield $\Pi\!\!\!/$, which projects onto the spheres $S_{t,u}$ (see Def. 3.39 and the remarks following it), has the following nonzero rectangular components (see (3.40)), $(i,j=1,2,3)$:

$$\Pi\!\!\!/_j^{\ i} = \delta_j^{\ i} - g_{ja}R^a R^i = (\underline{g}^{-1})^{ia} g\!\!\!/_{aj}, \tag{6.1}$$

where $\delta_j^{\ i}$ is the standard Kronecker delta, $\underline{g}^{-1}$ is the inverse first fundamental form from Def. 3.9, and we have used Lemmas 3.48 and 4.3.

DEFINITION 6.1 (**Rotation vectorfields**). For $l=1,2,3$, we define the *Euclidean rotation* $O_{(Flat;l)}$ to be the Σ_t-tangent vectorfield with the following rectangular spatial components, $(i=1,2,3)$:

$$O^i_{(Flat;l)} := \epsilon_{lai}x^a, \tag{6.2}$$

where ϵ_{ijk} is the fully antisymmetric symbol normalized by $\epsilon_{123}=1$.

We define the *geometric rotation* $O_{(l)}$ to be the $S_{t,u}$-tangent vectorfield with the following rectangular spatial components, $(i=1,2,3)$:

$$O^i_{(l)} := \Pi\!\!\!/_a^{\ i} O^a_{(Flat;l)}. \tag{6.3}$$

6.2. Basic properties of the rotation vectorfields

Some of our most delicate analysis involves the components of the Euclidean rotations in the direction of R. This component is captured in the next definition.

DEFINITION 6.2 (**The R component of $O_{(Flat;l)}$**). We define the following scalar-valued functions $\uprho_{(l)}$, $(l=1,2,3)$:

$$\uprho_{(l)} := g(O_{(Flat;l)}, R). \tag{6.4}$$

In the next lemma, we decompose $O_{(Flat;l)}$ into $O_{(l)}$ plus an error term in the radial direction R. We also provide an explicit expression for the radial component $\uprho_{(l)}$.

LEMMA 6.3 (**Decomposition $O_{(l)}$ into $O_{(Flat;l)}$ plus an error, and an expression for $\uprho_{(l)}$**). *The following identities hold, where $g_{ij}^{(Small)}$ and $R_{(Small)}^i$ are defined by* (2.4) *and* (4.1b)*:*

$$O_{(Flat;l)} = O_{(l)} + \uprho_{(l)} R, \tag{6.5}$$

$$\begin{aligned} \uprho_{(l)} &= g_{bc}\epsilon_{lab}x^a R^c \\ &= \epsilon_{lab}x^a R_{(Small)}^b + \epsilon_{lab} g_{bc}^{(Small)} x^a R_{(Small)}^c - \frac{1}{\varrho} g_{bc}^{(Small)} \epsilon_{lab} x^a x^c. \end{aligned} \tag{6.6}$$

PROOF. The identity (6.5) follows easily from definition (6.3), definition (6.4), and the fact that $g(R, R) = 1$ (see (3.20c)).

The first equality in (6.6) follows directly from expressing the right-hand side of (6.4) in rectangular spatial coordinates and from definition (6.2). To deduce the second equality, we insert the decompositions $g_{ab} = \delta_{ab} + g_{ab}^{(Small)}$ (see (2.4)) and $R^c = -\varrho^{-1}x^c + R_{(Small)}^c$ (see (4.1b)) into the first equality. By the anti-symmetry of $\epsilon_{\dots}$, the "large" term $\varrho^{-1}\delta_{bc}\epsilon_{lab}x^a x^c$ vanishes and hence the second equality follows. □

LEMMA 6.4 (**The rectangular components of $O_{(l)}$**). *The rectangular spatial components $O_{(l)}^i$, $(i = 1, 2, 3)$, can be expressed as follows:*

$$O_{(l)}^i = \epsilon_{lai}x^a - \uprho_{(l)} R^i = \epsilon_{lai}x^a + \uprho_{(l)} \frac{x^i}{\varrho} - \uprho_{(l)} R_{(Small)}^i. \tag{6.7}$$

PROOF. Lemma 6.4 follows easily from definition (6.3), the identity (6.5), and the decomposition (4.1b). □

LEMMA 6.5 (**An expression for the $S_{t,u}$ components of $O_{(l)}$**). *The following identities hold, where $g_{ij}^{(Small)}$ is defined by* (2.4)*,* ($A = 1, 2$ *and* $\not{d}^A x^b = (\not{g}^{-1})^{AB} \not{d}_B x^b$)*:*

$$O_{(l)}^A = \epsilon_{lca} g_{ab} x^c \not{d}^A x^b = \epsilon_{lab} x^a \not{d}^A x^b + \epsilon_{lca} g_{ab}^{(Small)} x^c \not{d}^A x^b. \tag{6.8}$$

PROOF. First, using definition (6.2), (6.5), and the identities $g(R, X_B) = 0$ and $X_B^i = \not{d}_B x^i$ (see (3.61)), we compute that $O_{(l)B} := \not{g}(O_{(l)}, X_B) = g(O_{(Flat;l)}, X_B) = g_{ab}\epsilon_{lca}x^c X_B^b = \epsilon_{lca} g_{ab} x^c \not{d}_B x^b$. The first equality in (6.8) now follows from contracting the previous identity against $(\not{g}^{-1})^{AB}$. The second then follows from the decomposition $g_{ab} = \delta_{ab} + g_{ab}^{(Small)}$ (see (2.4)). □

In the next lemma, we derive an expression for the Lie derivatives $\mathcal{L}_{O_{(l)}} O_{(m)}$. The most important aspect of the lemma is that *the expression does not involve the dangerous transversal derivative $R\Psi = \upmu^{-1}\breve{R}\Psi$.*

LEMMA 6.6 (**Expression for the commutators $[O_{(m)}, O_{(n)}]$**). *Let $\upchi^{(Small)}$ and $\Theta^{(Tan-\Psi)}$ be the type $\binom{0}{2}$ $S_{t,u}$ tensorfields defined in* (4.1c) *and* (4.11d)*, let $R_{(Small)}$ be the $\Sigma_t^{U_0}$-tangent vectorfield defined in* (4.1b)*, and let $\uprho_{(l)}$, $(l = 1, 2, 3)$, be the scalar-valued functions defined in* (6.4)*. Then the following identity holds:*

$$\mathcal{L}_{O_{(l)}} O_{(m)} = [O_{(l)}, O_{(m)}] = -\epsilon_{lmn} O_{(n)} + {}^{(l,m)}W^{\#}, \tag{6.9}$$

where ${}^{(l,m)}W$ *is the* $S_{t,u}$ *one-form given by*

$$
\begin{aligned}
{}^{(l,m)}W_A = & -\rho_{(l)} O_{(m)}^B \chi_{BA}^{(Small)} + \rho_{(m)} O_{(l)}^B \chi_{BA}^{(Small)} \\
& + \rho_{(l)} O_{(m)}^B \Theta_{BA}^{(Tan-\Psi)} - \rho_{(m)} O_{(l)}^B \Theta_{BA}^{(Tan-\Psi)} \\
& - \rho_{(l)} \epsilon_{mab} R_{(Small)}^a g_{bc} \not{d}_A x^c + \rho_{(m)} \epsilon_{lab} R_{(Small)}^a g_{bc} \not{d}_A x^c.
\end{aligned} \tag{6.10}
$$

PROOF. Using (6.5) and the decompositions (4.1b) and (6.7), we compute that

$$
\begin{aligned}
O_{(l)} O_{(m)}^i & = O_{(l)}^c \epsilon_{mci} - \rho_{(m)} O_{(l)} R^i - (O_{(l)} \rho_{(m)}) R^i \\
& = O_{(Flat;l)}^c \epsilon_{mci} + \varrho^{-1} \rho_{(l)} x^c \epsilon_{mci} - \rho_{(l)} R_{(Small)}^c \epsilon_{mci} + \varrho^{-1} \rho_{(m)} \epsilon_{lci} x^c \\
& \quad - \varrho^{-1} \rho_{(l)} \rho_{(m)} R^i - \rho_{(m)} O_{(l)} R_{(Small)}^i - (O_{(l)} \rho_{(m)}) R^i.
\end{aligned} \tag{6.11}
$$

Using equation (4.10d) to substitute for $O_{(l)} R_{(Small)}^i = O_{(l)}^A \not{d}_A R_{(Small)}^i$ on the right-hand side of (6.11), we deduce that

$$
\begin{aligned}
O_{(l)} O_{(m)}^i = & O_{(Flat;l)}^c \epsilon_{mci} - \rho_{(l)} R_{(Small)}^c \epsilon_{mci} + \varrho^{-1} \rho_{(l)} \epsilon_{mci} x^c + \varrho^{-1} \rho_{(m)} \epsilon_{lci} x^c \\
& + \rho_{(m)} (\not{g}^{-1})^{BC} \chi_{AB}^{(Small)} O_{(l)}^A \not{d}_C x^i - \rho_{(m)} (\not{g}^{-1})^{BC} \Theta_{AB}^{(Tan-\Psi)} O_{(l)}^A \not{d}_C x^i \\
& - \varrho^{-1} \rho_{(l)} \rho_{(m)} R^i - \rho_{(m)} \theta_A^{(Tan-\Psi)} O_{(l)}^A R^i - (O_{(l)} \rho_{(m)}) R^i.
\end{aligned} \tag{6.12}
$$

We now compute $[\mathcal{L}_{O_{(l)}} O_{(m)}]^i = O_{(l)} O_{(m)}^i - O_{(m)} O_{(l)}^i$ by interchanging the roles of l and m in (6.12) and then subtracting the two identities. It is straightforward to compute that the first term on the right-hand side of (6.12) generates the term $[O_{(Flat;l)}, O_{(Flat;m)}]^i = -\epsilon_{lmn} O_{(Flat;n)}^i = -\epsilon_{lmn} O_{(n)}^i - \rho_{(n)} \epsilon_{lmn} R^i$. Moreover, since $[O_{(l)}, O_{(m)}]$ is $S_{t,u}$-tangent, we can then apply the $S_{t,u}$ projection tensorfield $\not{\Pi}_i^{\ j}$ (6.1) to both sides of the resulting expression, which preserves the vector $[O_{(l)}, O_{(m)}]$ on the left-hand side and annihilates the product $\rho_{(n)} \epsilon_{lmn} R^i$ from the previous sentence as well as all terms on the last line of (6.12), which are proportional to the vector R. In total, we deduce from this line of reasoning that for $j = 1, 2, 3$, we have

$$
\begin{aligned}
[O_{(l)}, O_{(m)}]^j = & -\epsilon_{lmn} O_{(n)}^j - \rho_{(l)} R_{(Small)}^c \not{\Pi}_i^{\ j} \epsilon_{mci} + \rho_{(m)} R_{(Small)}^c \not{\Pi}_i^{\ j} \epsilon_{lci} \\
& + \rho_{(m)} (\not{g}^{-1})^{BC} \chi_{AB}^{(Small)} O_{(l)}^A \not{d}_C x^j - \rho_{(l)} (\not{g}^{-1})^{BC} \chi_{AB}^{(Small)} O_{(m)}^A \not{d}_C x^j \\
& - \rho_{(m)} (\not{g}^{-1})^{BC} \Theta_{AB}^{(Tan-\Psi)} O_{(l)}^A \not{d}_C x^j + \rho_{(l)} (\not{g}^{-1})^{BC} \Theta_{AB}^{(Tan-\Psi)} O_{(m)}^A \not{d}_C x^j.
\end{aligned} \tag{6.13}
$$

Clearly, the first term on the right-hand side of (6.13) is the first term on the right-hand side of (6.10) as desired. To show that the remaining terms on the right-hand side of (6.13) are (when viewed as $S_{t,u}$-tangent vectors with rectangular components $(\cdot)^j$) $\not{g}$-dual to ${}^{(l,m)}W$, we contract them against $g_{jk} \not{d}_D x^k$ and use the identities $g_{jk} (\not{d}_D x^k) \not{\Pi}_i^{\ j} = g_{ic} \not{d}_D x^c$ and $g_{jk} (\not{d}_C x^j) \not{d}_D x^k = \not{g}_{CD}$ (see (3.61)). □

CHAPTER 7

Definition of the Commutation Vectorfields and Deformation Tensor Calculations

In this chapter, we define the set $\mathscr{Z}$ of commutation vectorfields that we use to commute the wave equation $\mu\square_{g(\Psi)}\Psi = 0$. We also compute the components of the deformation tensors ${}^{(Z)}\pi$ of the vectorfields $Z \in \mathscr{Z}$ relative to the rescaled frame $\{L, \breve{R}, X_1, X_2\}$. In order to prove our sharp classical lifespan theorem, we must precisely understand the structure of some of these components. The reason is that the derivatives of the ${}^{(Z)}\pi$ appear as inhomogeneous terms in the commuted wave equation (see Lemma 9.4 and Prop. 9.6), and some of the corresponding frame components are difficult to analyze. In fact, some of them seem to be on the border of the kinds of terms that would allow the proof of our sharp classical lifespan theorem to go through.

7.1. The commutation vectorfields

To prove our sharp classical lifespan theorem, we commute the covariant wave equation $\mu\square_{g(\Psi)}\Psi = 0$ with the vectorfields belonging to the following set $\mathscr{Z}$.

DEFINITION 7.1 (**Commutation vectorfields**). We define the set $\mathscr{Z}$ of commutation vectorfields as follows:

$$\mathscr{Z} := \{\varrho L, \breve{R}, O_{(1)}, O_{(2)}, O_{(3)}\}, \tag{7.1}$$

where ϱ, L, $\breve{R}$, and $O_{(l)}$ are defined in (3.5), (3.12), (3.19a), and (6.3).

In our analysis, we also use the following subsets of $\mathscr{Z}$.

DEFINITION 7.2 (**Spatial and rotation commutation subsets**). We define the subsets $\mathscr{S}, \mathscr{O} \subset \mathscr{Z}$ of spatial commutation vectorfields and rotation commutation vectorfields as follows:

$$\mathscr{S} := \{\breve{R}, O_{(1)}, O_{(2)}, O_{(3)}\}, \tag{7.2a}$$

$$\mathscr{O} := \{O_{(1)}, O_{(2)}, O_{(3)}\}. \tag{7.2b}$$

REMARK 7.3 (**The meaning of O**). Throughout the monograph, we often use the symbol O to denote a generic element of $\mathscr{O}$.

On rare occasion, we commute the wave equation with the Euclidean rotations.

DEFINITION 7.4 (**Euclidean rotation commutation subset**). We define the subset $\mathscr{O}_{(Flat)}$ of Euclidean rotation vectorfields as follows:

$$\mathscr{O}_{(Flat)} := \{O_{(Flat;1)}, O_{(Flat;2)}, O_{(Flat;3)}\}, \tag{7.3}$$

where $O_{(Flat;l)}$ is defined in (6.2).

7.2. Deformation tensor calculations

We recall that the deformation tensor ${}^{(V)}\pi$ of a vectorfield V is defined in Def. 4.9. In this section, we calculate ${}^{(V)}\pi$ for various vectorfields V. We begin with the next lemma, which shows that for some important vectorfields V, the $S_{t,u}$ projection of ${}^{(V)}\pi$ is the same as $\not{\mathcal{L}}_V \not{g}$.

LEMMA 7.5 (**Connection between projected Lie derivatives of $\not{g}$ and $\not{\pi}$**). *Let $\not{g}$ be the first fundamental form of $S_{t,u}$ from Def. 3.9. If $V \in \{L, \varrho L, \breve{R}\}$ or V is an $S_{t,u}$-tangent vectorfield, then*

$$\not{\mathcal{L}}_V \not{g}_{AB} = {}^{(Z)}\not{\pi}_{AB}, \tag{7.4a}$$

$$(\not{\mathcal{L}}_V \not{g}^{-1})^{AB} = -{}^{(Z)}\not{\pi}^{AB}. \tag{7.4b}$$

PROOF. The identity (7.4a) follows from Cor. 4.12, Def. 4.9, and the identity $\not{g} = \not{\Pi} g$. (7.4b) then follows from (7.4a) and the identity $(\not{\mathcal{L}}_V \not{g}^{-1})^{AB} = -(\not{g}^{-1})^{AC}(\not{g}^{-1})^{BD}\mathcal{L}_V \not{g}_{CD} = -(\not{g}^{-1})^{AC}(\not{g}^{-1})^{BD}\not{\mathcal{L}}_V \not{g}_{CD}$, which is a simple consequence of the identity $\not{g}_{AC}(\not{g}^{-1})^{CB} = \delta_A^B$, the Leibniz rule, and the fact that $\not{\mathcal{L}}_V \delta_A^B = 0$ (see Lemma 4.11). □

We now calculate the frame components of the deformation tensors ${}^{(V)}\pi_{\mu\nu}$ of various vectorfields including the commutation vectorfields $\mathscr{Z}$. The main result is the following proposition.

REMARK 7.6 (**Some important structural features**). The following three aspects of the proposition are highly important.

- The dangerous transversal derivative $R\Psi = \upmu^{-1}\breve{R}\Psi$ is completely absent from the right-hand sides of the expressions.
- ${}^{(Z)}\pi_{LL} = 0$ for all $Z \in \mathscr{Z}$. This identity in particular implies the absence of the dangerous quadratic term $\upmu^{-1(Z)}\pi_{LL}\breve{R}\Psi$, whose derivatives would appear on the right-hand side of the identity (9.13). The presence of such a term could in principle destroy our proof of sharp classical lifespan theorem because we would have no obvious way to control the dangerous factor $\upmu^{-1}$.
- $\upmu^{-1}\left\{{}^{(Z)}\pi_{L\breve{R}} + Z\upmu\right\}$ is either 0 or -1 for all $Z \in \mathscr{Z}$. We elaborate upon the importance of this fact in Remark 9.2.

PROPOSITION 7.7 (**Expressions for the frame components of various deformation tensors**). *We have the following identities, where* $\chi^{(Small)}$ *is the symmetric type* $\binom{0}{2}$ $S_{t,u}$ *tensorfield defined in* (4.1c), *we are using the notation of Sects.* 3.11 *and* 3.20, *and the quantities* f *are smooth functions of* Ψ*:*

$$ {}^{(\breve{R})}\pi_{LL} = 0, \tag{7.5a} $$

$$ {}^{(\breve{R})}\pi_{\breve{R}\breve{R}} = 2\breve{R}\upmu, \tag{7.5b} $$

$$ {}^{(\breve{R})}\pi_{L\breve{R}} = -\breve{R}\upmu, \tag{7.5c} $$

$$ {}^{(\breve{R})}\not{\pi}_{LA} = -\not{d}_A\upmu + G_{(Frame)}\begin{pmatrix} \upmu L\Psi \\ \breve{R}\Psi \\ \upmu\not{d}\Psi \end{pmatrix}, \tag{7.5d} $$

$$ {}^{(\breve{R})}\not{\pi}_{\breve{R}A} = 0, \tag{7.5e} $$

$$ \begin{aligned} {}^{(\breve{R})}\not{\pi}_{AB} &= -2\upmu\hat{\chi}_{AB}^{(Small)} + 2\hat{\not{k}}_{AB}^{(Trans-\Psi)} + 2\upmu\hat{\not{k}}_{AB}^{(Tan-\Psi)} \\ &= -2\upmu\hat{\chi}_{AB}^{(Small)} + G_{(Frame)}\hat{\otimes}\begin{pmatrix} \upmu L\Psi \\ \breve{R}\Psi \\ \upmu\not{d}\Psi \end{pmatrix}, \end{aligned} \tag{7.5f} $$

$$ \begin{aligned} \mathrm{tr}_{\not{g}}{}^{(\breve{R})}\not{\pi} &= -\frac{4}{\varrho}\upmu - 2\upmu\mathrm{tr}_{\not{g}}\chi^{(Small)} + 2\mathrm{tr}_{\not{g}}\not{k}^{(Trans-\Psi)} + 2\upmu\mathrm{tr}_{\not{g}}\not{k}^{(Tan-\Psi)} \\ &= -\frac{4}{\varrho}\upmu - 2\upmu\mathrm{tr}_{\not{g}}\chi^{(Small)} + G_{(Frame)}^{\#}\begin{pmatrix} \upmu L\Psi \\ \breve{R}\Psi \\ \upmu\not{d}\Psi \end{pmatrix}, \end{aligned} \tag{7.5g} $$

$$ {}^{(L)}\pi_{LL} = 0, \tag{7.6a} $$

$$ {}^{(L)}\pi_{\breve{R}\breve{R}} = 2L\upmu = G_{(Frame)}\begin{pmatrix} \upmu L\Psi \\ \breve{R}\Psi \end{pmatrix}, \tag{7.6b} $$

$$ {}^{(L)}\pi_{L\breve{R}} = -L\upmu = G_{(Frame)}\begin{pmatrix} \upmu L\Psi \\ \breve{R}\Psi \end{pmatrix}, \tag{7.6c} $$

$$ {}^{(L)}\not{\pi}_{LA} = 0, \tag{7.6d} $$

$$ {}^{(L)}\not{\pi}_{\breve{R}A} = \not{d}_A\upmu + G_{(Frame)}\begin{pmatrix} \upmu L\Psi \\ \breve{R}\Psi \\ \upmu\not{d}\Psi \end{pmatrix}, \tag{7.6e} $$

$$ {}^{(L)}\not{\pi}_{AB} = 2\hat{\chi}_{AB}^{(Small)}, \tag{7.6f} $$

$$ \mathrm{tr}_{\not{g}}{}^{(L)}\not{\pi} = 2\mathrm{tr}_{\not{g}}\chi^{(Small)} + \frac{4}{\varrho}, \tag{7.6g} $$

$$ {}^{(\varrho L)}\pi_{LL} = 0, \tag{7.7a} $$

$$ {}^{(\varrho L)}\pi_{\breve{R}R} = 2\varrho L\upmu + 2\upmu = \varrho G_{(Frame)} \begin{pmatrix} \upmu L\Psi \\ \breve{R}\Psi \end{pmatrix} + 2\upmu, \tag{7.7b} $$

$$ {}^{(\varrho L)}\pi_{L\breve{R}} = -\varrho L\upmu - \upmu = \varrho G_{(Frame)} \begin{pmatrix} \upmu L\Psi \\ \breve{R}\Psi \end{pmatrix} - \upmu, \tag{7.7c} $$

$$ {}^{(\varrho L)}\not\pi_{LA} = 0, \tag{7.7d} $$

$$ {}^{(\varrho L)}\not\pi_{\breve{R}A} = \varrho \not d_A \upmu + \varrho G_{(Frame)} \begin{pmatrix} \upmu L\Psi \\ \breve{R}\Psi \\ \upmu \not d\Psi \end{pmatrix}, \tag{7.7e} $$

$$ {}^{(\varrho L)}\hat{\not\pi}_{AB} = 2\varrho \hat{\upchi}_{AB}^{(Small)}, \tag{7.7f} $$

$$ \mathrm{tr}_{\not g}{}^{(\varrho L)}\not\pi = 2\varrho \mathrm{tr}_{\not g}\upchi^{(Small)} + 4, \tag{7.7g} $$

$$ {}^{(O_{(l)})}\pi_{LL} = 0, \tag{7.8a} $$

$$ {}^{(O_{(l)})}\pi_{\breve{R}R} = 2O_{(l)}\upmu, \tag{7.8b} $$

$$ {}^{(O_{(l)})}\pi_{L\breve{R}} = -O_{(l)}\upmu, \tag{7.8c} $$

$$ {}^{(O_{(l)})}\not\pi_{LA} = -\upchi_{AB}^{(Small)} O_{(l)}^B + {}^{(O_{(l)};Error)}\not\pi_{LA}, \tag{7.8d} $$

$$ {}^{(O_{(l)})}\not\pi_{\breve{R}A} = \upmu \upchi_{AB}^{(Small)} O_{(l)}^B + \uprho_{(l)} \not d_A \upmu + {}^{(O_{(l)};Error)}\not\pi_{\breve{R}A}, \tag{7.8e} $$

$$ {}^{(O_{(l)})}\hat{\not\pi}_{AB} = 2\uprho_{(l)} \hat{\upchi}_{AB}^{(Small)} + {}^{(O_{(l)};Error)}\hat{\not\pi}_{AB}, \tag{7.8f} $$

$$ \mathrm{tr}_{\not g}{}^{(O_{(l)})}\not\pi = 2\uprho_{(l)} \mathrm{tr}_{\not g}\upchi^{(Small)} + \mathrm{tr}_{\not g}{}^{(O_{(l)};Error)}\not\pi, \tag{7.8g} $$

$$ {}^{(O_{(l)};Error)}\not\pi_L = G_{(Frame)} \begin{pmatrix} O_{(l)} \\ \uprho_{(l)} \end{pmatrix} \begin{pmatrix} L\Psi \\ \not d\Psi \end{pmatrix} + f(\Psi) L_{(Small)} \not d x, \tag{7.9a} $$

$$ {}^{(O_{(l)};Error)}\not\pi_{\breve{R}} = G_{(Frame)} \begin{pmatrix} \upmu O_{(l)} \\ \upmu \uprho_{(l)} \end{pmatrix} \begin{pmatrix} L\Psi \\ \not d\Psi \end{pmatrix} + G_{(Frame)} \uprho_{(l)} \breve{R}\Psi + \upmu f(\Psi) \begin{pmatrix} \Psi \\ L_{(Small)} \end{pmatrix} \not d x, \tag{7.9b} $$

$$ {}^{(O_{(l)};Error)}\hat{\not\pi} = G_{(Frame)} \hat{\otimes} \begin{pmatrix} O_{(l)} \\ \uprho_{(l)} \end{pmatrix} \begin{pmatrix} L\Psi \\ \not d\Psi \end{pmatrix} + f(\Psi)\Psi \not d x \hat{\otimes} \not d x, \tag{7.9c} $$

$$ \mathrm{tr}_{\not g}{}^{(O_{(l)};Error)}\not\pi = G_{(Frame)} \not g^{-1} \begin{pmatrix} O_{(l)} \\ \uprho_{(l)} \end{pmatrix} \begin{pmatrix} L\Psi \\ \not d\Psi \end{pmatrix} + \frac{\uprho_{(l)}}{\varrho} + f(\Psi)\Psi(\not d^{\#} x)\not d x. \tag{7.9d} $$

REMARK 7.8 (**Clarification of the meaning of $O_{(l)}$ in the above schematic relations**). In (7.9a)-(7.9d), the vectorfield $O_{(l)}$ appearing in the first array is not acting as a differential operator, but rather as a tensorfield that is being contracted against other tensorfields such as $G_{(Frame)}$ and $\not d\Psi$.

Before proving the proposition, we first compute some covariant derivatives of the $S_{t,u}$ projection tensorfield $\not\Pi$ relative to the frame vectorfields $\{L, \breve{R}, X_1, X_2\}$.

LEMMA 7.9 (**Frame covariant derivatives of the spherical projection tensorfield $\Pi\!\!\!/$**). *Let $\Pi\!\!\!/$ be the type $\binom{1}{1}$ $S_{t,u}$ projection tensorfield defined in* (3.39b). *Then we have the following identities:*

(7.10a)
$$(\mathscr{D}_L \Pi\!\!\!/) \cdot R = \mu^{-1}\zeta^{(Trans-\Psi)A} X_A + \zeta^{(Tan-\Psi)A} X_A,$$
(7.10b)
$$(\mathscr{D}_L \Pi\!\!\!/) \cdot X_A = -\mu^{-1}\zeta_A^{(Trans-\Psi)} L - \zeta_A^{(Tan-\Psi)} L,$$
(7.10c)
$$(\mathscr{D}_{\breve{R}} \Pi\!\!\!/) \cdot R = (\nabla\!\!\!/^A \mu) X_A,$$
(7.10d)
$$(\mathscr{D}_{\breve{R}} \Pi\!\!\!/) X_A = \zeta_A^{(Trans-\Psi)} L + \mu\zeta_A^{(Tan-\Psi)} L + \zeta_A^{(Trans-\Psi)} R + \mu\zeta_A^{(Tan-\Psi)} R + (\nabla\!\!\!/_A \mu) R,$$
(7.10e)
$$(\mathscr{D}_A \Pi\!\!\!/) \cdot R = \chi_A^{\ B} X_B - \mu^{-1} k\!\!\!/_A^{\ (Trans-\Psi)B} X_B - k\!\!\!/_A^{\ (Tan-\Psi)B} X_B,$$
(7.10f)
$$(\mathscr{D}_A \Pi\!\!\!/) \cdot X_B = \mu^{-1} k\!\!\!/_{AB}^{(Trans-\Psi)} L + k\!\!\!/_{AB}^{(Tan-\Psi)} L + \chi_{AB} R.$$

In the above expressions, the $S_{t,u}$ tensorfields χ, $\zeta^{(Trans-\Psi)}$, $k\!\!\!/^{(Trans-\Psi)}$, $\zeta^{(Tan-\Psi)}$, and $k\!\!\!/^{(Tan-\Psi)}$ are defined by (3.70), (5.4a), (5.4b), (5.5a), *and* (5.5b).

PROOF. The main idea of the proof is to use the decompositions provided by Lemma 5.1. As examples, we prove (7.10a) and (7.10d). The remaining identities in (7.10a)-(7.10f) can be proved using similar arguments. To prove (7.10a), we differentiate the identity $\Pi\!\!\!/ \cdot \breve{R} = 0$ and use the identity $\breve{R} = \mu R$ to deduce that $(\mathscr{D}_L \Pi\!\!\!/) \cdot R = \mu^{-1}(\mathscr{D}_L \Pi\!\!\!/) \cdot \breve{R} = -\mu^{-1}\Pi\!\!\!/ \cdot \mathscr{D}_L \breve{R}$. The desired identity (7.10a) now follows easily from the previous identity, (5.2d), and (5.3a).

To prove (7.10d), we differentiate the identity $\Pi\!\!\!/ X_A = X_A$ to deduce $(\mathscr{D}_{\breve{R}} \Pi\!\!\!/) \cdot X_A = \mathscr{D}_{\breve{R}} X_A - \Pi\!\!\!/ \cdot \mathscr{D}_{\breve{R}} X_A$. Since $\mathscr{D}_{\breve{R}} X_A - \mathscr{D}_A \breve{R} = [\breve{R}, X_A]$ is $S_{t,u}$-tangent (see (4.21c)), it follows that $(\mathscr{D}_{\breve{R}} \Pi\!\!\!/) \cdot X_A = \mathscr{D}_A \breve{R} - \Pi\!\!\!/ \cdot \mathscr{D}_A \breve{R}$. The desired identity (7.10d) now follows easily from the previous identity, (5.2f), and (5.3a). □

PROOF OF PROP. 7.7. The main idea of the proof is to use the decompositions provided by Lemmas 5.1 and 7.9. We give four examples, which by far involve the most difficult computations.

We begin by proving (7.8f) and (7.8g). These identities and (7.8d) are the only ones in the proposition for which it requires some effort to see the cancellation of some terms involving the dangerous transversal derivative $R\Psi = \mu^{-1}\breve{R}\Psi$. To proceed, we use Defs. 4.9 and 6.1 to deduce that

$$\begin{aligned} {}^{(O_{(l)})}\pi\!\!\!/_{AB} &= g(\mathscr{D}_A[\Pi\!\!\!/ \cdot O_{(Flat;l)}], X_B) + g(\mathscr{D}_B[\Pi\!\!\!/ \cdot O_{(Flat;l)}], X_A) \\ &= g([\mathscr{D}_A \Pi\!\!\!/] \cdot O_{(Flat;l)}, X_B) + g([\mathscr{D}_B \Pi\!\!\!/] \cdot O_{(Flat;l)}, X_A) \\ &\quad - g(\mathscr{D}_A O_{(Flat;l)}, X_B) + g(\mathscr{D}_B O_{(Flat;l)}, X_A). \end{aligned} \tag{7.11}$$

From (6.5), (7.10e), (7.10f), (5.3b), (5.4b), and (5.5b), we deduce that

(7.12)

$$g([\mathscr{D}_A \not{\Pi}] \cdot O_{(Flat;l)}, X_B) + g([\mathscr{D}_B \not{\Pi}] \cdot O_{(Flat;l)}, X_A)$$
$$= 2\rho_{(l)}\chi_{AB} - 2\mu^{-1}\rho_{(l)}\not{k}_{AB}^{(Trans-\Psi)} - 2\rho_{(l)}\not{k}_{AB}^{(Tan-\Psi)}$$
$$= 2\rho_{(l)}\chi_{AB} - \mu^{-1}\rho_{(l)}\not{G}_{AB}\,\breve{R}\Psi + \rho_{(l)}G_{(Frame)}\begin{pmatrix} L\Psi \\ \not{d}\Psi \end{pmatrix}.$$

Next, we use (4.4a), (4.4b), Def. 6.1, (6.5), the identity $R = \mu^{-1}\breve{R}$, and the identity $\not{d}_A x^i = X_A^i$ (see (3.61)) to compute (relative to rectangular coordinates) that

(7.13)

$$g(\mathscr{D}_A O_{(Flat;l)}, X_B) + g(\mathscr{D}_B O_{(Flat;l)}, X_A)$$
$$= \epsilon_{lca} g_{ab} \not{d}_A x^c \not{d}_B x^b + \epsilon_{lca} g_{ab} \not{d}_B x^c \not{d}_A x^b + \not{G}_{AB}\, O_{(Flat;l)}\Psi.$$

We now insert the decompositions $O_{(Flat;l)} = O_{(l)} + \rho_{(l)} R$ and $g_{ab} = \delta_{ab} + g_{ab}^{(Small)}$ (see (2.4)) into (7.13) and observe that by the antisymmetry of $\epsilon_{\dots}$, the δ_{ab} part cancels from the first two terms on the right-hand side of (7.13). Hence, we have

$$(7.14)\qquad g(\mathscr{D}_A O_{(Flat;l)}, X_B) + g(\mathscr{D}_B O_{(Flat;l)}, X_A)$$
$$= \epsilon_{lca} g_{ab}^{(Small)} (\not{d}_A x^c \not{d}_B x^b + \not{d}_B x^c \not{d}_A x^b)$$
$$+ \mu^{-1}\rho_{(l)}\not{G}_{AB}\,\breve{R}\Psi + \not{G}_{AB}\, O_{(l)}\Psi.$$

We now add (7.12) and (7.14) and note the cancellation of the dangerous product $\mu^{-1}\rho_{(l)}\not{G}_{AB}\,\breve{R}\Psi$. Also using (7.11) and the fact that $g_{ab}^{(Small)}$ is a smooth function of Ψ that vanishes at $\Psi = 0$ (see (2.6)), we conclude that

$$(7.15)\qquad {}^{(O_{(l)})}\not{\pi}_{AB} = 2\rho_{(l)}\chi_{AB} + G_{(Frame)}\begin{pmatrix} O_{(l)} \\ \rho_{(l)} \end{pmatrix}\begin{pmatrix} L\Psi \\ \not{d}\Psi \end{pmatrix} + f(\Psi)\Psi \not{d}x \otimes \not{d}x.$$

The desired identities (7.8f) and (7.8g) now follow from (7.15) and (4.1c).

As a third example, we now prove (7.8d). To proceed, we use Defs. 6.1 and 4.9 to deduce

(7.16)

$$ {}^{(O_{(l)})}\not{\pi}_{LA} = g(\mathscr{D}_L[\not{\Pi} \cdot O_{(Flat;l)}], X_A) + g(\mathscr{D}_A[\not{\Pi} \cdot O_{(Flat;l)}], L)$$
$$= g([\mathscr{D}_L \not{\Pi}] \cdot O_{(Flat;l)}, X_A) + g([\mathscr{D}_A \not{\Pi}] \cdot O_{(Flat;l)}, L) + g(\mathscr{D}_L O_{(Flat;l)}, X_A).$$

From (4.1c), (6.5), (6.8), (7.10a), (7.10b), (7.10e), (7.10f), (5.3a), (5.4a), and (5.5a), we deduce that

(7.17)

$$g([\mathscr{D}_L \not{\Pi}] \cdot O_{(Flat;l)}, X_A) + g([\mathscr{D}_A \not{\Pi}] \cdot O_{(Flat;l)}, L)$$
$$= -\chi_{AB} O_{(l)}^B + \mu^{-1}\rho_{(l)}\zeta_A^{(Trans-\Psi)} + \rho_{(l)}\zeta_A^{(Tan-\Psi)}$$
$$= -\frac{1}{\varrho}\epsilon_{lca} g_{ab} x^c \not{d}_A x^b - \chi_{AB}^{(Small)} O_{(l)}^B - \frac{1}{2}\mu^{-1}\rho_{(l)}\not{G}_{LA}\,\breve{R}\Psi$$
$$+ \rho_{(l)} G_{(Frame)}\begin{pmatrix} L\Psi \\ \not{d}\Psi \end{pmatrix}.$$

Next, we use (4.4a), (4.4b), Def. 6.1, (6.5), the identity $R = \upmu^{-1}\breve{R}$, and the identity $\not{d}_A x^i = X_A^i$ (see (3.61)) to compute (relative to rectangular coordinates) that

$$g(\mathscr{D}_L O_{(Flat;l)}, X_A) = \epsilon_{lca} g_{ab} L^c \not{d}_A x^b + \frac{1}{2} \not{G}_{LA} O_{(Flat;l)} \Psi - \frac{1}{2} \not{G}_{O_{(Flat;l)} A} L\Psi - \frac{1}{2} G_{L O_{(Flat;l)}} \not{d}_A \Psi. \tag{7.18}$$

We now insert the decompositions $O_{(Flat;l)} = O_{(l)} + \uprho_{(l)} R$ and $L^c = \frac{x^c}{\varrho} + L^c_{(Small)}$ (see (4.1a)) into (7.18) and deduce that

(7.19)
$$g(\mathscr{D}_L O_{(Flat;l)}, X_A) = \frac{1}{\varrho} \epsilon_{lca} g_{ab} x^c \not{d}_A x^b + \epsilon_{lca} g_{ab} L^c_{(Small)} \not{d}_A x^b + \frac{1}{2} \upmu^{-1} \uprho_{(l)} \not{G}_{LA} \breve{R}\Psi + G_{(Frame)} \begin{pmatrix} O_{(l)} \\ \uprho_{(l)} \end{pmatrix} \begin{pmatrix} L\Psi \\ \not{d}\Psi \end{pmatrix}.$$

We now add (7.17) and (7.19) and observe that the first and third terms on the right-hand sides cancel. The desired identity (7.8d) thus follows.

As a final example, we prove (7.8e). This example is somewhat simpler than the previous examples in the sense that we do not have to observe the cancellation of any dangerous $\upmu^{-1}\breve{R}\Psi$-containing products (although we still have to observe some other cancellations). To proceed, we use Defs. 4.9 and 6.1 to deduce

(7.20)
$$\begin{aligned} {}^{(O_{(l)})}\not{\pi}_{\breve{R}A} &= g(\mathscr{D}_{\breve{R}}[\Pi \cdot O_{(Flat;l)}], X_A) + g(\mathscr{D}_A[\Pi \cdot O_{(Flat;l)}], \breve{R}) \\ &= g([\mathscr{D}_{\breve{R}}\Pi] \cdot O_{(Flat;l)}, X_A) + g([\mathscr{D}_A \Pi] \cdot O_{(Flat;l)}, \breve{R}) + g(\mathscr{D}_{\breve{R}} O_{(Flat;l)}, X_A). \end{aligned}$$

Next, using (4.1c), (6.5), (6.8), (7.10c), (7.10d), (7.10e), (7.10f), (5.3b), (5.4b), (5.5b), and the identity $\breve{R} = \upmu R$, we deduce that

(7.21)
$$\begin{aligned} g([\mathscr{D}_{\breve{R}}\Pi] \cdot O_{(Flat;l)}, X_A) &+ g([\mathscr{D}_A \Pi] \cdot O_{(Flat;l)}, \breve{R}) \\ &= \uprho_{(l)} \not{d}_A \upmu - \not{k}^{(Trans-\Psi)}_{AB} O^B_{(l)} - \upmu \not{k}^{(Tan-\Psi)}_{AB} O^B_{(l)} + \upmu \chi_{AB} O^B_{(l)} \\ &= \frac{1}{\varrho} \upmu \epsilon_{lca} g_{ab} x^c \not{d}_A x^b + \upmu \chi^{(Small)}_{AB} O^B_{(l)} + \uprho_{(l)} \not{d}_A \upmu \\ &\quad - \frac{1}{2} \not{G}_{AB} O^B_{(l)} \breve{R}\Psi + G_{(Frame)} \begin{pmatrix} \upmu L\Psi \\ \upmu \not{d}\Psi \end{pmatrix} O_{(l)}. \end{aligned}$$

Next, from (4.4a), (4.4b), Def. 6.1, the identity $\breve{R} = \upmu R$, and the identity $\not{d}_A x^b = X_A^b$ (see (3.61)), we compute (relative to rectangular coordinates) that

$$g(\mathscr{D}_{\breve{R}} O_{(Flat;l)}, X_A) = \upmu \epsilon_{lca} g_{ab} R^c \not{d}_A x^b + \frac{1}{2} \not{G}_{O_{(Flat;l)} A} \breve{R}\Psi + \frac{1}{2} \upmu \not{G}_{RA} O_{(Flat;l)} \Psi - \frac{1}{2} \upmu G_{R O_{(Flat;l)}} \not{d}_A \Psi. \tag{7.22}$$

We now insert the decompositions $O_{(Flat;l)} = O_{(l)} + \rho_{(l)} R$ and $R^c = -\frac{x^c}{\varrho} + R^c_{(Small)}$ (see (6.5) and (4.1b)) into (7.22) and deduce that

(7.23)

$$g(\mathscr{D}_{\breve{R}} O_{(Flat;l)}, X_A) = -\frac{1}{\varrho}\mu\epsilon_{lca} g_{ab} x^c \not{d}_A x^b + \mu\epsilon_{lca} g_{bc} R^c_{(Small)} \not{d}_A x^b + \frac{1}{2}\not{G}_{AB}\, O^B_{(l)} \breve{R}\Psi$$
$$+ G_{(Frame)} \left(\mu \not{d}\Psi\right) O_{(l)} + G_{(Frame)} \rho_{(l)} \breve{R}\Psi + \mu G_{(Frame)} \rho_{(l)} \not{d}\Psi.$$

Adding (7.21) and (7.23), noting that the first and third terms on the right-hand side of (7.23) are exactly canceled by two terms on the right-hand side of (7.21), and using the fact that $R^i = -L^i_{(Small)}$ plus a smooth function of Ψ that vanishes at $\Psi = 0$ (that is, (4.3)), we arrive at (7.8e).

The remaining identities in the proposition can be proved in a similar fashion with the help of Lemma 4.10 and the torsion-free property $\mathscr{D}_V W - \mathscr{D}_W V = [V, W]$, valid for arbitrary vectorfields V and W. The computations are much simpler and we leave the details to the reader. □

COROLLARY 7.10 (**Expression for the $S_{t,u}$ covariant derivatives of the rotation vectorfields**). *The type* $\binom{1}{1}$ $S_{t,u}$ *tensorfield* $\not{\nabla} O_{(l)}$ *can be expressed as follows, where* χ *is the symmetric type* $\binom{0}{2}$ $S_{t,u}$ *tensorfield defined in* (3.70), *we are using the notation of Sects.* 3.11 *and* 3.20 *and Remark* 7.8, *and* f *is a smooth function of* Ψ*:*

(7.24)
$$\not{\nabla}_A O^B_{(l)} = \epsilon_{lab} (\not{d}_A x^a) \not{d}^B x^b + \rho_{(l)} \chi_A^{\ B}$$
$$+ G_{(Frame)} \not{g}^{-1} \begin{pmatrix} O_{(l)} \\ \rho_{(l)} \end{pmatrix} \begin{pmatrix} L\Psi \\ \not{d}\Psi \end{pmatrix} + f(\Psi)\Psi \not{d}x \otimes \not{d}^{\#} x.$$

PROOF. Since $\not{\nabla}_A O^B_{(l)} = (\not{g}^{-1})^{BC} g(\mathscr{D}_A O_{(l)}, X_B)$, the proof is essentially the same as the proof of (7.15). The main difference is that in the present case, we do not have the cancellation of the "δ_{ab} part" noted below equation (7.13) because the quantity under consideration is not symmetrized over the indices A, B. This results in the presence of the "nonsmall" term $\epsilon_{lab} (\not{d}_A x^a) \not{d}^B x^b$ on the right-hand side of (7.24). □

CHAPTER 8

Geometric Operator Commutator Formulas and Schematic Notation for Repeated Differentiation

In this chapter, we provide geometric expressions for the commutators of some operators that we encounter in our analysis. We also define some schematic notation connected to repeated differentiation with respect to commutation vectorfields $Z \in \mathscr{Z}$.

8.1. Definitions of various differential operators

In this section, we define various differential operators that play a role in our analysis.

DEFINITION 8.1 (**Angular divergence**). If Y is an $S_{t,u}$-tangent vectorfield, then we define $\mathrm{div\!\!\!/}\, Y$ to be the following function:

$$\mathrm{div\!\!\!/}\, Y := \nabla\!\!\!\!/_A Y^A. \tag{8.1a}$$

If ξ is an $S_{t,u}$ one-form, then we define $\mathrm{div\!\!\!/}\, \xi$ to be the following function:

$$\mathrm{div\!\!\!/}\, \xi := (g\!\!\!/^{-1})^{AB} \nabla\!\!\!\!/_A \xi_B. \tag{8.1b}$$

If ξ is a type $\binom{1}{1}$ $S_{t,u}$ tensorfield, then we define $\mathrm{div\!\!\!/}\, \xi$ to be the following $S_{t,u}$ one-form:

$$(\mathrm{div\!\!\!/}\, \xi)_A := \nabla\!\!\!\!/_B \xi^B_{\ A}. \tag{8.1c}$$

If ξ is a symmetric type $\binom{0}{2}$ $S_{t,u}$ tensorfield, then we define $\mathrm{div\!\!\!/}\, \xi$ to be the following $S_{t,u}$ one-form:

$$(\mathrm{div\!\!\!/}\, \xi)_A := (g\!\!\!/^{-1})^{BC} \nabla\!\!\!\!/_B \xi_{AC}. \tag{8.1d}$$

If ξ is a symmetric type $\binom{2}{0}$ $S_{t,u}$ tensorfield, then we define $\mathrm{div\!\!\!/}\, \xi$ to be the following $S_{t,u}$-tangent vectorfield:

$$(\mathrm{div\!\!\!/}\, \xi)^A := \nabla\!\!\!\!/_B \xi^{AB}. \tag{8.1e}$$

DEFINITION 8.2 (**Trace-free $S_{t,u}$-projected Lie derivatives**). If ξ is a type $\binom{0}{2}$ $S_{t,u}$ tensorfield and V is a spacetime vectorfield, then we define the trace-free type $\binom{0}{2}$ $S_{t,u}$ tensorfield $\hat{\mathcal{L}\!\!\!/}_V \xi$ as follows:

$$\hat{\mathcal{L}\!\!\!/}_V \xi_{AB} := \mathcal{L}\!\!\!/_V \xi_{AB} - \frac{1}{2}(\mathrm{tr}_{g\!\!\!/} \mathcal{L}\!\!\!/_V \xi) g\!\!\!/_{AB}. \tag{8.2}$$

DEFINITION 8.3 (**Trace-free part of $\nabla\!\!\!/^2$**). If f is a function, then we define the trace-free type $\binom{0}{2}$ $S_{t,u}$ tensorfield $\hat{\nabla\!\!\!/}^2 f$ as follows:

$$\hat{\nabla\!\!\!/}^2_{AB} f := \nabla\!\!\!/^2_{AB} f - \frac{1}{2}(\Delta\!\!\!/ f) g\!\!\!/_{AB}. \tag{8.3}$$

DEFINITION 8.4 (**The operator $\breve{\nabla\!\!\!/}$**). If ξ is a type $\binom{0}{2}$ $S_{t,u}$ tensorfield, then we define the type $\binom{0}{3}$ $S_{t,u}$ tensorfield $\breve{\nabla\!\!\!/}\xi$ as follows:

$$\breve{\nabla\!\!\!/}_A \xi_{BC} := \frac{1}{2}\left\{\nabla\!\!\!/_A \xi_{BC} + \nabla\!\!\!/_B \xi_{AC} - \nabla\!\!\!/_C \xi_{AB}\right\}. \tag{8.4}$$

8.2. Operator commutator identities

In this section, we provide a collection of operator commutator identities that we use in our analysis.

DEFINITION 8.5 (**Commutator of two operators**). If P and Q are two operators, then $[P, Q] := PQ - QP$ denotes their commutator.

LEMMA 8.6 (L, ϱL, $\breve{R}$, O **commute with** $d\!\!\!/$). *If f is a function and $V \in \{L, \varrho L, \breve{R}, O_{(1)}, O_{(2)}, O_{(3)}\}$, then*

$$\mathcal{L}\!\!\!/_V d\!\!\!/ f = d\!\!\!/ V f. \tag{8.5}$$

PROOF. We prove (8.5) only when $V = \breve{R}$ since the other vectorfields V can be treated by using a similar argument and the fact that $d\!\!\!/ \varrho = 0$. To proceed, we set $V = \breve{R}$, contract the right-hand side of (8.5) against the vector X_A, and use the Leibniz rule together with the fact that $[X_A, \breve{R}]$ is $S_{t,u}$-tangent (which follows from (4.21c)) in order to deduce that

$$d\!\!\!/_A \breve{R} f = X_A(\breve{R} f) = \breve{R}(X_A f) + d\!\!\!/ f \cdot [X_A, \breve{R}]. \tag{8.6}$$

On the other hand, we contract the left-hand side of (8.5) against X_A and use the Leibniz rule together with (4.21c) to deduce that

$$(\mathcal{L}\!\!\!/_{\breve{R}} d\!\!\!/ f) \cdot X_A = (\mathcal{L}_{\breve{R}} d\!\!\!/ f) \cdot X_A = \breve{R}(d\!\!\!/ f \cdot X_A) + d\!\!\!/ f \cdot [X_A, \breve{R}]. \tag{8.7}$$

We now note that the right-hand sides of (8.6) and (8.7) are equal, which yields the desired identity (8.5). □

LEMMA 8.7 (**Leibniz rules with an angular differential commutation**). *If X is an $S_{t,u}$-tangent vectorfield, $V \in \{L, \varrho L, \breve{R}, O_{(1)}, O_{(2)}, O_{(3)},\}$, and f is a function, then the following commutator identity holds:*

$$V(X \cdot d\!\!\!/ f) = X \cdot d\!\!\!/ L f + \mathcal{L}\!\!\!/_V X \cdot d\!\!\!/ f. \tag{8.8}$$

PROOF. Lemma 8.7 follows from the Leibniz rule and Lemma 8.6. □

LEMMA 8.8 (**Commutator of the Lie derivative of the trace-free part and the trace-free Lie derivative**). *If ξ is a symmetric type $\binom{0}{2}$ $S_{t,u}$ tensorfield and $Z \in \mathscr{Z}$, then*

$$\mathcal{L}_Z \hat{\xi}_{AB} - \hat{\mathcal{L}\!\!\!/}_Z \xi_{AB} = \frac{1}{2} {}^{(Z)}\pi\!\!\!/^{CD} \hat{\xi}_{CD} g\!\!\!/_{AB} - \frac{1}{2} \mathrm{tr}_{g\!\!\!/} \xi \, {}^{(Z)}\pi\!\!\!/_{AB}. \tag{8.9a}$$

If in addition ξ is trace-free, then

$$\mathcal{L}\!\!\!/_Z \xi_{AB} - \hat{\mathcal{L}\!\!\!/}_Z \xi_{AB} = \frac{1}{2} {}^{(Z)}\pi\!\!\!/^{CD} \xi_{CD} g\!\!\!/_{AB}. \tag{8.9b}$$

PROOF. From the decomposition $\xi = \frac{1}{2}\mathrm{tr}_{\not g}\xi \not g + \hat{\xi}$, the Leibniz rule, and the identities $\not{\mathcal{L}} \not g_{AB} = {}^{(Z)}\not\pi_{AB}$ and $(\not{\mathcal{L}}_Z \not g^{-1})^{AB} = -{}^{(Z)}\not\pi^{AB}$ (see Lemma 7.5), we compute that

$$Z\mathrm{tr}_{\not g}\xi = -{}^{(Z)}\not\pi^{AB}\xi_{AB} + \mathrm{tr}_{\not g}\not{\mathcal{L}}_Z\xi, \tag{8.10}$$

$$\begin{aligned} \not{\mathcal{L}}_Z\hat{\xi}_{AB} &= \not{\mathcal{L}}_Z\xi_{AB} - \frac{1}{2}\mathrm{tr}_{\not g}\not{\mathcal{L}}_Z\xi \not g_{AB} + \frac{1}{2}{}^{(Z)}\not\pi^{CD}\xi_{CD}\not g_{AB} - \frac{1}{2}\mathrm{tr}_{\not g}\xi^{(Z)}\not\pi_{AB} \\ &= \hat{\not{\mathcal{L}}}_Z\xi_{AB} + \frac{1}{2}{}^{(Z)}\not\pi^{CD}\xi_{CD}\not g_{AB} - \frac{1}{2}\mathrm{tr}_{\not g}\xi^{(Z)}\not\pi_{AB} \\ &= \hat{\not{\mathcal{L}}}_Z\xi_{AB} + \frac{1}{2}{}^{(Z)}\not\pi^{CD}\hat{\xi}_{CD}\not g_{AB} - \frac{1}{2}\mathrm{tr}_{\not g}\xi^{(Z)}\hat{\not\pi}_{AB}, \end{aligned} \tag{8.11}$$

where to deduce the second equality in (8.11) we used definition (8.2), and to deduce the last equality in (8.11) we used the decompositions ${}^{(Z)}\not\pi_{AB} = \frac{1}{2}\mathrm{tr}_{\not g}{}^{(Z)}\not\pi \not g_{AB} + {}^{(Z)}\hat{\not\pi}_{AB}$ and ${}^{(Z)}\not\pi^{CD} = \frac{1}{2}\mathrm{tr}_{\not g}{}^{(Z)}\not\pi(\not g^{-1})^{CD} + {}^{(Z)}\hat{\not\pi}^{CD}$. We have thus proved (8.9a). The identity (8.9b) follows trivially from (8.9a). □

LEMMA 8.9 (**Commuting ϱL, $\breve{R}$, O with $\not\nabla$**). *If ξ is an $S_{t,u}$ one-form and $V \in \{L, \varrho L, \breve{R}, O_{(1)}, O_{(2)}, O_{(3)}\}$, then we have the following commutator identity, where the operator $\check{\not\nabla}$ is defined in Def. 8.4:*

$$([\not\nabla_A, \not{\mathcal{L}}_V]\xi)_B = (\check{\not\nabla}_A {}^{(V)}\not\pi_B^{\ C})\xi_C. \tag{8.12}$$

Similarly, if ξ is a symmetric type $\binom{0}{2}$ $S_{t,u}$ tensorfield and $V \in \{L, \varrho L, \breve{R}, O_{(1)}, O_{(2)}, O_{(3)}\}$, then we have the following commutator identities:

$$([\not\nabla_A, \not{\mathcal{L}}_V]\xi)_{BC} = (\check{\not\nabla}_A {}^{(V)}\not\pi_B^{\ D})\xi_{CD} + (\check{\not\nabla}_A {}^{(V)}\not\pi_C^{\ D})\xi_{BD}, \tag{8.13}$$

$$([\not\nabla_A, \not{\mathcal{L}}_V]\not\nabla\xi)_{BCD} = (\check{\not\nabla}_A {}^{(V)}\not\pi_B^{\ E})\not\nabla_E\xi_{CD} + (\check{\not\nabla}_A {}^{(V)}\not\pi_C^{\ E})\not\nabla_B\xi_{ED} + (\check{\not\nabla}_A {}^{(V)}\not\pi_D^{\ E})\not\nabla_B\xi_{CE}. \tag{8.14}$$

PROOF. We first prove (8.13) in the case $V = L$. We compute relative to the geometric coordinates, and we use the fact that the vectorfield $L = \frac{\partial}{\partial t}$ commutes with the coordinate vectorfields $(\frac{\partial}{\partial \vartheta^1}, \frac{\partial}{\partial \vartheta^2}) = (X_1, X_2)$. Also using Lemma 7.5, we deduce the following identities: $\frac{\partial}{\partial t}\not g_{AB} = \not{\mathcal{L}}_L \not g_{AB} = {}^{(L)}\not\pi_{AB}$, $\frac{\partial}{\partial t}(\not g^{-1})^{AB} = (\not{\mathcal{L}}_L \not g^{-1})^{AB} = -{}^{(L)}\not\pi^{AB}$. To proceed, we note that the Christoffel symbols $\not\Gamma_{A\ B}^{\ C}$ of $\not g$ (relative to the local coordinates $(\vartheta^1, \vartheta^2)$) can be expressed as

$$\not\Gamma_{A\ B}^{\ C} = \frac{1}{2}(\not g^{-1})^{CD}\left\{\frac{\partial}{\partial\vartheta^A}\not g_{DB} + \frac{\partial}{\partial\vartheta^B}\not g_{AD} - \frac{\partial}{\partial\vartheta^D}\not g_{AB}\right\}. \tag{8.15}$$

Next, we state the following standard covariant derivative identities:

$$\not\nabla_A\xi_{BC} = \frac{\partial}{\partial\vartheta^A}\xi_{BC} - \not\Gamma_{A\ B}^{\ D}\xi_{DC} - \not\Gamma_{A\ C}^{\ D}\xi_{BD}, \tag{8.16}$$

$$\not\nabla_A\not{\mathcal{L}}_L\xi_{BC} = \not\nabla_A\frac{\partial}{\partial t}\xi_{BC} = \frac{\partial}{\partial t}\frac{\partial}{\partial\vartheta^A}\xi_{BC} - \not\Gamma_{A\ B}^{\ D}\frac{\partial}{\partial t}\xi_{DC} - \not\Gamma_{A\ C}^{\ D}\frac{\partial}{\partial t}\xi_{BD}. \tag{8.17}$$

We now set $\not{\Gamma}_{ADB} := \not{g}_{CD}\not{\Gamma}_{A\ B}^{\ C} = \left\{\frac{\partial}{\partial\vartheta^A}\not{g}_{DB} + \frac{\partial}{\partial\vartheta^B}\not{g}_{AD} - \frac{\partial}{\partial\vartheta^D}\not{g}_{AB}\right\}$, refer to definition (8.4), and use the above calculations and the symmetry property $\not{\Gamma}_{ADB} = \not{\Gamma}_{BDA}$ to compute that

(8.18)

$$\begin{aligned}\frac{\partial}{\partial t}\not{\Gamma}_{A\ B}^{\ C} &= -{}^{(L)}\not{\pi}^{CD}\not{\Gamma}_{ADB} + \frac{1}{2}(\not{g}^{-1})^{CD}\left\{\frac{\partial}{\partial\vartheta^A}{}^{(L)}\not{\pi}_{DB} + \frac{\partial}{\partial\vartheta^B}{}^{(L)}\not{\pi}_{AD} - \frac{\partial}{\partial\vartheta^D}{}^{(L)}\not{\pi}_{AB}\right\} \\ &= (\not{g}^{-1})^{CD}\check{\not{\nabla}}_A{}^{(L)}\not{\pi}_{BD}.\end{aligned}$$

Using (8.16) and (8.18), we compute that

(8.19)

$$\begin{aligned}\not{\mathcal{L}}_L\not{\nabla}_A\xi_{BC} = \frac{\partial}{\partial t}\not{\nabla}_A\xi_{BC} &= \frac{\partial}{\partial\vartheta^A}\frac{\partial}{\partial t}\xi_{BC} - \not{\Gamma}_{A\ B}^{\ D}\frac{\partial}{\partial t}\xi_{DC} - \not{\Gamma}_{A\ C}^{\ D}\frac{\partial}{\partial t}\xi_{BD} \\ &\quad - (\not{g}^{-1})^{DE}(\check{\not{\nabla}}_A{}^{(L)}\not{\pi}_{BE})\xi_{DC} - (\not{g}^{-1})^{DE}(\check{\not{\nabla}}_A{}^{(L)}\not{\pi}_{CE})\xi_{BD}.\end{aligned}$$

Finally, subtracting (8.19) from (8.17) and noting that the right-hand side of (8.17) cancels the first three terms on the right-hand side of (8.19), we arrive at the desired identity (8.13) in the case $V = L$.

Since ${}^{(\varrho L)}\not{\pi} = \varrho^{(L)}\not{\pi}$, the identity (8.13) for ϱL follows directly from the identity for L.

A similar argument involving one minor adjustment yields (8.13) when $V = \breve{R}$. The minor adjustment is that we first endow $\Sigma_t^{U^0}$ with local coordinates $(u, \widetilde{\vartheta}^1, \widetilde{\vartheta}^2)$ (where u is the eikonal function) in such a way that $\breve{R} = \frac{\partial}{\partial u}|_{\widetilde{\vartheta}^1,\widetilde{\vartheta}^2}$. We already carried out such a construction in our proof of Cor. 3.47. We can then proceed as in the proof of (8.13) in the case $V = L$, but using the coordinates $\widetilde{\vartheta}^A$ in place of ϑ^A.

To prove (8.13) in the case $V = O$, we let $\varphi_{(\lambda)} : \mathcal{M}_{t,u} \to \mathcal{M}_{t,u}$ be the flow map (see Remark 3.56) of the $S_{t,u}$-tangent vectorfield O. Let $\varphi_{(\lambda)}^*(\not{\nabla}\xi)$, $\varphi_{(\lambda)}^*\not{g}$, and $\varphi_{(\lambda)}^*\xi$ respectively denote the pullbacks (see equation (3.65)) of the tensorfields $\not{\nabla}\xi$, $\not{g}$ and ξ by $\varphi_{(\lambda)}$, and let $\varphi_{(\lambda)}^*\not{\nabla}$ denote the Levi-Civita connection of $\varphi_{(\lambda)}^*\not{g}$. Then by covariance, we have the identity $\varphi_{(\lambda)}^*(\not{\nabla}\xi) = (\varphi_{(\lambda)}^*\not{\nabla})\varphi_{(\lambda)}^*\xi$.

To simplify the notation, we denote ${}^{(\lambda)}\not{\nabla} := \varphi_{(\lambda)}^*\not{\nabla}$, ${}^{(\lambda)}\not{g} := \varphi_{(\lambda)}^*\not{g}$, and ${}^{(\lambda)}\xi := \varphi_{(\lambda)}^*\xi$. We also let ${}^{(\lambda)}\not{\Gamma}_{A\ B}^{\ C}$ denote the Christoffel symbols of ${}^{(\lambda)}\not{g}$ relative to the local coordinates $(\vartheta^1, \vartheta^2)$. In particular, the above covariance identity can be written as

$$\varphi_{(\lambda)}^*(\not{\nabla}\xi) = {}^{(\lambda)}\not{\nabla}{}^{(\lambda)}\xi. \tag{8.20}$$

Using the fact that the Lie derivative of a tensorfield with respect to a vectorfield is the derivative (with respect to the flow parameter λ) of the pullback of the tensorfield by the flow map of the vectorfield, the fact that O and $\not{g} = \not{\Pi}g$ (see Remark 3.43) are $S_{t,u}$ tensorfields, and Cor. 4.12, we deduce the tensorial identities $\frac{d}{d\lambda}|_{\lambda=0}{}^{(\lambda)}\not{g}_{AB} = \not{\mathcal{L}}_O\not{g}_{AB} = {}^{(O)}\not{\pi}_{AB}$ and $\frac{d}{d\lambda}|_{\lambda=0}({}^{(\lambda)}\not{g}^{-1})^{AB} = -{}^{(O)}\not{\pi}^{AB}$, where ${}^{(\lambda)}\not{g}^{-1}$ denotes the inverse of ${}^{(\lambda)}\not{g}$. From these identities, we deduce that the identity (8.18) holds with the left-hand side replaced by $\frac{d}{d\lambda}|_{\lambda=0}{}^{(\lambda)}\not{\Gamma}_{A\ B}^{\ C}$ and all terms ${}^{(L)}\not{\pi}$ on the right-hand side replaced by ${}^{(O)}\not\pi$. That is, we have

$$\frac{d}{d\lambda}|_{\lambda=0}{}^{(\lambda)}\not\Gamma_{A\ B}^{\ C} = (\not g^{-1})^{CD}\breve{\not\nabla}_A {}^{(O)}\not\pi_{BD}. \tag{8.21}$$

We now consider the ABC component of the right-hand side of (8.20), that is ${}^{(\lambda)}\not\nabla_A {}^{(\lambda)}\xi_{BC}$. This component is equal to the right-hand side of (8.16) except with ${}^{(\lambda)}\not\nabla$, ${}^{(\lambda)}\xi$, and ${}^{(\lambda)}\not\Gamma$ respectively in place of $\not\nabla$, ξ, and $\not\Gamma$. Hence, applying the operator $\frac{d}{d\lambda}|_{\lambda=0}$ to the right-hand side of (8.20) and using the identity (8.21), we deduce that

$$\left(\frac{d}{d\lambda}|_{\lambda=0}\left\{{}^{(\lambda)}\not\nabla {}^{(\lambda)}\xi\right\}\right)_{ABC} = (\not\nabla \not{\mathcal{L}}_O\xi)_{ABC} + (\breve{\not\nabla}_A {}^{(O)}\not\pi_B^{\ D})\xi_{DC} + (\breve{\not\nabla}_A {}^{(O)}\not\pi_C^{\ D})\xi_{BD}. \tag{8.22}$$

On the other hand, applying the operator $\frac{d}{d\lambda}|_{\lambda=0}$ to the left-hand side of (8.20) and considering the ABC component, we deduce (from the representation of Lie differentiation in terms of the derivative of the pullback by the flow map) that

$$\left(\frac{d}{d\lambda}|_{\lambda=0}\left\{\varphi_{(\lambda)}^*(\not\nabla\xi)\right\}\right)_{ABC} = (\not{\mathcal{L}}_O\not\nabla\xi)_{ABC}. \tag{8.23}$$

Subtracting (8.23) from (8.22), we conclude (8.13) in the case $V = O$. We have thus proved (8.13) in all cases.

The identities (8.12) and (8.14) can be proved in an analogous fashion, and we omit the details. □

Lemma 8.10 (**$[\{\not{\mathcal{L}}_L + \mathrm{tr}_{\not g}\chi\}, \not{\mathrm{div}}] = 0$ for $S_{t,u}$-tangent vectorfields**). *Let Y be an $S_{t,u}$-tangent vectorfield and let χ be the $S_{t,u}$ tensorfield defined in (3.70). Then we have the following identity:*

$$\{L + \mathrm{tr}_{\not g}\chi\}\not{\mathrm{div}} Y = \not{\mathrm{div}}\left\{\not{\mathcal{L}}_L Y + \mathrm{tr}_{\not g}\chi Y\right\}. \tag{8.24}$$

Proof. Let ξ be the $S_{t,u}$ one-form that is $\not g$-dual to Y. Using the Leibniz rule and the identity $(\not{\mathcal{L}}_L\not g^{-1})^{AB} = -{}^{(L)}\not\pi^{AB}$ (see (7.4b)) and commuting the operators $\not{\mathcal{L}}_L$ and $\not\nabla$, we deduce that $L\not{\mathrm{div}} Y = L\left\{(\not g^{-1})^{AB}\not\nabla_A\xi_B)\right\} = -{}^{(L)}\not\pi^{AB}\not\nabla_A Y_B + (\not g^{-1})^{AB}\not\nabla_A\not{\mathcal{L}}_L\xi_B + (\not g^{-1})^{AB}[\not{\mathcal{L}}_L, \not\nabla]_A Y_B$. Using the identity ${}^{(L)}\not\pi = 2\chi$ (see (7.4a) and (3.72)) and (8.12) with L in the role of V, we deduce that $(\not g^{-1})^{AB}[\not{\mathcal{L}}_L, \not\nabla]_A\xi_B = Y\mathrm{tr}_{\not g}\chi - 2(\not{\mathrm{div}}\chi)\cdot Y$. Furthermore, since $\xi_B = \not g_{BC}Y^C$, we deduce from the Leibniz rule and (7.4a) that $\not{\mathcal{L}}_L\xi_B = 2\chi_{BC}Y^C + \not g_{BC}\not{\mathcal{L}}_L Y^C$. Hence, we deduce that $(\not g^{-1})^{AB}\not\nabla_A\not{\mathcal{L}}_L\xi_B = 2(\not{\mathrm{div}}\chi)\cdot Y + 2\chi^{AB}\not\nabla_A Y_B + \not{\mathrm{div}}\not{\mathcal{L}}_L Y$. Combining these identities, we deduce that

$$L\not{\mathrm{div}} Y = \not{\mathrm{div}}\not{\mathcal{L}}_L Y + Y\mathrm{tr}_{\not g}\chi. \tag{8.25}$$

The desired identity (8.24) now follows from (8.25) and the Leibniz rule identity $\not{\mathrm{div}}(\mathrm{tr}_{\not g}\chi Y) = \mathrm{tr}_{\not g}\chi\not{\mathrm{div}} Y + Y\mathrm{tr}_{\not g}\chi$. □

Lemma 8.11 (**Commuting ϱL, $\breve{R}$, O with $\not\nabla^2$**). *If f is a scalar-valued function and $V \in \{L, \varrho L, \breve{R}, O_{(1)}, O_{(2)}, O_{(3)}\}$, then we have the following commutator identities:*

$$([\not\nabla^2, \not{\mathcal{L}}_V]f)_{AB} = (\breve{\not\nabla}_A {}^{(V)}\not\pi_B^{\ C})\not d_C f, \tag{8.26a}$$

$$[\not\Delta, \not{\mathcal{L}}_V]f = {}^{(V)}\not\pi^{AB}\not\nabla_{AB}^2 f + (\breve{\not\nabla}_A {}^{(V)}\not\pi^{AB})\not d_B f. \tag{8.26b}$$

PROOF. The identity (8.26a) follows from (8.12) with $\nabla\!\!\!/ f$ in the role of ξ and Lemma 8.6.

To deduce (8.26b), we take the trace over the AB indices in (8.26a) and use the operator commutator identity $(g\!\!\!/^{-1})^{AB}\mathcal{L}\!\!\!/_V \nabla\!\!\!/^2_{AB} f = V\Delta\!\!\!/ f + {}^{(V)}\pi\!\!\!/^{AB}\nabla\!\!\!/^2_{AB} f$, which is a simple consequence of (7.4b) and the Leibniz rule. □

8.3. Notation for repeated differentiation

In this section, we introduce some notation that allows for a compact presentation of expressions involving repeated differentiation.

DEFINITION 8.12 (**Notation for repeated differentiation**). We fix a labeling $\mathscr{Z} = \{Z_{(i)}\}_{i=1}^5$ of the five commutation vectorfields (see Def. 7.1). We use the following notation.

- If $\vec{I} = (\iota_1, \iota_2, \cdots, \iota_N)$ is a multi-index of order $|\vec{I}| := N$ with $\iota_1, \iota_2, \cdots, \iota_N \in \{1, 2, 3, 4, 5\}$, then $\mathscr{Z}^{\vec{I}} := Z_{(\iota_1)} Z_{(\iota_2)} \cdots Z_{(\iota_N)}$ denotes the corresponding N^{th} order differential operator.
- When we are not concerned with the precise structure of the multi-index $\vec{I}$, we abbreviate $\mathscr{Z}^N := Z_{(\iota_1)} Z_{(\iota_2)} \cdots Z_{(\iota_N)}$.
- Similarly, $\mathcal{L}\!\!\!/^N_{\mathscr{Z}} := \mathcal{L}\!\!\!/_{Z_{(\iota_1)}} \mathcal{L}\!\!\!/_{(Z_{\iota_2})} \cdots \mathcal{L}\!\!\!/_{Z_{(\iota_N)}}$ denotes an Nth order $S_{t,u}$-projected Lie derivative operator (see Def. 3.57).
- Similarly, $\hat{\mathcal{L}\!\!\!/}^N_{\mathscr{Z}} := \mathcal{L}\!\!\!/_{Z_{(\iota_1)}} \mathcal{L}\!\!\!/_{(Z_{\iota_2})} \cdots \mathcal{L}\!\!\!/_{Z_{(\iota_N)}}$ denotes an Nth order trace-free $S_{t,u}$-projected Lie derivative operator (see Def. 8.2).
- We use similar notation with $\mathscr{S}$, $\mathscr{O}$, or $\mathscr{O}_{(Flat)}$ in place of $\mathscr{Z}$ when the derivatives are all, respectively, spatial derivatives, geometric rotation derivatives, or Euclidean rotation derivatives (see Defs. 7.2 and 7.4).

DEFINITION 8.13 (**Schematic notation for a pointwise norm of a quantity and its derivatives**). We use the following schematic notation:

- If f is a function, then $|\mathscr{Z}^{\leq N} f|$ denotes any term that is $\leq \sum_{|\vec{I}| \leq N} c_{\vec{I}} \left| \mathscr{Z}^{\vec{I}} f \right|$, where the $c_{\vec{I}}$ are nonnegative constants that verify $\sum_{|\vec{I}| \leq N} c_{\vec{I}} \leq 1$. Similarly, $|\mathscr{Z}^N f|$ denotes any term that is $\leq \sum_{|\vec{I}| = N} c_{\vec{I}} \left| \mathscr{Z}^{\vec{I}} f \right|$, where the $c_{\vec{I}}$ are nonnegative constants that verify $\sum_{|\vec{I}| = N} c_{\vec{I}} \leq 1$. Similarly, if V is a vectorfield, then $|V \mathscr{Z}^{\leq N} f|$ denotes any term that is $\leq \sum_{|\vec{I}| \leq N} c_{\vec{I}} \left| V \mathscr{Z}^{\vec{I}} f \right|$, where the $c_{\vec{I}}$ are nonnegative constants that verify $\sum_{|\vec{I}| \leq N} c_{\vec{I}} \leq 1$. Similarly, $|V \mathscr{Z}^N f|$ denotes any term that is $\leq \sum_{|\vec{I}| = N} c_{\vec{I}} \left| V \mathscr{Z}^{\vec{I}} f \right|$, where the $c_{\vec{I}}$ are nonnegative constants that verify $\sum_{|\vec{I}| = N} c_{\vec{I}} \leq 1$. Similarly, $|d\!\!\!/ \mathscr{Z}^{\leq N} f|$ denotes any term that is $\leq \sum_{|\vec{I}| \leq N} c_{\vec{I}} \left| d\!\!\!/ \mathscr{Z}^{\vec{I}} f \right|$, where the $c_{\vec{I}}$ are nonnegative constants that verify $\sum_{|\vec{I}| \leq N} c_{\vec{I}} \leq 1$. We use similar notation for other expressions.
- If ξ is an $S_{t,u}$ tensor, then $\left| \mathcal{L}\!\!\!/^{\leq N}_{\mathscr{Z}} \xi \right|$ denotes any term that is less than or equal to $\sum_{|\vec{I}| \leq N} c_{\vec{I}} \left| \mathcal{L}\!\!\!/^{\vec{I}}_{\mathscr{Z}} \xi \right|$, where the $c_{\vec{I}}$ are nonnegative constants that verify $\sum_{|\vec{I}| \leq N} c_{\vec{I}} \leq 1$. Similarly, $\left| \mathcal{L}\!\!\!/^N_{\mathscr{Z}} \xi \right|$ denotes any term that is

$\leq \sum_{|\vec{I}|=N} c_{\vec{I}} \left| \not{\mathcal{L}}_{\mathscr{Z}}^{\vec{I}} \xi \right|$, where the $c_{\vec{I}}$ are nonnegative constants that verify $\sum_{|\vec{I}|=N} c_{\vec{I}} \leq 1$. Similarly, if V is a vectorfield, then $\left| \not{\mathcal{L}}_V \not{\mathcal{L}}_{\mathscr{Z}}^{\leq N} \xi \right|$ denotes any term that is $\leq \sum_{|\vec{I}|\leq N} c_{\vec{I}} \left| \not{\mathcal{L}}_V \not{\mathcal{L}}_{\mathscr{Z}}^{\vec{I}} \xi \right|$, where the $c_{\vec{I}}$ are nonnegative constants that verify $\sum_{|\vec{I}|\leq N} c_{\vec{I}} \leq 1$. Similarly, $\left| \not{\mathcal{L}}_V \not{\mathcal{L}}_{\mathscr{Z}}^{N} \xi \right|$ denotes any term that is $\leq \sum_{|\vec{I}|=N} c_{\vec{I}} \left| \not{\mathcal{L}}_V \not{\mathcal{L}}_{\mathscr{Z}}^{\vec{I}} \xi \right|$, where the $c_{\vec{I}}$ are nonnegative constants that verify $\sum_{|\vec{I}=N} c_{\vec{I}} \leq 1$. We use similar notation for other expressions, including the case in which trace-free $S_{t,u}$-projected (see Def. 8.2) Lie derivatives $\hat{\not{\mathcal{L}}}$ are present instead of ordinary $S_{t,u}$-projected Lie derivatives $\not{\mathcal{L}}$.

- We use similar notation with $\mathscr{S}$, $\mathscr{O}$, or $\mathscr{O}_{(Flat)}$ in place of $\mathscr{Z}$ when the derivatives are all, respectively, spatial derivatives, geometric rotation derivatives, or Euclidean rotation derivatives (see Defs. 7.2 and 7.4).

CHAPTER 9

The Structure of the Wave Equation Inhomogeneous Terms After One Commutation

In this chapter, we commute the wave equation $\upmu\square_{g(\Psi)}\Psi = 0$ with a commutation vectorfield $Z \in \mathscr{Z}$ (see Def. 7.1) and decompose the corresponding inhomogeneous terms relative to the frame $\{L, \breve{R}, X_1, X_2\}$. In order to prove our sharp classical lifespan theorem, we need to know the precise structure, including numerical constants, of some of the terms. Furthermore, the special properties of Z lead to some exact cancellations that are essential for the proof of the theorem. We provide the main result in Proposition 9.6.

9.1. Preliminary calculations

We start with the following lemma, which provides a preliminary decomposition of the commutator term $[Z, \upmu\square_{g(\Psi)}]\Psi$.

LEMMA 9.1 (**Preliminary vectorfield-covariant wave operator commutator expression**). *Let Z be any spacetime vectorfield and let ${}^{(Z)}\pi_{\mu\nu}$ be its deformation tensor (see Def. 4.9). Assume that ${}^{(Z)}\pi_{LL} := {}^{(Z)}\pi_{\alpha\beta}L^{\alpha}L^{\beta} = 0$. Then the following commutation identity holds:*

(9.1)

$$\begin{aligned}\upmu\square_{g(\Psi)}(Z\Psi) &= \upmu\mathscr{D}_{\alpha}\left({}^{(Z)}\pi^{\alpha\beta}\mathscr{D}_{\beta}\Psi - \frac{1}{2}\mathrm{tr}_g\,{}^{(Z)}\pi\,\mathscr{D}^{\alpha}\Psi\right)\\ &\quad + Z(\upmu\square_{g(\Psi)}\Psi) \underbrace{-\upmu^{-1}\left\{{}^{(Z)}\pi_{L\breve{R}} + Z\upmu\right\}(\upmu\square_{g(\Psi)}\Psi)}_{\textit{potentially dangerous factor}} + \frac{1}{2}\mathrm{tr}_{\not g}\,{}^{(Z)}\not\pi(\upmu\square_{g(\Psi)}\Psi).\end{aligned}$$

PROOF. Using the Leibniz rule for $\mathscr{D}$, the symmetry property $\mathscr{D}_{\alpha}\mathscr{D}_{\beta}\Psi = \mathscr{D}_{\beta}\mathscr{D}_{\alpha}\Psi$, the fact that $\mathscr{D}g = 0$, and the decomposition $\mathscr{D}_{\alpha}Z_{\beta} = \frac{1}{2}(\mathscr{D}_{\alpha}Z_{\beta} - \mathscr{D}_{\beta}Z_{\alpha}) + \frac{1}{2}{}^{(Z)}\pi_{\alpha\beta}$, we compute that

$$\begin{aligned}\upmu\square_{g(\Psi)}(Z\Psi) &= Z(\upmu\square_{g(\Psi)}\Psi) - (Z\upmu)\square_{g(\Psi)}\Psi\\ &\quad - \upmu Z^{\alpha}\mathscr{D}^{\beta}\mathscr{D}_{\beta}\mathscr{D}_{\alpha}\Psi - \upmu Z^{\alpha}\mathscr{D}_{\alpha}\mathscr{D}^{\beta}\mathscr{D}_{\beta}\Psi + \upmu\,{}^{(Z)}\pi^{\alpha\beta}\mathscr{D}_{\alpha}\mathscr{D}_{\beta}\Psi\\ &\quad - \frac{1}{2}\upmu(\mathscr{D}_{\alpha}{}^{(Z)}\pi^{\alpha\beta})\mathscr{D}_{\beta}\Psi + \frac{1}{2}\upmu\left\{\mathscr{D}^{\alpha}(\mathscr{D}_{\alpha}Z^{\beta} - \mathscr{D}^{\beta}Z_{\alpha})\right\}\mathscr{D}_{\beta}\Psi.\end{aligned} \tag{9.2}$$

Furthermore, with $\mathscr{R}_{\alpha\beta} := (g^{-1})^{\kappa\lambda}\mathscr{R}_{\alpha\kappa\beta\lambda}$ denoting the Ricci curvature tensor[1] of g, we compute that the following commutator identities hold:

$$Z^{\alpha}\mathscr{D}^{\beta}\mathscr{D}_{\beta}\mathscr{D}_{\alpha}\Psi - Z^{\alpha}\mathscr{D}_{\alpha}\mathscr{D}^{\beta}\mathscr{D}_{\beta}\Psi = \mathscr{R}_{\alpha}^{\ \beta}Z^{\alpha}\mathscr{D}_{\beta}\Psi, \tag{9.3}$$

$$\frac{1}{2}\left\{\mathscr{D}^{\alpha}\underbrace{(\mathscr{D}_{\alpha}Z^{\beta} - \mathscr{D}^{\beta}Z_{\alpha})}_{{}^{(Z)}\pi_{\alpha}^{\ \beta} - 2\mathscr{D}^{\beta}Z_{\alpha}}\right\}\mathscr{D}_{\beta}\Psi = \frac{1}{2}(\mathscr{D}_{\alpha}{}^{(Z)}\pi^{\alpha\beta})\mathscr{D}_{\beta}\Psi - \frac{1}{2}(\mathscr{D}^{\alpha}\mathrm{tr}_g{}^{(Z)}\pi)\mathscr{D}_{\alpha}\Psi - \mathscr{R}_{\alpha}^{\ \beta}Z^{\alpha}\mathscr{D}_{\beta}\Psi. \tag{9.4}$$

The desired identity (9.1) now follows from (9.2), (9.3), (9.4), Lemma 3.48, the decomposition $\frac{1}{2}\mathrm{tr}_g{}^{(Z)}\pi = -\mu^{-1(Z)}\pi_{L\breve{R}} + \frac{1}{2}\mathrm{tr}_{g\!\!/}{}^{(Z)}\pi\!\!\!/$ (which is based in part on the assumption ${}^{(Z)}\pi_{LL} = 0$), and straightforward computations. □

REMARK 9.2 (**Some good properties of the commutation vectorfields**). For commutation vectorfields $Z \in \mathscr{Z}$ (see Def. 7.1), we have ${}^{(Z)}\pi_{LL} = 0$. Furthermore, for $Z \in \mathscr{Z}$, the "potentially dangerous factor" on the right-hand side of (9.1) contains no powers of μ^{-1} and therefore, despite superficial appearances, is not actually dangerous. Specifically, Prop. 7.7 implies that for each $Z \in \mathscr{Z}$, $\mu^{-1}\left\{{}^{(Z)}\pi_{L\breve{R}} + Z\mu\right\}$ is equal to either 0 or -1. These properties of $\mathscr{Z}$ are essential for closing our L^2 estimates.

We now define a family of commutation current vectorfields that will facilitate our analysis of the structure of the inhomogeneous terms on the right-hand side of (9.1).

DEFINITION 9.3 (**Commutation currents**). Let Z be any vectorfield, and let ${}^{(Z)}\pi_{\mu\nu}$ be its deformation tensor (see Def. 4.9). We define the corresponding commutation current (vectorfield) ${}^{(Z)}\mathscr{J}^{\alpha}[\Psi]$ as follows:

$${}^{(Z)}\mathscr{J}^{\alpha}[\Psi] := {}^{(Z)}\pi^{\alpha\beta}\mathscr{D}_{\beta}\Psi - \frac{1}{2}\mathrm{tr}_g{}^{(Z)}\pi\mathscr{D}^{\alpha}\Psi. \tag{9.5}$$

We now use commutation currents to provide an alternate description of the inhomogeneous terms on the right-hand side of (9.1). The advantage of the alternate description is that it is easy to decompose the corresponding expression relative to the frame $\{L, \breve{R}, X_1, X_2\}$.

LEMMA 9.4 (**Basic structure of the inhomogeneous terms in the once-commuted wave equation**). *Given any $Z \in \mathscr{Z}$ (note that ${}^{(Z)}\pi_{LL} = 0$ by Prop. 7.7), let ${}^{(Z)}\mathscr{J}[\Psi]$ be the commutation current* (9.5). *Then $Z\Psi$ verifies the inhomogeneous wave equation*

$$\mu\square_{g(\Psi)}(Z\Psi) = \mu\mathscr{D}_{\alpha}{}^{(Z)}\mathscr{J}^{\alpha}[\Psi] + Z(\mu\square_{g(\Psi)}\Psi) \underbrace{-\mu^{-1}\left\{{}^{(Z)}\pi_{L\breve{R}} + Z\mu\right\}(\mu\square_{g(\Psi)}\Psi)}_{\textit{potentially dangerous factor}} + \frac{1}{2}\mathrm{tr}_{g\!\!/}{}^{(Z)}\pi\!\!\!/(\mu\square_{g(\Psi)}\Psi). \tag{9.6}$$

[1] See Def. 11.1 for our sign conventions for the Riemann curvature tensor $\mathscr{R}_{\alpha\kappa\beta\lambda}$.

PROOF. The lemma follows easily from applying $\mathscr{D}_\alpha$ to the right-hand side of (9.5) and comparing the resulting expression to the right-hand side of (9.1). □

Lemma 9.4 shows that in order to estimate the inhomogeneous terms in the wave equation verified by $Z\Psi$, we have to analyze the structure of the commutation currents $\mu\mathscr{D}_\alpha{}^{(Z)}\mathscr{J}^\alpha[\Psi]$. Moreover, to obtain estimates for the higher-order derivatives of Ψ, we must analyze the derivatives of $\mu\mathscr{D}_\alpha{}^{(Z)}\mathscr{J}^\alpha[\Psi]$. As a first step, in Prop. 9.6, we decompose $\mu\mathscr{D}_\alpha^{(Z)}\mathscr{J}^\alpha[\Psi]$ relative to the frame $\{L, \breve{R}, X_1, X_2\}$. We use the following preliminary lemma in our proof of the proposition.

LEMMA 9.5 (**The divergence of $\mathscr{J}$ in terms of rescaled frame derivatives**). *Let $\mathscr{J}$ be any spacetime vectorfield and consider its decomposition relative to the rescaled frame $\{L, \breve{R}, X_1, X_2\}$ (see Remark 3.38):*

$$\mathscr{J} = -\mathscr{J}_L L - \mu^{-1}\mathscr{J}_{\breve{R}} L - \mu^{-1}\mathscr{J}_L \breve{R} + \not{\mathscr{J}}, \tag{9.7}$$

where $\not{\mathscr{J}}$ is the projection of $\mathscr{J}$ onto $S_{t,u}$. Then the following μ-weighted spacetime covariant divergence identity holds:

$$\begin{aligned}\mu\mathscr{D}_\alpha\mathscr{J}^\alpha = & -L(\mu\mathscr{J}_L) - L(\mathscr{J}_{\breve{R}}) - \breve{R}(\mathscr{J}_L) + \operatorname{d\not{i}v}(\mu\not{\mathscr{J}}) \\ & - \left\{\operatorname{tr}_{\not{g}}\not{k}^{(Trans-\Psi)} + \mu\operatorname{tr}_{\not{g}}\not{k}^{(Tan-\Psi)}\right\}\mathscr{J}_L - \operatorname{tr}_{\not{g}}\chi\mathscr{J}_{\breve{R}}.\end{aligned} \tag{9.8}$$

The $S_{t,u}$ tensorfields χ, $\not{k}^{(Trans-\Psi)}$, and $\not{k}^{(Tan-\Psi)}$ appearing in (9.8) are defined in (3.70), (5.4b), and (5.5b).

PROOF. We fix a hypersurface region $\Sigma_t^{U_0}$. We will show that (9.8) holds along $\Sigma_t^{U_0}$. Since the identity (9.8) is invariant with respect to changes of angular coordinates of the form $\widetilde{\vartheta}^1 = f^1(u, \vartheta^1, \vartheta^2)$, $\widetilde{\vartheta}^2 = f^2(u, \vartheta^1, \vartheta^2)$ (where the f^A are the coordinate transformation functions), it suffices to show that (3.50a) holds for a well-chosen version $(\widetilde{\vartheta}^1, \widetilde{\vartheta}^2)$ of such angular coordinates. To this end, we use the angular coordinates constructed in the proof of Cor. 3.47, which are such that along $\Sigma_t^{U_0}$, we have $\breve{R} = \frac{\partial}{\partial u}|_{t,\widetilde{\vartheta}^1,\widetilde{\vartheta}^2}$. By transporting $\widetilde{\vartheta}^1, \widetilde{\vartheta}^2$ via the ODEs $L\widetilde{\vartheta}^1 = L\widetilde{\vartheta}^2 = 0$, we may assume that $L = \frac{\partial}{\partial t}|_{u,\widetilde{\vartheta}^1,\widetilde{\vartheta}^2}$ along $\Sigma_t^{U_0}$. We now expand $\mathscr{J}$ relative to the coordinate frame $\left\{\frac{\partial}{\partial t}, \frac{\partial}{\partial u}, \frac{\partial}{\partial\widetilde{\vartheta}^1}, \frac{\partial}{\partial\widetilde{\vartheta}^2}\right\}$ as follows: $\mathscr{J} = \mathscr{J}^t\frac{\partial}{\partial t} + \mathscr{J}^u\frac{\partial}{\partial u} + \not{\mathscr{J}}^A\frac{\partial}{\partial\widetilde{\vartheta}^A}$, where $\not{\mathscr{J}}$ is $S_{t,u}$-tangent. Furthermore, from the identities $g(L, L) = g(L, \frac{\partial}{\partial\widetilde{\vartheta}^A}) = g(\breve{R}, \frac{\partial}{\partial\widetilde{\vartheta}^A}) = 0$, $g(L, \breve{R}) = -\mu$, and $g(\breve{R}, \frac{\partial}{\partial u}) = \mu^2$ (see (3.46) and Lemma 3.20), it is straightforward to compute that

$$\mathscr{J}^t = -\mathscr{J}_L - \mu^{-1}\mathscr{J}_{\breve{R}}, \qquad \mathscr{J}^u = -\mu^{-1}\mathscr{J}_L. \tag{9.9}$$

To complete the proof, we use the following standard formula for the divergence $\mathscr{D}_\alpha\mathscr{J}^\alpha$ of $\mathscr{J}$ relative to the coordinates $(t, u, \widetilde{\vartheta}^1, \widetilde{\vartheta}^2)$:

$$\mathscr{D}_\alpha\mathscr{J}^\alpha = \frac{1}{\sqrt{|\det g|}}\left\{\frac{\partial}{\partial t}(\sqrt{|\det g|}\mathscr{J}^t) + \frac{\partial}{\partial u}(\sqrt{|\det g|}\mathscr{J}^u) + \frac{\partial}{\partial\widetilde{\vartheta}^A}(\sqrt{|\det g|}\not{\mathscr{J}}^A)\right\}. \tag{9.10}$$

In (9.10), $\det g$ is taken relative to the coordinates $(t, u, \widetilde{\vartheta}^1, \widetilde{\vartheta}^2)$. We next note that the identity (3.50a) holds relative to the coordinates $(t, u, \widetilde{\vartheta}^1, \widetilde{\vartheta}^2)$, as was shown in the proof of (3.50a). Inserting the square root of the identity into (9.10), using the fact that

$$\frac{1}{\sqrt{|\det \not{g}|}} \frac{\partial}{\partial \widetilde{\vartheta}^A}(\mu \sqrt{|\det \not{g}|} \not{\mathscr{J}}^A) = \not{\mathrm{div}}(\mu \not{\mathscr{J}}) \tag{9.11}$$

(where the determinant in (9.11) is taken relative to the coordinates $(\widetilde{\vartheta}^1, \widetilde{\vartheta}^2)$ induced on $S_{t,u}$), and multiplying the resulting equation by μ, we deduce that

$$\begin{aligned} \mu \mathscr{D}_\alpha \mathscr{J}^\alpha = & \frac{\partial}{\partial t}(\mu \mathscr{J}^t) + \frac{\partial}{\partial u}(\mu \mathscr{J}^u) + \not{\mathrm{div}}(\mu \not{\mathscr{J}}) \\ & + \left(\frac{\partial}{\partial t} \ln \sqrt{\det \not{g}}\right) \mu \mathscr{J}^t + \left(\frac{\partial}{\partial u} \ln \sqrt{\det \not{g}}\right) \mu \mathscr{J}^u. \end{aligned} \tag{9.12}$$

Next, from the standard matrix identity $\frac{\partial}{\partial t} \ln \sqrt{\det \not{g}} = \frac{1}{2}(\not{g}^{-1})^{AB} \frac{\partial}{\partial t} \not{g}_{AB}$ and the fact that $\frac{\partial}{\partial t} \not{g}_{AB} = \not{\mathcal{L}}_L \not{g}_{AB} = {}^{(L)}\not{\pi}_{AB} = 2\chi_{AB}$ (see (7.4a) and (3.72)), we deduce that $\frac{\partial}{\partial t} \ln \sqrt{\det \not{g}} = \mathrm{tr}_{\not{g}} \chi$. Similarly, with the help of (7.5g) and (4.1c), we deduce that $\frac{\partial}{\partial u} \ln \sqrt{\det \not{g}} = \frac{1}{2} \mathrm{tr}_{\not{g}} {}^{(\breve{R})}\not{\pi} = -\mu \mathrm{tr}_{\not{g}} \chi + \mathrm{tr}_{\not{g}} \not{k}^{(Trans-\Psi)} + \mu \mathrm{tr}_{\not{g}} \not{k}^{(Tan-\Psi)}$. Inserting these identities, the identities $\frac{\partial}{\partial t}|_{u, \widetilde{\vartheta}^1, \widetilde{\vartheta}^2} = L$ and $\frac{\partial}{\partial u}|_{t, \widetilde{\vartheta}^1, \widetilde{\vartheta}^2} = \breve{R}$ (along $\Sigma_t^{U_0}$), and (9.9) into (9.12), we arrive at (9.8). □

9.2. Frame decomposition of the commutation current

We now use Lemma 9.5 to derive a detailed frame decomposition of the term $\mu \mathscr{D}_\alpha {}^{(Z)}\mathscr{J}^\alpha[\Psi]$ on the right-hand side of (9.6). When we derive a priori L^2 estimates, the top-order derivatives of this term are by far the most difficult quantities that we have to bound. In fact, in order to close our top-order L^2 estimates, we must understand the precise structure of some of the terms in $\mu \mathscr{D}_\alpha {}^{(Z)}\mathscr{J}^\alpha[\Psi]$ including the exact numerical constants that appear. In order to close our below-top-order L^2 estimates, we need to know much less about their structure.

PROPOSITION 9.6 (**Frame decomposition of the divergence of the commutation current**). *Let Z be any vectorfield verifying ${}^{(Z)}\pi_{LL} = 0$, and let ${}^{(Z)}\mathscr{J}[\Psi]$ be the commutation current* (9.5) *associated to Z. Then*

(9.13)
$$\begin{aligned} \mu \mathscr{D}_\alpha {}^{(Z)}\mathscr{J}^\alpha[\Psi] = & \mathscr{K}_{(\pi-Danger)}^{(Z)}[\Psi] + \mathscr{K}_{(\pi-Cancel-1)}^{(Z)}[\Psi] + \mathscr{K}_{(\pi-Cancel-2)}^{(Z)}[\Psi] \\ & + \mathscr{K}_{(\pi-Elliptic)}^{(Z)}[\Psi] + \mathscr{K}_{(\pi-Good)}^{(Z)}[\Psi] + \mathscr{K}_{(\Psi)}^{(Z)}[\Psi] + \mathscr{K}_{(Low)}^{(Z)}[\Psi], \end{aligned}$$

where

(9.14a) $\mathscr{K}_{(\pi-Danger)}^{(Z)}[\Psi] := -(\mathrm{d}\!\!/\mathrm{iv}\, {}^{(Z)}\!\not\pi_L^{\#})\breve{R}\Psi,$

(9.14b)
$$\mathscr{K}_{(\pi-Cancel-1)}^{(Z)}[\Psi] := \left\{\frac{1}{2}\breve{R}\mathrm{tr}_{\not g}{}^{(Z)}\!\not\pi - \mathrm{d}\!\!/\mathrm{iv}\, {}^{(Z)}\!\not\pi_{\breve{R}}^{\#} - \upmu \mathrm{d}\!\!/\mathrm{iv}\, {}^{(Z)}\!\not\pi_L^{\#}\right\} L\Psi,$$

(9.14c)
$$\mathscr{K}_{(\pi-Cancel-2)}^{(Z)}[\Psi] := \left\{-\not\mathcal{L}_{\breve{R}}{}^{(Z)}\!\not\pi_L^{\#} + \not d^{\#}\,{}^{(Z)}\pi_{L\breve{R}}\right\}\cdot \not d\Psi,$$

(9.14d) $\mathscr{K}_{(\pi-Elliptic)}^{(Z)}[\Psi] := \upmu(\mathrm{d}\!\!/\mathrm{iv}\, {}^{(Z)}\!\not\pi^{\#\#})\cdot \not d\Psi,$

(9.14e)
$$\begin{aligned}\mathscr{K}_{(\pi-Good)}^{(Z)}[\Psi] := &\frac{1}{2}\upmu(L\mathrm{tr}_{\not g}{}^{(Z)}\!\not\pi)L\Psi + (L\,{}^{(Z)}\pi_{L\breve{R}})L\Psi + (L\,{}^{(Z)}\pi_{\breve{R}R})L\Psi \\ &+ \frac{1}{2}(L\mathrm{tr}_{\not g}{}^{(Z)}\!\not\pi)\breve{R}\Psi - \upmu(\not\mathcal{L}_L{}^{(Z)}\!\not\pi_L^{\#})\cdot \not d\Psi - (\not\mathcal{L}_L{}^{(Z)}\!\not\pi_{\breve{R}}^{\#})\cdot \not d\Psi,\end{aligned}$$

(9.15)
$$\begin{aligned}\mathscr{K}_{(\Psi)}^{(Z)}[\Psi] := &\varrho^{-1}\left\{\frac{1}{2}\upmu\mathrm{tr}_{\not g}{}^{(Z)}\!\not\pi + {}^{(Z)}\pi_{L\breve{R}} + {}^{(Z)}\pi_{\breve{R}R}\right\}\left\{L(\varrho L\Psi) + \frac{1}{2}\mathrm{tr}_{\not g}\chi(\varrho L\Psi)\right\} \\ &+ \mathrm{tr}_{\not g}{}^{(Z)}\!\not\pi\left\{L\breve{R}\Psi + \frac{1}{2}\mathrm{tr}_{\not g}\chi\breve{R}\Psi\right\} \\ &- 2\varrho^{-1}\upmu\,{}^{(Z)}\!\not\pi_L^{\#}\cdot \not d(\varrho L\Psi) - 2\,{}^{(Z)}\!\not\pi_L^{\#}\cdot \not d\breve{R}\Psi - 2\varrho^{-1}\,{}^{(Z)}\!\not\pi_{\breve{R}}^{\#}\cdot \not d(\varrho L\Psi) \\ &+ {}^{(Z)}\pi_{L\breve{R}}\not\Delta\Psi + \upmu\,{}^{(Z)}\!\hat{\not\pi}^{\#\#}\cdot \hat{\not\nabla}^2\Psi,\end{aligned}$$

and

(9.16)
$$\begin{aligned}\mathscr{K}_{(Low)}^{(Z)}[\Psi] := &\frac{1}{2}\mathrm{tr}_{\not g}\chi^{(Small)}\left\{{}^{(Z)}\pi_{L\breve{R}} + {}^{(Z)}\pi_{\breve{R}R} - \frac{1}{2}\upmu\mathrm{tr}_{\not g}{}^{(Z)}\!\not\pi\right\}L\Psi \\ &+ \frac{1}{2}\left\{L\upmu + \mathrm{tr}_{\not g}\not k^{(Trans-\Psi)} + \upmu\mathrm{tr}_{\not g}\not k^{(Tan-\Psi)} - 2\varrho^{-1}\upmu\right\}\mathrm{tr}_{\not g}{}^{(Z)}\!\not\pi L\Psi \\ &- ({}^{(Z)}\!\not\pi_L^{\#}\cdot \not d\upmu)L\Psi \\ &- \left\{L\upmu + \mathrm{tr}_{\not g}\not k^{(Trans-\Psi)} + \upmu\mathrm{tr}_{\not g}\not k^{(Tan-\Psi)}\right\}{}^{(Z)}\!\not\pi_L^{\#}\cdot \not d\Psi \\ &- \left\{2\varrho^{-1} + \mathrm{tr}_{\not g}\chi^{(Small)}\right\}{}^{(Z)}\!\not\pi_{\breve{R}}^{\#}\cdot \not d\Psi + \frac{1}{2}\mathrm{tr}_{\not g}{}^{(Z)}\!\not\pi(\not d^{\#}\upmu)\cdot \not d\Psi \\ &+ \mathrm{tr}_{\not g}{}^{(Z)}\!\not\pi\left\{\zeta^{(Trans-\Psi)\#} + \upmu\zeta^{(Tan-\Psi)\#}\right\}\cdot \not d\Psi + (\not d\upmu)\cdot {}^{(Z)}\!\hat{\not\pi}^{\#\#}\cdot \not d\Psi.\end{aligned}$$

In the above expressions the $S_{t,u}$ tensorfields χ, $\chi^{(Small)}$, $\zeta^{(Trans-\Psi)}$, $\not k^{(Trans-\Psi)}$, $\zeta^{(Tan-\Psi)}$, and $\not k^{(Tan-\Psi)}$ are defined by (3.70), (4.1c), (5.4a), (5.4b), (5.5a), *and* (5.5b). *In addition, the type $\binom{0}{2}$ $S_{t,u}$ tensor $\hat{\not\nabla}^2\Psi$ is defined by* (8.3). *Moreover, $\varrho = 1 - u + t$ is the geometric radial variable defined in* (3.5).

REMARK 9.7 (**The importance of ${}^{(Z)}\pi_{LL} = 0$**). The assumption ${}^{(Z)}\pi_{LL} = 0$ implies the absence, on the right-hand sides of (9.14a)-(9.16), of the dangerous term mentioned in Remark 7.3.

REMARK 9.8 (**Favorable combinations**). Note that we have carefully combined various terms to create sums of the form $LZ\Psi + \frac{1}{2}\mathrm{tr}_{\not g}\chi Z\Psi$ (where $Z \in \mathscr{Z}$) and placed them on the right-hand side of in (9.15). These combinations have better t-decay properties than the individual summands; see, for example, Lemma 12.58.

PROOF. With the help of Lemma 3.48 and the decomposition $\frac{1}{2}\mathrm{tr}_g {}^{(Z)}\pi = -\upmu^{-1(Z)}\pi_{L\breve{R}} + \frac{1}{2}\mathrm{tr}_{\not g} {}^{(Z)}\not\pi$ (which relies in part on the assumption ${}^{(Z)}\pi_{LL} = 0$), we compute that the components of ${}^{(Z)}\mathscr{J}[\Psi]$ relative to the frame $\{L, \breve{R}, X_1, X_2\}$ are:

$$ {}^{(Z)}\mathscr{J}_L[\Psi] := g({}^{(Z)}\mathscr{J}, L) = -\frac{1}{2}\mathrm{tr}_{\not g} {}^{(Z)}\not\pi L\Psi + {}^{(Z)}\not\pi_L^{\#} \cdot \not d\Psi, \tag{9.17} $$

$$ \begin{aligned} {}^{(Z)}\mathscr{J}_{\breve{R}}[\Psi] := g({}^{(Z)}\mathscr{J}, \breve{R}) = -{}^{(Z)}\pi_{L\breve{R}}L\Psi - {}^{(Z)}\pi_{\breve{R}R}L\Psi + {}^{(Z)}\not\pi_{\breve{R}}^{\#} \cdot \not d\Psi \\ - \frac{1}{2}\mathrm{tr}_{\not g} {}^{(Z)}\not\pi \breve{R}\Psi, \end{aligned} \tag{9.18} $$

(9.19)

$$ \begin{aligned} \upmu \not\mathscr{J}_A[\Psi] := \upmu g({}^{(Z)}\mathscr{J}, X_A) = -\upmu^{(Z)}\not\pi_{LA}L\Psi - {}^{(Z)}\not\pi_{\breve{R}A}L\Psi - {}^{(Z)}\not\pi_{LA}\breve{R}\Psi \\ + {}^{(Z)}\pi_{L\breve{R}}\not d_A\Psi + \upmu^{(Z)}\not\pi_A^{\ B}\not d_B\Psi. \end{aligned} $$

The identity (9.13) now follows from applying Lemma 9.5 to the frame components (9.17)-(9.19) and carrying out straightforward computations with the help of the commutator identity (5.9), Lemma 8.6, the identity $\mathrm{tr}_{\not g}\chi = 2\varrho^{-1} + \mathrm{tr}_{\not g}\chi^{(Small)}$ (see (4.1c)), and the identity $L\varrho = 1$. □

REMARK 9.9 (**Explanation of the various types of error terms**). We now make some remarks concerning the terms on the right-hand side of (9.13) for the commutation vectorfields $Z \in \mathscr{Z}$ defined in Def. 7.1. When we derive top-order L^2 estimates for the error terms in the commuted wave equation, we have to be very careful in how we estimate the top-order derivatives of $\mathscr{K}_{(\pi-Danger)}^{(Z)}[\Psi]$, $\mathscr{K}_{(\pi-Cancel-1)}^{(Z)}[\Psi]$, $\mathscr{K}_{(\pi-Cancel-2)}^{(Z)}[\Psi]$, $\mathscr{K}_{(\pi-Elliptic)}^{(Z)}[\Psi]$, and $\mathscr{K}_{(\pi-Good)}^{(Z)}[\Psi]$, for it is precisely these terms that lead to the presence of the top-order derivatives of the eikonal function quantities; as we outlined in Sect. 2.10.4, such terms are difficult to estimate in L^2 and lead to $\upmu^{-1}$-degenerate top-order L^2 estimates.

- The most difficult terms are generated by the top-order derivatives of $\mathscr{K}_{(\pi-Danger)}^{(Z)}[\Psi]$. The reason is that the factor $\breve{R}\Psi$ decays at the nonintegrable rate $\varepsilon(1+t)^{-1}$ (see Sect. 12.4.1). In closing our top-order L^2 estimates for this term, we must uncover some special structure that allows us to connect the product $G_{LL}\breve{R}\Psi$ back to $L\upmu$ via the transport equation (4.5); see, for example, the proof of Prop. 17.7.
- We must exploit some critically important algebraic cancellation that occurs in $\mathscr{K}_{(\pi-Cancel-1)}^{(Z)}[\Psi]$ and $\mathscr{K}_{(\pi-Cancel-2)}^{(Z)}[\Psi]$; see just below equation (16.52). The cancellation is based in part on the special structure of the deformation tensors of the vectorfields belonging to the commutation set $\mathscr{Z}$. In particular, if not for the cancellation present in $\mathscr{K}_{(\pi-Cancel-2)}^{(Z)}[\Psi]$, our error terms would involve certain top-order derivatives of the eikonal function quantities that we would have no means of controlling (that is, our estimates would lose derivatives); see Remark 2.14.

- The top-order derivatives of $\mathscr{K}_{(\pi-Elliptic)}^{(Z)}[\Psi]$ are relatively easy to bound, even though we must treat some of the terms with elliptic estimates on the spheres $S_{t,u}$ in order to avoid losing derivatives.
- The top-order derivatives of $\mathscr{K}_{(\pi-Good)}^{(Z)}[\Psi]$ are relatively easy to bound because all of the top-order derivatives of the eikonal function quantities contain at least one L differentiation; for such terms, there is no danger of losing derivatives.
- The top-order derivatives of $\mathscr{K}_{(\Psi)}^{(Z)}[\Psi]$ generate terms involving the top-order derivatives of Ψ, but *not of the eikonal function quantities*; such terms are relatively easy to bound.
- The term $\mathscr{K}_{(Low)}^{(Z)}[\Psi]$ consists of products that are lower-order in the sense of number of derivatives. Hence, $\mathscr{K}_{(Low)}^{(Z)}[\Psi]$ is relatively easy to bound at all derivative levels.

CHAPTER 10

Energy and Cone Flux Definitions and the Fundamental Divergence Identities

In this chapter, we define energies and cone fluxes, which are L^2-type quantities that we use to control solutions to $\square_{g(\Psi)}\Psi = 0$. We then use the vectorfield multiplier method to derive divergence theorem-type identities verified by the energies and cone fluxes. To derive the identities, we use two distinct vectorfield multipliers, a timelike multiplier and a Morawetz-type multiplier, which in the present context is a weighted version of L. In the case of the identity generated by the Morawetz multiplier, we use the correction current (10.18c) to generate some additional terms that lead to favorable cancellations. We use the identities in Chapter 20 in order to derive, via a lengthy Gronwall argument, a priori L^2 estimates for Ψ and its up-to-top-order derivatives.

10.1. Preliminary calculations

REMARK 10.1 (**The size of the $S_{t,u}$ area form**). Throughout this monograph, the geometric spherical area form $d\upsilon_{\not g}$ on $S_{t,u}$ implicitly contains a factor of magnitude $\approx \varrho^2$ (see inequality (12.152)) compared to the standard Euclidean area form on the Euclidean-unit sphere $(\mathbb{S}^2, \not e)$, where $\not e$ is the Riemannian metric induced on the Euclidean unit sphere by the Euclidean metric e on Σ_t. Note that relative to the rectangular spatial coordinates on Σ_t, we have $e_{ij} = \delta_{ij}$.

We start with the following standard lemma, which we often use in our analysis.

LEMMA 10.2 (**Derivatives of $S_{t,u}$ integrals**). *We have the following identities:*

(10.1a)
$$\frac{\partial}{\partial t}\left(\int_{S_{t,u}} f \, d\upsilon_{\not g}\right) = \int_{S_{t,u}} Lf + \mathrm{tr}_{\not g}\chi f \, d\upsilon_{\not g},$$

(10.1b)
$$\begin{aligned}
&\frac{\partial}{\partial u}\left(\int_{S_{t,u}} f \, d\upsilon_{\not g}\right) \\
&= \int_{S_{t,u}} \breve{R} f \, d\upsilon_{\not g} \\
&\quad + \int_{S_{t,u}} \left\{-\frac{2}{\varrho} - \frac{2}{\varrho}(\upmu - 1) - \upmu \mathrm{tr}_{\not g}\chi^{(Small)} + \mathrm{tr}_{\not g}\not k^{(Trans-\Psi)} + \upmu \mathrm{tr}_{\not g}\not k^{(Tan-\Psi)}\right\} f \, d\upsilon_{\not g}.
\end{aligned}$$

In the above expressions, the $S_{t,u}$ tensorfields χ, $\chi^{(Small)}$, $\not k^{(Trans-\Psi)}$, and $\not k^{(Tan-\Psi)}$ are defined by (3.70), (4.1c), (5.4b), *and* (5.5b).

PROOF. For each real number λ belonging to a small neighborhood of 0, let $\varphi_{(\lambda)} : \mathcal{M}_{t,u} \to \mathcal{M}_{t,u}$ be the flow map of L (see Remark 3.56). Let $\varphi^*_{(\lambda)}$ denote pullback by $\varphi_{(\lambda)}$ (see equation (3.65)) so that in particular, $\varphi^*_{(\lambda)} f = f \circ \varphi_{(\lambda)}$ for functions f. Note that since $Lt = 1$, it follows that $\varphi_{(\lambda)}$ bijectively maps the sphere $S_{t,u}$ onto the sphere $S_{t+\lambda,u}$. Hence, by covariance, we have

$$\int_{S_{t+\lambda,u}} f \, d\upsilon_{\not g} = \int_{S_{t,u}} \varphi^*_{(\lambda)} f \, d\upsilon_{\varphi^*_{(\lambda)} \not g}. \tag{10.2}$$

We now claim that

$$\frac{d}{d\lambda}|_{\lambda=0} \, d\upsilon_{\varphi^*_{(\lambda)} \not g} = \not{\mathcal{L}}_L d\upsilon_{\not g} = \frac{1}{2} \mathrm{tr}_{\not g} {}^{(L)}\not\pi d\upsilon_{\not g}. \tag{10.3}$$

The identity (10.3) is straightforward to verify in the local $S_{t,u}$ coordinates $(\vartheta^1, \vartheta^2)$, in which $d\upsilon_{\varphi^*_{(\lambda)} \not g} = \sqrt{\det \varphi^*_{(\lambda)} \not g} \, d\vartheta^1 \wedge d\vartheta^2$. In particular, one uses the identity

(10.4)
$$\frac{d}{d\lambda}|_{\lambda=0} \ln \det \varphi^*_{(\lambda)} \not g = ([\varphi^*_{(\lambda)} \not g]^{-1})^{AB} |_{\lambda=0} \frac{d}{d\lambda}|_{\lambda=0} \varphi^*_{(\lambda)} \not g_{AB} = (\not g^{-1})^{AB} \not{\mathcal{L}}_L \not g_{AB},$$

(7.4a), and the identity $\not{\mathcal{L}}_L \not g_{AB} = {}^{(L)}\not\pi_{AB}$ (see (7.4a)). Thus, differentiating each side of (10.2) with $\frac{d}{d\lambda}$, setting $\lambda = 0$, and using (10.3), we deduce

$$\begin{aligned} \frac{\partial}{\partial t} \left(\int_{S_{t,u}} f \, d\upsilon_{\not g} \right) &= \frac{d}{d\lambda}|_{\lambda=0} \int_{S_{t+\lambda,u}} f \, d\upsilon_{\not g} = \int_{S_{t,u}} (Lf) \, d\upsilon_{\not g} + \int_{S_{t,u}} f \not{\mathcal{L}}_L d\upsilon_{\not g} \\ &= \int_{S_{t,u}} (Lf) \, d\upsilon_{\not g} + \frac{1}{2} \int_{S_{t,u}} f \mathrm{tr}_{\not g} {}^{(L)}\not\pi \, d\upsilon_{\not g}. \end{aligned} \tag{10.5}$$

Similarly, since $\breve{R} u = 1$, the identity (10.5) holds with $\frac{\partial}{\partial t}$ replaced by $\frac{\partial}{\partial u}$ and L replaced by $\breve{R}$. Lemma 10.2 thus follows from the identity (10.5) and the identities

$$\frac{1}{2} \mathrm{tr}_{\not g} {}^{(L)}\not\pi = \mathrm{tr}_{\not g} \chi, \tag{10.6}$$

$$\frac{1}{2} \mathrm{tr}_{\not g} {}^{(\breve{R})}\not\pi = \left\{ -\frac{2}{\varrho} \upmu - \upmu \mathrm{tr}_{\not g} \chi^{(Small)} + \mathrm{tr}_{\not g} \not k^{(Trans-\Psi)} + \upmu \mathrm{tr}_{\not g} \not k^{(Tan-\Psi)} \right\}, \tag{10.7}$$

which follow from Prop. 7.7. □

The following tensorfield is a core ingredient in our derivation of L^2-type estimates for solutions to the inhomogeneous wave equation $\upmu \square_{g(\Psi)} \Psi = \mathfrak{F}$.

DEFINITION 10.3 (**Energy-momentum tensorfield**). We define $Q[\Psi]$, the energy-momentum tensorfield associated to Ψ, to be the following type $\binom{0}{2}$ tensorfield:

$$Q_{\mu\nu} = Q_{\mu\nu}[\Psi] := \mathscr{D}_\mu \Psi \mathscr{D}_\nu \Psi - \frac{1}{2} g_{\mu\nu} (g^{-1})^{\alpha\beta} \mathscr{D}_\alpha \Psi \mathscr{D}_\beta \Psi. \tag{10.8}$$

In the next lemma, we provide the fundamental divergence property of $Q[\Psi]$ for solutions to the inhomogeneous wave equation.

LEMMA 10.4 (**Fundamental divergence property of** $Q[\Psi]$). *If Ψ verifies the $\upmu$-weighted equation*

$$\upmu \square_g \Psi = \mathfrak{F}, \tag{10.9}$$

then

$$\mu \mathscr{D}_\alpha Q^\alpha{}_\nu[\Psi] = \mathfrak{F} \mathscr{D}_\nu \Psi. \tag{10.10}$$

PROOF. To derive (10.10), we take the divergence of the right-hand side of (10.8), multiply by μ, and replace the factor $\mu \square_{g(\Psi)} \Psi = \mu (g^{-1})^{\alpha\beta} \mathscr{D}^2_{\alpha\beta} \Psi$ with $\mathfrak{F}$ when it arises. □

REMARK 10.5 (**The role of the μ-weighted wave operator**). We use the μ-weighted product $\mu \square_{g(\Psi)} \Psi$ in (10.9) because this is the product that will appear when we apply the divergence theorem. The reason that this product appears rather than $\square_{g(\Psi)} \Psi$ is that our spacetime integrals are defined in terms of a rescaled spacetime volume $d\varpi$ in which the term μ has been factored out (see Sect. 3.19).

We use the following two multiplier vectorfields to construct the L^2-type quantities that control Ψ.

DEFINITION 10.6 (**Multiplier vectorfields**). We define the timelike multiplier vectorfield T and the Morawetz multiplier vectorfield $\widetilde{K}$ as follows:

$$T := (1 + \mu) L + \underline{\breve{L}} = (1 + 2\mu) L + 2\breve{R}, \tag{10.11a}$$

$$\widetilde{K} := \varrho^2 L, \tag{10.11b}$$

where ϱ is defined in (3.5).

We note that T is g-future-directed and g-timelike (in particular, we have $g(T,T) = -4\mu(1+\mu) < 0$). Moreover, $\widetilde{K}$ is g-future-directed and g-null. T is an analog of the Minkowskian vectorfield $2\partial_t + L_{(Flat)} = 3\partial_t + \partial_r$, but it has been carefully constructed so as to be useful both in regions where μ is large and in regions where μ is small. $\widetilde{K}$ is an analog of the Morawetz multiplier (1.70) that we encountered in our earlier review of the linear wave equation. We note that neither T nor $\widetilde{K}$ are Killing fields, even for the background solution $\Psi \equiv 0$. That is, the deformation tensors ${}^{(T)}\pi$ and ${}^{(\widetilde{K})}\pi$ are both nonzero. In fact, some of the error terms generated by ${}^{(\widetilde{K})}\pi$ contain weights that are large in t; see (10.50a). To handle such error terms when deriving L^2-type estimates, we exploit the U_0-thinness (relative to the eikonal function u) of the spacetime region.

We now derive expressions for the components of $Q[\Psi]$ relative to the rescaled null frame.

LEMMA 10.7 (**Null components of $Q_{\mu\nu}$**). *Let $Q[\Psi]$ be the energy-momentum tensorfield defined in* (10.8). *The components of $Q[\Psi]$ relative to the rescaled null frame $\{L, \underline{\breve{L}}, X_1, X_2\}$ are:*

$$Q_{LL}[\Psi] = (L\Psi)^2, \tag{10.12a}$$

$$Q_{\underline{\breve{L}}\underline{\breve{L}}}[\Psi] = (\underline{\breve{L}}\Psi)^2, \tag{10.12b}$$

$$Q_{L\underline{\breve{L}}}[\Psi] = \mu |\not{d}\Psi|^2, \tag{10.12c}$$

$$Q_{LA}[\Psi] = (L\Psi) \not{d}_A \Psi, \tag{10.12d}$$

$$Q_{\underline{\breve{L}}A}[\Psi] = (\underline{\breve{L}}\Psi) \not{d}_A \Psi, \tag{10.12e}$$

$$Q_{AB}[\Psi] = (\not{d}_A \Psi)(\not{d}_B \Psi) - \frac{1}{2} \not{g}_{AB} \left\{ -\mu^{-1} (L\Psi)(\underline{\breve{L}}\Psi) + |\not{d}\Psi|^2 \right\}. \tag{10.12f}$$

PROOF. The proof is a series of computations that can be carried out with the help of the decomposition (3.51b) of g^{-1} relative to the null frame. For example, using the identity $g(L,\breve{\underline{L}}) = -2\upmu$ (see (3.13)), we compute that $Q_{L\breve{\underline{L}}}[\Psi] = (L\Psi)(\breve{\underline{L}}\Psi) - \frac{1}{2}g(L,\breve{\underline{L}})\left\{-\upmu^{-1}(L\Psi)(\breve{\underline{L}}\Psi) + |\nabla\!\!\!/\Psi|^2\right\} = \upmu|\nabla\!\!\!/\Psi|^2$. We have thus proved (10.12c). We leave the proofs of the remaining identities to the reader. □

LEMMA 10.8 (**Null components of $^{(T)}\pi$ and $^{(\widetilde{K})}\pi - \varrho^2\mathrm{tr}_{g\!\!\!/}\chi g$**). *Let T and $\widetilde{K}$ be the multiplier vectorfields from Def.* 10.6. *Let $^{(T)}\pi$ and $^{(\widetilde{K})}\pi$ denote the corresponding deformation tensors as defined in Def.* 4.9. *Then relative to the rescaled null frame $\{L,\breve{\underline{L}},X_1,X_2\}$, the components of $^{(T)}\pi$ are:*

(10.13a) $$^{(T)}\pi_{LL} = 0,$$

(10.13b) $$^{(T)}\pi_{\breve{\underline{L}}\breve{\underline{L}}} = -4\upmu\left\{\breve{\underline{L}}\upmu - (1+\upmu)L\upmu\right\},$$

(10.13c) $$^{(T)}\pi_{L\breve{\underline{L}}} = -2\upmu\left\{\upmu^{-1}(\breve{\underline{L}}\upmu + L\upmu) + 3L\upmu\right\},$$

(10.13d) $$^{(T)}\pi_{LA} = -2\nabla\!\!\!/_A\upmu - 4\zeta_A^{(Trans-\Psi)} - 4\upmu\zeta_A^{(Tan-\Psi)},$$

(10.13e) $$^{(T)}\pi_{\breve{\underline{L}}A} = 2(1-\upmu)\nabla\!\!\!/_A\upmu + 4(1+\upmu)\left\{\zeta_A^{(Trans-\Psi)} + \upmu\zeta_A^{(Tan-\Psi)}\right\},$$

(10.13f) $$^{(T)}\pi\!\!\!/_{AB} = 2\left\{\hat{\chi}_{AB} + 2\hat{k\!\!\!/}_{AB}^{(Trans-\Psi)} + 2\upmu\hat{k\!\!\!/}_{AB}^{(Tan-\Psi)}\right\},$$

(10.13g) $$\mathrm{tr}_{g\!\!\!/}{}^{(T)}\pi\!\!\!/ = 2\left\{\mathrm{tr}_{g\!\!\!/}\chi + 2\mathrm{tr}_{g\!\!\!/}k\!\!\!/^{(Trans-\Psi)} + 2\upmu\mathrm{tr}_{g\!\!\!/}k\!\!\!/^{(Tan-\Psi)}\right\}.$$

Furthermore, the components of $^{(\widetilde{K})}\pi - \varrho^2\mathrm{tr}_{g\!\!\!/}\chi g$ are:

(10.14a) $$^{(\widetilde{K})}\pi_{LL} - \varrho^2\mathrm{tr}_{g\!\!\!/}\chi\underbrace{g(L,L)}_{=0} = 0,$$

(10.14b) $$^{(\widetilde{K})}\pi_{\breve{\underline{L}}\breve{\underline{L}}} - \frac{1}{2}\varrho^2\mathrm{tr}_{g\!\!\!/}\chi\underbrace{g(\breve{\underline{L}},\breve{\underline{L}})}_{=0} = -4\upmu\left\{2\varrho\breve{\underline{L}}\varrho - \varrho^2L\upmu\right\},$$

(10.14c) $$^{(\widetilde{K})}\pi_{L\breve{\underline{L}}} - \varrho^2\mathrm{tr}_{g\!\!\!/}\chi\underbrace{g(L,\breve{\underline{L}})}_{=-2\upmu} = -2\varrho^2\upmu\left\{\frac{L\upmu}{\upmu} - \mathrm{tr}_{g\!\!\!/}\chi^{(Small)}\right\},$$

(10.14d) $$^{(\widetilde{K})}\pi_{LA} - \varrho^2\mathrm{tr}_{g\!\!\!/}\chi\underbrace{g(L,X_A)}_{=0} = 0,$$

(10.14e) $$^{(\widetilde{K})}\pi_{\breve{\underline{L}}A} - \varrho^2\mathrm{tr}_{g\!\!\!/}\chi\underbrace{g(\breve{\underline{L}},X_A)}_{=0} = 2\varrho^2\left\{\nabla\!\!\!/_A\upmu + 2\zeta_A^{(Trans-\Psi)} + 2\upmu\zeta_A^{(Tan-\Psi)}\right\},$$

(10.14f) $$^{(\widetilde{K})}\pi\!\!\!/_{AB} - \varrho^2\mathrm{tr}_{g\!\!\!/}\chi\underbrace{\hat{g\!\!\!/}_{AB}}_{=0} = 2\varrho^2\hat{\chi}_{AB}^{(Small)},$$

(10.14g) $$\mathrm{tr}_{g\!\!\!/}{}^{(\widetilde{K})}\pi\!\!\!/ - 2\varrho^2\mathrm{tr}_{g\!\!\!/}\chi = 0.$$

In the above expressions, the $S_{t,u}$ tensorfields χ, $\chi^{(Small)}$, $\zeta^{(Trans-\Psi)}$, $k\!\!\!/^{(Trans-\Psi)}$, $\zeta^{(Tan-\Psi)}$, and $k\!\!\!/^{(Tan-\Psi)}$ are defined by (3.70), (4.1c), (5.4a), (5.4b), (5.5a), *and* (5.5b).

PROOF. The proof is a series of computations based on the rescaled null frame connection coefficient identities of Lemma 5.3 and the ideas used in the proof of Lemma 10.7. We give two examples. We use the notation $\langle V, W\rangle := g(V,W)$ to simplify the presentation. As a first example, we prove (10.13f) and (10.13g). Using (4.20) and (10.11a), we compute that

(10.15)
$$
\begin{aligned}
{}^{(T)}\not{\pi}_{AB} &= \left\langle \mathscr{D}_A \left\{(1+\upmu)L + \breve{\underline{L}}\right\}, X_B\right\rangle + \left\langle \mathscr{D}_B \left\{(1+\upmu)L + \breve{\underline{L}}\right\}, X_A\right\rangle \\
&= (1+\upmu)\langle \mathscr{D}_A L, X_B\rangle + (1+\upmu)\langle \mathscr{D}_B L, X_A\rangle + \langle \mathscr{D}_A \breve{\underline{L}}, X_B\rangle + \langle \mathscr{D}_B \breve{\underline{L}}, X_A\rangle.
\end{aligned}
$$

Also using Lemma 5.3 and the decomposition (5.3b), we deduce that the right-hand side of (10.15) is equal to $2\chi_{AB} + 4\not{k}_{AB}^{(Trans-\Psi)} + 4\upmu\not{k}_{AB}^{(Tan-\Psi)}$. The desired identities (10.13f) and (10.13g) now follow from splitting this expression into its trace and trace-free parts.

As a second example, we prove (10.14c). Using (4.20), (10.11b), and the identities $L\varrho = 1$ and $\breve{\underline{L}}\varrho = \upmu - 2$, we compute that

$$
\begin{aligned}
{}^{(\widetilde{K})}\pi_{L\breve{\underline{L}}} &= \langle \mathscr{D}_L(\varrho^2 L), \breve{\underline{L}}\rangle + \langle \mathscr{D}_{\breve{\underline{L}}}(\varrho^2 L), L\rangle \\
&= -4\varrho\upmu + \varrho^2\langle \mathscr{D}_L L, \breve{\underline{L}}\rangle + \varrho^2\langle \mathscr{D}_{\breve{\underline{L}}} L, L\rangle.
\end{aligned}
\tag{10.16}
$$

Also using Lemma 5.3, we deduce that the right-hand side of (10.16) is equal to $-4\varrho\upmu - 2\varrho^2 L\upmu$, which is the first product on the left-hand side of (10.14c). Furthermore, using the decomposition $\mathrm{tr}_{\not{g}}\chi = 2\varrho^{-1} + \mathrm{tr}_{\not{g}}\chi^{(Small)}$, we compute that the second term on the left-hand side of (10.14c) is

$$
-\varrho^2 \mathrm{tr}_{\not{g}}\chi g(L, \breve{\underline{L}}) = 4\varrho\upmu + 2\upmu\varrho^2\mathrm{tr}_{\not{g}}\chi^{(Small)}.
\tag{10.17}
$$

Adding the previous term $-4\varrho\upmu - \varrho^2 L\upmu$ to (10.17) yields $-\varrho^2 L\upmu + 2\upmu\varrho^2\mathrm{tr}_{\not{g}}\chi^{(Small)}$, from which we easily conclude the desired identity (10.14c). □

We now define vectorfields that we use for bookkeeping when we integrate by parts to derive energy-cone flux identities.

DEFINITION 10.9 (**Compatible currents**). Let Ψ be a function and let $Q[\Psi]$ be the energy-momentum tensorfield defined in (10.8). Let T and $\widetilde{K}$ be the multiplier vectorfields from Def. 10.6. We associate to Ψ the following four "compatible current" vectorfields:

$$
{}^{(T)}J^{\nu}[\Psi] := Q^{\nu}{}_{\alpha}[\Psi]T^{\alpha},
\tag{10.18a}
$$

$$
{}^{(\widetilde{K})}J^{\nu}[\Psi] := Q^{\nu}{}_{\alpha}[\Psi]\widetilde{K}^{\alpha},
\tag{10.18b}
$$

$$
{}^{(Correction)}J^{\nu}[\Psi] := \frac{1}{2}\left\{\varrho^2\mathrm{tr}_{\not{g}}\chi\Psi\mathscr{D}^{\nu}\Psi - \frac{1}{2}\Psi^2\mathscr{D}^{\nu}[\varrho^2\mathrm{tr}_{\not{g}}\chi]\right\},
\tag{10.18c}
$$

$$
{}^{(\widetilde{K}+Correction)}J^{\nu}[\Psi] := {}^{(\widetilde{K})}J^{\nu}[\Psi] + {}^{(Correction)}J^{\nu}[\Psi].
\tag{10.18d}
$$

In the next lemma, as a precursor to applying the divergence theorem, we compute the divergence of the compatible currents.

LEMMA 10.10 (**Preliminary expressions for the $\upmu$-weighted divergence of the currents**). *If Ψ verifies the $\upmu$-weighted inhomogeneous wave equation*

$$
\upmu\square_{g(\Psi)}\Psi = \mathfrak{F},
\tag{10.19}
$$

then the following divergence identities hold for the compatible currents from Def. 10.9*:*

$$\upmu \mathscr{D}_\alpha {}^{(T)}J^\alpha[\Psi] = \left\{(1+2\upmu)L\Psi + 2\breve{R}\Psi\right\}\mathfrak{F} + \frac{1}{2}\upmu Q^{\alpha\beta}[\Psi]{}^{(T)}\pi_{\alpha\beta}, \tag{10.20a}$$

$$\upmu \mathscr{D}_\alpha {}^{(\widetilde{K})}J^\alpha[\Psi] = \varrho^2(L\Psi)\mathfrak{F} + \frac{1}{2}\upmu Q^{\alpha\beta}[\Psi]{}^{(\widetilde{K})}\pi_{\alpha\beta}, \tag{10.20b}$$

$$\begin{aligned}\upmu \mathscr{D}_\alpha {}^{(Correction)}J^\alpha[\Psi] &= \frac{1}{2}\varrho^2 \mathrm{tr}_{\not g}\chi\Psi\mathfrak{F} + \frac{1}{2}\varrho^2\mathrm{tr}_{\not g}\chi\upmu(g^{-1})^{\alpha\beta}\mathscr{D}_\alpha\Psi\mathscr{D}_\beta\Psi \\ &\quad - \frac{1}{4}\upmu\left\{\square_{g(\Psi)}[\varrho^2\mathrm{tr}_{\not g}\chi]\right\}\Psi^2,\end{aligned} \tag{10.20c}$$

$$\begin{aligned}\upmu \mathscr{D}_\alpha {}^{(\widetilde{K}+Correction)}J^\alpha[\Psi] &= \varrho^2\mathfrak{F}\left\{L\Psi + \frac{1}{2}\mathrm{tr}_{\not g}\chi\Psi\right\} \\ &\quad + \frac{1}{2}\upmu Q^{\alpha\beta}[\Psi]\left\{{}^{(\widetilde{K})}\pi_{\alpha\beta} - \varrho^2\mathrm{tr}_{\not g}\chi g_{\alpha\beta}\right\} \\ &\quad - \frac{1}{4}\upmu\left\{\square_{g(\Psi)}[\varrho^2\mathrm{tr}_{\not g}\chi]\right\}\Psi^2,\end{aligned} \tag{10.20d}$$

where χ *is the* $S_{t,u}$ *tensorfield defined in* (3.70) *and* ${}^{(V)}\pi$ *is the deformation tensor of* V *(see Def.* 4.9*).*

PROOF. Equation (10.20a) follows from applying $\mathscr{D}_\nu$ to (10.18a), using the Leibniz rule and Lemma 10.4, definition (4.20), definition (10.11b), and the symmetry of $Q_{\mu\nu}[\Psi]$ and ${}^{(T)}\pi_{\mu\nu}$. The proof of (10.20b) is based on definition (10.18b) and is similar; we omit the details.

Equation (10.20c) follows from applying $\mathscr{D}_\nu$ to (10.18c), using the Leibniz rule, and from using equation (10.19) to substitute for the term $\upmu\square_{g(\Psi)}\Psi$ when it arises.

Equation (10.20d) follows from definition (10.18d), equations (10.20b) and (10.20c), and the identity $g_{\alpha\beta}Q^{\alpha\beta}[\Psi] = -(g^{-1})^{\alpha\beta}\mathscr{D}_\alpha\Psi\mathscr{D}_\beta\Psi$, which easily follows from definition (10.8). □

10.2. The energy-cone flux integral identities

We are now ready to define the energies and cone fluxes that we use in our L^2 analysis of solutions. We establish their coerciveness properties in Lemma 14.1.

DEFINITION 10.11 (**Energies and cone fluxes**). Let Ψ be a scalar function and let $N = \frac{1}{2}L + \frac{1}{2}\upmu^{-1}\breve{\underline{L}} = L + R$ denote the future-directed unit normal to Σ_t (see Lemma 3.20). We define the energies $\mathbb{E}[\Psi](t,u)$, $\widetilde{\mathbb{E}}[\Psi](t,u)$ and the cone fluxes $\mathbb{F}[\Psi](t,u)$, $\widetilde{\mathbb{F}}[\Psi](t,u)$ as follows, where the rescaled volume forms $d\underline{\varpi}$ and $d\overline{\varpi}$ are defined in Def. 3.68, and ${}^{(T)}J^\alpha[\Psi]$, ${}^{(\widetilde{K})}J^\alpha[\Psi]$, ${}^{(Correction)}J^\alpha[\Psi]$, and ${}^{(\widetilde{K}+Correction)}J^\alpha[\Psi]$ are defined in Def. 10.9:

$$\mathbb{E}[\Psi](t,u) := \int_{\Sigma_t^u}\upmu\,{}^{(T)}J_N[\Psi]\,d\underline{\varpi}, \tag{10.21a}$$

$$\begin{aligned}\widetilde{\mathbb{E}}[\Psi](t,u) &:= \int_{\Sigma_t^u}\upmu\,{}^{(\widetilde{K}+Correction)}J_N[\Psi]\,d\underline{\varpi} - \frac{1}{2}\int_{\Sigma_t^u}\varrho^2\mathrm{tr}_{\not g}\chi\Psi\breve{R}\Psi\,d\underline{\varpi} \\ &\quad + \frac{1}{4}\int_{\Sigma_t^u}\left\{\upmu L[\varrho^2\mathrm{tr}_{\not g}\chi] + \breve{R}[\varrho^2\mathrm{tr}_{\not g}\chi] + \frac{1}{2}\upmu\varrho^2(\mathrm{tr}_{\not g}\chi)^2\right\}\Psi^2\,d\underline{\varpi},\end{aligned} \tag{10.21b}$$

(10.22a)

$$\mathbb{F}[\Psi](t,u) := \int_{\mathcal{C}_u^t} {}^{(T)}J_L[\Psi] \, d\overline{\varpi},$$

(10.22b)

$$\widetilde{\mathbb{F}}[\Psi](t,u) := \int_{\mathcal{C}_u^t} {}^{(\widetilde{K}+Correction)}J_L[\Psi] \, d\overline{\varpi} + \frac{1}{4}\int_{\mathcal{C}_u^t} \left\{ L[\varrho^2 \mathrm{tr}_{\not g}\chi \Psi^2] + \varrho^2(\mathrm{tr}_{\not g}\chi)^2\Psi^2 \right\} d\overline{\varpi}.$$

In the next proposition, we derive our main energy-cone flux identities, which are the starting points of our L^2 analysis of Ψ and its derivatives. As a preliminary step, we first prove the following lemma, which provides a version of the divergence theorem on the spacetime region $\mathcal{M}_{t,u}$. *Note that we explicitly indicate the factors of* $\upmu$ *in the lemma*; there are no "hidden factors" in the volume forms.

LEMMA 10.12 (**The divergence theorem on the region $\mathcal{M}_{t,u}$**). *Let J be any spacetime vectorfield. Let $N = L + R$ (see* (3.19c) *and Lemma* 3.20*) be the future-directed unit normal to $\Sigma_t^{U_0}$. Then the following integral identity holds, where the rescaled volume forms $d\underline{\varpi}$, $d\overline{\varpi}$, and $d\varpi$ are defined in Def.* 3.68*:*

(10.23)

$$\int_{\Sigma_t^u} \upmu J_N \, d\underline{\varpi} + \int_{\mathcal{C}_u^t} J_L \, d\overline{\varpi} - \int_{\Sigma_0^u} \upmu J_N \, d\underline{\varpi} - \int_{\mathcal{C}_0^t} J_L \, d\overline{\varpi} = - \int_{\mathcal{M}_{t,u}} \upmu \mathscr{D}_\alpha J^\alpha \, d\varpi.$$

PROOF. We can expand J relative to the geometric coordinate frame as follows: $J = J^t \frac{\partial}{\partial t} + J^u \frac{\partial}{\partial u} + \not J^A \frac{\partial}{\partial \vartheta^A}$, where $\not J$ is the projection of J onto $S_{t,u}$. Furthermore, as we obtained in the proof of (9.9), we have

$$J^t = -J_N, \qquad J^u = -\upmu^{-1} J_L. \tag{10.24}$$

We will use the following standard formula for the divergence $\mathscr{D}_\alpha J^\alpha$ of J:

(10.25)

$$\mathscr{D}_\alpha J^\alpha = \frac{1}{\sqrt{|\mathrm{det} g|}} \left\{ \frac{\partial}{\partial t}(\sqrt{|\mathrm{det} g|} J^t) + \frac{\partial}{\partial u}(\sqrt{|\mathrm{det} g|} J^u) + \frac{\partial}{\partial \vartheta^A}(\sqrt{|\mathrm{det} g|} \not J^A) \right\},$$

where the determinant $\mathrm{det} g$ is taken relative to the geometric coordinates (t, u, ϑ). Inserting the square root of (3.50a) into (10.25), we deduce that

(10.26)

$$\frac{\partial}{\partial t}(\upmu \sqrt{\mathrm{det} \not g} J^t) + \frac{\partial}{\partial u}(\upmu \sqrt{\mathrm{det} \not g} J^u) + \frac{\partial}{\partial \vartheta^A}(\upmu \sqrt{|\mathrm{det} \not g|} \not J^A) = (\upmu \mathscr{D}_\alpha J^\alpha) \sqrt{\mathrm{det} \not g}.$$

Integrating (10.26) over $\mathbb{S}^2$ with respect to $d\vartheta$ and noting that the integral of the last term on the left-hand side of (10.26) vanishes, we deduce that

$$\frac{\partial}{\partial t}\left(\int_{S_{t,u}} \upmu J^t \, d\upsilon_{\not g} \right) + \frac{\partial}{\partial u}\left(\int_{S_{t,u}} \upmu J^u \, d\upsilon_{\not g} \right) = \int_{S_{t,u}} \upmu \mathscr{D}_\alpha J^\alpha \, d\upsilon_{\not g}. \tag{10.27}$$

Integrating (10.27) with respect to $du'dt'$ over the coordinate rectangle $[0,t] \times [0,u]$, we deduce that

(10.28)

$$\int_{\Sigma_t^u} \upmu J^t \, d\underline{\varpi} - \int_{\Sigma_0^u} \upmu J^t \, d\underline{\varpi} + \int_{\mathcal{C}_u^t} \upmu J^u \, d\overline{\varpi} - \int_{\mathcal{C}_0^t} \upmu J^u \, d\overline{\varpi} = \int_{\mathcal{M}_{t,u}} \upmu \mathscr{D}_\alpha J^\alpha \, d\varpi.$$

Inserting the identities (10.24) into (10.28), we arrive at (10.23). □

We now derive our main energy-cone flux identities.

PROPOSITION 10.13 (**Energy-cone flux identities via the divergence theorem with important cancellations**). *Consider the energies and fluxes of Def. 10.11. For solutions Ψ to*

$$\upmu \square_{g(\Psi)} \Psi = \mathfrak{F}$$

that vanish along the outer null cone[1] $\mathcal{C}_0$, *we have the following integral identities:*

$$\mathbb{E}[\Psi](t,u) + \mathbb{F}[\Psi](t,u) = \mathbb{E}[\Psi](0,u) - \int_{\mathcal{M}_{t,u}} \left\{(1+2\upmu)(L\Psi) + 2\breve{R}\Psi\right\} \mathfrak{F}\, d\varpi - \frac{1}{2}\int_{\mathcal{M}_{t,u}} \upmu Q^{\alpha\beta}[\Psi]\,{}^{(T)}\pi_{\alpha\beta}\, d\varpi, \tag{10.29a}$$

$$\begin{aligned}
&\widetilde{\mathbb{E}}[\Psi](t,u) + \widetilde{\mathbb{F}}[\Psi](t,u) \\
&= \widetilde{\mathbb{E}}[\Psi](0,u) \\
&\quad - \frac{1}{4}\int_{\Sigma_0^u} \left\{ \breve{\underline{L}}[\varrho^2 \mathrm{tr}_{\not g}\chi] - \frac{1}{2}\varrho^2\upmu(\mathrm{tr}_{\not g}\chi)^2 \right\} \Psi^2\, d\varpi \\
&\quad - \frac{1}{4}\int_{\Sigma_0^u} \left\{ \varrho^2 \mathrm{tr}_{\not g}\chi \mathrm{tr}_{\not g}\not k^{(Trans-\Psi)} + \varrho^2\upmu \mathrm{tr}_{\not g}\chi \mathrm{tr}_{\not g}\not k^{(Tan-\Psi)} \right\} \Psi^2\, d\varpi \\
&\quad + \frac{1}{4}\int_{\Sigma_t^u} \left\{ \breve{\underline{L}}[\varrho^2 \mathrm{tr}_{\not g}\chi] - \frac{1}{2}\varrho^2\upmu(\mathrm{tr}_{\not g}\chi)^2 \right\} \Psi^2\, d\varpi \\
&\quad + \frac{1}{4}\int_{\Sigma_t^u} \left\{ \varrho^2 \mathrm{tr}_{\not g}\chi \mathrm{tr}_{\not g}\not k^{(Trans-\Psi)} + \varrho^2\upmu \mathrm{tr}_{\not g}\chi \mathrm{tr}_{\not g}\not k^{(Tan-\Psi)} \right\} \Psi^2\, d\varpi \\
&\quad - \int_{\mathcal{M}_{t,u}} \varrho^2 \left\{ L\Psi + \frac{1}{2}\mathrm{tr}_{\not g}\chi\Psi \right\} \mathfrak{F}\, d\varpi \\
&\quad - \frac{1}{2}\int_{\mathcal{M}_{t,u}} \upmu Q^{\alpha\beta}[\Psi] \left\{ {}^{(\widetilde{K})}\pi_{\alpha\beta} - \varrho^2 \mathrm{tr}_{\not g}\chi g_{\alpha\beta} \right\} d\varpi \\
&\quad + \frac{1}{4}\int_{\mathcal{M}_{t,u}} \upmu \left\{ \square_{g(\Psi)}[\varrho^2 \mathrm{tr}_{\not g}\chi] \right\} \Psi^2\, d\varpi.
\end{aligned} \tag{10.29b}$$

In the above expressions, the $S_{t,u}$ tensorfields χ, $\not k^{(Trans-\Psi)}$, and $\not k^{(Tan-\Psi)}$ are defined by (3.70), (5.4b), *and* (5.5b). *Moreover, the rescaled volume form $d\varpi$ is defined in Def.* 3.68.

PROOF. The overall strategy is to apply Lemma 10.12 with various currents J from Def. 10.9. Specifically, (10.29a) follows from applying Lemma 10.12 with the current ${}^{(T)}J[\Psi]$ defined in (10.18a). The proof is simpler than the proof of (10.29b), which we provide below. Hence, we omit the details of the proof of (10.29a).

[1] Recall that this vanishing is a simple consequence of our assumption that the data are supported in the Euclidean unit ball Σ_0^1.

To prove (10.29b), we first apply Lemma 10.12 with the current ${}^{(\widetilde{K}+Correction)}J[\Psi]$ defined in (10.18d). Using the fact that $\Psi \equiv 0$ along $\mathcal{C}_0^t$, we deduce the following identity:

$$(10.30) \quad \int_{\Sigma_t^u} \upmu\, {}^{(\widetilde{K}+Correction)}J_N[\Psi]\, d\underline{\varpi} + \int_{\mathcal{C}_u^t} {}^{(\widetilde{K}+Correction)}J_L[\Psi]\, d\overline{\varpi}$$
$$= \int_{\Sigma_0^u} \upmu\, {}^{(\widetilde{K}+Correction)}J_N[\Psi]\, d\underline{\varpi} - \int_{\mathcal{M}_{t,u}} \upmu \mathscr{D}_\alpha {}^{(\widetilde{K}-Correction)}J^\alpha[\Psi]\, d\varpi.$$

Examining (10.20d), (10.21b), (10.22b), (10.29b), and (10.30), we see that the desired identity (10.29b) will follow once we show that the last two Σ_t^u integrals on the right-hand side of (10.21b) and the last $\mathcal{C}_u^t$ integral on the right-hand side of (10.22b) sum to

$$(10.31) \quad \frac{1}{4}\int_{\Sigma_t^u} \left\{ \breve{\underline{L}}[\varrho^2 \mathrm{tr}_{\not g}\chi] - \frac{1}{2}\upmu\varrho^2(\mathrm{tr}_{\not g}\chi)^2 \right\} \Psi^2\, d\varpi$$
$$+ \frac{1}{4}\int_{\Sigma_t^u} \left\{ \varrho^2 \mathrm{tr}_{\not g}\chi \mathrm{tr}_{\not g}\not k^{(Trans-\Psi)} + \upmu\varrho^2 \mathrm{tr}_{\not g}\chi \mathrm{tr}_{\not g}\not k^{(Tan-\Psi)} \right\} \Psi^2\, d\varpi$$
$$- \frac{1}{4}\int_{S_{0,u}} \varrho^2 \mathrm{tr}_{\not g}\chi \Psi^2\, d\upsilon_{\not g}.$$

Note that the integral $-\frac{1}{4}\int_{S_{0,u}} \varrho^2 \mathrm{tr}_{\not g}\chi\Psi^2\, d\upsilon_{\not g}$ on the right-hand side of (10.31) does not depend on t and hence cancels out of the expression (10.29b). To deduce (10.31), we first analyze the last two Σ_t^u integrals on the right-hand side of (10.21b). To this end, we set $f := \varrho^2 \mathrm{tr}_{\not g}\chi\Psi^2$ and integrate (10.1b) in u to compute that

$$(10.32) \quad \int_{S_{t,u}} \varrho^2 \mathrm{tr}_{\not g}\chi\Psi^2\, d\upsilon_{\not g} = \int_0^u \frac{\partial}{\partial u'}\left(\int_{S_{t,u'}} \varrho^2 \mathrm{tr}_{\not g}\chi\Psi^2\, d\upsilon_{\not g} \right) du'$$
$$= \int_0^u \int_{S_{t,u'}} \varrho^2 \mathrm{tr}_{\not g}\chi \breve{R}(\Psi^2)\, d\upsilon_{\not g}\, du'$$
$$+ \int_0^u \int_{S_{t,u'}} \left\{ \breve{R}[\varrho^2 \mathrm{tr}_{\not g}\chi] - \upmu\varrho^2(\mathrm{tr}_{\not g}\chi)^2 \right\} \Psi^2\, d\upsilon_{\not g}\, du'$$
$$+ \int_0^u \int_{S_{t,u'}} \left\{ \varrho^2 \mathrm{tr}_{\not g}\chi \mathrm{tr}_{\not g}\not k^{(Trans-\Psi)} + \upmu\varrho^2 \mathrm{tr}_{\not g}\chi \mathrm{tr}_{\not g}\not k^{(Tan-\Psi)} \right\} \Psi^2\, d\upsilon_{\not g}\, du'.$$

We now use (10.32), the identity $\upmu L = \breve{\underline{L}} - 2\breve{R}$ (see (3.19a)), and straightforward computations to express the last two Σ_t^u integrals on the right-hand side of (10.21b) as follows:

$$
\begin{aligned}
(10.33) \qquad & -\frac{1}{4}\int_{\Sigma_t^u} \varrho^2 \mathrm{tr}_{\not g}\chi \breve{R}(\Psi^2)\, d\underline{\varpi} \\
& + \frac{1}{4}\int_{\Sigma_t^u} \left\{ \upmu L[\varrho^2 \mathrm{tr}_{\not g}\chi] + \breve{R}[\varrho^2 \mathrm{tr}_{\not g}\chi] + \frac{1}{2}\upmu \varrho^2 (\mathrm{tr}_{\not g}\chi)^2 \right\} \Psi^2\, d\underline{\varpi} \\
& = -\frac{1}{4}\int_{S_{t,u}} \varrho^2 \mathrm{tr}_{\not g}\chi \Psi^2\, d\upsilon_{\not g} \\
& + \frac{1}{4}\int_{\Sigma_t^u} \left\{ \breve{\underline{L}}[\varrho^2 \mathrm{tr}_{\not g}\chi] - \frac{1}{2}\upmu \varrho^2 (\mathrm{tr}_{\not g}\chi)^2 \right\} \Psi^2\, d\varpi \\
& + \frac{1}{4}\int_{\Sigma_t^u} \left\{ \varrho^2 \mathrm{tr}_{\not g}\chi \mathrm{tr}_{\not g}\not k^{(Trans-\Psi)} + \upmu \varrho^2 \mathrm{tr}_{\not g}\chi \mathrm{tr}_{\not g}\not k^{(Tan-\Psi)} \right\} \Psi^2\, d\varpi.
\end{aligned}
$$

Similarly, we use (10.1a) with $f := \varrho^2 \mathrm{tr}_{\not g}\chi \Psi^2$ to express the last $\mathcal{C}_u^t$ integral on the right-hand side of (10.22b) as follows:

$$
\begin{aligned}
(10.34) \qquad & \frac{1}{4}\int_{\mathcal{C}_u^t} \left\{ L[\varrho^2 \mathrm{tr}_{\not g}\chi \Psi^2] + \varrho^2 (\mathrm{tr}_{\not g}\chi)^2 \Psi^2 \right\} d\overline{\varpi} \\
& = \frac{1}{4}\int_{S_{t,u}} \varrho^2 \mathrm{tr}_{\not g}\chi \Psi^2\, d\upsilon_{\not g} - \frac{1}{4}\int_{S_{0,u}} \varrho^2 \mathrm{tr}_{\not g}\chi \Psi^2\, d\upsilon_{\not g}.
\end{aligned}
$$

We now add (10.33) and (10.34) and note the cancellation of

$$\frac{1}{4}\int_{S_{t,u}} \varrho^2 \mathrm{tr}_{\not g}\chi \Psi^2\, d\upsilon_{\not g},$$

which yields the desired identity (10.31). □

10.3. Integration by parts identities for the top-order square integral estimates

In order to estimate some of the top-order $\mathfrak{F}$-containing integrals on the right-hand side of (10.29b), we need to derive alternate expressions via integration by parts. The alternate expressions are based on the following two simple lemmas.

LEMMA 10.14 (**Integration by parts on $\mathcal{M}_{t,u}$**). *If Ψ and f are functions, then we have the following integration by parts identity:*

$$
\begin{aligned}
(10.35) \qquad \int_{\mathcal{M}_{t,u}} f \left\{ L\Psi + \frac{1}{2}\mathrm{tr}_{\not g}\chi \Psi \right\} d\varpi = & -\int_{\mathcal{M}_{t,u}} \Psi \left\{ Lf + \frac{1}{2}\mathrm{tr}_{\not g}\chi f \right\} d\varpi \\
& - \int_{\Sigma_0^u} \Psi f\, d\underline{\varpi} + \int_{\Sigma_t^u} \Psi f\, d\underline{\varpi},
\end{aligned}
$$

where χ is the $S_{t,u}$ tensorfield defined in (3.70).

PROOF. From (10.1a), we deduce that

(10.36)

$$\frac{\partial}{\partial t}\int_{\Sigma_t^u} \Psi f\, d\underline{\varpi} = \int_{\Sigma_t^u} \Psi \left\{ Lf + \frac{1}{2}\mathrm{tr}_{\not g}\chi f \right\} d\underline{\varpi} + \int_{\Sigma_t^u} f \left\{ L\Psi + \frac{1}{2}\mathrm{tr}_{\not g}\chi \Psi \right\} d\underline{\varpi}.$$

The desired identity (10.35) now follows from integrating (10.36) in time from 0 to t. □

LEMMA 10.15 (**Integration by parts on $S_{t,u}$**). *If Y is an $S_{t,u}$-tangent vectorfield and Ψ and f are functions, then we have the following integration by parts identity:*

$$\int_{S_{t,u}} (Y\Psi)f \, d\upsilon_{\not g} = -\int_{S_{t,u}} (Yf)\Psi \, d\upsilon_{\not g} - \frac{1}{2}\int_{S_{t,u}} \mathrm{tr}_{\not g}{}^{(Y)}\not\pi \Psi f \, d\upsilon_{\not g}, \tag{10.37}$$

where ${}^{(Y)}\not\pi$ is the projection of ${}^{(Y)}\pi$ (see Def. 4.9) onto $S_{t,u}$.

PROOF. Let $\not\mathrm{div}$ be the operator defined in (8.1a). By the Leibniz rule, we have

$$\begin{aligned} \not\mathrm{div}(f\Psi Y) &= (Y\Psi)f + (Yf)\Psi + \frac{1}{2}f\Psi(\not g^{-1})^{AB}\{\not\nabla_A Y_B + \not\nabla_B Y_A\} \\ &= (Y\Psi)f + (Yf)\Psi + \frac{1}{2}f\Psi\mathrm{tr}_{\not g}{}^{(Y)}\not\pi. \end{aligned} \tag{10.38}$$

The desired identity (10.37) now follows from integrating (10.38) over $S_{t,u}$ and noting that the $S_{t,u}$ integral of the angular divergence of an $S_{t,u}$-tangent vectorfield must vanish. □

As we mentioned above, some of the integrands on the right-hand side of (10.29b) are not quite in a form that allows us to close our top-order L^2 estimates. Hence, in order to bound certain top-order error integrals, we need to carry out some additional integrations by parts; see Sect. 20.4. We carry out these integrations by parts in the next two lemmas. The proofs involve lengthy computations but are conceptually very simple.

REMARK 10.16 (**Important integrals vs. negligible integrals**). In the next two lemmas, the explicitly written integrals on the right-hand sides are the difficult ones that contribute to the degeneracy of our $\widetilde{\text{top-order}}$ L^2 estimates, while the integrals featuring the integrands $\mathrm{Error}_{(i)}$ and $\widetilde{\mathrm{Error}}_{(i)}$ have a negligible effect on the dynamics.

LEMMA 10.17 (**Integration by parts identity useful for bounding some top-order error integrals**). *Let η be a function and let w be a "weight" function. Let $O \in \{O_{(1)}, O_{(2)}, O_{(3)}\}$ be a geometric rotation vectorfield (see (6.3)). Then we have the following integration by parts identity:*

$$\begin{aligned} &-\int_{\mathcal{M}_{t,u}} w(\breve{R}\Psi)\left\{L\mathscr{Z}^N\Psi + \frac{1}{2}\mathrm{tr}_{\not g}\chi\mathscr{Z}^N\Psi\right\} O\eta \, d\varpi \\ &= -\int_{\mathcal{M}_{t,u}} w(\breve{R}\Psi)(L\eta)O\mathscr{Z}^N\Psi \, d\varpi \\ &+ \int_{\Sigma_t^u} w(\breve{R}\Psi)\eta C\mathscr{Z}^N\Psi \, d\underline{\varpi} \\ &+ \int_{\mathcal{M}_{t,u}} \mathrm{Error}_{(1)}[\mathscr{Z}^N\Psi; w; \eta] \, d\varpi + \int_{\Sigma_t^u} \mathrm{Error}_{(2)}[\mathscr{Z}^N\Psi; w; \eta] \, d\underline{\varpi} \\ &+ \int_{\Sigma_0^u} \mathrm{Error}_{(3)}[\mathscr{Z}^N\Psi; w; \eta] \, d\underline{\varpi}, \end{aligned} \tag{10.39}$$

where

(10.40a)

$$\begin{aligned}\mathrm{Error}_{(1)}[\mathscr{Z}^N\Psi; w; \eta] := & -w(O\breve{R}\Psi)(L\eta)\mathscr{Z}^N\Psi - (Ow)(\breve{R}\Psi)(L\eta)\mathscr{Z}^N\Psi \\ & - \frac{1}{2}w\mathrm{tr}_{\not g}{}^{(O)}\not\pi(\breve{R}\Psi)(L\eta)\mathscr{Z}^N\Psi \\ & - (Lw)(O\breve{R}\Psi)\eta\mathscr{Z}^N\Psi - (OLw)(\breve{R}\Psi)\eta\mathscr{Z}^N\Psi \\ & - \frac{1}{2}(Lw)\mathrm{tr}_{\not g}{}^{(O)}\not\pi(\breve{R}\Psi)\eta\mathscr{Z}^N\Psi \\ & - w(\breve{R}\Psi)\eta({}^{(O)}\not\pi_L^{\#}\cdot \not d\mathscr{Z}^N\Psi) - w({}^{(O)}\not\pi_L^{\#}\cdot \not d\breve{R}\Psi)(L\eta)\mathscr{Z}^N\Psi \\ & - ({}^{(O)}\not\pi_L^{\#}\cdot \not dw)(\breve{R}\Psi)\eta\mathscr{Z}^N\Psi \\ & - w(\mathrm{d}\not{\mathrm{i}}\mathrm{v}\,{}^{(O)}\not\pi_L^{\#})(\breve{R}\Psi)\eta\mathscr{Z}^N\Psi \\ & - w\left\{L\breve{R}\Psi + \frac{1}{2}\mathrm{tr}_{\not g}\chi\breve{R}\Psi\right\}\eta O\mathscr{Z}^N\Psi \\ & - (Lw)(\breve{R}\Psi)\eta O\mathscr{Z}^N\Psi \\ & - w\left(O\left\{L\breve{R}\Psi + \frac{1}{2}\mathrm{tr}_{\not g}\chi\breve{R}\Psi\right\}\right)\eta\mathscr{Z}^N\Psi \\ & - (Ow)\left\{L\breve{R}\Psi + \frac{1}{2}\mathrm{tr}_{\not g}\chi\breve{R}\Psi\right\}\eta\mathscr{Z}^N\Psi \\ & - \frac{1}{2}w\mathrm{tr}_{\not g}{}^{(O)}\not\pi\left\{L\breve{R}\Psi + \frac{1}{2}\mathrm{tr}_{\not g}\chi\breve{R}\Psi\right\}\eta\mathscr{Z}^N\Psi,\end{aligned}$$

(10.40b)

$$\begin{aligned}\mathrm{Error}_{(2)}[\mathscr{Z}^N\Psi; w; \eta] := & \, w(O\breve{R}\Psi)\eta\mathscr{Z}^N\Psi + (Ow)(\breve{R}\Psi)\eta\mathscr{Z}^N\Psi \\ & + \frac{1}{2}w\mathrm{tr}_{\not g}{}^{(O)}\not\pi(\breve{R}\Psi)\eta\mathscr{Z}^N\Psi,\end{aligned}$$

(10.40c)

$$\begin{aligned}\mathrm{Error}_{(3)}[\mathscr{Z}^N\Psi; w; \eta] := & -w(\breve{R}\Psi)\eta O\mathscr{Z}^N\Psi - w(O\breve{R}\Psi)\eta\mathscr{Z}^N\Psi - (Ow)(\breve{R}\Psi)\eta\mathscr{Z}^N\Psi \\ & - \frac{1}{2}w\mathrm{tr}_{\not g}{}^{(O)}\not\pi(\breve{R}\Psi)\eta\mathscr{Z}^N\Psi.\end{aligned}$$

PROOF. We first use the identity (10.35) with $\mathscr{Z}^N\Psi$ in the role of Ψ and $w(\breve{R}\Psi)O\eta$ in the role of f. We then split

(10.41)
$$\begin{aligned}\left\{L + \frac{1}{2}\mathrm{tr}_{\not g}\chi\right\}\left\{w(\breve{R}\Psi)O\eta\right\} = & \, w\left\{L\breve{R}\Psi + \frac{1}{2}\mathrm{tr}_{\not g}\chi\breve{R}\Psi\right\}O\eta \\ & + w(\breve{R}\Psi)LO\eta + (Lw)(\breve{R}\Psi)O\eta.\end{aligned}$$

Furthermore, we commute the operators L and O in the product $w(\breve{R}\Psi)LO\eta$ on the right-hand side of (10.41) and then integrate by parts on the $S_{t,u}$ via Lemma 10.15 (note that $\frac{1}{2}\mathrm{tr}_{\not g}{}^{(Y)}\not\pi = \mathrm{d}\not{\mathrm{i}}\mathrm{v}\,Y$ for $S_{t,u}$-tangent vectorfields Y) to pull, in all products, all instances of O and the commutator $[L, O] = {}^{(O)}\not\pi_L^{\#}$ (see Lemma 4.10) off of η. In total, these steps lead to the lemma. □

LEMMA 10.18 (**A second integration by parts identity useful for bounding some top-order error integrals**). *Let Y be an $S_{t,u}$-tangent vectorfield and let $w = w(t,u)$ be a "weight" function. Then we have the following integration by parts identity:*

$$(10.42)\qquad \begin{aligned} &-\int_{\mathcal{M}_{t,u}} w(\breve{R}\Psi)\left\{L\mathscr{Z}^N\Psi + \frac{1}{2}\mathrm{tr}_{g\!\!/}\chi\mathscr{Z}^N\Psi\right\}\mathrm{di}\!\!/\mathrm{v}\, Y\, d\varpi \\ &= -\int_{\mathcal{M}_{t,u}} w(\breve{R}\Psi)\left(\{\not{\mathcal{L}}_L Y + \mathrm{tr}_{g\!\!/}\chi Y\}\cdot d\!\!/\mathscr{Z}^N\Psi\right) d\varpi \\ &\quad + \int_{\Sigma_t^u} w(\breve{R}\Psi)(Y\cdot d\!\!/\mathscr{Z}^N\Psi)\, d\underline{\varpi} \\ &\quad + \int_{\mathcal{M}_{t,u}} \widetilde{\mathrm{Error}}_{(1)}[\mathscr{Z}^N\Psi; w; Y]\, d\varpi + \int_{\Sigma_t^u} \widetilde{\mathrm{Error}}_{(2)}[\mathscr{Z}^N\Psi; w; Y]\, d\underline{\varpi} \\ &\quad + \int_{\Sigma_0^u} \widetilde{\mathrm{Error}}_{(3)}[\mathscr{Z}^N\Psi; w; Y]\, d\underline{\varpi}, \end{aligned}$$

where

(10.43a)

$$\begin{aligned} \widetilde{\mathrm{Error}}_{(1)}[\mathscr{Z}^N\Psi; w; Y] := &-w\left(\{\not{\mathcal{L}}_L Y + \mathrm{tr}_{g\!\!/}\chi Y\}\cdot d\!\!/\breve{R}\Psi\right)\mathscr{Z}^N\Psi \\ &- (\breve{R}\Psi)\left(\{\not{\mathcal{L}}_L Y + \mathrm{tr}_{g\!\!/}\chi Y\}\cdot d\!\!/ w\right)\mathscr{Z}^N\Psi \\ &- \left\{Lw - \mathrm{tr}_{g\!\!/}\chi w\right\}(\breve{R}\Psi)(Y\cdot d\!\!/\mathscr{Z}^N\Psi) \\ &- \left\{Lw - \mathrm{tr}_{g\!\!/}\chi w\right\}(Y\cdot d\!\!/\breve{R}\Psi)\mathscr{Z}^N\Psi \\ &- (\breve{R}\Psi)(Y\cdot d\!\!/ Lw)\mathscr{Z}^N\Psi + (\breve{R}\Psi)\mathrm{tr}_{g\!\!/}\chi(Y\cdot d\!\!/ w)\mathscr{Z}^N\Psi \\ &+ w(\breve{R}\Psi)(Y\cdot d\!\!/\mathrm{tr}_{g\!\!/}\chi^{(Small)})\mathscr{Z}^N\Psi \\ &- w\left\{L\breve{R}\Psi + \frac{1}{2}\mathrm{tr}_{g\!\!/}\chi\breve{R}\Psi\right\}(Y\cdot d\!\!/\mathscr{Z}^N\Psi) \\ &- w\left(Y\cdot d\!\!/\left\{L\breve{R}\Psi + \frac{1}{2}\mathrm{tr}_{g\!\!/}\chi\breve{R}\Psi\right\}\right)\mathscr{Z}^N\Psi \\ &- (Y\cdot d\!\!/ w)\left\{L\breve{R}\Psi + \frac{1}{2}\mathrm{tr}_{g\!\!/}\chi\breve{R}\Psi\right\}\mathscr{Z}^N\Psi, \end{aligned}$$

(10.43b)

$$\widetilde{\mathrm{Error}}_{(2)}[\mathscr{Z}^N\Psi; w; Y] := w(Y\cdot d\!\!/\breve{R}\Psi)\mathscr{Z}^N\Psi + (Y\cdot d\!\!/ w)(\breve{R}\Psi)\mathscr{Z}^N\Psi,$$

(10.43c)

$$\begin{aligned} \widetilde{\mathrm{Error}}_{(3)}[\mathscr{Z}^N\Psi; w; Y] := &-w(\breve{R}\Psi)(Y\cdot d\!\!/\mathscr{Z}^N\Psi) - w(Y\cdot d\!\!/\breve{R}\Psi)\mathscr{Z}^N\Psi \\ &- (Y\cdot d\!\!/ w)(\breve{R}\Psi)\mathscr{Z}^N\Psi. \end{aligned}$$

PROOF. We first use (10.35) with $\mathscr{Z}^N\Psi$ in the role of Ψ and $f = f_1 f_2 f_3 := w(\breve{R}\Psi)\mathrm{di}\!\!/\mathrm{v}\, Y$ to remove the operator $\left\{L + \frac{1}{2}\mathrm{tr}_{g\!\!/}\chi\right\}$ off of $\mathscr{Z}^N\Psi$ on the left-hand side of (10.42). After removal, we have the spacetime integrand factor

$$(10.44)\quad (\mathscr{Z}^N\Psi)\left\{L+\frac{1}{2}\mathrm{tr}_{\not g}\chi\right\}(f_1f_2f_3)$$
$$= w(\breve{R}\Psi)(\mathscr{Z}^N\Psi)\left\{L+\mathrm{tr}_{\not g}\chi\right\}\mathrm{d\not iv}Y + (\breve{R}\Psi)(\mathscr{Z}^N\Psi)\left\{Lw-\mathrm{tr}_{\not g}\chi w\right\}\mathrm{d\not iv}Y$$
$$+ w\left\{L\breve{R}\Psi+\frac{1}{2}\mathrm{tr}_{\not g}\chi\breve{R}\Psi\right\}(\mathscr{Z}^N\Psi)\mathrm{d\not iv}Y.$$

We use Lemma 8.10 to rewrite the first product on the right-hand side of (10.44) as $w(\breve{R}\Psi)(\mathscr{Z}^N\Psi)\mathrm{d\not iv}\left\{\not{\mathcal{L}}_L Y+\mathrm{tr}_{\not g}\chi Y\right\}$. Finally, we integrate by parts on the $S_{t,u}$ to remove all covariant angular derivatives off of $\not{\mathcal{L}}_L Y+\mathrm{tr}_{\not g}\chi Y$ and Y and use the fact that by (4.1c), we have $\not d\mathrm{tr}_{\not g}\chi = \not d(2\varrho^{-1}+\mathrm{tr}_{\not g}\chi^{(Small)}) = \not d\mathrm{tr}_{\not g}\chi^{(Small)}$. In total, these steps lead to the identity (10.42). □

10.4. Error integrands arising from the deformation tensors of the multiplier vectorfields

In the next definition, we give names to the energy error integrands corresponding to the deformation tensors of the multiplier vectorfields (that is, the last integrand in (10.29a) and the next-to-last integrand in (10.29b)).

Definition 10.19 (**The error integrands corresponding to T and $\widetilde{K}$**). Given a function Ψ, we associate to it the following quantities ${}^{(T)}\mathfrak{P}[\Psi]$ and ${}^{(\widetilde{K})}\mathfrak{P}[\Psi]$, where $Q[\Psi]$ is the energy-momentum tensorfield of Def. 10.3 and T and $\widetilde{K}$ are the multiplier vectorfields from Def. 10.6:

$$(10.45a)\qquad {}^{(T)}\mathfrak{P}[\Psi] := -\frac{1}{2}\mu Q^{\alpha\beta}[\Psi]\,{}^{(T)}\pi_{\alpha\beta},$$

$$(10.45b)\qquad {}^{(\widetilde{K})}\mathfrak{P}[\Psi] := -\frac{1}{2}\mu Q^{\alpha\beta}[\Psi]\left\{{}^{(\widetilde{K})}\pi_{\alpha\beta}-\varrho^2\mathrm{tr}_{\not g}\chi g_{\alpha\beta}\right\}.$$

To derive our main results, we separately analyze the positive and negative parts of $L\mu$. This motivates the following definition.

Definition 10.20 (f_+ **and** f_-). Given any real-valued function f, we decompose it into its positive part f_+ and its negative part f_- as follows:

$$(10.46)\qquad f = f_+ - f_-,$$

where

$$(10.47)\qquad f_+ := \max\{f,0\},\qquad f_- := \max\{-f,0\}.$$

In the next lemma, decompose the quantities ${}^{(T)}\mathfrak{P}[\Psi]$ and ${}^{(\widetilde{K})}\mathfrak{P}[\Psi]$ from Def. 10.19 relative to the rescaled null frame $\{L,\breve{\underline{L}},X_1,X_2\}$.

Lemma 10.21 (**Null decomposition of the error integrands corresponding to T and $\widetilde{K}$**). *Let $[L\mu]_-$ and $[L\mu]_+$ denote the negative and positive parts of $L\mu$ (see Def. 10.20). Then the quantities ${}^{(T)}\mathfrak{P}[\Psi]$ and ${}^{(\widetilde{K})}\mathfrak{P}[\Psi]$ from Def. 10.19 can* *be decomposed relative to the rescaled null frame $\{L, \breve{\underline{L}}, X_1, X_2\}$ as follows:*

$$
{}^{(T)}\mathfrak{P}[\Psi] = \sum_{i=1}^{6} {}^{(T)}\mathfrak{P}_{(i)}[\Psi], \tag{10.48a}
$$

$$
{}^{(\widetilde{K})}\mathfrak{P}[\Psi] = -\frac{1}{2}\varrho^2|\not{d}\Psi|^2[L\upmu]_- + \sum_{i=1}^{4} {}^{(\widetilde{K})}\mathfrak{P}_{(i)}, \tag{10.48b}
$$

where

(10.49a)
$$
{}^{(T)}\mathfrak{P}_{(1)}[\Psi] := \frac{1}{2}(L\Psi)^2\left\{\breve{\underline{L}}\upmu - (1+\upmu)L\upmu\right\},
$$

(10.49b)
$$
{}^{(T)}\mathfrak{P}_{(2)}[\Psi] := \frac{1}{2}|\not{d}\Psi|^2\left\{(\breve{\underline{L}}\upmu + L\upmu) + 3\upmu L\upmu\right\},
$$

(10.49c)
$$
{}^{(T)}\mathfrak{P}_{(3)}[\Psi] := -(\breve{\underline{L}}\Psi)(\not{d}^{\#}\Psi)\cdot\left\{\not{d}\upmu + 2\zeta^{(Trans-\Psi)} + 2\upmu\zeta^{(Tan-\Psi)}\right\},
$$

(10.49d)
$$
{}^{(T)}\mathfrak{P}_{(4)}[\Psi] := (L\Psi)(\not{d}^{\#}\Psi)\cdot\left\{(1-\upmu)\not{d}\upmu + 2(1+\upmu)\zeta^{(Trans-\Psi)} + 2\upmu(1+\upmu)\zeta^{(Tan-\Psi)}\right\},
$$

(10.49e)
$$
{}^{(T)}\mathfrak{P}_{(5)}[\Psi] := -(\upmu\not{d}^{\#}\Psi\hat{\otimes}\not{d}^{\#}\Psi)\cdot\left\{\hat{\chi}^{(Small)} + 2\hat{\not{k}}^{(Trans-\Psi)} + 2\upmu\hat{\not{k}}^{(Tan-\Psi)}\right\},
$$

(10.49f)
$$
{}^{(T)}\mathfrak{P}_{(6)}[\Psi] := -\frac{1}{2}(L\Psi)(\breve{\underline{L}}\Psi)\left\{\mathrm{tr}_{\not{g}}\chi + 2\mathrm{tr}_{\not{g}}\not{k}^{(Trans-\Psi)} + 2\upmu\mathrm{tr}_{\not{g}}\not{k}^{(Tan-\Psi)}\right\},
$$

and

$$
{}^{(\widetilde{K})}\mathfrak{P}_{(1)}[\Psi] := (L\Psi)^2\left\{2\varrho\breve{\underline{L}}\varrho - \varrho^2 L\upmu\right\}, \tag{10.50a}
$$

$$
{}^{(\widetilde{K})}\mathfrak{P}_{(2)}[\Psi] := \frac{1}{2}\varrho^2(|\not{d}\Psi|^2)\left\{\frac{[L\upmu]_+}{\upmu} - \mathrm{tr}_{\not{g}}\chi^{(Small)}\right\}, \tag{10.50b}
$$

$$
{}^{(\widetilde{K})}\mathfrak{P}_{(3)}[\Psi] := \varrho^2(L\Psi)(\not{d}^{\#}\Psi)\cdot\left\{\not{d}\upmu + 2\zeta^{(Trans-\Psi)} + 2\upmu\zeta^{(Tan-\Psi)}\right\}, \tag{10.50c}
$$

$$
{}^{(\widetilde{K})}\mathfrak{P}_{(4)}[\Psi] := -\varrho^2(\upmu\not{d}^{\#}\Psi\hat{\otimes}\not{d}^{\#}\Psi)\cdot\hat{\chi}^{(Small)}. \tag{10.50d}
$$

In the above expressions, the $S_{t,u}$ tensorfields χ, $\chi^{(Small)}$, $\zeta^{(Trans-\Psi)}$, $\not{k}^{(Trans-\Psi)}$, $\zeta^{(Tan-\Psi)}$, and $\not{k}^{(Tan-\Psi)}$ are defined by (3.70), (4.1c), (5.4a), (5.4b), (5.5a), *and* (5.5b).

REMARK 10.22 (**Isolating the Morawetz term**). Note that in (10.48b), we have singled out the nonpositive term $-\frac{1}{2}\varrho^2[L\upmu]_-|\not{d}\Psi|^2$, by "removing it" from ${}^{(\widetilde{K})}\mathfrak{P}_{(2)}[\Psi]$. The reason is that it generates an important Morawetz-type spacetime integral that plays a fundamental role in our analysis; see the discussion surrounding equation (2.53) and Lemma 14.5.

PROOF. The basic idea is to use the null decomposition (3.51b) of g^{-1} and the null decompositions of Lemmas 10.7 and 10.8 to expand the products (10.45a) and

(10.45b) relative to the frame $\{L, \breve{\underline{L}}, X_1, X_2\}$. For example, to compute (10.49a)-(10.49f), we first use (3.51b) to deduce that the left-hand side of (10.45a) can be expressed as

$$-\frac{1}{2}\upmu\left\{-\frac{1}{2}\upmu^{-1}\left(L^{\alpha}\breve{\underline{L}}^{\kappa}+L^{\kappa}\breve{\underline{L}}^{\alpha}\right)+(\mathscr{g}^{-1})^{AB}X_A^{\alpha}X_B^{\kappa}\right\}$$
$$\times\left\{-\frac{1}{2}\upmu^{-1}\left(L^{\beta}\breve{\underline{L}}^{\lambda}+L^{\lambda}\breve{\underline{L}}^{\beta}\right)+(\mathscr{g}^{-1})^{AB}X_A^{\beta}X_B^{\lambda}\right\}Q_{\alpha\beta}[\Psi]{}^{(T)}\pi_{\kappa\lambda}.$$

We then fully expand the above product in terms of the null components of $Q[\Psi]$ and ${}^{(T)}\pi$ and use Lemmas 10.7 and 10.8 to substitute for these null components. Carrying out straightforward computations, we arrive at (10.49a)-(10.49f).

The proofs of (10.50a)-(10.50d) are similar. As we noted in Remark 10.22, we removed the important term $-\frac{1}{2}\varrho^2[L\upmu]_-|\mathscr{d}\Psi|^2$, which appears in the product

$$-\frac{1}{4}\upmu^{-1}Q_{L\breve{\underline{L}}}[\Psi]\left\{{}^{(\widetilde{K})}\pi_{L\breve{\underline{L}}}-\varrho^2\mathrm{tr}_{\mathscr{g}}\chi g(L,\breve{\underline{L}})\right\},$$

and explicitly placed it on the right-hand side of (10.48b). □

CHAPTER 11

Avoiding Derivative Loss and Other Difficulties via Modified Quantities

We use the results of Chapter 11 only to close our top-order L^2 estimates. Specifically, in this chapter, we derive transport equations for several "modified" versions of the eikonal function quantities $\mathrm{tr}_{\not g}\chi^{(Small)}$ and $\not d\mu$. When combined with elliptic estimates on the $S_{t,u}$, the modified quantities allow us to estimate some important top-order derivatives of $\mathrm{tr}_{\not g}\chi^{(Small)}$ *without losing derivatives.* These steps are necessary even for proving an up-to-top-order local well-posedness result relative to the geometric coordinates. The idea that one should work with modified quantities in order to close the L^2 estimates for the eikonal function u first appeared in Christodoulou-Klainerman's proof of the global stability of the Minkowski spacetime as a solution to Einstein's equations [**19**]. A similar idea was later employed[1] by Klainerman-Rodnianski [**50**] in their proof of low regularity local well-posedness for quasilinear wave equations of the form $-\partial_t^2\Psi + (g^{-1})^{ab}(\Psi)\partial_a\partial_b\Psi = \mathcal{N}(\Psi, \partial\Psi)$; their work was also heavily based on using an eikonal function to prove sharp estimates. A similar idea was later used by Christodoulou in his shock formation monograph [**17**]. In the latter two works, the key step is analogous to Cor. 11.4 in the present work.

The main idea in avoiding the derivative loss is that for solutions to $\square_g\Psi = 0$, a certain component of the Ricci curvature tensor of g, specifically $(\not g^{-1})^{AB}\mathscr{R}_{LALB}$ (where $\mathscr{R}$ is the Riemann curvature tensor of g) can be written as a perfect L derivative of the first derivatives of Ψ plus some error terms. The main point is that the Ψ-dependent error terms involve only Ψ and its first derivatives, and *not second-order derivatives of* Ψ, which are of principal order from the point of view of differentiability. The perfect L derivative structure is what allows us to construct suitably modified versions of $\mathrm{tr}_{\not g}\chi^{(Small)}$ that do not lose derivatives.

We now provide more details concerning this structure. Schematically, we have $L\mathrm{tr}_{\not g}\chi^{(Small)} = \mathscr{D}^2\Psi + \mathrm{Error}(\Psi, \mathscr{D}\Psi)$. Integrating this identity along the integral curves of L, we find only that $\mathrm{tr}_{\not g}\chi^{(Small)}$ has the same degree of differentiability as $\mathscr{D}^2\Psi$. We can commute this identity with vectorfields $Z \in \mathcal{Z}$ to obtain a similar result for the higher-order derivatives of $\mathrm{tr}_{\not g}\chi^{(Small)}$. The problem is that the top-order result is $\mathscr{Z}^N\mathrm{tr}_{\not g}\chi^{(Small)} \sim \mathscr{D}^2\mathscr{Z}^N\Psi$, where the right-hand side involves one more derivative of Ψ than our top-order energy-cone flux quantities allow us to control. Moreover, to close the top-order L^2 estimates corresponding to the timelike multiplier T, we actually *do need* to be able to control $\mathscr{Z}^N\mathrm{tr}_{\not g}\chi^{(Small)}$ since it appears as an error term. To overcome the apparent derivative loss, we construct

[1]Although the authors in [**19**] and [**50**] had to use modified quantities to avoid losing a derivative, they did not have to face the difficulty of obtaining estimates in regions where μ is near 0.

a "fully modified" version of $\mathrm{tr}_{\not g}\chi^{(Small)}$ (to be distinguished from the "partially modified" quantities discussed below) by multiplying it by μ and then adding to it certain terms connected to the perfect L derivative structure mentioned above. The new quantity verifies $L(\mathsf{Modified}) = \mathrm{Error}(\Psi, \mathscr{D}\Psi)$, and we can integrate this identity to conclude that $\mathsf{Modified}$ has the same regularity as $\mathscr{D}\Psi$, a gain of one derivative. Actually, these schematic expressions are somewhat misleading in the following sense: one of the error terms on the right-hand side of the equation $L(\mathsf{Modified}) = \cdots$ is roughly of the form $|\hat{\chi}|^2$ and cannot be eliminated through modification. At the top-order, the corresponding term would lead to derivative loss. To control the top-order derivatives of $|\hat{\chi}|^2$ we will, as we alluded to above, derive a family of elliptic estimates on the $S_{t,u}$ (see Chapter 18). The elliptic estimates allow us to control the top-order derivatives of $\hat{\chi}$ in terms of quantities that do not lose derivatives. This strategy was first employed in [**19**] and later in [**50**] and [**17**]. We also stress that in [**17**] and in the present work, *the structure of the top-order and the lower-order terms are both important*. In particular, we have to take care in keeping track of factors of μ and μ^{-1}, which can greatly affect the dynamics near the shock. We derive the transport equation verified by the lowest-order modified quantity in Cor. 11.7. We provide the higher-order versions of this equation in Lemma 11.8 and Prop. 11.10.

In [**17**], Christodoulou also encountered a derivative loss difficulty for the term $\not\Delta\mu$, similar to the one for $\mathrm{tr}_{\not g}\chi^{(Small)}$ described above. To circumvent it, he derived an independent transport equation and corresponding estimates for a fully modified version of $\not\Delta\mu$, in analogy with his treatment of $\mathrm{tr}_{\not g}\chi^{(Small)}$. Although his strategy solved the derivative loss problem, implementing it required a large number of complicated calculations and estimates. In the present monograph, we adopt an alternate strategy to handle this term. Specifically, in Chapter 15, we use commutation estimates to control the relevant top-order derivatives of $\not\Delta\mu$ in terms of related top-order derivatives of $\mathrm{tr}_{\not g}\chi^{(Small)}$. In doing so, we avoid having to derive an analog of Christodoulou's transport equation for a modified version of $\not\Delta\mu$, which saves a large amount of work. Our strategy, when supplemented with elliptic estimates for $\not\Delta\mu$, allows us to bound all top-order derivatives of μ without loss, *as long as the derivative operator involves an L derivative or two angular derivatives*; see Remark 2.14. Amazingly, the structure of the right-hand side of (9.13) together with the identities of Prop. 7.7 imply that, thanks to the special properties of the vectorfields $Z \in \mathscr{Z}$, all top-order derivatives of μ that appear in our equations involve an L derivative or two angular derivatives.

In order to close the top-order L^2 estimates corresponding to the Morawetz multiplier $\widetilde{K}$, we also need to derive transport equations for "partially modified" versions of $\mathrm{tr}_{\not g}\chi^{(Small)}$ and $\not d\mu$. The partial modifications are not connected to avoiding derivative loss, but rather to avoiding certain error integrals that have unfavorable time-growth properties. We derive these equations in Sects. 11.3 and 11.4.

11.1. Preliminary structural identities

In order to prove our sharp classical lifespan theorem, we must understand the sharp structure of the transport equations verified by the modified quantities. To derive this sharp structure, we rely on identities verified by the Riemann curvature tensor of g. We now define this tensor, and, for later use, the Riemann curvature tensor of $\not{g}$ as well.

DEFINITION 11.1 (**The Riemann curvature tensors of** g **and** $\not{g}$). The Riemann curvature tensor $\mathscr{R}$ of the spacetime metric g is the type $\binom{0}{4}$ spacetime tensorfield defined by

$$\mathscr{R}(W,X,Y,Z) := g(-\mathscr{D}^2_{WX}Y + \mathscr{D}^2_{XW}Y, Z), \tag{11.1}$$

where W, X, Y, and Z are arbitrary spacetime vectors and $\mathscr{D}^2_{WX}Y := W^\alpha X^\beta \mathscr{D}_\alpha \mathscr{D}_\beta Y$, as in Remark 3.13.

Similarly, the Riemann curvature tensor $\mathfrak{R}$ of the Riemannian metric $\not{g}$ on $S_{t,u}$ induced by g (see Defs. 3.9 and 3.11) is the type $\binom{0}{4}$ $S_{t,u}$ tensorfield defined by

$$\mathfrak{R}(W,X,Y,Z) := \not{g}(-\nabla\!\!\!/^2_{WX}Y + \nabla\!\!\!/^2_{XW}Y, Z), \tag{11.2}$$

where W, X, Y, and Z are arbitrary $S_{t,u}$-tangent vectors.

It is a standard fact (see, for example, [**67**, Proposition 4]) that the following symmetry properties hold:

$$\mathscr{R}(W,X,Y,Z) = -\mathscr{R}(X,W,Y,Z) = -\mathscr{R}(W,X,Z,Y) = \mathscr{R}(Y,Z,W,X), \tag{11.3}$$

and similarly for $\mathfrak{R}(W,X,Y,Z)$.

In the next lemma, we compute the components of $\mathscr{R}$ relative to the rectangular spacetime coordinates. This is a preliminary calculation that will help us compute some of the components of $\mathscr{R}$ relative to the rescaled frame $\{L, \breve{R}, X_1, X_2\}$.

LEMMA 11.2 (**Rectangular components of the Riemann curvature tensor of** g). *Relative to the* **rectangular** *coordinate system, the components of* $\mathscr{R}$ *can be expressed as follows:*

(11.4)

$$\begin{aligned}
\mathscr{R}_{\mu\nu\alpha\beta} = & \frac{1}{2}\left\{G_{\beta\mu}\mathscr{D}^2_{\alpha\nu}\Psi + G_{\alpha\nu}\mathscr{D}^2_{\beta\mu}\Psi - G_{\beta\nu}\mathscr{D}^2_{\alpha\mu}\Psi - G_{\alpha\mu}\mathscr{D}^2_{\beta\nu}\Psi\right\} \\
& + \frac{1}{4}G_{\mu\alpha}G_{\nu\beta}(g^{-1})^{\kappa\lambda}(\partial_\kappa\Psi)(\partial_\lambda\Psi) - \frac{1}{4}G_{\mu\beta}G_{\nu\alpha}(g^{-1})^{\kappa\lambda}(\partial_\kappa\Psi)(\partial_\lambda\Psi) \\
& + \frac{1}{4}(g^{-1})^{\kappa\lambda}[G_{\kappa(\nu}\partial_{\alpha)}\Psi][G_{\lambda(\mu}\partial_{\beta)}\Psi] - \frac{1}{4}(g^{-1})^{\kappa\lambda}[G_{\kappa(\mu}\partial_{\alpha)}\Psi][G_{\lambda(\nu}\partial_{\beta)}\Psi] \\
& + \frac{1}{2}\left\{G'_{\beta\mu}(\partial_\alpha\Psi)(\partial_\nu\Psi) + G'_{\alpha\nu}(\partial_\beta\Psi)(\partial_\mu\Psi)\right\} \\
& - \frac{1}{2}\left\{G'_{\beta\nu}(\partial_\alpha\Psi)(\partial_\mu\Psi) + G'_{\alpha\mu}(\partial_\beta\Psi)(\partial_\nu\Psi)\right\}.
\end{aligned}$$

In (11.4), $\mathscr{D}$ *denotes the Levi-Civita connection of* g, $\mathscr{D}^2\Psi$ *(which is a symmetric type* $\binom{0}{2}$ *tensorfield) denotes the second covariant derivative of* Ψ, $G_{\alpha\beta} = G_{\alpha\beta}(\Psi) = \dfrac{dg_{\alpha\beta}(\Psi)}{d\Psi}$, $G'_{\alpha\beta}(\Psi) = \dfrac{dG_{\alpha\beta}(\Psi)}{d\Psi}$, *and* $G_{\mu(\nu}\partial_{\lambda)}\Psi := G_{\mu\nu}\partial_\lambda\Psi + G_{\mu\lambda}\partial_\nu\Psi$.

PROOF. It is a standard fact (see, for example, [**17**, Chapter 4], and note, in view of (11.3), that the sign conventions for $\mathscr{R}_{\mu\nu\alpha\beta}$ are the same as those used here but that the index conventions for the Christoffel symbols are different), based on definition (11.1) and equation (4.4a), that relative to an arbitrary coordinate system, the components of the Riemann curvature tensor of g can be expressed as

(11.5)

$$\begin{aligned}\mathscr{R}_{\mu\nu\alpha\beta} &= \frac{1}{2}\Big\{\mathscr{D}^2_{\alpha\nu}g_{\beta\mu} + \mathscr{D}^2_{\beta\mu}g_{\alpha\nu} - \mathscr{D}^2_{\alpha\mu}g_{\beta\nu} - \mathscr{D}^2_{\beta\nu}g_{\alpha\mu}\Big\} \\ &+ \frac{1}{8}(g^{-1})^{\kappa\lambda}\Big\{\partial_{(\kappa}g_{\beta\mu)}[\partial_{(\alpha}g_{\nu)\lambda} - \partial_\lambda g_{\alpha\nu}] + \partial_{(\kappa}g_{\alpha\nu)}[\partial_{(\beta}g_{\mu)\lambda} - \partial_\lambda g_{\beta\mu}]\Big\} \\ &- \frac{1}{8}(g^{-1})^{\kappa\lambda}\Big\{\partial_{(\kappa}g_{\beta\nu)}[\partial_{(\alpha}g_{\mu)\lambda} - \partial_\lambda g_{\alpha\mu}] + \partial_{(\kappa}g_{\alpha\mu)}[\partial_{(\beta}g_{\nu)\lambda} - \partial_\lambda g_{\beta\nu}]\Big\},\end{aligned}$$

where $\mathscr{D}$ denotes the Levi-Civita connection of g, the components $g_{\mu\nu}$ are treated as **scalar-valued functions** for the purpose of covariant differentiation, $\partial_{(\lambda}g_{\mu\nu)} := \partial_\lambda g_{\mu\nu} + \partial_\nu g_{\lambda\mu} + \partial_\mu g_{\nu\lambda}$, and $\partial_{(\lambda}g_{\mu)\nu} := \partial_\lambda g_{\mu\nu} + \partial_\mu g_{\lambda\nu}$. The desired expression (11.4) now follows readily from straightforward computations. □

We now compute the curvature component $\mathscr{R}_{LALB}$.

COROLLARY 11.3 (**A $\upmu^{-1}$-regular expression for $\mathscr{R}_{LALB}$**). *The curvature component $\mathscr{R}_{LALB}$ can be expressed as follows, where we are using the notation of Sect.* 3.20 *to schematically denote the terms on the last line:*

$$\begin{aligned}\mathscr{R}_{LALB} &= \frac{1}{2}\Big\{-\not{G}_{AB}\,L(L\Psi) + \not{G}_{LA}\,\not{d}_B L\Psi + \not{G}_{LB}\,\not{d}_A L\Psi - G_{LL}\not{\nabla}^2_{AB}\Psi\Big\} \\ &+ \frac{1}{2}\upmu^{-1}G_{LL}\chi_{AB}\breve{R}\Psi - \frac{1}{2}\not{G}_{LA}\,\chi_B^{\ C}\not{d}_C\Psi - \frac{1}{2}\not{G}_{LB}\,\chi_A^{\ C}\not{d}_C\Psi \\ &+ \begin{pmatrix} G^2_{(Frame)} \\ G'_{(Frame)} \end{pmatrix}\begin{pmatrix} L\Psi \\ \not{d}\Psi \end{pmatrix}^2 + G^2_{(Frame)}\not{g}^{-1}\begin{pmatrix} L\Psi \\ \not{d}\Psi \end{pmatrix}^2.\end{aligned} \tag{11.6}$$

In (11.6), χ *is the* $S_{t,u}$ *tensorfield defined in* (3.70).

PROOF. We contract each side of (11.4) against $L^\mu X_A^\nu L^\alpha X_B^\beta$ and use the decompositions (see Lemma 3.48)

$$\begin{aligned}(g^{-1})^{\kappa\lambda} &= (\not{g}^{-1})^{\kappa\lambda} - L^\kappa L^\lambda - L^\kappa R^\lambda - R^\kappa L^\lambda \\ &= (\not{g}^{-1})^{\kappa\lambda} - L^\kappa L^\lambda - \upmu^{-1}L^\kappa \breve{R}^\lambda - \upmu^{-1}\breve{R}^\kappa L^\lambda\end{aligned} \tag{11.7}$$

to deduce that

(11.8)

$$\begin{aligned}\mathscr{R}_{\mu\nu\alpha\beta}L^\mu X_A^\nu L^\alpha X_B^\beta &= \frac{1}{2}\{-\not{G}_{AB}\,\mathscr{D}^2_{LL}\Psi + \not{G}_{LA}\,\mathscr{D}^2_{LB}\Psi + \not{G}_{LB}\,\mathscr{D}^2_{LA}\Psi - G_{LL}\mathscr{D}^2_{AB}\Psi\} \\ &+ \frac{1}{2}\upmu^{-1}\Big\{-G_{LL}\not{G}_{AB}\,(L\Psi)\breve{R}\Psi + \not{G}_{LA}\,\not{G}_{LB}\,(L\Psi)\breve{R}\Psi\Big\} \\ &+ \begin{pmatrix} G^2_{(Frame)} \\ G'_{(Frame)} \end{pmatrix}\begin{pmatrix} L\Psi \\ \not{d}\Psi \end{pmatrix}^2 + G^2_{(Frame)}\not{g}^{-1}\begin{pmatrix} L\Psi \\ \not{d}\Psi \end{pmatrix}^2.\end{aligned}$$

In deriving (11.8), we used the fact that the terms arising from the last three lines on the right-hand side of (11.4) generate harmless lower-order terms that can be incorporated into the terms on the last line on the right-hand side of (11.8). This fact is easy to see for the terms generated by the terms on the last two lines on the right-hand side of (11.4). To obtain this fact for the terms generated by the terms on the third-from-last line on the right-hand side of (11.4), we have exploited that in these terms, the λ, κ indices are contractions against G and *not differentiations* of

Ψ. Hence, we can use the decomposition $(g^{-1})^{\kappa\lambda} = (\not{g}^{-1})^{\kappa\lambda} - L^{\kappa}L^{\lambda} - L^{\kappa}R^{\lambda} - R^{\kappa}L^{\lambda}$ to substitute for $(g^{-1})^{\kappa\lambda}$ in these terms.

Our next goal, which is the most important part of the proof, is to show that the μ^{-1}-containing terms on the second line of (11.8) are canceled by corresponding terms in first line, so that the sum of the first two lines contains no such terms. To this end, we use Lemmas 4.6 and 5.1 to compute

(11.9)
$$\mathscr{D}^2_{LL}\Psi = LL\Psi - \frac{1}{2}\mu^{-1}G_{LL}(L\Psi)\breve{R}\Psi + G_{(Frame)}(L\Psi)^2,$$

(11.10)
$$\mathscr{D}^2_{LA}\Psi = \not{d}_A L\Psi - \frac{1}{2}\mu^{-1}\not{G}_{LA}(L\Psi)\breve{R}\Psi - \chi_A^{\ B}\not{d}_B\Psi + G_{(Frame)}\begin{pmatrix} L\Psi \\ \not{d}\Psi \end{pmatrix} L\Psi,$$

(11.11)
$$\mathscr{D}^2_{AB}\Psi = \not{\nabla}^2_{AB}\Psi - \frac{1}{2}\mu^{-1}\not{G}_{AB}(L\Psi)\breve{R}\Psi - \mu^{-1}\chi_{AB}\breve{R}\Psi + G_{(Frame)}\begin{pmatrix} L\Psi \\ \not{d}\Psi \end{pmatrix} L\Psi.$$

Inserting these identities into (11.8), we observe the desired cancellations and thus arrive at (11.6). □

We now show that up to lower-order error terms, $(\not{g}^{-1})^{AB}\mathscr{R}_{LALB}$ can be written as a perfect L derivative of the first derivatives of Ψ. This is the key step that will allow us to avoid losing derivatives in our estimates of the top-order derivatives of the eikonal function u.

COROLLARY 11.4 (**The key identity verified by the curvature component** $(\not{g}^{-1})^{AB}\mathscr{R}_{LALB}$). *Assume that* $\square_{g(\Psi)}\Psi = 0$. *Then the curvature component* $\mu(\not{g}^{-1})^{AB}\mathscr{R}_{LALB}$ *can be expressed as a perfect* L *derivative plus lower-order terms* $\mathfrak{A}$ *as follows:*

(11.12)
$$\mu(\not{g}^{-1})^{AB}\mathscr{R}_{LALB} = L\left\{-G_{LL}\breve{R}\Psi - \frac{1}{2}\mu\not{G}_A^{\ A}L\Psi - \frac{1}{2}\mu G_{LL}L\Psi + \mu\not{G}_L^{\ A}\not{d}_A\Psi\right\} + \mathfrak{A},$$

where $\mathfrak{A}$ *has the following schematic structure (see Sect. 3.20):*

$$\mathfrak{A} = \sum_{p=0}^{1}\begin{pmatrix} G^2_{(Frame)}\not{g}^{-1} \\ G'_{(Frame)} \end{pmatrix}(\not{g}^{-1})^p\begin{pmatrix} \mu L\Psi \\ \breve{R}\Psi \\ \mu\not{d}\Psi \end{pmatrix}\begin{pmatrix} L\Psi \\ \not{d}\Psi \end{pmatrix}. \tag{11.13}$$

Furthermore, without assuming $\square_{g(\Psi)}\Psi = 0$, *we have*

$$\begin{aligned}(\not{g}^{-1})^{AB}\mathscr{R}_{LALB} = {} & \frac{(L\mu)}{\mu}\mathrm{tr}_{\not{g}}\chi + L\left\{-\frac{1}{2}\not{G}_A^{\ A}L\Psi - \frac{1}{2}G_{LL}L\Psi + \not{G}_L^{\ A}\not{d}_A\Psi\right\} \\ & - \frac{1}{2}G_{LL}\not{\Delta}\Psi + \mathfrak{B},\end{aligned} \tag{11.14}$$

where $\mathfrak{B}$ *has the following schematic structure:*

$$\mathfrak{B} = G_{(Frame)}\mathrm{tr}_{\not{g}}\chi L\Psi + \sum_{p=0}^{1}\begin{pmatrix} G^2_{(Frame)}\not{g}^{-1} \\ G'_{(Frame)} \end{pmatrix}(\not{g}^{-1})^p\begin{pmatrix} L\Psi \\ \not{d}\Psi \end{pmatrix}^2. \tag{11.15}$$

PROOF. We first prove (11.12). To proceed, we contract (11.6) against $\upmu(\not{g}^{-1})^{AB}$ to deduce that

$$\begin{aligned}(11.16)\quad \upmu(\not{g}^{-1})^{AB}\mathscr{R}_{LALB} &= -\frac{1}{2}\upmu\not{G}_A^{\ A}\, L(L\Psi) + \upmu\not{G}_L^{\ A}\, \not{d}_A L\Psi - \frac{1}{2}\upmu G_{LL}\not{\Delta}\Psi \\ &+ \frac{1}{2}G_{LL}\mathrm{tr}_{\not{g}}\chi\breve{R}\Psi - \upmu\not{G}_L^{\ A}\chi_A^{\ B}\not{d}_B\Psi \\ &+ \sum_{p=0}^{1}\begin{pmatrix} G^2_{(Frame)}(\not{g}^{-1})^p \\ G'_{(Frame)} \end{pmatrix}\not{g}^{-1}\begin{pmatrix} \upmu L\Psi \\ \upmu\not{d}\Psi \end{pmatrix}\begin{pmatrix} L\Psi \\ \not{d}\Psi \end{pmatrix}.\end{aligned}$$

Using Cor. 4.15, the identity $L\upmu = G_{(Frame)}\begin{pmatrix} \upmu L\Psi \\ \breve{R}\Psi \end{pmatrix}$ (that is, Lemma 4.6), and the identities $\not{\mathcal{L}}_L\not{g}_{AB} = 2\chi_{AB}$ (see (3.72)) and $(\not{\mathcal{L}}_L\not{g}^{-1})^{AB} = -2\chi^{AB}$ (see also (7.4a)-(7.4b)), we rewrite the first two terms on the right-hand side of (11.16) as

$$\begin{aligned}(11.17)\quad -\frac{1}{2}\upmu\not{G}_A^{\ A}\, L(L\Psi) &= -\frac{1}{2}L\left\{\upmu\not{G}_A^{\ A}\, L\Psi\right\} \\ &+ \sum_{p=0}^{1}\begin{pmatrix} G^2_{(Frame)}(\not{g}^{-1})^p \\ G'_{(Frame)} \end{pmatrix}\not{g}^{-1}\begin{pmatrix} \upmu L\Psi \\ \breve{R}\Psi \\ \upmu\not{d}\Psi \end{pmatrix}\begin{pmatrix} L\Psi \\ \not{d}\Psi \end{pmatrix},\end{aligned}$$

$$\begin{aligned}(11.18)\quad \upmu\not{G}_L^{\ A}\, \not{d}_A L\Psi &= L\left\{\upmu\not{G}_L^{\ A}\, \not{d}_A\Psi\right\} + \upmu\not{G}_L^{\ A}\chi_A^{\ B}\not{d}_B L\Psi \\ &+ \sum_{p=0}^{1}\begin{pmatrix} G^2_{(Frame)}(\not{g}^{-1})^p \\ G'_{(Frame)} \end{pmatrix}\not{g}^{-1}\begin{pmatrix} \upmu L\Psi \\ \breve{R}\Psi \\ \upmu\not{d}\Psi \end{pmatrix}\begin{pmatrix} L\Psi \\ \not{d}\Psi \end{pmatrix}.\end{aligned}$$

Furthermore, from Cor. 4.15, the identity $L\upmu = G_{(Frame)}\begin{pmatrix} \upmu L\Psi \\ \breve{R}\Psi \end{pmatrix}$ mentioned above, (5.12a), and the assumption[2] $\square_{g(\Psi)}\Psi = 0$, we deduce that the third term on the right-hand side of (11.16) can be rewritten as

$$\begin{aligned}(11.19)\quad -\frac{1}{2}\upmu G_{LL}\not{\Delta}\Psi &= -\frac{1}{2}L\left\{G_{LL}(\upmu L\Psi + 2\breve{R}\Psi)\right\} - \frac{1}{2}G_{LL}\mathrm{tr}_{\not{g}}\chi\breve{R}\Psi \\ &+ \begin{pmatrix} G^2_{(Frame)}\not{g}^{-1} \\ G'_{(Frame)} \end{pmatrix}\begin{pmatrix} \upmu L\Psi \\ \breve{R}\Psi \\ \upmu\not{d}\Psi \end{pmatrix}\begin{pmatrix} L\Psi \\ \not{d}\Psi \end{pmatrix}.\end{aligned}$$

Inserting (11.17), (11.18), and (11.19) into (11.16), we deduce the desired identities (11.12) and (11.13).

To prove (11.14), we repeat the above argument, but we do not use equation (11.19) to substitute for the $\not{\Delta}\Psi$ term on the right-hand side of (11.16). This results in the presence of the term $\frac{1}{2}\upmu^{-1}G_{LL}\mathrm{tr}_{\not{g}}\chi\breve{R}\Psi$, arising from the first term on the second line of (11.16), on the right-hand side of (11.14). Using Lemma 4.6, we can rewrite this term as $\upmu^{-1}(L\upmu)\mathrm{tr}_{\not{g}}\chi$ plus a term that is schematically of the form of the first term on the right-hand side of (11.15). This explains the origin of the first term on the right-hand side of (11.14). □

[2]This is the critical step in the proof where we use the wave equation.

11.2. Full modification of the trace of the re-centered null second fundamental form

We recall that we defined the re-centered null second fundamental form $\chi^{(Small)}$ in (4.1c). We now define our fully modified version of $\mathrm{tr}_{\not g}\chi^{(Small)}$.

DEFINITION 11.5 (**Lowest-order fully modified version of $\mathrm{tr}_{\not g}\chi^{(Small)}$**). We define

$$\mathscr{X} := \upmu \mathrm{tr}_{\not g}\chi^{(Small)} + \mathfrak{X}, \tag{11.20a}$$

$$\mathfrak{X} := -G_{LL}\breve{R}\Psi - \frac{1}{2}\upmu \not G_A^{\ A} L\Psi - \frac{1}{2}\upmu G_{LL} L\Psi + \upmu \not G_L^{\ A} \not d_A \Psi. \tag{11.20b}$$

REMARK 11.6 (**Motivation for the definition of $\mathfrak{X}$**). Note that $\mathfrak{X}$ is the term in braces on the right-hand side of (11.12).

In the next corollary, we derive the transport equation verified by $\mathscr{X}$. The equation is an analog of the well-known *Raychaudhuri equation* [**70**] in General Relativity.

COROLLARY 11.7 (**The key "Raychaudhuri-type" transport equation satisfied by the fully modified version of $\mathrm{tr}_{\not g}\chi^{(Small)}$**). *Assume that* $\square_{g(\Psi)}\Psi = 0$ *Let $\mathscr{X}$ be the fully modified quantity defined in* (11.20a) *and let $\mathfrak{A}$ be the inhomogeneous term appearing on the right-hand side of* (11.12). *Then $\mathscr{X}$ satisfies the following transport equation:*

$$L\mathscr{X} = 2(L\upmu)\mathrm{tr}_{\not g}\chi - \frac{1}{2}\upmu(\mathrm{tr}_{\not g}\chi)^2 - 2\frac{1}{\varrho}L\upmu + 2\frac{1}{\varrho^2}\upmu - \upmu\left|\hat{\chi}^{(Small)}\right|^2 - \mathfrak{A}, \tag{11.21}$$

where $\varrho = 1 - u + t$ is the geometric radial variable defined in (3.5).

PROOF. From the identity $\chi_{AB} = g(\mathscr{D}_A L, X_B)$ (see (3.73)), the fact that $[L, X_A] = 0$, the Def. 11.1 of the Riemann curvature tensor, the torsion-free property $\mathscr{D}_L X_B = \mathscr{D}_B L$ of $\mathscr{D}$, the fact that $\mathscr{D}g = 0$, and Lemma 5.1, we deduce

(11.22)

$$\begin{aligned}\not{\mathcal{L}}_L\chi_{AB} &= L(\chi_{AB}) = L[g(\mathscr{D}_A L, X_B)] = g(\mathscr{D}_L(\mathscr{D}_A L), X_B) + g(\mathscr{D}_A L, \mathscr{D}_L X_B)\\ &= g(\mathscr{D}_A(\mathscr{D}_L L), X_B) - \mathscr{R}_{LALB} + g(\mathscr{D}_A L, \mathscr{D}_B L)\\ &= \frac{(L\upmu)}{\upmu}\chi_{AB} + \chi_A^{\ C}\chi_{BC} - \mathscr{R}_{LALB}.\end{aligned}$$

Contracting the left-hand and right-hand sides of (11.22) with $(\not g^{-1})^{AB}$, multiplying by $\upmu$, and using the identity $(\not{\mathcal{L}}_L \not g^{-1})^{AB} = -2\chi^{AB}$ (see (3.72)), we deduce

$$\upmu L\mathrm{tr}_{\not g}\chi = (L\upmu)\mathrm{tr}_{\not g}\chi - \upmu|\chi|^2 - \upmu(\not g^{-1})^{AB}\mathscr{R}_{LALB}. \tag{11.23}$$

Using (11.12), we see that the last term on the right-hand side of (11.23) can be expressed as $\upmu(\not g^{-1})^{AB}\mathscr{R}_{LALB} = L\mathfrak{X} + \mathfrak{A}$. Equation (11.21) now follows from the decompositions $\chi = \varrho^{-1}\not g + \chi^{(Small)}$ (see (4.1c)), $|\chi|^2 = \frac{1}{2}(\mathrm{tr}_{\not g}\chi)^2 + \left|\hat{\chi}^{(Small)}\right|^2$, $\mathrm{tr}_{\not g}\chi = 2\varrho^{-1} + \mathrm{tr}_{\not g}\chi^{(Small)}$, the identity $L\varrho = 1$, and straightforward calculations. $\square$

In the next lemma, we reveal the structure of the equation that arises after commuting the modified equation (11.21) with one spatial commutation vectorfield $S \in \mathscr{S}$. The proof is somewhat delicate because we have to carefully identify some important cancellations.

LEMMA 11.8 (**The transport equation satisfied by the fully modified version of** $S\mathrm{tr}_{\not g}\chi^{(Small)}$). *Assume that* $\square_{g(\Psi)}\Psi = 0$. *Let* $S \in \mathscr{S}$ *be a spatial commutation vectorfield (see Def. 7.2) and let* $\mathfrak{X}$ *be the quantity defined in* (11.20b). *We define the fully modified quantity* ${}^{(S)}\mathscr{X}$ *as follows:*

$$ {}^{(S)}\mathscr{X} := \upmu S\mathrm{tr}_{\not g}\chi^{(Small)} + S\mathfrak{X}. \tag{11.24} $$

Then ${}^{(S)}\mathscr{X}$ *satisfies the following transport equation:*

$$ \begin{aligned} L{}^{(S)}\mathscr{X} &= -\mathrm{tr}_{\not g}\chi{}^{(S)}\mathscr{X} + \upmu[L,S]\mathrm{tr}_{\not g}\chi^{(Small)} + 2(L\upmu)S\mathrm{tr}_{\not g}\chi^{(Small)} + \frac{1}{2}\mathrm{tr}_{\not g}\chi S\mathfrak{X} \\ &\quad - S\left(\upmu\left|\hat{\chi}^{(Small)}\right|^2\right) + {}^{(S)}\mathfrak{J}, \end{aligned} \tag{11.25} $$

where the inhomogeneous term ${}^{(S)}\mathfrak{J}$ *is equal to*

$$ \begin{aligned} {}^{(S)}\mathfrak{J} &= -S\mathfrak{A} + \frac{1}{2}\mathrm{tr}_{\not g}\chi S \overbrace{\left\{-\frac{3}{2}\upmu G_{LL}L\Psi - 2\upmu G_{L\breve{R}}L\Psi - \frac{1}{2}\upmu \not G_A^{\ A} L\Psi + \upmu \not G_L^{\ A} \not d_A\Psi\right\}}^{=2L\upmu+\mathfrak{X}} \\ &\quad - (S\upmu)\left\{L\mathrm{tr}_{\not g}\chi^{(Small)} + 2\frac{1}{\varrho}\mathrm{tr}_{\not g}\chi^{(Small)} + \frac{1}{2}(\mathrm{tr}_{\not g}\chi^{(Small)})^2\right\} \\ &\quad + [L,S]\mathfrak{X} - 2\frac{1}{\varrho^2}(L\upmu)S\varrho + 2\upmu\frac{1}{\varrho^2}\mathrm{tr}_{\not g}\chi^{(Small)}S\varrho, \end{aligned} \tag{11.26} $$

and $\mathfrak{A}$ *is the inhomogeneous term appearing on the right-hand side of* (11.12).

PROOF. We apply S to both sides of (11.21). Clearly, the last term on the right-hand side of (11.21) gives rise to the first term on the right-hand side of (11.26). In our analysis of the remaining terms, we will draw boxes around the important terms that appear explicitly on either the left-hand or right-hand sides of equation (11.25); the remaining terms are either canceled by other terms or are by definition part of ${}^{(S)}\mathfrak{J}$.

Referring to definitions (11.20a) and (11.24), we compute that the term $SL\mathscr{X}$ arising from the left-hand side of (11.21) can be written as

$$ \boxed{L{}^{(S)}\mathscr{X}} + \boxed{\upmu[S,L]\mathrm{tr}_{\not g}\chi^{(Small)}} + [S,L]\mathfrak{X} + (S\upmu)L\mathrm{tr}_{\not g}\chi^{(Small)} + (SL\upmu)\mathrm{tr}_{\not g}\chi^{(Small)}. \tag{11.27} $$

Using the decomposition $\mathrm{tr}_{\not g}\chi = 2\varrho^{-1} + \mathrm{tr}_{\not g}\chi^{(Small)}$ (see (4.1c)), we compute that the term $2S\{(L\upmu)\mathrm{tr}_{\not g}\chi\}$ arising from the right-hand side of (11.21) can be written as

$$ \boxed{2(L\upmu)S\mathrm{tr}_{\not g}\chi^{(Small)}} - \frac{4}{\varrho^2}(L\upmu)S\varrho + 2(SL\upmu)\mathrm{tr}_{\not g}\chi. \tag{11.28} $$

Referring to definition (11.24) and using the decomposition $\mathrm{tr}_{\not g}\chi = 2\varrho^{-1} + \mathrm{tr}_{\not g}\chi^{(Small)}$, we compute that the term $-\frac{1}{2}S\{\upmu(\mathrm{tr}_{\not g}\chi)^2\}$ arising from the right-hand side of (11.21) can be written as

$$\boxed{-\mathrm{tr}_{\not g}\chi^{(S)}\mathscr{X}} + \boxed{\mathrm{tr}_{\not g}\chi S\mathfrak{X}} + \frac{4}{\varrho^3}\upmu S\varrho + \frac{2}{\varrho^2}\upmu\mathrm{tr}_{\not g}\chi^{(Small)}S\varrho - 2\frac{1}{\varrho^2}S\upmu - 2\frac{1}{\varrho}\mathrm{tr}_{\not g}\chi^{(Small)}S\upmu - \frac{1}{2}(S\upmu)(\mathrm{tr}_{\not g}\chi^{(Small)})^2. \tag{11.29}$$

We now explicitly place exactly half of the second term $\mathrm{tr}_{\not g}\chi S\mathfrak{X}$ in (11.29) on the right-hand side of (11.25). We then add the other half to half of the last term in (11.28) (that is, to $(SL\upmu)\mathrm{tr}_{\not g}\chi$) and we place the sum on the right-hand side of (11.26) as the second term. We then use equations (4.5) and (11.20b) to put this sum into the form that appears on the right-hand side of (11.26) (the important point is the cancellation of the products $G_{LL}\breve{R}\Psi$). The remaining part $(SL\upmu)\mathrm{tr}_{\not g}\chi$ of the last term in (11.28) is completely canceled by the last term in (11.27) (which gets multiplied by -1 when it is moved over to the right-hand side of (11.25)) and the last term in (11.30) below (the cancellation occurs because of the identity $\mathrm{tr}_{\not g}\chi = 2\varrho^{-1} + \mathrm{tr}_{\not g}\chi^{(Small)}$).

We compute that the term $-2S\left(\frac{1}{\varrho}L\upmu\right)$ arising from the right-hand side of (11.21) can be written as

$$2\frac{1}{\varrho^2}(L\upmu)S\varrho - 2\frac{1}{\varrho}SL\upmu. \tag{11.30}$$

We compute that the term $2S\left(\frac{1}{\varrho^2}\upmu\right)$ arising from the right-hand side of (11.21) can be written as

$$2\frac{1}{\varrho^2}S\upmu - 4\frac{1}{\varrho^3}\upmu S\varrho. \tag{11.31}$$

The term $-S\left(\upmu\left|\hat{\chi}^{(Small)}\right|^2\right)$ arising from the right-hand side of (11.21) appears manifestly on the right-hand side of (11.25).

Combining (11.27), (11.28), (11.29), (11.30), and (11.31), taking into account the remarks made just below (11.29) and in the previous sentence, and carrying out straightforward computations, we arrive at the desired equations (11.25) and (11.26). □

We now define higher-order versions of the fully modified quantity $\mathscr{X}$. We use them to close our top-order L^2 estimates corresponding to the timelike multiplier T.

DEFINITION 11.9 (**Fully modified version of the pure spatial derivatives of $\mathrm{tr}_{\not g}\chi^{(Small)}$**). Let $\mathscr{S}^N$ be an Nth order pure spatial commutation vectorfield operator (see Def. 7.2) and let $\mathfrak{X}$ be the quantity defined in (11.20b). We define the fully modified function ${}^{(\mathscr{S}^N)}\mathscr{X}$ as follows:

$${}^{(\mathscr{S}^N)}\mathscr{X} := \upmu\mathscr{S}^N\mathrm{tr}_{\not g}\chi^{(Small)} + \mathscr{S}^N\Xi. \tag{11.32}$$

In the next proposition, we derive the transport equation verified by ${}^{(\mathscr{S}^N)}\mathscr{X}$.

PROPOSITION 11.10 (**The transport equation satisfied by the fully modified version of $\mathscr{S}^N\mathrm{tr}_{\not g}\chi^{(Small)}$**). *Assume that $\square_{g(\Psi)}\Psi = 0$. Consider an Nth order pure spatial commutation vectorfield operator of the form $\mathscr{S}^N = \mathscr{S}^{N-1}S$ (see*

Def. 7.2) and let $^{(\mathscr{S}^N)}\mathscr{X}$ *and* $\mathfrak{X}$ *be the corresponding quantities defined in* (11.32) *and* (11.20b). *Then* $^{(\mathscr{S}^N)}\mathscr{X}$ *satisfies the following transport equation:*

(11.33)
$$L^{(\mathscr{S}^N)}\mathscr{X} - \left\{2\frac{L\mu}{\mu} - \mathrm{tr}_{\not g}\chi\right\}^{(\mathscr{S}^N)}\mathscr{X} = \left\{\frac{1}{2}\mathrm{tr}_{\not g}\chi - 2\frac{L\mu}{\mu}\right\}\mathscr{S}^N\mathfrak{X} + \mu[L,\mathscr{S}^N]\mathrm{tr}_{\not g}\chi^{(Small)} - 2\mu\hat{\chi}^{(Small)\#\#}\cdot \not{\mathcal{L}}_{\mathscr{S}}^N\hat{\chi}^{(Small)} + {}^{(\mathscr{S}^N)}\mathfrak{I},$$

where $^{(\mathscr{S}^N)}\mathfrak{I}$ *is the inhomogeneous term*

(11.34)
$$\begin{aligned} {}^{(\mathscr{S}^N)}\mathfrak{I} := {} & \mathscr{S}^{N-1(S)}\mathfrak{I} + [L,\mathscr{S}^{N-1}]S\mathfrak{X} \\ & - [\mathscr{S}^{N-1},\mu\mathrm{tr}_{\not g}\chi]S\mathrm{tr}_{\not g}\chi^{(Small)} + \frac{1}{2}[\mathscr{S}^{N-1},\mathrm{tr}_{\not g}\chi]S\mathfrak{X} \\ & + [\mathscr{S}^{N-1},L\mu]S\mathrm{tr}_{\not g}\chi^{(Small)} - [\mathscr{S}^{N-1},\mu]SL\mathrm{tr}_{\not g}\chi^{(Small)} \\ & + \left\{2\mu\hat{\chi}^{(Small)\#\#}\cdot\not{\mathcal{L}}_{\mathscr{S}}^N\hat{\chi}^{(Small)} - \mathscr{S}^N\left(\mu\hat{\chi}^{(Small)\#\#}\cdot\hat{\chi}^{(Small)}\right)\right\}, \end{aligned}$$

and $^{(S)}\mathfrak{I}$ *is the term given by* (11.26).

Proof. We apply $\mathscr{S}^{N-1}$ to the transport equation (11.25) for $^{(S)}\mathscr{X}$. Clearly, the last term on the right-hand side of (11.25) gives rise to the first term on the right-hand side of (11.34). In our analysis of the remaining terms, we draw boxes around the important terms that appear explicitly on either the left-hand or right-hand sides of equation (11.33); the remaining terms are either canceled by other terms or are error terms that are by definition part of $^{(\mathscr{S}^N)}\mathfrak{I}$. To begin, we consider definitions (11.24) and (11.32) and compute that the corresponding left-hand side of $\mathscr{S}^{N-1}$ applied to (11.25) (that is, $\mathscr{S}^{N-1}L^{(S)}\mathscr{X}$) can be written as

(11.35)
$$\boxed{L^{(\mathscr{S}^N)}\mathscr{X}} + \boxed{\mu[\mathscr{S}^N,L]\mathrm{tr}_{\not g}\chi^{(Small)}} + [\mathscr{S}^{N-1},L\mu]S\mathrm{tr}_{\not g}\chi^{(Small)} + [\mathscr{S}^{N-1},\mu]LS\mathrm{tr}_{\not g}\chi^{(Small)} + \mu\mathscr{S}^{N-1}[L,S]\mathrm{tr}_{\not g}\chi^{(Small)} + [\mathscr{S}^{N-1},L]S\mathfrak{X}.$$

The term $-\mathscr{S}^{N-1}(\mathrm{tr}_{\not g}\chi^{(S)}\mathscr{X})$ arising from the right-hand side of (11.25) can be written as

(11.36)
$$-\boxed{\mathrm{tr}_{\not g}\chi^{(\mathscr{S}^N)}\mathscr{X}} - [\mathscr{S}^{N-1},\mu\mathrm{tr}_{\not g}\chi]S\mathrm{tr}_{\not g}\chi^{(Small)}.$$

The term $\mathscr{S}^{N-1}\left\{\mu[L,S]\mathrm{tr}_{\not g}\chi^{(Small)}\right\}$ arising from the right-hand side of (11.25) can be written as

(11.37)
$$\mu\mathscr{S}^{N-1}\left([L,S]\mathrm{tr}_{\not g}\chi^{(Small)}\right) + [\mathscr{S}^{N-1},\mu]\left([L,S]\mathrm{tr}_{\not g}\chi^{(Small)}\right).$$

The term $2\mathscr{S}^{N-1}\{(L\mu)S\mathrm{tr}_{\not g}\chi^{(Small)}\}$ arising from the right-hand side of (11.25) can be written as

(11.38)
$$\boxed{2\frac{L\mu}{\mu}{}^{(\mathscr{S}^N)}\mathscr{X}} - \boxed{2\frac{L\mu}{\mu}\mathscr{S}^N\mathfrak{X}} + 2[\mathscr{S}^{N-1},L\mu]S\mathrm{tr}_{\not g}\chi^{(Small)}.$$

The term $\frac{1}{2}\mathscr{S}^{N-1}\left(\mathrm{tr}_{\not g}\chi S\mathfrak{X}\right)$ arising from the right-hand side of (11.25) can be written as

(11.39)
$$\boxed{\frac{1}{2}\mathrm{tr}_{\not g}\chi\mathscr{S}^N\mathfrak{X}} + \frac{1}{2}[\mathscr{S}^{N-1},\mathrm{tr}_{\not g}\chi]S\mathfrak{X}.$$

The term $-\mathscr{S}^{N-1}S\left(\upmu\left|\hat{\upchi}^{(Small)}\right|^2\right)$ arising from the right-hand side of (11.25) can be written as

$$-\boxed{2\upmu\hat{\upchi}^{(Small)\#\#}\cdot\not{\mathcal{L}}_{\mathscr{S}}^{N}\hat{\upchi}^{(Small)}} + \left\{2\upmu\hat{\upchi}^{(Small)\#\#}\cdot\not{\mathcal{L}}_{\mathscr{S}}^{N}\hat{\upchi}^{(Small)} - \mathscr{S}^{N}(\upmu\hat{\upchi}^{(Small)\#\#}\cdot\hat{\upchi}^{(Small)})\right\}. \tag{11.40}$$

Combining (11.35), (11.36), (11.37), (11.38), (11.39), and (11.40), we deduce the desired equations (11.33) and (11.34). $\square$

11.3. Partial modification of the trace of the re-centered null second fundamental form

To close our top-order L^2 estimates involving the Morawetz multiplier $\widetilde{K}$, we need to use the partially modified version of ${\rm tr}_{\not{g}}\upchi^{(Small)}$ given in the next definition. The partially modified quantity allows us to avoid certain spacetime error integrals with damaging time growth.

Definition 11.11 (**Lowest-order partially modified version of ${\rm tr}_{\not{g}}\upchi^{(Small)}$**). We define the scalar-valued functions $\widetilde{\mathscr{X}}$ and $\widetilde{\mathfrak{X}}$ as follows:

$$\widetilde{\mathscr{X}} := {\rm tr}_{\not{g}}\upchi^{(Small)} + \widetilde{\mathfrak{X}}, \tag{11.41a}$$

$$\widetilde{\mathfrak{X}} := -\frac{1}{2}\not{G}_A^{\ A}L\Psi - \frac{1}{2}G_{LL}L\Psi + \not{G}_L^{\ A}\not{d}_A\Psi. \tag{11.41b}$$

Remark 11.12 (**Motivation for the definition of $\widetilde{\mathfrak{X}}$**). Note that $\widetilde{\mathfrak{X}}$ is the term in braces on the right-hand side of (11.14).

In the next lemma, we derive the transport equation verified by $\widetilde{\mathscr{X}}$.

Lemma 11.13 (**The transport equation satisfied by the partially modified version of ${\rm tr}_{\not{g}}\upchi^{(Small)}$**). *Let $\widetilde{\mathscr{X}}$ be the partially modified quantity defined in* (11.41a) *and let $\varrho = 1 - u + t$ be as defined in* (3.5). *Then $\varrho^2\widetilde{\mathscr{X}}$ satisfies the following transport equation:*

$$L(\varrho^2\widetilde{\mathscr{X}}) = \frac{1}{2}\varrho^2 G_{LL}\not{\Delta}\Psi - \varrho^2|\upchi^{(Small)}|^2 + \widetilde{\mathfrak{B}}, \tag{11.42}$$

where

$$\widetilde{\mathfrak{B}} := -\varrho^2\mathfrak{B} + 2\varrho\widetilde{\mathfrak{X}}, \tag{11.43}$$

$\mathfrak{B}$ is the term appearing on the right-hand side of (11.14), *and $\widetilde{\mathfrak{X}}$ is defined in* (11.41b).

Proof. From equations (11.14) and (11.23), we deduce that

$$L\left\{{\rm tr}_{\not{g}}\upchi + \widetilde{\mathfrak{X}}\right\} = \frac{1}{2}G_{LL}\not{\Delta}\Psi - \mathfrak{B} - |\upchi|^2. \tag{11.44}$$

The desired equation (11.42) now follows from (11.44), the decompositions $\upchi = \varrho^{-1}\not{g} + \upchi^{(Small)}$ (see (4.1c)), $|\upchi|^2 = \frac{1}{2}({\rm tr}_{\not{g}}\upchi)^2 + \left|\hat{\upchi}^{(Small)}\right|^2$, ${\rm tr}_{\not{g}}\upchi = 2\varrho^{-1} + {\rm tr}_{\not{g}}\upchi^{(Small)}$, the fact that $L\varrho = 1$, and from straightforward computations. $\square$

We now define higher-order versions of $\widetilde{\mathscr{X}}$ and $\widetilde{\mathfrak{X}}$.

DEFINITION 11.14 (**Partially modified version of $\mathscr{Z}^N \mathrm{tr}_{\not g}\upchi^{(Small)}$**). We define the partially modified function ${}^{(\mathscr{Z}^N)}\widetilde{\mathscr{X}}$ as follows:

$$(11.45a)\qquad {}^{(\mathscr{Z}^N)}\widetilde{\mathscr{X}} := \mathscr{Z}^N \mathrm{tr}_{\not g}\upchi^{(Small)} + {}^{(\mathscr{Z}^N)}\widetilde{\mathfrak{X}},$$

$$(11.45b)\qquad {}^{(\mathscr{Z}^N)}\widetilde{\mathfrak{X}} := -\frac{1}{2}\not G_A^{\ A} L\mathscr{Z}^N\Psi - \frac{1}{2}G_{LL}L\mathscr{Z}^N\Psi + \not G_L^{\ A}\not d_A\mathscr{Z}^N\Psi.$$

In the next lemma, we derive the transport equation verified by ${}^{(\mathscr{S}^{N-1})}\widetilde{\mathscr{X}}$.

LEMMA 11.15 (**The transport equation satisfied by the partially modified version of $\mathscr{S}^{N-1}\mathrm{tr}_{\not g}\upchi^{(Small)}$**). *Let $\mathscr{S}^{N-1}$ be an $(N-1)^{st}$ order pure spatial commutation vectorfield operator (see Def. 7.2) and let ${}^{(\mathscr{S}^{N-1})}\widetilde{\mathscr{X}}$ be the partially modified quantity defined in* (11.45a). *Let $\varrho = 1-u+t$ be as defined in* (3.5). *Then $\varrho^2\widetilde{\mathscr{X}}$ satisfies the following transport equation:*

$$(11.46)\qquad L(\varrho^{2\,(\mathscr{S}^{N-1})}\widetilde{\mathscr{X}}) = \frac{1}{2}\varrho^2 G_{LL}\not\Delta\mathscr{S}^{N-1}\Psi + {}^{(\mathscr{S}^{N-1})}\widetilde{\mathfrak{B}},$$

where the inhomogeneous term ${}^{(\mathscr{S}^{N-1})}\widetilde{\mathfrak{B}}$ is given by

(11.47)

$$\begin{aligned}
{}^{(\mathscr{S}^{N-1})}\widetilde{\mathfrak{B}} &= \mathscr{S}^{N-1}\widetilde{\mathfrak{B}} - \mathscr{S}^{N-1}(\varrho^2|\upchi^{(Small)}|^2) \\
&\quad + \frac{1}{2}[\mathscr{S}^{N-1}, \varrho^2 G_{LL}]\not\Delta\Psi + \frac{1}{2}\varrho^2 G_{LL}[\mathscr{S}^{N-1}, \not\Delta]\Psi \\
&\quad + [L, \mathscr{S}^{N-1}](\varrho^2\mathrm{tr}_{\not g}\upchi^{(Small)}) + [L, \mathscr{S}^{N-1}](\varrho^2\widetilde{\mathfrak{X}}) \\
&\quad + L\left\{[\varrho^2, \mathscr{S}^{N-1}]\mathrm{tr}_{\not g}\upchi^{(Small)}\right\} + L\left\{\varrho^{2\,(\mathscr{S}^{N-1})}\widetilde{\mathfrak{X}} - \mathscr{S}^{N-1}(\varrho^2\widetilde{\mathfrak{X}})\right\},
\end{aligned}$$

$\widetilde{\mathfrak{B}}$ is defined in (11.43), *$\widetilde{\mathfrak{X}}$ is defined in* (11.41b), *and ${}^{(\mathscr{S}^{N-1})}\widetilde{\mathfrak{X}}$ is defined in* (11.45b).

PROOF. We apply $\mathscr{S}^{N-1}$ to each side of equation (11.42) and perform several straightforward operator commutations. □

11.4. Partial modification of the angular gradient of the inverse foliation density

This section complements Sect. 11.3. Specifically, we define partially modified versions of $\not d\upmu$ in order to close our top-order L^2 estimates corresponding to the multiplier $\widetilde{K}$. We start by providing the definition of the lowest-order partially modified version of $\not d\upmu$.

DEFINITION 11.16 (**Lowest-order partially modified version of $\not d\upmu$**). We define the $S_{t,u}$ one-forms $\widetilde{\mathscr{M}}$ and $\widetilde{\mathfrak{M}}$ as follows:

$$(11.48a)\qquad \widetilde{\mathscr{M}} := \not d\upmu + \widetilde{\mathfrak{M}},$$

$$(11.48b)\qquad \widetilde{\mathfrak{M}} := \frac{1}{2}\upmu G_{LL}\not d\Psi + \upmu G_{LR}\not d\Psi.$$

In the next lemma, we derive the transport equation verified by $\widetilde{\mathscr{M}}$.

LEMMA 11.17 (**The transport equation satisfied by the partially modified version of $\not d\upmu$**). *Let $\widetilde{\mathscr{M}}$ be the partially modified $S_{t,u}$ one-form defined in*

(11.48a). *Then* $\widetilde{\mathscr{M}}$ *satisfies the following transport equation:*

$$\not{\mathcal{L}}_L \widetilde{\mathscr{M}} = \frac{1}{2} G_{LL} \not{d} \breve{R}\Psi + \mathfrak{J}, \tag{11.49}$$

where the $S_{t,u}$ *one-form* $\mathfrak{J}$ *has the following schematic form (see Sect. 3.20):*

$$\begin{aligned} \mathfrak{J} = & \begin{pmatrix} \upmu L\Psi \\ \breve{R}\Psi \end{pmatrix} \not{d} G_{(Frame)} + \upmu (L G_{(Frame)}) \not{d}\Psi \\ & + G_{(Frame)}^2 \begin{pmatrix} \upmu L\Psi \\ \breve{R}\Psi \end{pmatrix} \not{d}\Psi + G_{(Frame)} (L\Psi) \not{d}\upmu, \end{aligned} \tag{11.50}$$

and no $S_{t,u}$ *tensors such as* $\not{G}_L$ *are present in the* $G_{(Frame)}$ *term on the right-hand side of* (11.50) *(that is, only scalar-valued functions such as* G_{LR} *are present).*

PROOF. We apply $\not{d}$ to equation (4.5) and use Lemma 8.6 to rewrite the left-hand side as $\not{d}L\upmu = \not{\mathcal{L}}_L \not{d}\upmu$. We now address the terms that arise when $\not{d}$ falls on the term $\frac{1}{2} G_{LL} \breve{R}\Psi$ from the right-hand side of (4.5). We explicitly place the term $\frac{1}{2} G_{LL} \not{d}\breve{R}\Psi$ on the right-hand side of (11.49), while the term $\frac{1}{2}(\not{d}G_{LL})\breve{R}\Psi$ is part of the inhomogeneous term (11.50).

We now address the terms that arise when $\not{d}$ falls on the terms $-\frac{1}{2}\upmu G_{LL} L\Psi$ and $-\upmu G_{LR} L\Psi$ from the right-hand side of (4.5). When $\not{d}$ falls on $\upmu$ or $G_{(Frame)}$, we consider these terms to be part of the inhomogeneous term (11.50). When $\not{d}$ falls on $L\Psi$, we use Lemma 8.6 to commute the L derivative all the way out, and then move the terms $-\frac{1}{2}\not{\mathcal{L}}_L(\upmu G_{LL}\not{d}\Psi)$ and $-\not{\mathcal{L}}_L(\upmu G_{LR}\not{d}\Psi)$ to the left-hand side; the products underneath the $\not{\mathcal{L}}_L$ differentiation are part of the quantity $\widetilde{\mathscr{M}}$. The commutator terms are included in the inhomogeneous term (11.50), and we use the equation (4.5) to replace all instances of $L\upmu$ with the right-hand side of (4.5). Lemma 11.17 thus follows. □

We now define higher-order versions of $\widetilde{\mathscr{M}}$ and $\widetilde{\mathfrak{M}}$.

DEFINITION 11.18 (**Partially modified version of** $\not{d}\mathscr{Z}^N\upmu$). We define the partially modified $S_{t,u}$ one-form ${}^{(\mathscr{Z}^N)}\widetilde{\mathscr{M}}$ as follows:

$${}^{(\mathscr{Z}^N)}\widetilde{\mathscr{M}} := \not{d}\mathscr{Z}^N\upmu + {}^{(\mathscr{Z}^N)}\widetilde{\mathfrak{M}}, \tag{11.51a}$$

$${}^{(\mathscr{Z}^N)}\widetilde{\mathfrak{M}} := \frac{1}{2}\upmu G_{LL}\not{d}\mathscr{Z}^N\Psi + \upmu G_{LR}\not{d}\mathscr{Z}^N\Psi. \tag{11.51b}$$

In the next lemma, we derive the transport equation verified by ${}^{(\mathscr{S}^{N-1})}\widetilde{\mathscr{M}}$.

LEMMA 11.19 (**The transport equation satisfied by the partially modified version of** $\not{d}\mathscr{S}^{N-1}\upmu$). *Let* $\mathscr{S}^{N-1}$ *be an* $(N-1)^{st}$ *order pure spatial commutation vectorfield operator (see Def. 7.2) and let* ${}^{(\mathscr{S}^{N-1})}\widetilde{\mathscr{M}}$ *be the partially modified* $S_{t,u}$ *one-form defined in* (11.51a). *Then* ${}^{(\mathscr{S}^{N-1})}\widetilde{\mathscr{M}}$ *satisfies the following transport equation:*

$$\not{\mathcal{L}}_L {}^{(\mathscr{S}^{N-1})}\widetilde{\mathscr{M}} = \frac{1}{2} G_{LL}\not{d}\mathscr{S}^{N-1}\breve{R}\Psi + {}^{(\mathscr{S}^{N-1})}\mathfrak{J}, \tag{11.52}$$

where the $S_{t,u}$ one-form $^{(\mathscr{S}^{N-1})}\mathfrak{J}$ *is given by*

$$^{(\mathscr{S}^{N-1})}\mathfrak{J} = \not{\mathcal{L}}_{\mathscr{S}}^{N-1}\mathfrak{J} + \frac{1}{2}[\not{\mathcal{L}}_{\mathscr{S}}^{N-1}, G_{LL}]\not{d}\breve{R}\Psi + [\not{\mathcal{L}}_L, \not{\mathcal{L}}_{\mathscr{S}}^{N-1}]\not{d}\upmu + [\not{\mathcal{L}}_L, \not{\mathcal{L}}_{\mathscr{S}}^{N-1}]\widetilde{\mathfrak{M}} + \not{\mathcal{L}}_L \left\{ ^{(\mathscr{S}^{N-1})}\widetilde{\mathfrak{M}} - \not{\mathcal{L}}_{\mathscr{S}}^{N-1}\widetilde{\mathfrak{M}} \right\}, \tag{11.53}$$

the $S_{t,u}$ one-form $\widetilde{\mathfrak{M}}$ is defined in (11.48b), *the $S_{t,u}$ one-form $\mathfrak{J}$ is defined in* (11.50), *and the $S_{t,u}$ one-form* $^{(\mathscr{S}^{N-1})}\widetilde{\mathfrak{M}}$ *is defined in* (11.51b).

PROOF. We apply $\not{\mathcal{L}}_{\mathscr{S}}^{N-1}$ to each side of equation (11.49) and perform several straightforward operator commutations. □

CHAPTER 12

Small Data, Sup-Norm Bootstrap Assumptions, and First Pointwise Estimates

In this chapter, we define the size of the data for solutions to the system

$$\square_{g(\Psi)}\Psi = 0, \qquad (g^{-1})^{\alpha\beta}(\Psi)\partial_\alpha u \partial_\beta u = 0.$$

We then make suitable bootstrap assumptions for solutions on a spacetime region of the form $\mathcal{M}_{T_{(Bootstrap)},U_0}$. Our bootstrap assumptions include fundamental positivity bootstrap assumption for $\upmu$, C^0 bootstrap assumptions for Ψ and its lower-order derivatives, and auxiliary C^0 bootstrap assumptions for $\upmu$, $L^i_{(Small)}$, and $\chi^{(Small)}$ and their lower-order derivatives. Our main goal in this chapter is to use the bootstrap assumptions to derive C^0 and pointwise estimates on $\mathcal{M}_{T_{(Bootstrap)},U_0}$ for various quantities and their derivatives with respect to the commutation vectorfields $Z \in \mathscr{Z}$. The estimates that we derive here are tedious but not too difficult; we derive related but more precise (and significantly more difficult) estimates in Chapters 13 and 15-17.

REMARK 12.1 (**Suppression of the independent variables**). Throughout the remainder of the monograph, we state many of our pointwise estimates in the form

$$|f_1| \lesssim h(t,u)|f_2| \tag{12.1}$$

for some function h. To avoid cluttering the notation, we use the convention that unless we indicate otherwise, in such inequalities, both f_1 and f_2 are evaluated at the point with geometric coordinates (t, u, ϑ).

12.1. Restricting the analysis to solutions of the evolution equations

From now until Appendix A, we assume that u is an outgoing solution to the eikonal equation (3.4a) with initial conditions (3.4b) and that there exists a time $T_{(Bootstrap)} > 0$ such that on $\mathcal{M}_{T_{(Bootstrap)},U_0}$ (see Def. 3.7), Ψ is a solution to the covariant wave equation (2.1a). That is, we assume that

$$(g^{-1})^{\alpha\beta}(\Psi)\partial_\alpha u \partial_\beta u = 0, \qquad \partial_t u > 0, \qquad u|_{\Sigma_0} = 1 - r$$

and that

$$\square_{g(\Psi)}\Psi = 0, \qquad (\Psi|_{\Sigma_0}, \partial_t\Psi|_{\Sigma_0}) = (\mathring{\Psi}, \mathring{\Psi}_0),$$

where $(\mathring{\Psi}, \mathring{\Psi}_0)$ are compactly supported in Σ_0^1, the Euclidean unit ball centered at the origin in Σ_0. We also assume that $g_{\mu\nu}(\Psi) = m_{\mu\nu} + g_{\mu\nu}^{(Small)}(\Psi)$ and $g_{\mu\nu}^{(Small)}(0) = 0$, as stated in (2.4) and (2.6). We also recall that until Sect. 23.2, U_0 denotes a fixed parameter verifying $0 < U_0 < 1$ and that $\mathcal{M}_{T_{(Bootstrap)},U_0}$ is the nontrivial portion of the future development of the portion of the data lying in the exterior of S_{0,U_0}, the Euclidean sphere of radius $1 - U_0$ centered at the origin in Σ_0.

REMARK 12.2 (**The constants can depend on U_0**). Throughout our analysis, some of the explicit constants "C" and the implicit constants tied to the notation "$\lesssim$" and "$\mathcal{O}(\cdots)$" are allowed to depend on U_0.

12.2. Small data

In this section, we define the size of the data for the covariant wave equation $\square_{g(\Psi)}\Psi = 0$.

DEFINITION 12.3 (**Definition of the size of the data**). Let $(\mathring{\Psi} := \Psi|_{\Sigma_0}, \mathring{\Psi}_0 := \partial_t\Psi|_{\Sigma_0})$ be initial data for the covariant wave equation (2.1a) that are compactly supported in the Euclidean unit ball Σ_0^1 (see (3.4b) and Def. 3.7). We define the size of the data as follows:

$$\mathring{\epsilon} := \|\mathring{\Psi}\|_{H_e^{25}(\Sigma_0^1)} + \|\mathring{\Psi}_0\|_{H_e^{24}(\Sigma_0^1)}. \tag{12.2}$$

In (12.2), $H_e^N(\Sigma_0^1)$ is the standard Euclidean Sobolev space involving rectangular spatial coordinate partial derivatives along Σ_0^1.

Preliminary small data assumption

The main results of this monograph are valid whenever $\mathring{\epsilon}$ is sufficiently small. We make the following assumption throughout our analysis:

$$\mathring{\epsilon} \leq \varepsilon, \tag{12.3}$$

where $\varepsilon > 0$ is the small "bootstrap parameter" appearing in bootstrap assumptions of Sects. 12.4.1 and 12.4.2.

12.3. Fundamental positivity bootstrap assumption for the inverse foliation density

We make the following bootstrap assumption for the solution on $\mathcal{M}_{T_{(Bootstrap)},U_0}$:

$$\mu > 0 \text{ on } \mathcal{M}_{T_{(Bootstrap)},U_0}. \tag{$\mathbf{BA}\mu > 0$}$$

The assumption $(\mathbf{BA}\mu > 0)$ implies that no shock is present in $\mathcal{M}_{T_{(Bootstrap)},U_0}$.

12.4. Sup-norm bootstrap assumptions

Our quantitative bootstrap assumptions involve a small parameter $\varepsilon > 0$. Throughout our analysis, we adjust the smallness of ε as necessary.

12.4.1. Fundamental sup-norm bootstrap assumptions. Our fundamental C^0 bootstrap assumptions for Ψ are that for $(t,u) \in [0, T_{(Bootstrap)}) \times [0, U_0]$, we have:

$$\left\|\mathscr{Z}^{\leq 13}\Psi\right\|_{C^0(\Sigma_t^u)}, \left\|\varrho L\mathscr{Z}^{\leq 12}\Psi\right\|_{C^0(\Sigma_t^u)}, \left\|\varrho{\not d}\mathscr{Z}^{\leq 12}\Psi\right\|_{C^0(\Sigma_t^u)} \leq \frac{\varepsilon}{1+t}, \tag{$\mathbf{BA}\Psi$}$$

$$\int_{t'=0}^{t}\left\|\begin{pmatrix} \mathscr{Z}^{\leq 14}\Psi \\ \varrho L\mathscr{Z}^{\leq 13}\Psi \\ \varrho{\not d}\mathscr{Z}^{\leq 13}\Psi \end{pmatrix}\right\|_{C^0(\Sigma_{t'}^u)} dt' \leq \varepsilon\ln(e+t), \tag{$\mathbf{BA}'\Psi$}$$

where the norm $\|\cdot\|_{C^0(\Sigma_t^u)}$ is defined in Def. 3.70.

We use the bootstrap assumptions $(\mathbf{BA}\Psi)$ many times throughout the monograph, while we use $(\mathbf{BA}'\Psi)$ only on a few key occasions (see Remark 12.7).

REMARK 12.4 (**Dispersive decay rates**). Since $\varrho(t,u) \approx 1+t$ in the spacetime region under study (see (12.5)), the bootstrap assumptions (**BA**Ψ) capture that Ψ decays, with respect to t, at the same rate as a solution to the linear wave equation (see Sect. 1.2.3). This is one manifestation of how the geometric coordinates allow us to transform the shock formation problem into a more traditional one in which we derive global existence-type estimates.

REMARK 12.5 (**The role of Sobolev embedding**). Much later in the monograph, we use Sobolev embedding-type estimates to derive an improvement of (**BA**Ψ)-(**BA**$'\Psi$). The improvement is based on Cor. 18.11.

REMARK 12.6 (**The need for the bootstrap assumption** (**BA**$'\Psi$)). Our main theorem, Theorem 22.1, does *not* yield an estimate that is consistent with a bootstrap assumption of the form

$$\left\| \begin{pmatrix} \mathscr{Z}^{\leq 14}\Psi \\ \varrho L \mathscr{Z}^{\leq 13}\Psi \\ \varrho {\not d} \mathscr{Z}^{\leq 13}\Psi \end{pmatrix} \right\|_{C^0(\Sigma_t^u)} \leq \frac{\varepsilon}{1+t}, \tag{12.4}$$

even though it *does* yield one that is compatible with (**BA**$'\Psi$).

REMARK 12.7 (**Further remarks on** (**BA**$'\Psi$)). As we will see in the proof of Prop. 12.46, the bootstrap assumptions (**BA**$'\Psi$) allows us to control one extra derivative of $\upmu$, $L_{(Small)}^i$, and $\chi^{(Small)}$ in the norm $\|\cdot\|_{C^0(\Sigma_t^u)}$ compared to the number of derivatives that we would be able to control using only the bootstrap assumptions (**BA**Ψ). In Chapter 17, we use this extra derivative to control some error terms that arise when we derive pointwise estimates for the top-order derivatives of $\chi^{(Small)}$; see Remark 17.2. The bootstrap assumptions (**BA**$'\Psi$) could be eliminated from our arguments in Chapter 17 by imposing additional Sobolev regularity requirements on the initial data; the additional regularity would allow us to increase the number of derivatives in our bootstrap assumptions (**BA**Ψ), which would render (**BA**$'\Psi$) unnecessary.

12.4.2. Auxiliary bootstrap assumptions. Let $\upmu$, $L_{(Small)}^i$, $R_{(Small)}^i$, and $\chi^{(Small)}$ be the quantities from Defs. 3.15 and 4.1. Our auxiliary C^0 bootstrap assumptions for these quantities are that for $(t,u) \in [0, T_{(Bootstrap)}) \times [0, U_0]$, we have:

(**AUX**$\upmu$)
$$\left\| \mathscr{Z}^{\leq 13}(\upmu - 1) \right\|_{C^0(\Sigma_t^u)} \leq \varepsilon^{1/2}\ln(e+t),$$

(**AUX**$LR_{(Small)}$)
$$\left\| \mathscr{Z}^{\leq 13} L_{(Small)}^i \right\|_{C^0(\Sigma_t^u)}, \left\| \mathscr{Z}^{\leq 13} R_{(Small)}^i \right\|_{C^0(\Sigma_t^u)} \leq \varepsilon^{1/2}\frac{\ln(e+t)}{1+t},$$

(**AUX**$\chi^{(Small)}$)
$$\left\| {\not{\mathcal{L}}}_{\mathscr{Z}}^{\leq 12} \chi^{(Small)} \right\|_{C^0(\Sigma_t^u)} \leq \varepsilon^{1/2}\frac{\ln(e+t)}{(1+t)^2}.$$

REMARK 12.8 (**Improving the auxiliary bootstrap assumptions**). Already within Chapter 12, we derive estimates showing that the bootstrap assumptions (**AUX**$\upmu$), (**AUX**$LR_{(Small)}$), and (**AUX**$\chi^{(Small)}$) in fact hold on $\mathcal{M}_{T_{(Bootstrap)},U_0}$ with $\sqrt{\varepsilon}$ replaced by $C\varepsilon$, which is an improvement for sufficiently small ε; see Cor. 12.47. In this sense, we consider these bootstrap assumptions to be "auxiliary."

REMARK 12.9 (**Conventions for repeated differentiation**). In this section, we often use the conventions for repeated differentiation described in Sect. 8.3.

12.5. Basic estimates for the geometric radial variable

In this section, we derive some simple pointwise and commutator estimates involving the geometric radial variable $\varrho = 1 - u + t$.

LEMMA 12.10 (**Basic estimates for the geometric radial variable** $\varrho = 1 - u + t$). *There exists a (U_0-dependent) constant $C > 1$ such that for $0 \leq u \leq U_0$, we have the following comparison estimates:*

$$C^{-1}(1+t) \leq \varrho(t,u) \leq 1+t. \tag{12.5}$$

Furthermore, if $0 \leq N \leq 24$ is an integer, then the following identities hold, where $\mathscr{Z}$ is the set of commutation vectorfields from definition (7.1) *(see also Remark* 7.3*):*

$$L\varrho = -\breve{R}\varrho = 1, \qquad O\varrho = 0, \tag{12.6a}$$

$$\mathscr{Z}^N \varrho \in \{0, -1, \varrho\}. \tag{12.6b}$$

Furthermore, if ξ is any $S_{t,u}$ tensorfield, M is an integer (not necessarily positive), and $1 \leq N \leq 24$ is an integer, then we have the following commutator estimates:

$$\left|[\not{\mathcal{L}}_{\mathscr{Z}}^N, \varrho]\xi\right| \lesssim \varrho \left|\not{\mathcal{L}}_{\mathscr{Z}}^{\leq N-1}\xi\right|, \tag{12.7a}$$

$$\left|\not{\mathcal{L}}_L \left\{[\not{\mathcal{L}}_{\mathscr{Z}}^N, \varrho^M]\xi\right\}\right| \overset{M}{\lesssim} \left|\not{\mathcal{L}}_L \left\{\varrho^M \not{\mathcal{L}}_{\mathscr{Z}}^{\leq N-1}\xi\right\}\right| + \frac{1}{\varrho^2}\left|\varrho^M \not{\mathcal{L}}_{\mathscr{Z}}^{\leq N-1}\xi\right|. \tag{12.7b}$$

PROOF. The first two identities in (12.6a) follow from Lemmas 3.18 and 3.20. The identity $O\varrho = 0$ is trivial because O is $S_{t,u}$-tangent. The identity (12.6b) then follows from the definition (7.1) of $\mathscr{Z}$. Inequalities (12.7a) and (12.7b) follow easily from (12.6a) and (12.6b). □

REMARK 12.11 (**Silent use of** $\varrho(t,u) \approx 1+t$). Throughout the remainder of the monograph, we often use the estimate (12.5) without explicitly mentioning it.

12.6. Basic estimates for the rectangular spatial coordinate functions

In this section, we derive some simple pointwise estimates involving the rectangular spatial coordinate functions x^i and the Euclidean radial coordinate r.

LEMMA 12.12 (**Basic estimates for the rectangular spatial coordinate functions** x^i **and** r). *Let x^i, $(i = 1,2,3)$, be the rectangular spatial coordinate functions, let $r = \sqrt{\sum_{a=1}^3 (x^a)^2}$ be the standard Euclidean radial coordinate, and let $\varrho(t,u) = 1 - u + t$ be the geometric radial variable. Under the small-data and bootstrap assumptions of Sects.* 12.1-12.4*, if ε is sufficiently small, then the following pointwise estimates hold on $\mathcal{M}_{T_{(Bootstrap)},U_0}$:*

$$|x^i| \leq \left(1 + C\varepsilon^{1/2}\frac{\ln(e+t)}{1+t}\right)\varrho, \tag{12.8a}$$

$$\left|\frac{r}{\varrho} - 1\right| \leq C\varepsilon^{1/2}\frac{\ln(e+t)}{1+t}. \tag{12.8b}$$

Furthermore, the following estimates hold on $\mathcal{M}_{T_{(Bootstrap)},U_0}$:

$$|\nabla\!\!\!/ x^i| \leq 1 + C\varepsilon\frac{1}{1+t}. \tag{12.9}$$

PROOF. To prove (12.8b), we first use (2.4), (3.20c), and (4.1b) to deduce the identity

$$\frac{r}{\varrho} = \sqrt{1 - 2\delta_{ab}R^a R^b_{(Small)} - g_{ab}^{(Small)} R^a R^b + \delta_{cb} R^a_{(Small)} R^b_{(Small)}}. \tag{12.10}$$

Using $|x^i| \leq r$ and the bootstrap assumptions (**BA**Ψ) and (**AUX**$LR_{(Small)}$), we deduce that

$$\frac{r}{\varrho} = \sqrt{1 + \mathcal{O}\left(\varepsilon^{1/2}\frac{\ln(e+t)}{1+t}\frac{r}{\varrho}\right) + \mathcal{O}\left(\varepsilon\frac{1}{1+t}\frac{r^2}{\varrho^2}\right) + \mathcal{O}\left(\varepsilon\frac{\ln^2(e+t)}{(1+t)^2}\right)}. \tag{12.11}$$

The desired estimate (12.8b) now follows easily from (12.11).

The estimate (12.8a) then follows from (12.8b) and the fact that $|x^i| \leq r$.

To prove (12.9), we first use (**BA**Ψ) to deduce (see (2.4)-(2.6) and Def. 3.9) that $|\underline{g}_{ij} - \delta_{ij}| = \left|g_{ij}^{(Small)}\right| \lesssim \varepsilon(1+t)^{-1}$. Hence, Taylor expanding the components of $\underline{g}^{-1}$ in terms of the components of $\underline{g}$, we deduce that $|(\underline{g}^{-1})^{ij} - \delta^{ij}| \lesssim \varepsilon(1+t)^{-1}$. Using this estimate and the fact that the magnitude of the angular differential of a function as measured by $g\!\!\!/$ is no larger than the magnitude of its spatial differential as measured by $\underline{g}$, we conclude the desired estimate as follows (note that there is no summation over i in (12.12)):

(12.12)

$$|\nabla\!\!\!/ x^i|^2 := |(g\!\!\!/^{-1})^{ab}\partial_a x^i \partial_b x^i|^2 \leq |(\underline{g}^{-1})^{ab}\partial_a x^i \partial_b x^i| = |(\underline{g}^{-1})^{ii}| \leq 1 + C\varepsilon\frac{1}{1+t}.$$

□

12.7. Estimates for the rectangular components of the metrics and the spherical projection tensorfield

In this section, we derive pointwise estimates for the rectangular components of the metrics and the $S_{t,u}$ projection tensorfield.

LEMMA 12.13 (**Pointwise estimates for the rectangular components of the metrics and the $S_{t,u}$ projection tensorfield**). *Let $g\!\!\!/$, $g\!\!\!/^{-1}$, and $\Pi\!\!\!/$ be the $S_{t,u}$ tensors from Defs. 3.9 and 3.39 and let $\underline{g}$ be the $\Sigma_t^{U_0}$ tensor from Def. 3.9. Let $0 \leq N \leq 24$ be an integer. Under the small-data and bootstrap assumptions of Sects. 12.1-12.4, if ε is sufficiently small, then the following pointwise estimates hold on the spacetime domain $\mathcal{M}_{T_{(Bootstrap)},U_0}$ for their nonzero rectangular components*

($\mu, \nu = 0,1,2,3$ and $i,j = 1,2,3$):

(12.13a)
$$\left| \mathscr{Z}^N \overbrace{\{g_{\mu\nu} - m_{\mu\nu}\}}^{g_{\mu\nu}^{(Small)}} \right|, \left| \mathscr{Z}^N \left\{ (g^{-1})^{\mu\nu} - (m^{-1})^{\mu\nu} \right\} \right| \lesssim \left| \mathscr{Z}^{\leq N} \Psi \right|,$$

(12.13b)
$$\left| \mathscr{Z}^N \left\{ (\underline{g}^{-1})^{ij} - \delta^{ij} \right\} \right| \lesssim \left| \mathscr{Z}^{\leq N} \Psi \right|,$$

(12.13c)
$$\left| \mathscr{Z}^N \left\{ g\!\!\!/_{ij} - \left(\delta_{ij} - \frac{x^i x^j}{\varrho^2} \right) \right\} \right| \lesssim \left| \mathscr{Z}^{\leq N} \Psi \right| + \sum_{a=1}^{3} \left| \mathscr{Z}^{\leq N} L_{(Small)}^a \right|,$$

(12.13d)
$$\left| \mathscr{Z}^N \left\{ g\!\!\!/_{ij} - (\delta_{ij} - R_i R_j) \right\} \right| \lesssim \left| \mathscr{Z}^{\leq N} \Psi \right| + \sum_{a=1}^{3} \left| \mathscr{Z}^{\leq N} L_{(Small)}^a \right|,$$

(12.13e)
$$\left| \mathscr{Z}^N \left\{ (g\!\!\!/^{-1})^{ij} - \left(\delta^{ij} - \frac{x^i x^j}{\varrho^2} \right) \right\} \right| \lesssim \left| \mathscr{Z}^{\leq N} \Psi \right| + \sum_{a=1}^{3} \left| \mathscr{Z}^{\leq N} L_{(Small)}^a \right|,$$

(12.13f)
$$\left| \mathscr{Z}^N \left\{ (g\!\!\!/^{-1})^{ij} - (\delta^{ij} - R^i R^j) \right\} \right| \lesssim \left| \mathscr{Z}^{\leq N} \Psi \right| + \sum_{a=1}^{3} \left| \mathscr{Z}^{\leq N} L_{(Small)}^a \right|,$$

(12.13g)
$$\left| \mathscr{Z}^N \left\{ \Pi\!\!\!/_j^{\ i} - \left(\delta_j^{\ i} - \frac{x^i x^j}{\varrho^2} \right) \right\} \right| \lesssim \left| \mathscr{Z}^{\leq N} \Psi \right| + \sum_{a=1}^{3} \left| \mathscr{Z}^{\leq N} L_{(Small)}^a \right|.$$

Furthermore, the following estimates hold:

(12.14a)
$$\left\| \mathscr{Z}^{\leq 13} \overbrace{\{g_{\mu\nu} - m_{\mu\nu}\}}^{g_{\mu\nu}^{(Small)}} \right\|_{C^0(\Sigma_t^u)}, \left\| \mathscr{Z}^{\leq 13} \left\{ (g^{-1})^{\mu\nu} - (m^{-1})^{\mu\nu} \right\} \right\|_{C^0(\Sigma_t^u)} \lesssim \varepsilon \frac{1}{1+t},$$

(12.14b)
$$\left\| \mathscr{Z}^{\leq 13} \left\{ (\underline{g}^{-1})^{ij} - \delta^{ij} \right\} \right\|_{C^0(\Sigma_t^u)} \lesssim \varepsilon \frac{1}{1+t},$$

(12.14c)
$$\left\| \mathscr{Z}^{\leq 12} \left\{ g\!\!\!/_{ij} - \left(\delta_{ij} - \frac{x^i x^j}{\varrho^2} \right) \right\} \right\|_{C^0(\Sigma_t^u)} \lesssim \varepsilon^{1/2} \frac{\ln(e+t)}{1+t},$$

(12.14d)
$$\left\| \mathscr{Z}^{\leq 12} \left\{ g\!\!\!/_{ij} - (\delta_{ij} - R_i R_j) \right\} \right\|_{C^0(\Sigma_t^u)} \lesssim \varepsilon^{1/2} \frac{\ln(e+t)}{1+t},$$

(12.14e)
$$\left\| \mathscr{Z}^{\leq 12} \left\{ (g\!\!\!/^{-1})^{ij} - \left(\delta^{ij} - \frac{x^i x^j}{\varrho^2} \right) \right\} \right\|_{C^0(\Sigma_t^u)} \lesssim \varepsilon^{1/2} \frac{\ln(e+t)}{1+t},$$

(12.14f)
$$\left\| \mathscr{Z}^{\leq 12} \left\{ (g\!\!\!/^{-1})^{ij} - (\delta^{ij} - R^i R^j) \right\} \right\|_{C^0(\Sigma_t^u)} \lesssim \varepsilon^{1/2} \frac{\ln(e+t)}{1+t},$$

(12.14g)
$$\left\| \mathscr{Z}^{\leq 12} \left\{ \Pi\!\!\!/_j^{\ i} - \left(\delta_j^{\ i} - \frac{x^i x^j}{\varrho^2} \right) \right\} \right\|_{C^0(\Sigma_t^u)} \lesssim \varepsilon^{1/2} \frac{\ln(e+t)}{1+t}.$$

Finally, the following pointwise estimates hold on $\mathcal{M}_{T_{(Bootstrap)},U_0}$:

(12.15)
$$\left| g\!\!\!/_{ij} \right|, \left| (g\!\!\!/^{-1})^{ij} \right|, \left| \Pi\!\!\!/_j^{\ i} \right| \leq 1 + C\varepsilon^{1/2}.$$

In the above estimates, δ_{ij}, $\delta_j^{\ i}$, and δ^{ij} are all standard Kronecker deltas.

PROOF. Recall that $g_{\mu\nu} = \delta_{\mu\nu} + g_{\mu\nu}^{(Small)}(\Psi)$, where $g_{\mu\nu}^{(Small)}(\cdot)$ is smooth and verifies $g_{\mu\nu}^{(Small)}(0) = 0$ (see (2.4)-(2.6)). The inequalities in (12.13a) thus follow easily. (12.14a) then follows from (12.13a) and (**BA**Ψ). The proofs of (12.13b) and (12.14b) are similar.

To prove (12.13e) we first use (3.52c) to expand $(\not{g}^{-1})^{ij} = (g^{-1})^{ij} + L^i L^j + L^i R^j + R^i L^j$. We next use Def. 4.1 and (4.3) to expand $L^i = \varrho^{-1}x^i + L^i_{(Small)}$ and $R^i = -\varrho^{-1}x^i - L^i_{(Small)} - (g^{-1})^{0i}$. Inequality (12.13e) now easily follows from these expansions and the estimates (12.6b), (12.8a), (12.13a), and (12.13b). (12.14e) then follows from (12.13e), (**BA**Ψ), and (**AUX**$LR_{(Small)}$).

The proofs of the remaining inequalities are very similar, and we omit the details. $\square$

COROLLARY 12.14 (**Comparison between the norms of $S_{t,u}$ tensors and the size of their rectangular components**). *Let be ξ an $S_{t,u}$ tensor. Under the small-data and bootstrap assumptions of Sects.* 12.1-12.4, *if ε is sufficiently small, then the following comparison estimates hold on $\mathcal{M}_{T_{(Bootstrap)},U_0}$:*

$$|\xi| \approx \sum |\xi_{Rectangular}|, \tag{12.16}$$

where the left-hand side is as in Def. 3.54 *and the sum on the right-hand side is taken over the components of ξ relative to the rectangular spatial coordinates.*

PROOF. We give the proof in the case that ξ is an $S_{t,u}$ one-form. The proof will generalize in an obvious fashion to the case of general $S_{t,u}$ tenors. We first recall that $|\xi|^2 = (\not{g}^{-1})^{ab}\xi_a\xi_b$. Hence, by the estimate (12.15) for the rectangular components of $\not{g}^{-1}$, we have $|\xi|^2 \lesssim \sum_{a=1}^3 |\xi_a|^2$, which easily implies that the left-hand side of (12.16) is $\lesssim$ the right-hand side as desired.

To prove the reverse inequality, we first note that $\sum_{a=1}^3 |\xi_a|^2 = \delta^{ab}\xi_a\xi_b$. Using (12.14f), we see that we can replace δ^{ab} with $(\not{g}^{-1})^{ab}$ plus an error term that is in magnitude $\lesssim \varepsilon^{1/2}\ln(e+t)(1+t)^{-1}$ plus tensorial products of the vector R that are completely annihilated because they are paired with copies of ξ. We therefore deduce that $\left|\delta^{ab}\xi_a\xi_b - |\xi|^2\right| \lesssim \delta^{ab}\xi_a\xi_b\varepsilon^{1/2}\ln(e+t)(1+t)^{-1}$, from which we easily conclude that the right-hand side of (12.16) is $\lesssim$ the left-hand side as desired. $\square$

12.8. The behavior of quantities along the initial data hypersurface

In this section, we provide quantitative estimates showing that when $\mathring{\epsilon}$ is small (see (12.2)), many other quantities such as $\upmu - 1$, $L^i_{(Small)}$, etc. are also small along Σ_0^1.

LEMMA 12.15 (**All rectangular derivatives of Ψ are small along Σ_0^1**). *Assume that $\square_{g(\Psi)}\Psi = 0$ and let $\mathring{\epsilon}$ be the size of the data as defined in Def.* 12.3. *Then if $\mathring{\epsilon}$ is sufficiently small, we have*

$$\sum_{M=0}^{25} \|\partial_t^M \Psi\|_{H_e^{25-M}(\Sigma_0^1)} \lesssim \mathring{\epsilon}. \tag{12.17}$$

In (12.17), $H_e^N(\Sigma_0^1)$ is the standard Euclidean Sobolev space involving order $\leq N$ rectangular spatial coordinate partial derivatives along Σ_0^1.

PROOF. The cases $M = 0, 1$ in (12.17) follow directly from the definition of the size of the data. To deduce (12.17) in the case $M = 2$, we use (2.3) and the wave equation (2.17) to solve for $\partial_t^2 \Psi$ as follows:

(12.18)
$$\partial_t^2 \Psi = (g^{-1})^{ab} \partial_a \partial_b \Psi + 2(g^{-1})^{0a} \partial_a \partial_t \Psi - \frac{1}{2}(g^{-1})^{\alpha\beta}(g^{-1})^{\kappa\lambda} \left\{ 2G_{\alpha\kappa} \partial_\beta \Psi - G_{\alpha\beta} \partial_\kappa \Psi \right\}.$$

The desired estimate (12.17) then follows from setting $t = 0$ in (12.18) and applying the standard Sobolev calculus. To deduce (12.17) in the cases $M \geq 3$, we repeatedly differentiate (12.18) with respect to t and use the equations to express all higher-order time derivatives in terms of spatial derivatives of Ψ and $\partial_t \Psi$. The desired estimate (12.17) then follows from the standard Sobolev calculus at $t = 0$ as in the case $M = 2$. □

LEMMA 12.16 ($\mu - 1$, $L_{(Small)}^i$, **and** Ξ^i **are small along** Σ_0^1). *Let*

(12.19)
$$z := \sqrt{(\underline{g}^{-1})^{ab} \frac{x^a x^b}{r^2}} = \sqrt{1 + \left\{ (\underline{g}^{-1})^{ab} - \delta^{ab} \right\} \frac{x^a x^b}{r^2}},$$

where $\underline{g}$ *is defined in Def. 3.9. Then along the initial data hypersurface region* Σ_0^1, *the following identities hold,* $(i = 1, 2, 3)$:

(12.20a)
$$\mu^{-2} = 1 + \left\{ (\underline{g}^{-1})^{ab} - \delta^{ab} \right\} \frac{x^a x^b}{r^2},$$

(12.20b)
$$L_{(Small)}^i = \frac{1}{z} \left\{ (\underline{g}^{-1})^{ia} - \delta^{ia} \right\} \frac{x^a}{r} + \left\{ \frac{1}{z} - 1 \right\} \frac{x^i}{r} - (g^{-1})^{0i},$$

(12.20c)
$$\Xi^i = \left\{ \frac{1}{z^2} - 1 \right\} (\underline{g}^{-1})^{ia} \frac{x^a}{r} + \left\{ (\underline{g}^{-1})^{ia} - \delta^{ia} \right\} \frac{x^a}{r}.$$

Above, Ξ *is the* $S_{t,u}$*-tangent vectorfield defined by (3.46).*

In addition, under the hypotheses of Lemma 12.15, the following estimates hold:

(12.21a)
$$\sum_{M=0}^{25} \left\| \partial_t^M (\mu - 1) \right\|_{H_e^{25-M}(\Sigma_0^1)} \lesssim \mathring{\epsilon},$$

(12.21b)
$$\sum_{M=0}^{25} \left\| \partial_t^M L_{(Small)}^i \right\|_{H_e^{25-M}(\Sigma_0^1)} \lesssim \mathring{\epsilon},$$

(12.21c)
$$\sum_{M=0}^{25} \left\| \partial_t^M \Xi^i \right\|_{H_e^{25-M}(\Sigma_0^1)} \lesssim \mathring{\epsilon}.$$

Above, $H_e^N(\Sigma_0^1)$ *is the standard Euclidean Sobolev space involving order* $\leq N$ *rectangular spatial coordinate partial derivatives along* Σ_0^1.

PROOF. In this proof, the identities that we state hold along Σ_0^1. In particular, we have $u = 1 - r$ (see (3.4b)), where r is the standard Euclidean radial variable on $\mathbb{R}^3$, and hence $\partial_i u = -\frac{x^i}{r}$. From this identity and (3.24), we deduce (12.20a). Next, to prove (12.20b), we first use definition (4.1b), the identities (3.25) and (4.3), the above calculations, and the fact that $\varrho = r$ along Σ_0^1 to deduce that $L_{(Small)}^i = -R_{(Small)}^i - (g^{-1})^{0i} = z^{-1} (\underline{g}^{-1})^{ia} \frac{x^a}{r} - \frac{x^i}{r} - (g^{-1})^{0i}$. The desired identity (12.20b) follows easily from the previous equation. Next, to prove (12.20c), we note that the geometric coordinates are constructed (see Sect. 3.5) in such a way that along Σ_0^1, we have $\frac{\partial}{\partial u} = -\partial_r$, where $\partial_r = \frac{x^a}{r}\partial_a$ is the standard Euclidean radial vectorfield. Also using (3.25) and (3.46), we deduce that along Σ_0^1, we have

$$\Xi^i = \Xi x^i = -\partial_r x^i - \breve{R}^i = -\frac{x^i}{r} - \upmu R^i = -\frac{x^i}{r} + z^{-2}(\underline{g}^{-1})^{ia}\frac{x^a}{r}.$$

The desired identity (12.20c) thus follows.

The estimates (12.21a)-(12.21c) in the case $M = 0$ then follow from the identities (12.20a)-(12.20c), the fact that $(\underline{g}^{-1})^{ij} - \delta^{ij}$ and $(g^{-1})^{0i}$ are smooth scalar-valued functions of Ψ that vanish at $\Psi = 0$ (see (2.4)-(2.6)), the estimate (12.17), and the standard Sobolev calculus. To obtain the desired estimates in the cases $M \geq 1$, we inductively use the evolution equations (3.47) (this equation needs to be rewritten relative to rectangular coordinates[1]), (4.5), and (4.9b) and equations (12.20a)-(12.20c) to solve for the ∂_t^K derivatives of $\upmu$, $L_{(Small)}^i$, and Ξ^i (for $1 \leq K \leq 25$) in terms of the rectangular coordinate derivatives of Ψ and the rectangular coordinate derivatives of the eikonal function u. Furthermore, along Σ_0^1, u is a smooth function of the spatial coordinates, while by using the eikonal equation (3.4a), we can solve for the time derivatives of u in terms of the rectangular spacetime derivatives of Ψ and the rectangular spatial derivatives of u. The desired estimates thus follow from the standard Sobolev calculus. $\square$

LEMMA 12.17 (**Various Sobolev norms are initially small**). *Under the hypotheses of Lemma* 12.15, *we have the following estimates,* $(i = 1, 2, 3)$*:*

(12.22a)

$$\left\|\mathscr{Z}^{\leq 25}\Psi\right\|_{L^2(\Sigma_0^1)}, \left\|\mathscr{Z}^{\leq 25}(\upmu - 1)\right\|_{L^2(\Sigma_0^1)}, \left\|\varrho\mathscr{Z}^{\leq 25}L_{(Small)}^i\right\|_{L^2(\Sigma_0^1)}, \left\|\mathscr{Z}^{\leq 25}\Xi^i\right\|_{L^2(\Sigma_0^1)} \lesssim \mathring{\epsilon},$$

(12.22b)

$$\left\|\mathscr{Z}^{\leq 23}\Psi\right\|_{C^0(\Sigma_0^1)}, \left\|\mathscr{Z}^{\leq 23}(\upmu - 1)\right\|_{C^0(\Sigma_0^1)}, \left\|\varrho\mathscr{Z}^{\leq 23}L_{(Small)}^i\right\|_{C^0(\Sigma_0^1)}, \left\|\mathscr{Z}^{\leq 23}\Xi^i\right\|_{C^0(\Sigma_0^1)} \lesssim \mathring{\epsilon}.$$

Above, $\|\cdot\|_{L^2(\Sigma_0^1)}$ *is the Sobolev norm from* (3.85b).

Furthermore, we have the following estimates, where Ξ *is the* $S_{t,u}$*-tangent vectorfield defined by* (3.46) *and* $\Xi_\flat$ *denotes the* $\mathscr{g}$*-dual of* Ξ*:*

$$\left\|\not{\mathcal{L}}_{\mathscr{Z}}^{\leq 25}\Xi\right\|_{L^2(\Sigma_0^1)}, \left\|\not{\mathcal{L}}_{\mathscr{Z}}^{\leq 25}\Xi_\flat\right\|_{L^2(\Sigma_0^1)} \lesssim \mathring{\epsilon}, \tag{12.23a}$$

$$\left\|\not{\mathcal{L}}_{\mathscr{Z}}^{\leq 23}\Xi\right\|_{C^0(\Sigma_0^1)}, \left\|\not{\mathcal{L}}_{\mathscr{Z}}^{\leq 23}\Xi_\flat\right\|_{C^0(\Sigma_0^1)} \lesssim \mathring{\epsilon}. \tag{12.23b}$$

PROOF. Since $d\underline{\varpi}$ (see Def. 3.68) is equal to $\upmu^{-1}$ times the volume form induced by g on Σ_0 (see the discussion at the beginning of Sect. 3.19), it follows that $d\underline{\varpi} = \upmu^{-1}\mathrm{det}\underline{g}dx^1dx^2dx^3$, where $\underline{g}$ is as in Def. 3.9 and det is taken relative to rectangular coordinates. From (12.21a) and (2.4)-(2.6), we deduce that along Σ_0, $\upmu^{-1}\mathrm{det}\underline{g}dx^1dx^2dx^3$ is near-Euclidean in the sense that

$$1 - C\mathring{\epsilon} \leq \upmu^{-1}\mathrm{det}\underline{g} \leq 1 + C\mathring{\epsilon}. \tag{12.24}$$

---

[1]With the help of (4.21a), we see that in rectangular coordinates, equation (3.47) reads $L\Xi^i = \Xi^a\partial_a L^i - {}^{(\breve{R})}\not{\pi}_L^{\ i}$. Note that equation (7.5d) provides an expression for ${}^{(\breve{R})}\not{\pi}_{LA}$.

It follows that the norm $\|\cdot\|_{L^2(\Sigma_0^1)}$ is comparable to the standard Euclidean Lebesgue norm $\|\cdot\|_{L^2_e(\Sigma_0^1)}$. Hence, to deduce (12.22a), it suffices to prove the same inequalities using the standard Euclidean Lebesgue norm $\|\cdot\|_{L^2_e(\Sigma_0^1)}$ in place of $\|\cdot\|_{L^2(\Sigma_0^1)}$. To this end, we completely expand $\mathscr{Z}^{\leq 25}(\upmu-1)$ in terms of rectangular coordinate derivatives and use the standard Sobolev calculus to deduce that

$$\|\mathscr{Z}^{\leq 25}(\upmu-1)\|_{L^2_e(\Sigma_0^1)} \lesssim \left(\sum_{M=0}^{25}\|\partial_t^M(\upmu-1)\|_{H_e^{25-M}(\Sigma_0^1)}\right) \times \left\{1+\left(\sum_{M=0}^{24}\sum_{Z\in\mathscr{Z}}\sum_{\alpha=0}^{3}\|\partial_t^M Z^\alpha\|_{H_e^{24-M}(\Sigma_0^1)}\right)^{25}\right\}. \tag{12.25}$$

Inequality (12.21a) implies that $\|\partial_t^M(\upmu-1)\|_{H_e^{25-M}(\Sigma_0^1)} \lesssim \mathring{\epsilon}$. Furthermore, from the identities $\varrho L^0=\varrho$, $\varrho L^i = x^i + \varrho L^i_{(Small)}$ (see (4.1a)), (4.3), and (6.7), it follows that all of the rectangular components Z^α, $(\alpha=0,1,2,3)$, are smooth functions of the rectangular coordinate functions, Ψ, u, $\upmu$, and the rectangular components $L^i_{(Small)}$. It thus follows from the remarks made in the proof of Lemma 12.16 about how to estimate the rectangular derivatives of u, the standard Sobolev calculus, (12.17), and (12.21a)-(12.21b) that $\|\partial_t^M Z^\alpha\|_{H_e^{25-M}(\Sigma_0^1)} \lesssim 1$. We have thus proved (12.22a) for $\upmu-1$. The remaining three inequalities in (12.22a) can be proved using a similar argument with the help of Lemma 12.16; we omit the details. The estimates (12.22b) follow in a similar fashion thanks to the standard Sobolev embedding estimate $\|f\|_{C^0(\Sigma_0^1)} \lesssim \|f\|_{H_e^2(\Sigma_0^1)}$. The estimates (12.23a) and (12.23b) for Ξ follow similarly. In particular, by Lemma 4.10 with $Y=\Xi$, the $S_{t,u}$-projected Lie derivatives of Ξ agree with standard Lie derivatives. Hence, $|\not{\mathcal{L}}_{\mathscr{Z}}^N\Xi|^2 = \not{g}_{ab}(\mathcal{L}_{\mathscr{Z}}^N\Xi^a)\mathcal{L}_{\mathscr{Z}}^N\Xi^b$ and the computations can be carried out relative to the rectangular coordinates. To derive the desired estimates (12.23a) and (12.23b) for $\Xi_\flat$, we note the identity $|\not{\mathcal{L}}_{\mathscr{Z}}^N\Xi_\flat|^2 = (\not{g}^{-1})^{ab}(\not{\mathcal{L}}_{\mathscr{Z}}^N(\not{g}_{ac}\Xi^c))\not{\mathcal{L}}_{\mathscr{Z}}^N(\not{g}_{bd}\Xi^d)$, which, in view of the identity $(\not{g}^{-1})^{ij} = (g^{-1})^{ij} + L^iL^j + L^iR^j + R^iL^j$ (see (3.52c)), allows us to carry out the computations relative to the rectangular coordinates. □

12.9. Estimates for the derivatives of rectangular components of various vectorfields and the radial component of the Euclidean rotations

In this section, we derive pointwise estimates for the derivatives of the rectangular components of various vectorfields that we use in our analysis. We also derive pointwise estimates for the scalar-valued functions $\uprho_{(l)}$, $(l=1,2,3)$, defined in (6.6).

LEMMA 12.18 (**First pointwise estimates for the spatial rectangular components of various vectorfields and the functions $\uprho_{(l)}$**). *Let $0\leq N\leq 24$ be an integer, and let L^i and R^i, $(i=1,2,3)$, be the (scalar-valued) rectangular spatial components of L and R. Under the small-data and bootstrap assumptions of Sects.* 12.1-12.4, *if ε is sufficiently small, then the following pointwise estimates* *hold on* $\mathcal{M}_{T_{(Bootstrap)},U_0}$*:*

$$\left|\begin{pmatrix} \mathscr{Z}^N L^i \\ \mathscr{Z}^N R^i \end{pmatrix}\right| \lesssim \left|\mathscr{Z}^{\leq N}\Psi\right| + \frac{1}{1+t}\left|\begin{pmatrix} \mathscr{Z}^{\leq N-1}(\upmu - 1) \\ \sum_{a=1}^3 \varrho \left|\mathscr{Z}^{\leq N} L^a_{(Small)}\right| \end{pmatrix}\right| + 1, \tag{12.26a}$$

(12.26b)

$$\left\|\begin{pmatrix} \mathscr{Z}^{\leq 12} L^i \\ \mathscr{Z}^{\leq 12} R^i \end{pmatrix}\right\|_{C^0(\Sigma_t^u)} \lesssim 1.$$

Furthermore, for each $Z \in \mathscr{Z} = \{\varrho L, \breve{R}, O_{(1)}, O_{(2)}, O_{(3)}\}$*, we have the following estimates for the rectangular spatial components* Z^i*,* $(i=1,2,3)$*, on* $\mathcal{M}_{T_{(Bootstrap)},U_0}$*:*

$$\left|\mathscr{Z}^N Z^i\right| \lesssim (1+t)\left|\mathscr{Z}^{\leq N}\Psi\right| + \left|\begin{pmatrix} \mathscr{Z}^{\leq N}(\upmu - 1) \\ \sum_{a=1}^3 \varrho \left|\mathscr{Z}^{\leq N} L^a_{(Small)}\right| \end{pmatrix}\right| + 1 + t, \tag{12.27a}$$

(12.27b)

$$\left\|\mathscr{Z}^{\leq 12} Z^i\right\|_{C^0(\Sigma_t^u)} \lesssim 1 + t.$$

In addition, we have the following estimates for the scalar-valued functions $\uprho_{(l)}$*,* $(l = 1, 2, 3)$*, defined in* (6.6) *on* $\mathcal{M}_{T_{(Bootstrap)},U_0}$*:*

$$\left|\mathscr{Z}^N \uprho_{(l)}\right| \lesssim (1+t)\left|\mathscr{Z}^{\leq N}\Psi\right| + \sum_{a=1}^3 \varrho\left|\mathscr{Z}^{\leq N} L^a_{(Small)}\right|, \tag{12.28a}$$

$$\left\|\mathscr{Z}^{\leq 12}\uprho_{(l)}\right\|_{C^0(\Sigma_t^u)} \lesssim \varepsilon^{1/2}\ln(e+t). \tag{12.28b}$$

PROOF. Throughout this proof, we use the bootstrap assumptions (**BA**Ψ), (**AUX**$\upmu$), and (**AUX**$LP_{(Small)}$).

We first prove (12.27a) by induction. In the base case $N = 0$, we have to bound the magnitudes of $\varrho L^i = x^i + \varrho L^i_{(Small)}$, $\breve{R}^i = -\upmu \frac{x^i}{\varrho} + \upmu R^i_{(Small)}$ (see Lemma 4.1), and $O^i_{(l)} = \epsilon_{lai}x^a + \uprho_{(l)}\frac{x^i}{\varrho} - \uprho_{(l)} R^i_{(Small)}$ (see (6.7)) by the right-hand side of (12.27a). Recall that an expression for $\uprho_{(l)}$ is given by (6.6). We bound all factors of x^i via (12.8a). Furthermore, from (2.4)-(2.6) and (4.3), we see that $g_{ab}^{(Small)}$ is a smooth function of Ψ that vanishes at $\Psi = 0$ and that $R^i_{(Small)} = -L^i_{(Small)}$ plus a smooth function of Ψ that vanishes at $\Psi = 0$. Thus, we have $|R^i_{(Small)}| \lesssim |L^i_{(Small)}| + |\Psi|$. From these facts and the bootstrap assumptions, we deduce the desired estimate (12.27a) when $N = 0$.

To carry out the induction, we apply $\mathscr{Z}^N$ to the right-hand side of the above identities for ϱL^i, $\breve{R}^i$, and $O^i_{(l)}$. Using the Leibniz rule, we can easily bound most terms that arise via the bootstrap assumptions, the estimate (12.8a), and the estimate (12.6b) (which we use to bound factors involving derivatives of ϱ) as in the previous paragraph (without the need for induction). The factors that require the induction hypotheses are those of the form $\mathscr{Z}^M x^i$, $1 \leq M \leq N$. To bound these terms, we schematically write $\mathscr{Z}^M = \mathscr{Z}^{M-1}Z$ and hence $\mathscr{Z}^M x^i = \mathscr{Z}^{M-1}Z^i$, which can be bounded by the induction hypotheses. We have thus proved (12.27a). (12.27b) then follows from (12.27a) and the bootstrap assumptions.

The estimate (12.28a) then follows from (2.6), (6.6), the fact that $R^i = -L^i_{(Small)}$ plus a smooth function of Ψ that vanishes at $\Psi = 0$, the estimates (12.6b), (12.8a), and (12.27a), and the bootstrap assumptions. (12.28b) then follows from (12.28a) and the bootstrap assumptions.

The estimate (12.26a) for L^i follows similarly from the decomposition $L^i = \frac{x^i}{\varrho} + L^i_{(Small)}$, the estimates (12.6b), (12.8a), and (12.27a), and the bootstrap assumptions. The estimate (12.26a) for R^i then follows from the fact that $R^i = -L^i$ plus a smooth function of Ψ that vanishes at $\Psi = 0$ (see (2.4)-(2.6) and (4.2b)), the estimates for $\mathscr{Z}^{\leq N} L^i$, and the bootstrap assumptions. Inequality (12.26b) then follows from (12.26a) and the bootstrap assumptions. □

12.10. Estimates for the rectangular components of the metric dual of the unit-length radial vectorfield

We now derive some simple pointwise estimates for $R_i + \frac{x^i}{\varrho}$ and $R_i^{(Small)}$. We recall that we lower and raise indices with, respectively, g and g^{-1}, *and not with the Minkowski metric.*

We start with the following lemma, in which we derive higher-order analogs of inequality (12.8b)

LEMMA 12.19 (**Estimates for** $\frac{r}{\varrho}$). *Let* $r = \sqrt{\sum_{a=1}^{3}(x^a)^2}$ *be the Euclidean radial coordinate and let* $\varrho(t,u) = 1 - u + t$ *be the geometric radial variable. Under the small-data and bootstrap assumptions of Sects.* 12.1-12.4, *if* ε *is sufficiently small, then the following estimates hold for* $(t,u) \in [0, T_{(Bootstrap)}) \times [0, U_0]$*:*

$$\left\| \mathscr{Z}^{\leq 12}\left(\frac{r}{\varrho} - 1\right)\right\|_{C^0(\Sigma_t^u)} \lesssim \varepsilon^{1/2}\frac{\ln(e+t)}{1+t}. \tag{12.29}$$

PROOF. Inequality (12.29) follows from applying $\mathscr{Z}^{\leq 12}$ to (12.10), using the estimates (12.14a) and (12.26b), and using the bootstrap assumptions (**AUX**$LR_{(Small)}$). □

LEMMA 12.20 (**Estimates for** $R_i + \frac{x^i}{\varrho}$ and $R_i^{(Small)}$). *Let* $0 \leq N \leq 24$ *be an integer. Under the small-data and bootstrap assumptions of Sects.* 12.1-12.4, *if* ε *is sufficiently small, then the following pointwise estimates hold on the spacetime domain* $\mathcal{M}_{T_{(Bootstrap)},U_0}$, $(i = 1,2,3)$*:*

$$\left| \mathscr{Z}^N\left(R_i + \frac{x^i}{\varrho}\right)\right| \lesssim \left|\mathscr{Z}^{\leq N}\Psi\right| + \sum_{a=1}^{3}\left|\mathscr{Z}^{\leq N} L^a_{(Small)}\right|, \tag{12.30a}$$

$$\left| \mathscr{Z}^N R_i^{(Small)}\right| \lesssim \left|\mathscr{Z}^{\leq N}\Psi\right| + \sum_{a=1}^{3}\left|\mathscr{Z}^{\leq N} L^a_{(Small)}\right|. \tag{12.30b}$$

Furthermore, the following estimates hold:

$$\left\| \mathscr{Z}^{\leq 12} \left(R_i + \frac{x^i}{\varrho} \right) \right\|_{C^0(\Sigma_t^u)} \lesssim \varepsilon^{1/2} \frac{\ln(e+t)}{1+t}, \tag{12.31a}$$

$$\left\| \mathscr{Z}^{\leq 12} R_i^{(Small)} \right\|_{C^0(\Sigma_t^u)} \lesssim \varepsilon^{1/2} \frac{\ln(e+t)}{1+t}. \tag{12.31b}$$

In addition,

(12.32) *inequality* (12.31a) *holds with* $\frac{x^i}{\varrho}$ *replaced by* $\frac{x^i}{r}$, *where* $r = \sqrt{\sum_{a=1}^{3} (x^a)^2}$.

PROOF. To prove (12.30a), we first use (2.4), (4.1a), and (4.2b), to deduce the identity

$$R_i + \frac{x^i}{\varrho} = -g_{ia}^{(Small)} \frac{x^a}{\varrho} - g_{ia} L^a_{(Small)} - g_{ia} (g^{-1})^{0a}. \tag{12.33}$$

Applying $\mathscr{Z}^N$ to (12.33), using the identity $Zx^i = Z^i$, using the estimates (12.6b), (12.8a), (12.13a), (12.14a), (12.27a), and (12.27b), and using the bootstrap assumptions (**BA**Ψ), and (**AUX**$LR_{(Small)}$), we conclude the desired estimate (12.30a). The estimate (12.31a) then follows from (12.30a) and the bootstrap assumptions (**BA**Ψ), and (**AUX**$LR_{(Small)}$).

The proofs of (12.30b) and (12.31b) are based on the identity $R_i^{(Small)} = R_i + g_{ia} \frac{x^a}{\varrho}$ (see (4.1b)) and are similar; we omit the details.

Inequality (12.32) follows from expanding $\frac{x^i}{\varrho} = \frac{x^i}{r} + \frac{x^i}{r}\left(\frac{r}{\varrho} - 1\right)$ and using the identity $Zx^i = Z^i$ and the estimates (12.6b), (12.8a), (12.27b), (12.29), and (12.31a). ☐

12.11. Precise pointwise estimates for the rotation vectorfields

In this section, we derive some sharp pointwise estimates for the rotation vectorfields and their spherical covariant derivatives. We carefully estimate the non-small constants such as the "1" on the right-hand side of (12.34a). The reason that we are interested in precise constants is that the same constants appear in the differential operator comparison estimates of Sect. 12.12, and these comparison estimates ultimately affect the number of derivatives we need to close our top-order L^2 estimates.

LEMMA 12.21 (**Precise pointwise estimates for the rotation vectorfields**). *Let* $O_{(l)}$, $(l = 1, 2, 3)$, *be the geometric rotation vectorfield from Def.* 6.1. *Under the small-data and bootstrap assumptions of Sects.* 12.1-12.4, *if* ε *is sufficiently small, then the following pointwise estimates hold on* $\mathcal{M}_{T_{(Bootstrap)}, U_0}$:

$$|O_{(l)}| \leq (1 + C\varepsilon^{1/2})\varrho, \tag{12.34a}$$

$$|[O_{(l)}, O_{(m)}]| \leq (1 + C\varepsilon^{1/2})\varrho, \tag{12.34b}$$

$$|\nabla\!\!\!/ O_{(l)}|^2 \leq 2(1 + C\varepsilon^{1/2}). \tag{12.34c}$$

Furthermore, let $\not{g}$, $\not{g}^{-1}$, *and* $\not{\Pi}$ *be the* $S_{t,u}$ *tensorfields from Defs.* 3.9 *and* 3.39. *Then the following pointwise estimates for rectangular components hold on* $\mathcal{M}_{T_{(Bootstrap)},U_0}$, $(a, b, i, j, m, n = 1, 2, 3)$*:*

$$\left| (\not{g}^{-1})^{mn} - \varrho^{-2} \sum_{l=1}^{3} O_{(l)}^m O_{(l)}^n \right| \lesssim \varepsilon^{1/2} \frac{\ln(e+t)}{1+t}, \tag{12.35a}$$

$$\left| \not{g}_{ij} (\not{g}^{-1})^{mn} - \not{\Pi}_i{}^n \not{\Pi}_j{}^m - \sum_{l=1}^{3} (\not{\nabla}_i O_{(l)}^m)(\not{\nabla}_j O_{(l)}^n) \right| \lesssim \varepsilon^{1/2} \frac{\ln(e+t)}{1+t}, \tag{12.35b}$$

$$\left| (\not{g}^{-1})^{mn} - \sum_{l=1}^{3} (\not{g}^{-1})^{ab} (\not{\nabla}_a O_{(l)}^m)(\not{\nabla}_b O_{(l)}^n) \right| \lesssim \varepsilon^{1/2} \frac{\ln(e+t)}{1+t}, \tag{12.35c}$$

$$\left| \sum_{l=1}^{3} O_{(l)}^m \not{\nabla}_b O_{(l)}^n \right| \lesssim \varepsilon^{1/2} \ln(e+t). \tag{12.35d}$$

PROOF. We first prove (12.34a). Since $O_{(l)}$ is $S_{t,u}$-tangent, the following identities hold relative to the rectangular spatial coordinates: $|O_{(l)}|^2 = \not{g}_{ab} O_{(l)}^a O_{(l)}^b = g_{ab} O_{(l)}^a O_{(l)}^b$. Using this equation, equations (2.4) and (6.5), and the fact that $g_{ab} R^a R^b = 1$ (see (3.20c)), we deduce that

$$|O_{(l)}|^2 \leq \delta_{ab} O_{(Flat;l)}^a O_{(Flat;l)}^b + C \left| g_{ab}^{(Small)} O_{(Flat;l)}^a O_{(Flat;l)}^b \right| + C \mathrm{p}_{(l)}^2.$$

Next, using (6.2) and the estimate (12.8b), we deduce that $\delta_{ab} O_{(Flat;l)}^a O_{(Flat;l)}^b \leq \sum_{a=1}^{3} (x^a)^2 = r^2 \leq (1 + C\varepsilon^{1/2})\varrho^2$. Inequality (12.34a) therefore easily follows once we show that $\left| g_{ab}^{(Small)} O_{(Flat;l)}^a O_{(Flat;l)}^b \right| \leq C\varepsilon^{1/2}\varrho$ and $|\mathrm{p}_{(l)}| \leq C\varepsilon^{1/2}\varrho$. The first of these two estimates follows from definition (6.2) and the estimates (12.8a) and (12.14a). The second estimate follows from (12.28b).

We now prove (12.34b). We take the norm of each term on the right-hand side of (6.9). By (12.34a), the first term is in magnitude $\leq (1 + C\varepsilon^{1/2})\varrho$. It remains for us to show that the terms on the right-hand side of (6.10) are in magnitude $\lesssim \varepsilon^{1/2}\varrho$. To see that the terms on the first line of the right-hand side of (6.10) are in fact $\lesssim \varepsilon \ln^2(e+t)(1+t)^{-1}$, we use the bootstrap assumption ($\mathbf{AUX}\chi^{(Small)}$) and the estimates (12.28b) and (12.34a). To see that the terms on the third line of the right-hand side of (6.10) are in fact $\lesssim \varepsilon \ln^2(e+t)(1+t)^{-1}$, we treat all lowercase Latin-indexed quantities as scalar valued functions and the $\not{d}x^c$ as $S_{t,u}$ one-forms. The desired bound follows from the bootstrap assumption ($\mathbf{AUX}LR_{(Small)}$) and the estimates (12.9), (12.14a), and (12.28b). To see that the terms on the second line of the right-hand side of (6.10) are in fact $\lesssim \varepsilon \ln(e+t)(1+t)^{-1}$, we use the estimates (12.28b), (12.34a), and the estimate $|\Theta^{(Tan-\Psi)}| \lesssim \varepsilon(1+t)^{-2}$. To prove this latter bound, we first use (4.11d) to write $\Theta^{(Tan-\Psi)}$ in the schematic form $\Theta^{(Tan-\Psi)} = G_{(Frame)} \begin{pmatrix} L\Psi \\ \not{d}\Psi \end{pmatrix}$. By the bootstrap assumptions ($\mathbf{BA}\Psi$), the terms in the array of Ψ derivatives are $\lesssim \varepsilon(1+t)^{-2}$ in magnitude. It remains for us to show that $|G_{(Frame)}| \lesssim 1$. To this end, we note that the rectangular components of the $S_{t,u}$ tensorfields belonging to the array $G_{(Frame)}$ (see Def. 3.62) can be viewed as smooth functions of Ψ, of the rectangular components of the vectorfields L and R, and of the rectangular components of the $S_{t,u}$ projection tensorfield $\not{\Pi}$. Hence, from inequality (12.16), the estimate (12.15), and the bootstrap assumptions ($\mathbf{BA}\Psi$) and ($\mathbf{AUX}LR_{(Small)}$), we deduce that $|G_{(Frame)}| \lesssim 1$ as desired.

We now prove (12.35a). To begin, we first expand

(12.36)
$$(\not g^{-1})^{mn} = \left\{\delta^{mn} - \frac{x^m x^n}{r^2}\right\} + \left\{\frac{1}{r^2} - \frac{1}{\varrho^2}\right\} x^m x^n + \left\{(\not g^{-1})^{mn} - \left(\delta^{mn} - \frac{x^m x^n}{\varrho^2}\right)\right\}.$$

Using (6.2), it is straightforward to verify that the first difference on the right-hand side of (12.36) can be expressed as

$$\left\{\delta^{mn} - \frac{x^m x^n}{r^2}\right\} = \sum_{l=1}^{3} r^{-2} O_{(Flat;l)}^m O_{(Flat;l)}^n. \tag{12.37}$$

From equation (6.5) and the estimates (12.26b) and (12.28b), we deduce that $|O_{(Flat;l)}^m - O_{(l)}^m| = |\mathrm{p}_{(l)}||R^m| \lesssim \varepsilon^{1/2}\ln(e+t)$. Also using the estimates (12.8a) and (12.8b), we deduce from (12.37) that $\left\{\delta^{mn} - \frac{x^m x^n}{\varrho^2}\right\} = \sum_{l=1}^{3} \varrho^{-2} O_{(l)}^m O_{(l)}^n$ plus an error term that is in magnitude $\lesssim \varepsilon^{1/2}\ln(e+t)(1+t)^{-1}$.

To bound the second difference $\left\{\frac{1}{r^2} - \frac{1}{\varrho^2}\right\} x^m x^n$ on the right-hand side of (12.36), we use the estimates (12.8a) and (12.8b) to deduce that it is in magnitude $\lesssim \varepsilon^{1/2}\ln(e+t)(1+t)^{-1}$ as desired.

To bound the final difference on the right-hand side of (12.36), we use the estimate (12.14e) to deduce that its magnitude is $\lesssim \varepsilon^{1/2}\ln(e+t)(1+t)^{-1}$ as desired

We now prove (12.35b). We use the identity (7.24) and as above, we treat all uppercase Latin indices as tensorial $S_{t,u}$ indices, and all lowercase Latin-indexed quantities as scalar-valued functions. The decomposition $\chi = \frac{\not g}{\varrho} + \chi^{(Small)}$ (see (4.1c)) and same arguments used above imply that the norms of the last three products on the right-hand side of (7.24) are (viewed as $S_{t,u}$ tensors) in magnitude $\lesssim \varepsilon^{1/2}\ln(e+t)(1+t)^{-1}$. Furthermore, the first term $\epsilon_{lab}\not d x^a \otimes \not d^{\#} x^b$ is (viewed as an $S_{t,u}$ tensor) in magnitude by $\lesssim 1$. By Cor. 12.14, the same estimates hold for the rectangular components of these tensors. Furthermore, using the identities $\not d_i x^a = \not\Pi_i^{\ a}$ and $\not d^{\#} x^j = (\not g^{-1})^{ij}$, we note that the i,m rectangular components of the type $\binom{1}{1}$ $S_{t,u}$ tensor $\epsilon_{lab}\not d x^a \otimes \not d^{\#} x^b$ are $\epsilon_{lab}\not\Pi_i^{\ a}(\not g^{-1})^{mb}$. Hence, the main contribution to the sum $\sum_{l=1}^{3}(\not\nabla_i O_{(l)}^m)(\not\nabla_j O_{(l)}^n)$ comes from the product term $\sum_{l=1}^{3}\epsilon_{lab}\epsilon_{lcd}\not\Pi_i^{\ a}(\not g^{-1})^{mb}\not\Pi_j^{\ c}(\not g^{-1})^{nd} = \delta^{kl}\epsilon_{kab}\epsilon_{lcd}\not\Pi_i^{\ a}(\not g^{-1})^{mb}\not\Pi_j^{\ c}(\not g^{-1})^{nd}$, and the desired estimate (12.35b) will follow once we prove the following estimate:

(12.38)
$$\left|\not g_{ij}(\not g^{-1})^{mn} - \not\Pi_i^{\ n}\not\Pi_j^{\ m} - \delta^{kl}\epsilon_{kab}\epsilon_{lcd}\not\Pi_i^{\ a}(\not g^{-1})^{mb}\not\Pi_j^{\ c}(\not g^{-1})^{nd}\right| \lesssim \varepsilon^{1/2}\frac{\ln(e+t)}{1+t}.$$

To proceed, we first use the identity $\delta^{kl}\epsilon_{kab}\epsilon_{lcd} = \delta_{ac}\delta_{bd} - \delta_{ad}\delta_{bc}$ to deduce that the third (final) product on the left-hand side of (12.38) is equal to

$$\{\delta_{ac}\delta_{bd} - \delta_{ad}\delta_{bc}\}\,\not\Pi_i^{\ a}(\not g^{-1})^{mb}\not\Pi_j^{\ c}(\not g^{-1})^{nd}. \tag{12.39}$$

Thus, to conclude (12.35b), we have only to show that (12.39) is equal to $\not g_{ij}(\not g^{-1})^{mn} - \not\Pi_i^{\ n}\not\Pi_j^{\ m}$ plus error terms that are in magnitude $\lesssim \varepsilon^{1/2}\ln(e+t)(1+t)^{-1}$.

To this end, we use (12.14d) and (12.15) to replace $\delta_{ac}\delta_{bd} - \delta_{ad}\delta_{bc}$ with $\not{g}_{ac}\not{g}_{bd} - \not{g}_{ad}\not{g}_{bc}$ plus an error term that is in magnitude $\lesssim \varepsilon^{1/2}\ln(e+t)(1+t)^{-1}$ plus tensorial products of one-forms that are g-dual to R and hence are annihilated by the factor $\not{\Pi}_i{}^a(\not{g}^{-1})^{mb}\not{\Pi}_j{}^c(\not{g}^{-1})^{nd}$ in (12.39). Finally, after this replacement-up-to-errors, we compute that

$$\{\not{g}_{ac}\not{g}_{bd} - \not{g}_{ad}\not{g}_{bc}\}\not{\Pi}_i{}^a(\not{g}^{-1})^{mb}\not{\Pi}_j{}^c(\not{g}^{-1})^{nd} = \not{g}_{ij}(\not{g}^{-1})^{mn} - \not{\Pi}_i{}^n\not{\Pi}_j{}^m \tag{12.40}$$

as desired.

We now prove (12.35c). To this end, we use the identities $(\not{g}^{-1})^{ij}\not{g}_{ij} = 2$, $(\not{g}^{-1})^{ij}\not{\Pi}_i{}^n = \not{\Pi}^{jn}$, and $\not{\Pi}^{jn}\not{\Pi}_j{}^m = (\not{g}^{-1})^{mn}$ to deduce that the contraction against $(\not{g}^{-1})^{ij}$ of the quantity under the norm on the left-hand side of (12.35b) is exactly the quantity under the norm on the left-hand side of (12.35c). The desired estimate (12.35c) now follows from the estimates (12.15) and (12.35b).

We now prove (12.34c) with the help of the identity (7.24). In our proof of (12.35b), we showed that the norms of the last three products on the right-hand side of (7.24) are (viewed as $S_{t,u}$ tensors) in magnitude $\lesssim \varepsilon^{1/2}\ln(e+t)(1+t)^{-1}$. Using the identity $\not{d}^A x^a \not{d}_A x^c = (\not{g}^{-1})^{ij}\partial_i x^a \partial_j x^c = (\not{g}^{-1})^{ac}$, we deduce that the square of the norm of the first product on the right-hand side of (7.24) is (viewed as a type $\binom{1}{1}$ $S_{t,u}$ tensor) equal to

$$|\epsilon_{lab}\epsilon_{lcd}(\not{g}^{-1})^{ac}(\not{g}^{-1})^{bd}|, \tag{12.41}$$

where there is no summation over l in (12.41). The expression (12.41) can be viewed as the square of the norm of the projection of the type $\binom{0}{2}$ Σ_t tensor $\epsilon_{l\cdots}$ onto $S_{t,u}$. Using the fact that the norm of the projection of such a tensor is no larger than the norm of the tensor itself, we deduce that (12.41) is $\leq \epsilon_{lab}\epsilon_{lcd}(\underline{g}^{-1})^{ac}(\underline{g}^{-1})^{bd}$, where $\underline{g}^{-1}$ is as in Def. 3.9. Using (12.14b), we see that we can replace $(\underline{g}^{-1})^{ac}$ and $(\underline{g}^{-1})^{bd}$ respectively with δ^{ac} and δ^{bd} up to an error term that is in magnitude $\lesssim \varepsilon(1+t)^{-1}$. The term arising after the replacement verifies

$$|\epsilon_{lab}\epsilon_{lcd}\delta^{ac}\delta^{bd}| = 2. \tag{12.42}$$

Combining the above estimates, we conclude the desired inequality (12.34c).

Finally, we prove (12.35d). Using Cor. 12.14, we see that it suffices to prove the following type $\binom{2}{1}$ $S_{t,u}$ tensor bound:

$$\left|\sum_{l=1}^{3} O_{(l)} \otimes \not{\nabla} O_{(l)}\right| \lesssim \varepsilon^{1/2}\ln(e+t). \tag{12.43}$$

Using the identities (6.8) and (7.24), the estimates (12.14a) and (12.34a), and the line of reasoning that we used at the beginning of the proof of (12.35b), we see that the main contribution to the tensor on the left-hand side of (12.43) is $\delta^{kl}\epsilon_{kab}\epsilon_{lcd}x^a\not{d}^{\#}x^b \otimes \not{d}x^c \otimes \not{d}^{\#}x^d$ and that all remaining error terms are in magnitude $\lesssim \varepsilon^{1/2}\ln(e+t)$. Hence, it suffices to prove the following $S_{t,u}$ tensor bound:

$$\left|\delta^{kl}\epsilon_{kab}\epsilon_{lcd}x^a\not{d}^{\#}x^b \otimes \not{d}x^c \otimes \not{d}^{\#}x^d\right| \lesssim \varepsilon^{1/2}\ln(e+t). \tag{12.44}$$

Using Cor. 12.14, we see that it suffices to prove the same bound for the rectangular m, n, p components of the $S_{t,u}$ tensor on the left-hand side of (12.44). Using the identity $\delta^{kl}\epsilon_{kab}\epsilon_{lcd} = \delta_{ac}\delta_{bd} - \delta_{ad}\delta_{bc}$, we see that the rectangular components of interest can be expressed as

$$\begin{aligned}(12.45)\qquad &\{\delta_{ac}\delta_{bd} - \delta_{ad}\delta_{bc}\}\, x^a (\not{g}^{-1})^{mb} \not{\Pi}_p^c (\not{g}^{-1})^{nd} \\ &= x^c \delta_{bd} (\not{g}^{-1})^{mb} \not{\Pi}_p^c (\not{g}^{-1})^{nd} - x^d \delta_{bc} (\not{g}^{-1})^{mb} \not{\Pi}_p^c (\not{g}^{-1})^{nd}.\end{aligned}$$

To prove the desired estimate for the rectangular components, we first use (12.31a) to replace x^c with $-\varrho R_c$ and x^d with $-\varrho R_d$ plus an error term with magnitude $\lesssim \varepsilon^{1/2}\ln(e+t)$. Furthermore, (12.15) implies that the remaining "non-x" factors on the right-hand side of (12.45) are in magnitude $\lesssim 1$, so that the total product generated by the error term is in magnitude $\lesssim \varepsilon^{1/2}\ln(e+t)$ as desired. We now note that the ϱR_c and ϱR_d terms can be viewed as the components of one-forms that are $S_{t,u}$-orthogonal. Since the c and d indices in (12.45) are paired with the rectangular components of $S_{t,u}$ tensors, these terms lead to products that completely vanish. We have thus proved (12.35d). □

12.12. Precise pointwise differential operator comparison estimates

In this section, we derive some pointwise differential operator comparison estimates. Most of the estimates are sharp in the sense that the explicit order-unity constants are the same constants that our proofs would yield in the case of Minkowski spacetime.

LEMMA 12.22 (**$\nabla\!\!\!/$ in terms of $\not{\mathcal{L}}_{\mathscr{O}}$**). *Let f be a scalar-valued function and let $\{O_{(1)}, O_{(2)}, O_{(3)}\}$ be the geometric rotation vectorfields from Def. 6.1. Under the small-data and bootstrap assumptions of Sects. 12.1-12.4, if ε is sufficiently small, then the following pointwise estimates hold on $\mathcal{M}_{T_{(Bootstrap)},U_0}$:*

$$(12.46a)\qquad \varrho^2 |\not{d} f|^2 \leq (1 + C\varepsilon^{1/2}) \sum_{l=1}^{3} |O_{(l)} f|^2,$$

$$(12.46b)\qquad \varrho^2 \left|\nabla\!\!\!/^2 f\right|^2 + |\not{d} f|^2 \leq (1 + C\varepsilon^{1/2}) \sum_{l=1}^{3} |\not{d} O_{(l)} f|^2,$$

$$(12.46c)\qquad \varrho^2 |\Delta\!\!\!/ f|^2 \leq 2(1 + C\varepsilon^{1/2}) \sum_{l=1}^{3} |\not{d} O_{(l)} f|^2.$$

Similarly, if ξ is an $S_{t,u}$ one-form, then

$$(12.47)\qquad \varrho^2 |\nabla\!\!\!/ \xi|^2 + |\xi|^2 \leq (1 + C\varepsilon^{1/2}) \sum_{l=1}^{3} \left|\not{\mathcal{L}}_{O_{(l)}} \xi\right|^2.$$

In addition, if ξ is a symmetric type $\binom{0}{2}$ $S_{t,u}$ tensorfield, then

$$(12.48)\qquad \varrho^2 |\nabla\!\!\!/ \xi|^2 + 4\left|\hat{\xi}\right|^2 \leq (1 + C\varepsilon^{1/2}) \sum_{l=1}^{3} \left|\not{\mathcal{L}}_{O_{(l)}} \xi\right|^2 + C\varepsilon^{1/2} (\mathrm{tr}_{\not{g}} \xi)^2.$$

Finally, if ξ is a symmetric type $\binom{0}{2}$ $S_{t,u}$ tensorfield, then

$$(12.49)\qquad |\not{\mathcal{L}}_O \xi| \leq C\varrho |\nabla\!\!\!/ \xi| + C|\xi|.$$

PROOF. First, we note that (12.46c) follows from (12.46b) and the fact that $|\mathrm{tr}_{\not g}\xi|^2 \leq 2|\xi|^2$ for type $\binom{0}{2}$ $S_{t,u}$ tensors. Next, we note that (12.46b) follows from (12.47) with $\xi := \not d f$ and the identity $\not\mathcal{L}_O \not d f = \not d \not\mathcal{L}_O f = \not d O f$, (that is, Lemma 8.6). Hence, to complete the proof of the lemma, we have to prove (12.46a), (12.47), (12.48), and (12.49).

To prove these inequalities, we first use Lemma 3.58 to deduce that the following identity holds for any type $\binom{0}{k}$ $S_{t,u}$ tensorfield ξ with rectangular components $\xi_{j_1 \cdots j_k}$:

$$\not\mathcal{L}_{O_{(l)}} \xi_{j_1 \cdots j_k} = \not\nabla_{O_{(l)}} \xi_{j_1 \cdots j_k} + \sum_{i=1}^{k} \xi_{\cdots j_{i-1} m j_{i+1} \cdots j_k} \not\nabla_{j_i} O_{(l)}^m. \tag{12.50}$$

We now note that the desired estimate (12.49) follows easily from (12.50), (12.34a), and (12.34c).

To prove the remaining estimates, we first square the norms of both sides of (12.50) (viewed as $S_{t,u}$ tensors) and then sum over the index l to deduce that (where we suppress the j. indices from (12.50))

$$\begin{aligned} \sum_{l=1}^{3} \left| \not\mathcal{L}_{O_{(l)}} \xi \right|^2 = \sum_{l=1}^{3} \left| O_{(l)}^m \not\nabla_m \xi \right|^2 + 2 \sum_{l=1}^{3} \sum_{i=1}^{k} (O_{(l)}^m \not\nabla_m \xi)^{\# \cdots \#} \cdot (\not\nabla O_{(l)})^n \xi_{\cdots n \cdots} \\ + \sum_{l=1}^{3} \left| \sum_{i=1}^{k} (\not\nabla O_{(l)})^m \xi_{\cdots m \cdots} \right|^2 . \end{aligned} \tag{12.51}$$

From (12.35a) and Cor. 12.14, we deduce that the first sum on the right-hand side of (12.51) is equal to $\varrho^2 |\not\nabla \xi|^2$ plus an error term that is in magnitude $\lesssim \varepsilon^{1/2} \varrho^2 |\not\nabla \xi|^2$. Furthermore, using (12.35d) and Cor. 12.14, we deduce that the second sum (which is a double sum) is in magnitude $\lesssim \varepsilon^{1/2} \varrho^2 |\not\nabla \xi|^2 + \varepsilon^{1/2} |\xi|^2$.

We split the analysis of the last (double) sum on the right-hand side of (12.51) into the cases where ξ is a function, type $\binom{0}{1}$, and symmetric type $\binom{0}{2}$ respectively. If ξ is a function f, then the left-hand side of (12.50) is equal to $\sum_{l=1}^{3} |O_{(l)} f|^2$ while the estimates in the previous paragraph imply that the right-hand side is equal to $\varrho^2 |\not d f|^2$ plus an error term that is in magnitude $\lesssim \varepsilon^{1/2} \varrho^2 |\not d f|^2$ (note that the second and third sums on the right-hand side of (12.51) are absent in this case). The desired estimate (12.46a) easily follows from these facts.

If ξ is type $\binom{0}{1}$, then the last (double) sum on the right-hand side of (12.51) is equal to

$$\sum_{l=1}^{3} (\not g^{-1})^{jj'} (\not\nabla_j O_{(l)}^m)(\not\nabla_{j'} O_{(l)}^n) \xi_m \xi_n. \tag{12.52}$$

Hence, using (12.35c) and Cor. 12.14, we deduce that the last sum is equal to $|\xi|^2$ plus an error term that is in magnitude $\lesssim \varepsilon^{1/2} |\xi|^2$.

If ξ is symmetric type $\binom{0}{2}$, then the last (double) sum on the right-hand side of (12.51) is equal to

$$2\sum_{l=1}^{3}(\not g^{-1})^{jj'}(\not g^{-1})^{kk'}\xi_{mj}\xi_{nj'}(\not\nabla_k O_{(l)}^m)\not\nabla_{k'}O_{(l)}^n + 2\sum_{l=1}^{3}(\not g^{-1})^{jj'}(\not g^{-1})^{kk'}\xi_{mj}\xi_{nk'}(\not\nabla_k O_{(l)}^m)\not\nabla_{j'}O_{(l)}^n. \tag{12.53}$$

Inequality (12.35c) and Cor. 12.14 imply that the first sum in (12.53) is equal to $2|\xi|^2$ plus an error term that is in magnitude $\lesssim \varepsilon^{1/2}|\xi|^2$. Furthermore, using (12.35b) and Cor. 12.14, we deduce that the second sum in (12.53) is equal to $2|\xi|^2 - 2(\mathrm{tr}_{\not g}\xi)^2$ plus an error term that is in magnitude $\lesssim \varepsilon^{1/2}|\xi|^2$. The desired estimate (12.48) now follows easily from these two estimates for the sums in (12.53), our prior analysis of the first two sums on the right-hand side of (12.51), and the identity $|\xi|^2 = \frac{1}{2}(\mathrm{tr}_{\not g}\xi)^2 + |\hat{\xi}|^2$. This completes the proof of Lemma 12.22. □

12.13. Useful estimates for avoiding detailed commutators

In this section, we provide some nonoptimal commutator estimates for vectorfields acting on functions. The estimates are useful when precision is not required.

LEMMA 12.23 (**Estimates for avoiding detailed commutators**). *Let f be a function, and let $1 \leq N \leq 24$ be an integer. Under the small-data and bootstrap assumptions of Secs. 12.1-12.4, if ε is sufficiently small, then the following pointwise estimates hold on $\mathcal{M}_{T_{(Bootstrap)},U_0}$:*

$$\left|\begin{pmatrix} \varrho\mathscr{Z}^N L f \\ \mathscr{Z}^N \breve{R} f \\ \varrho\not{\mathcal{L}}_{\mathscr{Z}}^N \not d f\end{pmatrix}\right| \lesssim \left|\begin{pmatrix} \varrho L\mathscr{Z}^{\leq N} f \\ \breve{R}\mathscr{Z}^{\leq N} f \\ \varrho\not d\mathscr{Z}^{\leq N} f\end{pmatrix}\right|, \tag{12.54}$$

$$\left|\mathscr{Z}^N f\right| \lesssim \left|\begin{pmatrix} \varrho L\mathscr{Z}^{\leq N-1} f \\ \breve{R}\mathscr{Z}^{\leq N-1} f \\ \varrho\not d\mathscr{Z}^{\leq N-1} f \\ \mathscr{Z}^{N-1} f\end{pmatrix}\right|, \tag{12.55}$$

$$\left|\begin{pmatrix} \varrho L\mathscr{Z}^{\leq N-1} f \\ \breve{R}\mathscr{Z}^{\leq N-1} f \\ \varrho\not d\mathscr{Z}^{\leq N-1} f \\ \mathscr{Z}^{N-1} f\end{pmatrix}\right| \lesssim \left|\mathscr{Z}^{\leq N} f\right|. \tag{12.56}$$

PROOF. We first prove (12.54). The estimate for $\varrho\not{\mathcal{L}}_{\mathscr{Z}}^N \not d f$ follows trivially from Lemma 8.6. To deduce the estimate for $\varrho\mathscr{Z}^N L f$, we express $Lf = \varrho^{-1}Zf$, where $Z := \varrho L \in \mathscr{Z}$. Then using (12.6b), we deduce that $\left|\varrho\mathscr{Z}^N Lf\right| \lesssim \left|\mathscr{Z}^{\leq N+1} f\right|$. If $\mathscr{Z}^{\leq N+1} = \varrho L\mathscr{Z}^{\leq N}$, or $\mathscr{Z}^{\leq N+1} = \breve{R}\mathscr{Z}^{\leq N}$, then the desired bound is obvious. If $\mathscr{Z}^{\leq N+1} = O\mathscr{Z}^{\leq N}$, then we use (12.34a) to deduce that $|O\mathscr{Z}^{\leq N} f| \lesssim |O|\left|\not d\mathscr{Z}^{\leq N} f\right| \lesssim \varrho\left|\not d\mathscr{Z}^{\leq N} f\right|$. In remains for us to prove the estimate (12.54) for $\mathscr{Z}^N \breve{R} f$. We denote $\mathscr{Z}^N \breve{R} := \mathscr{Z}^{N+1}$. If $\mathscr{Z}^{N+1} = \varrho L\mathscr{Z}^N$ or $\mathscr{Z}^{N+1} = \breve{R}\mathscr{Z}^N$, then the desired bound is obvious. If $\mathscr{Z}^{N+1} = O\mathscr{Z}^N$, then we argue as we did in our analysis of $O\mathscr{Z}^{\leq N}$. We have thus proved (12.54).

Inequalities (12.55) and (12.56) can be proved using similar arguments and, in the case of the estimate (12.56) for $\varrho \not{d} \mathscr{Z}^{\leq N-1} f$, inequality (12.46a); we omit the details. □

12.14. Estimates for the derivatives of the angular differential of the rectangular spatial coordinate functions

In this section, we derive pointwise estimates for the $S_{t,u}$-projected Lie derivatives of $\not{d} x^i$, $(i = 1, 2, 3)$

LEMMA 12.24 (**Pointwise estimates for** $\not{d} x^i$). *Let $0 \leq N \leq 24$ be an integer. Under the small-data and bootstrap assumptions of Sects. 12.1-12.4, if ε is sufficiently small, then the following pointwise estimates hold on $\mathcal{M}_{T_{(Bootstrap)}, U_0}$:*

(12.57a)

$$\left| \not{\mathcal{L}}_{\mathscr{Z}}^N \not{d} x^i \right|, \left| \not{d} \mathscr{Z}^N x^i \right| \lesssim \left| \mathscr{Z}^{\leq N} \Psi \right| + \frac{1}{1+t} \left\| \begin{pmatrix} \mathscr{Z}^{\leq N}(\upmu - 1) \\ \sum_{a=1}^3 \varrho \left| \mathscr{Z}^{\leq N} L_{(Small)}^a \right| \end{pmatrix} \right\| + 1,$$

(12.57b)

$$\left\| \begin{pmatrix} \not{\mathcal{L}}_{\mathscr{Z}}^{\leq 12} \not{d} x^i \\ \not{d} \mathscr{Z}^{\leq 12} x^i \end{pmatrix} \right\|_{C^0(\Sigma_t^u)} \lesssim 1.$$

PROOF. To prove (12.57a) and (12.57b), we use Lemma 8.6 and (12.46a) to deduce that $|\not{\mathcal{L}}_{\mathscr{Z}}^N \not{d} x^i| \lesssim (1+t)^{-1} |\mathscr{Z}^{N+1} x^i| = (1+t)^{-1} |\mathscr{Z}^N Z^i|$. The desired estimates thus follow from (12.26a) and (12.27b). □

12.15. Pointwise estimates for the Lie derivatives of the frame components of the derivative of the rectangular components of the metric with respect to the solution

In this section, we derive pointwise estimates for the $S_{t,u}$-projected Lie derivatives of the arrays of $S_{t,u}$ tensorfields $G_{(Frame)}$ and $G'_{(Frame)}$.

LEMMA 12.25 (**Pointwise estimates for the Lie derivatives of** $G_{(Frame)}$ **and** $G'_{(Frame)}$). *Let $G_{(Frame)}$ and $G'_{(Frame)}$ be the arrays of $S_{t,u}$ tensorfields given by Defs. 2.7 and 3.62. Let $0 \leq N \leq 24$ be an integer. Under the small-data and bootstrap assumptions of Sects. 12.1-12.4, if ε is sufficiently small, then the following pointwise estimates hold on $\mathcal{M}_{T_{(Bootstrap)}, U_0}$:*

$$\left| \not{\mathcal{L}}_{\mathscr{Z}}^{\leq N} G_{(Frame)} \right| \lesssim \left| \mathscr{Z}^{\leq N} \Psi \right| + \frac{1}{1+t} \left\| \begin{pmatrix} \mathscr{Z}^{\leq N}(\upmu - 1) \\ \sum_{a=1}^3 \varrho \left| \mathscr{Z}^{\leq N} L_{(Small)}^a \right| \end{pmatrix} \right\| + 1, \tag{12.58a}$$

(12.58b)

$$\left\| \not{\mathcal{L}}_{\mathscr{Z}}^{\leq 12} G_{(Frame)} \right\|_{C^0(\Sigma_t^u)} \lesssim 1.$$

Furthermore, the same estimates hold for $G'_{(Frame)}$.

PROOF. We prove only the desired estimate for the $S_{t,u}$ one-form $\not{G}_L$; the proofs of the estimates for the remaining elements of $G_{(Frame)}$ and for $G'_{(Frame)}$ are essentially the same. We use the notation of Lemma 4.14. Since the rectangular components $G_{\mu\nu} = G_{\mu\nu}(\Psi)$ are smooth functions of Ψ, the bootstrap assumptions $(\mathbf{BA}\Psi)$ imply that $|G_{\mu\nu}| \lesssim 1$. Using in addition the estimates (12.26a) and the estimates (12.15) for the rectangular components of $\not g^{-1}$ and the $S_{t,u}$ projections $\not\Pi$ we conclude that the scalar functions G_{La} and the $S_{t,u}$ one-forms $\not G_L$ are $\lesssim 1$ in magnitude. For the same reasons, the same estimates hold with G' in place of G and also with any higher derivative $\frac{d^k}{d\Psi^k}G$, $k \geq 2$, in place of G. Furthermore, Lemma 8.6, inequality (12.46a), and the estimates (12.26b) and (12.27b) imply that $\left\|\mathscr{Z}^{\leq 12} L^i\right\|_{C^0(\Sigma_t^u)}$ and that for any $Z \in \mathscr{Z}$, $\left\|\not{\mathcal{L}}_{\mathscr{Z}}^{\leq 11} \not d Z^i\right\|_{C^0(\Sigma_t^u)}$, $\left\|\not{\mathcal{L}}_{\mathscr{Z}}^{\leq 11} \not d Z^i\right\|_{C^0(\Sigma_t^u)} \lesssim 1$. Using all of these estimates and the bootstrap assumptions (**BA**Ψ), we now repeatedly $\not{\mathcal{L}}$-differentiate the identity (4.31b) (with $V = L$ in the identity) with respect to vectorfields $Z \in \mathscr{Z}$ and use Lemma 8.6 and inequality (12.46a) to deduce that

$$
\begin{aligned}
\left|\not{\mathcal{L}}_{\mathscr{Z}}^N \not G_L\right| &\lesssim \left|\mathscr{Z}^{\leq N}\Psi\right| + \sum_{a=1}^{3}\left|\mathscr{Z}^{\leq N} L^a\right| \\
&\quad + \sum_{Z\in\mathscr{Z}}\sum_{a=1}^{3}\left|\not d \mathscr{Z}^{\leq N-1} Z^a\right| \\
&\lesssim \left|\mathscr{Z}^{\leq N}\Psi\right| + \sum_{a=1}^{3}\left|\mathscr{Z}^{\leq N} L^a\right| \\
&\quad + \frac{1}{\varrho}\sum_{Z\in\mathscr{Z}}\sum_{a=1}^{3}\left|\mathscr{Z}^{\leq N} Z^a\right|.
\end{aligned}
\tag{12.59}
$$

The desired estimate (12.58a) for $\not G_L$ now follows from (12.59), (12.26a) and (12.27a). Inequality (12.58b) for $\not G_L$ then follows from (12.58a) and the bootstrap assumptions. This completes our proof of the estimates for $\not G_L$. We remark that the proofs of the estimates involving contractions against R (such as the estimates for $\not G_R$) are similar but also rely on the estimates (12.26a)-(12.26b) for R^i. This completes our proof of the lemma. □

12.16. Crude pointwise estimates for the Lie derivatives of the angular components of the deformation tensors

In this section, for various vectorfields V, we provide some crude pointwise estimates for the $S_{t,u}$-projected Lie derivatives of the $S_{t,u}$ tensorfields ${}^{(V)}\not\pi$ and the Lie derivatives of some related tensorfields.

LEMMA 12.26 (**Crude pointwise estimates for the Lie derivatives of the angular components of the deformation tensors**). *Let* χ *and* $\chi^{(Small)}$ *be the* $S_{t,u}$ *tensorfields defined in* (3.70) *and* (4.1c). *Let* $0 \leq N \leq 23$ *be an integer and let* $Z \in \mathscr{Z}$ *be a commutation vectorfield. Under the small-data and bootstrap assumptions of Sects.* 12.1-12.4, *if* ε *is sufficiently small, then the following pointwise* *estimates hold on* $\mathcal{M}_{T_{(Bootstrap)},U_0}$:

$$\left\| \begin{pmatrix} \varrho \not{\mathcal{L}}_{\mathscr{Z}}^N \chi \\ \varrho \not{\mathcal{L}}_{\mathscr{Z}}^N \chi^{\#} \\ \varrho \mathscr{Z}^N \mathrm{tr}_{\not{g}} \chi \\ \not{\mathcal{L}}_{\mathscr{Z}}^N {}^{(Z)}\not{\pi} \\ \not{\mathcal{L}}_{\mathscr{Z}}^N {}^{(Z)}\not{\pi}^{\#} \\ \not{\mathcal{L}}_{\mathscr{Z}}^N {}^{(Z)}\not{\pi}^{\#\#} \\ \mathscr{Z}^N \mathrm{tr}_{\not{g}} {}^{(Z)}\not{\pi} \\ \not{\mathcal{L}}_{\mathscr{Z}}^{N+1} \not{g} \\ \not{\mathcal{L}}_{\mathscr{Z}}^{N+1} \not{g}^{-1} \end{pmatrix} \right\| \lesssim \left| \begin{pmatrix} \varrho L \mathscr{Z}^{\leq N} \Psi \\ \breve{R} \mathscr{Z}^{\leq N} \Psi \\ \varrho \not{d} \mathscr{Z}^{\leq N} \Psi \\ \mathscr{Z}^{\leq N} \Psi \end{pmatrix} \right| + \frac{1}{1+t} \left| \begin{pmatrix} \mathscr{Z}^{\leq N+1}(\upmu - 1) \\ \varrho \sum_{a=1}^3 |\mathscr{Z}^{\leq N+1} L_{(Small)}^a| \end{pmatrix} \right| + 1, \tag{12.60a}$$

(12.60b)

$$\left\| \begin{pmatrix} \varrho \not{\mathcal{L}}_{\mathscr{Z}}^{\leq 11} \chi \\ \varrho \not{\mathcal{L}}_{\mathscr{Z}}^{\leq 11} \chi^{\#} \\ \varrho \mathscr{Z}^{\leq 11} \mathrm{tr}_{\not{g}} \chi \\ \not{\mathcal{L}}_{\mathscr{Z}}^{\leq 11} {}^{(Z)}\not{\pi} \\ \not{\mathcal{L}}_{\mathscr{Z}}^{\leq 11} {}^{(Z)}\not{\pi}^{\#} \\ \not{\mathcal{L}}_{\mathscr{Z}}^{\leq 11} {}^{(Z)}\not{\pi}^{\#\#} \\ \mathscr{Z}^{\leq 11} \mathrm{tr}_{\not{g}} {}^{(Z)}\not{\pi} \\ \not{\mathcal{L}}_{\mathscr{Z}}^{\leq 12} \not{g} \\ \not{\mathcal{L}}_{\mathscr{Z}}^{\leq 12} \not{g}^{-1} \end{pmatrix} \right\|_{C^0(\Sigma_t^u)} \lesssim 1.$$

PROOF. Throughout this proof, we use the bootstrap assumptions $(\mathbf{BA}\Psi)$, $(\mathbf{AUX}\upmu)$, and $(\mathbf{AUX}LR_{(Small)})$. We first prove (12.60a) for $\not{g}$. We differentiate both sides of the identity $\not{g}_{AB} = g_{ab} \not{d}_A x^a \not{d}_B x^b$ with the operator $\not{\mathcal{L}}_{\mathscr{Z}}^N$ and apply the Leibniz rule to the right-hand side. We treat all uppercase Latin indices as tensorial $S_{t,u}$ indices and all lowercase Latin-indexed quantities as scalar-valued functions. We use (8.5) to commute the operator $\not{\mathcal{L}}_{\mathscr{Z}}^N$ under $\not{d}$. It follows that

$$\left| \not{\mathcal{L}}_{\mathscr{Z}}^{N+1} \not{g} \right| \lesssim \sum_{N_1+N_2+N_3=N+1} \left| \mathscr{Z}^{N_1} g_{ab} \right| \left| \not{d} \mathscr{Z}^{N_2} x^a \right| \left| \not{d} \mathscr{Z}^{N_3} x^b \right|. \tag{12.61}$$

The desired estimate (12.60a) for $\not{\mathcal{L}}_{\mathscr{Z}}^{N+1} \not{g}$ now follows from (12.61) and the estimates (12.57a), (12.57b), (12.13a), (12.14a), and (12.55). The desired estimate (12.60b) for $\not{\mathcal{L}}_{\mathscr{Z}}^{N+1} \not{g}$ then follows from the estimate (12.60a) for $\not{\mathcal{L}}_{\mathscr{Z}}^{N+1} \not{g}$ and the bootstrap assumptions.

To prove the estimate (12.60a) for ${}^{(Z)}\not{\pi}$, we simply note that by Lemma 7.5, we have ${}^{(Z)}\not{\pi} = \not{\mathcal{L}}_Z \not{g}$. Hence the desired estimate for $\not{\mathcal{L}}_{\mathscr{Z}}^N {}^{(Z)}\not{\pi}$ follows from the previously proven estimate for $\not{\mathcal{L}}_{\mathscr{Z}}^{N+1} \not{g}$. The desired estimate (12.60b) for ${}^{(Z)}\not{\pi}$ then follows from the estimate (12.60a) for ${}^{(Z)}\not{\pi}$ and the bootstrap assumptions.

To prove the estimate (12.60a) for $\not{\mathcal{L}}_{\mathscr{Z}}^N \chi$, we set $Z = \varrho L$ and note that by Lemma 7.5, (7.7f), and (7.7g), we have $2\varrho\chi = \not{\mathcal{L}}_Z \not{g}$. Using Lemma 12.10, we see that the desired estimate follows from the previously proven estimate for $\not{\mathcal{L}}_{\mathscr{Z}}^{N+1} \not{g}$.

The desired estimate (12.60b) for $\not{\mathcal{L}}_{\mathscr{Z}}^{N}\chi$ then follows from the estimate (12.60a) for $\not{\mathcal{L}}_{\mathscr{Z}}^{N}\chi$ and the bootstrap assumptions.

To prove the estimate (12.60a) for $\not{g}^{-1}$, we start with the identity $\not{\mathcal{L}}_{Z}\not{g}^{-1} = -{}^{(Z)}\not{\pi}^{\#\#}$ (see Lemma 7.5). Applying $\not{\mathcal{L}}_{\mathscr{Z}}^{N}$ to this identity and arguing inductively, we see that the resulting expression involves only $S_{t,u}$-tensorial products of $\not{g}^{-1}$ and the Lie derivatives of the $\not{\pi}$. From the previously proven estimate (12.60b) for the lower-order derivatives of the $\not{\pi}$, we infer that all terms that are quadratic in the derivatives of the $\not{\pi}$ can be pointwise bounded by a constant times a term that is linear in the derivatives of the $\not{\pi}$. It thus follows that

$$\left|\not{\mathcal{L}}_{\mathscr{Z}}^{N+1}\not{g}^{-1}\right| \lesssim \max_{Z\in\mathscr{Z}}\left|\not{\mathcal{L}}_{\mathscr{Z}}^{\leq N}{}^{(Z)}\not{\pi}\right|, \tag{12.62}$$

and hence the desired estimate (12.60a) for $\not{\mathcal{L}}_{\mathscr{Z}}^{N+1}\not{g}^{-1}$ follows from the previously proven estimate (12.60a) for ${}^{(Z)}\not{\pi}$. The desired estimate (12.60b) for $\not{\mathcal{L}}_{\mathscr{Z}}^{N+1}\not{g}^{-1}$ then follows from the estimate (12.60a) for $\not{\mathcal{L}}_{\mathscr{Z}}^{N+1}\not{g}^{-1}$ and the bootstrap assumptions. Finally, in view of the schematic identities $\chi^{\#} = \not{g}^{-1}\chi$, ${}^{(Z)}\not{\pi}^{\#} = \not{g}^{-1(Z)}\not{\pi}$, and ${}^{(Z)}\not{\pi}^{\#\#} = (\not{g}^{-1})^{2(Z)}\not{\pi}$, we see that the estimates (12.60a) and (12.60b) for $\not{\mathcal{L}}_{\mathscr{Z}}^{N}\chi^{\#}$, $\not{\mathcal{L}}_{\mathscr{Z}}^{N}{}^{(Z)}\not{\pi}^{\#}$, and $\not{\mathcal{L}}_{\mathscr{Z}}^{N}{}^{(Z)}\not{\pi}^{\#\#}$ follow easily from the Leibniz rule, the previously proven estimates for $\not{g}^{-1}$, χ, and ${}^{(Z)}\not{\pi}$, and the bootstrap assumptions. □

COROLLARY 12.27 (**Pointwise estimates for the Lie derivatives of $G_{(Frame)}^{\#}$**). *Under the assumptions of Lemma 12.25, the estimates* (12.58a) *and* (12.58b) *also hold with $G_{(Frame)}^{\#}$ (see* (3.78b) *and Def.* 3.64*) in place of $G_{(Frame)}$.*

PROOF. We have the schematic identity $G_{(Frame)}^{\#} = \not{g}^{-1}G_{(Frame)}$ for the non-scalar elements of $G_{(Frame)}$. We now differentiate this identity with $\not{\mathcal{L}}_{\mathscr{Z}}^{N}$, apply the Leibniz rule, and use the estimates (12.58a) and (12.58b) for $G_{(Frame)}$ and the estimates (12.60a) and (12.60b) for $\not{g}^{-1}$. The corollary thus follows. □

12.17. Two additional crude differential operator comparison estimates

In this section, we provide two additional differential operator comparison estimates. We do not bother to derive sharp constants in the estimates.

LEMMA 12.28 (**A crude differential operator comparison estimate involving $\not{\mathcal{L}}_X$ in terms of $\not{\mathcal{L}}_{\mathscr{O}}$**). *Let X be an $S_{t,u}$-tangent vectorfield and let ξ be an $S_{t,u}$ one-form or a symmetric type $\binom{0}{2}$ $S_{t,u}$ tensorfield. Under the small-data and bootstrap assumptions of Sects. 12.1-12.4, if ε is sufficiently small, then the following pointwise estimate holds on $\mathcal{M}_{T_{(Bootstrap)},U_0}$:*

$$|\not{\mathcal{L}}_X\xi| \leq C\frac{1}{1+t}\left(|X||\not{\mathcal{L}}_{\mathscr{O}}^{\leq 1}\xi| + |\xi||\not{\mathcal{L}}_{\mathscr{O}}^{\leq 1}X\right). \tag{12.63}$$

PROOF. Let $X_\flat$ be the $S_{t,u}$ one-form that is $\not{g}$-dual to X. Using the identity (12.50) with X in place of $O_{(l)}$ to express $S_{t,u}$-projected Lie derivatives in terms of

$S_{t,u}$-covariant derivatives, we deduce the schematic identity $\not{\mathcal{L}}_X \xi = X \cdot \nabla\!\!\!/ \xi + \xi^{\#} \cdot \nabla\!\!\!/ X_{\flat}$. From this identity and inequalities (12.47) and (12.48), we deduce that $|\not{\mathcal{L}}_X \xi| \lesssim (1+t)^{-1} \left(|X| |\not{\mathcal{L}}_{\mathscr{O}}^{\leq 1} \xi| + |\xi| |\not{\mathcal{L}}_{\mathscr{O}}^{1} X_{\flat}| \right)$. To bound the Lie derivatives of $X_{\flat}$, we first use (7.4a) to obtain, for any $O \in \mathscr{O}$, $|\not{\mathcal{L}}_O X_{\flat}| = |\not{\mathcal{L}}_O (\not{g} \cdot X)| \lesssim |\not{\mathcal{L}}_O X| + |{}^{(O)}\not{\pi}||X|$. Using this inequality and the estimate (12.60b) (to bound the magnitude of ${}^{(O)}\not{\pi}$), we deduce that $|\not{\mathcal{L}}_O X_{\flat}| \lesssim |\not{\mathcal{L}}_{\mathscr{O}}^{\leq 1} X|$, and the desired estimate (12.63) readily follows. □

LEMMA 12.29 (**A crude differential operator comparison estimate involving second-order $S_{t,u}$ covariant derivatives**). *Let ξ be a symmetric type $\binom{0}{2}$ $S_{t,u}$ tensorfield. Under the small-data and bootstrap assumptions of Sects.* 12.1-12.4, *if ε is sufficiently small, then the following pointwise estimate holds on $\mathcal{M}_{T_{(Bootstrap)}, U_0}$:*

$$\varrho^4 |\nabla\!\!\!/^2 \xi|^2 \leq C |\not{\mathcal{L}}_{\mathscr{O}}^{\leq 2} \xi|^2. \tag{12.64}$$

PROOF. We first consider the identity (12.51) with $\nabla\!\!\!/ \xi$ in the role of ξ, multiply it by ϱ^2, and use the line of reasoning just below it together with inequality (12.34c) (to bound the final product on the right-hand side of (12.51)) to deduce that

$$\varrho^4 |\nabla\!\!\!/^2 \xi|^2 \lesssim \varrho^2 \sum_{l=1}^{3} |\not{\mathcal{L}}_{O_{(l)}} \nabla\!\!\!/ \xi|^2 + \varrho^2 |\nabla\!\!\!/ \xi|^2. \tag{12.65}$$

Using the commutator identity (8.13), the estimate (12.48), the simple bound $|\mathrm{tr}_{\not{g}} \xi| \lesssim |\xi|$, and the estimate (12.60b) (to bound the magnitude of the Lie derivatives of the ${}^{(O_{(l)})}\not{\pi}$), we deduce that the right-hand side of (12.65) is

$$\begin{aligned} &\lesssim \varrho^2 \sum_{l=1}^{3} |\nabla\!\!\!/ \not{\mathcal{L}}_{O_{(l)}} \xi|^2 + \varrho^2 \sum_{l=1}^{3} |\nabla\!\!\!/ {}^{(O_{(l)})}\not{\pi}|^2 |\xi|^2 + \varrho^2 |\nabla\!\!\!/ \xi|^2 \\ &\lesssim |\not{\mathcal{L}}_{\mathscr{O}}^{\leq 2} \xi|^2 + \sum_{l=1}^{3} |\not{\mathcal{L}}_{\mathscr{O}}^{\leq 1} {}^{(O_{(l)})}\not{\pi}|^2 |\xi|^2 \lesssim |\not{\mathcal{L}}_{\mathscr{O}}^{\leq 2} \xi|^2. \end{aligned} \tag{12.66}$$

□

12.18. Pointwise estimates for the derivatives of the re-centered null second fundamental form in terms of other quantities

In this section, we derive pointwise estimates for the $S_{t,u}$-projected Lie derivatives of the tensorfield $\chi^{(Small)}$.

LEMMA 12.30 (**Pointwise estimates for $\chi^{(Small)}$ in terms of other quantities**). *Let $\chi^{(Small)}$ be the $S_{t,u}$ tensorfield defined in* (4.1c) *and let $0 \leq N \leq 23$ be an integer. Under the small-data and bootstrap assumptions of Sects.* 12.1-12.4, *if ε*

is sufficiently small, then the following pointwise estimates hold on $\mathcal{M}_{T_{(Bootstrap)},U_0}$:

$$\left| \begin{pmatrix} \not{\mathcal{L}}_{\mathscr{Z}}^N \chi^{(Small)} \\ \mathscr{Z}^N \mathrm{tr}_{\not{g}} \chi^{(Small)} \\ \not{\mathcal{L}}_{\mathscr{Z}}^N \hat{\chi}^{(Small)} \\ \not{\mathcal{L}}_{\mathscr{Z}}^N \chi^{(Small)\#} \\ \not{\mathcal{L}}_{\mathscr{Z}}^N \hat{\chi}^{(Small)\#} \\ \not{\mathcal{L}}_{\mathscr{Z}}^N \chi^{(Small)\#\#} \\ \not{\mathcal{L}}_{\mathscr{Z}}^N \hat{\chi}^{(Small)\#\#} \end{pmatrix} \right| \lesssim \frac{1}{1+t} \left| \begin{pmatrix} \varrho L \mathscr{Z}^{\leq N} \Psi \\ \breve{R} \mathscr{Z}^{\leq N} \Psi \\ \varrho \not{d} \mathscr{Z}^{\leq N} \Psi \\ \mathscr{Z}^{\leq N} \Psi \end{pmatrix} \right| + \frac{1}{(1+t)^2} \left| \begin{pmatrix} \mathscr{Z}^{\leq N}(\upmu - 1) \\ \sum_{a=1}^3 \varrho |\mathscr{Z}^{\leq N+1} L_{(Small)}^a| \end{pmatrix} \right|, \tag{12.67a}$$

$$\left\| \begin{pmatrix} \not{\mathcal{L}}_{\mathscr{Z}}^N \chi^{(Small)} \\ \mathscr{Z}^N \mathrm{tr}_{\not{g}} \chi^{(Small)} \\ \not{\mathcal{L}}_{\mathscr{Z}}^N \hat{\chi}^{(Small)} \\ \not{\mathcal{L}}_{\mathscr{Z}}^N \chi^{(Small)\#} \\ \not{\mathcal{L}}_{\mathscr{Z}}^N \hat{\chi}^{(Small)\#} \\ \not{\mathcal{L}}_{\mathscr{Z}}^N \chi^{(Small)\#\#} \\ \not{\mathcal{L}}_{\mathscr{Z}}^N \hat{\chi}^{(Small)\#\#} \end{pmatrix} \right\|_{C^0(\Sigma_t^u)} \lesssim \varepsilon^{1/2} \frac{\ln(e+t)}{(1+t)^2}. \tag{12.67b}$$

PROOF. By (4.19a), we have the following identity, where the last product on the right-hand side is written schematically:

$$\chi_{AB}^{(Small)} = g_{ab}(\not{d}_A x^a)\not{d}_B L_{(Small)}^b + G_{(Frame)} \begin{pmatrix} L\Psi \\ \not{d}\Psi \end{pmatrix}. \tag{12.68}$$

We now apply $\not{\mathcal{L}}_{\mathscr{Z}}^N$ to both sides of (12.68). As in our proof of (12.60a), we treat all uppercase Latin indices in the term $g_{ab}(\not{d}_A x^a)\not{d}_B L_{(Small)}^b$ as tensorial $S_{t,u}$ indices, and all lowercase Latin-indexed quantities as scalar-valued functions. The Leibniz rule, the identity (8.5), and inequality (12.46a) thus yield

$$\left| \not{\mathcal{L}}_{\mathscr{Z}}^N \chi^{(Small)} \right| \lesssim \frac{1}{1+t} \sum_{\substack{N_1+N_2+N_3 \leq N+1 \\ N_1, N_2 \leq N}} \left| \mathscr{Z}^{N_1} g_{ab} \right| \left| \not{d} \mathscr{Z}^{N_2} x^a \right| \left| \mathscr{Z}^{N_3} L_{(Small)}^b \right| + \sum_{N_1+N_2=N} \left| \not{\mathcal{L}}_{\mathscr{Z}}^{N_1} G_{(Frame)} \right| \left| \begin{pmatrix} \mathscr{Z}^{N_2} L\Psi \\ \not{d}\mathscr{Z}^{N_2}\Psi \end{pmatrix} \right|. \tag{12.69}$$

The desired estimate (12.67a) for $\not{\mathcal{L}}_{\mathscr{Z}}^N \chi^{(Small)}$ now follows from inequality (12.69), the estimates (12.13a), (12.14a), (12.58a), (12.58b), (12.57a), (12.57b), Lemma 12.23, and the bootstrap assumptions. The desired estimate (12.67b) for $\not{\mathcal{L}}_{\mathscr{Z}}^N \chi^{(Small)}$ then follows from the estimate (12.67a) for $\not{\mathcal{L}}_{\mathscr{Z}}^N \chi^{(Small)}$ and the bootstrap assumptions.

The desired estimates (12.67a) and (12.67b) for $\chi^{(Small)\#}$ and $\chi^{(Small)\#\#}$ then follow from the estimates for $\chi^{(Small)}$ in the same way that the estimates for # quantities in Lemma 12.26 follow from the estimates for the non-# quantities (see the proof of the lemma).

The desired estimates (12.67a) and (12.67b) for $\mathrm{tr}_{\not{g}}\chi^{(Small)}$ then follow from the corresponding estimates for $\chi^{(Small)\#}$, the fact that $\mathscr{Z}^{\leq N}\mathrm{tr}_{\not{g}}\chi^{(Small)}$ is the pure trace of the type $\binom{1}{1}$ tensorfield $\not{\mathcal{L}}_{\mathscr{Z}}^{N} \chi^{(Small)\#}$, and the fact that magnitude of the pure trace of a type $\binom{1}{1}$ tensorfield is $\lesssim$ the magnitude of the type $\binom{1}{1}$ tensorfield itself.

The desired estimates (12.67a) and (12.67b) for $\hat{\chi}^{(Small)}$ then follow from the identity $\hat{\chi}^{(Small)} = \chi^{(Small)} - \frac{1}{2}\mathrm{tr}_{\not{g}}\chi^{(Small)}\not{g}$, the estimates (12.67a) and (12.67b) for $\chi^{(Small)}$ and $\mathrm{tr}_{\not{g}}\chi^{(Small)}$, and the estimates of Lemma 12.26 for the quantities $\not{\mathcal{L}}_{\mathscr{Z}}^{M}\not{g}$.

The desired estimates (12.67b) and (12.67a) for $\hat{\chi}^{(Small)\#}$ and $\hat{\chi}^{(Small)\#\#}$ then follow from the estimates for $\hat{\chi}^{(Small)}$ in the same way that the estimates for # quantities in Lemma 12.26 follow from the estimates for the non-# quantities. □

12.19. Pointwise estimates for the Lie derivatives of the rotation vectorfields

In this section, we derive pointwise estimates for the $S_{t,u}$-projected Lie derivatives of the rotation vectorfields $\{O_{(1)}, O_{(2)}, O_{(3)}\}$.

LEMMA 12.31 (**Pointwise estimates for the Lie derivatives of the rotation vectorfields**). *Let $0 \leq N \leq 24$ be an integer. Let $O \in \{O_{(1)}, O_{(2)}, O_{(3)}\}$ be a geometric rotation vectorfield from Def.* 6.1 *and let $O_\flat$ denote its $\not{g}$-dual. Under the small-data and bootstrap assumptions of Sects.* 12.1-12.4, *if ε is sufficiently small, then the following pointwise estimates hold on $\mathcal{M}_{T_{(Bootstrap)},U_0}$:*

(12.70a)
$$\left|\begin{pmatrix} \not{\mathcal{L}}_{\mathscr{Z}}^{\leq N} O \\ \not{\mathcal{L}}_{\mathscr{Z}}^{\leq N} O_\flat \end{pmatrix}\right| \lesssim (1+t)\left|\mathscr{Z}^{\leq N}\Psi\right| + \left|\begin{pmatrix} \mathscr{Z}^{\leq N}(\upmu - 1) \\ \sum_{a=1}^{3} \varrho\left|\mathscr{Z}^{\leq N} L_{(Small)}^{a}\right| \end{pmatrix}\right| + 1 + t,$$

(12.70b)
$$\left\|\begin{pmatrix} \not{\mathcal{L}}_{\mathscr{Z}}^{\leq 12} O \\ \not{\mathcal{L}}_{\mathscr{Z}}^{\leq 12} O_\flat \end{pmatrix}\right\|_{C^0(\Sigma_t^u)} \lesssim 1 + t.$$

PROOF. We first decompose, with the help of (3.61),

$$O^A = g_{ab}(\not{g}^{-1})^{AB}(\not{d}_B x^a)O^b.$$

We now apply $\not{\mathcal{L}}_{\mathscr{Z}}^{N}$ to both sides of this identity. Arguing as in our proof of (12.60a), we apply the Leibniz rule to the right-hand side and treat all uppercase Latin indices as tensorial $S_{t,u}$ indices and all lowercase Latin-indexed quantities as scalar-valued functions. Also using Lemma 8.6, we see that

$$\left|\not{\mathcal{L}}_{\mathscr{Z}}^{N} O\right| \lesssim \sum_{N_1+N_2+N_3+N_4 \leq N} \left|\mathscr{Z}^{N_1} g_{ab}\right| \left|\not{\mathcal{L}}_{\mathscr{Z}}^{N_2}\not{g}^{-1}\right| \left|\not{d}\mathscr{Z}^{N_3} x^a\right| \left|\mathscr{Z}^{N_4} O^b\right|. \tag{12.71}$$

The desired bounds (12.70a) and (12.70b) for O now follow from the estimates (12.71), (12.13a), (12.14a), (12.57a), (12.57b), (12.27a), (12.27b), (12.60a), (12.60b), and (12.56), and the bootstrap assumptions (**BA**Ψ), (**AUX**$\upmu$), and (**AUX**$LR_{(Small)}$).

To obtain the bounds (12.70a) and (12.70b) for $O_\flat = \not{g} \cdot O$, we use a similar argument based on the decomposition $O_A = g_{ab}(\not{d}_A x^a)O^b$. □

12.20. Pointwise estimates for the angular one-forms and vectorfields corresponding to the commutation vectorfield deformation tensors

In this section, for various vectorfields V, we derive pointwise estimates for the $S_{t,u}$-projected Lie derivatives of the $S_{t,u}$ tensorfields ${}^{(V)}\not\pi_L$ and ${}^{(V)}\not\pi_{\breve{R}}$ and their $\not g$-duals.

LEMMA 12.32 (**Pointwise estimates for the angular one-forms and vectorfields corresponding to the commutation vectorfield deformation tensors**). *Let $0 \leq N \leq 23$ be an integer and let $O \in \{O_{(1)}, O_{(2)}, O_{(3)}\}$. Under the small-data and bootstrap assumptions of Sects. 12.1-12.4, if ε is sufficiently small, then the following pointwise estimates hold on $\mathcal{M}_{T_{(Bootstrap)},U_0}$:*

$$\left| \begin{pmatrix} \not{\mathcal{L}}_{\mathscr{Z}}^N {}^{(\breve{R})}\not\pi_L \\ \not{\mathcal{L}}_{\mathscr{Z}}^N {}^{(\breve{R})}\not\pi_L^{\#} \end{pmatrix} \right| \lesssim \left| \begin{pmatrix} \varrho L \mathscr{Z}^{\leq N}\Psi \\ \breve{R}\mathscr{Z}^{\leq N}\Psi \\ \varrho \not{d} \mathscr{Z}^{\leq N}\Psi \\ \mathscr{Z}^{\leq N}\Psi \end{pmatrix} \right| + \frac{1}{1+t} \left| \begin{pmatrix} \mathscr{Z}^{\leq N+1}(\upmu - 1) \\ \sum_{a=1}^{3} \varrho \left| \mathscr{Z}^{\leq N} L_{(Small)}^a \right| \end{pmatrix} \right|, \tag{12.72a}$$

$$\left\| \begin{pmatrix} \not{\mathcal{L}}_{\mathscr{Z}}^{\leq 11} {}^{(\breve{R})}\not\pi_L \\ \not{\mathcal{L}}_{\mathscr{Z}}^{\leq 11} {}^{(\breve{R})}\not\pi_L^{\#} \end{pmatrix} \right\|_{C^0(\Sigma_t^u)} \lesssim \varepsilon^{1/2} \frac{\ln(e+t)}{1+t}, \tag{12.72b}$$

$$\left| \begin{pmatrix} \not{\mathcal{L}}_{\mathscr{Z}}^N {}^{(\varrho L)}\not\pi_{\breve{R}} \\ \not{\mathcal{L}}_{\mathscr{Z}}^N {}^{(\varrho L)}\not\pi_{\breve{R}}^{\#} \end{pmatrix} \right| \lesssim (1+t) \left| \begin{pmatrix} \varrho L \mathscr{Z}^{\leq N}\Psi \\ \breve{R}\mathscr{Z}^{\leq N}\Psi \\ \varrho \not{d} \mathscr{Z}^{\leq N}\Psi \\ \mathscr{Z}^{\leq N}\Psi \end{pmatrix} \right| + \left| \begin{pmatrix} \mathscr{Z}^{\leq N+1}(\upmu - 1) \\ \sum_{a=1}^{3} \varrho \left| \mathscr{Z}^{\leq N} L_{(Small)}^a \right| \end{pmatrix} \right|, \tag{12.73a}$$

$$\left\| \begin{pmatrix} \not{\mathcal{L}}_{\mathscr{Z}}^{\leq 11} {}^{(\varrho L)}\not\pi_{\breve{R}} \\ \not{\mathcal{L}}_{\mathscr{Z}}^{\leq 11} {}^{(\varrho L)}\not\pi_{\breve{R}}^{\#} \end{pmatrix} \right\|_{C^0(\Sigma_t^u)} \lesssim \varepsilon^{1/2} \ln(e+t), \tag{12.73b}$$

$$\left| \begin{pmatrix} \not{\mathcal{L}}_{\mathscr{Z}}^N {}^{(O)}\not\pi_L \\ \not{\mathcal{L}}_{\mathscr{Z}}^N {}^{(O)}\not\pi_L^{\#} \end{pmatrix} \right| \lesssim \left| \begin{pmatrix} \varrho L \mathscr{Z}^{\leq N}\Psi \\ \breve{R}\mathscr{Z}^{\leq N}\Psi \\ \varrho \not{d} \mathscr{Z}^{\leq N}\Psi \\ \mathscr{Z}^{\leq N}\Psi \end{pmatrix} \right| + \frac{1}{1+t} \left| \begin{pmatrix} \mathscr{Z}^{\leq N}(\upmu - 1) \\ \sum_{a=1}^{3} \varrho \left| \mathscr{Z}^{\leq N+1} L_{(Small)}^a \right| \end{pmatrix} \right|, \tag{12.74a}$$

$$\left\| \begin{pmatrix} \not{\mathcal{L}}_{\mathscr{Z}}^{\leq 11} {}^{(O)}\not\pi_L \\ \not{\mathcal{L}}_{\mathscr{Z}}^{\leq 11} {}^{(O)}\not\pi_L^{\#} \end{pmatrix} \right\|_{C^0(\Sigma_t^u)} \lesssim \varepsilon^{1/2} \frac{\ln(e+t)}{1+t}, \tag{12.74b}$$

$$\left| \begin{pmatrix} \not{\mathcal{L}}_{\mathscr{Z}}^{N} {}^{(O)}\not\pi_{\breve{R}} \\ \not{\mathcal{L}}_{\mathscr{Z}}^{N} {}^{(O)}\not\pi_{\breve{R}}^{\#} \end{pmatrix} \right| \lesssim \ln(e+t) \left| \begin{pmatrix} \varrho L \mathscr{Z}^{\leq N}\Psi \\ \breve{R}\mathscr{Z}^{\leq N}\Psi \\ \varrho \not{d} \mathscr{Z}^{\leq N}\Psi \\ \mathscr{Z}^{\leq N}\Psi \end{pmatrix} \right| + \frac{\ln(e+t)}{1+t} \left| \begin{pmatrix} \mathscr{Z}^{\leq N+1}(\upmu - 1) \\ \sum_{a=1}^{3} \varrho |\mathscr{Z}^{\leq N+1} L_{(Small)}^a| \end{pmatrix} \right|, \tag{12.75a}$$

$$\left\| \begin{pmatrix} \not{\mathcal{L}}_{\mathscr{Z}}^{\leq 11} {}^{(O)}\not\pi_{\breve{R}} \\ \not{\mathcal{L}}_{\mathscr{Z}}^{\leq 11} {}^{(O)}\not\pi_{\breve{R}}^{\#} \end{pmatrix} \right\|_{C^0(\Sigma_t^u)} \lesssim \varepsilon^{1/2} \frac{\ln^2(e+t)}{1+t}. \tag{12.75b}$$

PROOF. To prove (12.75a) for $\not{\mathcal{L}}_{\mathscr{Z}}^{N} {}^{(O)}\not\pi_{\breve{R}}$, we first use (7.8e) and (7.9b) to decompose (see Remark 7.8)

$$\begin{aligned} {}^{(O_{(l)})}\not\pi_{\breve{R}A} = \upmu \chi_{AB}^{(Small)} O_{(l)}^B + \uprho_{(l)} \not{d}_A \upmu + G_{(Frame)} \begin{pmatrix} \upmu O_{(l)} \\ \upmu \uprho_{(l)} \end{pmatrix} \begin{pmatrix} L\Psi \\ \not{d}\Psi \end{pmatrix} \\ + G_{(Frame)} \uprho_{(l)} \breve{R}\Psi + \upmu f(\Psi) \begin{pmatrix} \Psi \\ L_{(Small)} \end{pmatrix} \not{d}x. \end{aligned} \tag{12.76}$$

We now apply $\not{\mathcal{L}}_{\mathscr{Z}}^{N}$ to the terms on the right-hand side of (12.76) and apply the Leibniz rule. To bound $\mathscr{Z}^M(\upmu - 1)$ for $M \leq 12$, we use the bootstrap assumptions (**AUX**$\upmu$), while the higher-order derivatives of $\upmu - 1$ appear explicitly on the right-hand side of (12.75a). To bound $\not{\mathcal{L}}_{\mathscr{Z}}^{M} \not{d}\upmu$ for $M \leq 11$, we use Lemma 8.6, inequality (12.46a), and the bootstrap assumptions (**AUX**$\upmu$), while we bound the higher-order derivatives by $\left|\not{\mathcal{L}}_{\mathscr{Z}}^{M} \not{d}\upmu\right| \lesssim (1+t)^{-1} \mathscr{Z}^{M+1}\upmu$, and the latter term appears explicitly on the right-hand side of (12.75a). We similarly bound $\not{\mathcal{L}}_{\mathscr{Z}}^{M} L_{(Small)}$ with the help of the bootstrap assumptions (**AUX**$LR_{(Small)}$). To bound $\not{\mathcal{L}}_{\mathscr{Z}}^{M} \chi^{(Small)}$, we use Lemma 12.30. To bound $\not{\mathcal{L}}_{\mathscr{Z}}^{M} O_{(l)}$, we use Lemma 12.31. To bound $\not{\mathcal{L}}_{\mathscr{Z}}^{M} \uprho_{(l)}$, we use inequalities (12.28a) and (12.28b). We bound the terms $\not{\mathcal{L}}_{\mathscr{Z}}^{M} G_{(Frame)}$ with Lemma 12.25. To bound the terms $\begin{pmatrix} \mathscr{Z}^M L\Psi \\ \not{\mathcal{L}}_{\mathscr{Z}}^{M} \not{d}\Psi \end{pmatrix}$, $\not{\mathcal{L}}_{\mathscr{Z}}^{M} f(\Psi)$, and $\mathscr{Z}^M\Psi$, we use Lemma 12.23 and the bootstrap assumptions (**BA**Ψ). To bound the terms $\not{\mathcal{L}}_{\mathscr{Z}}^{M} \not{d}x$, we use the estimates (12.57a) and (12.57b). In total, these estimates yield inequality (12.75a) for $\not{\mathcal{L}}_{\mathscr{Z}}^{\leq N} {}^{(O)}\not\pi_{\breve{R}}$. We can prove the estimate (12.75a) for $\not{\mathcal{L}}_{\mathscr{Z}}^{\leq N} {}^{(O)}\not\pi_{\breve{R}}^{\#} = \not{\mathcal{L}}_{\mathscr{Z}}^{\leq N} (\not{g}^{-1} \cdot {}^{(O)}\not\pi_{\breve{R}})$ in a similar fashion, but we also need to use the estimates (12.60a) and (12.60b) to bound the factors of $\not{\mathcal{L}}_{\mathscr{Z}}^{M} \not{g}^{-1}$ that arise in the estimates.

Inequality (12.75b) then follows from (12.75a) and the bootstrap assumptions (**BA**Ψ), (**AUX**$\upmu$), and (**AUX**$LR_{(Small)}$).

Using the identities (7.8d) and (7.9a), we similarly deduce inequalities (12.74a) and (12.74b).

A similar but simpler proof based on the identity (7.5d) yields inequalities (12.72a) and (12.72b).

A similar but simpler proof based on the identity (7.7e) yields inequalities (12.73a) and (12.73b). □

12.21. Preliminary Lie derivative commutator estimates

In this section, we derive some preliminary commutator estimates involving $S_{t,u}$-projected Lie derivatives.

LEMMA 12.33 (**Preliminary quantitative estimate involving Lie derivative commutators**). *Let $1 \leq N \leq 23$ be an integer and let ξ be an $S_{t,u}$ one-form or a symmetric type $\binom{0}{2}$ $S_{t,u}$ tensorfield. Let $O \in \{O_{(1)}, O_{(2)}, O_{(3)}\}$. Under the small-data and bootstrap assumptions of Sects. 12.1-12.4, if ε is sufficiently small, then the following pointwise estimates hold on $\mathcal{M}_{T_{(Bootstrap)},U_0}$:*

(12.77a)
$$\left|[\not{\mathcal{L}}_{\varrho L}, \not{\mathcal{L}}_{\mathscr{Z}}^N]\xi\right| \lesssim \left|\not{\mathcal{L}}_L \not{\mathcal{L}}_{\mathscr{Z}}^{\leq N-1}\xi\right| + \frac{1}{1+t} \sum_{N_1+N_2 \leq N} \max_{l=1,2,3} \left|\not{\mathcal{L}}_{\mathscr{Z}}^{N_1} \begin{pmatrix} \varrho^{(\breve{R})}\not{\pi}_L^{\#} \\ \varrho^{(O_{(l)})}\not{\pi}_L^{\#} \end{pmatrix}\right| \left|\not{\mathcal{L}}_{\mathscr{Z}}^{N_2}\xi\right|,$$

(12.77b)
$$\left|[\not{\mathcal{L}}_{\breve{R}}, \not{\mathcal{L}}_{\mathscr{Z}}^N]\xi\right| \lesssim \left|\not{\mathcal{L}}_L \not{\mathcal{L}}_{\mathscr{Z}}^{\leq N-1}\xi\right| + \frac{1}{1+t} \sum_{N_1+N_2 \leq N} \max_{l=1,2,3} \left|\not{\mathcal{L}}_{\mathscr{Z}}^{N_1} \begin{pmatrix} \varrho^{(\breve{R})}\not{\pi}_L^{\#} \\ ^{(O_{(l)})}\not{\pi}_L^{\#} \\ ^{(O_{(l)})}\not{\pi}_{\breve{R}}^{\#} \end{pmatrix}\right| \left|\not{\mathcal{L}}_{\mathscr{Z}}^{N_2}\xi\right|,$$

(12.77c)
$$\left|[\not{\mathcal{L}}_{O}, \not{\mathcal{L}}_{\mathscr{Z}}^N]\xi\right| \lesssim \frac{1}{1+t} \sum_{N_1+N_2 \leq N} \max_{l,m=1,2,3} \left|\not{\mathcal{L}}_{\mathscr{Z}}^{N_1} \begin{pmatrix} \varrho^{(O_{(l)})}\not{\pi}_L^{\#} \\ ^{(O_{(l)})}\not{\pi}_{\breve{R}}^{\#} \\ \not{\mathcal{L}}_{O_{(l)}} O_{(m)} \end{pmatrix}\right| \left|\not{\mathcal{L}}_{\mathscr{Z}}^{N_2}\xi\right|,$$

(12.77d)
$$\left|[\not{\mathcal{L}}_{L}, \not{\mathcal{L}}_{\mathscr{Z}}^N]\xi\right| \lesssim \left|\not{\mathcal{L}}_L \not{\mathcal{L}}_{\mathscr{Z}}^{\leq N-1}\xi\right| + \frac{1}{1+t} \sum_{N_1+N_2 \leq N} \max_{l=1,2,3} \left|\not{\mathcal{L}}_{\mathscr{Z}}^{N_1} \begin{pmatrix} ^{(\breve{R})}\not{\pi}_L^{\#} \\ ^{(O_{(l)})}\not{\pi}_L^{\#} \end{pmatrix}\right| \left|\not{\mathcal{L}}_{\mathscr{Z}}^{N_2}\xi\right|,$$

(12.77e)
$$\left|[\not{\mathcal{L}}_{L}, \not{\mathcal{L}}_{\mathscr{S}}^N]\xi\right| \lesssim \frac{1}{1+t} \sum_{N_1+N_2 \leq N} \max_{l=1,2,3} \left|\not{\mathcal{L}}_{\mathscr{Z}}^{N_1} \begin{pmatrix} ^{(\breve{R})}\not{\pi}_L^{\#} \\ ^{(O_{(l)})}\not{\pi}_L^{\#} \end{pmatrix}\right| \left|\not{\mathcal{L}}_{\mathscr{Z}}^{N_2}\xi\right|.$$

Furthermore, if $1 \leq N \leq 24$ and f is a scalar-valued function, then

(12.78)
$$\text{the estimates (12.77a)} - \text{(12.77e) also hold with } \xi \text{ replaced by } f, \text{ but with the sums } \sum_{N_1+N_2 \leq N} \text{ on the right-hand sides replaced by } \sum_{\substack{N_1+N_2 \leq N \\ N_1 \leq N-1}}.$$

PROOF. In this proof, we sometimes use Lemma 4.10 and Cor. 4.13 silently. We begin by proving (12.77d). As a first step, let $Z \in \mathscr{Z}$. Then either $Z = \varrho L$, in which case $[L, Z] = L$, or $Z = S \in \mathscr{S}$ is a spatial commutation vectorfield, in which case $[L, S]$ is $S_{t,u}$-tangent by Lemma 4.10. In the former case, we have $[\not{\mathcal{L}}_L, \not{\mathcal{L}}_Z]\xi = \not{\mathcal{L}}_L \xi$, while in the latter case, we have $[\not{\mathcal{L}}_L, \not{\mathcal{L}}_S]\xi = \not{\mathcal{L}}_{[L,S]}\xi = \not{\mathcal{L}}_{{}^{(S)}\not{\pi}_L^{\#}}\xi$. Iterating these identities, we deduce the following schematic identity, where some terms on the right-hand side may be absent:

$$[\not{\mathcal{L}}_L, \not{\mathcal{L}}_{\mathscr{Z}}^N]\xi = \sum_{M \leq N-1} \not{\mathcal{L}}_L \not{\mathcal{L}}_{\mathscr{Z}}^M \xi + \sum_{S \in \mathscr{S}} \sum_{N_1 + N_2 \leq N-1} \not{\mathcal{L}}_{\not{\mathcal{L}}_{\mathscr{Z}}^{N_1}\, {}^{(S)}\not{\pi}_L^{\#}} \not{\mathcal{L}}_{\mathscr{Z}}^{N_2} \xi. \tag{12.79}$$

The desired estimate (12.77d) now follows from (12.79) and inequality (12.63).

To prove the estimate (12.78) in the case of $[L, \mathscr{Z}^N]f$, we first use an argument similar to the one used to derive (12.79) together with Lemma 8.6 in order to deduce that

$$[L, \mathscr{Z}^N]f = \sum_{M \leq N-1} L \mathscr{Z}^M f + \sum_{S \in \mathscr{S}} \sum_{N_1 + N_2 \leq N-1} (\not{\mathcal{L}}_{\mathscr{Z}}^{N_1}\, {}^{(S)}\not{\pi}_L^{\#}) \cdot \not{d} \mathscr{Z}^{N_2} f. \tag{12.80}$$

The desired estimate (12.78) in this case now follows from (12.80) and inequality (12.46a). Note that inequality (12.80) involves one fewer derivatives of ${}^{(S)}\not{\pi}_L^{\#}$ compared to (12.77d) because we do not need to use inequality (12.63) in the proof of (12.80).

The proofs of (12.77a)-(12.77c) and (12.77e) are similar to the proof of (12.77d). To proceed, we first note that by Lemma 4.10 and (12.6b), if $Z \in \mathscr{Z}$, then $[\varrho L, Z]$ is either 0, $\varrho^{(\breve{R})}\not{\pi}_L^{\#} + L$, or $\varrho^{(O_{(l)})}\not{\pi}_L^{\#}$; $[\breve{R}, Z]$ is either 0, $-(\varrho^{(\breve{R})}\not{\pi}_L^{\#} + L)$, or ${}^{(O_{(l)})}\not{\pi}_{\breve{R}}^{\#}$; $[O, Z]$ is either $\pm\varrho^{(O)}\not{\pi}_L^{\#}$, $-{}^{(O)}\not{\pi}_{\breve{R}}^{\#}$, or $\not{\mathcal{L}}_O O_{(m)}$; and if $S \in \mathscr{S}$, then $[L, S] = {}^{(S)}\not{\pi}_L^{\#}$. Based on these commutator identities and (12.6b), the remainder of the proofs of (12.77a)-(12.77c) and (12.77e) essentially mirrors the proof of (12.77d).

The proofs of the remaining estimates in (12.78) are similarly connected to the proof in the case of $[L, \mathscr{Z}^N]f$. □

12.22. Commutator estimates for vectorfields acting on functions and spherical covariant tensorfields

In this section, we derive pointwise commutator estimates for vectorfields acting on functions and $S_{t,u}$ covariant tensorfields.

REMARK 12.34 (**Floor and ceiling functions**). In what follows, $\lfloor \cdot \rfloor$ and $\lceil \cdot \rceil$ respectively denote the floor and ceiling functions. That is, if M is a nonnegative integer, then $\lfloor M/2 \rfloor = M/2$ for M even and $\lfloor M/2 \rfloor = (M-1)/2$ for M odd, while $\lceil M/2 \rceil = M/2$ for M even and $\lceil M/2 \rceil = (M+1)/2$ for M odd.

LEMMA 12.35 (**Commutator estimates for vectorfields acting on functions**). *Let $1 \leq N \leq 24$ be an integer and let f be a scalar-valued function. Let $O \in \{O_{(1)}, O_{(2)}, O_{(3)}\}$. Under the small-data and bootstrap assumptions of Sects. 12.1-12.4, if ε is sufficiently small, then the following pointwise estimates* *hold on* $\mathcal{M}_{T_{(Bootstrap)},U_0}$:

(12.81a)

$$\left| \begin{pmatrix} [\mathscr{Z}^N, \varrho L]f \\ [\mathscr{Z}^N, \breve{R}]f \end{pmatrix} \right| \lesssim \left| L\mathscr{Z}^{\leq N-1} f \right| + \varepsilon^{1/2} \frac{\ln(e+t)}{1+t} \left| \begin{pmatrix} \varrho L \mathscr{Z}^{\leq N-1} f \\ \breve{R}\mathscr{Z}^{\leq N-1} f \\ \varrho {\not d} \mathscr{Z}^{\leq N-1} f \\ \mathscr{Z}^{\leq N-1} f \end{pmatrix} \right|$$
$$+ \left\| \mathscr{Z}^{\leq \lfloor N/2 \rfloor} f \right\|_{C^0(\Sigma_t^u)} \left| \begin{pmatrix} \varrho L \mathscr{Z}^{\leq N-1} \Psi \\ \breve{R}\mathscr{Z}^{\leq N-1} \Psi \\ \varrho {\not d} \mathscr{Z}^{\leq N-1} \Psi \\ \mathscr{Z}^{\leq N-1} \Psi \end{pmatrix} \right|$$
$$+ \frac{1}{1-t} \left\| \mathscr{Z}^{\leq \lfloor N/2 \rfloor} f \right\|_{C^0(\Sigma_t^u)} \left| \begin{pmatrix} \mathscr{Z}^{\leq N}(\upmu - 1) \\ \sum_{a=1}^3 \varrho \left| \mathscr{Z}^{\leq N} L_{(Small)}^a \right| \end{pmatrix} \right|,$$

(12.81b)

$$\left| [\mathscr{Z}^N, O] f \right| \lesssim \left| \begin{pmatrix} \varrho L \mathscr{Z}^{\leq N-1} f \\ \breve{R}\mathscr{Z}^{\leq N-1} f \\ \varrho {\not d} \mathscr{Z}^{\leq N-1} f \\ \mathscr{Z}^{\leq N-1} f \end{pmatrix} \right|$$
$$+ (1+t) \left\| \mathscr{Z}^{\leq \lfloor N/2 \rfloor} f \right\|_{C^0(\Sigma_t^u)} \left| \begin{pmatrix} \varrho L \mathscr{Z}^{\leq N-1} \Psi \\ \breve{R}\mathscr{Z}^{\leq N-1} \Psi \\ \varrho {\not d} \mathscr{Z}^{\leq N-1} \Psi \\ \mathscr{Z}^{\leq N-1} \Psi \end{pmatrix} \right|$$
$$+ \left\| \mathscr{Z}^{\leq \lfloor N/2 \rfloor} f \right\|_{C^0(\Sigma_t^u)} \left| \begin{pmatrix} \mathscr{Z}^{\leq N}(\upmu - 1) \\ \sum_{a=1}^3 \varrho \left| \mathscr{Z}^{\leq N} L_{(Small)}^a \right| \end{pmatrix} \right|,$$

(12.81c)

$$\left| [\mathscr{Z}^N, L] f \right| \lesssim \left| L\mathscr{Z}^{\leq N-1} f \right| + \varepsilon^{1/2} \frac{\ln(e+t)}{(1+t)^2} \left| \begin{pmatrix} \varrho L \mathscr{Z}^{\leq N-1} f \\ \breve{R}\mathscr{Z}^{\leq N-1} f \\ \varrho {\not d} \mathscr{Z}^{\leq N-1} f \\ \mathscr{Z}^{\leq N-1} f \end{pmatrix} \right|$$
$$+ \frac{1}{1-t} \left\| \mathscr{Z}^{\leq \lfloor N/2 \rfloor} f \right\|_{C^0(\Sigma_t^u)} \left| \begin{pmatrix} \varrho L \mathscr{Z}^{\leq N-1} \Psi \\ \breve{R}\mathscr{Z}^{\leq N-1} \Psi \\ \varrho {\not d} \mathscr{Z}^{\leq N-1} \Psi \\ \mathscr{Z}^{\leq N-1} \Psi \end{pmatrix} \right|$$
$$+ \frac{1}{(1+t)^2} \left\| \mathscr{Z}^{\leq \lfloor N/2 \rfloor} f \right\|_{C^0(\Sigma_t^u)} \left| \begin{pmatrix} \mathscr{Z}^{\leq N}(\upmu - 1) \\ \sum_{a=1}^3 \varrho \left| \mathscr{Z}^{\leq N} L_{(Small)}^a \right| \end{pmatrix} \right|.$$

In addition,

(12.82)

(12.81c) *also holds with the first term on the right replaced by* $\left| \mathscr{Z}^{\leq N-1} L f \right|$.

Finally,

(12.83)

(12.81c) *also holds without the first term on the right*

if the left-hand side is equal to $\left|[\mathscr{S}^N, L]f\right|$,

where $\mathscr{S}^N$ *is an* N^{th} *order pure spatial commutation vectorfield operator.*

PROOF. The two estimates in (12.81a) and the estimate (12.81c) follow from (12.77a), (12.77b), (12.77d), (12.78), (12.6b), (12.46a), (12.55), (12.72a), (12.72b) (12.73a), (12.73b), (12.74a), (12.74b), (12.75a), (12.75b), and the identity $\varrho^{(\breve{R})}\not\pi_L = -^{(\varrho L)}\not\pi_{\breve{R}}$, which is straightforward to verify using the arguments that we used to prove (4.21a). The estimate (12.82) then follows inductively from (12.81c).

The proof of (12.83) is similar, but relies on inequalities (12.77e) and (12.78).

The proof of (12.81b) is similar, but we also need use the estimates (12.70a) and (12.70b), which results in a right-hand side with larger factors of $1+t$. □

LEMMA 12.36 (**Pointwise estimates involving commutations with the operator** $L + \frac{1}{2}\mathrm{tr}_{\not g}\chi$). *Let* $1 \leq N \leq 23$ *be an integer, let* χ *be the* $S_{t,u}$ *tensorfield defined in* (3.70), *and let* f *be a scalar-valued function. Under the small-data and bootstrap assumptions of Sects.* 12.1-12.4, *if* ε *is sufficiently small, then the following pointwise estimates hold on* $\mathcal{M}_{T_{(Bootstrap)},U_0}$:

$$
\begin{aligned}
(12.84)\qquad & \left|\left[\mathscr{Z}^N, \left\{L + \frac{1}{2}\mathrm{tr}_{\not g}\chi\right\}\right] f\right| \\
& \lesssim \left|\left\{L + \frac{1}{2}\mathrm{tr}_{\not g}\chi\right\} \mathscr{Z}^{\leq N-1} f\right| \\
& \quad + \frac{\ln(e+t)}{(1+t)^2}\left|\begin{pmatrix} \varrho \not d \mathscr{Z}^{\leq N-1} f \\ \mathscr{Z}^{\leq N-1} f \end{pmatrix}\right| \\
& \quad + \frac{1}{1+t}\left\|\mathscr{Z}^{\leq \lfloor N/2 \rfloor} f\right\|_{C^0(\Sigma_t^u)} \left|\begin{pmatrix} \varrho L \mathscr{Z}^{\leq N}\Psi \\ \breve{R}\mathscr{Z}^{\leq N}\Psi \\ \varrho \not d \mathscr{Z}^{\leq N}\Psi \\ \mathscr{Z}^{\leq N}\Psi \end{pmatrix}\right| \\
& \quad + \frac{1}{(1+t)^2}\left\|\mathscr{Z}^{\leq \lfloor N/2 \rfloor} f\right\|_{C^0(\Sigma_t^u)} \left|\begin{pmatrix} \mathscr{Z}^{\leq N+1}(\upmu - 1) \\ \sum_{a=1}^3 \varrho |\mathscr{Z}^{\leq N+1} L_{(Small)}^a| \end{pmatrix}\right|.
\end{aligned}
$$

PROOF. We first recall the decomposition $\frac{1}{2}\mathrm{tr}_{\not g}\chi = \varrho^{-1} + \frac{1}{2}\mathrm{tr}_{\not g}\chi^{(Small)}$ (see (4.1c)). We separate the proof into case **i)** $\mathscr{Z}^N = \mathscr{Z}^{N-1}S$, where $S \in \{\breve{R}, O_{(1)}, O_{(2)}, O_{(3)}\}$ is a spatial commutation vectorfield and case **ii)** $\mathscr{Z}^N = \mathscr{Z}^{N-1}Z$, where $Z = \varrho L$. Using the aforementioned decomposition, the identity $L\varrho = 1$, and the fact that $[L, S] = {}^{(S)}\not\pi_L^{\#}$ (see Lemma 7.5), we compute that in cases **i)** and **ii)**, we respectively have

(12.85)

$$\mathscr{Z}^N\left\{Lf+\frac{1}{2}\mathrm{tr}_{\not g}\chi f\right\}=\mathscr{Z}^{N-1}\left\{LSf+\frac{1}{2}\mathrm{tr}_{\not g}\chi Sf\right\}$$
$$-\mathscr{Z}^{N-1}({}^{(S)}\not\pi_L^{\#}\cdot\not d f)+\frac{1}{2}\mathscr{Z}^{N-1}\left\{(S\varrho^{-1}+S\mathrm{tr}_{\not g}\chi^{(Small)})f\right\}$$

(12.86)

$$\mathscr{Z}^N\left\{Lf+\frac{1}{2}\mathrm{tr}_{\not g}\chi f\right\}=\mathscr{Z}^{N-1}\left\{LZf+\frac{1}{2}\mathrm{tr}_{\not g}\chi Zf\right\}$$
$$-\mathscr{Z}^{N-1}\left\{Lf+\frac{1}{2}\mathrm{tr}_{\not g}\chi f\right\}$$
$$+\frac{1}{2}\mathscr{Z}^{N-1}\left\{\mathrm{tr}_{\not g}\chi^{(Small)}f\right\}+\frac{1}{2}\mathscr{Z}^{N-1}\left\{(Z\mathrm{tr}_{\not g}\chi^{(Small)})f\right\}.$$

We now use the identities (12.85) and (12.86) to inductively commute the operators $\mathscr{Z}^M$ through the operator $L+\frac{1}{2}\mathrm{tr}_{\not g}\chi$. Using also the Leibniz rule, Lemma 8.6, the fact that $S\varrho^{-1}$ is equal to either ϱ^{-2} or 0, and (12.6b), we deduce that

(12.87)
$$\left|\mathscr{Z}^N\left\{Lf+\frac{1}{2}\mathrm{tr}_{\not g}\chi f\right\}-\left\{L+\frac{1}{2}\mathrm{tr}_{\not g}\chi\right\}\mathscr{Z}^N f\right|$$
$$\lesssim\left|\left\{L+\frac{1}{2}\mathrm{tr}_{\not g}\chi\right\}\mathscr{Z}^{\leq N-1}f\right|+\sum_{N_1+N_2\leq N-1}\sum_{S\in\mathscr{S}}\left|\not{\mathcal{L}}_{\mathscr{Z}}^{N_1}\,{}^{(S)}\not\pi_L^{\#}\right|\left|\not d\mathscr{Z}^{N_2}f\right|$$
$$+\sum_{\substack{N_1+N_2\leq N\\ N_2\leq N-1}}\left|\mathscr{Z}^{N_1}\mathrm{tr}_{\not g}\chi^{(Small)}\right|\left|\mathscr{Z}^{N_2}f\right|+\frac{1}{(1+t)^2}\left|\mathscr{Z}^{\leq N-1}f\right|.$$

The desired bound (12.84) now follows from (12.87), the estimates (12.67a), (12.72a), (12.72b), (12.74a), (12.74b), and (12.128a), and the bootstrap assumptions (**BA**Ψ). ☐

COROLLARY 12.37 (**Pointwise estimates involving commutations with the operator** $L+\frac{1}{2}\mathrm{tr}_{\not g}\chi$). *Let $1\leq N\leq 24$ be an integer and let $Z\in\mathscr{Z}$. Under the small-data and bootstrap assumptions of Sects. 12.1-12.4, if ε is sufficiently small, then the following pointwise estimates hold on $\mathcal{M}_{T_{(Bootstrap)},U_0}$:*

(12.88a)
$$\left|\mathscr{Z}^{N-1}\left\{LZ\Psi+\frac{1}{2}\mathrm{tr}_{\not g}\chi Z\Psi\right\}\right|\lesssim\left|\left\{L+\frac{1}{2}\mathrm{tr}_{\not g}\chi\right\}\mathscr{Z}^{\leq N}\Psi\right|$$
$$+\frac{\ln(e+t)}{(1+t)^2}\left|\begin{pmatrix}\varrho\not d\mathscr{Z}^{\leq N-1}\Psi\\ \mathscr{Z}^{\leq N-1}\Psi\end{pmatrix}\right|$$
$$+\varepsilon\frac{1}{(1+t)^3}\left|\begin{pmatrix}\mathscr{Z}^{\leq N}(\upmu-1)\\ \sum_{a=1}^3\varrho\left|\mathscr{Z}^{\leq N}L_{(Small)}^a\right|\end{pmatrix}\right|,$$

(12.88b)
$$\left\|\mathscr{Z}^{\leq 11}\left\{LZ\Psi+\frac{1}{2}\mathrm{tr}_{\not g}\chi Z\Psi\right\}\right\|_{C^0(\Sigma_t^u)}\lesssim\varepsilon\frac{1}{(1+t)^2}.$$

PROOF. We simply use Lemma 12.36 with $Z\Psi$ in the role of f and $N-1$ in the role of N and the bootstrap assumptions (**BA**Ψ). □

LEMMA 12.38 (**Commutator estimates for vectorfields acting on tensorfields**). *Let $1 \leq N \leq 23$ be an integer and let ξ be an $S_{t,u}$ one-form or a symmetric type $\binom{0}{2}$ $S_{t,u}$ tensorfield. Let $Z \in \mathscr{Z}$ be a commutation vectorfield. Under the small-data and bootstrap assumptions of Sects. 12.1-12.4, if ε is sufficiently small, then the following commutator estimates hold on $\mathcal{M}_{T_{(Bootstrap)},U_0}$:*

(12.89a)

$$\left|\begin{pmatrix} [\not{\mathcal{L}}_{\mathscr{Z}}^N, \not{\mathcal{L}}_{\varrho L}]\xi \\ [\not{\mathcal{L}}_{\mathscr{Z}}^N, \not{\mathcal{L}}_{\breve{R}}]\xi \end{pmatrix}\right| \lesssim \left|\not{\mathcal{L}}_L \not{\mathcal{L}}_{\mathscr{Z}}^{N-1}\xi\right| + \varepsilon^{1/2}\frac{\ln(e+t)}{1+t}\left|\not{\mathcal{L}}_{\mathscr{Z}}^{\leq N}\xi\right|$$
$$+ \left\|\not{\mathcal{L}}_{\mathscr{Z}}^{\leq \lfloor N/2 \rfloor}\xi\right\|_{C^0(\Sigma_t^u)}\left|\begin{pmatrix} \varrho L\mathscr{Z}^{\leq N}\Psi \\ \breve{R}\mathscr{Z}^{\leq N}\Psi \\ \varrho \not{d}\mathscr{Z}^{\leq N}\Psi \\ \mathscr{Z}^{\leq N}\Psi \end{pmatrix}\right|$$
$$+ \frac{1}{1+t}\left\|\not{\mathcal{L}}_{\mathscr{Z}}^{\leq \lfloor N/2 \rfloor}\xi\right\|_{C^0(\Sigma_t^u)}\left|\begin{pmatrix} \mathscr{Z}^{\leq N+1}(\upmu - 1) \\ \sum_{a=1}^3 \varrho|\mathscr{Z}^{\leq N+1}L_{(Small)}^a| \end{pmatrix}\right|,$$

(12.89b)

$$\left|[\not{\mathcal{L}}_{\mathscr{Z}}^N, \not{\mathcal{L}}_O]\xi\right| \lesssim \left|\not{\mathcal{L}}_{\mathscr{Z}}^{\leq N}\xi\right| + (1+t)\left\|\not{\mathcal{L}}_{\mathscr{Z}}^{\leq \lfloor N/2 \rfloor}\xi\right\|_{C^0(\Sigma_t^u)}\left|\begin{pmatrix} \varrho L\mathscr{Z}^{\leq N}\Psi \\ \breve{R}\mathscr{Z}^{\leq N}\Psi \\ \varrho \not{d}\mathscr{Z}^{\leq N}\Psi \\ \mathscr{Z}^{\leq N}\Psi \end{pmatrix}\right|$$
$$+ \left\|\not{\mathcal{L}}_{\mathscr{Z}}^{\leq \lfloor N/2 \rfloor}\xi\right\|_{C^0(\Sigma_t^u)}\left|\begin{pmatrix} \mathscr{Z}^{\leq N+1}(\upmu - 1) \\ \sum_{a=1}^3 \varrho|\mathscr{Z}^{\leq N+1}L_{(Small)}^a| \end{pmatrix}\right|,$$

(12.89c)

$$\left|[\not{\mathcal{L}}_{\mathscr{Z}}^N, \not{\mathcal{L}}_L]\xi\right| \lesssim \left|\not{\mathcal{L}}_L \not{\mathcal{L}}_{\mathscr{Z}}^{N-1}\xi\right| + \varepsilon^{1/2}\frac{\ln(e+t)}{(1+t)^2}\left|\not{\mathcal{L}}_{\mathscr{Z}}^{\leq N}\xi\right|$$
$$+ \frac{1}{1+t}\left\|\not{\mathcal{L}}_{\mathscr{Z}}^{\leq \lfloor N/2 \rfloor}\xi\right\|_{C^0(\Sigma_t^u)}\left|\begin{pmatrix} \varrho L\mathscr{Z}^{\leq N}\Psi \\ \breve{R}\mathscr{Z}^{\leq N}\Psi \\ \varrho \not{d}\mathscr{Z}^{\leq N}\Psi \\ \mathscr{Z}^{\leq N}\Psi \end{pmatrix}\right|$$
$$+ \frac{1}{(1+t)^2}\left\|\not{\mathcal{L}}_{\mathscr{Z}}^{\leq \lfloor N/2 \rfloor}\xi\right\|_{C^0(\Sigma_t^u)}\left|\begin{pmatrix} \mathscr{Z}^{\leq N+1}(\upmu - 1) \\ \sum_{a=1}^3 \varrho|\mathscr{Z}^{\leq N+1}L_{(Small)}^a| \end{pmatrix}\right|.$$

In addition,

(12.90)

(12.89c) *also holds with the first term on the right replaced by* $\left|\not{\mathcal{L}}_{\mathscr{Z}}^{\leq N-1}\not{\mathcal{L}}_L\xi\right|$.

Finally,

(12.91) (12.89c) *also holds without the first term on the right*

if the left-hand side is equal to $\left|[\not{\mathcal{L}}_{\mathscr{S}}^N, \not{\mathcal{L}}_L]\xi\right|$,

where $\not{\mathcal{L}}_{\mathscr{S}}^N$ *is an* N^{th} *order pure spatial commutation vectorfield operator.*

PROOF. The proof of Lemma 12.38 is essentially the same as the proof of Lemma 12.35, but we use the estimates (12.77a)-(12.77e) in place of (12.78); note that the former estimates involve one additional derivative of deformation tensors compared to the latter, and this fact is reflected in the estimates stated in the present lemma. □

12.23. Commutator estimates for vectorfields acting on the covariant angular derivative of a spherical tensorfield

In this section, we derive pointwise commutator estimates for vectorfields acting on $\nabla\xi$, where ξ is a covariant $S_{t,u}$ tensorfield.

LEMMA 12.39 (**Commutator estimates for vectorfields acting on $\nabla\xi$, where ξ is an $S_{t,u}$ tensorfield**). *Let $1 \leq N \leq 23$ be an integer and let ξ be an $S_{t,u}$ one-form or a symmetric type $\binom{0}{2}$ $S_{t,u}$ tensorfield. Under the small-data and bootstrap assumptions of Sects. 12.1-12.4, if ε is sufficiently small, then the following commutator estimates hold on $\mathcal{M}_{T_{(Bootstrap)},U_0}$:*

(12.92)

$$\left|[\nabla, \not{\mathcal{L}}_{\mathscr{Z}}^N]\xi\right| \lesssim \frac{1}{1+t}\left|\not{\mathcal{L}}_{\mathscr{Z}}^{\leq N-1}\xi\right| - \frac{1}{1+t}\left\|\not{\mathcal{L}}_{\mathscr{Z}}^{\leq \lfloor N/2 \rfloor}\xi\right\|_{C^0(\Sigma_t^u)}\left|\begin{pmatrix} \varrho L\mathscr{Z}^{\leq N}\Psi \\ \breve{R}\mathscr{Z}^{\leq N}\Psi \\ \varrho\not{d}\mathscr{Z}^{\leq N}\Psi \\ \mathscr{Z}^{\leq N}\Psi \end{pmatrix}\right|$$
$$+ \frac{1}{(1+t)^2}\left\|\not{\mathcal{L}}_{\mathscr{Z}}^{\leq \lfloor N/2 \rfloor}\xi\right\|_{C^0(\Sigma_t^u)}\left|\begin{pmatrix} \mathscr{Z}^{\leq N+1}(\upmu - 1) \\ \sum_{a=1}^3 \varrho|\mathscr{Z}^{\leq N+1}L_{(Small)}^a| \end{pmatrix}\right|.$$

PROOF. We first prove (12.92) in the case that ξ is a symmetric type $\binom{0}{2}$ $S_{t,u}$ tensor. Let $Z \in \mathscr{Z}$. From (8.13), we deduce the schematic identity

$$[\nabla, \not{\mathcal{L}}_Z]\xi = \not{g}^{-1}(\nabla^{(Z)}\not{\pi})\xi. \tag{12.93}$$

Iterating (12.93), using the schematic identity $\not{\mathcal{L}}_Z\not{g}^{-1} = -(\not{g}^{-1})^{2(Z)}\not{\pi}$ (see Lemma 7.5), and using the fact that ${}^{(Z)}\not{\pi}$ is also a symmetric type $\binom{0}{2}$ $S_{t,u}$ tensor in order to commute Lie derivatives with the operator ∇ in $\nabla^{(Z)}\not{\pi}$, we deduce the schematic identity

(12.94)

$$[\nabla, \not{\mathcal{L}}_{\mathscr{Z}}^N]\xi = \sum_{M=1}^{N} \sum_{N_1+N_2+\cdots+N_{M+2}=N-M}$$
$$(\not{\mathcal{L}}_{\mathscr{Z}}^{N_1}\not{g}^{-1})\cdots(\not{\mathcal{L}}_{\mathscr{Z}}^{N_k}\not{g}^{-1})(\nabla\not{\mathcal{L}}_{\mathscr{Z}}^{N_{k+1}}\not{\pi})(\not{\mathcal{L}}_{\mathscr{Z}}^{N_{k+2}}\not{\pi})\cdots(\not{\mathcal{L}}_{\mathscr{Z}}^{N_{M+1}}\not{\pi})\not{\mathcal{L}}_{\mathscr{Z}}^{N_{M+2}}\xi,$$

where the $\not{\pi}$ are the $S_{t,u}$ projections of deformation tensors of vectorfields in $\mathscr{Z}$. Inequality (12.92) now follows from the identity (12.94) and inequalities (12.60a), (12.60b), and (12.48).

Thanks to the identity (8.12), the proof of (12.92) in the case that ξ is an $S_{t,u}$ one-form is essentially the same; we omit the details. □

12.24. Commutator estimates for vectorfields acting on the angular Hessian of a function

In this section, we derive pointwise commutator estimates for vectorfields acting on $\nabla^2 f$, where f is a function.

LEMMA 12.40 (**Commutator estimates for vectorfields acting on $\nabla\!\!\!/^2 f$**). *Let $1 \leq N \leq 23$ be an integer and let f be a scalar-valued function. Under the small-data and bootstrap assumptions of Sects. 12.1-12.4, if ε is sufficiently small, then the following commutator estimates hold on $\mathcal{M}_{T_{(Bootstrap)},U_0}$:*

(12.95)

$$\left|[\nabla\!\!\!/^2, \mathcal{L}\!\!\!/_{\mathscr{Z}}^N]f\right| \lesssim \frac{1}{(1+t)^2}\left|\mathscr{Z}^{\leq N} f\right| + \frac{1}{(1+t)^2}\left\|\mathscr{Z}^{\leq \lceil N/2 \rceil} f\right\|_{C^0(\Sigma_t^u)}\left|\begin{pmatrix} \varrho L \mathscr{Z}^{\leq N}\Psi \\ \breve{R}\mathscr{Z}^{\leq N}\Psi \\ \varrho d\!\!\!/ \mathscr{Z}^{\leq N}\Psi \\ \mathscr{Z}^{\leq N}\Psi \end{pmatrix}\right| + \frac{1}{(1+t)^3}\left\|\mathscr{Z}^{\leq \lceil N/2 \rceil} f\right\|_{C^0(\Sigma_t^u)}\left|\begin{pmatrix} \mathscr{Z}^{\leq N+1}(\upmu - 1) \\ \sum_{a=1}^3 \varrho|\mathscr{Z}^{\leq N+1} L_{(Small)}^a| \end{pmatrix}\right|,$$

(12.96)

$$\left|[\hat{\nabla\!\!\!/}^2, \mathcal{L}\!\!\!/_{\mathscr{Z}}^N]f\right| \lesssim \frac{1}{(1+t)^2}\left|\mathscr{Z}^{\leq N+1} f\right| + \frac{1}{(1+t)^2}\left\|\mathscr{Z}^{\leq \lceil N/2 \rceil} f\right\|_{C^0(\Sigma_t^u)}\left|\begin{pmatrix} \varrho L \mathscr{Z}^{\leq N}\Psi \\ \breve{R}\mathscr{Z}^{\leq N}\Psi \\ \varrho d\!\!\!/ \mathscr{Z}^{\leq N}\Psi \\ \mathscr{Z}^{\leq N}\Psi \end{pmatrix}\right| + \frac{1}{(1+t)^3}\left\|\mathscr{Z}^{\leq \lceil N/2 \rceil} f\right\|_{C^0(\Sigma_t^u)}\left|\begin{pmatrix} \mathscr{Z}^{\leq N+1}(\upmu - 1) \\ \sum_{a=1}^3 \varrho|\mathscr{Z}^{\leq N+1} L_{(Small)}^a| \end{pmatrix}\right|,$$

(12.97)

$$\left|[\Delta\!\!\!/, \mathscr{Z}^N]f\right| \lesssim \frac{1}{(1+t)^2}\left|\mathscr{Z}^{\leq N+1} f\right| + \frac{1}{(1+t)^2}\left\|\mathscr{Z}^{\leq \lceil N/2 \rceil} f\right\|_{C^0(\Sigma_t^u)}\left|\begin{pmatrix} \varrho L \mathscr{Z}^{\leq N}\Psi \\ \breve{R}\mathscr{Z}^{\leq N}\Psi \\ \varrho d\!\!\!/ \mathscr{Z}^{\leq N}\Psi \\ \mathscr{Z}^{\leq N}\Psi \end{pmatrix}\right| + \frac{1}{(1+t)^3}\left\|\mathscr{Z}^{\leq \lceil N/2 \rceil} f\right\|_{C^0(\Sigma_t^u)}\left|\begin{pmatrix} \mathscr{Z}^{\leq N+1}(\upmu - 1) \\ \sum_{a=1}^3 \varrho|\mathscr{Z}^{\leq N+1} L_{(Small)}^a| \end{pmatrix}\right|.$$

PROOF. We first prove (12.95). To this end, let $Z \in \mathscr{Z}$. From (8.26a), we deduce the schematic identity

$$[\nabla\!\!\!/^2, \mathcal{L}\!\!\!/_Z]f = g\!\!\!/^{-1}(\nabla\!\!\!/^{(Z)}\pi\!\!\!/)d\!\!\!/ f. \tag{12.98}$$

Iterating (12.98), using the schematic identity $\mathcal{L}\!\!\!/_Z g\!\!\!/^{-1} = -(g\!\!\!/^{-1})^{2(Z)}\pi\!\!\!/$ (see Lemma 7.5), commuting Lie derivatives with the operator $\nabla\!\!\!/$ in $\nabla\!\!\!/^{(Z)}\pi\!\!\!/$ as in our proof of

(12.94), and using Lemma 8.6, we deduce the schematic identity

(12.99)

$$[\nabla\!\!\!/^2, \mathcal{L}\!\!\!/_{\mathscr{Z}}^N]f = \sum_{M=1}^{N} \sum_{N_1+N_2+\cdots+N_{M+2}=N-M} (\mathcal{L}\!\!\!/_{\mathscr{Z}}^{N_1} g\!\!\!/^{-1}) \cdots (\mathcal{L}\!\!\!/_{\mathscr{Z}}^{N_k} g\!\!\!/^{-1})(\nabla\!\!\!/ \mathcal{L}\!\!\!/_{\mathscr{Z}}^{N_{k+1}} \pi\!\!\!/)(\mathcal{L}\!\!\!/_{\mathscr{Z}}^{N_{k+2}} \pi\!\!\!/) \cdots (\mathcal{L}\!\!\!/_{\mathscr{Z}}^{N_{M+1}} \pi\!\!\!/) d\!\!\!/ \mathcal{L}\!\!\!/_{\mathscr{Z}}^{N_{M+2}} f,$$

where the $\pi\!\!\!/$ are the $S_{t,u}$ projections of deformation tensors of vectorfields in $\mathscr{Z}$. Inequality (12.95) now follows from the identity (12.99) and inequalities (12.60a), (12.60b), (12.46a), and (12.48).

Inequality (12.97) can be proved in a similar fashion with the help of (8.26b) and (12.46b). Note that the right-hand side of (8.26b) depends on one extra derivative of f compared to (8.26a) and hence the same is true for (12.97) compared to (12.95).

To prove (12.96), we first use the decomposition $\hat{\nabla\!\!\!/}^2 f = \nabla\!\!\!/^2 f - \frac{1}{2}(\Delta\!\!\!/ f) g\!\!\!/$ and the Leibniz rule to deduce the following identity:

(12.100)

$$\begin{aligned}[\hat{\nabla\!\!\!/}^2, \mathcal{L}\!\!\!/_{\mathscr{Z}}^N]f &= [\nabla\!\!\!/^2, \mathcal{L}\!\!\!/_{\mathscr{Z}}^N]f - \frac{1}{2}([\Delta\!\!\!/, \mathcal{L}\!\!\!/_{\mathscr{Z}}^N]f) g\!\!\!/ \\ &+ \sum_{\substack{N_1+N_2 \leq N \\ N_1 \leq N-1}} (\Delta\!\!\!/ \mathcal{L}\!\!\!/_{\mathscr{Z}}^{N_1} f) \mathcal{L}\!\!\!/_{\mathscr{Z}}^{N_2} g\!\!\!/ \\ &+ \sum_{\substack{N_1+N_2 \leq N \\ N_1 \leq N-1}} ([\mathcal{L}\!\!\!/_{\mathscr{Z}}^{N_1}, \Delta\!\!\!/] f) \mathcal{L}\!\!\!/_{\mathscr{Z}}^{N_2} g\!\!\!/.\end{aligned}$$

To bound the magnitude first term on the right-hand side of (12.100) by the right-hand side of (12.96), we simply use the estimate (12.95). Similarly, to bound the second term, we use (12.97). To bound the magnitude of the third term on the right-hand side of (12.100) by the right-hand side of (12.96), we use (12.60a), (12.60b), (12.46a), and (12.46c). To bound the magnitude of the last term on the right-hand side of (12.100) by the right-hand side of (12.96), we use (12.60a), (12.60b), (12.97), and the bootstrap assumptions (**BA**Ψ), (**AUX**$\upmu$), and (**AUX**$LR_{(Small)}$). □

COROLLARY 12.41 (**Estimates for vectorfields acting on $\nabla\!\!\!/^2\Psi$**). *Let $0 \leq N \leq 23$ be an integer. Under the small-data and bootstrap assumptions of Sects. 12.1-12.4, if ε is sufficiently small, then the following estimates hold on $\mathcal{M}_{T_{(Bootstrap)},U_0}$:*

(12.101a)

$$\left| \begin{pmatrix} \mathcal{L}\!\!\!/_{\mathscr{Z}}^N \nabla\!\!\!/^2 \Psi \\ \mathscr{Z}^N \Delta\!\!\!/ \Psi \\ \mathcal{L}\!\!\!/_{\mathscr{Z}}^N \hat{\nabla\!\!\!/}^2 \Psi \end{pmatrix} \right| \lesssim \frac{1}{(1+t)^2} \left| \begin{pmatrix} \varrho L \mathscr{Z}^{\leq N+1} \Psi \\ \breve{R} \mathscr{Z}^{\leq N+1} \Psi \\ \varrho d\!\!\!/ \mathscr{Z}^{\leq N+1} \Psi \\ \mathscr{Z}^{\leq N+1} \Psi \end{pmatrix} \right| + \varepsilon \frac{1}{(1+t)^4} \left| \begin{pmatrix} \mathscr{Z}^{\leq N+1}(\upmu - 1) \\ \sum_{a=1}^3 \varrho |\mathscr{Z}^{\leq N+1} L_{(Small)}^a| \end{pmatrix} \right|,$$

(12.101b)

$$\left\| \begin{pmatrix} \mathcal{L}\!\!\!/_{\mathscr{Z}}^{\leq 11} \nabla\!\!\!/^2 \Psi \\ \mathscr{Z}^{\leq 11} \Delta\!\!\!/ \Psi \\ \mathcal{L}\!\!\!/_{\mathscr{Z}}^{\leq 11} \hat{\nabla\!\!\!/}^2 \Psi \end{pmatrix} \right\|_{C^0(\Sigma_t^u)} \lesssim \varepsilon \frac{1}{(1+t)^3}.$$

PROOF. The estimates (12.101a) and (12.101b) follow easily from Lemma 12.40 with Ψ in the role of f and the bootstrap assumptions (**BA**Ψ). $\square$

12.25. Commutator estimates involving the trace and trace-free parts

In this section, we derive pointwise commutator estimates involving the trace and trace-free parts of symmetric type $\binom{0}{2}$ $S_{t,u}$ tensorfields. Some of the estimates involve the trace-free projected Lie derivative operators $\hat{\not{\mathcal{L}}}_V$ of Def. 8.2.

REMARK 12.42 (**Clarification of the meaning of $\mathscr{Z}^{N-1}$**). It is understood that on the *left-hand* sides of the inequalities of Lemma 12.43, the string of vectorfields $\mathscr{Z}^N$ is the same each time it occurs. For example, in the expression $\left|\not{\mathcal{L}}_{\mathscr{Z}}^N \hat{\xi} - \hat{\not{\mathcal{L}}}_{\mathscr{Z}}^N \hat{\xi}\right|$ from the left-hand side of (12.103), the operators $\not{\mathcal{L}}_{\mathscr{Z}}^N$ and $\hat{\not{\mathcal{L}}}_{\mathscr{Z}}^N$ correspond to the same N^{th} order string of commutation vectorfields. Furthermore, this remark applies to many other inequalities proved in the remainder of the monograph.

LEMMA 12.43 (**Commutator estimates involving the trace and trace-free parts of symmetric type $\binom{0}{2}$ $S_{t,u}$ tensorfields**). *Let $0 \leq N \leq 24$ be an integer and let ξ be a symmetric type $\binom{0}{2}$ $S_{t,u}$ tensorfield. Under the small-data and bootstrap assumptions of Sects. 12.1-12.4, if ε is sufficiently small, then the following pointwise estimates hold on $\mathcal{M}_{T_{(Bootstrap)},U_0}$:*

$$\left|\not{\mathcal{L}}_{\mathscr{Z}}^N \hat{\xi}\right| \lesssim \left|\not{\mathcal{L}}_{\mathscr{Z}}^{\leq N} \xi\right| + \left\|\not{\mathcal{L}}_{\mathscr{Z}}^{\leq \lfloor N/2 \rfloor} \xi\right\|_{C^0(\Sigma_t^u)} \left|\begin{pmatrix} \varrho L \mathscr{Z}^{\leq N-1}\Psi \\ \breve{R} \mathscr{Z}^{\leq N-1}\Psi \\ \varrho \not{d} \mathscr{Z}^{\leq N-1}\Psi \\ \mathscr{Z}^{\leq N-1}\Psi \end{pmatrix}\right| + \frac{1}{1+t}\left\|\not{\mathcal{L}}_{\mathscr{Z}}^{\leq \lfloor N/2 \rfloor} \xi\right\|_{C^0(\Sigma_t^u)} \left|\begin{pmatrix} \mathscr{Z}^{\leq N}(\upmu - 1) \\ \sum_{a=1}^3 \varrho \left|\mathscr{Z}^{\leq N} L_{(Small)}^a\right| \end{pmatrix}\right|. \tag{12.102}$$

Furthermore, if $1 \leq N \leq 24$ is an integer, then the following pointwise estimates hold on $\mathcal{M}_{T_{(Bootstrap)},U_0}$:

(12.103)

$$\left|\not{\mathcal{L}}_{\mathscr{Z}}^N \hat{\xi} - \hat{\not{\mathcal{L}}}_{\mathscr{Z}}^N \hat{\xi}\right| \lesssim \left|\not{\mathcal{L}}_{\mathscr{Z}}^{\leq N-1} \xi\right| + \left\|\not{\mathcal{L}}_{\mathscr{Z}}^{\leq \lfloor N/2 \rfloor} \xi\right\|_{C^0(\Sigma_t^u)} \left|\begin{pmatrix} \varrho L \mathscr{Z}^{\leq N-1}\Psi \\ \breve{R} \mathscr{Z}^{\leq N-1}\Psi \\ \varrho \not{d} \mathscr{Z}^{\leq N-1}\Psi \\ \mathscr{Z}^{\leq N-1}\Psi \end{pmatrix}\right| + \frac{1}{1+t}\left\|\not{\mathcal{L}}_{\mathscr{Z}}^{\leq \lfloor N/2 \rfloor} \xi\right\|_{C^0(\Sigma_t^u)} \left|\begin{pmatrix} \mathscr{Z}^{\leq N}(\upmu - 1) \\ \sum_{a=1}^3 \varrho \left|\mathscr{Z}^{\leq N} L_{(Small)}^a\right| \end{pmatrix}\right|,$$

(12.104)

$$\left|\mathrm{tr}_{\not{g}}\not{\mathcal{L}}_{\mathscr{Z}}^{N}\xi - \mathscr{Z}^{N}\mathrm{tr}_{\not{g}}\xi\right| \lesssim \left|\not{\mathcal{L}}_{\mathscr{Z}}^{\leq N-1}\xi\right| + \left\|\not{\mathcal{L}}_{\mathscr{Z}}^{\leq \lfloor N/2\rfloor}\xi\right\|_{C^0(\Sigma_t^u)}\left|\begin{pmatrix} \varrho L\mathscr{Z}^{\leq N-1}\Psi \\ \breve{R}\mathscr{Z}^{\leq N-1}\Psi \\ \varrho\not{d}\mathscr{Z}^{\leq N-1}\Psi \\ \mathscr{Z}^{\leq N-1}\Psi \end{pmatrix}\right|$$
$$+ \frac{1}{1+t}\left\|\not{\mathcal{L}}_{\mathscr{Z}}^{\leq \lfloor N/2\rfloor}\xi\right\|_{C^0(\Sigma_t^u)}\left|\begin{pmatrix} \mathscr{Z}^{\leq N}(\upmu-1) \\ \sum_{a=1}^{3}\varrho\left|\mathscr{Z}^{\leq N}L_{(Small)}^{a}\right| \end{pmatrix}\right|,$$

(12.105)

$$\left|\not{\mathcal{L}}_{\mathscr{Z}}^{N}\hat{\xi} - \widehat{(\not{\mathcal{L}}_{\mathscr{Z}}^{N}\xi)}\right| \lesssim \left|\not{\mathcal{L}}_{\mathscr{Z}}^{\leq N-1}\xi\right| + \left\|\not{\mathcal{L}}_{\mathscr{Z}}^{\leq \lfloor N/2\rfloor}\xi\right\|_{C^0(\Sigma_t^u)}\left|\begin{pmatrix} \varrho L\mathscr{Z}^{\leq N-1}\Psi \\ \breve{R}\mathscr{Z}^{\leq N-1}\Psi \\ \varrho\not{d}\mathscr{Z}^{\leq N-1}\Psi \\ \mathscr{Z}^{\leq N-1}\Psi \end{pmatrix}\right|$$
$$+ \frac{1}{1+t}\left\|\not{\mathcal{L}}_{\mathscr{Z}}^{\leq \lfloor N/2\rfloor}\xi\right\|_{C^0(\Sigma_t^u)}\left|\begin{pmatrix} \mathscr{Z}^{\leq N}(\upmu-1) \\ \sum_{a=1}^{3}\varrho\left|\mathscr{Z}^{\leq N}L_{(Small)}^{a}\right| \end{pmatrix}\right|,$$

(12.106)

$$\left|\not{\mathcal{L}}_{\mathscr{Z}}^{N}\xi - \left\{\not{\mathcal{L}}_{\mathscr{Z}}^{N}\hat{\xi} + \frac{1}{2}(\mathscr{Z}^{N}\mathrm{tr}_{\not{g}}\xi)\not{g}\right\}\right|$$
$$\lesssim \left|\not{\mathcal{L}}_{\mathscr{Z}}^{\leq N-1}\xi\right| + \left\|\not{\mathcal{L}}_{\mathscr{Z}}^{\leq \lfloor N/2\rfloor}\xi\right\|_{C^0(\Sigma_t^u)}\left|\begin{pmatrix} \varrho L\mathscr{Z}^{\leq N-1}\Psi \\ \breve{R}\mathscr{Z}^{\leq N-1}\Psi \\ \varrho\not{d}\mathscr{Z}^{\leq N-1}\Psi \\ \mathscr{Z}^{\leq N-1}\Psi \end{pmatrix}\right|$$
$$+ \frac{1}{1+t}\left\|\not{\mathcal{L}}_{\mathscr{Z}}^{\leq \lfloor N/2\rfloor}\xi\right\|_{C^0(\Sigma_t^u)}\left|\begin{pmatrix} \mathscr{Z}^{\leq N}(\upmu-1) \\ \sum_{a=1}^{3}\varrho\left|\mathscr{Z}^{\leq N}L_{(Small)}^{a}\right| \end{pmatrix}\right|.$$

Furthermore, if $0 \leq N \leq 23$ is an integer, then the following estimates hold on $\mathcal{M}_{T_{(Bootstrap)},U_0}$:

(12.107)

$$\left|\not\nabla\not{\mathcal{L}}_{\mathscr{Z}}^{N}\xi - \left\{\not\nabla\not{\mathcal{L}}_{\mathscr{Z}}^{N}\hat{\xi} + \frac{1}{2}(\not{d}\mathscr{Z}^{N}\mathrm{tr}_{\not{g}}\xi)\not{g}\right\}\right|$$
$$\lesssim \frac{1}{1+t}\left|\not{\mathcal{L}}_{\mathscr{Z}}^{\leq N}\xi\right| + \frac{1}{1+t}\left\|\not{\mathcal{L}}_{\mathscr{Z}}^{\leq \lfloor N/2\rfloor}\xi\right\|_{C^0(\Sigma_t^u)}\left|\begin{pmatrix} \varrho L\mathscr{Z}^{\leq N}\Psi \\ \breve{R}\mathscr{Z}^{\leq N}\Psi \\ \varrho\not{d}\mathscr{Z}^{\leq N}\Psi \\ \mathscr{Z}^{\leq N}\Psi \end{pmatrix}\right|$$
$$+ \frac{1}{(1+t)^2}\left\|\not{\mathcal{L}}_{\mathscr{Z}}^{\leq \lfloor N/2\rfloor}\xi\right\|_{C^0(\Sigma_t^u)}\left|\begin{pmatrix} \mathscr{Z}^{\leq N+1}(\upmu-1) \\ \varrho\sum_{a=1}^{3}\left|\mathscr{Z}^{\leq N+1}L_{(Small)}^{a}\right| \end{pmatrix}\right|,$$

(12.108)

$$\left|\nabla\!\!\!/ \mathcal{L}\!\!\!/_{\mathscr{Z}}^N \hat{\xi} - \nabla\!\!\!/ \hat{\mathcal{L}\!\!\!/}_{\mathscr{Z}}^N \hat{\xi}\right| \lesssim \frac{1}{1+t}\left|\mathcal{L}\!\!\!/_{\mathscr{Z}}^{\leq N}\xi\right| + \frac{1}{1+t}\left\|\mathcal{L}\!\!\!/_{\mathscr{Z}}^{\leq \lfloor N/2 \rfloor}\xi\right\|_{C^0(\Sigma_t^u)}\left|\begin{pmatrix} \varrho L \mathscr{Z}^{\leq N}\Psi \\ \breve{R}\mathscr{Z}^{\leq N}\Psi \\ \varrho d\!\!\!/ \mathscr{Z}^{\leq N}\Psi \\ \mathscr{Z}^{\leq N}\Psi \end{pmatrix}\right| + \frac{1}{(1+t)^2}\left\|\mathcal{L}\!\!\!/_{\mathscr{Z}}^{\leq \lfloor N/2 \rfloor}\xi\right\|_{C^0(\Sigma_t^u)}\left|\begin{pmatrix} \mathscr{Z}^{\leq N+1}(\upmu - 1) \\ \varrho\sum_{a=1}^3 |\mathscr{Z}^{\leq N+1}L^a_{(Small)}| \end{pmatrix}\right|,$$

(12.109)

$$\left|\mathcal{L}\!\!\!/_{\mathscr{Z}}^N \mathrm{div}\!\!\!/\, \xi - \mathrm{div}\!\!\!/\, \mathcal{L}\!\!\!/_{\mathscr{Z}}^N \xi\right| \lesssim \frac{1}{1+t}\left|\mathcal{L}\!\!\!/_{\mathscr{Z}}^{\leq N}\xi\right| + \frac{1}{1+t}\left\|\mathcal{L}\!\!\!/_{\mathscr{Z}}^{\leq \lfloor N/2 \rfloor}\xi\right\|_{C^0(\Sigma_t^u)}\left|\begin{pmatrix} \varrho L \mathscr{Z}^{\leq N}\Psi \\ \breve{R}\mathscr{Z}^{\leq N}\Psi \\ \varrho d\!\!\!/ \mathscr{Z}^{\leq N}\Psi \\ \mathscr{Z}^{\leq N}\Psi \end{pmatrix}\right| + \frac{1}{(1+t)^2}\left\|\mathcal{L}\!\!\!/_{\mathscr{Z}}^{\leq \lfloor N/2 \rfloor}\xi\right\|_{C^0(\Sigma_t^u)}\left|\begin{pmatrix} \mathscr{Z}^{\leq N+1}(\upmu - 1) \\ \varrho\sum_{a=1}^3 |\mathscr{Z}^{\leq N+1}L^a_{(Small)}| \end{pmatrix}\right|,$$

(12.110)

$$\left|\mathcal{L}\!\!\!/_{\mathscr{Z}}^N \mathrm{div}\!\!\!/\, \hat{\xi} - \mathrm{div}\!\!\!/\, \hat{\mathcal{L}\!\!\!/}_{\mathscr{Z}}^N \hat{\xi}\right| \lesssim \frac{1}{1+t}\left|\mathcal{L}\!\!\!/_{\mathscr{Z}}^{\leq N}\xi\right| + \frac{1}{1+t}\left\|\mathcal{L}\!\!\!/_{\mathscr{Z}}^{\leq \lfloor N/2 \rfloor}\xi\right\|_{C^0(\Sigma_t^u)}\left|\begin{pmatrix} \varrho L \mathscr{Z}^{\leq N}\Psi \\ \breve{R}\mathscr{Z}^{\leq N}\Psi \\ \varrho d\!\!\!/ \mathscr{Z}^{\leq N}\Psi \\ \mathscr{Z}^{\leq N}\Psi \end{pmatrix}\right| + \frac{1}{(1+t)^2}\left\|\mathcal{L}\!\!\!/_{\mathscr{Z}}^{\leq \lfloor N/2 \rfloor}\xi\right\|_{C^0(\Sigma_t^u)}\left|\begin{pmatrix} \mathscr{Z}^{\leq N+1}(\upmu - 1) \\ \varrho\sum_{a=1}^3 |\mathscr{Z}^{\leq N+1}L^a_{(Small)}| \end{pmatrix}\right|.$$

PROOF. To prove (12.102), we first decompose $\hat{\xi}_{AB} = \xi_{AB} - \frac{1}{2}(g\!\!\!/^{-1})^{CD}\xi_{CD} g\!\!\!/_{AB}$ (see (3.56b)). Applying $\mathcal{L}\!\!\!/_{\mathscr{Z}}^N$ to both sides of this identity and applying the Leibniz rule to the product on the right-hand side, we deduce that

$$\left|\mathcal{L}\!\!\!/_{\mathscr{Z}}^N \hat{\xi}\right| \lesssim \left|\mathcal{L}\!\!\!/_{\mathscr{Z}}^N \xi\right| + \sum_{N_1+N_2+N_3=N} \left|\mathcal{L}\!\!\!/_{\mathscr{Z}}^{N_1}\xi\right| \left|\mathcal{L}\!\!\!/_{\mathscr{Z}}^{N_2} g\!\!\!/\right| \left|\mathcal{L}\!\!\!/_{\mathscr{Z}}^{N_3} g\!\!\!/^{-1}\right|. \tag{12.111}$$

The desired estimate (12.102) now follows from (12.111) and the estimates (12.60a) and (12.60b).

To prove (12.104), we first deduce from the definition of the operator $\mathrm{tr}_{g\!\!\!/}$ (see (3.56a)) and the Leibniz rule that

$$\left|\mathrm{tr}_{g\!\!\!/}\mathcal{L}\!\!\!/_{\mathscr{Z}}^N \xi - \mathscr{Z}^N \mathrm{tr}_{g\!\!\!/}\xi\right| \lesssim \sum_{\substack{N_1+N_2=N \\ N_2 \leq N-1}} \left|\mathcal{L}\!\!\!/_{\mathscr{Z}}^{N_1} g\!\!\!/^{-1}\right| \left|\mathcal{L}\!\!\!/_{\mathscr{Z}}^{N_2}\xi\right|. \tag{12.112}$$

The desired estimate (12.104) now follows from (12.112), (12.60a) and (12.60b).

To prove (12.105), we first decompose $\hat{\xi}_{AB} = \xi_{AB} - \frac{1}{2}(g\!\!\!/^{-1})^{CD}\xi_{CD} g\!\!\!/_{AB}$. Applying $\mathcal{L}\!\!\!/_{\mathscr{Z}}^N$ to both sides of this identity and applying the Leibniz rule to the product on the right-hand side, we deduce that

$$\left|\mathcal{L}\!\!\!/_{\mathscr{Z}}^N \hat{\xi} - \left\{\mathcal{L}\!\!\!/_{\mathscr{Z}}^N \xi - \frac{1}{2}(\mathrm{tr}_{g\!\!\!/}\mathcal{L}\!\!\!/_{\mathscr{Z}}^N \xi) g\!\!\!/\right\}\right| \lesssim \sum_{\substack{N_1+N_2+N_3=N \\ N_1 \leq N-1}} \left|\mathcal{L}\!\!\!/_{\mathscr{Z}}^{N_1}\xi\right| \left|\mathcal{L}\!\!\!/_{\mathscr{Z}}^{N_2} g\!\!\!/\right| \left|\mathcal{L}\!\!\!/_{\mathscr{Z}}^{N_3} g\!\!\!/^{-1}\right|. \tag{12.113}$$

We now conclude the desired estimate (12.105) by noting that the term in braces on the left-hand side of (12.113) is precisely $\widehat{(\not{\mathcal{L}}_{\mathscr{Z}}^N \xi)}$ and by using the estimates (12.60a) and (12.60b) to bound the magnitude of the right-hand side of (12.113) by the right-hand side of (12.105).

To prove (12.103), we let $Z \in \mathscr{Z}$. From (8.9b) and the definition (3.56b) of the trace-free part of a tensor we deduce the schematic identity

$$\mathcal{L}_Z \hat{\xi} - \hat{\not{\mathcal{L}}}_Z \hat{\xi} = \left\{ (\not{g}^{-1})^{2(Z)}\not{\pi}\xi + (\not{g}^{-1})^{(Z)}\not{\pi}\not{g}^{-1}\xi \right\} \not{g}. \tag{12.114}$$

Iterating (12.114), using the schematic identities $\not{\mathcal{L}}_Z \not{g} = {}^{(Z)}\not{\pi}$ and $\not{\mathcal{L}}_Z \not{g}^{-1} = -(\not{g}^{-1})^{2(Z)}\not{\pi}$ (see Lemma 7.5), and using (8.2) to express the right-hand side of (12.115) in terms of the Lie derivatives of ξ, $\not{g}$ and $\not{g}^{-1}$, we deduce the following schematic identity, in which not all factors on the right-hand side appear in the "correct order" (the order is irrelevant from the point of view of our estimates):

(12.115)

$$\mathcal{L}_{\mathscr{Z}}^N \hat{\xi} - \hat{\not{\mathcal{L}}}_{\mathscr{Z}}^N \hat{\xi} = \sum_{M=1}^{N} \sum_{N_1+N_2+\cdots+N_k=N-M} (\not{\mathcal{L}}_{\mathscr{Z}}^{N_1} \not{g}^{-1}) \cdots (\not{\mathcal{L}}_{\mathscr{Z}}^{N_a} \not{g}^{-1})(\not{\mathcal{L}}_{\mathscr{Z}}^{N_{a+1}} \not{g}) \cdots (\not{\mathcal{L}}_{\mathscr{Z}}^{N_{k-2}} \not{g})(\not{\mathcal{L}}_{\mathscr{Z}}^{N_{k-1}} \not{\pi})(\not{\mathcal{L}}_{\mathscr{Z}}^{N_k} \xi).$$

Above, the $\not{\pi}$ are the $S_{t,u}$ projections of deformation tensors of vectorfields in $\mathscr{Z}$. We now conclude the desired estimate (12.103) by using the estimates (12.60a) and (12.60b) to bound the magnitude of the right-hand sides of (12.115) by the right-hand side of (12.103).

To prove (12.108), we first use the schematic identity $\not{\mathcal{L}}_Z \not{g}^{-1} = -(\not{g}^{-1})^2 \not{\mathcal{L}}_Z \not{g}$ (see Lemma 7.5) to replace all $S_{t,u}$-projected Lie derivatives of $\not{g}^{-1}$ on the right-hand side of (12.115) with products consisting of powers of $\not{g}^{-1}$ and $S_{t,u}$-projected Lie derivatives of $\not{g}$. We then apply $\not{\nabla}$ to both sides of the resulting identity. Finally, we use inequality (12.48) and the same reasoning used just after equation (12.115) to bound the magnitude of the right-hand side of the resulting equation by the right-hand side of (12.108) as desired.

To prove (12.106), we first decompose $\xi_{AB} = \hat{\xi}_{AB} + \frac{1}{2}(\not{g}^{-1})^{CD}\xi_{CD}\not{g}_{AB}$. Applying $\not{\mathcal{L}}_{\mathscr{Z}}^N$ to this identity and using the Leibniz rule, we deduce that

(12.116)

$$\left| \not{\mathcal{L}}_{\mathscr{Z}}^N \xi - \left\{ \not{\mathcal{L}}_{\mathscr{Z}}^N \hat{\xi} + \frac{1}{2}(\mathscr{Z}^N \mathrm{tr}_{\not{g}} \xi)\not{g} \right\} \right| \lesssim \sum_{\substack{N_1+N_2+N_3 \leq N \\ N_2 \leq N-1}} \left| \not{\mathcal{L}}_{\mathscr{Z}}^{N_1} \not{g}^{-1} \right| \left| \not{\mathcal{L}}_{\mathscr{Z}}^{N_2} \xi \right| \left| \not{\mathcal{L}}_{\mathscr{Z}}^{N_3} \not{g} \right|.$$

We then use the estimates (12.60a) and (12.60b) to bound the magnitude of the right-hand side of (12.116) by the right-hand side of (12.106) as desired. The desired estimate (12.106) now follows from using the already proven estimate (12.103) to replace the operator $\not{\mathcal{L}}_{\mathscr{Z}}^N \hat{\xi}$ on the left-hand side of (12.116) with $\hat{\not{\mathcal{L}}}_{\mathscr{Z}}^N \hat{\xi}$ up to error terms that are bounded in magnitude by the right-hand side of (12.106).

To prove (12.107), we first decompose $\xi_{AB} = \hat{\xi}_{AB} + \frac{1}{2}(\not{g}^{-1})^{CD}\xi_{CD}\not{g}_{AB}$. Applying $\not{\nabla}\not{\mathcal{L}}_{\mathscr{Z}}^N$ to this identity and using the Leibniz rule, we deduce that

$$\left|\not{\nabla}\not{\mathcal{L}}_{\mathscr{Z}}^N\xi - \left\{\not{\nabla}\not{\mathcal{L}}_{\mathscr{Z}}^N\hat{\xi} + \frac{1}{2}(\not{d}\mathscr{Z}^N\mathrm{tr}_{\not{g}}\xi)\not{g}\right\}\right| \lesssim \sum_{\substack{N_1+N_2+N_3\leq N \\ N_2\leq N-1}} \sum_{i_1+i_2+i_3=1} \left|\not{\nabla}^{i_1}\not{\mathcal{L}}_{\mathscr{Z}}^{N_1}\not{g}^{-1}\right| \left|\not{\nabla}^{i_2}\not{\mathcal{L}}_{\mathscr{Z}}^{N_2}\xi\right| \left|\not{\nabla}^{i_3}\not{\mathcal{L}}_{\mathscr{Z}}^{N_3}\not{g}\right|. \tag{12.117}$$

Next, we use the schematic identity $\not{\mathcal{L}}_Z\not{g}^{-1} = -(\not{g}^{-1})^2\not{\mathcal{L}}_Z\not{g}$ mentioned above to replace the factors $\not{\mathcal{L}}_{\mathscr{Z}}^{N_1}\not{g}^{-1}$ on the right-hand side of (12.117) with products consisting of powers of $\not{g}^{-1}$ and $S_{t,u}$-projected Lie derivatives of $\not{g}$. Also using inequality (12.48) and the estimates (12.60a) and (12.60b), we bound the magnitude of the right-hand side of (12.117) by the right-hand side of (12.107) as desired. Finally, we use the already proven estimate (12.108) to replace the term $\not{\nabla}\not{\mathcal{L}}_{\mathscr{Z}}^N\hat{\xi}$ on the left-hand side of (12.117) with the term $\not{\nabla}\hat{\not{\mathcal{L}}}_{\mathscr{Z}}^N\hat{\xi}$ up to error terms that are bounded by in magnitude by the right-hand side of (12.107).

To prove (12.110), we view the quantity inside the norm on the left-hand side of (12.108) as the difference of the two type $\binom{0}{3}$ $S_{t,u}$ tensors. We consider the ABC components of these tensors, where the A index corresponds to $\not{\nabla}$ in each tensor. Clearly, the AB trace of this difference of tensors is an $S_{t,u}$ one-form that is bounded in magnitude by $\lesssim$ the right-hand side of (12.108). Furthermore, the AB trace of the second tensor on the left-hand side of (12.108) is precisely $\mathrm{d\acute{\imath}v}\hat{\not{\mathcal{L}}}_{\mathscr{Z}}^N\hat{\xi}$. Hence, the desired estimate (12.110) will follow once we show that the difference between the AB trace of the first tensor on the left-hand side of (12.108) and $\not{\mathcal{L}}_{\mathscr{Z}}^N\mathrm{d\acute{\imath}v}\hat{\xi}$ is in magnitude $\lesssim$ the right-hand side of (12.110). To this end, we use the Leibniz rule to commute the operators $\not{\nabla}$ and $\not{g}^{-1}\not{\mathcal{L}}_{\mathscr{Z}}^{N_2}$, thereby deducing that the difference under consideration can be bounded as follows:

$$\left|\not{\mathcal{L}}_{\mathscr{Z}}^N\left\{(\not{g}^{-1})^{AB}\not{\nabla}_A\hat{\xi}_{B\cdot})\right\} - (\not{g}^{-1})^{AB}\not{\mathcal{L}}_{\mathscr{Z}}^N\not{\nabla}_A\hat{\xi}_{B\cdot}\right| \lesssim \sum_{\substack{N_1+N_2\leq N \\ N_2\leq N-1}} \left|\not{\mathcal{L}}_{\mathscr{Z}}^{N_1}\not{g}^{-1}\right|\left|\not{\nabla}\not{\mathcal{L}}_{\mathscr{Z}}^{N_2}\hat{\xi}\right| + \sum_{\substack{N_1+N_2\leq N \\ N_2\leq N-1}} \left|\not{\mathcal{L}}_{\mathscr{Z}}^{N_1}\not{g}^{-1}\right|\left|[\not{\nabla},\not{\mathcal{L}}_{\mathscr{Z}}^{N_2}]\hat{\xi}\right|, \tag{12.118}$$

where the $\cdot$'s in (12.118) signify that the contracted quantities on the left-hand side are $S_{t,u}$ one-forms. To bound the first sum on the right-hand side of (12.118) by the right-hand side of (12.110), we use the estimates (12.60a), (12.60b), (12.48), and (12.102), and the bootstrap assumptions (**BA**$\boldsymbol{\Psi}$), (**AUX**$\boldsymbol{\upmu}$), and (**AUX**$\boldsymbol{LR_{(Small)}}$). To bound the second sum on the right-hand side of (12.118) by the right-hand side of (12.110), we use the estimates (12.60a), (12.60b), (12.92), and (12.102), and the bootstrap assumptions (**BA**$\boldsymbol{\Psi}$), (**AUX**$\boldsymbol{\upmu}$), and (**AUX**$\boldsymbol{LR_{(Small)}}$). We have thus proved inequality (12.110). Inequality (12.109) can be proved by using essentially the same argument, and we omit the details. □

12.26. Pointwise estimates, in terms of other quantities, for the Lie derivatives of the re-centered null second fundamental form involving an outgoing null differentiation

In this section, we derive pointwise estimates for $\not{\mathcal{L}}_L \not{\mathcal{L}}_{\mathscr{Z}}^N \chi_{(Small)}$ and for related quantities. We start with the following lemma, which provides an expression for $\not{\mathcal{L}}_L \chi^{(Small)}$ in terms of Ψ and L^i, $(i = 1, 2, 3)$.

LEMMA 12.44 (**Expressions for $\not{\mathcal{L}}_L \chi_{(Small)}$ in terms of other quantities**). *The symmetric type $\binom{0}{2}$ $S_{t,u}$ tensorfield $\chi^{(Small)}$ defined in* (4.1c) *verifies the following transport equation:*

(12.119)

$$\not{\mathcal{L}}_L \chi_{AB}^{(Small)} = \varrho^{-1} g_{ab} (\not{d}_A x^a) \not{d}_B L(\varrho L_{(Small)}^b) + \varrho^{-2} g_{ab} (\not{d}_A (\varrho L_{(Small)}^a)) \not{d}_B (\varrho L_{(Small)}^b)$$
$$+ \varrho^{-1} G_{ab} (\not{d}_A x^a)(\not{d}_B (\varrho L_{(Small)}^b)) L\Psi - \not{\mathcal{L}}_L \Lambda_{AB}^{(Tan-\Psi)},$$

where $\varrho = 1 - u + t$ is the geometric radial variable and $\Lambda_{AB}^{(Tan-\Psi)}$ is the type $\binom{0}{2}$ $S_{t,u}$ tensorfield given by (4.11c).

PROOF. We first use the decomposition $L^a = \varrho^{-1} x^a + L^a_{(Small)}$ (that is, (4.1a)), the identity $Lx^a = L^a$, and Lemma 8.6 to deduce that

$$\not{\mathcal{L}}_L \not{d}_A x^a = \not{d}_A L^a = \varrho^{-1} \not{d}_A x^a + \not{d}_A L^a_{(Small)}. \tag{12.120}$$

We now apply $\not{\mathcal{L}}_L$ to both sides of equation (4.19a) and treat all uppercase Latin indices as tensorial $S_{t,u}$ indices, and all lowercase Latin-indexed quantities as scalar-valued functions. Also using (12.120), the identity $L\varrho = 1$, and the chain rule identity $Lg_{ab} = G_{ab} L\Psi$, we deduce that

$$\not{\mathcal{L}}_L \chi_{AB}^{(Small)} = \varrho^{-1} g_{ab} (\not{d}_A x^a) \not{d}_B L(\varrho L_{(Small)}^b) + g_{ab} (\not{d}_A L_{(Small)}^a) \not{d}_B L_{(Small)}^b$$
$$+ G_{ab} (\not{d}_A x^a)(\not{d}_B L_{(Small)}^b) L\Psi - \not{\mathcal{L}}_L \Lambda_{AB}^{(Tan-\Psi)}. \tag{12.121}$$

The desired identity (12.119) now easily follows from (12.121) and the fact that $\not{d}\varrho = 0$. ☐

LEMMA 12.45 (**Pointwise estimates for $\not{\mathcal{L}}_L \not{\mathcal{L}}_{\mathscr{Z}}^N \chi_{(Small)}$ in terms of other variables**). *Let $\chi^{(Small)}$ be the type $\binom{0}{2}$ $S_{t,u}$ tensorfield defined in* (4.1c). *Let $0 \leq N \leq 23$ be an integer and recall that $\varrho = 1 - u + t$. Under the small-data and bootstrap assumptions of Sects.* 12.1-12.4, *if ε is sufficiently small, then the following estimates hold on $\mathcal{M}_{T_{(Bootstrap)}, U_0}$:*

(12.122)

$$\left| \begin{pmatrix} \varrho^2 \not{\mathcal{L}}_L \not{\mathcal{L}}_{\mathscr{Z}}^N \chi^{(Small)} \\ \not{\mathcal{L}}_L (\varrho^2 \not{\mathcal{L}}_{\mathscr{Z}}^N \chi^{(Small)\#}) \\ L(\varrho^2 \mathscr{Z}^N \mathrm{tr}_{\not{g}} \chi^{(Small)}) \\ \varrho^2 \not{\mathcal{L}}_L \not{\mathcal{L}}_{\mathscr{Z}}^N \hat{\chi}^{(Small)} \\ \not{\mathcal{L}}_L (\varrho^2 \not{\mathcal{L}}_{\mathscr{Z}}^N \hat{\chi}^{(Small)\#}) \end{pmatrix} \right| \lesssim \left| \begin{pmatrix} \varrho L \mathscr{Z}^{\leq N+1} \Psi \\ \breve{R} \mathscr{Z}^{\leq N+1} \Psi \\ \varrho \not{d} \mathscr{Z}^{\leq N+1} \Psi \\ \mathscr{Z}^{\leq N+1} \Psi \end{pmatrix} \right| + \sum_{a=1}^{3} \left| L(\varrho \mathscr{Z}^{\leq N+1} L^a_{(Small)}) \right|$$
$$+ \frac{\ln(e+t)}{(1+t)^2} \left| \begin{pmatrix} \mathscr{Z}^{\leq N+1} (\upmu - 1) \\ \sum_{a=1}^{3} \varrho |\mathscr{Z}^{\leq N+1} L^a_{(Small)}| \end{pmatrix} \right|.$$

PROOF. We first prove the estimate (12.122) for $\varrho^2 \not{\mathcal{L}}_L \not{\mathcal{L}}_{\mathscr{Z}}^N \chi^{(Small)}$. As a first step, we bound $\varrho^2 \not{\mathcal{L}}_{\mathscr{Z}}^N \not{\mathcal{L}}_L \chi^{(Small)}$ by the right-hand side of (12.122). To this end, we write equation (12.119) in the following schematic form (recall that $\varrho L \in \mathscr{Z}$):

$$\begin{aligned} (12.123) \quad \not{\mathcal{L}}_L \chi^{(Small)} &= \frac{1}{\varrho} f(\Psi)(\not{d}x)\not{d}(L(\varrho L_{(Small)})) + \frac{1}{\varrho^2} f(\Psi)(\not{d}(\varrho L_{(Small)}))^2 \\ &+ \frac{1}{\varrho} f(\Psi)(\not{d}x)(\not{d}(\varrho L_{(Small)}))L\Psi \\ &+ \frac{1}{\varrho}(\not{\mathcal{L}}_{\varrho L} G_{(Frame)}) \begin{pmatrix} L\Psi \\ \not{d}\Psi \end{pmatrix} + \frac{1}{\varrho} G_{(Frame)} \not{\mathcal{L}}_{\varrho L} \begin{pmatrix} L\Psi \\ \not{d}\Psi \end{pmatrix}, \end{aligned}$$

where the f are smooth scalar-valued functions of Ψ. We then apply the operator $\not{\mathcal{L}}_{\mathscr{Z}}^N$ to the right-hand side of (12.123) and apply the Leibniz rule to the products on the right-hand side. We bound the terms $\mathscr{Z}^M \varrho$ with (12.6b). We bound the terms $\mathscr{Z}^M f(\Psi)$ with the bootstrap assumptions (**BA**Ψ). The factors of $\not{\mathcal{L}}_{\mathscr{Z}}^M \not{d}(L(\varrho L_{(Small)}))$ are the only ones that contribute to the second term on the right-hand side of (12.122). To bound them, we first use Lemma 8.6 and inequality (12.46a) to deduce that $\left|\not{\mathcal{L}}_{\mathscr{Z}}^M \not{d}(L(\varrho L_{(Small)}))\right| \lesssim \frac{1}{1+t}\left|\mathscr{Z}^{\leq M+1} L(\varrho L_{(Small)})\right|$. We then use the commutator estimate (12.81c) with $f = \varrho L_{(Small)}$, the fact that $L\varrho = 1$, (12.6b), and the bootstrap assumptions (**AUX**$LR_{(Small)}$) to bound the right-hand side of the previous inequality by

$$(12.124) \quad \begin{aligned} &\lesssim \frac{1}{1+t}\left|L(\varrho \mathscr{Z}^{\leq M+1} L_{(Small)})\right| \\ &+ \frac{\ln(e+t)}{(1+t)^2}\left|\begin{pmatrix} \varrho L \mathscr{Z}^{\leq M}\Psi \\ \breve{R}\mathscr{Z}^{\leq M}\Psi \\ \varrho\not{d}\mathscr{Z}^{\leq M}\Psi \\ \mathscr{Z}^{\leq M}\Psi \end{pmatrix}\right| + \frac{\ln(e+t)}{(1+t)^3}\left|\begin{pmatrix} \mathscr{Z}^{\leq M+1}(\upmu - 1) \\ \sum_{a=1}^3 \varrho|\mathscr{Z}^{\leq M+1} L^a_{(Small)}| \end{pmatrix}\right|. \end{aligned}$$

In particular, using (12.124), (12.6b), and the bootstrap assumptions (**BA**Ψ), (**AUX**$\upmu$), and (**AUX**$LR_{(Small)}$), we deduce that when $M \leq 10$, we have $\left|\not{\mathcal{L}}_{\mathscr{Z}}^M \not{d}(L(\varrho L_{(Small)}))\right| \lesssim \varepsilon^{1/2}\frac{\ln(e+t)}{(1+t)^2}$. We bound the terms $\not{\mathcal{L}}_{\mathscr{Z}}^M G_{(Frame)}$ and $\not{\mathcal{L}}_{\mathscr{Z}}^M(\not{\mathcal{L}}_{\varrho L} G_{(Frame)})$ with Lemma 12.25. We bound the terms $\not{\mathcal{L}}_{\mathscr{Z}}^M \not{d}x$ with the estimates (12.57a) and (12.57b). We bound the terms $\begin{pmatrix} \mathscr{Z}^M L\Psi \\ \not{\mathcal{L}}_{\mathscr{Z}}^M \not{d}\Psi \end{pmatrix}$ with Lemma 12.23 and the bootstrap assumptions (**BA**Ψ). In total, these estimates imply that ϱ^2 times the magnitude of $\not{\mathcal{L}}_{\mathscr{Z}}^N$ applied to the right-hand side of (12.123) is $\lesssim$ the right-hand side of (12.122) as desired.

To complete the proof of the estimate (12.122) for $\varrho^2 \not{\mathcal{L}}_L \not{\mathcal{L}}_{\mathscr{Z}}^N \chi^{(Small)}$ we must bound the commutator term $\varrho^2[\not{\mathcal{L}}_L, \not{\mathcal{L}}_{\mathscr{Z}}^N]\chi^{(Small)}$ by the right-hand side (12.122). To derive the desired estimate, we use inequality (12.89c) with $\chi^{(Small)}$ in the role of ξ (and the first term $\left|\not{\mathcal{L}}_L \not{\mathcal{L}}_{\mathscr{Z}}^{N-1}\chi^{(Small)}\right|$ on the right-hand side of (12.89c) is bounded by induction) and the estimates of Lemma 12.30.

To bound the magnitude of $\not{\mathcal{L}}_L(\varrho^2 \not{\mathcal{L}}_{\mathscr{Z}}^N \chi^{(Small)\#})$ by the right-hand side of (12.122), we first note the following identity, which holds for any symmetric type

$\binom{0}{2}$ $S_{t,u}$ tensorfield ξ:

$$\not{\mathcal{L}}_L(\varrho^2\xi^{\#}) = -2\varrho^2\chi_{AB}^{(Small)}\xi^{AB} + \varrho^2(\not{\mathcal{L}}_L\xi)^{\#}. \tag{12.125}$$

The identity (12.125) follows easily from the identity $(\not{\mathcal{L}}_L \not{g}^{-1})^{AB} = -2\chi^{AB}$ (see (3.72) and Lemma 7.5), (4.1c), and the identity $L\varrho = 1$. We now set $\xi = \not{\mathcal{L}}_{\mathscr{Z}}^N \chi^{(Small)}$ in (12.125). The second term on the right-hand side of (12.125) has already been suitably bounded by the previously proven estimate (12.122) for $\varrho^2 \not{\mathcal{L}}_L \not{\mathcal{L}}_{\mathscr{Z}}^N \chi^{(Small)}$. To bound the first term on the right-hand side of (12.125) in magnitude by the right-hand side of (12.122), we use the estimates of Lemma 12.30.

To bound the magnitude of $L(\varrho^2 \mathscr{Z}^N \mathrm{tr}_{\not{g}}\chi^{(Small)})$ by the right-hand side of (12.122), we note that $L(\varrho^2 \mathscr{Z}^N \mathrm{tr}_{\not{g}}\chi^{(Small)})$ is the pure trace of the type $\binom{1}{1}$ $S_{t,u}$ tensorfield $\not{\mathcal{L}}_L(\varrho^2 \not{\mathcal{L}}_{\mathscr{Z}}^N \chi^{(Small)\#})$. Since the magnitude of the pure trace of a type $\binom{1}{1}$ tensorfield is $\lesssim$ the magnitude of the type $\binom{1}{1}$ tensorfield itself, the desired estimate follows from the previously proven estimate (12.122) for $\not{\mathcal{L}}_L(\varrho^2 \not{\mathcal{L}}_{\mathscr{Z}}^N \chi^{(Small)\#})$.

To bound the magnitude of $\varrho^2 \not{\mathcal{L}}_L \not{\mathcal{L}}_{\mathscr{Z}}^N \hat{\chi}^{(Small)}$ by the right-hand side of (12.122), we first use the Leibniz rule to derive the following analog of (12.111):

(12.126)

$$\begin{aligned}\left|\not{\mathcal{L}}_L \not{\mathcal{L}}_{\mathscr{Z}}^N \hat{\chi}^{(Small)}\right| &\lesssim \left|\not{\mathcal{L}}_L \not{\mathcal{L}}_{\mathscr{Z}}^N \chi^{(Small)}\right| \\ &+ \sum_{i_1+i_2+i_3=1} \sum_{N_1+N_2+N_3=N} \left|\not{\mathcal{L}}_L^{i_1} \not{\mathcal{L}}_{\mathscr{Z}}^{N_1} \chi^{(Small)}\right| \left|\not{\mathcal{L}}_L^{i_2} \not{\mathcal{L}}_{\mathscr{Z}}^{N_2} \not{g}\right| \left|\not{\mathcal{L}}_L^{i_3} \not{\mathcal{L}}_{\mathscr{Z}}^{N_3} \not{g}^{-1}\right|.\end{aligned}$$

The desired bound now follows from (12.126), the previously proven estimate (12.122) for $\varrho^2 \not{\mathcal{L}}_L \not{\mathcal{L}}_{\mathscr{Z}}^N \chi^{(Small)}$, the estimates of Lemmas 12.26 and 12.30, and the fact that $\varrho L \in \mathscr{Z}$.

Finally, the estimate (12.122) for $\not{\mathcal{L}}_L(\varrho^2 \not{\mathcal{L}}_{\mathscr{Z}}^N \hat{\chi}^{(Small)\#})$ follows from the estimate (12.122) for $\varrho^2 \not{\mathcal{L}}_L \not{\mathcal{L}}_{\mathscr{Z}}^N \hat{\chi}^{(Small)}$ in the same way that the estimate for $\not{\mathcal{L}}_L(\varrho^2 \not{\mathcal{L}}_{\mathscr{Z}}^N \chi^{(Small)\#})$ followed from the estimate for $\varrho^2 \not{\mathcal{L}}_L \not{\mathcal{L}}_{\mathscr{Z}}^N \chi^{(Small)}$. □

12.27. Improvement of the auxiliary bootstrap assumptions

In this section, we use previously derived transport equations to derive improvements of the C^0 bootstrap assumptions for μ, $L_{(Small)}^i$, and their lower-order derivatives. At the same time, we derive pointwise estimates for these quantities and their derivatives. The pointwise estimates *involve a loss of one order of differentiability* relative to Ψ because the right-hand sides of the transport equations depend on the derivatives of Ψ. In Chapter 19, we will use the pointwise estimates to derive energy estimates for the below-top-order derivatives of the eikonal function quantities.

Proposition 12.46 (**Estimates for μ, $L_{(Small)}^i$, and $\chi^{(Small)}$ derived from transport equations in the direction L**). *Let χ and $\chi^{(Small)}$ be the type $\binom{0}{2}$ $S_{t,u}$ tensorfields defined in (3.70) and (4.1c). Let $0 \le N \le 24$ be an integer and recall that $\varrho(t,u) = 1 - u + t$. Under the small data and bootstrap assumptions of Sects. 12.1-12.4, if ε is sufficiently small, then the following pointwise estimates* *hold on* $\mathcal{M}_{T_{(Bootstrap)},U_0}$*:*

$$\left| \begin{pmatrix} \varrho^2 \not{\mathcal{L}}_{\mathscr{Z}}^{N-1} \upchi^{(Small)} \\ \varrho^2 \not{\mathcal{L}}_{\mathscr{Z}}^{N-1} \upchi^{(Small)\#} \\ \varrho^2 \mathscr{Z}^{N-1} \mathrm{tr}_{\not{g}} \upchi^{(Small)} \\ \varrho^2 \not{\mathcal{L}}_{\mathscr{Z}}^{N-1} \hat{\upchi}^{(Small)} \\ \varrho^2 \not{\mathcal{L}}_{\mathscr{Z}}^{N-1} \hat{\upchi}^{(Small)\#} \end{pmatrix} \right| \lesssim (1+t) \left| \begin{pmatrix} \varrho L \mathscr{Z}^{\leq N} \Psi \\ \breve{R} \mathscr{Z}^{\leq N} \Psi \\ \varrho \not{d} \mathscr{Z}^{\leq N} \Psi \\ \mathscr{Z}^{\leq N} \Psi \end{pmatrix} \right| + \left| \begin{pmatrix} \mathscr{Z}^{\leq N-1}(\upmu - 1) \\ \sum_{a=1}^3 \varrho \left| \mathscr{Z}^{\leq N} L^a_{(Small)} \right| \end{pmatrix} \right|, \tag{12.127a}$$

$$\left| \begin{pmatrix} L \mathscr{Z}^{\leq N} \upmu \\ \sum_{a=1}^3 \left| L(\varrho \mathscr{Z}^{\leq N} L^a_{(Small)}) \right| \\ \varrho^2 \not{\mathcal{L}}_L \not{\mathcal{L}}_{\mathscr{Z}}^{\leq N-1} \upchi^{(Small)} \\ \not{\mathcal{L}}_L (\varrho^2 \not{\mathcal{L}}_{\mathscr{Z}}^{\leq N-1} \upchi^{(Small)\#}) \\ L(\varrho^2 \mathscr{Z}^{\leq N-1} \mathrm{tr}_{\not{g}} \upchi^{(Small)}) \\ \varrho^2 \not{\mathcal{L}}_L \not{\mathcal{L}}_{\mathscr{Z}}^{\leq N-1} \hat{\upchi}^{(Small)} \\ \not{\mathcal{L}}_L (\varrho^2 \not{\mathcal{L}}_{\mathscr{Z}}^{\leq N-1} \hat{\upchi}^{(Small)\#}) \end{pmatrix} \right|, \left| \begin{pmatrix} \mathscr{Z}^{\leq N} L \upmu \\ \sum_{a=1}^3 \left| \mathscr{Z}^{\leq N} (L(\varrho L^a_{(Small)})) \right| \\ \not{\mathcal{L}}_{\mathscr{Z}}^{\leq N-1} (\varrho^2 \not{\mathcal{L}}_L \upchi^{(Small)}) \\ \not{\mathcal{L}}_{\mathscr{Z}}^{\leq N-1} \not{\mathcal{L}}_L (\varrho^2 \upchi^{(Small)\#}) \\ \mathscr{Z}^{\leq N-1} L(\varrho^2 \mathrm{tr}_{\not{g}} \upchi^{(Small)}) \\ \not{\mathcal{L}}_{\mathscr{Z}}^{\leq N-1} (\varrho^2 \not{\mathcal{L}}_L \hat{\upchi}^{(Small)}) \\ \not{\mathcal{L}}_{\mathscr{Z}}^{\leq N-1} \not{\mathcal{L}}_L (\varrho^2 \hat{\upchi}^{(Small)\#}) \end{pmatrix} \right| \lesssim \left| \begin{pmatrix} \varrho L \mathscr{Z}^{\leq N} \Psi \\ \breve{R} \mathscr{Z}^{\leq N} \Psi \\ \varrho \not{d} \mathscr{Z}^{\leq N} \Psi \\ \mathscr{Z}^{\leq N} \Psi \end{pmatrix} \right| + \frac{\ln(e+t)}{(1+t)^2} \left| \begin{pmatrix} \mathscr{Z}^{\leq N}(\upmu - 1) \\ \sum_{a=1}^3 \varrho \left| \mathscr{Z}^{\leq N} L^a_{(Small)} \right| \end{pmatrix} \right|. \tag{12.127b}$$

Furthermore, the following estimates hold for $t \in [0, T_{(Bootstrap)})$*:*

$$\left\| \begin{pmatrix} \mathscr{Z}^{\leq 13}(\upmu - 1) \\ \sum_{a=1}^3 \left| \varrho \mathscr{Z}^{\leq 13} L^a_{(Small)} \right| \\ \varrho^2 \not{\mathcal{L}}_{\mathscr{Z}}^{\leq 12} \upchi^{(Small)} \\ \varrho^2 \not{\mathcal{L}}_{\mathscr{Z}}^{\leq 12} \upchi^{(Small)\#} \\ \varrho^2 \mathscr{Z}^{\leq 12} \mathrm{tr}_{\not{g}} \upchi^{(Small)} \\ \varrho^2 \not{\mathcal{L}}_{\mathscr{Z}}^{\leq 12} \hat{\upchi}^{(Small)} \\ \varrho^2 \not{\mathcal{L}}_{\mathscr{Z}}^{\leq 12} \hat{\upchi}^{(Small)\#} \end{pmatrix} \right\|_{C^0(\Sigma_t^u)} \lesssim \varepsilon \ln(e+t). \tag{12.128a}$$

Finally, the following estimates hold for $t \in [0, T_{(Bootstrap)})$*:*

(12.128b)
$$\left\| \begin{pmatrix} L\mathscr{Z}^{\leq 12}\upmu \\ \sum_{a=1}^{3} \left| L(\varrho \mathscr{Z}^{\leq 12} L_{(Small)}^a) \right| \\ \varrho^2 \not{\mathcal{L}}_L \not{\mathcal{L}}_{\mathscr{Z}}^{\leq 11} \upchi^{(Small)} \\ \not{\mathcal{L}}_L (\varrho^2 \not{\mathcal{L}}_{\mathscr{Z}}^{\leq 11} \upchi^{(Small)\#}) \\ L(\varrho^2 \mathscr{Z}^{\leq 11} \mathrm{tr}_{\not{g}} \upchi^{(Small)}) \\ \varrho^2 \not{\mathcal{L}}_L \not{\mathcal{L}}_{\mathscr{Z}}^{\leq 11} \hat{\upchi}^{(Small)} \\ \not{\mathcal{L}}_L (\varrho^2 \not{\mathcal{L}}_{\mathscr{Z}}^{\leq 11} \hat{\upchi}^{(Small)\#}) \end{pmatrix} \right\|_{C^0(\Sigma_t^u)}, \left\| \begin{pmatrix} \mathscr{Z}^{\leq 12} L\upmu \\ \sum_{a=1}^{3} \left| \mathscr{Z}^{\leq 12}(L(\varrho L_{(Small)}^a)) \right| \\ \not{\mathcal{L}}_{\mathscr{Z}}^{\leq 11} (\varrho^2 \not{\mathcal{L}}_L \upchi^{(Small)}) \\ \not{\mathcal{L}}_{\mathscr{Z}}^{\leq 11} (\not{\mathcal{L}}_L (\varrho^2 \upchi^{(Small)\#})) \\ \mathscr{Z}^{\leq 11}(L(\varrho^2 \mathrm{tr}_{\not{g}} \upchi^{(Small)})) \\ \not{\mathcal{L}}_{\mathscr{Z}}^{\leq 11} (\varrho^2 \not{\mathcal{L}}_L \hat{\upchi}^{(Small)}) \\ \not{\mathcal{L}}_{\mathscr{Z}}^{\leq 11} (\not{\mathcal{L}}_L (\varrho^2 \hat{\upchi}^{(Small)\#})) \end{pmatrix} \right\|_{C^0(\Sigma_t^u)} \lesssim \varepsilon \frac{1}{1+t}.$$

PROOF. We first note that inequality (12.127a) follows easily from inequality (12.67a).

We now prove (12.127b). We prove only the estimates for the first array on the left-hand side because the proofs of the estimates for the second array do not involve commutations and hence are simpler. To proceed with our estimates for the first array, we note that by Lemma 12.45, the estimates for all $\upchi^{(Small)}$-related terms in the array will follow once we prove inequality (12.127b) for only the first two terms $L\mathscr{Z}^{\leq N}\upmu$ and $\sum_{a=1}^{3} \left| L(\varrho \mathscr{Z}^{\leq N} L_{(Small)}^a) \right|$ in the array.

Proof of the bound for $\left| L(\varrho \mathscr{Z}^{\leq N} L_{(Small)}^i) \right|$: We now use induction in N to prove (12.127b) for the term $\left| L(\varrho \mathscr{Z}^{\leq N} L_{(Small)}^i) \right|$ on the left-hand side. To proceed, we write equation (4.9b) in the schematic form

(12.129)
$$L(\varrho L_{(Small)}^i) = G_{(Frame)} \begin{pmatrix} L\Psi \\ \not{d}\Psi \end{pmatrix} \begin{pmatrix} \varrho L_{(Small)}^i \\ x^i \\ \varrho f^i(\Psi) \\ \varrho \not{d}^{\#} x^i \end{pmatrix} := \mathfrak{I}^i,$$

where $f^i(\Psi)$ is smooth function of Ψ that vanishes at $\Psi = 0$ (specifically, $f^i(\Psi) = (g^{-1})^{0i}$; see (2.4)-(2.6)). Commuting (12.129) with $\mathscr{Z}^N$, we deduce that

(12.130)
$$L(\varrho \mathscr{Z}^N L_{(Small)}^i) = \mathscr{Z}^N \mathfrak{I}^i + [L, \mathscr{Z}^N](\varrho L_{(Small)}^i) + L\left\{ [\varrho, \mathscr{Z}^N] L_{(Small)}^i \right\}.$$

We claim that the following bound holds for the first term $\mathscr{Z}^N\mathfrak{J}^i$ on the right-hand side of (12.130):

(12.131)

$$|\mathscr{Z}^N\mathfrak{J}^i| \lesssim \left|\begin{pmatrix} \varrho L\mathscr{Z}^{\leq N}\Psi \\ \varrho \not{d}\mathscr{Z}^{\leq N}\Psi \\ \breve{R}\mathscr{Z}^{\leq N}\Psi \\ \mathscr{Z}^{\leq N}\Psi \end{pmatrix}\right| + \frac{\ln(e+t)}{(1+t)^2}\left|\begin{pmatrix} \mathscr{Z}^{\leq N}(\upmu - 1) \\ \sum_{a=1}^{3}\varrho\left|\mathscr{Z}^{\leq N}L^a_{(Small)}\right| \end{pmatrix}\right|.$$

Let us accept (12.131) for the moment; we will independently prove (12.131) at the end of the argument without using induction. We note that the right-hand side of (12.131) is $\lesssim$ the right-hand side of (12.127b) as desired. This immediately yields the desired estimate in the base case $N = 0$.

To carry out the induction, we assume that the estimate (12.127b) for $\left|L(\varrho\mathscr{Z}^{\leq N-1}L^i_{(Small)})\right|$ has been proved. To bound the term $L\left\{[\varrho, \mathscr{Z}^N]L^i_{(Small)}\right\}$ on the right-hand side of (12.130), we use (12.7b) to deduce that

(12.132)

$$\left|L\left\{[\varrho, \mathscr{Z}^N]L^i_{(Small)}\right\}\right| \lesssim \left|L(\varrho\mathscr{Z}^{\leq N-1}L^i_{(Small)})\right| + \frac{1}{(1+t)^2}\left|\varrho\mathscr{Z}^{\leq N-1}L^i_{(Small)}\right|.$$

The second term on the right-hand side of (12.132) is manifestly $\lesssim$ the right-hand side of (12.127b) as desired, while to bound the first, we use the induction hypothesis.

To bound the term $[L, \mathscr{Z}^N](\varrho L^i_{(Small)})$ on the right-hand side of (12.130), we use the commutator estimate (12.82) with $f = \varrho L^i_{(Small)}$, inequalities (12.6b) and (12.56), and the bootstrap assumption $(\mathbf{AUX}LR_{(Small)})$ to deduce that

(12.133)

$$\begin{aligned}\left|[L, \mathscr{Z}^N](\varrho L^i_{(Small)})\right| &\lesssim \left|\mathscr{Z}^{\leq N-1}L(\varrho L^i_{(Small)})\right| + \frac{\ln(e+t)}{1+t}\left|\begin{pmatrix} \varrho L\mathscr{Z}^{\leq N}\Psi \\ \breve{R}\mathscr{Z}^{\leq N}\Psi \\ \varrho\not{d}\mathscr{Z}^{\leq N}\Psi \\ \mathscr{Z}^{\leq N}\Psi \end{pmatrix}\right| \\ &\quad + \frac{\ln(e+t)}{(1+t)^2}\left|\begin{pmatrix} \mathscr{Z}^{\leq N}(\upmu - 1) \\ \sum_{a=1}^{3}\varrho\left|\mathscr{Z}^{\leq N}L^a_{(Small)}\right| \end{pmatrix}\right|.\end{aligned}$$

The last two terms on the right-hand side of (12.133) are manifestly $\lesssim$ the right-hand side of (12.127b) as desired, while to bound the first, we use the estimate (12.131).

It remains for us to prove (12.131). We apply the operator $\not{\mathcal{L}}^N_{\mathscr{Z}}$ to the terms in (12.129) and use the Leibniz rule. We bound the terms $\not{\mathcal{L}}^M_{\mathscr{Z}}G_{(Frame)}$ with the estimates of Lemma 12.25. To bound $\begin{pmatrix} \mathscr{Z}^M L\Psi \\ \not{\mathcal{L}}^M_{\mathscr{Z}}\not{d}\Psi \end{pmatrix}$, we use Lemma 12.23 and the bootstrap assumptions $(\mathbf{BA}\Psi)$. To bound $\mathscr{Z}^M(\varrho L^i_{(Small)})$ when $M \leq 12$, we use inequality (12.6b) and the bootstrap assumptions $(\mathbf{AUX}LR_{(Small)})$ (when $M \geq 13$, the term $\varrho\mathscr{Z}^M L^i_{(Small)}$ appears on the right-hand side of (12.131)). To bound $\mathscr{Z}^M x^i$, we use inequalities (12.8a), (12.27a), and (12.27b) as well as the observation that $\mathscr{Z}^M x^i = \mathscr{Z}^{M-1}Zx^i = \mathscr{Z}^{M-1}Z^i$. To bound $\mathscr{Z}^M(\varrho f^i(\Psi))$, we use inequality (12.6b) and the bootstrap assumptions $(\mathbf{BA}\Psi)$, which in particular imply that $|\mathscr{Z}^M f^i(\Psi)| \lesssim |\mathscr{Z}^{\leq M}\Psi|$. To bound $\not{\mathcal{L}}_{\mathscr{Z}}^M(\varrho \not{d}^{\#} x^i)$, we use inequalities (12.6b), (12.57a), (12.57b) (12.60a), and (12.60b). In total, these estimates yield inequality (12.131), and our proof of the bound for $\left|L(\varrho \mathscr{Z}^{\leq N} L^i_{(Small)})\right|$ is complete.

Proof of the bound for $|L\mathscr{Z}^{\leq N}\upmu|$: We now use induction in N to prove the desired estimate for the term $|L\mathscr{Z}^{\leq N}\upmu|$ on the left-hand side of (12.127b). The proof is simpler than our proof of the bound for $\left|L(\varrho \mathscr{Z}^{\leq N} L^i_{(Small)})\right|$. To proceed, we write equation (4.5) in the schematic form

$$L\upmu = G_{(Frame)} \begin{pmatrix} \upmu L\Psi \\ \breve{R}\Psi \end{pmatrix} := \mathfrak{J}. \tag{12.134}$$

Commuting (12.134) with $\mathscr{Z}^N$, we have

$$|L\mathscr{Z}^N\upmu| \lesssim |[L, \mathscr{Z}^N]\upmu| + |\mathscr{Z}^N\mathfrak{J}|. \tag{12.135}$$

We claim that

$$|\mathscr{Z}^N\mathfrak{J}| \lesssim \left| \begin{pmatrix} \varrho \underline{L} \mathscr{Z}^{\leq N}\Psi \\ \breve{R}\mathscr{Z}^{\leq N}\Psi \\ \varrho \not{d} \mathscr{Z}^{\leq N}\Psi \\ \mathscr{Z}^{\leq N}\Psi \end{pmatrix} \right| + \frac{\ln(e+t)}{(1+t)^2} \left| \begin{pmatrix} \mathscr{Z}^{\leq N}(\upmu - 1) \\ \sum_{a=1}^3 \varrho \left|\mathscr{Z}^{\leq N} L^a_{(Small)}\right| \end{pmatrix} \right|. \tag{12.136}$$

To derive the estimate (12.136), we apply the Leibniz rule. We bound the terms $\mathscr{Z}^M G_{(Frame)}$ with the estimates of Lemma 12.25. To bound the terms $\not{\mathcal{L}}_{\mathscr{Z}}^M \begin{pmatrix} L\Psi \\ \breve{R}\Psi \end{pmatrix}$, we use Lemma 12.23 and the bootstrap assumptions (**BA**Ψ). To bound $\mathscr{Z}^M\upmu$ when $M \leq 12$, we use the bootstrap assumptions (**AUX**$\upmu$) (when $M \geq 13$, the term $\mathscr{Z}^M(\upmu - 1)$ appears on the right-hand side of (12.136)).

We now note that the right-hand side of (12.136) is manifestly $\lesssim$ the right-hand side of (12.127b) as desired. The estimate (12.136) therefore immediately yields the desired estimate for $|L\mathscr{Z}^{\leq N}\upmu|$ in the base case $N = 0$.

To carry out the induction, we assume that the estimate (12.127b) for $|L\mathscr{Z}^{\leq N-1}\upmu|$ has been proved. To bound the commutator term on the right-hand side of (12.135), we use the commutator estimate (12.81c) with $f = \upmu$, inequality (12.56), and the bootstrap assumptions (**AUX**$\upmu$) to deduce that

$$\begin{aligned} |[L, \mathscr{Z}^N]\upmu| \lesssim |L\mathscr{Z}^{\leq N-1}\upmu| &+ \frac{\ln(e+t)}{1+t}|\mathscr{Z}^{\leq N}\Psi| \\ &+ \frac{\ln(e+t)}{(1+t)^2} \left| \begin{pmatrix} \mathscr{Z}^{\leq N}(\upmu - 1) \\ \sum_{a=1}^3 \varrho \left|\mathscr{Z}^{\leq N} L^a_{(Small)}\right| \end{pmatrix} \right|. \end{aligned} \tag{12.137}$$

The last two terms on the right-hand side of (12.137) are manifestly bounded by the right-hand side of (12.127b) as desired, while to bound the first, we use the induction hypothesis.

Proof of (12.128a)-(12.128b): To prove (12.128a), we first simplify the notation by defining

$$q(t,u,\vartheta) := \left| \begin{pmatrix} \mathscr{Z}^{\leq 13}(\upmu - 1) \\ \sum_{a=1}^3 \left|\varrho \mathscr{Z}^{\leq 13} L^a_{(Small)}\right| \end{pmatrix} \right|, \qquad \mathring{q}(u,\vartheta) := q(0,u,\vartheta). \tag{12.138}$$

The bootstrap assumptions ($\mathbf{BA}'\Psi$) imply that the integral $\int_{t'=0}^{t} \cdot \, dt'$ of the array Ψ quantities on the right-hand side of (12.127b) (with $N = 13$ in (12.127b)) is bounded in the norm $\| \cdot \|_{C^0(\Sigma_t^u)}$ by $\lesssim \varepsilon \ln(e+t)$. Hence, integrating inequality (12.127b) in time along the integral curves of L, we deduce that

$$q(t,u,\vartheta) \leq \mathring{q}(u,\vartheta) + C\varepsilon \ln(e+t) + C \int_{t'=0}^{t} \frac{\ln(e+t')}{(1+t')^2} q(t',u,\vartheta) \, dt'. \tag{12.139}$$

Inequality (12.22b) and the small data assumption $\mathring{\epsilon} \leq \varepsilon$ together imply that $\mathring{q}(u,\vartheta) \lesssim \varepsilon$. Hence, applying Gronwall's inequality to (12.139), we deduce that $q(t,u,\vartheta) \lesssim \varepsilon \ln(e+t)$. This yields the desired inequalities for the first two terms in the array on the left-hand side of (12.128a). The estimates for the remaining quantities on the left-hand side of (12.128a) then follow from these estimates, inequality (12.67a), and the bootstrap assumptions ($\mathbf{BA}\Psi$).

Inequality (12.128b) then follows from inequalities (12.127b) and (12.128a) and the bootstrap assumptions ($\mathbf{BA}\Psi$). □

COROLLARY 12.47 (**Improvement of the auxiliary bootstrap assumptions**). *Under the small data and bootstrap assumptions of Sects.* 12.1-12.4, *if* ε *is sufficiently small, then the auxiliary bootstrap assumptions* ($\mathbf{AUX}\upmu$)-($\mathbf{AUX}\chi^{(Small)}$) *hold with* $\varepsilon^{1/2}$ *replaced by* $C\varepsilon$.

PROOF. The corollary follows directly from (12.128a). □

Using the estimates of Cor. 12.47 in place of the auxiliary bootstrap assumptions, we can repeat the proofs of all of the previous estimates in this chapter to arrive at the following corollary.

COROLLARY 12.48 ($\varepsilon^{1/2}$ **can be replaced with** $C\varepsilon$). *Under the small data and bootstrap assumptions of Sects.* 12.1-12.4, *all of the estimates in Chapter* 12 *that we proved before Cor.* 12.47 *and that involve an explicit factor of precisely* $\varepsilon^{1/2}$ *in fact hold with* $\varepsilon^{1/2}$ *replaced by* $C\varepsilon$.

12.28. Sharp pointwise estimates for a frame component of the derivative of the metric with respect to the solution

In this section, we provide sharp pointwise estimates for G_{LL} and LG_{LL}. In particular, we show that G_{LL} is well-approximated by the future null condition failure factor ${}^{(+)}\aleph$.

LEMMA 12.49 (**Bounds for** G_{LL}). *Let* ${}^{(+)}\aleph$ *be the future null condition failure factor from Def.* 3.32. *Under the small data and bootstrap assumptions of Sects.* 12.1-12.4, *if* ε *is sufficiently small, then the following pointwise estimates hold on* $\mathcal{M}_{T_{(Bootstrap)},U_0}$:

$$|LG_{LL}| \lesssim \varepsilon \frac{1}{(1+t)^2}, \tag{12.140}$$

$$\left| G_{LL} - {}^{(+)}\aleph \right| \lesssim \varepsilon \frac{\ln(e+t)}{1+t}. \tag{12.141}$$

Furthermore, if $0 \leq s \leq t$, *then*

$$|G_{LL}(t,u,\vartheta) - G_{LL}(s,u,\vartheta)| \lesssim \varepsilon \frac{t-s}{(1+s)(1+t)} \lesssim \varepsilon \frac{1}{1+s}. \tag{12.142}$$

PROOF. From (4.33a), the estimate (12.58b), and the bootstrap assumptions (**BA**Ψ), we deduce that

$$|LG_{LL}| \lesssim \varepsilon \frac{1}{(1+t)^2}, \tag{12.143}$$

which is the desired estimate (12.140).

Inequality (12.142) then follows from integrating (12.140) along the integral curves of L from time s to time t.

To prove (12.141), we first note that by the mean value theorem, the estimate (12.26b), and the bootstrap assumptions (**BA**Ψ), we have

$$\left|G_{LL} - G_{\alpha\beta}(\Psi = 0)L^\alpha L^\beta\right| \lesssim \frac{\varepsilon}{1+t}. \tag{12.144}$$

Next, we recall that $L_{(Flat)} = \partial_t + \frac{x^a}{r}\partial_a$. Hence, using definition (4.1a), the estimates (12.8b) and (12.128a), Cor. 12.48, and the fact that $L^0 = L^0_{(Flat)} = 1$, we deduce that the following estimate holds for $\nu = 0, 1, 2, 3$:

$$\left|L^\nu - L^\nu_{(Flat)}\right| \lesssim \frac{\varepsilon \ln(e+t)}{1+t}. \tag{12.145}$$

Using the estimates (12.26b), (12.144), and (12.145), we conclude the desired bound (12.141). □

12.29. Pointwise estimates for the angular Laplacian of the derivatives of the rectangular components of the re-centered version of the outgoing null vectorfield

In this section, we derive pointwise estimates for the scalar-valued functions $\not\Delta \mathscr{Z}^{\leq N-1} L^i_{(Small)}$.

REMARK 12.50 (**The estimates for $\not\Delta \mathscr{Z}^{\leq N-1} L^i_{(Small)}$ are "optional"**). We do not need the estimates of Lemma 12.51 to prove the most important aspects of our main sharp classical lifespan theorem. We provide these estimates only in an effort to give a complete description of the behavior of the functions $L^i_{(Small)}$.

LEMMA 12.51 (**Pointwise estimate for $\not\Delta \mathscr{Z}^{\leq N-1} L^i_{(Small)}$ in terms of other variables**). *Let $1 \leq N \leq 24$ be an integer and let $L^i_{(Small)}$, $(i = 1, 2, 3)$, be the scalar-valued functions defined in* (4.1a). *Under the small data and bootstrap assumptions of Sects. 12.1-12.4, if ε is sufficiently small, then the following pointwise estimate holds on $\mathcal{M}_{T_{(Bootstrap)}, U_0}$.*

$$\begin{aligned}
\left|\not\Delta \mathscr{Z}^{\leq N-1} L^i_{(Small)}\right| &\lesssim \frac{1}{(1+t)^2}\left|\begin{pmatrix} \varrho L \mathscr{Z}^{\leq N}\Psi \\ \breve{R}\mathscr{Z}^{\leq N}\Psi \\ \varrho \not{d}\mathscr{Z}^{\leq N}\Psi \\ \mathscr{Z}^{\leq N}\Psi \end{pmatrix}\right| \\
&\quad + \frac{1}{(1+t)^3}\left|\begin{pmatrix} \mathscr{Z}^{\leq N}(\upmu - 1) \\ \sum_{a=1}^3 \varrho \left|\mathscr{Z}^{\leq N} L^a_{(Small)}\right| \end{pmatrix}\right| \\
&\quad + \left|\text{d}\not{\text{i}}\text{v} \not{\mathcal{L}}_{\mathscr{Z}}^{N-1} \chi^{(Small)}\right|.
\end{aligned} \tag{12.146}$$

PROOF. We first apply $\not{\mathcal{L}}_{\mathscr{Z}}^{\leq N-1}$ and then $\nabla\!\!\!/^A$ to equation (4.10c). By Lemma 8.6, the left-hand side of the resulting expression is equal to $\Delta\!\!\!/ \mathscr{Z}^{\leq N-1} L^i_{(Small)}$. Hence, Lemma 12.51 will follow once we show that the right-hand side of the resulting expression is bounded in magnitude by $\lesssim$ the right-hand side of (12.146). We begin by addressing the term $\nabla\!\!\!/^A \not{\mathcal{L}}_{\mathscr{Z}}^{\leq N-1} \left\{ (g\!\!\!/^{-1})^{BC} \chi_{AB}^{(Small)} d\!\!\!/_C x^i \right\}$ generated by the first product on the right-hand side of (4.10c). When all derivatives fall on $\chi_{AB}^{(Small)}$, we use the estimate (12.57b) to bound this term by $\lesssim \left| \mathrm{div}\!\!\!/ \not{\mathcal{L}}_{\mathscr{Z}}^{\leq N-1} \chi^{(Small)} \right|$, which is manifestly bounded by the last term on the right-hand side of (12.146) as desired. Next, using the schematic identity $\not{\mathcal{L}}_Z g\!\!\!/^{-1} = -(g\!\!\!/^{-1})^2 {}^{(Z)}\pi\!\!\!/$ (see (7.4b)), Lemma 8.6, and inequalities (12.47) and (12.48), we see that the remaining terms that arise in the Leibniz expansion are bounded in magnitude by

$$\lesssim \frac{1}{1+t} \sum_{k=0}^{N} \sum_{\substack{N_1+\cdots+N_{k+2} \leq N-k \\ N_{k+1} \leq N-1}} \left| \not{\mathcal{L}}_{\mathscr{Z}}^{N_1} \pi\!\!\!/ \right| \left| \not{\mathcal{L}}_{\mathscr{Z}}^{N_2} \pi\!\!\!/ \right| \cdots \left| \not{\mathcal{L}}_{\mathscr{Z}}^{N_k} \pi\!\!\!/ \right| \left| \not{\mathcal{L}}_{\mathscr{Z}}^{N_{k+1}} \chi^{(Small)} \right| \left| d\!\!\!/ \mathscr{Z}^{N_{k+2}} x^i \right|, \tag{12.147}$$

where the $\pi\!\!\!/$ are the $S_{t,u}$-projections of deformation tensors of vectorfields belonging to $\mathscr{Z}$. From the estimates (12.57a), (12.57b), (12.60a), (12.60b), (12.67a), and (12.67b), we deduce that the right-hand side of (12.147) is $\lesssim$ the right-hand side of (12.146) as desired.

We now address the term $\nabla\!\!\!/^A \not{\mathcal{L}}_{\mathscr{Z}}^{\leq N-1} \left\{ \lambda_A^{(Tan-\Psi)} R^i_{(Small)} \right\}$ generated by the second product on the right-hand side of (4.10c). Using definition (4.11a), Lemma 8.6, inequality (12.47), and the Leibniz rule, we see that this term is bounded in magnitude by

$$\lesssim \frac{1}{1+t} \sum_{a=1}^{3} \sum_{N_1+N_2+N_3 \leq N} \left| \not{\mathcal{L}}_{\mathscr{Z}}^{N_1} G_{(Frame)} \right| \left| d\!\!\!/ \mathscr{Z}^{N_2} \Psi \right| \left| \mathscr{Z}^{N_3} R^a_{(Small)} \right|. \tag{12.148}$$

From the identity (4.3), the fact that $(g^{-1})^{0i}$ is smooth function of Ψ that vanishes at $\Psi = 0$ (see (2.4)-(2.6)), the bootstrap assumptions (**BAΨ**), and the estimates (12.58a), (12.58b), and (12.128a), we deduce that the right-hand side of (12.148) is $\lesssim$ the right-hand side of (12.146) as desired.

We now address the term $\nabla\!\!\!/^A \not{\mathcal{L}}_{\mathscr{Z}}^{\leq N-1} \left\{ \frac{x^i}{\varrho} \lambda_A^{(Tan-\Psi)} \right\}$ generated by the third product on the right-hand side of (4.10c). From definition (4.11a), Lemma 8.6, inequality (12.47), and the Leibniz rule, it follows that this term is bounded in magnitude by

$$\lesssim \frac{1}{1+t} \sum_{N_1+N_2+N_3 \leq N} \left| \mathscr{Z}^{N_1} \left(\frac{x^i}{\varrho} \right) \right| \left| \not{\mathcal{L}}_{\mathscr{Z}}^{N_2} G_{(Frame)} \right| \left| d\!\!\!/ \mathscr{Z}^{N_3} \Psi \right|. \tag{12.149}$$

From the bootstrap assumptions (**BAΨ**), (12.6b), the estimates (12.8a), (12.27a), (12.27b), (12.58a), (12.58b), and (12.128a), and the fact that $Zx^i = Z^i$, we deduce that the right-hand side of (12.149) is $\lesssim$ the right-hand side of (12.146) as desired.

We now address the term $\nabla\!\!\!/^A \not{\mathcal{L}}_{\mathscr{Z}}^{\leq N-1} \left\{ (g\!\!\!/^{-1})^{BC} \Lambda_{AB}^{(Tan-\Psi)} d\!\!\!/_C x^i \right\}$ generated by the final product on the right-hand side of (4.10c). Using definition (4.11c), Lemma 8.6, inequality (12.47), and the Leibniz rule, we argue as in the proof of (12.147) to bound this term in magnitude by

(12.150)

$$\lesssim \frac{1}{1+t}\sum_{k=0}^{N}\sum_{N_1+\cdots+N_{k+3}\leq N-k} \left|\not{\mathcal{L}}_{\mathscr{Z}}^{N_1}\not{\pi}\right|\left|\not{\mathcal{L}}_{\mathscr{Z}}^{N_2}\not{\pi}\right|\cdots\left|\not{\mathcal{L}}_{\mathscr{Z}}^{N_k}\not{\pi}\right.\left|\not{\mathcal{L}}_{\mathscr{Z}}^{N_{k+1}}G_{(Frame)}\right|\left|\not{\mathcal{L}}_{\mathscr{Z}}^{N_{k+2}}\begin{pmatrix} L\Psi \\ \not{d}\Psi \end{pmatrix}\right|\left|\not{d}\mathscr{Z}^{N_{k+3}}x^i\right|.$$

From the estimates (12.54), (12.57a), (12.57b), (12.58a), (12.58b), (12.60a), (12.60b), and (12.128a), and the bootstrap assumptions ($\mathbf{BA}\Psi$), we deduce that the right-hand side of (12.150) is $\lesssim$ the right-hand side of (12.146) as desired. $\square$

12.30. Estimates related to integrals over the spheres

In this section, we provide a collection of estimates that are connected to integrals over $S_{t,u}$. We start with a lemma that provides bounds for the error terms on the right-hand side of equation (10.1b).

LEMMA 12.52 (**Estimates for the error terms in $\not{\mathcal{L}}_{\breve{R}}dv_{\not{g}}$**). *Under the small data and bootstrap assumptions of Sects. 12.1-12.4, if ε is sufficiently small, then the last four terms in braces in equation* (10.1b) *verify the following pointwise estimate on $\mathcal{M}_{T_{(Bootstrap)},U_0}$:*

(12.151)

$$\left|-\frac{2}{\varrho}(\mu-1)-\mu\mathrm{tr}_{\not{g}}\chi^{(Small)}+\mathrm{tr}_{\not{g}}\not{k}^{(Trans-\Psi)}+\mu\mathrm{tr}_{\not{g}}\not{k}^{(Tan-\Psi)}\right|\lesssim\varepsilon\frac{\ln(e+t)}{(1+t)}.$$

PROOF. The last two terms on the left-hand side of (12.151) have the schematic form $G_{(Frame)}^{\#}\begin{pmatrix}\mu L\Psi \\ \breve{R}\Psi \\ \mu\not{d}\Psi\end{pmatrix}$. Hence, the desired estimate (12.151) follows from Cor. 12.27, the estimates (12.58b) and (12.128a), and the bootstrap assumptions ($\mathbf{BA}\Psi$). $\square$

LEMMA 12.53 (**Comparison of the area forms $dv_{\not{g}}$ and $dv_{\not{e}}$**). *Let $p=p(\vartheta)$ be a nonnegative function of the geometric angular coordinates. Under the small data and bootstrap assumptions of Sects. 12.1-12.4, if ε is sufficiently small, then the following estimate holds for $(t,u)\in[0,T_{(Bootstrap)})\times[0,U_0]$:*

$$\int_{S_{t,u}}p(\vartheta)dv_{\not{g}(t,u,\vartheta)}=\{1+\mathcal{O}(\varepsilon)\}\,\varrho^2(t,u)\int_{\vartheta\in\mathbb{S}^2}p(\vartheta)dv_{\not{e}(\vartheta)},\tag{12.152}$$

where $\varrho(t,u)=1-u+t$, $dv_{\not{g}(t,u,\vartheta)}$ is the area form defined in (3.84a), *and $dv_{\not{e}(\vartheta)}$ is the standard Euclidean area form on the* **Euclidean unit** *sphere $(\mathbb{S}^2,\not{e})$.*

PROOF. We set

$$P(t,u)=\int_{S_{t,u}}p(\vartheta)dv_{\not{g}(t,u,\vartheta)}.$$

From Lemma 10.2 and the decomposition $\mathrm{tr}_{\not{g}}\chi=2\varrho^{-1}+\mathrm{tr}_{\not{g}}\chi^{(Small)}$ (see (4.1c)), we have

$$\frac{\partial}{\partial t}P(t,u)=\int_{S_{t,u}}\left\{\frac{2}{\varrho}+\mathrm{tr}_{\not{g}}\chi^{(Small)}\right\}p\,dv_{\not{g}},\tag{12.153}$$

(12.154)
$$\frac{\partial}{\partial u}P(t,u) = \int_{S_{t,u}} \left\{ -\frac{2}{\varrho} - \frac{2}{\varrho}(\upmu - 1) - \upmu \mathrm{tr}_{\not{g}}\chi^{(Small)} + \mathrm{tr}_{\not{g}}\not{k}^{(Trans-\Psi)} + \upmu \mathrm{tr}_{\not{g}}\not{k}^{(Tan-\Psi)} \right\} p\, dv_{\not{g}}.$$

Inserting the estimates $|\mathrm{tr}_{\not{g}}\chi^{(Small)}| \lesssim \varepsilon \ln(e+t)(1+t)^{-2}$ (that is, (12.128a)) and (12.151) at $t = 0$ into (12.153) and (12.154) respectively, we deduce that

$$\left(1 - C\varepsilon\frac{\ln(e+t)}{1+t}\right)\frac{2}{\varrho(t,u)} \leq \frac{\partial}{\partial t}\ln P(t,u) \leq \left(1 + C\varepsilon\frac{\ln(e+t)}{1+t}\right)\frac{2}{\varrho(t,u)}, \tag{12.155}$$

$$-(1 + C\varepsilon)\frac{2}{\varrho(0,u)} \leq \frac{\partial}{\partial u}\ln P(0,u) \leq -(1 - C\varepsilon)\frac{2}{\varrho(0,u)}. \tag{12.156}$$

Respectively integrating (12.155) and (12.156) with respect to t and u, we deduce the following inequalities:

$$(1 - C\varepsilon)\frac{\varrho^2(t,u)}{\varrho^2(0,u)} \leq \frac{P(t,u)}{P(0,u)} \leq (1 + C\varepsilon)\frac{\varrho^2(t,u)}{\varrho^2(0,u)}, \tag{12.157}$$

$$(1 - C\varepsilon)\frac{\varrho^2(0,u)}{\varrho^2(0,0)} \leq \frac{P(0,u)}{P(0,0)} \leq (1 + C\varepsilon)\frac{\varrho^2(0,u)}{\varrho^2(0,0)}. \tag{12.158}$$

The desired estimate (12.152) now follows easily from (12.157) and (12.158) and the identities $\varrho(0,0) = 1$ and $P(0,0) = \int_{\vartheta \in \mathbb{S}^2} p(\vartheta) dv_{\not{e}(\vartheta)}$ (the latter identity is valid because in view of our assumption that the data are supported in Σ_0^1, we have that $\not{g}(0,0,\cdot) = \not{e}(\cdot)$, where $\not{e}$ is the standard round metric on $\mathbb{S}^2$). □

COROLLARY 12.54 (**Estimates for integrals of constants**). *Under the small data and bootstrap assumptions of Sects.* 12.1-12.4, *if ε is sufficiently small, then the following estimates hold for $(t,u) \in [0, T_{(Bootstrap)}) \times [0, U_0]$:*

$$4\pi(1 - C\varepsilon)\varrho^2(t,u) \leq \int_{S_{t,u}} 1\, dv_{\not{g}} \leq 4\pi(1 + C\varepsilon)\varrho^2(t,u), \tag{12.159a}$$

$$C^{-1}\varrho^2(t,u) \leq \int_{\Sigma_t^u} 1\, d\underline{\varpi} \leq C\varrho^2(t,u), \tag{12.159b}$$

$$C^{-1}\varrho(t,u) \leq \|1\|_{L^2(\Sigma_t^u)} \leq C\varrho(t,u). \tag{12.159c}$$

PROOF. Cor. 12.54 follows easily from the definitions of the quantities involved, Lemma 12.53, and the identity $\int_{\vartheta \in \mathbb{S}^2} 1 dv_{\not{e}(\vartheta)} = 4\pi$. □

LEMMA 12.55 (**Pointwise estimates for $\sqrt{\det \not{g}}$**). *Let $\sqrt{\det \not{g}}(t,u,\vartheta)$ denote the area form factor of $\not{g}$ relative to the geometric coordinates (and hence the determinant is taken relative to the geometric angular coordinates $(\vartheta^1, \vartheta^2)$) and let $\sqrt{\det \not{e}}(\vartheta)$ denote the area form factor of $\not{e}$ relative to the geometric coordinates. Here, $\not{e}$ is the Riemannian metric induced on the Euclidean unit sphere by the Euclidean metric[2] e on Σ_t. Recall that $\varrho(t,u) = 1 - u + t$. Under the small data and bootstrap assumptions of Sects.* 12.1-12.4, *if ε is sufficiently small, then the*

[2]Relative to the rectangular spatial coordinates on Σ_t, we have $e_{ij} = \delta_{ij}$.

following estimate holds on the spacetime domain $\mathcal{M}_{T_{(Bootstrap)},U_0}$:

$$\left| \frac{\varrho^{-2}\sqrt{\det \slashed{g}}}{\sqrt{\det \slashed{e}}} - 1 \right| \lesssim \varepsilon \frac{\ln(e+t)}{1+t}. \tag{12.160}$$

PROOF. Let $p \in \Sigma_t^{U_0}$ and let $\gamma : [0, U_0] \to \Sigma_t^{U_0}$ be the integral curve of $\breve{R}$ that passes through p and that is parametrized by the eikonal function u. From the proof of Lemma 10.2 and in particular (10.7), and the identities $\frac{d}{du}\varrho \circ \gamma = \breve{R}\varrho = -1$ and $\breve{R}u = 1$, we deduce that

$$\begin{aligned}\frac{d}{du}\left\{\ln[(\varrho^{-2}\sqrt{\det \slashed{g}})] \circ \gamma\right\} &= -\frac{2}{\varrho}(\upmu - 1) - \upmu \mathrm{tr}_{\slashed{g}}\chi^{(Small)} \\ &\quad + \mathrm{tr}_{\slashed{g}}\slashed{k}^{(Trans-\Psi)} + \upmu \mathrm{tr}_{\slashed{g}}\slashed{k}^{(Tan-\Psi)}.\end{aligned} \tag{12.161}$$

Inserting the estimate (12.151) into the right-hand side of (12.161), we deduce that

$$\left|\frac{d}{du}\left\{\ln[(\varrho^{-2}\sqrt{\det \slashed{g}})] \circ \gamma\right\}\right| \lesssim \varepsilon \frac{\ln(e+t)}{1+t}. \tag{12.162}$$

Integrating (12.162) du from the point p until the initial sphere $S_{t,0}$ (where $u = 0$) and using the identity $\varrho^{-2}\sqrt{\det \slashed{g}}|_{S_{t,0}} = \sqrt{\det \slashed{e}}$ (which holds because, in view of our assumption that the data are supported in Σ_0^1, we have that $\slashed{g}(0,0,\cdot) = \varrho^2(0,t)\slashed{e}(\cdot) = (1+t)^2\slashed{e}(\cdot)$, where $\slashed{e}$ is the standard round metric on $\mathbb{S}^2$), we conclude the desired estimate (12.160). □

LEMMA 12.56 (**Pointwise estimates for the time integral of** $\mathrm{tr}_{\slashed{g}}\chi$). *Let* χ *be the symmetric type* $\binom{0}{2}$ *tensorfield defined in* (3.70) *and recall that* $\varrho(t,u) = 1 - u + t$. *Under the small data and bootstrap assumptions of Sects.* 12.1-12.4, *if* ε *is sufficiently small, then the following estimate holds for* $(t,u) \in [0, T_{(Bootstrap)}) \times [0, U_0]$:

$$\ln\left(\frac{\varrho^2(t,u)}{\varrho^2(0,u)}\right) - C\varepsilon \leq \int_{s=0}^{t} \mathrm{tr}_{\slashed{g}}\chi(s,u,\vartheta)\, ds \leq \ln\left(\frac{\varrho^2(t,u)}{\varrho^2(0,u)}\right) + C\varepsilon. \tag{12.163}$$

PROOF. The inequalities in (12.163) follow easily from the decomposition $\mathrm{tr}_{\slashed{g}}\chi = 2\varrho^{-1} + \mathrm{tr}_{\slashed{g}}\chi^{(Small)}$ (see (4.1c)) and the pointwise estimate $|\mathrm{tr}_{\slashed{g}}\chi^{(Small)}| \leq C\varepsilon \ln(e+t)(1+t)^{-2}$ (that is, (12.128a)). □

LEMMA 12.57 (**Estimate for the norm** $\|\cdot\|_{L^2(\Sigma_t^u)}$ **of time-integrated functions**). *Let* f *be a function on spacetime and let*

$$F(t,u,\vartheta) := \int_{t'=0}^{t} f(t',u,\vartheta)\, dt'. \tag{12.164}$$

Recall that $\varrho(t,u) = 1 - u + t$. *Under the small data and bootstrap assumptions of Sects.* 12.1-12.4, *if* ε *is sufficiently small, then the following estimate holds for* $(t,u) \in [0, T_{(Bootstrap)}) \times [0, U_0]$:

$$\|F\|_{L^2(\Sigma_t^u)} \leq (1 + C\varepsilon)\varrho(t,u) \int_{t'=0}^{t} \frac{\|f\|_{L^2(\Sigma_{t'}^u)}}{\varrho(t',u)}\, dt'. \tag{12.165}$$

PROOF. From the definition (12.164) of F, the definition (3.85b) of $\|F\|^2_{L^2(\Sigma_t^u)}$, the estimate (12.152), Minkowski's inequality for integrals, and the fact that $\varrho(t',u') \geq \varrho(t',u)$ when $0 \leq u' \leq u$, we have

(12.166)

$$\begin{aligned}
&\|F\|^2_{L^2(\Sigma_t^u)} \\
&= \int_{u'=0}^{u} \int_{S_{t,u}} F^2(t,u',\vartheta)\, d\upsilon_{g\!\!/}\, du' \\
&\leq (1+C\varepsilon)\varrho^2(t,u) \int_{u'=0}^{u} \int_{\vartheta\in\mathbb{S}^2} F^2(t,u',\vartheta)\, d\upsilon_{e\!\!/}\, du' \\
&= (1+C\varepsilon)\varrho^2(t,u) \int_{u'=0}^{u} \int_{\vartheta\in\mathbb{S}^2} \left(\int_{t'=0}^{t} f(t',u',\vartheta)\, dt' \right)^2 d\upsilon_{e\!\!/}\, du' \\
&\leq (1+C\varepsilon)\varrho^2(t,u) \left\{ \int_{t'=0}^{t} \left(\int_{u'=0}^{u} \int_{\vartheta\in\mathbb{S}^2} f^2(t',u',\vartheta)\, d\upsilon_{e\!\!/}\, du' \right)^{1/2} dt' \right\}^2 \\
&\leq (1+C\varepsilon)\varrho^2(t,u) \left\{ \int_{t'=0}^{t} \left(\int_{u'=0}^{u} \int_{S_{t',u'}} f^2(t',u',\vartheta) \frac{d\upsilon_{g\!\!/}}{\varrho^2(t',u')}\, du' \right)^{1/2} dt' \right\}^2 \\
&\leq (1+C\varepsilon)\varrho^2(t,u) \left\{ \int_{t'=0}^{t} \left(\frac{1}{\varrho^2(t',u)} \int_{u'=0}^{u} \int_{S_{t',u'}} f^2(t',u',\vartheta)\, d\upsilon_{g\!\!/}\, du' \right)^{1/2} dt' \right\}^2 \\
&= (1+C\varepsilon)\varrho^2(t,u) \left\{ \int_{t'=0}^{t} \frac{1}{\varrho(t',u)} \|f\|_{L^2(\Sigma_{t'}^u)}\, dt' \right\}^2 .
\end{aligned}$$

Inequality (12.165) now follows from taking the square root of both sides of (12.166). □

12.31. Faster than expected decay for certain wave-variable-related quantities

We use the estimates in the following lemma in order to bound some of the error terms that appear in the integration by parts identities of Lemmas 10.17 and 10.18. The estimates also play a critical role in our sharp analysis of μ (see Chapter 13) and in our proof that μ can vanish in finite time (see Chapter 23).

LEMMA 12.58 (**Faster than expected decay for certain Ψ-related quantities**). *Recall that Ψ verifies the covariant wave equation $\mu\square_{g(\Psi)}\Psi = 0$ and that Ψ vanishes in the exterior of the flat outgoing null cone $\mathcal{C}_0$. Let χ be the symmetric type $\binom{0}{2}$ tensorfield defined in (3.70) and recall that $\varrho(t,u) = 1 - u + t$. Under the small data and bootstrap assumptions of Sects. 12.1-12.4, if ε is sufficiently small, then the following estimate holds for $t \in [0, T_{(Bootstrap)})$:*

$$\left\| L(\varrho \breve{R}\Psi) \right\|_{C^0(\Sigma_t^u)}, \ \|L(\varrho L\Psi)\|_{C^0(\Sigma_t^u)} \lesssim \varepsilon \frac{\ln(e+t)}{(1+t)^2}, \tag{12.167}$$

(12.168)
$$\left\| \mathscr{O}^{\leq 1}\left(L\Psi + \frac{1}{2}\mathrm{tr}_{\not g}\chi\Psi\right)\right\|_{C^0(\Sigma_t^u)}, \quad \left\| \mathscr{O}^{\leq 1}\left(L\breve{R}\Psi + \frac{1}{2}\mathrm{tr}_{\not g}\chi\breve{R}\Psi\right)\right\|_{C^0(\Sigma_t^u)} \lesssim \varepsilon\frac{\ln(e+t)}{(1+t)^3}.$$

PROOF. We first prove (12.168) for $L\Psi + \frac{1}{2}\mathrm{tr}_{\not g}\chi\Psi$. Writing the right-hand side of the wave operator decomposition (5.13b) (with $f = \Psi$) in schematic form, we have

(12.169)
$$\breve{R}\left\{\varrho\left[L\Psi + \frac{1}{2}\mathrm{tr}_{\not g}\chi\Psi\right]\right\} = \begin{pmatrix} \mu\underline{L}(\varrho L\Psi) \\ \mu\varrho\not\Delta\Psi \end{pmatrix} + \varrho G_{(Frame)}\not g^{-1}\begin{pmatrix} \mu L\Psi \\ \breve{R}\Psi \\ \mu\not d\Psi \end{pmatrix}\begin{pmatrix} L\Psi \\ \not d\Psi \end{pmatrix}$$
$$+ \begin{pmatrix} \mu \\ 1 \end{pmatrix} L\Psi + \varrho\not g^{-1}(\not d\mu)\not d\Psi + \begin{pmatrix} \varrho\breve{R}\mathrm{tr}_{\not g}\chi^{(Small)} \\ \mathrm{tr}_{\not g}\chi^{(Small)} \end{pmatrix}\Psi.$$

From (12.169), the inequalities (12.46a) and (12.46c), the estimates (12.58b) and (12.128a), and the bootstrap assumptions ($\mathbf{BA}\Psi$), we deduce that

(12.170)
$$\left|\breve{R}\left\{\varrho\left[L\Psi + \frac{1}{2}\mathrm{tr}_{\not g}\chi\Psi\right]\right\}\right| \lesssim \varepsilon\frac{\ln(e+t)}{(1+t)^2}.$$

Fixing t, integrating (12.170) along the integral curves of $-\breve{R}$ until reaching the sphere $S_{t,0}$, recalling that $\breve{R}u = 1$, and recalling that $\Psi \equiv 0$ in the exterior of the outgoing null cone portion $\mathcal{C}_0^{T_{(Bootstrap)}}$, we deduce that

(12.171)
$$\left|\varrho\left[L\Psi - \frac{1}{2}\mathrm{tr}_{\not g}\chi\Psi\right]\right| \lesssim \varepsilon\frac{\ln(e+t)}{(1+t)^2}.$$

The desired estimate (12.168) for $L\Psi + \frac{1}{2}\mathrm{tr}_{\not g}\chi\Psi$ now follows from dividing inequality (12.171) by ϱ.

To prove the desired estimate (12.168) for $O\left(L\Psi + \frac{1}{2}\mathrm{tr}_{\not g}\chi\Psi\right)$, where $O \in \mathscr{O}$ is a rotation, we note that $O\left(L\Psi + \frac{1}{2}\mathrm{tr}_{\not g}\chi\Psi\right)$ verifies an equation similar to the equation (12.169) verified by $L\Psi + \frac{1}{2}\mathrm{tr}_{\not g}\chi\Psi$, but with the right-hand side of (12.169) replaced by O applied to the right-hand side of (12.169), and with the additional commutator term (see (4.21c)) $\varrho\,{}^{(O)}\not\pi_{\breve{R}}^{\#} \cdot \not d\left(L\Psi + \frac{1}{2}\mathrm{tr}_{\not g}\chi\Psi\right)$ on the right-hand side. We now claim that the magnitudes of the terms arising from O applied to the right-hand side of (12.169) are $\lesssim \varepsilon\ln(e+t)(1+t)^{-2}$. To prove this claim, we use Lemma 8.6, the same estimates that we used to deduce (12.170) together with (12.60b) (to bound $\mathcal{L}_O\not g^{-1}$) and, to bound the term $O\not\Delta\Psi$, the estimate (12.101b).

Furthermore, using the decomposition $\mathrm{tr}_{\not g}\chi = 2\varrho^{-1} + \mathrm{tr}_{\not g}\chi^{(Small)}$ (see (4.1c)), inequality (12.46a), the estimates (12.75b) and (12.128a), and the bootstrap assumptions ($\mathbf{BA}\Psi$), we bound the commutator term as follows:

$$\begin{aligned}(12.172)\qquad & \varrho\left|{}^{(O)}\not\pi_{\breve{R}}^{\#}\cdot\not d\left(L\Psi+\frac{1}{2}\mathrm{tr}_{\not g}\chi\Psi\right)\right| \\ & \lesssim \left|{}^{(O)}\not\pi_{\breve{R}}\right|\left\{\sum_{l=1}^{3}\left(|O_{(l)}L\Psi| + |O_{(l)}\mathrm{tr}_{\not g}\chi^{(Small)}||\Psi| + \frac{1}{\varrho}|O_{(l)}\Psi|\right)\right\} \\ & \lesssim \varepsilon\frac{\ln^2(e+t)}{(1+t)^3}.\end{aligned}$$

Combining these estimates, we deduce that inequality (12.170) holds with $\varrho\left[L\Psi+\frac{1}{2}\mathrm{tr}_{\not g}\chi\Psi\right]$ replaced by $O\left\{\varrho\left[L\Psi+\frac{1}{2}\mathrm{tr}_{\not g}\chi\Psi\right]\right\}$. Thus, the desired estimate (12.168) for $O\left[L\Psi+\frac{1}{2}\mathrm{tr}_{\not g}\chi\Psi\right]$ follows from integrating along the integral curves of $-\breve{R}$, as we did above.

To prove estimate (12.168) for $L\breve{R}\Psi+\frac{1}{2}\mathrm{tr}_{\not g}\chi\breve{R}\Psi$, we first use the wave operator decomposition (5.12a) with $f=\Psi$, Lemma 4.6, and the simple identity $LL\Psi = \frac{1}{\varrho}L(\varrho L\Psi) - \frac{1}{\varrho}L\Psi$ to deduce the schematic equation

$$(12.173)\qquad L\breve{R}\Psi+\frac{1}{2}\mathrm{tr}_{\not g}\chi\breve{R}\Psi = \begin{pmatrix}\frac{1}{\varrho}\upmu L(\varrho L\Psi)\\ \upmu\not\Delta\Psi\end{pmatrix} + \frac{1}{\varrho}\upmu L\Psi + G_{(Frame)}\not g^{-1}\begin{pmatrix}\upmu L\Psi\\ \breve{R}\Psi\\ \upmu\not d\Psi\end{pmatrix}\begin{pmatrix}L\Psi\\ \not d\Psi\end{pmatrix}.$$

Using the same estimates we used to prove (12.170), we bound the right-hand side of (12.173) in magnitude by $\lesssim \varepsilon\ln(e+t)(1+t)^{-3}$. We have thus proved the desired estimate (12.168) for $L\breve{R}\Psi+\frac{1}{2}\mathrm{tr}_{\not g}\chi\breve{R}\Psi$.

To prove the estimate (12.168) for $O\left(L\breve{R}\Psi+\frac{1}{2}\mathrm{tr}_{\not g}\chi\breve{R}\Psi\right)$, we apply O to both sides of (12.173). Using Lemma 8.6, the same estimates we used to bound the right-hand side of (12.173) together with (12.60b) (to bound $\mathcal{L}_O\not g^{-1}$), and to bound the term $O\not\Delta\Psi$, the estimate (12.101b). We conclude that $\left|O\left(L\breve{R}\Psi+\frac{1}{2}\mathrm{tr}_{\not g}\chi\breve{R}\Psi\right)\right| \lesssim \varepsilon\ln(e+t)(1+t)^{-3}$ as desired.

The desired estimate (12.167) for $L(\varrho\breve{R}\Psi)$ follows from the identity $L(\varrho\breve{R}\Psi) = \varrho\left(L\breve{R}\Psi+\frac{1}{2}\mathrm{tr}_{\not g}\chi\breve{R}\Psi\right)-\frac{1}{2}\varrho\mathrm{tr}_{\not g}\chi^{(Small)}\breve{R}\Psi$, the estimate (12.168) for $L\breve{R}\Psi+\frac{1}{2}\mathrm{tr}_{\not g}\chi\breve{R}\Psi$, the estimate (12.128a) for $\mathrm{tr}_{\not g}\chi^{(Small)}$, and the bootstrap assumption $\left\|\breve{R}\Psi\right\|_{C^0(\Sigma_t^u)} \leq \varepsilon(1+t)^{-1}$.

The desired (nonoptimal) estimate (12.167) for $L(\varrho L\Psi)$ follows easily from the bootstrap assumption $|\varrho L(\varrho L\Psi)| \leq \varepsilon(1+t)^{-1}$ (that is, ($\mathbf{BA}\Psi$)). □

12.32. Pointwise estimates for the vectorfield Xi

In this section, we derive pointwise estimates for the vectorfield Ξ from the decomposition $\breve{R} = \frac{\partial}{\partial u} - \Xi$.

LEMMA 12.59 (**Estimates for** Ξ). *Let Ξ be the $S_{t,u}$-tangent vectorfield from (3.46) and let $\Xi_\flat$ be the corresponding $\not{g}$-dual one-form. Let $0 \leq N \leq 23$ be an integer. Under the small data and bootstrap assumptions of Sects.* 12.1-12.4, *if ε is sufficiently small, then the following pointwise estimates hold on the spacetime domain $\mathcal{M}_{T_{(Bootstrap)},U_0}$:*

$$\left|\not{\mathcal{L}}_L(\varrho^{-2}\not{\mathcal{L}}_{\mathscr{Z}}^N\Xi_\flat)\right| \lesssim \frac{1}{(1+t)^2}\left|\begin{pmatrix} \varrho L\mathscr{Z}^{\leq N}\Psi \\ \breve{R}\mathscr{Z}^{\leq N}\Psi \\ \varrho\not{d}\mathscr{Z}^{\leq N}\Psi \\ \mathscr{Z}^{\leq N}\Psi \end{pmatrix}\right| + \frac{1}{(1+t)^3}\left|\begin{pmatrix} \mathscr{Z}^{\leq N+1}(\upmu-1) \\ \sum_{a=1}^3 \varrho|\mathscr{Z}^{\leq N+1}L_{(Small)}^a| \end{pmatrix}\right| + \frac{\ln(e+t)}{(1+t)^2}\left|\varrho^{-2}\not{\mathcal{L}}_{\mathscr{Z}}^{\leq N}\Xi_\flat\right|. \tag{12.174a}$$

In addition, the following estimate holds for $(t,u) \in [0,T_{(Bootstrap)}) \times [0,U_0]$:

$$\left\|\not{\mathcal{L}}_{\mathscr{Z}}^{\leq 12}\Xi\right\|_{C^0(\Sigma_t^u)} \lesssim \varepsilon(1+t). \tag{12.174b}$$

Furthermore, the following estimate holds for the rectangular spatial components Ξ^i, $(i=1,2,3)$:

$$\left\|\mathscr{Z}^{\leq 12}\Xi^i\right\|_{C^0(\Sigma_t^u)} \lesssim \varepsilon(1+t). \tag{12.174c}$$

PROOF. We first lower the index in equation (3.47) with $\not{g}$, use the identities $\not{\mathcal{L}}_L\not{g} = 2\varrho^{-1}\not{g} + 2\chi^{(Small)}$ (see (3.72) and (4.1c)) and $L\varrho = 1$, use the fact that $[L,\breve{R}]_\flat = {}^{(\breve{R})}\not{\pi}_L$ (see (4.21a)), and commute with ϱ^{-2} followed by $\not{\mathcal{L}}_{\mathscr{Z}}^N$ to deduce that

$$\not{\mathcal{L}}_L(\varrho^{-2}\not{\mathcal{L}}_{\mathscr{Z}}^N\Xi_\flat) = 2\not{\mathcal{L}}_{\mathscr{Z}}^N\left\{\varrho^{-2}\chi^{(Small)\#}\cdot\Xi_\flat\right\} - \not{\mathcal{L}}_{\mathscr{Z}}^N(\varrho^{-2}{}^{(\breve{R})}\not{\pi}_L) + [\not{\mathcal{L}}_L,\not{\mathcal{L}}_{\mathscr{Z}}^N](\varrho^{-2}\Xi_\flat) + \not{\mathcal{L}}_L\left\{[\varrho^{-2},\not{\mathcal{L}}_{\mathscr{Z}}^N]\Xi_\flat\right\}. \tag{12.175}$$

We now derive bounds for the right-hand side of (12.175). To simplify the proof, we make the following temporary bootstrap assumptions (which are justified for short times by (12.25b)):

$$\textbf{(TEMP}-\textbf{BA}\Xi_\flat\textbf{)} \quad \left\|\not{\mathcal{L}}_{\mathscr{Z}}^{\leq 12}\Xi_\flat\right\|_{C^0(\Sigma_t^u)} \leq 1+t, \quad (t,u) \in [0,T_{(Bootstrap)}) \times [0,U_0].$$

From the Leibniz rule, the estimates (12.6b), (12.67a), and (12.128a), and the bootstrap assumptions ($\mathbf{BA\Psi}$) and ($\mathbf{TEMP} - \mathbf{BA}\Xi_\flat$), we deduce that

$$\left|\not{\mathcal{L}}_{\mathscr{Z}}^N\left\{\varrho^{-2}\chi^{(Small)\#}\cdot\Xi_\flat\right\}\right| \lesssim \frac{1}{(1+t)^2}\left|\begin{pmatrix}\varrho L\mathscr{Z}^{\leq N}\Psi\\ \breve{R}\mathscr{Z}^{\leq N}\Psi\\ \varrho\not{d}\mathscr{Z}^{\leq N}\Psi\\ \mathscr{Z}^{\leq N}\Psi\end{pmatrix}\right| + \frac{1}{(1+t)^3}\left|\begin{pmatrix}\mathscr{Z}^{\leq N}(\upmu-1)\\ \sum_{a=1}^3\varrho|\mathscr{Z}^{\leq N+1}L_{(Small)}^a|\end{pmatrix}\right| + \varepsilon\frac{\ln(e+t)}{(1+t)^2}\left|\varrho^{-2}\not{\mathcal{L}}_{\mathscr{Z}}^{\leq N}\Xi_\flat\right|. \tag{12.176}$$

To bound the second term on the right-hand side of (12.175), we use the Leibniz rule, the identity (7.5d), Lemma 8.6, inequalities (12.46a) and (12.54), the estimates (12.6b), (12.58a), (12.58b), and (12.128a), and the bootstrap assumptions ($\mathbf{BA\Psi}$) to deduce that

$$\left|\not{\mathcal{L}}_{\mathscr{Z}}^N(\varrho^{-2(\breve{R})}\not{\pi}_L)\right| \lesssim \frac{1}{(1+t)^2}\left|\begin{pmatrix}\varrho L\mathscr{Z}^{\leq N}\Psi\\ \breve{R}\mathscr{Z}^{\leq N}\Psi\\ \varrho\not{d}\mathscr{Z}^{\leq N}\Psi\\ \mathscr{Z}^{\leq N}\Psi\end{pmatrix}\right| + \frac{1}{(1+t)^3}\left|\begin{pmatrix}\mathscr{Z}^{\leq N+1}(\upmu-1)\\ \sum_{a=1}^3\varrho\left|\mathscr{Z}^{\leq N}L_{(Small)}^a\right|\end{pmatrix}\right|. \tag{12.177}$$

To bound the third term on the right-hand side of (12.175), we use inequality (12.7b), inequality (12.89c) with $\varrho^{-2}\Xi_\flat$ in the role of ξ (and the first term $\left|\not{\mathcal{L}}_L\not{\mathcal{L}}_{\mathscr{Z}}^{N-1}(\varrho^{-2}\Xi_\flat)\right|$ on the right-hand side of (12.89c) is bounded by induction), the estimates (12.6b) and (12.128a), and the bootstrap assumptions ($\mathbf{BA\Psi}$) and ($\mathbf{TEMP} - \mathbf{BA}\Xi_\flat$) to deduce that

$$\left|[\not{\mathcal{L}}_L,\not{\mathcal{L}}_{\mathscr{Z}}^N](\varrho^{-2}\Xi_\flat)\right| \lesssim \frac{1}{(1+t)^2}\left|\begin{pmatrix}\varrho L\mathscr{Z}^{\leq N}\Psi\\ \breve{R}\mathscr{Z}^{\leq N}\Psi\\ \varrho\not{d}\mathscr{Z}^{\leq N}\Psi\\ \mathscr{Z}^{\leq N}\Psi\end{pmatrix}\right| + \frac{1}{(1+t)^3}\left|\begin{pmatrix}\mathscr{Z}^{\leq N+1}(\upmu-1)\\ \sum_{a=1}^3\varrho\left|\mathscr{Z}^{\leq N+1}L_{(Small)}^a\right|\end{pmatrix}\right| + \left|\not{\mathcal{L}}_L(\varrho^{-2}\not{\mathcal{L}}_{\mathscr{Z}}^{\leq N-1}\Xi_\flat)\right| + \frac{\ln(e+t)}{(1+t)^2}\left|\varrho^{-2}\not{\mathcal{L}}_{\mathscr{Z}}^{\leq N}\Xi_\flat\right|. \tag{12.178}$$

To bound the last term on the right-hand side of (12.175), we use (12.7b) to deduce that

$$\left|\not{\mathcal{L}}_L\left\{[\varrho^{-2},\not{\mathcal{L}}_{\mathscr{Z}}^N]\Xi_\flat\right\}\right| \lesssim \left|\not{\mathcal{L}}_L(\varrho^{-2}\not{\mathcal{L}}_{\mathscr{Z}}^{\leq N-1}\Xi_\flat)\right| + \frac{1}{(1+t)^2}\left|\varrho^{-2}\not{\mathcal{L}}_{\mathscr{Z}}^{\leq N}\Xi_\flat\right|. \tag{12.179}$$

Combining (12.175), (12.176), (12.177), (12.178), and (12.179), and using induction in N to handle the next-to-last term on the right-hand side of (12.178) and the first term on the right-hand side of (12.179), we deduce that

(12.180)
$$\left|\not{\mathcal{L}}_L(\varrho^{-2}\not{\mathcal{L}}_{\mathscr{Z}}^N\Xi_\flat)\right| \lesssim \frac{1}{(1+t)^2}\left|\left(\begin{array}{c}\varrho L\mathscr{Z}^{\leq N}\Psi\\ \breve{R}\mathscr{Z}^{\leq N}\Psi\\ \varrho\not{d}\mathscr{Z}^{\leq N}\Psi\\ \mathscr{Z}^{\leq N}\Psi\end{array}\right)\right| - \frac{1}{(1+t)^3}\left|\left(\begin{array}{c}\mathscr{Z}^{\leq N+1}(\upmu-1)\\ \sum_{a=1}^3 \varrho|\mathscr{Z}^{\leq N+1}L_{(Small)}^a|\end{array}\right)\right| - \frac{\ln(e+t)}{(1+t)^2}\left|\varrho^{-2}\not{\mathcal{L}}_{\mathscr{Z}}^{\leq N}\Xi_\flat\right|.$$

In particular, we have proved the desired bound (12.174a).

We now use the identities $L\varrho = 1$ and $\not{\mathcal{L}}_L\not{g}^{-1} = -2\varrho^{-1}\not{g}^{-1} - 2\chi^{(Small)\#\#}$ (see (3.72), (4.1c), and Lemma 7.5) to deduce that for any $S_{t,u}$ one-form ξ, we have $L(|\varrho\xi|^2) = -2\varrho^2\chi_{AB}^{(Small)}\xi^A\xi^B + 2\varrho^2\xi^A\not{\mathcal{L}}_L\xi_A$. Using this identity with $\xi := \varrho^{-2}\not{\mathcal{L}}_{\mathscr{Z}}^N\Xi_\flat$, the Cauchy-Schwarz inequality, the estimates (12.128a) and (12.180), and the bootstrap assumptions (**BA**Ψ), we deduce that

(12.181)
$$\left|L\left|\varrho^{-1}\not{\mathcal{L}}_{\mathscr{Z}}^{\leq 12}\Xi_\flat\right|\right| \lesssim \varepsilon\frac{\ln(e+t)}{(1+t)^2} + \frac{\ln(e+t)}{(1+t)^2}\left|\varrho^{-1}\not{\mathcal{L}}_{\mathscr{Z}}^{\leq 12}\Xi_\flat\right|.$$

We now apply Gronwall's inequality to the quantity $\left|\varrho^{-1}\not{\mathcal{L}}_{\mathscr{Z}}^{\leq 12}\Xi_\flat\right|$ in (12.181) along the integral curves of L and use the small data estimate (12.23b), thereby concluding $\left\|\varrho^{-1}\not{\mathcal{L}}_{\mathscr{Z}}^{\leq 12}\Xi_\flat\right\|_{C^0(\Sigma_t^u)} \lesssim \varepsilon$. In particular, we have improved the bootstrap assumption (**TEMP** $-$ **BA**$\Xi_\flat$) and proved (12.174b) with $\Xi_\flat$ in place of Ξ. To conclude the desired bound (12.174b) for Ξ, we apply the Leibniz rule to the right-hand side of the identity $\not{\mathcal{L}}_{\mathscr{Z}}^N\Xi = \not{\mathcal{L}}_{\mathscr{Z}}^N(\not{g}^{-1}\cdot\Xi_\flat)$ and use the estimate for $\left\|\varrho^{-1}\not{\mathcal{L}}_{\mathscr{Z}}^{\leq 12}\Xi_\flat\right\|_{C^0(\Sigma_t^u)}$ together with (12.60b).

Inequality (12.174c) then follows from the identity $\mathscr{Z}^N\Xi^i = \not{\mathcal{L}}_{\mathscr{Z}}^N(\Xi\cdot\not{d}x^i)$, the Leibniz rule, and inequalities (12.57b) and (12.174b). □

12.33. Estimates for the components of the commutation vectorfields relative to the geometric coordinates

In this section, we derive estimates for the components of the commutation vectorfields $Z \in \mathscr{Z}$ relative to the geometric coordinates. As a simple corollary, we deduce that if f is a function, $0 \leq N \leq 13$, and $\mathscr{Z}^{\leq N}f$ is a continuous function of the geometric coordinates $(t, u, \vartheta^1, \vartheta^2)$, then f is also a C^N function of $(t, u, \vartheta^1, \vartheta^2)$.

LEMMA 12.60 (**Estimates for the components of the commutation vectorfields relative to the geometric coordinates**). *Let $0 \leq N \leq 12$ be an integer and let $Z \in \mathscr{Z}$. Let $\{(\mathbb{D}_i, \vartheta_i^1, \vartheta_i^2)\}_{i=1,2,3,4}$ be the standard atlas from Def.* 3.22. *Under the small-data and bootstrap assumptions of Sects.* 12.1-12.4, *if ε is sufficiently small, then for $\mathbb{D} \in \cup_{i=1}^4\{\mathbb{D}_i\}$ and $A = 1, 2$, the following pointwise estimates hold* *for* $(t,u) \in [0, T_{(Bootstrap)}) \times [0, U_0]$*:*

$$\left\| L \mathscr{Z}^{\leq N} Z \vartheta^A \right\|_{C^0(\{t\}\times[0,u]\times\mathbb{D})} \overset{\mathbb{D}}{\lesssim} \frac{\ln(e+t)}{(1+t)^2}, \tag{12.182a}$$

$$\left\| \mathscr{Z}^{\leq N} Z \vartheta^A \right\|_{C^0(\{t\}\times[0,u]\times\mathbb{D})} \overset{\mathbb{D}}{\lesssim} 1. \tag{12.182b}$$

Furthermore, the following estimates hold:

$$\left\| L \mathscr{Z}^{\leq N} Z t \right\|_{C^0(\Sigma_t^u)} \lesssim 1, \tag{12.183a}$$

$$\left\| \mathscr{Z}^{\leq N} Z t \right\|_{C^0(\Sigma_t^u)} \lesssim 1 + t, \tag{12.183b}$$

$$\left\| L \mathscr{Z}^{\leq N} Z u \right\|_{C^0(\Sigma_t^u)} = 0, \tag{12.184a}$$

$$\left\| \mathscr{Z}^{\leq N} Z u \right\|_{C^0(\Sigma_t^u)} \lesssim 1. \tag{12.184b}$$

PROOF. We first note that since $L\vartheta^A = 0$, it follows that

$$L \mathscr{Z}^N Z \vartheta^A = [L, \mathscr{Z}^N Z] \vartheta^A = [L, \mathscr{Z}^{N+1}] \vartheta^A. \tag{12.185}$$

Hence, by making straightforward changes to the proof of (12.81c), using the bootstrap assumptions (**BA**Ψ) and (**AUX**$\upmu$)-(**AUX**$\chi^{(Small)}$), and arguing inductively to bound the term analogous to the first term on the right-hand side of (12.81c), we deduce the following estimate on $[0, T_{(Bootstrap)}) \times [0, U_0] \times \mathbb{D}$:

$$\sum_{Z \in \mathscr{Z}} \left| L \mathscr{Z}^{\leq N} Z \vartheta^A \right| \lesssim \sum_{Z \in \mathscr{Z}} \frac{\ln(e+t)}{(1+t)^2} \left| \mathscr{Z}^{\leq N} Z \vartheta^A \right|. \tag{12.186}$$

From (12.186), the data estimate $\left| \mathscr{Z}^{\leq N+1} \vartheta^A \right| (0, u, \vartheta) \overset{\mathbb{D}}{\lesssim} 1$, which is straightforward to verify using the estimates of Sect. 12.8, and Gronwall's inequality (in the quantity $\sum_{Z \in \mathscr{Z}} \left| \mathscr{Z}^{\leq N} Z \vartheta^A \right|$), we conclude the desired estimate (12.182b). The estimate (12.182a) then follows from (12.186) and (12.182b). Next, since $Zt \in \{0, \varrho\}$, the estimates (12.183a)-(12.183b) follow easily from Lemma 12.10. Finally, since $Zu \in \{0, 1\}$, the desired estimates (12.184a)-(12.184b) follow easily. □

COROLLARY 12.61 (**Geometric coordinate regularity follows from vectorfield differential operator regularity**). *Let f be a scalar-valued function and let $0 \leq N \leq 13$ be an integer. Assume that for $0 \leq M \leq N$, $\mathscr{Z}^M f$ is a continuous function of (t, u, ϑ). Under the small-data and bootstrap assumptions of Sects.* 12.1-12.4, *if ε is sufficiently small, then f is N-times continuously differentiable with respect to the geometric coordinates $(t, u, \vartheta^1, \vartheta^2)$ on $\mathcal{M}_{T_{(Bootstrap)}, U_0}$.*

PROOF. We first show that if $Z \in \mathscr{Z}$, then the components of Z relative to the geometric coordinates are C^{12} functions of (t, u, ϑ). To this end, we first note that the commutation set $\mathscr{Z}$ has span equal to the span of the geometric coordinate partial derivative vectorfields $\left\{ \frac{\partial}{\partial t}, \frac{\partial}{\partial u}, X_1, X_2 \right\}$. From Lemma 12.60 with $N = 0$, it follows that each geometric coordinate vectorfield can (locally) be written as a linear combination of the vectorfields in $\mathscr{Z}$ with coefficients that are C^0 functions of $(t, u, \vartheta^1, \vartheta^2)$. Hence, we can use Lemma 12.60 with $N = 1$ to conclude that these coefficients are in fact C^1 functions of $(t, u, \vartheta^1, \vartheta^2)$ and that the same regularity statement holds for the components $Zt, Zu, Z\vartheta^1, Z\vartheta^2$ of the vectorfields $Z \in \mathscr{Z}$ relative to the geometric coordinates. Continuing by induction with the help of Lemma 12.60, we conclude that the components are C^{12} as desired. Cor. 12.61 now follows from expressing the geometric coordinate derivatives of f in terms of the aforementioned linear combinations of vectorfields $Z \in \mathscr{Z}$ and noting that in the resulting expressions, no more than 12 derivatives fall on the components of the Z. □

12.34. Estimates for the rectangular spatial derivatives of the eikonal function

In this section, we derive some simple estimates for the first rectangular spatial derivatives of the eikonal function.

LEMMA 12.62 (**Estimates for the first rectangular spatial derivatives of u**). *Under the small-data and bootstrap assumptions of Sects.* 12.1-12.4, *if ε is sufficiently small, then for $i = 1, 2, 3$ the following estimates hold for the rectangular spatial derivatives of the eikonal function for $(t, u) \in [0, T_{(Bootstrap)}) \times [0, U_0]$:*

$$\left\| \mathscr{Z}^{\leq 12} \left(\mu \partial_i u + \frac{x^i}{r} \right) \right\|_{C^0(\Sigma_t^u)} \leq C\varepsilon \frac{\ln(e+t)}{1+t}. \tag{12.187}$$

PROOF. Inequality (12.187) follows from the identity (3.25), the estimate (12.32), and Cor. 12.48. □

CHAPTER 13

Sharp Estimates for the Inverse Foliation Density

In this chapter, we derive a variety of sharp pointwise estimates for the inverse foliation density $\upmu$ and its derivatives. We then derive estimates for some related time integrals involving $\upmu$. The time integral estimates play a critical role in our proof of the main Gronwall-type lemma (see Lemma 20.10), which we use to deduce our main a priori L^2 estimates for Ψ. The time integral estimates are sensitive and affect the degree of degeneracy-in-$\upmu^{-1}$ of our L^2 estimates for the top derivatives of Ψ, which in turn affects the number of derivatives we need to close the proof of the sharp classical lifespan theorem. The most difficult estimates in this chapter are based on the fact that $L(\varrho L\upmu)$ is integrable in time along the integral curves of L. This fact leads to the approximate monotonicity of $\upmu$ along the integral curves, where along a fixed integral curve the sign of the monotonicity *up to* a "late" time t is determined by the sign of $L\upmu$ *exactly at* time t; see Sect 2.10.5 for an overview.

REMARK 13.1 (**The role of a posteriori estimates**). The most difficult estimates involving $\upmu$ in Chapter 13 are not a priori nature, but rather a posteriori. Hence, we derive many of the estimates on time intervals of the form $0 \leq s \leq t$, and our arguments are based, on a case-by-case basis, on the possible (but not known in advance) behavior of various quantities at time t.

13.1. Basic ingredients in the analysis

We begin by recalling that $\upmu$ verifies the transport equation (see (4.5))

$$L\upmu = \omega. \tag{13.1}$$

In Chapter 13, we often view $\upmu = \upmu(s, u, \vartheta)$ and similarly for other quantities. In doing so, we are referring to the moving time coordinate as "s" in order to distinguish it from a "fixed later time" t verifying $t \geq s$. We recall that relative to these coordinates, we have (see (3 31)) $L = \frac{\partial}{\partial s} := \frac{\partial}{\partial s}|_{u,\vartheta}$. In summary, we have

$$\frac{\partial}{\partial s}\upmu(s, u, \vartheta) = \omega(s, u, \vartheta). \tag{13.2}$$

If f is a function that depends on the geometric coordinates (s, u, ϑ) as well as the late time $t \geq s$, then we indicate this by writing $f(s, u, \vartheta; t)$.

We now define some auxiliary quantities that we use to analyze $\upmu$.

DEFINITION 13.2 (**Auxiliary quantities used to analyze** $\upmu$). We define the following quantities, where we assume that $0 \leq s \leq t$ for those quantities that depend on both s and t:

$$\Omega(s,u,\vartheta) := \varrho(s,u)\omega(s,u,\vartheta) = \varrho(s,u)L\upmu(s,u,\vartheta), \tag{13.3a}$$

$$m(s,u,\vartheta;t) := \frac{\Omega(t,u,\vartheta) - \Omega(s,u,\vartheta)}{\varrho(s,u)}, \tag{13.3b}$$

$$M(s,u,\vartheta;t) := \int_{s'=s}^{s'=t} m(s',u,\vartheta;t)\,ds', \tag{13.3c}$$

$$\mathring{\upmu}(u,\vartheta) := \upmu(s=0,u,\vartheta), \tag{13.3d}$$

$$\widetilde{\Omega}(s,u,\vartheta;t) := \frac{\Omega(s,u,\vartheta)}{\mathring{\upmu}(u,\vartheta) - M(0,u,\vartheta;t)}, \tag{13.3e}$$

$$\widetilde{m}(s,u,\vartheta;t) := \frac{m(s,u,\vartheta;t)}{\mathring{\upmu}(u,\vartheta) - M(0,u,\vartheta;t)}, \tag{13.3f}$$

$$\widetilde{M}(s,u,\vartheta;t) := \frac{M(s,u,\vartheta;t)}{\mathring{\upmu}(u,\vartheta) - M(0,u,\vartheta;t)}, \tag{13.3g}$$

$$\upmu_{(Approx)}(s,u,\vartheta;t) := 1 + \widetilde{\Omega}(t,u,\vartheta;t)\ln\left(\frac{\varrho(s,u)}{\varrho(0,u)}\right) + \widetilde{M}(s,u,\vartheta;t). \tag{13.3h}$$

REMARK 13.3 (**The role of $\upmu_{(Approx)}(s,u,\vartheta;t)$**). The main point of Def. 13.2 is that $\upmu_{(Approx)}(s,u,\vartheta;t)$ is a good approximation to $\upmu(s,u,\vartheta)$ that is monotonic along each integral curve of L for $0 \leq s \leq t$, up to the error term $\widetilde{M}(s,u,\vartheta;t)$. Furthermore, the sign of the approximate monotonicity is determined by $\widetilde{\Omega}(t,u,\vartheta;t)$.

The identities stated in the next lemma follow easily from Def. 13.2; we omit the simple proofs.

LEMMA 13.4 (**Some identities verified by the auxiliary quantities**). *The following identities hold, where we assume that $0 \leq s \leq t$ for those quantities that depend on both s and t:*

$$\frac{\partial}{\partial s}\upmu(s,u,\vartheta) = \frac{\Omega(s,u,\vartheta)}{\varrho(s,u)} = \frac{\Omega(t,u,\vartheta)}{\varrho(s,u)} - m(s,u,\vartheta;t), \tag{13.4a}$$

$$\frac{\partial}{\partial s}M(s,u,\vartheta;t) = -m(s,u,\vartheta;t), \tag{13.4b}$$

$$\frac{\partial}{\partial s}\upmu_{(Approx)}(s,u,\vartheta;t) = \frac{\widetilde{\Omega}(t,u,\vartheta;t)}{\varrho(s,u)} - \widetilde{m}(s,u,\vartheta;t), \tag{13.4c}$$

$$\frac{\frac{\partial}{\partial s}\upmu_{(Approx)}(s,u,\vartheta;t)}{\upmu_{(Approx)}(s,u,\vartheta;t)} = \frac{\frac{\partial}{\partial s}\upmu(s,u,\vartheta)}{\upmu(s,u,\vartheta)}, \tag{13.4d}$$

(13.4e)
$$\frac{\partial}{\partial s}\upmu_{(Approx)}(s,u,\vartheta;t) = \frac{\upmu_{(Approx)}(s,u,\vartheta;t) - 1}{\varrho(s,u)\left\{1 + \ln\left(\frac{\varrho(s,u)}{\varrho(0,u)}\right)\right\}} + \frac{\widetilde{\Omega}(t,u,\vartheta;t) - \widetilde{M}(s,u,\vartheta;t)}{\varrho(s,u)\left\{1 + \ln\left(\frac{\varrho(s,u)}{\varrho(0,u)}\right)\right\}} - \widetilde{m}(s,u,\vartheta;t),$$

$$\frac{\partial}{\partial s}\upmu(s,u,\vartheta) = \frac{\Omega(t,u,\vartheta)}{\varrho(s,u)} - m(s,u,\vartheta;t), \tag{13.4f}$$

$$\frac{\partial}{\partial s}\upmu(s,u,\vartheta) = \{\mathring{\upmu}(u,\vartheta) - M(0,u,\vartheta;t)\}\left\{\frac{\widetilde{\Omega}(t,u,\vartheta;t)}{\varrho(s,u)} - \widetilde{m}(s,u,\vartheta;t)\right\}, \tag{13.4g}$$

$$\upmu(s,u,\vartheta) = \{\mathring{\upmu}(u,\vartheta) - M(0,u,\vartheta;t)\}\,\upmu_{(Approx)}(s,u,\vartheta;t). \tag{13.4h}$$

□

We now define some additional quantities that play a role in our analysis of $\upmu$.

DEFINITION 13.5 (**$\upmu_{(Min)}$, $\upmu_\star$, and $\widetilde{\Omega}_{(Min)}$**). We define the following quantities, where we assume that $0 \leq s \leq t$ in (13.7):

$$\upmu_{(Min)}(s,u) := \min_{\Sigma_s^u} \upmu = \min_{u' \in [0,u], \vartheta \in \mathbb{S}^2} \upmu(s,u',\vartheta), \tag{13.5}$$

$$\upmu_\star(s,u) := \min\left\{1, \min_{\Sigma_s^u} \upmu\right\} = \min\left\{1, \upmu_{(Min)}(s,u)\right\}, \tag{13.6}$$

$$\widetilde{\Omega}_{(Min)}(s,u;t) := \min_{u' \in [0,u], \vartheta \in \mathbb{S}^2} \widetilde{\Omega}(s,u',\vartheta;t). \tag{13.7}$$

The function $\widetilde{\Omega}$ appearing on the right-hand side of (13.7) is defined by (13.3e).

REMARK 13.6 (**Some redundancy in definition** (13.6)). Since $\upmu \equiv 1$ along $\mathcal{C}_0$, there is redundancy in definition (13.6); we have taken the min against 1 only to emphasize that $\upmu_\star(t,u) \leq 1$.

In the next lemma, we provide some basic estimates for the auxiliary quantities. Below, we use these basic estimates to prove more intricate estimates (see Prop. 13.9).

LEMMA 13.7 (**First estimates for the auxiliary quantities used to analyze $\upmu$**). *Recall that $\varrho(t,u) = 1 - u + t$ is the geometric radial variable. Under the small data and bootstrap assumptions of Sects. 12.1-12.4, if ε is sufficiently small, then the following pointwise estimates hold for $(t,u,\vartheta) \in [0, T_{(Bootstrap)}) \times [0,U_0] \times \mathbb{S}^2$ and $0 \leq s \leq t$:*

$$\left|\mathring{\upmu}(u,\vartheta) - 1\right| \lesssim \varepsilon, \tag{13.8}$$

$$\left|\mathring{\upmu}(u,\vartheta) - M(0,u,\vartheta;t) - 1\right| \lesssim \varepsilon, \tag{13.9}$$

$$\left|\Omega(s,u,\vartheta)\right|, \left|\widetilde{\Omega}(s,u,\vartheta;t)\right| \lesssim \varepsilon, \tag{13.10}$$

$$\left|\Omega(t,u,\vartheta) - \frac{1}{2}[\varrho G_{LL} \breve{R}\Psi](t,u,\vartheta)\right| \lesssim \varepsilon \frac{1}{1+t}. \tag{13.11}$$

In addition, the following pointwise estimates hold:

(13.12)
$$\left|\Omega(s,u,\vartheta) - \Omega(t,u,\vartheta)\right|, \left|\widetilde{\Omega}(s,u,\vartheta;t) - \widetilde{\Omega}(t,u,\vartheta;t)\right| \lesssim \varepsilon \frac{\ln(e+s)}{1+s}\left(\frac{t-s}{1+t}\right),$$

$$\left|m(s,u,\vartheta;t)\right|, \left|\widetilde{m}(s,u,\vartheta;t)\right| \lesssim \varepsilon \frac{\ln(e+s)}{(1+s)^2}\left(\frac{t-s}{1+t}\right), \tag{13.13}$$

$$\left|M(s,u,\vartheta;t)\right|, \left|\widetilde{M}(s,u,\vartheta;t)\right| \lesssim \varepsilon \frac{\ln(e+s)}{1+s}\left(\frac{t-s}{1+t}\right), \tag{13.14}$$

$$\left|\upmu_{(Approx)}(s,u,\vartheta;t) - 1\right| \lesssim \varepsilon \ln(e+s), \tag{13.15}$$

(13.16)

$$(1 - C\varepsilon)\upmu_{(Approx)}(s, u, \vartheta; t) \leq \upmu(s, u, \vartheta) \leq (1 + C\varepsilon)\upmu_{(Approx)}(s, u, \vartheta; t),$$

(13.17)

$$(1 - C\varepsilon)\frac{\partial}{\partial s}\upmu_{(Approx)}(s, u, \vartheta; t) \leq \frac{\partial}{\partial s}\upmu(s, u, \vartheta) \leq (1 + C\varepsilon)\frac{\partial}{\partial s}\upmu_{(Approx)}(s, u, \vartheta; t),$$

$$\left|\upmu(s, u, \vartheta) - \left\{\mathring{\upmu}(u, \vartheta) + \frac{1}{2}\ln\left(\frac{\varrho(s, u)}{\varrho(0, u)}\right)[\varrho G_{LL}\breve{R}\Psi](t, u, \vartheta)\right\}\right| \lesssim \varepsilon. \tag{13.18}$$

Proof. The estimate (13.8) follows from (12.22b) and (12.3).

The estimate (13.10) for Ω follows from definition (13.3a) and the estimate (12.128b).

To prove (13.11), we first use equation (4.5) and definition (13.3a) to deduce that

$$\Omega - \frac{1}{2}\varrho G_{LL}\breve{R}\Psi = \varrho G_{(Frame)}L\Psi. \tag{13.19}$$

The desired estimate (13.11) now follows from (13.19), the estimate (12.58b), and the bootstrap assumptions (**BA**Ψ).

We now prove the estimate (13.12) for Ω. We first use equation (4.5) and definition (13.3a) to deduce that

$$L\Omega = L\left\{\frac{1}{2}G_{LL}\varrho\breve{R}\Psi - \frac{1}{2}\upmu G_{LL}\varrho L\Psi - \upmu G_{LR}\varrho L\Psi\right\}, \tag{13.20}$$

where we view the left-hand and right-hand sides of (13.20) as functions of (s', u, ϑ) and $L = \frac{\partial}{\partial s'}$. From the estimates (12.58b), (12.128a), (12.128b), (12.140), and (12.167) and the bootstrap assumptions (**BA**Ψ), it follows that the magnitude of the right-hand side of (13.20) is $\lesssim \varepsilon\ln(e + s')(e + s')^{-2}$. Hence, we have

$$\begin{aligned}
|\Omega(s, u, \vartheta) - \Omega(t, u, \vartheta)| &\lesssim \varepsilon\int_{s'=s}^{t}\frac{\ln(e + s')}{(e + s')^2}\,ds' \\
&= \varepsilon\left\{\frac{\ln(e + s) + 1}{e + s}\right\} - \varepsilon\left\{\frac{\ln(e + t) + 1}{e + t}\right\} \\
&\leq \varepsilon\left\{\frac{\ln(e + s) + 1}{e + s}\right\} - \varepsilon\left\{\frac{\ln(e + s) + 1}{e + t}\right\} \\
&\leq C\varepsilon\frac{\ln(e + s)}{e + s}\left(\frac{t - s}{e + t}\right) \leq C\varepsilon\frac{\ln(e + s)}{1 + s}\left(\frac{t - s}{1 + t}\right).
\end{aligned} \tag{13.21}$$

We have thus proved the desired inequality (13.12) for the first term on the left-hand side.

The estimate (13.13) for m then follows easily from definition (13.3b) and the inequality (13.12) for Ω.

The estimate (13.14) for M then follows from definition (13.3c), the inequality (13.13) for m, and the reasoning that we used to obtain (13.21).

The estimate (13.9) then follows from (13.8) and the estimate (13.14) for $|M(0, u, \vartheta; t)|$.

The estimates (13.10), (13.12), (13.13), and (13.14) for $\widetilde{\Omega}$, $\widetilde{m}$, and $\widetilde{M}$ then follow from definitions (13.3e), (13.3f), (13.3g), the corresponding estimates for the nontilded quantities, and the estimate (13.9).

The estimate (13.15) then follows from definition (13.3h) and the estimates (13.10) and (13.14) involving $\widetilde{\Omega}$ and $\widetilde{M}$.

The estimate (13.16) follows from the identity (13.4h) and the estimate (13.9).

The estimate (13.17) follows from the identity (13.4d) and the estimate (13.16).

To prove (13.18), we first use definition (13.3a) and the estimates (13.11) and (13.12) to deduce that for $0 \leq s' \leq t$, we have

$$\left| L\mu(s', u, \vartheta) - \frac{1}{2}\frac{1}{\varrho(s', u)}[\varrho G_{LL}\breve{R}\Psi](t, u, \vartheta) \right| \lesssim \varepsilon \frac{\ln(e + s')}{(1 + s')^2}. \tag{13.22}$$

Integrating (13.22) along the integral curves of $L = \frac{\partial}{\partial s'}$ from $s' = 0$ to $s' = s$, we conclude (13.18). □

13.2. Sharp pointwise estimates for the inverse foliation density

When we derive our a priori L^2 estimates, it will be essential that we treat the region where μ is not appreciably shrinking in a different fashion than we treat the region where μ is appreciably shrinking. The estimates of Lemma 13.7 (see in particular definition (13.3h) and the estimates (13.14) and (13.16)) imply that along the portion of the integral curve of L corresponding to $0 \leq s \leq t$ and fixed (u, ϑ), the behavior of μ is essentially determined by $\widetilde{\Omega}(t, u, \vartheta; t)$. This discussion motivates the following definitions

DEFINITION 13.8 (**Regions of distinct μ behavior**). Let $\widetilde{\Omega}(s, u, \vartheta; t)$ be the function defined in (13.3e). For each $t \in [0, T_{(Bootstrap)})$, $s \in [0, t]$, and $u \in [0, U_0]$, we partition

$$[0, u] \times \mathbb{S}^2 = {}^{(+)}\mathcal{V}_t^u \cup {}^{(-)}\mathcal{V}_t^u, \tag{13.23a}$$

$$\Sigma_s^u = {}^{(+)}\Sigma_{s;t}^u \cup {}^{(-)}\Sigma_{s;t}^u, \tag{13.23b}$$

where

$${}^{(+)}\mathcal{V}_t^u := \left\{(u', \vartheta) \in [0, u] \times \mathbb{S}^2 \mid \widetilde{\Omega}(t, u', \vartheta; t) \geq 0\right\}, \tag{13.24a}$$

$${}^{(-)}\mathcal{V}_t^u := \left\{(u', \vartheta) \in [0, u] \times \mathbb{S}^2 \mid \widetilde{\Omega}(t, u', \vartheta; t) < 0\right\}, \tag{13.24b}$$

$${}^{(+)}\Sigma_{s;t}^u := \left\{(s, u', \vartheta) \in \Sigma_s^u \mid \widetilde{\Omega}(t, u', \vartheta; t) \geq 0\right\}, \tag{13.24c}$$

$${}^{(-)}\Sigma_{s;t}^u := \left\{(s, u', \vartheta) \in \Sigma_s^u \mid \widetilde{\Omega}(t, u', \vartheta; t) < 0\right\}. \tag{13.24d}$$

In the next proposition, we provide the sharp pointwise estimates for the inverse foliation density μ and some of its derivatives.

PROPOSITION 13.9 (**Sharp pointwise estimates for μ, $L\mu$, and $\breve{R}\mu$**). *Assume that the small data and bootstrap assumptions of Sects.* 12.1-12.4 *hold for* $(t, u, \vartheta) \in \mathcal{M}_{T_{(Bootstrap)}, U_0}$. *If ε is sufficiently small, $0 \leq s \leq t < T_{(Bootstrap)}$, and $0 \leq u \leq U_0$, then the following estimates hold, where $\varrho(t, u) = 1 - u + t$.*

Simple upper bound for $L\mu$ under all conditions.

$$\|L\mu\|_{C^0(\Sigma_s^u)} \leq C\varepsilon \frac{1}{1 + s}. \tag{13.25}$$

Upper bounds for $\frac{[L\upmu]_+}{\upmu}$ ***under all conditions.***

$$\left\| \frac{[L\upmu]_+}{\upmu} \right\|_{C^0(\Sigma_s^u)} \leq (1 + C\varepsilon) \frac{1}{\varrho(s,u)\left\{1 + \ln\left(\frac{\varrho(s,u)}{\varrho(0,u)}\right)\right\}}. \tag{13.26}$$

Furthermore,

$$\left\| \frac{[L\upmu]_+}{\upmu} \right\|_{C^0(\Sigma_s^u)} \leq C\varepsilon \frac{1}{1+s}. \tag{13.27}$$

Small $\upmu$ ***implies*** $L\upmu$ ***is negative.***

$$\upmu(s,u,\vartheta) \leq \frac{1}{4} \implies L\upmu(s,u,\vartheta) \leq \frac{-2}{\varrho(s,u)\left\{1 + \ln\left(\frac{\varrho(s,u)}{\varrho(0,u)}\right)\right\}}. \tag{13.28}$$

Upper bound for $\frac{[\breve{R}\upmu]_+}{\upmu}$ ***under all conditions.***

$$\left\| \frac{[\breve{R}\upmu]_+}{\upmu} \right\|_{C^0(\Sigma_s^u)} \leq C\varepsilon^{1/2} \frac{\ln(e+s)}{\sqrt{\ln(e+t) - \ln(e+s)}} + C\varepsilon \ln(e+s). \tag{13.29}$$

Upper bound for $\frac{[L\upmu + \breve{\underline{L}}\upmu]_+}{\upmu}$ ***under all conditions.***

$$\left\| \frac{[L\upmu + \breve{\underline{L}}\upmu]_+}{\upmu} \right\|_{C^0(\Sigma_s^u)} \leq C\varepsilon^{1/2} \frac{\ln(e+s)}{\sqrt{\ln(e+t) - \ln(e+s)}} + C\varepsilon \ln(e+s). \tag{13.30}$$

Sharp spatially uniform estimates when $\upmu$ ***is not decaying.*** *Consider a time interval* $s \in [0,t]$ *and assume the right endpoint inequality*

$$\widetilde{\Omega}_{(Min)}(t,u;t) \geq 0, \tag{13.31}$$

where $\widetilde{\Omega}_{(Min)}$ *is defined in* (13.7). *Let* $\upmu_\star(t,u)$ *be the quantity defined in* (13.6) *and let*

$$B > 0$$

be a real number. Then there exists a constant $C > 0$ *such that if* $B\sqrt{\varepsilon} < 1$, *then the following estimates hold for* $s \in [0,t]$:

$$1 - C\varepsilon \leq \upmu_\star(s,u) \leq 1, \tag{13.32a}$$

$$1 - C\sqrt{\varepsilon} \leq \upmu_\star^B(s,u) \leq 1, \tag{13.32b}$$

$$\|\varrho[L\upmu]_-\|_{C^0(\Sigma_s^u)} \leq C\varepsilon \frac{\ln(e+s)}{1+s}. \tag{13.32c}$$

Sharp spatially uniform estimates when $\upmu$ ***is decaying.*** *Consider a time interval* $s \in [0,t]$ *and assume the right endpoint inequality*

$$\delta := -\widetilde{\Omega}_{(Min)}(t,u;t) > 0, \tag{13.33}$$

where $\widetilde{\Omega}_{(Min)}$ *is defined in* (13.7). *Then the following small-time and large-time estimates hold.*

Small-time estimates. *Let* $\upmu_{\star}(t,u)$ *be the quantity defined in* (13.6) *and let*

$$B > 0$$

be a real number. There exists a constant $C > 0$ *such that if* $B\sqrt{\varepsilon}|\ln\varepsilon| < 1$ *and* $s \in \left[0, (1-u)\frac{\varepsilon}{\delta^2} - (1-u)\right)$, *then*

$$1 - C\varepsilon|\ln\varepsilon| \leq \upmu_{\star}(s,u) \leq 1, \tag{13.34a}$$

$$1 - C\sqrt{\varepsilon} \leq \upmu_{\star}^{B}(s,u) \leq 1, \tag{13.34b}$$

$$\|\varrho[L\upmu]_-\|_{C^0(\Sigma_s^u)} = \|\Omega_-\|_{C^0(\Sigma_s^u)} \leq (1+C\varepsilon)\left\{\delta + C\varepsilon\frac{\ln(e+s)}{1+s}\right\}. \tag{13.34c}$$

Large-time estimates. *Let* $\upmu_{\star}(t,u)$ *be the quantity defined in* (13.6). *Under the hypotheses of the small-time estimates, there exists a constant* $C > 0$ *such that if* $B\sqrt{\varepsilon}|\ln\varepsilon| < 1$ *and* $s \in \left[(1-u)\frac{\varepsilon}{\delta^2} - (1-u), t\right]$, *then*

(13.35a)

$$(1-C\varepsilon|\ln\varepsilon|)\left\{1-\delta\ln\left(\frac{\varrho(s,u)}{\varrho(0,u)}\right)\right\} \leq \upmu_{\star}(s,u) \leq (1+C\varepsilon|\ln\varepsilon|)\left\{1-\delta\ln\left(\frac{\varrho(s,u)}{\varrho(0,u)}\right)\right\},$$

(13.35b)

$$(1-C\sqrt{\varepsilon})\left\{1-\delta\ln\left(\frac{\varrho(s,u)}{\varrho(0,u)}\right)\right\}^{B} \leq \upmu_{\star}^{B}(s,u) \leq (1+C\sqrt{\varepsilon})\left\{1-\delta\ln\left(\frac{\varrho(s,u)}{\varrho(0,u)}\right)\right\}^{B},$$

(13.35c)

$$\|\varrho[L\upmu]_-\|_{C^0(\Sigma_s^u)} = \|\Omega_-\|_{C^0(\Sigma_s^u)} \leq (1+C\sqrt{\varepsilon})\,\delta.$$

<u>***Sharp estimates when*** $(u',\vartheta) \in {}^{(+)}\mathcal{V}_t^u$</u>. *For each* $t \geq 0$, *we define the* ($t-$*dependent) constant* $\gamma \geq 0$ *by*

$$\gamma := \sup_{(u',\vartheta)\in{}^{(+)}\mathcal{V}_t^u} \widetilde{\Omega}(t,u',\vartheta;t), \tag{13.36}$$

where $\widetilde{\Omega}$ *is defined in* (13.3e) *and* ${}^{(+)}\mathcal{V}_t^u$ *is defined in* (13.24a).

Then the following estimate holds:

$$\gamma \leq C\varepsilon. \tag{13.37}$$

In addition, if $0 \leq s_1 \leq s_2 \leq t$, *then the following estimates hold:*

$$\sup_{(u',\vartheta)\in{}^{(+)}\mathcal{V}_t^u} \frac{\upmu(s_2,u',\vartheta)}{\upmu(s_1,u',\vartheta)} \leq (1+C\varepsilon)\frac{1+\gamma\left\{1+\ln\left(\frac{\varrho(s_2,u')}{\varrho(0,u')}\right)\right\}}{1+\gamma\left\{1+\ln\left(\frac{\varrho(s_1,u')}{\varrho(0,u')}\right)\right\}}, \tag{13.38a}$$

$$\sup_{(u',\vartheta)\in{}^{(+)}\mathcal{V}_t^u} \frac{\upmu(s_2,u',\vartheta)}{\upmu(s_1,u',\vartheta)} \leq (1+C\varepsilon)\ln(e+s_2). \tag{13.38b}$$

Furthermore, there exists a constant $C > 0$ *such that if* $0 < a \leq 1$ *is any constant, then*

$$\sup_{\substack{1\leq s_2^a\leq s_1\leq s_2\leq t\\(u',\vartheta)\in{}^{(+)}\mathcal{V}_t^u}} \frac{\upmu(s_2,u',\vartheta)}{\upmu(s_1,u',\vartheta)} \leq (1+C\varepsilon)\frac{1}{a}. \tag{13.39}$$

In addition, if $s \in [0,t]$ and ${}^{(+)}\Sigma_{s;t}^u$ is as defined in (13.24c), then we have

(13.40)
$$\left\| \frac{L\upmu}{\upmu} \right\|_{C^0({}^{(+)}\Sigma_{s;t}^u)} \leq (1+C\varepsilon)\frac{\upgamma}{\varrho(s,u)\left[1+\upgamma\left\{1+\ln\left(\frac{\varrho(s,u)}{\varrho(0,u)}\right)\right\}\right]} + C\varepsilon\frac{\ln(e+s)}{(1+s)^2}.$$

Furthermore, if $s \in [0,t]$, then we have

$$\left\| \frac{[L\upmu]_-}{\upmu} \right\|_{C^0({}^{(+)}\Sigma_{s;t}^u)} \leq C\varepsilon\frac{\ln(e+s)}{(1+s)^2}. \tag{13.41}$$

***Sharp estimates when** $(u',\vartheta) \in {}^{(-)}\mathcal{V}_t^u$.* *Assume that the set ${}^{(-)}\mathcal{V}_t^u$ defined in (13.24b) is nonempty and consider a time interval $s \in [0,t]$. Let $\updelta > 0$ be as in (13.33) and note that since ${}^{(-)}\mathcal{V}_t^u$ is nonempty, we have $\updelta = -\min_{(u',\vartheta)\in{}^{(-)}\mathcal{V}_t^u}\widetilde{\Omega}(t,u',\vartheta;t)$. Then the following estimate holds:*

$$\sup_{\substack{0\leq s_1\leq s_2\leq t \\ (u',\vartheta)\in{}^{(-)}\mathcal{V}_t^u}} \frac{\upmu(s_2,u',\vartheta)}{\upmu(s_1,u',\vartheta)} \leq 1+C\sqrt{\varepsilon}. \tag{13.42}$$

Furthermore, if $s \in [0,t]$ and ${}^{(-)}\Sigma_{s;t}^u$ is as defined in (13.24d), then the following estimate holds:

$$\|[L\upmu]_+\|_{C^0({}^{(-)}\Sigma_{s;t}^u)} \leq C\varepsilon\frac{\ln(e+s)}{(1+s)^2}. \tag{13.43}$$

In addition, there exists a constant $C>0$ such that if $0 \leq s \leq (1-u)\frac{\varepsilon}{\updelta^2}-(1-u) \leq t$, then

$$\|\varrho[L\upmu]_-\|_{C^0({}^{(-)}\Sigma_{s;t}^u)} \leq (1+C\varepsilon)\left\{\updelta + C\varepsilon\frac{\ln(e+s)}{1+s}\right\}. \tag{13.44a}$$

Finally, there exists a constant $C>0$ such that if $(1-u)\frac{\varepsilon}{\updelta^2}-(1-u) \leq s \leq t$, then

$$\|\varrho[L\upmu]_-\|_{C^0({}^{(-)}\Sigma_{s;t}^u)} \leq \left(1+C\sqrt{\varepsilon}\right)\updelta. \tag{13.44b}$$

***Approximate time-monotonicity of** $\upmu_\star^{-B}(s,u)$.* *Let $\upmu_\star(t,u)$ be the quantity defined in (13.6) and let*

$$B > 0$$

be a real number. There exists a constant $C>0$ such that if $B\sqrt{\varepsilon}|\ln\varepsilon| \leq 1$ and $0 \leq s_1 \leq s_2 \leq t$, then

$$\upmu_\star^{-B}(s_1,u) \leq (1+C\sqrt{\varepsilon})\upmu_\star^{-B}(s_2,u). \tag{13.45}$$

REMARK 13.10 (**A choice made out of convenience**). There is nothing special about the value $1/4$ in inequality (13.28); we could prove a similar estimate with "$1/4$" replaced by any positive constant less than 1 (but the smallness of ε would have to be adapted to that constant).

PROOF.
Proof of (13.25)**:** The estimate (13.25) has already been proved as (12.128b).

Proof of (13.26)-(13.27)**:** To prove (13.26), we note that by (13.4d), it suffices to bound the ratio

$$\frac{[\frac{\partial}{\partial s}\upmu_{(Approx)}(s,u,\vartheta;t)]_+}{\upmu_{(Approx)}(s,u,\vartheta;t)} \tag{13.46}$$

by the right-hand side of (13.26). To bound the ratio (13.46), we first divide both sides of the identity (13.4e) by $\upmu_{(Approx)}(s,u,\vartheta;t)$ to deduce

$$\begin{aligned} \frac{\frac{\partial}{\partial s}\upmu_{(Approx)}(s,u,\vartheta;t)}{\upmu_{(Approx)}(s,u,\vartheta;t)} &= \frac{1}{\varrho(s,u)\left\{1+\ln\left(\frac{\varrho(s,u)}{\varrho(0,u)}\right)\right\}} \\ &\quad - \frac{1}{\upmu_{(Approx)}(s,u,\vartheta;t)\varrho(s,u)\left\{1+\ln\left(\frac{\varrho(s,u)}{\varrho(0,u)}\right)\right\}} \\ &\quad + \frac{\widetilde{\Omega}(t,u,\vartheta;t)-\widetilde{M}(s,u,\vartheta;t)}{\upmu_{(Approx)}(s,u,\vartheta;t)\varrho(s,u)\left\{1+\ln\left(\frac{\varrho(s,u)}{\varrho(0,u)}\right)\right\}} \\ &\quad - \frac{\widetilde{m}(s,u,\vartheta;t)}{\upmu_{(Approx)}(s,u,\vartheta;t)}. \end{aligned} \tag{13.47}$$

We now observe that the right-hand side of (13.47) is negative (and hence the ratio (13.46) is 0) unless

$$\begin{aligned} \upmu_{(Approx)}(s,u,\vartheta;t)-1 \geq &\frac{-\widetilde{\Omega}(t,u,\vartheta;t)+\widetilde{M}(s,u,\vartheta;t)}{\varrho(s,u)\left\{1+\ln\left(\frac{\varrho(s,u)}{\varrho(0,u)}\right)\right\}} \\ &+ \widetilde{m}(s,u,\vartheta;t)\varrho(s,u)\left\{1+\ln\left(\frac{\varrho(s,u)}{\varrho(0,u)}\right)\right\}. \end{aligned} \tag{13.48}$$

We may thus assume that inequality (13.48) holds, in which case the estimates (13.10), (13.13), and (13.14) imply that

$$\frac{1}{\upmu_{(Approx)}(s,u,\vartheta;t)} \leq 1 + C\varepsilon\frac{\left\{1+\ln\left(\frac{\varrho(s,u)}{\varrho(0,u)}\right)\right\}^2}{\varrho(s,u)}. \tag{13.49}$$

Now since the positive part of the left-hand side of (13.47) is bounded by the sum of the first term on the right-hand side and the sum of the magnitudes of the last three terms on the right-hand side, the estimates (13.10), (13.13), (13.14), and (13.49) thus yield that (13.46) is bounded by the right-hand side of (13.26) as desired.

To prove (13.27), we use a slightly different argument that takes into account the bound (13.15). Inserting this bound and the estimates (13.10), (13.13), (13.14), and (13.49) into (13.47), we deduce that

$$
\begin{aligned}
&\frac{[\frac{\partial}{\partial s}\upmu_{(Approx)}(s,u,\vartheta;t)]_+}{\upmu_{(Approx)}(s,u,\vartheta;t)} \\
&\leq \frac{1}{\varrho(s,u)\left\{1+\ln\left(\frac{\varrho(s,u)}{\varrho(0,u)}\right)\right\}}\frac{|\upmu_{(Approx)}(s,u,\vartheta;t)-1|}{\upmu_{(Approx)}(s,u,\vartheta;t)} \\
&\quad + \frac{1}{\varrho(s,u)\left\{1+\ln\left(\frac{\varrho(s,u)}{\varrho(0,u)}\right)\right\}}\frac{1}{\upmu_{(Approx)}(s,u,\vartheta;t)}\left|\widetilde{\Omega}(t,u,\vartheta;t)-\widetilde{M}(s,u,\vartheta;t)\right| \\
&\quad + \frac{1}{\upmu_{(Approx)}(s,u,\vartheta;t)}|\widetilde{m}(s,u,\vartheta;t)| \\
&\leq C\varepsilon\frac{1}{1+s}.
\end{aligned}
\tag{13.50}
$$

The desired estimate (13.27) now follows from (13.4d) and (13.50).

Proof of (13.28)**:** To prove (13.28), we first note that the identity (13.47) and the estimates (13.10), (13.13), and (13.14) imply that whenever

$$
\upmu_{(Approx)}(s,u,\vartheta;t) \leq \frac{1}{3.5}, \tag{13.51}
$$

the following inequality necessarily holds:

$$
\frac{\partial}{\partial s}\upmu_{(Approx)}(s,u,\vartheta;t) \leq \frac{1-3.5+C\varepsilon}{\varrho(s,u)\left\{1+\ln\left(\frac{\varrho(s,u)}{\varrho(0,u)}\right)\right\}}. \tag{13.52}
$$

The desired conclusion (13.28) follows from the estimates (13.9), (13.16), and (13.52), and the following simple consequence of (13.4h):

$$
L\upmu(s,u,\vartheta) = \{\mathring{\upmu}(u,\vartheta) - M(0,u,\vartheta;t)\}\frac{\partial}{\partial s}\upmu_{(Approx)}(s,u,\vartheta;t). \tag{13.53}
$$

Proof of (13.32a)-(13.32c)**:** To prove (13.32a) and (13.32b), we first note that since $\widetilde{\Omega}(t,u,\vartheta;t) \geq \widetilde{\Omega}_{(Min)}(t,u;t) \geq 0$, (13.3h) and (13.14) imply that the following estimate holds for $s \in [0,t]$:

$$
\upmu_{(Approx)}(s,u,\vartheta;t) \geq 1 - C\varepsilon. \tag{13.54}
$$

Furthermore, (13.4h), (13.6), (13.9), and the estimate (13.54) imply that the following estimate holds for $s \in [0,t]$:

$$
\upmu_\star^B(s,u) \geq (1-C\varepsilon)^B. \tag{13.55}
$$

In particular, we have proved (13.32a) (the upper bound is trivial). Furthermore, since the function $h(y) := (1-y)^{1/y}$ monotonically increases to e^{-1} as y decreases to 0, we can set $y = C\varepsilon$ and use (13.55) and the trivial bound $\exp(-CB\varepsilon) \geq 1 - CB\varepsilon$ to conclude the desired estimate (13.32b) (the upper bound is trivial).

To prove (13.32c), we first decompose

$$
\widetilde{\Omega}(s,u,\vartheta;t) = \widetilde{\Omega}(t,u,\vartheta;t) + \left\{\widetilde{\Omega}(s,u,\vartheta;t) - \widetilde{\Omega}(t,u,\vartheta;t)\right\}. \tag{13.56}
$$

Then since $\widetilde{\Omega}(t,u,\vartheta;t) \geq 0$, it follows from (13.12) and (13.56) that

(13.57)
$$\widetilde{\Omega}_-(s,u,\vartheta;t) \leq \left|\widetilde{\Omega}(s,u,\vartheta;t) - \widetilde{\Omega}(t,u,\vartheta;t)\right| \leq C\varepsilon\frac{\ln(e+s)}{1+s}\left(\frac{t-s}{1+t}\right) \leq C\varepsilon\frac{\ln(e+s)}{1+s}.$$

The desired estimate (13.32c) now easily follows from definitions (13.3a) and (13.3e) and inequalities (13.9) and (13.57).

Proof of (13.34a)-(13.34c): We first prove (13.34a). In order to simplify the presentation, we define

(13.58)
$$t_1 := (1-u)\frac{\varepsilon}{\delta^2} - (1-u),$$

where δ is defined in (13.33). Note that by (13.10), the following simple inequality holds:

(13.59)
$$\delta \leq C\varepsilon.$$

Hence, it follows from (13.58) and (13.59) that if ε is sufficiently small, then

(13.60)
$$t_1 \geq \frac{1}{C\varepsilon}.$$

We may thus assume that (13.60) holds.

For the estimates under consideration, we are assuming that

(13.61)
$$0 \leq s \leq t_1.$$

From (13.58), (13.59), and the relations $\ln\left(\frac{\varrho(t_1,u)}{\varrho(0,u)}\right) = \ln\left(\frac{\varepsilon}{\delta^2}\right) \leq |\ln\delta| + \left|\ln\frac{\delta}{\varepsilon}\right|$, we deduce that for $0 \leq s \leq t_1$, we have

(13.62)
$$\begin{aligned} 1 - \delta\ln\left(\frac{\varrho(s,u)}{\varrho(0,u)}\right) &\geq 1 - \delta\ln\left(\frac{\varrho(t_1,u)}{\varrho(0,u)}\right) \geq 1 - \delta\left\{|\ln\delta| + \left|\ln\frac{\delta}{\varepsilon}\right|\right\} \\ &\geq 1 - C\varepsilon|\ln\varepsilon|. \end{aligned}$$

It then follows from (13.3h), (13.4h), (13.9), (13.14), (13.33), and (13.62) that for $0 \leq s \leq t_1$, we have

(13.63)
$$\mu(s,u,\vartheta) \geq 1 - C\varepsilon|\ln\varepsilon|.$$

Since inequality (13.63) is independent of u and ϑ, the same estimate holds for $\mu_\star(s,u)$. Hence, we have

(13.64)
$$\mu_\star^B(s,u) \geq (1 - C\varepsilon|\ln\varepsilon|)^B.$$

In particular, we have proved (13.34a) (the upper bound is trivial). Furthermore, by arguing as in our proof of (13.32b) (see the discussion below (13.55)) and in particular using (13.64), we deduce that

(13.65)
$$\mu_\star^B(s,u) \geq \exp(-CB\varepsilon|\ln\varepsilon|) \geq 1 - CB\varepsilon|\ln\varepsilon| \geq 1 - C\sqrt{\varepsilon}.$$

We have thus shown that the desired bounds (13.34b) hold for $0 \leq s \leq t_1$ (the upper bound is trivial).

To prove (13.34c), we first use the splitting (13.56) and the estimate (13.12) to deduce that

(13.66)
$$\widetilde{\Omega}_-(s,u,\vartheta;t) \leq \delta + C\varepsilon\frac{\ln(e+s)}{1+s}.$$

The desired bound (13.34c) now follows from definitions (13.3a) and (13.3e) and inequalities (13.9) and (13.66).

Proof of (13.35a)-(13.35c)**:** Let t_1 be as in (13.58). We introduce the logarithmic variables

$$\tau := \ln\left(\frac{\varrho(t,u)}{\varrho(0,u)}\right), \tag{13.67}$$

$$\tau_1 := \ln\left(\frac{\varrho(t_1,u)}{\varrho(0,u)}\right) = \ln\left(\frac{\varepsilon}{\delta^2}\right), \tag{13.68}$$

$$\sigma := \ln\left(\frac{\varrho(s,u)}{\varrho(0,u)}\right). \tag{13.69}$$

To prove (13.35a), we first note that (13.3h) implies that for $(t,u) \in [0, T_{(Bootstrap)}) \times [0, U_0]$, we have

$$\mu_{(Approx)}(t,u,\vartheta;t) = 1 - \delta \ln\left(\frac{\varrho(t,u)}{\varrho(0,u)}\right). \tag{13.70}$$

Since $(\mathbf{BA}\mu > 0)$, (13.4h), and (13.9) imply that $0 < \mu_{(Approx)}$, it follows from (13.70) that

$$\delta \ln\left(\frac{\varrho(t,u)}{\varrho(0,u)}\right) < 1. \tag{13.71}$$

We now use (13.3h), (13.4h), and (13.14) to deduce that for $s \in [t_1, t]$, we have

(13.72)

$$\mu(s,u,\vartheta) \geq \{\mathring{\mu}(u,\vartheta) - M(0,u,\vartheta;t)\}\left\{1 - \delta \ln\left(\frac{\varrho(s,u)}{\varrho(0,u)}\right) - C\varepsilon\frac{\ln(e+s)}{1+s}\left(\frac{t-s}{1+t}\right)\right\}.$$

Using (13.72) and the inequalities

$$\frac{\ln(e+s)}{1+s} \leq C\frac{1+\sigma}{\exp(\sigma)}, \tag{13.73}$$

$$\frac{t-s}{1+t} \leq \frac{\varrho(t,u) - \varrho(s,u)}{\varrho(t,u)}, \tag{13.74}$$

$$0 \leq \frac{\varrho(t,u) - \varrho(s,u)}{\varrho(t,u)} = \frac{\exp(\tau) - \exp(\sigma)}{\exp(\tau)} = 1 - \exp(\sigma - \tau) \leq \tau - \sigma \tag{13.75}$$

(where the last inequality in (13.75) follows from the fact that $1 - \exp(-y) \leq y$ whenever $y \geq 0$), we deduce that for $s \in [t_1, t]$, we have

(13.76)

$$\begin{aligned}\mu(s,u,\vartheta) &\geq \{\mathring{\mu}(u,\vartheta) - M(0,u,\vartheta;t)\}\left\{1 - \delta\sigma - C\varepsilon\frac{1+\sigma}{\exp(\sigma)}(\tau - \sigma)\right\} \\ &= \{\mathring{\mu}(u,\vartheta) - M(0,u,\vartheta;t)\}\left\{1 - \delta\tau + \delta\left(1 - C\frac{\varepsilon}{\delta}\frac{1+\sigma}{\exp(\sigma)}\right)(\tau - \sigma)\right\}.\end{aligned}$$

We now estimate the crucially important term $\frac{\varepsilon}{\delta}\frac{1+\sigma}{\exp(\sigma)}$ appearing in the second line of (13.76). To this end, we use the fact that $s \in [t_1, t]$, inequalities (13.59) and

(13.60), and the fact that the function $h(y) := (1+y)\exp(-y)$ is decreasing on the domain $[0, \infty)$ to deduce that

$$\frac{\varepsilon}{\delta}\frac{1+\sigma}{\exp(\sigma)} \leq \frac{\varepsilon}{\delta}\frac{1+\tau_1}{\exp(\tau_1)} = \delta\left\{1 + \ln\left(\frac{\varepsilon}{\delta^2}\right)\right\} \leq C\varepsilon|\ln\varepsilon|. \tag{13.77}$$

Inserting the bound (13.77) into inequality (13.76) and using the bound (13.9), we deduce the following lower bound for $s \in [t_1, t]$:

(13.78)

$$\begin{aligned}
\mu(s, u, \vartheta) &\geq \{\mathring{\mu}(u, \vartheta) - M(0, u, \vartheta; t)\}\left\{1 - \delta\tau + \delta\left(1 - C\frac{\varepsilon}{\delta}\frac{1+\sigma}{\exp(\sigma)}\right)(\tau - \sigma)\right\} \\
&\geq \{\mathring{\mu}(u, \vartheta) - M(0, u, \vartheta; t)\} \\
&\quad \times \{(1 - C\varepsilon|\ln\varepsilon|)(1 - \delta\tau) + \delta(1 - C\varepsilon|\ln\varepsilon|)(\tau - \sigma)\} \\
&= \{\mathring{\mu}(u, \vartheta) - M(0, u, \vartheta; t)\}(1 - C\varepsilon|\ln\varepsilon|)(1 - \delta\sigma) \\
&\geq (1 - C\varepsilon|\ln\varepsilon|)(1 - \delta\sigma).
\end{aligned}$$

Note that by (13.71), the right-hand side of (13.78) is positive. Since the final inequality in (13.78) is independent of ϑ and since $-\delta\sigma = -\delta\sigma(s, u)$ is decreasing in u, the same estimate holds for $\mu_\star(s, u)$. We have thus deduced the desired lower bound in (13.35a). A similar argument yields the upper bound. Then by using an argument similar to the one we used just after inequality (13.55), we also deduce from the bounds (13.35a) that the upper and lower bounds in (13.35b) hold.

To prove (13.35c), we first decompose $\widetilde{\Omega}(s, u, \vartheta; t)$ as in (13.56) and use definition (13.7), (13.33), and inequalities (13.12) and (13.73) to deduce

$$\widetilde{\Omega}_-(s, u, \vartheta; t) \leq \delta + C\varepsilon\frac{1+\sigma}{\exp(\sigma)} = \delta\left\{1 + C\frac{\varepsilon}{\delta}\frac{1+\sigma}{\exp(\sigma)}\right\}. \tag{13.79}$$

Arguing as in our proof of (13.77), we deduce from inequality (13.79) that

$$\widetilde{\Omega}_-(s, u, \vartheta; t) \leq (1 + C\varepsilon|\ln\varepsilon|)\,\delta \leq \left(1 + C\sqrt{\varepsilon}\right)\delta. \tag{13.80}$$

From definition (13.3e) and the estimates (13.9) and (13.80), we deduce the following bound for $s \in [t_1, t]$:

$$\Omega_-(s, u, \vartheta) \leq \left(1 + C\sqrt{\varepsilon}\right)\delta. \tag{13.81}$$

Since inequality (13.81) is independent of u and ϑ, the desired bound (13.35c) now follows from definition (13.3a).

Proof of (13.37)**:** Inequality (13.37) follows from (13.10).

Proof of (13.38a)-(13.39)**:** To prove (13.38a), we first note that by (13.4h), it suffices to show that for each $(u', \vartheta) \in {}^{(+)}\mathcal{V}_t^u$ and each $0 \leq s_1 \leq s_2 \leq t$, the ratio $\frac{\mu_{(Approx)}(s_2, u', \vartheta)}{\mu_{(Approx)}(s_1, u', \vartheta)}$ is $\leq$ the right-hand side of (13.38a). To this end, we use (13.3h) and the estimates (13.10) and (13.14) to deduce that

$$\frac{\upmu_{(Approx)}(s_2,u',\vartheta)}{\upmu_{(Approx)}(s_1,u',\vartheta)} = \frac{1+\widetilde{\Omega}(t,u',\vartheta;t)\left\{1+\ln\left(\frac{\varrho(s_2,u')}{\varrho(0,u')}\right)\right\}+\mathcal{O}(\varepsilon)}{1+\widetilde{\Omega}(t,u',\vartheta;t)\left\{1+\ln\left(\frac{\varrho(s_1,u')}{\varrho(0,u')}\right)\right\}+\mathcal{O}(\varepsilon)} \leq (1+C\varepsilon)\frac{1+\widetilde{\Omega}(t,u',\vartheta;t)\left\{1+\ln\left(\frac{\varrho(s_2,u')}{\varrho(0,u')}\right)\right\}}{1+\widetilde{\Omega}(t,u',\vartheta;t)\left\{1+\ln\left(\frac{\varrho(s_1,u')}{\varrho(0,u')}\right)\right\}}. \tag{13.82}$$

Since the right-hand side of (13.82) is increasing when viewed as a function of $\widetilde{\Omega}(t,u',\vartheta;t)$, it follows that the right-hand side of (13.82) is

$$\leq (1+C\varepsilon)\frac{1+\upgamma\left\{1+\ln\left(\frac{\varrho(s_2,u')}{\varrho(0,u')}\right)\right\}}{1+\upgamma\left\{1+\ln\left(\frac{\varrho(s_1,u')}{\varrho(0,u')}\right)\right\}} \tag{13.83}$$

as desired. We have thus proved (13.38a).

The estimate (13.38b) follows from (13.37) and (13.38a).

To deduce (13.39), we first note that (13.37) and (13.38a) imply that

$$\sup_{(u',\vartheta)\in {}^{(+)}\mathcal{V}_t^u}\frac{\upmu(s_2,u',\vartheta)}{\upmu(s_1,u',\vartheta)} \leq (1+C\varepsilon)\frac{1+\upgamma\ln(e+s_2)}{1+\upgamma\ln(e+s_1)}. \tag{13.84}$$

The desired estimate (13.39) now follows from (13.84) and the following simple estimate, which holds when $0<a\leq 1$ and $1\leq s_2^a\leq s_1\leq s_2$: $\ln(e+s_1)\geq \ln(e+s_2^a)\geq a\ln(e+s_2)$.

Proof of (13.40)**:** To prove (13.40), we first note that by (13.4d), it suffices to bound the ratio on the left-hand side of (13.47) on the domain $(s,u',\vartheta)\in {}^{(+)}\Sigma_{s;t}^u$ (see definition (13.23b)) by the right-hand side of (13.40). To this end, for each $(u',\vartheta)\in {}^{(+)}\mathcal{V}_t^u$, we use (13.3h) and (13.4c) and the estimates (13.10), (13.13), and (13.14) to deduce that

$$\frac{\frac{\partial}{\partial s}\upmu_{(Approx)}(s,u',\vartheta;t)}{\upmu_{(Approx)}(s,u',\vartheta;t)} \leq (1+C\varepsilon)\frac{1}{\varrho(s,u')}\frac{\widetilde{\Omega}(t,u',\vartheta;t)}{\left[1+\widetilde{\Omega}(t,u',\vartheta;t)\left\{1+\ln\left(\frac{\varrho(s,u')}{\varrho(0,u')}\right)\right\}\right]} + \mathcal{O}\left(\varepsilon\frac{\ln(e+s)}{(1+s)^2}\right). \tag{13.85}$$

Since the ratio $\dfrac{\widetilde{\Omega}(t,u',\vartheta;t)}{\left[1+\widetilde{\Omega}(t,u',\vartheta;t)\left\{1+\ln\left(\frac{\varrho(s,u')}{\varrho(0,u')}\right)\right\}\right]}$ is increasing when viewed as a function of $\widetilde{\Omega}(t,u',\vartheta;t)$, it follows that the right-hand side of (13.85) is bounded by

$$\leq (1+C\varepsilon)\frac{1}{\varrho(s,u')}\frac{\upgamma}{\left[1+\upgamma\left\{1+\ln\left(\frac{\varrho(s,u')}{\varrho(0,u')}\right)\right\}\right]} + \mathcal{O}\left(\varepsilon\frac{\ln(e+s)}{(1+s)^2}\right). \tag{13.86}$$

The desired estimate (13.40) now follows easily from (13.86).

Proof of (13.41)**:** To prove (13.41), we first use the splitting (13.56) and the estimate (13.12) to deduce that for $(s, u', \vartheta) \in {}^{(+)}\Sigma_{s;t}^{u}$ (see definition (13.24c)), we have

$$\widetilde{\Omega}_{-}(s, u', \vartheta; t) \leq C\varepsilon \frac{\ln(e+s)}{1+s}. \tag{13.87}$$

It thus follows from (13.3a), (13.3e), (13.9), and (13.87) that

$$[L\upmu]_{-}(s, u', \vartheta) \leq C\varepsilon \frac{\ln(e+s)}{(1+s)^2}, \tag{13.88}$$

$$L\upmu(s, u', \vartheta) \geq -C\varepsilon \frac{\ln(e+s)}{(1+s)^2}. \tag{13.89}$$

Integrating (13.89) along the integral curves of $L = \frac{\partial}{\partial s}$ emanating from Σ_0 and using the small-data estimate (13.8), we deduce that

$$\upmu(s, u, \vartheta) \geq 1 - C\varepsilon. \tag{13.90}$$

The desired estimate (13.41) now follows from (13.88) and (13.90).

Proof of (13.42): Note that this estimate concerns only points (s, u', ϑ) such that $(u', \vartheta) \in {}^{(-)}\mathcal{V}_t^u$ (see definition (13.24b)). We start by repeating the proof of inequality (13.63), but with $\updelta$ in the proof everywhere replaced by $-\widetilde{\Omega}(t, u, \vartheta; t)$, which changes the value of t_1 as defined in (13.58). The net effect is that for this new value of t_1, if $s \in [0, t_1]$, then we have

$$\upmu(s, u, \vartheta) \geq 1 - C\sqrt{\varepsilon}. \tag{13.91}$$

In addition, by making straightforward modifications to the proof of (13.91), we find that if $s \in [0, t_1]$, then

$$\upmu(s, u, \vartheta) \leq 1 + C\sqrt{\varepsilon}. \tag{13.92}$$

Similarly, we repeat the proof of inequality (13.78), but with $\updelta$ in the proof everywhere replaced by $-\widetilde{\Omega}(t, u, \vartheta; t)$. As above, this replacement changes the value of t_1 as defined in (13.58) and $\uptau_1$ as defined in (13.68). In total, we find that for this new value of t_1, if $s \in [t_1, t]$, then we have

$$\upmu(s, u', \vartheta) \geq (1 - C\sqrt{\varepsilon}) \left\{ 1 + \widetilde{\Omega}(t, u', \vartheta; t) \ln\left(\frac{\varrho(s, u')}{\varrho(0, u')}\right) \right\}. \tag{13.93}$$

In addition, by making straightforward modifications to the proof of (13.93), we find that for $s \in [t_1, t]$, we have

$$\upmu(s, u', \vartheta) \leq (1 + C\sqrt{\varepsilon}) \left\{ 1 + \widetilde{\Omega}(t, u', \vartheta; t) \ln\left(\frac{\varrho(s, u')}{\varrho(0, u')}\right) \right\}. \tag{13.94}$$

We now separately investigate the small-time regime $s_2 \leq t_1$ and the large-time regime $s_2 \geq t_1$, where we are referring to the new value of t_1 defined above. In the small-time regime, we easily bound the ratio $\frac{\upmu(s_2, u', \vartheta)}{\upmu(s_1, u', \vartheta)}$ by $\leq$ the right-hand side of (13.42) with the help of (13.91) and (13.92).

We now consider the large-time regime $s_2 \geq t_1$. We split this regime into two sub-cases: $s_1 \geq t_1$ and $s_1 \leq t_1$. In the first sub-case, we bound the ratio $\frac{\upmu(s_2, u', \vartheta)}{\upmu(s_1, u', \vartheta)}$ by $\leq$ the right-hand side of (13.42) by using the estimates (13.93) and (13.94), the assumption $\widetilde{\Omega}(t,u',\vartheta;t) < 0$, and the fact that $\ln\left(\frac{\varrho(s,u')}{\varrho(0,u')}\right)$ is increasing in s. In the second sub-case $s_1 \leq t_1$, we bound the ratio $\frac{\upmu(s_2,u',\vartheta)}{\upmu(s_1,u',\vartheta)}$ by $\leq$ the right-hand side of (13.42) by using the estimates (13.91) and (13.94) and the assumption $\widetilde{\Omega}(t,u',\vartheta;t) < 0$.

Proof of (13.43)**:** To prove (13.43), we first use the splitting (13.56) and the estimate (13.12) to deduce that for $(s,u',\vartheta) \in {}^{(-)}\Sigma_{s;t}^{u}$ (see definition (13.24d)), we have

$$\widetilde{\Omega}_{+}(s,u',\vartheta;t) \leq C\varepsilon\frac{\ln(e+s)}{1+s}. \tag{13.95}$$

From definitions (13.3a) and (13.3e) and the estimates (13.9) and (13.95), we conclude the desired estimate (13.43).

Proof of (13.44a)-(13.44b)**:** Inequality (13.44a) can be proved by using essentially the same argument that we used to prove (13.34c); we omit the details.

Similarly, inequality (13.44b) can be proved by using essentially the same argument that we used to prove (13.35c), and we omit the details.

Proof of (13.45)**:** We separately investigate the cases $\widetilde{\Omega}_{(Min)}(t,u;t) \geq 0$ and $\widetilde{\Omega}_{(Min)}(t,u;t) < 0$. In the former case, the desired estimate (13.45) follows from (13.32b).

We now investigate the remaining case in which (as in (13.33)) $\delta := -\widetilde{\Omega}_{(Min)}(t,u;t) > 0$. We separately investigate the small-time regime $s_2 \leq t_1$ and the large-time regime $s_2 \geq t_1$, where t_1 is defined in (13.58). In the small-time regime $s_2 \leq t_1$, the desired estimate (13.45) follows from (13.34b).

We now consider the large-time regime $s_2 \geq t_1$. This regime splits into two sub-cases: $s_1 \geq t_1$ and $s_1 \leq t_1$. In the first sub-case, the desired estimate (13.45) follows from (13.35b) and the fact that $1 - \delta\ln\left(\frac{\varrho(s,u)}{\varrho(0,u)}\right)$ is decreasing in s. In the second sub-case, the desired estimate (13.45) follows from (13.34b) and (13.35b).

Proof of (13.29)**:** We split the argument into four cases that exhaust all possibilities.

Case 1: $\widetilde{\Omega}_{(Min)}(t,U_0;t) \geq 0$*:* In this (easy) case, definitions (13.3h) and (13.7), the identity (13.4h), and the estimates (13.9) and (13.14) imply that

$$\upmu(s,u,\vartheta) \geq 1 - C\varepsilon. \tag{13.96}$$

Furthermore, from the estimate (12.128a), we have

$$\left\| \breve{R}\upmu \right\|_{C^0(\Sigma_s^u)} \leq C\varepsilon\ln(e+s). \tag{13.97}$$

Hence, in this case, we can use (13.96) and (13.97) to deduce that $\frac{[\breve{R}\upmu(s,u,\vartheta)]_{+}}{\upmu(s,u,\vartheta)}$ is $\leq$ the second term on the right-hand side of (13.29) as desired.

Case 2: Mild decay rate: In this case, we assume that $\widetilde{\Omega}_{(Min)}(t, U_0; t) < 0$. Using the notation $\delta := -\widetilde{\Omega}_{(Min)}(t, U_0; t) > 0$ much as in (13.33), we define the sets

$$\mathcal{V}(U_0; t) := \left\{(u, \vartheta) \in [0, U_0] \times \mathbb{S}^2 \mid \widetilde{\Omega}(t, u, \vartheta; t) < -\frac{\delta}{2}\right\}, \tag{13.98}$$

$$\mathcal{U}(s, U_0; t) := \left\{(s, u, \vartheta) \in \Sigma_s^{U_0} \mid (u, \vartheta) \in \mathcal{V}(U_0; t)\right\}. \tag{13.99}$$

Roughly, $\mathcal{U}(s, U_0; t)$ is the "dangerous" subset of $\Sigma_s^{U_0}$ where μ can decay at a relatively rapid rate. Note that $\mathcal{V}(U_0; t)$ is an open subset of $[0, U_0] \times \mathbb{S}^2$ while $\mathcal{U}(s, U_0; t)$ is an open subset of $\Sigma_s^{U_0}$.

Now if $(s, u, \vartheta) \in \Sigma_s^{U_0} \backslash \mathcal{U}(s, U_0; t)$, then we can slightly modify our proof of (13.96) (which was based in part on (13.3h)) and also use the bound (13.71) to deduce the following estimate:

(13.100)

$$\mu(s, u, \vartheta) \geq (1 - C\varepsilon)\left\{1 - \frac{1}{2}\delta \ln\left(\frac{\varrho(s, u)}{\varrho(0, u)}\right)\right\} \geq (1 - C\varepsilon)\left\{1 - \frac{1}{2}\delta \ln\left(\frac{\varrho(t, u)}{\varrho(0, u)}\right)\right\}$$
$$\geq \frac{1}{2}(1 - C\varepsilon).$$

From (13.97) and (13.100), we see that $\dfrac{[\breve{R}\mu(s, u, \vartheta)]_+}{\mu(s, u, \vartheta)}$ is $\leq$ the second term on the right-hand side of (13.29) as desired, as in Case 1.

Case 3: Dangerous decay rate, but short times: In this case, we assume that $\delta := -\widetilde{\Omega}_{(Min)}(t, U_0; t) > 0$, that $(s, u, \vartheta) \in \mathcal{U}(s, U_0; t)$, and that $s \leq t_1$, where we set

$$t_1 := (1 - u)\exp(\tau_1) - (1 - u), \tag{13.101}$$

$$\tau_1 := \frac{1}{2\delta}. \tag{13.102}$$

Using the estimate (13.59), we conclude that if ε is sufficiently small, then

$$\tau_1 \geq \frac{1}{C\varepsilon}. \tag{13.103}$$

As in our proof of (13.100), we can slightly modify our proof of (13.96) (which was based in part on (13.3h)) and also use the assumption $s \leq t_1$ to deduce the following estimate:

$$\mu(s, u, \vartheta) \geq (1 - C\varepsilon)\left\{1 - \delta \ln\left(\frac{\varrho(s, u)}{\varrho(0, u)}\right)\right\} \geq (1 - C\varepsilon)\{1 - \delta\tau_1\} \tag{13.104}$$
$$= \frac{1}{2}(1 - C\varepsilon).$$

Also using the bound (13.97), we see that $\dfrac{[\breve{R}\mu(s, u, \vartheta)]_+}{\mu(s, u, \vartheta)}$ is $\leq$ the second term on the right-hand side of (13.29) as desired, as in Cases 1 and 2.

Case 4: The most difficult case, involving dangerous decay rates and large times: In this case, we assume that $\delta := -\widetilde{\Omega}_{(Min)}(t, U_0; t) > 0$, that $(s, u, \vartheta) \in \mathcal{U}(s, U_0; t)$, and that $t_1 \leq s \leq t$, where t_1 is defined in (13.101). In particular, in terms of the logarithmic change of variable $\sigma := \ln\left(\frac{\varrho(s,u)}{\varrho(0,u)}\right)$, the following estimate holds for $t_1 \leq s \leq t$:

$$2\delta \geq \frac{1}{\sigma} \geq \frac{1}{\sigma+1}. \tag{13.105}$$

Furthermore, with $\tau := \ln\left(\frac{\varrho(t,u)}{\varrho(0,u)}\right)$, then as in (13.71), the following inequality holds:

$$\delta\tau < 1. \tag{13.106}$$

We may assume that $\breve{R}\mu(s,u,\vartheta) > 0$, since otherwise the estimate (13.29) is trivial. To prove (13.29) under the assumption $\breve{R}\mu(s,u,\vartheta) > 0$, we study the integral curves of $\breve{R}$. Thus, let

$$\gamma : [0, U_0] \to \Sigma_s^{U_0} \tag{13.107}$$

denote the integral curve of $\breve{R}$ passing through the point $p := (s,u,\vartheta)$. We parametrize γ via the eikonal function u', that is, $\gamma = \gamma(u')$, where $u' \in [0,U_0]$, and u is the value of u' that corresponds to the point p of interest. We use notation such as

$$\dot{\gamma}(u') := \frac{d}{du'}\gamma(u'), \qquad \ddot{\gamma}(u') := \frac{d^2}{d(u')^2}\gamma(u'). \tag{13.108}$$

We also set

$$F(u') := \mu \circ \gamma(u'), \tag{13.109}$$

which implies that

$$\dot{F}(u') = \breve{R}\mu|_{\gamma(u')}, \tag{13.110}$$

$$\ddot{F}(u') = \breve{R}\breve{R}\mu|_{\gamma(u')}. \tag{13.111}$$

Note that by assumption, we have

$$\dot{F}(u' = u) > 0. \tag{13.112}$$

We now set

$$u_* := \inf\left\{u' \in [0,u] \ : \ \gamma(u') \in \mathcal{U}(s,U_0;t)\right\}. \tag{13.113}$$

Note that $\gamma(u) \in \mathcal{U}(s,U_0;t)$ by assumption, while $\gamma(0) \notin \mathcal{U}(s,U_0;t)$ because $\widetilde{\Omega}(t,0,\vartheta;t) \equiv 0$ (recall that $u' = 0$ corresponds to a point on the null cone $\mathcal{C}_0$, where the solution is trivial). Also using the fact that $\mathcal{U}(s,U_0;t)$ is open, we deduce that

$$0 < u_* < u \leq U_0, \tag{13.114}$$

$$\gamma(u_*) \in \Sigma_s^{U_0} \backslash \mathcal{U}(s,U_0;t). \tag{13.115}$$

We consider two sub-cases. In the first sub-case, we assume that

$$\dot{F}(u') > 0 \text{ for } u' \in [u_*, u]. \tag{13.116}$$

In this first sub-case, it follows from (13.116) that

$$F(u_*) < F(u). \tag{13.117}$$

Furthermore, the relation (13.115) and the argument used in Case 2 imply that the bound (13.100) holds for $F(u_*) = \upmu|_{\gamma(u_*)}$. Also using inequality (13.117), we deduce that

$$\upmu(s,u,\vartheta) = F(u) > F(u_*) \geq \frac{1}{2}(1 - C\varepsilon). \tag{13.118}$$

Also using the bound (13.97), we see that $\dfrac{[\breve{R}\upmu(s,u,\vartheta)]_+}{\upmu(s,u,\vartheta)}$ is $\leq$ the second term on the right-hand side of (13.29) as desired, as in the other cases we have previously considered.

In the final (most difficult) sub-case, we assume that the statement (13.116) is false. Hence, there exists a point $u_{**} \in (u_*, u]$ such that

$$\dot{F}(u_{**}) = 0, \tag{13.119}$$

$$\dot{F}(u') > 0 \text{ for } u' \in (u_{**}, u]. \tag{13.120}$$

In this case, our analysis involves the following function $H(\cdot)$ of a single real variable:

$$H(s) := \sup_{\Sigma_s^{U_0}} \breve{R}\breve{R}\upmu. \tag{13.121}$$

By (12.128a), we have the following bound:

$$H(s) \leq C\varepsilon \ln(e + s). \tag{13.122}$$

By carefully studying the behavior of $\upmu$ along the integral curves of $\breve{R}$, we will prove that the following two inequalities also hold in this case:

$$H(s) > 0, \tag{13.123}$$

$$\frac{[\breve{R}\upmu(s,u,\vartheta)]_+}{\upmu(s,u,\vartheta)} \leq \sqrt{\frac{H(s)}{\upmu_{(Min)}(s,u)}}, \tag{13.124}$$

where $\upmu_{(Min)}(s,u)$ is defined in (13.5). Let us accept inequalities (13.123) and (13.124); we will derive them at the end of the proof.

In order to derive an upper bound for the right-hand side of (13.124), we now derive a lower bound for $\upmu_{(Min)}(s,u)$. We use the logarithmic change of variables $\sigma := \ln\left(\dfrac{\varrho(s,u)}{\varrho(0,u)}\right)$, $\tau := \ln\left(\dfrac{\varrho(t,u)}{\varrho(0,u)}\right)$, and $\tau_1 := \ln\left(\dfrac{\varrho(t_1,u)}{\varrho(0,u)}\right) = \dfrac{1}{2\delta}$. Using essentially the same reasoning that led to the first line of (13.76) together with the estimates (13.9) and (13.106), we deduce that

$$\begin{aligned} \upmu_{(Min)}(s,u) &\geq (1 - C\varepsilon)\left\{1 - \delta\sigma - C\varepsilon\frac{1+\sigma}{\exp(\sigma)}(\tau - \sigma)\right\} \\ &\geq (1 - C\varepsilon)\left\{\delta(\tau - \sigma) - C\varepsilon\frac{1+\sigma}{\exp(\sigma)}(\tau - \sigma)\right\} \\ &\geq \frac{1}{2}\delta\left\{1 - C\frac{\varepsilon}{\delta}\frac{1+\sigma}{\exp(\sigma)}\right\}(\tau - \sigma). \end{aligned} \tag{13.125}$$

We now estimate the crucially important term $C\dfrac{\varepsilon}{\delta}\dfrac{1+\sigma}{\exp(\sigma)}$ on the right-hand side of (13.125). Since $\sigma \geq \tau_1$ and since the function $h(y) := (1 + y)\exp(-y)$ is decreasing on $[0,\infty)$, we have

$$C\frac{\varepsilon}{\delta}\frac{1+\sigma}{\exp(\sigma)} \leq C\frac{\varepsilon}{\delta}\frac{1+\tau_1}{\exp(\tau_1)} = 2C\varepsilon\frac{1}{2\delta}\left\{1+\frac{1}{2\delta}\right\}\exp\left(-\frac{1}{2\delta}\right). \tag{13.126}$$

Since

$$\lim_{y\to\infty} y(1+y)\exp(-y) = 0, \tag{13.127}$$

it follows from (13.125), (13.126) (where we view $y = \frac{1}{2\delta}$), the estimate (13.105), and the simple bound (13.59) that if ε is sufficiently small, then the quantity in braces on the right-hand side of (13.125) is $\geq 1/2$ and hence

$$\mu_{(Min)}(s,u) \geq \frac{1}{8}\frac{(\tau-\sigma)}{1+\sigma}. \tag{13.128}$$

Inserting the bounds (13.122) and (13.128) into the right-hand side of inequality (13.124), we deduce that

$$\frac{[\breve{R}\mu(s,u,\vartheta)]_+}{\mu(s,u,\vartheta)} \leq C\varepsilon^{1/2}\frac{(1+\sigma)}{\sqrt{\tau-\sigma}}. \tag{13.129}$$

It is straightforward to check using the basic properties of the function $f(x) = \ln(x)$ that the right-hand side of (13.129) is $\leq$ the first term on the right-hand side of (13.29) as desired. Hence, in order to conclude the desired estimate (13.29) in this final sub-case, it remains only for us to prove inequalities (13.123) and (13.124).

To prove (13.123), we first use (13.112) and (13.119) to deduce that

$$\int_{u'=u_{**}}^{u'=u} \ddot{F}(u')\,du' = \dot{F}(u) > 0. \tag{13.130}$$

Hence, it follows from the definition (13.121) of $H(s)$, (13.111), and (13.130) that $H(s) > 0$ as desired.

We now derive inequality (13.124). To this end, we first use the fundamental theorem of calculus to deduce that if $\hat{u} \in [u_{**}, u]$, then

$$\dot{F}(u) - \dot{F}(\hat{u}) = \int_{u'=\hat{u}}^{u'=u} \ddot{F}(u')\,du' \leq H(s)(u-\hat{u}). \tag{13.131}$$

From inequality (13.131), it follows that whenever $\hat{u} \in [u_{**}, u]$ and

$$|\hat{u}-u| \leq \frac{1}{2}\frac{\dot{F}(u)}{H(s)}, \tag{13.132}$$

we have

$$\dot{F}(\hat{u}) \geq \frac{1}{2}\dot{F}(u) > 0. \tag{13.133}$$

In particular, by choosing $\hat{u} = u_1 := u - \frac{1}{2}\frac{\dot{F}(u)}{H(s)}$ and using the lower bound (13.133), we deduce that

$$\begin{aligned} F(u) - F(u_1) = \int_{u'=u_1}^{u'=u} \dot{F}(u')\,du' &\geq \frac{1}{2}\dot{F}(u)(u-u_1) \\ &= \frac{1}{4}\frac{\dot{F}^2(u)}{H(s)}. \end{aligned} \tag{13.134}$$

Moreover, since $\dot{F}(u_{**}) = 0$, the bound (13.133) for the domain of $\hat{u}$ values defined by (13.132) implies that u_{**} cannot belong to the domain, that is, that

$$u_{**} < u_1. \tag{13.135}$$

Since (13.120) implies that $\dot{F}(u') > 0$ for $u' \in (u_{**}, u_1]$, we also have

$$F(u_{**}) < F(u_1). \tag{13.136}$$

Hence, using the trivial bound $F(u_1) \geq \mu_{(Min)}(s,u)$ (which follows from the definitions of $F(u_1)$ and $\mu_{(Min)}(s,u)$) and the estimate (13.134), we deduce that

(13.137)

$$\begin{aligned}\mu(s,u,\vartheta) - \mu_{(Min)}(s,u) = F(u) - \mu_{(Min)}(s,u) &\geq F(u) - F(u_1) \\ &\geq \frac{1}{4}\frac{\dot{F}^2(u)}{H(s)} = \frac{1}{4}\frac{[\breve{R}\mu(s,u,\vartheta)]_+^2}{H(s)}.\end{aligned}$$

Solving for $\mu(s,u,\vartheta)$ in inequality (13.137) and then using the resulting bound to deduce a bound for $\dfrac{[\breve{R}\mu(s,u,\vartheta)]_-}{\mu(s,u,\vartheta)}$, we find that

$$\frac{[\breve{R}\mu(s,u,\vartheta)]_+}{\mu(s,u,\vartheta)} \leq \frac{[\breve{R}\mu(s,u,\vartheta)]_+}{\frac{1}{4}\frac{[\breve{R}\mu(s,u,\vartheta)]_+^2}{H(s)} + \mu_{(Min)}(s,u)}. \tag{13.138}$$

We now deduce an "algebraic" bound from inequality (13.138). To this end, we note that a standard calculus exercise yields that for any constant $a > 0$, the function

$$h(x,y) := \frac{x}{a^{-1}x^2 + y} \tag{13.139}$$

on the domain $[0,\infty) \times [0,\infty)$ is bounded as follows:

$$|h(x,y)| \leq \frac{\sqrt{a}}{2\sqrt{y}}. \tag{13.140}$$

Applying inequality (13.140) to the right-hand side of (13.138) with $a = 4H(s)$, $x = [\breve{R}\mu(s,u,\vartheta)]_+$, and $y = \mu_{(Min)}(s,u)$, we arrive at the desired inequality (13.124). This completes our proof of (13.29) in this final sub-case.

Proof of (13.30)**:** From the identity $\breve{\underline{L}}\mu = \mu L + 2\breve{R}$ (see (3.19a)) and the estimates (13.25), (13.27), and (13.29), we deduce the desired estimate as follows:

(13.141)

$$\begin{aligned}\frac{[L\mu(s,u,\vartheta) + \breve{\underline{L}}\mu(s,u,\vartheta)]_+}{\mu(s,u,\vartheta)} &\leq |L\mu(s,u,\vartheta)| + \frac{[L\mu(s,u,\vartheta)]_+}{\mu(s,u,\vartheta)} + 2\frac{[\breve{R}\mu(s,u,\vartheta)]_+}{\mu(s,u,\vartheta)} \\ &\leq C\varepsilon^{1/2}\frac{\ln(e+s)}{\sqrt{\ln(e+t) - \ln(e+s)}} + C\varepsilon\frac{1}{1+s}.\end{aligned}$$

□

The following simple corollary plays a role in our analysis of the transport inequalities of Sect. 17.3.

COROLLARY 13.11 (**Analysis of $\frac{1}{2}\mathrm{tr}_{\not g}\chi - 2\frac{L\mu}{\mu}$**). *Let χ be the symmetric type $\binom{0}{2}$ $S_{t,u}$ tensorfield defined in (3.70) and let $\varrho(t,u) = 1-u+t$ be the geometric radial variable. Under the small-data and bootstrap assumptions of Sects. 12.1-12.4, if ε is sufficiently small, then the following pointwise estimate holds on $\mathcal{M}_{T_{(Bootstrap)},U_0}$:*

$$(1 - C\varepsilon)\varrho^{-1} \leq \frac{1}{2}\mathrm{tr}_{\not g}\chi - 2\frac{L\mu}{\mu} \leq \frac{1}{2}\mathrm{tr}_{\not g}\chi + 2\frac{[L\mu]_-}{\mu}. \tag{13.142}$$

PROOF. From the decomposition $\mathrm{tr}_{\not g}\chi = 2\varrho^{-1} + \mathrm{tr}_{\not g}\chi^{(Small)}$ (see (4.1c)) and inequality (12.128a), we have that $\frac{1}{2}\mathrm{tr}_{\not g}\chi \geq (1 - C\varepsilon)\varrho^{-1}$. Furthermore, by (13.27), we have $-2\frac{L\mu}{\mu} \geq -C\varepsilon\varrho^{-1}$. Adding these two estimates, we conclude (13.142). □

13.3. Fundamental estimates for time integrals involving the foliation density

We now use the estimates of Prop. 13.9 to derive estimates for time integrals that involve powers of μ^{-1}. These estimates play a crucial role in our derivation of a priori L^2 estimates for Ψ. Specifically, the time integrals appear in Chapter 20, when we bound the error integrals of Prop. 10.13.

PROPOSITION 13.12 (**Fundamental estimates for time integrals involving μ^{-1}**). *Let $\mu_\star(t,u)$ be as defined in (13.6). Let*

$$B > 1$$

be a real number. Under the small-data and bootstrap assumptions of Sects. 12.1-12.4, if ε is sufficiently small, then the following estimates hold for $0 \leq t < T_{(Bootstrap)}$ and $0 \leq u \leq U_0$.

Estimates relevant for borderline top-order spacetime integrals. *There exists a constant $C > 0$ such that if $B\sqrt{\varepsilon}|\ln\varepsilon| \leq 1$, then*

$$\int_{s=0}^{t} \frac{\|[L\mu]_-\|_{C^0(\Sigma_s^u)}}{\mu_\star^B(s,u)}\, ds \leq \frac{1 + C\sqrt{\varepsilon}}{B-1}\mu_\star^{1-B}(t,u). \tag{13.143}$$

In addition, there exists a constant $C > 0$ such that if $\sqrt{\varepsilon} < b \leq 1/2$ is a constant, then

$$\frac{1}{\varrho(t,u)}\int_{s=0}^{t} \left\| \left(\frac{\mu(t,\cdot)}{\mu}\right)^2 \right\|_{C^0(\Sigma_s^u)} ds \leq 1 + Cb + C\frac{\ln^2(e+t)}{(1+t)^b}. \tag{13.144}$$

Estimates relevant for borderline top-order hypersurface integrals. *Let ${}^{(+)}\Sigma_{s;t}^u \subset \Sigma_s^u$ and ${}^{(-)}\Sigma_{s;t}^u \subset \Sigma_s^u$ be as defined in (13.24c) and (13.24d). There exists a constant $C > 0$ such that if $B\sqrt{\varepsilon}|\ln\varepsilon| \leq 1$ and if ${}^{(-)}\Sigma_{t;t}^u$ is nonempty, then*

$$\|\varrho L\mu\|_{C^0({}^{(-)}\Sigma_{t;t}^u)} \int_{s=0}^{t} \frac{1}{\varrho(s,u)\mu_\star^B(s,u)}\, ds \leq \frac{1 + C\sqrt{\varepsilon}}{B-1}\mu_\star^{1-B}(t,u). \tag{13.145a}$$

In addition, if $A \geq 0$ is a real number and ${}^{(+)}\Sigma_{t;t}^{u}$ is nonempty, then

(13.145b)

$$\left\| \frac{\varrho L \upmu}{\upmu} \right\|_{C^0({}^{(+)}\Sigma_{t;t}^u)} \int_{s=0}^{t} \left\| \sqrt{\frac{\upmu(t,\cdot)}{\upmu}} \right\|_{C^0({}^{(+)}\Sigma_{s;t}^u)} \frac{1}{\varrho(s,u)} \left\{ 1 + \ln \left(\frac{\varrho(s,u)}{\varrho(0,u)} \right) \right\}^A ds \leq \left\{ \frac{1}{A+1/2} + C\varepsilon \right\} \left\{ 1 + \ln \left(\frac{\varrho(t,u)}{\varrho(0,u)} \right) \right\}^A.$$

Estimates relevant for less dangerous top-order spacetime integrals. *There exists a constant $C > 0$ depending on* **upper** *bounds for a and A but* ***independent of*** *B such that if $a \geq \varepsilon^{1/4}$, $A\varepsilon^{1/4} \leq 1$, and $B\sqrt{\varepsilon}|\ln \varepsilon| \leq 1$, then*

$$\int_{s=0}^{t} \frac{\ln^A(e+s)}{(1+s)^{1+a}\upmu_\star^B(s,u)} ds \leq C \left\{ 1 + \frac{1}{B-1} \right\} \frac{1}{a^{A+1}} \upmu_\star^{1-B}(t,u). \tag{13.146}$$

In addition, there exists a constant $C > 0$ such that if $B\sqrt{\varepsilon}\ \ln \varepsilon| \leq 1$, then

$$\int_{s=0}^{t} \frac{1}{(1+s)\upmu_\star^B(s,u)} ds \leq C \left\{ 1 + \frac{1}{B-1} \right\} \ln(e+t) \upmu_\star^{1-B}(t,u). \tag{13.147}$$

Estimates for integrals that lead to only $\ln \upmu_\star^{-1}$ degeneracy. *There exists a constant $C > 0$ such that*

$$\int_{s=0}^{t} \frac{\| [L\upmu]_- \|_{C^0(\Sigma_s^u)}}{\upmu_\star(s,u)} ds \leq (1 + C\sqrt{\varepsilon}) \ln \upmu_\star^{-1}(t,u) + C\sqrt{\varepsilon}. \tag{13.148}$$

Furthermore, there exists a constant $C > 0$ depending on **upper** *bounds for a and A such that if $a \geq \varepsilon^{1/4}$ and $A\varepsilon \leq 1$, then*

$$\int_{s=0}^{t} \frac{\ln^A(e+s)}{(1+s)^{1+a}\upmu_\star(s,u)} ds \leq C \frac{1}{a^{A+1}} \left\{ \ln \upmu_\star^{-1}(t,u) + 1 \right\}. \tag{13.149}$$

In addition, there exists a constant $C > 0$ such that

$$\int_{s=0}^{t} \frac{1}{(1+s)} \frac{1}{\upmu_\star(s,u)} ds \leq C \ln(e+t) \left\{ \ln \upmu_\star^{-1}(t,u) + 1 \right\}. \tag{13.150}$$

Estimates for integrals that break the $\upmu_\star^{-1}$ degeneracy. *There exists a constant $C > 0$ depending on* **upper** *bounds for a and A such that if $a \geq \varepsilon^{1/4}$ and $A\varepsilon \leq 1$, then*

$$\int_{s=0}^{t} \frac{\ln^A(e+s)}{(1+s)^{1+a}\upmu_\star^{3/4}(s,u)} ds \leq \frac{C}{a^A}. \tag{13.151}$$

In addition, there exists a constant $C > 0$ such that

$$\int_{s=0}^{t} \frac{1}{(1+s)\upmu_\star^{3/4}(s,u)} ds \leq C \ln(e+t). \tag{13.152}$$

PROOF. **Proof of** (13.143) **and** (13.148)**:** We prove only (13.143). Inequality (13.148) can be proved by making simple modifications to our proof of (13.143). To prove (13.143), we first recall that

$$[L\upmu(s,u,\vartheta)]_- = \frac{[\Omega(s,u,\vartheta)]_-}{\varrho(s,u)}. \tag{13.153}$$

We consider separately the cases $\widetilde{\Omega}_{(Min)}(t,u;t) \geq 0$ and $\widetilde{\Omega}_{(Min)}(t,u;t) < 0$, where $\widetilde{\Omega}_{(Min)}$ is defined by (13.7). In the case $\widetilde{\Omega}_{(Min)}(t,u;t) \geq 0$, the estimates (13.32b) and (13.32c) imply that

$$\text{(13.154)} \quad \begin{aligned} \int_{s=0}^{t} \frac{\|[L\mu]_-\|_{C^0(\Sigma_s^u)}}{\mu_\star^B(s,u)}\, ds &\leq \int_{s=0}^{t} \frac{\|\Omega_-\|_{C^0(\Sigma_s^u)}}{\varrho(s,u)\mu_\star^B(s,u)}\, ds \\ &\leq C\varepsilon \int_{s=0}^{t} \frac{\ln(e+s)}{(1+s)^2}\, ds \leq C\varepsilon\mu_\star^{1-B}(s,u). \end{aligned}$$

We have thus proved (13.143) in this case.

In the remaining case, which is $\delta := -\widetilde{\Omega}_{(Min)}(t,u;t) > 0$, we split the integral under consideration into the two portions $\int_{s=0}^{t_1} \cdots\, ds$ and $\int_{s=t_1}^{t} \cdots\, ds$, where as in (13.58),

$$\begin{aligned} t_1 &:= (1-u)\frac{\varepsilon}{\delta^2} - (1-u), \\ \tau_1 &:= \ln\left(\frac{\varrho(t_1,u)}{\varrho(0,u)}\right) = \ln\left(\frac{\varepsilon}{\delta^2}\right). \end{aligned}$$

We recall that if ε is sufficiently small, then we have $t_1 > 0$ (see (13.60)). To estimate the integral portions, we use the change of variables (13.67)-(13.69). In particular, $d\sigma = \varrho^{-1}(s,u)\, ds$. We first estimate the integral portion $\int_{s=0}^{t_1} \cdots\, ds$. Using the change of variables, the small-time estimates (13.34b) and (13.34c), the estimate (13.59), and the estimate $\int_{\sigma=0}^{\infty}(1+\sigma)\exp(-\sigma)\, d\sigma \leq C$, we bound the integral of interest as follows:

$$\text{(13.155)} \quad \begin{aligned} \int_{s=0}^{t_1} \frac{\|[L\mu]_-\|_{C^0(\Sigma_s^u)}}{\mu_\star^B(s,u)}\, ds &\leq \int_{s=0}^{t_1} \frac{\|\Omega_-\|_{C^0(\Sigma_s^u)}}{\mu_\star^B(s,u)} \frac{ds}{\varrho(s,u)} \\ &\leq (1+C\sqrt{\varepsilon}) \int_{\sigma=0}^{\tau_1} \{\delta + C\varepsilon(1+\sigma)\exp(-\sigma)\}\, d\sigma \\ &\leq (1+C\sqrt{\varepsilon})\{\tau_1\delta + C\varepsilon\} \\ &\leq (1+C\sqrt{\varepsilon})\{2\delta|\ln\delta| + \delta|\ln\varepsilon| + C\varepsilon\} \\ &\leq C\varepsilon|\ln\varepsilon| \\ &\leq C\frac{1}{B-1}(B-1)\varepsilon|\ln\varepsilon|\mu_\star^{1-B}(t,u). \end{aligned}$$

Using our upper bound assumptions on the size of B in terms of ε, it is straightforward to see that the right-hand side of (13.155) is $\leq$ the term $C\frac{\sqrt{\varepsilon}}{B-1}\mu_\star^{1-B}(t,u)$ on the right-hand side of (13.143) as desired.

In the final step, we bound the integral portion $\int_{s=t_1}^{t} \cdots\, ds$ by

$$\leq \frac{1+C\sqrt{\varepsilon}}{B-1}\mu_\star^{1-B}(t,u).$$

To this end, we use the large-time estimates (13.35b) and (13.35c) and the change of variables (13.67) and (13.69) to derive the desired estimate as follows:

$$\begin{aligned}
(13.156) \qquad \int_{s=t_1}^{t} \frac{\|[L\mu]_-\|_{C^0(\Sigma_s^u)}}{\mu_\star^B(s,u)}\, ds &\leq \int_{s=t_1}^{t} \frac{\|\Omega_-\|_{C^0(\Sigma_s^u)}}{\mu_\star^B(s,u)} \frac{ds}{\varrho(s,u)} \\
&\leq (1+C\sqrt{\varepsilon})\delta \int_{\sigma=\tau_1}^{\tau} \frac{1}{(1-\delta\sigma)^B}\, d\sigma \\
&\leq \frac{1}{B-1}(1+C\sqrt{\varepsilon})\frac{1}{(1-\delta\tau)^{B-1}} \\
&\leq \frac{1}{B-1}(1+C\sqrt{\varepsilon})\mu_\star^{1-B}(t,u).
\end{aligned}$$

Combining (13.154) and the sum of (13.155) and (13.156), we conclude the desired estimate (13.143) in all cases.

Proof of (13.146), (13.149), **and** (13.151)**:** We prove only (13.146); the estimates (13.149) and (13.151) can be proved by making straightforward modifications to our proof of (13.146). To prove (13.146), we modify our proof of (13.143). We use the fact that the function

$$(13.157) \qquad f(\sigma) := (1+\sigma)^A \exp(-a(1+\sigma))$$

defined on the domain $[-1,\infty)$ achieves its maximum value at $\sigma_{(Max)} := \frac{A}{a} - 1$ and that

$$(13.158) \qquad f'(\sigma) \geq 0, \qquad \sigma \in [-1, \sigma_{(Max)}),$$

$$(13.159) \qquad f(\sigma_{(Max)}) = \left(\frac{A}{e}\right)^A \frac{1}{a^A},$$

$$(13.160) \qquad f'(\sigma) < 0, \qquad \sigma \in (\sigma_{(Max)}, \infty).$$

Furthermore, a straightforward integration by parts calculation yields that f verifies the integral bound

$$(13.161) \qquad \int_{\sigma=0}^{\infty} f(\sigma)\, d\sigma \leq \frac{A!}{a^{A+1}}.$$

We consider separately the cases $\widetilde{\Omega}_{(Min)}(t,u;t) \geq 0$ and $\widetilde{\Omega}_{(Min)}(t,u;t) < 0$, where $\widetilde{\Omega}_{(Min)}$ is defined by (13.7). In the case $\widetilde{\Omega}_{(Min)}(t,u;t) \geq 0$, we use the estimates (13.32b) and (13.161) and the change of variables $\sigma := \ln\left(\frac{\varrho(s,u)}{\varrho(0,u)}\right)$ to deduce

(13.162)
$$\begin{aligned}\int_{s=0}^{t} \frac{\ln^A(e+s)}{(1+s)^{1+a}\upmu_{\star}^B(s,u)}\,ds &\leq C \int_{s=0}^{t} \frac{\left\{1+\ln\left(\frac{\varrho(s,u)}{\varrho(0,u)}\right)\right\}^A}{\left\{1+\frac{\varrho(s,u)}{\varrho(0,u)}\right\}^a \upmu_{\star}^B(s,u)} \frac{ds}{\varrho(s,u)} \\ &\leq C \int_{\sigma=0}^{\infty} (1+\sigma)^A \exp\{-a(1+\sigma)\}\,d\sigma \leq C\frac{1}{a^{A+1}} \\ &\leq C\frac{1}{a^{A+1}}\upmu_{\star}^{1-B}(t,u).\end{aligned}$$

We have thus proved the desired bound (13.146) in this case.

In the remaining case, which is $\delta := -\widetilde{\Omega}_{(Min)}(t,u;t) > 0$, we split the time integration into the two portions $\int_{s=0}^{t_1}\cdots\,ds$ and $\int_{s=t_1}^{t}\cdots\,ds$, where we now set $t_1 := (1-u)\exp\left(\frac{1}{2B\delta}\right) - (1-u)$, $\tau_1 := \ln\left(\frac{\varrho(t_1,u)}{\varrho(0,u)}\right) = \frac{1}{2B\delta}$, $\tau := \ln\left(\frac{\varrho(t,u)}{\varrho(0,u)}\right)$. We first bound the integral portion $\int_{s=0}^{t_1}\cdots\,ds$. To this end, we first use the estimates (13.34b) and (13.35b), the fact that $1-\delta\ln\left(\frac{\varrho(s,u)}{\varrho(0,u)}\right) \geq 1-\frac{1}{2B}$ for $0 \leq s \leq t_1$, and the fact that there exists a uniform constant $C > 1$ such that $C^{-1} \leq \left\{1-\frac{1}{2B}\right\}^B \leq C$ for $B > 1$ to deduce that there exists a uniform constant $C > 1$ *independent of* $B > 1$ such that the following estimates hold:

$$C^{-1} \leq \upmu_{\star}^B(s,u) \leq C, \qquad s \in [0,t_1], \tag{13.163}$$
$$C^{-1} \leq \upmu_{\star}^{1-B}(s,u) \leq C, \qquad s \in [0,t_1]. \tag{13.164}$$

Also using (13.161), we can estimate the integral portion of interest by using essentially the same argument that we used in deriving inequality (13.162):

$$\int_{s=0}^{t_1} \frac{\ln^A(e+s)}{(1+s)^{1+a}\upmu_{\star}^B(s,u)}\,ds \leq C\frac{1}{a^{A+1}}\upmu_{\star}^{1-B}(t,u). \tag{13.165}$$

Clearly, the right-hand side of (13.165) is $\leq$ the right-hand side of (13.146) as desired.

We now bound the integral portion $\int_{s=t_1}^{t}\cdots\,ds$. We first note that (13.59) and our assumptions on A, B, and a imply that $\tau_1 \geq \frac{|\ln\varepsilon|}{C\varepsilon^{1/2}}$ and $1+\sigma_{(Max)} < \frac{1}{\varepsilon^{1/2}}$. Therefore, if ε is sufficiently small, then we have $\tau_1 > 1+\sigma_{(Max)}$, and, by (13.160), we see that $f(\sigma)$ is decreasing for $\sigma \geq \tau_1 - 1$. Hence, using the estimate (13.35b), the fact that $f(\sigma) < f(\sigma_{(Max)})$ over the integration range, and the definition of τ_1, we deduce that

(13.166)
$$\begin{aligned}\int_{s=t_1}^{t} \frac{\ln^A(e+s)}{(1+s)^{1+a}\mu_\star^B(s,u)}\,ds &\leq C\int_{\sigma=\tau_1}^{\tau} \frac{(1+\sigma)^A\exp(-a(1+\sigma))}{\mu_\star^B}\,d\sigma \\ &\leq C\tau_1^A\exp(-a\tau_1)\int_{\sigma=\tau_1}^{\tau}\frac{1}{\{1-\delta\sigma\}^B}\,d\sigma \\ &\leq C\tau_1^A\exp(-a\tau_1)\frac{1}{\delta}\frac{1}{(B-1)}(1-\delta\tau)^{1-B} \\ &\leq C\frac{B}{B-1}\left(\frac{a}{2B\delta}\right)^{A+1}\exp\left(-\frac{a}{2B\delta}\right)\frac{1}{a^{A+1}}\mu_\star^{1-B}(t,u).\end{aligned}$$

Using the fact that the function $h(y) := y^{A+1}\exp(-y)$ is $\leq \{(A+1)/e\}^{A+1}$ on the domain $y \in [0,\infty)$, we deduce (with $y := \frac{a}{2B\delta}$) that the product $\left(\frac{a}{2B\delta}\right)^{A+1}\exp\left(-\frac{a}{2B\delta}\right)$ on the right-hand side of (13.166) is $\leq \{(A+1)/e\}^{A+1}$. The desired estimate (13.146) thus follows in this case.

Proof of (13.147)**,** (13.150)**, and** (13.152)**:** The proof of (13.147) is similar to the proof of (13.143) but requires a few simple changes. To prove (13.147), we consider separately the cases $\widetilde{\Omega}_{(Min)}(t,u;t) \geq 0$ and $\widetilde{\Omega}_{(Min)}(t,u;t) < 0$, where $\widetilde{\Omega}_{(Min)}$ is defined by (13.7). In the case $\widetilde{\Omega}_{(Min)}(t,u;t) \geq 0$, we use the estimate (13.32b) to deduce that

$$\int_{s=0}^{t} \frac{1}{(1+s)\mu_\star^B(s,u)}\,ds \leq C\int_{s=0}^{t}\frac{1}{1+s}\,ds \leq C\ln(e+t)\mu_\star^{1-B}(t,u)$$

as desired.

In the remaining case, which is $\delta := -\widetilde{\Omega}_{(Min)}(t,u;t) > 0$, we set $t_1 := (1-u)\exp\left(\frac{1}{2B\delta}\right) - (1-u)$, $\tau_1 := \ln\left(\frac{\varrho(t_1,u)}{\varrho(0,u)}\right) = \frac{1}{2B\delta}$. We use the change of variables (13.67) and (13.69). We first consider the sub-case $t \leq t_1$. In this sub-case, we use the estimates (13.163) and (13.164) to bound the integral of interest as follows:

(13.167)
$$\begin{aligned}\int_{s=0}^{t}\frac{1}{(1+s)\mu_\star^B(s,u)}\,ds &\leq C\int_{s=0}^{t}\frac{1}{\mu_\star^B(s,u)}\frac{ds}{\varrho(s,u)} \\ &\leq C\int_{\sigma=0}^{\tau} 1\,d\sigma \leq C\tau\mu_\star^{1-B}(t,u) \\ &\leq C\ln(e+t)\mu_\star^{1-B}(t,u).\end{aligned}$$

We now consider the sub-case $\delta := -\widetilde{\Omega}_{(Min)}(t,u;t) > 0$, $t_1 < t$. We split the integral under consideration into the two portions $\int_{s=0}^{t_1}\cdots\,ds$ and $\int_{s=t_1}^{t}\cdots\,ds$. We first estimate the integral portion $\int_{s=0}^{t_1}\cdots\,ds$. We use the same argument used to prove (13.167) together with the approximate monotonicity inequality (13.45) to bound the integral of interest as follows:

$$\int_{s=0}^{t_1} \frac{1}{(1+s)\upmu_\star^B(s,u)}\,ds \leq C\ln(e+t_1)\upmu_\star^{1-B}(t_1,u) \leq C\ln(e+t)\upmu_\star^{1-B}(t,u). \tag{13.168}$$

It remains for us to bound the integral portion $\int_{s=t_1}^t \cdots\, ds$ by

$$\leq C\left\{1+\frac{1}{B-1}\right\}\ln(e+t)\upmu_\star^{1-B}(t,u).$$

To this end, we use the small-time estimate (13.34b), the large-time estimate (13.35b), the change of variables (13.67) and (13.69), and the inequalities

$$\frac{1}{(B-1)\delta} = \frac{2B}{B-1}\tau_1 \leq C\left\{1+\frac{1}{B-1}\right\}\tau \leq C\left\{1+\frac{1}{B-1}\right\}\ln(e+t) \tag{13.169}$$

to bound the integral of interest as follows:

$$\begin{aligned}\int_{s=t_1}^{t} \frac{1}{(1+s)\upmu_\star^B(s,u)}\,ds &\leq C\int_{s=t_1}^{t} \frac{1}{\upmu_\star^B(s,u)}\frac{ds}{\varrho(s,u)} \\ &\leq C\int_{\sigma=\tau_1}^{\tau} \frac{1}{\{1-\delta\sigma\}^B}\,d\sigma \\ &\leq \frac{C}{(B-1)\delta}\{1-\delta\tau\}^{1-B} \\ &\leq C\left\{1+\frac{1}{B-1}\right\}\ln(e+t)\upmu_\star^{1-B}(t,u).\end{aligned} \tag{13.170}$$

We have thus proved the desired estimate (13.147) in all cases.

The estimate (13.150) can be proved in a similar fashion by choosing $t_1 := (1-u)\exp\left(\frac{1}{2\delta}\right) - (1-u)$.

The estimate (13.152) can be proved in a similar fashion by choosing $t_1 := (1-u)\exp\left(\frac{1}{(3/2)\delta}\right) - (1-u)$.

Proof of (13.144)**:** To prove (13.144), we first consider the case $0 \leq t \leq 1$. In this case, we use the bound $\sup_{s\in[0,t]}\left\|\frac{\upmu(t,\cdot)}{\upmu}\right\|_{C^0(\Sigma_s^u)} \leq 1 + C\sqrt{\varepsilon}$ implied by Def. 13.8 and the estimates (13.38b) and (13.42) (with $s_1 := s$ and $s_2 := t$) in order to deduce that

$$\frac{1}{1-u+t}\int_{s=0}^{t}\left\|\left(\frac{\upmu(t,\cdot)}{\upmu}\right)^2\right\|_{C^0(\Sigma_s^u)} ds \leq (1+C\sqrt{\varepsilon})\frac{t}{1-u+t} \leq C\frac{\ln^2(e+t)}{(1+t)^b} \tag{13.171}$$

as desired.

We now consider the case $t \geq 1$. To proceed, we split the integral into the two portions $\int_{s=0}^{t^{1-b}}\cdots\, ds$ and $\int_{s=t^{1-b}}^{t}\cdots\, ds$. To bound the first portion, we use the bound $\sup_{s\in[0,t^{1-b}]}\left\|\frac{\upmu(t,\cdot)}{\upmu}\right\|_{C^0(\Sigma_s^u)} \leq (1+C\sqrt{\varepsilon})\ln(e+t)$ implied by Def. 13.8 and the estimates (13.38b) and (13.42) (with $s_1 := s$ and $s_2 := t$) in order to deduce that

$$\frac{1}{1-u+t}\int_{s=0}^{t^{1-b}}\left\|\left(\frac{\mu(t,\cdot)}{\mu}\right)^2\right\|_{C^0(\Sigma_s^u)} ds \leq C\frac{\ln^2(e+t)}{1+t}(t^{1-b}) \leq C\frac{\ln^2(e+t)}{(1+t)^b} \tag{13.172}$$

as desired.

To bound the second portion, we use the bound $\sup_{s\in[t^{1-b},t]}\left\|\frac{\mu(t,\cdot)}{\mu}\right\|_{C^0(\Sigma_s^u)} \leq 1 + C\sqrt{\varepsilon} + Cb$ implied by Def. 13.8 and the estimates (13.39) and (13.42) (with $a := 1-b$, $s_1 := s$, and $s_2 := t$) in order to deduce that

$$\begin{aligned}\frac{1}{1-u+t}\int_{s=t^{1-b}}^{t}\left\|\left(\frac{\mu(t,\cdot)}{\mu}\right)^2\right\|_{C^0(\Sigma_s^u)} ds &\leq (1+C\sqrt{\varepsilon}+Cb)\frac{t(1-t^{-b})}{1-u+t} \\ &\leq 1+C\sqrt{\varepsilon}+Cb+C\frac{1}{1+t}.\end{aligned} \tag{13.173}$$

We now observe that since $\sqrt{\varepsilon} < b \leq \frac{1}{2}$ by assumption, it follows that the sum of (13.172) and (13.173) is $\leq$ the right-hand side of (13.144) as desired.

Proof of (13.145b)**:** For convenience, we abbreviate $f(t,u) := 1 + \ln\left(\frac{\varrho(t,u)}{\varrho(0,u)}\right)$ throughout this proof. From inequality (13.40) with $s = t$ and the estimate (13.37), we deduce that

$$\left\|\frac{\varrho L\mu}{\mu}\right\|_{C^0((+)\Sigma_{t;u}^u)} \leq (1+C\varepsilon)\frac{\gamma}{\left[1+\gamma\left\{1+\ln\left(\frac{\varrho(t,u)}{\varrho(0,u)}\right)\right\}\right]} + C\varepsilon\frac{\ln(e+t)}{1+t}. \tag{13.174}$$

We now derive a suitable bound for the term generated by the first term on the right-hand side of (13.174). To this end, we use (13.38a) with $s_1 = s$ and $s_2 = t$, Def. 13.8, and the change of variables $\sigma(s) = f(s,u)$ (with u held fixed), $d\sigma = \varrho^{-1}(s,u)\,ds$, to deduce that the term under consideration is

$$\begin{aligned}&\leq (1+C\varepsilon)\frac{\gamma}{\sqrt{1-\gamma f(t,u)}}\int_{s=0}^{t}\frac{f^A(s,u)}{\varrho(s,u)\sqrt{1+\gamma f(s,u)}}\,ds \\ &= (1+C\varepsilon)\frac{\gamma}{\sqrt{1-\gamma f(t,u)}}\int_{\sigma=1}^{f(t,u)}\frac{\sigma^A}{\sqrt{1+\gamma\sigma}}\,d\sigma.\end{aligned} \tag{13.175}$$

Using the trivial bound $\sigma^A/\sqrt{1+\gamma\sigma} \leq \sigma^{A-1/2}/\sqrt{\gamma}$ to pointwise bound the integrand, we deduce that the right-hand side of (13.175) is

$$\begin{aligned}&\leq (1+C\varepsilon)\frac{\gamma}{\sqrt{1+\gamma f(t,u)}}\frac{1}{\sqrt{\gamma}}\int_{\sigma=1}^{f(t,u)}\sigma^{A-1/2}\,d\sigma \\ &\leq (1+C\varepsilon)\frac{\sqrt{\gamma}}{\sqrt{1+\gamma f(t,u)}}\frac{f^{A+1/2}(t,u)}{(A+1/2)}.\end{aligned} \tag{13.176}$$

It now easily follows that the right-hand side of (13.176) is $\leq$ the right-hand side of (13.145b) as desired.

It remains for us to address the term generated by the second term on the right-hand side of (13.174). Using (13.38b) with $s_1 = s$ and $s_2 = t$, we deduce that the product under consideration is

$$\leq C\varepsilon \frac{f(t,u)}{\varrho(t,u)} \int_{s=0}^{t} \frac{f^{A+1/2}(s,u)}{\varrho(s,u)}\, ds \leq C\varepsilon \frac{f^{A+5/2}(t,u)}{1+t} \leq C\varepsilon f^{A}(t,u), \tag{13.177}$$

which is $\leq$ the $C\varepsilon \cdots$ product on the right-hand side of (13.145b) as desired.

Proof of (13.145a)**:** The proof of (13.145a) is very similar to that of (13.143), so we are somewhat terse concerning the details. We begin with the trivial estimate

$$\|\varrho L\upmu\|_{C^0({}^{(-)}\Sigma_{t;t}^{u})} \leq \|\varrho[L\upmu]_-\|_{C^0({}^{(-)}\Sigma_{t;t}^{u})} + \|\varrho[L\upmu]_+\|_{C^0({}^{(-)}\Sigma_{t;t}^{u})} \,. \tag{13.178}$$

We separately bound the terms corresponding to the two terms on the right-hand side of (13.178). We first prove the easier bound involving $\|\varrho[L\upmu]_+\|_{C^0({}^{(-)}\Sigma_{t;t}^{u})}$. To this end, we use the estimates (13.43) (with $s = t$) and (13.147) to deduce the bound

(13.179)

$$\begin{aligned}\|\varrho[L\upmu]_+\|_{C^0({}^{(-)}\Sigma_{t;t}^{u})} \int_{s=0}^{t} \frac{1}{\varrho(s,u)\upmu_\star^{B}(s,u)}\, ds &\leq C\varepsilon \frac{\ln(e+t)}{1+t} \int_{s=0}^{t} \frac{1}{(e+s)\upmu_\star^{B}(s,u)}\, ds \\ &\leq C\varepsilon \frac{1}{B-1} \frac{\ln^2(e+t)}{1+t} \upmu_\star^{1-B}(t,u).\end{aligned}$$

Clearly, the right-hand side of (13.179) is $\leq$ the right-hand side of (13.145a) as desired.

To prove the bound corresponding to the second term $\|\varrho[L\upmu]_-\|_{C^0({}^{(-)}\Sigma_{t;t}^{u})}$ on the right-hand side of (13.179), we first note that since ${}^{(-)}\Sigma_{t;t}^{u}$ is nonempty, the positivity assumption on $\updelta$ stated in (13.33) holds. We separately consider the cases $0 \leq t \leq t_1$ and $t \geq t_1$, where $t_1 := (1-u)\frac{\varepsilon}{\updelta^2} - (1-u)$. In the first case $0 \leq t \leq t_1$, we use the estimates (13.34b), (13.44a), and (13.59) and the simple bounds $\int_{s=0}^{t} \frac{1}{e+s}\, ds \leq \ln\left(e + \frac{\varepsilon}{\updelta^2}\right)$ and $\int_{s=0}^{t} \frac{1}{e+s}\, ds \leq \ln(e+t)$ to deduce the desired estimate as follows:

$$\begin{aligned}&\|\varrho[L\upmu]_-\|_{C^0({}^{(-)}\Sigma_{t;t}^{u})} \int_{s=0}^{t} \frac{1}{\varrho(s,u)\upmu_\star^{B}(s,u)}\, ds \\ &\leq C\left\{\updelta + C\varepsilon\frac{\ln(e+t)}{1+t}\right\} \int_{s=0}^{t} \frac{1}{e+s}\, ds \\ &\leq C\left\{\updelta \ln\left(e+\frac{\varepsilon}{\updelta^2}\right) + C\varepsilon \frac{\ln^2(e+t)}{1+t}\right\} \\ &\leq C\varepsilon|\ln\varepsilon| \leq C\frac{1}{B-1}(B-1)\varepsilon|\ln\varepsilon|\upmu_\star^{1-B}(t,u) \\ &\leq C\sqrt{\varepsilon}\frac{1}{B-1}\upmu_\star^{1-B}(t,u).\end{aligned} \tag{13.180}$$

In the remaining case $t \geq t_1$, we use the estimates (13.34b), (13.35b), (13.44b), and (13.59) and the simple bound $\int_{s=0}^{t_1} \frac{1}{e+s}\, ds \leq \ln\left(e+\frac{\varepsilon}{\updelta^2}\right)$ to deduce the desired bound as follows:

(13.181)

$$\begin{aligned}
& \|\varrho[L\mu]_-\|_{C^0((-)\Sigma^u_{t;t})} \int_{s=0}^{t} \frac{1}{\varrho(s,u)\mu_\star^B(s,u)}\,ds \\
& \leq (1+C\sqrt{\varepsilon})\,\delta \int_{s=0}^{t_1} \frac{1}{e+s}\,ds + (1+C\sqrt{\varepsilon})\,\delta \int_{s=t_1}^{t} \frac{1}{\varrho(s,u)\left\{1-\delta\ln\left(\frac{\varrho(s,u)}{\varrho(0,u)}\right)\right\}^B}\,ds \\
& \leq C\varepsilon|\ln\varepsilon| + \frac{1+C\sqrt{\varepsilon}}{B-1}\left\{1-\delta\ln\left(\frac{\varrho(t,u)}{\varrho(0,u)}\right)\right\}^{1-B} \\
& \leq \frac{1+C\sqrt{\varepsilon}}{B-1}\mu_\star^{1-B}(t,u).
\end{aligned}$$

We have thus proved the desired estimate (13.145a). □

CHAPTER 14

Square Integral Coerciveness and the Fundamental Square-Integral-Controlling Quantities

Our first goal in this chapter is to reveal the coercive nature of the energies and cone fluxes defined in Def. 10.11; see Sect. 14.1. Then, in Sect. 14.2, we define a related family of coercive L^2-based quantities that we later use (see Chapter 20) to derive a priori L^2 estimates for Ψ and its derivatives. Finally, in Sect. 14.3, we quantify the coercive nature of these "fundamental L^2-controlling quantities."

14.1. Coerciveness of the energies and cone fluxes

In this section, we show that the energy-cone flux quantities from Def. 10.11 are coercive.

LEMMA 14.1 (**Coerciveness of the energies and cone fluxes**). *Under the small data and bootstrap assumptions of Sects. 12.1-12.4, if ε is sufficiently small, then the energies $\mathbb{E}[\Psi](t,u)$, $\widetilde{\mathbb{E}}[\Psi](t,u)$ and the cone fluxes $\mathbb{F}[\Psi](t,u)$, $\widetilde{\mathbb{F}}[\Psi](t,u)$ from Def. 10.11 have the following coerciveness properties (see Remark 10.1), where χ is the symmetric type $\binom{0}{2}$ $S_{t,u}$ tensorfield defined in (3.70):*

$$\begin{aligned}(14.1a)\quad \mathbb{E}[\Psi](t,u) &= \frac{1}{2}\int_{\Sigma_t^u}\left\{\upmu(1+\upmu)(L\Psi)^2 + (\breve{\underline{L}}\Psi)^2 + \upmu(1+2\upmu)|{\not{d}}\Psi|^2\right\} d\underline{\varpi} \\ &\geq \max\Big\{C^{-1}\|\Psi\|^2_{L^2(S_{t,u})}, C^{-1}\|\Psi\|^2_{L^2(\Sigma_t^u)}, \\ &\qquad \frac{1}{2}\|\sqrt{\upmu}L\Psi\|^2_{L^2(\Sigma_t^u)}, \frac{1}{2}\|\upmu L\Psi\|^2_{L^2(\Sigma_t^u)}, \left\|\breve{R}\Psi\right\|^2_{L^2(\Sigma_t^u)}\Big\} \\ &\quad + \frac{1}{2}\|\sqrt{\upmu}{\not{d}}\Psi\|^2_{L^2(\Sigma_t^u)} + \|\upmu{\not{d}}\Psi\|^2_{L^2(\Sigma_t^u)},\end{aligned}$$

$$\begin{aligned}(14.1b)\quad \mathbb{F}[\Psi](t,u) &= \int_{\mathcal{C}_u^t}\left\{(1+\upmu)(L\Psi)^2 + \upmu|{\not{d}}\Psi|^2\right\} d\overline{\varpi} \\ &= \|L\Psi\|^2_{L^2(\mathcal{C}_u^t)} + \|\sqrt{\upmu}L\Psi\|^2_{L^2(\mathcal{C}_u^t)} + \|\sqrt{\upmu}{\not{d}}\Psi\|^2_{L^2(\mathcal{C}_u^t)},\end{aligned}$$

(14.2a)

$$\widetilde{\mathbb{E}}[\Psi](t,u) = \frac{1}{2}\int_{\Sigma_t^u}\left\{\varrho^2\upmu\left(L\Psi + \frac{1}{2}\mathrm{tr}_{\not g}\chi\Psi\right)^2 + \varrho^2\upmu|\not d\Psi|^2\right\}d\underline{\varpi}$$
$$\approx (1+t)^2\left\|\sqrt{\upmu}\left(L\Psi + \frac{1}{2}\mathrm{tr}_{\not g}\chi\Psi\right)\right\|_{L^2(\Sigma_t^u)}^2 + (1+t)^2\left\|\sqrt{\upmu}\not d\Psi\right\|_{L^2(\Sigma_t^u)}^2,$$

(14.2b)

$$\widetilde{\mathbb{F}}[\Psi](t,u) = \int_{\mathcal{C}_u^t}\varrho^2\left(L\Psi + \frac{1}{2}\mathrm{tr}_{\not g}\chi\Psi\right)^2 d\overline{\varpi}$$
$$\approx \left\|(1+t')\left(L\Psi + \frac{1}{2}\mathrm{tr}_{\not g}\chi\Psi\right)\right\|_{L^2(\mathcal{C}_u^t)}^2.$$

PROOF. To prove (14.2a), we first use Lemma 10.7 and definitions (10.11b) and (10.18d) to compute that (recall that by Def. 3.19, we have $N = \frac{1}{2}L + \frac{1}{2}\underline{L} = L + \upmu^{-1}\breve{R}$)

$$(14.3)\qquad \upmu\,{}^{(\widetilde{K}+Correction)}J_N[\Psi] = \frac{1}{2}\upmu\varrho^2\left\{L\Psi + \frac{1}{2}\mathrm{tr}_{\not g}\chi\Psi\right\}^2 + \frac{1}{2}\upmu\varrho^2|\not d\Psi|^2$$
$$+ \frac{1}{2}\varrho^2\mathrm{tr}_{\not g}\chi\Psi\breve{R}\Psi - \frac{1}{8}\upmu\varrho^2(\mathrm{tr}_{\not g}\chi)^2\Psi^2$$
$$- \frac{1}{4}\upmu\Psi^2 L[\varrho^2\mathrm{tr}_{\not g}\chi] - \frac{1}{4}\Psi^2\breve{R}[\varrho^2\mathrm{tr}_{\not g}\chi].$$

From (14.3), it follows that the right-hand side of (10.21b) is equal to the integral lying to the right of the equal sign in (14.2a) as desired. The final inequality in (14.2a) follows trivially from the estimate (12.5).

To prove (14.2b), we first use Lemma 10.7 and definitions (10.11b) and (10.18d) to compute that

(14.4)

$${}^{(\widetilde{K}+Correction)}J_L[\Psi] = \varrho^2\left\{L\Psi + \frac{1}{2}\mathrm{tr}_{\not g}\chi\Psi\right\}^2 - \frac{1}{4}\varrho^2(\mathrm{tr}_{\not g}\chi)^2\Psi^2$$
$$- \frac{1}{2}\varrho^2\mathrm{tr}_{\not g}\chi\Psi L\Psi - \frac{1}{4}\Psi^2 L[\varrho^2\mathrm{tr}_{\not g}\chi]$$
$$= \varrho^2\left\{L\Psi + \frac{1}{2}\mathrm{tr}_{\not g}\chi\Psi\right\}^2 - \frac{1}{4}\varrho^2(\mathrm{tr}_{\not g}\chi)^2\Psi^2 - \frac{1}{4}\Psi^2 L[\varrho^2\mathrm{tr}_{\not g}\chi\Psi^2].$$

From (14.4), it follows that the right-hand side of (10.22b) is equal to the integral lying to the right of the equal sign in (14.2b) as desired. The final inequality in (14.2b) follows trivially from the estimate (12.5).

The proofs of estimates (14.1a) and (14.1b) are based on definitions (10.11a), (10.18a), and Lemma 10.7 and, with only two exceptions, are similar to the proofs of (14.2a)-(14.2b). We therefore omit the details aside from those for the two exceptions, which are the estimates for the terms $C^{-1}\|\Psi\|_{L^2(S_{t,u})}^2$ and $C^{-1}\|\Psi\|_{L^2(\Sigma_t^u)}^2$ in inequality (14.1a). In view of definitions (3.85a)-(3.85b) and the assumption $u \in [0, U_0]$, we see that the latter follows from the former by integrating in u. Hence, we prove only the former estimate. To this end, we first use the identity

(10.1b) and the estimate (12.151) to deduce that

(14.5)
$$\left| \frac{\partial}{\partial u} \int_{S_{t,u}} \Psi^2 \, d\upsilon_{\not g(t,u,\vartheta)} \right| \leq \int_{S_{t,u}} 2 \left| \Psi \breve{R} \Psi \right| + \left\{ \frac{2}{\varrho} + C\varepsilon \frac{\ln(e+t)}{1+t} \right\} \Psi^2 \, d\upsilon_{\not g(t,u,\vartheta)}.$$

From (14.5) and Cauchy-Schwarz, we deduce that

$$\left| \frac{\partial}{\partial u} \left\| \Psi \right\|_{L^2(S_{t,u})} \right| \leq \left\| \breve{R} \Psi \right\|_{L^2(S_{t,u})} + C \left\| \Psi \right\|_{L^2(S_{t,u})}. \tag{14.6}$$

Applying Gronwall's inequality to (14.6), using the fact that $\|\Psi\|_{L^2(S_{t,0})} = 0$ (in view of our assumption that the data are supported in Σ_0^1), using the fact that $u \in [0, U_0]$, and using Cauchy-Schwarz, we deduce that

$$\begin{aligned} \|\Psi\|_{L^2(S_{t,u})} &\leq C \int_{u'=0}^{u} \left\| \breve{R}\Psi \right\|_{L^2(S_{t,u'})} \, du' \\ &\leq C \left(\int_{u'=0}^{u} \left\| \breve{R}\Psi \right\|_{L^2(S_{t,u'})}^2 \, du' \right)^{1/2} = C \left\| \breve{R}\Psi \right\|_{L^2(\Sigma_t^u)}. \end{aligned} \tag{14.7}$$

Since the already proven inequality (14.1a) for $\left\| \breve{R}\Psi \right\|_{L^2(\Sigma_t^u)}$ implies that the right-hand side of (14.7) is $\lesssim \mathbb{E}^{1/2}[\Psi](t,u)$, the desired estimate (14.1a) for $\|\Psi\|_{L^2(S_{t,u})}$ thus follows. □

14.2. Definitions of the fundamental square-integral-controlling quantities

In this section, we define the "fundamental L^2-controlling quantities." These are the energy-type quantities for which we derive a priori estimates (see Chapter 20) in order to obtain control of Ψ and its derivatives in L^2. In particular, in this section, we introduce a family of coercive Morawetz spacetime integrals.

Definition 14.2 (**The main coercive quantities used for deriving L^2 estimates**). In terms of the energies and fluxes of Def. 10.11, we define

$$\mathbb{Q}_{(N)}(t,u) := \max_{|\vec{I}|=N} \sup_{(t',u') \in [0,t] \times [0,u]} \left\{ \mathbb{E}[\mathscr{Z}^{\vec{I}}\Psi](t',u') + \mathbb{F}[\mathscr{Z}^{\vec{I}}\Psi](t',u') \right\}, \tag{14.8a}$$

$$\widetilde{\mathbb{Q}}_{(N)}(t,u) := \max_{|\vec{I}|=N} \sup_{(t',u') \in [0,t] \times [0,u]} \left\{ \widetilde{\mathbb{E}}[\mathscr{Z}^{\vec{I}}\Psi](t',u') + \widetilde{\mathbb{F}}[\mathscr{Z}^{\vec{I}}\Psi](t',u') \right\}, \tag{14.8b}$$

where in (14.8a)-(14.8b), we are using the shorthand notation for repeated differentiation introduced in Sect. 8.3.

Moreover, we define

$$\mathbb{Q}_{(\leq N)}(t,u) := \max_{0 \leq M \leq N} \mathbb{Q}_{(M)}(t,u), \tag{14.8c}$$

$$\widetilde{\mathbb{Q}}_{(\leq N)}(t,u) := \max_{0 \leq M \leq N} \widetilde{\mathbb{Q}}_{(M)}(t,u). \tag{14.8d}$$

The following coercive spacetime integrals play a fundamental role in controlling some error integrals involving the angular derivatives of Ψ.

DEFINITION 14.3 (**The coercive Morawetz spacetime integral**). To a function Ψ, we associate the following coercive quantities:

$$\widetilde{\mathbb{K}}[\Psi](t,u) := \frac{1}{2}\int_{\mathcal{M}_{t,u}} \varrho^2 [L\upmu]_- |\slashed{d}\Psi|^2 \, d\varpi, \tag{14.9a}$$

$$\widetilde{\mathbb{K}}_{(N)}(t,u) := \max_{|\vec{I}|=N} \widetilde{\mathbb{K}}[\mathscr{Z}^{\vec{I}}\Psi](t,u), \tag{14.9b}$$

$$\widetilde{\mathbb{K}}_{(\leq N)}(t,u) := \max_{0\leq M\leq N} \widetilde{\mathbb{K}}_{(M)}(t,u), \tag{14.9c}$$

where in (14.9b), we are using the shorthand notation for repeated differentiation introduced in Sect. 8.3.

REMARK 14.4 (**The origin of the Morawetz integral**). Recall that the integral (14.9a) appears in the energy-flux identities for Ψ generated by the Morawetz multiplier $\widetilde{K}$; see (10.48b).

14.3. Coerciveness of the fundamental square-integral-controlling quantities

In this section, we exhibit the key coerciveness properties enjoyed by the fundamental L^2-controlling quantities defined in Sect. 14.2.

We start by quantifying the L^2-coerciveness of the Morawetz integrals.

LEMMA 14.5 (**Quantification of the L^2-coerciveness of $\widetilde{\mathbb{K}}[\Psi](t,u)$**). *Let $\mathbf{1}_{\{\upmu\leq 1/4\}}$ denote the characteristic function of the spacetime set $\{(t,u,\vartheta) \mid \upmu(t,u,\vartheta)\leq 1/4\}$. Under the small-data and bootstrap assumptions of Sects.* 12.1-12.4, *if ε is sufficiently small, then the following inequality holds for the Morawetz spacetime integral* (14.9a) *whenever $(t,u)\in[0,T_{(Bootstrap)})\times[0,U_0]$:*

$$\widetilde{\mathbb{K}}[\Psi](t,u) \geq \int_{\mathcal{M}_{t,u}} \mathbf{1}_{\{\upmu\leq 1/4\}} \frac{\varrho(t',u')}{1+\ln\left(\frac{\varrho(t',u')}{\varrho(0,u')}\right)} |\slashed{d}\Psi|^2 \, d\varpi. \tag{14.10}$$

PROOF. Inequality (14.10) follows from the estimate (13.28). $\square$

In the next proposition, we explicitly quantify the L^2-coerciveness of $\mathbb{Q}_{(N)}(t,u)$ and $\widetilde{\mathbb{Q}}_{(N)}(t,u)$ in a manner that will help us estimate the error integrals on the right-hand side of the identities of Prop. 10.13.

REMARK 14.6 (**The importance of some sharp constants**). Some of the sharp constants, such as the (implicit) factor 1 on the right-hand side of (14.11e) and the factor $\sqrt{2}$ on the right-hand side of (14.12d), are important because they affect the number of derivatives we need to close our estimates; see, for example, inequalities (20.51), (20.89), and (20.111).

PROPOSITION 14.7 (**Quantitative L^2 coerciveness of $\mathbb{Q}_{(N)}(t,u)$ and $\widetilde{\mathbb{Q}}_{(N)}(t,u)$**). *Let $0\leq N\leq 24$ be an integer. Under the small data and bootstrap assumptions of Sects.* 12.1-12.4, *if ε is sufficiently small, then the quantities $\mathbb{Q}_{(N)}(t,u)$ and $\widetilde{\mathbb{Q}}_{(N)}(t,u)$ from Def.* 14.2 *have the following coerciveness properties.*

Coerciveness along Σ_t^u

(14.11a) $$\left\|\mathscr{Z}^N\Psi\right\|_{L^2(S_{t,u})} \leq C\mathbb{Q}_{(N)}^{1/2}(t,u),$$

(14.11b) $$\left\|\mathscr{Z}^N\Psi\right\|_{L^2(\Sigma_t^u)} \leq C\mathbb{Q}_{(N)}^{1/2}(t,u),$$

(14.11c) $$\left\|\sqrt{\mu}L\mathscr{Z}^N\Psi\right\|_{L^2(\Sigma_t^u)} \leq \sqrt{2}\mathbb{Q}_{(N)}^{1/2}(t,u),$$

(14.11d) $$\left\|\mu L\mathscr{Z}^N\Psi\right\|_{L^2(\Sigma_t^u)} \leq \sqrt{2}\mathbb{Q}_{(N)}^{1/2}(t,u),$$

(14.11e) $$\left\|\breve{R}\mathscr{Z}^N\Psi\right\|_{L^2(\Sigma_t^u)} \leq \mathbb{Q}_{(N)}^{1/2}(t,u),$$

(14.11f) $$\left\|\sqrt{\mu}{d\!\!/}\mathscr{Z}^N\Psi\right\|_{L^2(\Sigma_t^u)} \leq \sqrt{2}\mathbb{Q}_{(N)}^{1/2}(t,u),$$

(14.11g) $$\left|\mu{d\!\!/}\mathscr{Z}^N\Psi\right\|_{L^2(\Sigma_t^u)} \leq \mathbb{Q}_{(N)}^{1/2}(t,u),$$

(14.12a) $$(1+t)\left\|L\mathscr{Z}^N\Psi\right\|_{L^2(\Sigma_t^u)} \leq C\mathbb{Q}_{(N)}^{1/2}(t,u) + C\frac{1}{\mu_\star^{1/2}(t,u)}\widetilde{\mathbb{Q}}_{(N)}^{1/2}(t,u),$$

(14.12b) $$(1+t)\left\|\sqrt{\mu}L\mathscr{Z}^N\Psi\right\|_{L^2(\Sigma_t^u)} \leq C\ln^{1/2}(e+t)\mathbb{Q}_{(N)}^{1/2}(t,u) + C\widetilde{\mathbb{Q}}_{(N)}^{1/2}(t,u),$$

(14.12c)
$$\left\|\varrho\sqrt{\mu}\left\{L+\frac{1}{2}\mathrm{tr}_{g\!\!/}\chi\right\}\mathscr{Z}^N\Psi\right\|_{L^2(\Sigma_t^u)} \leq \sqrt{2}\widetilde{\mathbb{Q}}_{(N)}^{1/2}(t,u),$$

(14.12d) $$\left\|\varrho\sqrt{\mu}{d\!\!/}\mathscr{Z}^N\Psi\right\|_{L^2(\Sigma_t^u)} \leq \sqrt{2}\widetilde{\mathbb{Q}}_{(N)}^{1/2}(t,u),$$

(14.13) $$\left\|\mathscr{Z}^N\Psi\right\|_{L^2(\Sigma_t^u)} \leq C\mathbb{Q}_{(N-1)}^{1/2}(t,u) + C\frac{1}{\mu_\star^{1/2}(t,u)}\widetilde{\mathbb{Q}}_{(N-1)}^{1/2}(t,u),$$

(14.14a) $$(1+t)\left\|L\mathscr{Z}^N\Psi\right\|_{L^2(\Sigma_t^u)} \leq C\mathbb{Q}_{(N+1)}^{1/2}(t,u),$$

(14.14b) $$(1+t)\left\|{d\!\!/}\mathscr{Z}^N\Psi\right\|_{L^2(\Sigma_t^u)} \leq C\mathbb{Q}_{(N+1)}^{1/2}(t,u).$$

Coerciveness along $\mathcal{C}_u^t$.

(14.15a) $$\left\|L\mathscr{Z}^N\Psi\right\|_{L^2(\mathcal{C}_u^t)} \leq \mathbb{Q}_{(N)}^{1/2}(t,u),$$

(14.15b) $$\left\|\sqrt{\mu}L\mathscr{Z}^N\Psi\right\|_{L^2(\mathcal{C}_u^t)} \leq \mathbb{Q}_{(N)}^{1/2}(t,u),$$

(14.15c) $$\left\|\sqrt{\mu}{d\!\!/}\mathscr{Z}^N\Psi\right\|_{L^2(\mathcal{C}_u^t)} \leq \mathbb{Q}_{(N)}^{1/2}(t,u),$$

(14.16) $$\left\|(1+t')\left\{L+\frac{1}{2}\mathrm{tr}_{g\!\!/}\chi\right\}\mathscr{Z}^N\Psi\right\|_{L^2(\mathcal{C}_u^t)} \leq C\widetilde{\mathbb{Q}}_{(N)}^{1/2}(t,u).$$

PROOF. Most of the inequalities in Prop. 14.7 follow directly from Lemma 14.1 and Def. 14.2. We sketch the proofs of the estimates that require some additional ingredients. To deduce (14.12a), we split $L\mathscr{Z}^N\Psi = \left\{L+\frac{1}{2}\mathrm{tr}_{g\!\!/}\chi\right\}\mathscr{Z}^N\Psi - \frac{1}{2}\mathrm{tr}_{g\!\!/}\chi\mathscr{Z}^N\Psi$ and use the estimate $\varrho|\mathrm{tr}_{g\!\!/}\chi| \lesssim 1$ (that is, (12.60b)). To deduce (14.12b), we also use the estimate $\mu \lesssim \ln(e+t)$ (that is, (12.128a)). The estimate (14.13) follows from separately considering the three cases $\mathscr{Z}^N = \varrho L \mathscr{Z}^{N-1}$, $\mathscr{Z}^N = \breve{R}\mathscr{Z}^{N-1}$, and $\mathscr{Z}^N = O\mathscr{Z}^{N-1}$, where in last case, $O \in \mathscr{O}$ is a rotation and we use the estimate (12.34a) to bound $\left|O\mathscr{Z}^{N-1}\Psi\right| \lesssim C(1+t)\left|\not{d}\mathscr{Z}^{N-1}\Psi\right|$. To deduce (14.14a), we need only to note that since $\varrho L \in \mathscr{Z}$, we have $L\mathscr{Z}^N = \varrho^{-1}\mathscr{Z}^{N+1}$. To deduce (14.14b), we need only to use inequality (12.46a) to deduce that $\left|\not{d}\mathscr{Z}^N\Psi\right| \lesssim \varrho^{-1}\sum_{l=1}^{3}\left|O_{(l)}\mathscr{Z}^N\Psi\right|$. □

In the next lemma, we quantify the smallness of $\mathbb{Q}_{(N)}$ and $\widetilde{\mathbb{Q}}_{(N)}$ along Σ_0^1 under our small data and bootstrap assumptions.

LEMMA 14.8 (**Smallness of $\mathbb{Q}_{(N)}$ and $\widetilde{\mathbb{Q}}_{(N)}$ along Σ_0^1**). *Let $0 \leq N \leq 24$ be an integer and let $\mathring{\epsilon}$ be the size of the data as defined by (12.2). Then if $\mathring{\epsilon}$ is sufficiently small and $u \in [0, U_0]$, we have*

$$\widetilde{\mathbb{Q}}_{(N)}^{1/2}(0,u) \leq C\mathbb{Q}_{(N)}^{1/2}(0,u), \tag{14.17a}$$

$$\mathbb{Q}_{(N)}^{1/2}(0,u) \leq C\mathring{\epsilon}. \tag{14.17b}$$

PROOF. Lemma 14.8 follows easily from Lemma 12.17, the *equalities* in Lemma 14.1, Def. 14.2, and the fact that $\varrho|\mathrm{tr}_{\not{g}}\chi| \lesssim 1$ (that is, (12.60b)). □

CHAPTER 15

Top-Order Pointwise Commutator Estimates Involving the Eikonal Function

In Chapter 15, we derive a collection of pointwise commutator estimates for various quantities that are constructed out of the eikonal function. In Chapter 20, we use these commutator estimates to help us derive top-order L^2 estimates without losing derivatives.

15.1. Top-order pointwise commutator estimates connecting the angular Hessian of the inverse foliation density to the radial Lie derivative of the re-centered null second fundamental form

Our main goal in this section is to show that all top-order derivatives of $\nabla\!\!\!/^2 \mu$ and $\Delta\!\!\!/ \mu$ can be controlled in terms of the top-order derivatives of $\mathcal{L}\!\!\!/_{\breve{R}} \chi^{(Small)}$ plus harmless error terms. This will save us a great deal of redundant effort by allowing us to avoid deriving a separate fully modified equation for $\Delta\!\!\!/ \mu$, as we did for $\mathrm{tr}_{g\!\!\!/} \chi^{(Small)}$ in Chapter 11. Specifically, by the fourth inequality in (15.1), the top-order derivatives $\Delta\!\!\!/ \mathscr{Z}^{N-1} \mu$ are determined by the top-order derivatives $\mathscr{Z}^{N-1} \breve{R} \mathrm{tr}_{g\!\!\!/} \chi^{(Small)}$ up to error terms; the error terms can be controlled without the need to use modified quantities, and furthermore, we already derived the necessary fully modified equation for $\mathscr{Z}^{N-1} \breve{R} \mathrm{tr}_{g\!\!\!/} \chi^{(Small)}$ in Prop. 11.10 (in the case $\mathscr{Z}^{N-1} = \mathscr{S}^{N-1}$, where full modification is needed).

REMARK 15.1 (**Clarification of the meaning of $\mathscr{Z}^{N-1}$**). It is understood that on the left-hand sides of the inequalities of Prop. 15.2, the string of vectorfields $\mathscr{Z}^{N-1}$ is the same each time it occurs. For example, in the expression $\left| \nabla\!\!\!/^2 \mathscr{Z}^{N-1} \mu - \mathcal{L}\!\!\!/_{\mathscr{Z}}^{N-1} \mathcal{L}\!\!\!/_{\breve{R}} \chi^{(Small)} \right|$ from the left-hand side of (15.28), the operators $\mathscr{Z}^{N-1}$ and $\mathcal{L}\!\!\!/_{\mathscr{Z}}^{N-1}$ correspond to the same $(N-1)^{st}$ order string of commutation vectorfields. Furthermore, this remark applies to many other inequalities proved in the remainder of the monograph.

PROPOSITION 15.2 (**Top-order pointwise commutator estimates connecting $\nabla\!\!\!/^2 \mu$ to $\mathcal{L}\!\!\!/_{\breve{R}} \chi^{(Small)}$**). *Let $1 \leq N \leq 24$ be an integer and let $\chi_{(Small)}$ be the symmetric type $\binom{0}{2}$ tensorfield defined in* (4.1c). *Under the small-data and bootstrap assumptions of Sects.* 12.1-12.4, *if ε is sufficiently small, then the following* *pointwise estimates hold on* $\mathcal{M}_{T_{(Bootstrap)},U_0}$:

$$
\begin{aligned}
(15.1)\quad & \left|\nabla\!\!\!/^2 \mathscr{Z}^{N-1}\upmu - \mathcal{L}\!\!\!/_{\mathscr{Z}}^{N-1}\mathcal{L}\!\!\!/_{\breve{R}}\chi^{(Small)}\right|, \\
& \left|\nabla\!\!\!/^2 \mathscr{Z}^{N-1}\upmu - \mathcal{L}\!\!\!/_{\breve{R}}\mathcal{L}\!\!\!/_{\mathscr{Z}}^{N-1}\chi^{(Small)}\right|, \\
& \left|\hat{\nabla\!\!\!/}^2 \mathscr{Z}^{N-1}\upmu - \mathcal{L}\!\!\!/_{\breve{R}}\mathcal{L}\!\!\!/_{\mathscr{Z}}^{N-1}\hat{\chi}^{(Small)}\right|, \\
& \left|\Delta\!\!\!/ \mathscr{Z}^{N-1}\upmu - \mathscr{Z}^{N-1}\breve{R}\mathrm{tr}_{g\!\!\!/}\chi^{(Small)}\right|, \\
& \left|\Delta\!\!\!/ \mathscr{Z}^{N-1}\upmu - \breve{R}\mathscr{Z}^{N-1}\mathrm{tr}_{g\!\!\!/}\chi^{(Small)}\right| \\
& \lesssim \frac{1}{1+t}\left|\begin{pmatrix} \varrho L\mathscr{Z}^{\leq N}\Psi \\ \breve{R}\mathscr{Z}^{\leq N}\Psi \\ \varrho d\!\!\!/\mathscr{Z}^{\leq N}\Psi \\ \mathscr{Z}^{\leq N}\Psi \end{pmatrix}\right| + \frac{1}{(1+t)^2}\left|\begin{pmatrix} \mathscr{Z}^{\leq N}(\upmu-1) \\ \sum_{a=1}^{3}\varrho\left|\mathscr{Z}^{\leq N}L_{(Small)}^a\right| \end{pmatrix}\right|.
\end{aligned}
$$

Our proof of Prop. 15.2 is based on the following commutator-type lemma.

LEMMA 15.3 (**An expression for $\mathcal{L}\!\!\!/_{\breve{R}}\chi^{(Small)}$ in terms of other variables**). *The symmetric type* $\binom{0}{2}$ $S_{t,u}$ *tensorfield* $\chi^{(Small)}$ *defined in* (4.1c) *verifies the following transport equation, where the capital-Latin-indexed-containing products are exact and the remaining ones are schematic:*

(15.2)

$$
\begin{aligned}
\mathcal{L}\!\!\!/_{\breve{R}}\chi_{AB}^{(Small)} &= \nabla\!\!\!/_{AB}^2\upmu + \frac{\upmu-1}{\varrho^2}g\!\!\!/_{AB} \\
&+ G_{(Frame)}\begin{pmatrix} \upmu L\breve{R}\Psi \\ d\!\!\!/\breve{R}\Psi \\ \upmu\nabla\!\!\!/^2\Psi \end{pmatrix} \\
&+ \upmu\chi^{(Small)\#}\chi^{(Small)} + \sum_{p=0}^{1}(g\!\!\!/^{-1})^p G_{(Frame)}\chi^{(Small)}\begin{pmatrix} \upmu L\Psi \\ \breve{R}\Psi \\ \upmu d\!\!\!/\Psi \end{pmatrix} \\
&+ G_{(Frame)}(d\!\!\!/\upmu)\begin{pmatrix} L\Psi \\ d\!\!\!/\Psi \end{pmatrix} \\
&+ \sum_{p=0}^{1}\frac{1}{\varrho}(g\!\!\!/^{-1})^p G_{(Frame)}\begin{pmatrix} \upmu L\Psi \\ \breve{R}\Psi \\ \upmu d\!\!\!/\Psi \end{pmatrix} \\
&+ (\nabla\!\!\!/ G_{(Frame)})\begin{pmatrix} \upmu L\Psi \\ \breve{R}\Psi \\ \upmu d\!\!\!/\Psi \end{pmatrix} \\
&+ \sum_{p=0}^{1}\begin{pmatrix} (g\!\!\!/^{-1})^p G_{(Frame)}^2 \\ G'_{(Frame)} \end{pmatrix}\begin{pmatrix} \upmu L\Psi \\ \breve{R}\Psi \\ \upmu d\!\!\!/\Psi \end{pmatrix}\begin{pmatrix} L\Psi \\ d\!\!\!/\Psi \end{pmatrix}.
\end{aligned}
$$

Before proving Lemma 15.3, we first use it to prove the proposition.

PROOF OF PROP. 15.2. We first prove the bound for the first term

$$\not\nabla^2 \mathscr{Z}^{N-1}\upmu - \not\mathcal{L}_{\mathscr{Z}}^{N-1}\not\mathcal{L}_{\breve{R}}\chi^{(Small)}\Big|$$

on the left-hand side of (15.1). To this end, we write equation (15.2) in the form

$$\not\mathcal{L}_{\breve{R}}\chi_{AB}^{(Small)} = \not\nabla_{AB}^2\upmu + \mathfrak{I}_{AB}, \tag{15.3}$$

where $\mathfrak{I}$ denotes all of the terms on the right-hand side of (15.2) except for the first one. Applying $\not\mathcal{L}_{\mathscr{Z}}^{N-1}$ to equation (15.3), we deduce that

$$\not\mathcal{L}_{\mathscr{Z}}^{N-1}\not\mathcal{L}_{\breve{R}}\chi_{AB}^{(Small)} - \not\nabla_{AB}^2\mathscr{Z}^{N-1}\upmu = -([\not\nabla^2, \not\mathcal{L}_{\mathscr{Z}}^{N-1}]\upmu)_{AB} + \not\mathcal{L}_{\mathscr{Z}}^{N-1}\mathfrak{I}_{AB}. \tag{15.4}$$

The desired estimate will follow once we show that the terms on the right-hand side of (15.4) are in magnitude $\lesssim$ the right-hand side of (15.1). To bound the term $([\not\nabla^2, \not\mathcal{L}_{\mathscr{Z}}^{N-1}]\upmu)_{AB}$, we use inequality (12.95) with $N-1$ in the role of N and $\upmu - 1$ in the role of f and inequality (12.128a) to deduce that

$$\left|[\not\nabla^2, \not\mathcal{L}_{\mathscr{Z}}^{N-1}](\upmu - 1)\right| \lesssim \frac{\ln(e+t)}{(1+t)^2}\left|\begin{pmatrix} \varrho L\mathscr{Z}^{\leq N-1}\Psi \\ \breve{R}\mathscr{Z}^{\leq N-1}\Psi \\ \varrho \not{d}\mathscr{Z}^{\leq N-1}\Psi \\ \mathscr{Z}^{\leq N-1}\Psi \end{pmatrix}\right| + \frac{1}{(1+t)^2}\left|\begin{pmatrix} \mathscr{Z}^{\leq N}(\upmu - 1) \\ \sum_{a=1}^{3}\varrho\left|\mathscr{Z}^{\leq N}L_{(Small)}^a\right| \end{pmatrix}\right|. \tag{15.5}$$

It remains for us to bound the magnitude of $\not\mathcal{L}_{\mathscr{Z}}^{N-1}\mathfrak{I}$ by the right-hand side of (15.1). To this end, we apply $\not\mathcal{L}_{\mathscr{Z}}^{N-1}$ to (15.2) and apply the Leibniz rule to the terms on the right-hand side of (15.2) (except for $\not\nabla^2\upmu$, which we have already handled). We bound the terms $\mathscr{Z}^M\varrho$ with (12.6b). We bound the terms $\mathscr{Z}^M\upmu$, $\mathscr{Z}^M(\upmu-1)$, and $\not\mathcal{L}_{\mathscr{Z}}^M\not{d}\upmu$ with Lemma 8.6, (12.46a), and (12.128a). We bound the terms $\not\mathcal{L}_{\mathscr{Z}}^M\not{g}$ and $\not\mathcal{L}_{\mathscr{Z}}^M\not{g}^{-1}$ with Lemma 12.26. We bound the terms $\not\mathcal{L}_{\mathscr{Z}}^M G_{(Frame)}$ and $\not\mathcal{L}_{\mathscr{Z}}^M G'_{(Frame)}$ with Lemma 12.25. We bound the terms $\not\mathcal{L}_{\mathscr{Z}}^M\not\nabla G_{(Frame)}$ with Lemma 8.6, (12.46a), (12.47), (12.48), (12.92), Lemma 12.25, (12.128a), and the bootstrap assumptions (**BA**Ψ). We bound the terms $\not\mathcal{L}_{\mathscr{Z}}^M\chi^{(Small)}$ and $\not\mathcal{L}_{\mathscr{Z}}^M\chi^{(Small)\#}$ with (12.67a) and (12.128a). We bound the terms $\begin{pmatrix} \mathscr{Z}^M L\Psi \\ \mathscr{Z}^M\breve{R}\Psi \\ \not\mathcal{L}_{\mathscr{Z}}^M\not{d}\Psi \end{pmatrix}$ with Lemma 12.23 and the bootstrap assumptions (**BA**Ψ). In total, these estimates imply that $\left|\not\mathcal{L}_{\mathscr{Z}}^{N-1}\mathfrak{I}\right|$ is $\lesssim$ the right-hand side of (15.1) as desired, which completes the proof of the bound for $\left|\not\nabla^2\mathscr{Z}^{N-1}\upmu - \not\mathcal{L}_{\mathscr{Z}}^{N-1}\not\mathcal{L}_{\breve{R}}\chi^{(Small)}\right|$.

To prove (15.1) for $\left|\not\Delta\mathscr{Z}^{N-1}\upmu - \breve{R}\mathscr{Z}^{N-1}\mathrm{tr}_{\not{g}}\chi^{(Small)}\right|$, we first note that it follows trivially from the previously proven estimate for $\left|\not\nabla^2\mathscr{Z}^{N-1}\upmu - \not\mathcal{L}_{\mathscr{Z}}^{N-1}\not\mathcal{L}_{\breve{R}}\chi^{(Small)}\right|$ that $\left|\not\Delta\mathscr{Z}^{N-1}\upmu - \mathrm{tr}_{\not{g}}\not\mathcal{L}_{\mathscr{Z}}^{N-1}\not\mathcal{L}_{\breve{R}}\chi^{(Small)}\right|$ is $\lesssim$ the right-hand side of (15.1). Hence, to prove the desired estimate, it suffices to show that

$$(15.6)\quad \begin{gathered}\left|[\not{\mathcal{L}}_{\breve{R}}, \not{\mathcal{L}}_{\mathscr{Z}}^{N-1}]\chi^{(Small)}\right|, \\ \left|\mathrm{tr}_{\not{g}}\not{\mathcal{L}}_{\breve{R}}\not{\mathcal{L}}_{\mathscr{Z}}^{N-1}\chi^{(Small)} - \breve{R}\mathscr{Z}^{N-1}\mathrm{tr}_{\not{g}}\chi^{(Small)}\right| \\ \lesssim \frac{1}{1+t}\left|\begin{pmatrix} \varrho L\mathscr{Z}^{\leq N-1}\Psi \\ \breve{R}\mathscr{Z}^{\leq N-1}\Psi \\ \varrho\not{d}\mathscr{Z}^{\leq N-1}\Psi \\ \mathscr{Z}^{\leq N-1}\Psi \end{pmatrix}\right| + \frac{1}{(1+t)^2}\left|\begin{pmatrix} \mathscr{Z}^{\leq N}(\upmu - 1) \\ \sum_{a=1}^3 \varrho\left|\mathscr{Z}^{\leq N}L_{(Small)}^a\right| \end{pmatrix}\right|.\end{gathered}$$

To deduce the desired estimate (15.6) for the first term on the left-hand side of (15.6), we use inequality (12.89a) with $N-1$ in the role of N and $\chi^{(Small)}$ in the role of ξ, inequality (12.67a) with $N-1$ in the role of N, and inequality (12.128a). To deduce the desired estimate (15.6) for the second term on the left-hand side of (15.6), we use inequality (12.104) with $\chi^{(Small)}$ in the role of ξ, (12.128a), and (12.67a) with $N-1$ in the role of N (to bound the first term on the right-hand side of (12.104)). We have thus proved the desired bound (15.1) for $\left|\not{\Delta}\mathscr{Z}^{N-1}\upmu - \breve{R}\mathscr{Z}^{N-1}\mathrm{tr}_{\not{g}}\chi^{(Small)}\right|$. The proof of the estimate (15.1) for $\left|\not{\Delta}\mathscr{Z}^{N-1}\upmu - \mathscr{Z}^{N-1}\breve{R}\mathrm{tr}_{\not{g}}\chi^{(Small)}\right|$ is similar; we omit the details. The estimate (15.1) for $\left|\not{\nabla}^2\mathscr{Z}^{N-1}\upmu - \not{\mathcal{L}}_{\breve{R}}\not{\mathcal{L}}_{\mathscr{Z}}^{N-1}\chi^{(Small)}\right|$ follows from the estimate (15.1) for the first term on the left-hand side and the estimate (15.6) for the first term on the left-hand side.

Finally, we note that the trace-free part of a tensor is in magnitude $\lesssim$ the magnitude of the tensor itself. Hence, we infer from the already proven estimate for $\left|\not{\nabla}^2\mathscr{Z}^{N-1}\upmu - \not{\mathcal{L}}_{\mathscr{Z}}^{N-1}\not{\mathcal{L}}_{\breve{R}}\chi^{(Small)}\right|$ and the estimate (15.6) for the first term on the left-hand side that in order to prove that $\left|\hat{\not{\nabla}}^2\mathscr{Z}^{N-1}\upmu - \not{\mathcal{L}}_{\breve{R}}\not{\mathcal{L}}_{\mathscr{Z}}^{N-1}\hat{\chi}^{(Small)}\right|$ is $\lesssim$ the right-hand side of (15.1), it suffices to show that

$$(15.7)\quad \begin{gathered}\left|\not{\mathcal{L}}_{\breve{R}}\not{\mathcal{L}}_{\mathscr{Z}}^{N-1}\hat{\chi}^{(Small)} - \left\{(\not{\mathcal{L}}_{\breve{R}}\not{\mathcal{L}}_{\mathscr{Z}}^{N-1}\chi^{(Small)}) - \frac{1}{2}\mathrm{tr}_{\not{g}}(\not{\mathcal{L}}_{\breve{R}}\not{\mathcal{L}}_{\mathscr{Z}}^{N-1}\chi^{(Small)})\not{g}\right\}\right| \\ \lesssim \frac{1}{1+t}\left|\begin{pmatrix} \varrho L\mathscr{Z}^{\leq N}\Psi \\ \breve{R}\mathscr{Z}^{\leq N}\Psi \\ \varrho\not{d}\mathscr{Z}^{\leq N}\Psi \\ \mathscr{Z}^{\leq N}\Psi \end{pmatrix}\right| + \frac{1}{(1+t)^2}\left|\begin{pmatrix} \mathscr{Z}^{\leq N}(\upmu - 1) \\ \sum_{a=1}^3 \varrho\left|\mathscr{Z}^{\leq N}L_{(Small)}^a\right| \end{pmatrix}\right|.\end{gathered}$$

The estimate (15.7) follows from inequality (12.105) with $\chi^{(Small)}$ in the role of ξ, (12.128a), and (12.67a) with $N-1$ in the role of N (to bound the first term on the right-hand side of (12.105)). □

It remains for us to prove Lemma 15.3. Our proof is based in part on the following lemma, which is an analog of Cor. 11.3. Among the most important aspects of the lemma is our identification of the $\upmu^{-1}$-singular terms, which are on the second line of the right-hand side of (15.8).

LEMMA 15.4 (**An expression for the curvature component $\mathscr{R}_{\breve{R}ALB}$**). *Let $\mathscr{R}$ be the Riemann curvature tensor of the spacetime metric g from Def. 11.1. Then the curvature component $\mathscr{R}_{\breve{R}ALB}$ can be expressed as follows, where the terms on* *the first three lines are exact and the remaining ones are schematic:*

$$\begin{aligned}(15.8)\qquad \mathscr{R}_{\breve{R}ALB} &= \frac{1}{2}\left\{\mu \mathscr{G}_{RA}\,\not{d}_B L\Psi + \mathscr{G}_{LA}\,\not{d}_B \breve{R}\Psi - \mathscr{G}_{AB}\, L\breve{R}\Psi - \mu G_{LR}\nabla^2_{AB}\Psi\right\}\\ &+ \frac{1}{4}\mu^{-1}\mathscr{G}_{LA}\,\mathscr{G}_{LB}\,(\breve{R}\Psi)^2 - \frac{1}{2}\mu^{-1}\mathscr{G}_{LA}\,(\not{d}_B\mu)\breve{R}\Psi\\ &+ \frac{1}{2}\left\{G_{LR}\chi_{AB}\breve{R}\Psi + \mu\mathscr{G}_{LA}\,\chi_B^{\ C}\not{d}_C\Psi - \mu\mathscr{G}_{RB}\,\chi_A^{\ C}\not{d}_C\Psi\right\}\\ &+ \sum_{p=0}^{1}\begin{pmatrix} (\not{g}^{-1})^p G^2_{(Frame)}\\ G'_{(Frame)}\end{pmatrix}\begin{pmatrix}\mu L\Psi\\ \breve{R}\Psi\\ \mu\not{d}\Psi\end{pmatrix}\begin{pmatrix} L\Psi\\ \not{d}\Psi\end{pmatrix}.\end{aligned}$$

PROOF. We claim that by contracting (11.4) against $\breve{R}^\mu X_A^\nu L^\alpha X_B^\beta$, we can deduce the following identity:

$$\begin{aligned}(15.9)\qquad \mathscr{R}_{\breve{R}ALB} &= \frac{1}{2}\{\mathscr{G}_{LA}\,\mathscr{D}^2_{\breve{R}B}\Psi + \mu\mathscr{G}_{RB}\,\mathscr{D}^2_{LA}\Psi - \mathscr{G}_{AB}\,\mathscr{D}^2_{L\breve{R}}\Psi - \mu G_{LR}\mathscr{D}^2_{AB}\Psi\}\\ &+ \begin{pmatrix} G^2_{(Frame)}\\ G'_{(Frame)}\end{pmatrix}\begin{pmatrix}\mu L\Psi\\ \breve{R}\Psi\\ \mu\not{d}\Psi\end{pmatrix}\begin{pmatrix} L\Psi\\ \not{d}\Psi\end{pmatrix}\\ &+ G^2_{(Frame)}\not{g}^{-1}\begin{pmatrix}\mu L\Psi\\ \mu\not{d}\Psi\end{pmatrix}\begin{pmatrix} L\Psi\\ \not{d}\Psi\end{pmatrix}.\end{aligned}$$

Clearly, the terms on the first line of the right-hand side of (15.9) arise from the terms in the first line on the right-hand side of (11.4). To analyze the terms that arise from the terms in the second line of the right-hand side of (11.4), we first factor the contraction vector $\breve{R}$ as follows: $\breve{R}^\mu = \mu R^\mu$. We then pair this factor μ with the factor $(g^{-1})^{\kappa\lambda}$ and use the decomposition (11.7) to deduce that the terms of interest are a product of $G_{(Frame)}G^\#_{(Frame)}$ and $\mu(g^{-1})^{\kappa\lambda}(\partial_\kappa\Psi)(\partial_\lambda\Psi) = \mu|\not{d}\Psi|^2 - \mu(L\Psi)^2 - 2(L\Psi)(\breve{R}\Psi)$. Hence, these terms are of the form of the terms on the second and third lines of the right-hand side of (15.9). In the terms that arise from the terms on the third line of the right-hand side of (11.4), the λ,κ indices are contractions and not differentiations of Ψ. Therefore, we can use the decomposition $(g^{-1})^{\kappa\lambda} = (\not{g}^{-1})^{\kappa\lambda} - L^\kappa L^\lambda - L^\kappa R^\lambda - R^\kappa L^\lambda$ from (11.7) to deduce that the terms that arise are a product of $\mu G^2_{(Frame)}$ and two distinct elements of the set $\{R\Psi, L\Psi, \not{d}_A\Psi, \not{d}_B\Psi\}$. Hence, these terms are of the form of the terms on the second and thirds lines of the right-hand side of (15.9). Similarly, the terms that arise from the terms on the last line of the right-hand side of (11.4) are a product of $\mu G'_{(Frame)}$ and two distinct elements of the set $\{R\Psi, L\Psi, \not{d}_A\Psi, \not{d}_B\Psi\}$ and thus are of the form of the terms on the second and third lines of the right-hand side of (15.9).

We now further analyze the terms on the first line of the right-hand side of (15.9). To this end, we use Lemma 5.1, the decompositions (5.3a) and (5.3b), and the identity $L\mu = G_{(Frame)}\begin{pmatrix}\mu L\Psi\\ \breve{R}\Psi\end{pmatrix}$ (which follows from Lemma 4.6) to deduce that

$$\mathscr{D}^2_{\breve{R}B}\Psi = \not{d}_B \breve{R}\Psi + \frac{1}{2}\upmu^{-1}\not{G}_{LB}(\breve{R}\Psi)^2 - \upmu^{-1}(\not{d}_B\upmu)\breve{R}\Psi + \upmu\chi_B^{\ C}\not{d}_C\Psi + G_{(Frame)}\begin{pmatrix}\upmu L\Psi \\ \breve{R}\Psi\end{pmatrix}\begin{pmatrix}L\Psi \\ \not{d}\Psi\end{pmatrix} + G_{(Frame)}\not{g}^{-1}\begin{pmatrix}\upmu L\Psi \\ \breve{R}\Psi \\ \upmu\not{d}\Psi\end{pmatrix}\not{d}\Psi, \tag{15.10}$$

$$\mathscr{D}^2_{L\breve{R}}\Psi = L\breve{R}\Psi + G_{(Frame)}\begin{pmatrix}\upmu L\Psi \\ \breve{R}\Psi\end{pmatrix}L\Psi + G_{(Frame)}\not{g}^{-1}\begin{pmatrix}\upmu L\Psi \\ \breve{R}\Psi \\ \upmu\not{d}\Psi\end{pmatrix}\not{d}\Psi. \tag{15.11}$$

Substituting (15.10)-(15.11) and also the identities (11.10)-(11.11) into the terms on the first line of the right-hand side of (15.9), we arrive at the desired decomposition (15.8). □

The next lemma provides an analog of the identity (11.22) but with the derivative $\not{\mathcal{L}}_{\breve{R}}$ in place of $\not{\mathcal{L}}_L$. Among the most important aspects of the lemma is our identification of the $\upmu^{-1}$-singular terms, which are on the second line of the right-hand side of (15.12).

LEMMA 15.5 (**A preliminary expression for $\not{\mathcal{L}}_{\breve{R}}\chi$ in terms of other variables**). *The symmetric type $\binom{0}{2}$ $S_{t,u}$ tensorfield χ verifies the following transport equation, where the terms on the first five lines are exact and the remaining ones are schematic:*

$$\begin{aligned}
\not{\mathcal{L}}_{\breve{R}}\chi_{AB} &= \not{\nabla}^2_{AB}\upmu + \frac{1}{2}\left\{\not{\nabla}_A(\upmu\zeta)_B + \not{\nabla}_B(\upmu\zeta)_A\right\} - \frac{1}{2}\left\{\mathscr{R}_{\breve{R}ALB} + \mathscr{R}_{\breve{R}BLA}\right\} \\
&+ \frac{1}{4}\upmu^{-1}\not{G}_{LA}\not{G}_{LB}(\breve{R}\Psi)^2 - \frac{1}{4}\upmu^{-1}\not{G}_{LA}\not{d}_B\upmu - \frac{1}{4}\upmu^{-1}\not{G}_{LB}\not{d}_A\upmu \\
&- \upmu\chi_A^{\ C}\chi_{BC} \\
&+ \left\{-\frac{1}{2}G_{LL}\breve{R}\Psi + \frac{1}{2}\upmu G_{LL}L\Psi + \upmu G_{LR}L\Psi\right\}\chi_{AB} \\
&+ \frac{1}{2}\left\{\upmu\chi_A^{\ C}\not{k}_{BC} + \upmu\chi_B^{\ C}\not{k}_{AC}\right\} \\
&+ \begin{pmatrix}G^2_{(Frame)} \\ G'_{(Frame)}\end{pmatrix}\begin{pmatrix}\upmu L\Psi \\ \breve{R}\Psi \\ \upmu\not{d}\Psi\end{pmatrix}\begin{pmatrix}L\Psi \\ \not{d}\Psi\end{pmatrix} + G^2_{(Frame)}\not{g}^{-1}\begin{pmatrix}\upmu L\Psi \\ \breve{R}\Psi \\ \upmu\not{d}\Psi\end{pmatrix}\begin{pmatrix}L\Psi \\ \not{d}\Psi\end{pmatrix} \\
&+ G_{(Frame)}(\not{d}\upmu)\begin{pmatrix}L\Psi \\ \not{d}\Psi\end{pmatrix}.
\end{aligned} \tag{15.12}$$

PROOF. Using the torsion-free property $\mathcal{L}_{\breve{R}}X_A = [\breve{R}, X_A] = \mathscr{D}_{\breve{R}}X_A - \mathscr{D}_A\breve{R}$, we compute that

$$\begin{aligned}
\not{\mathcal{L}}_{\breve{R}}\chi_{AB} = \mathcal{L}_{\breve{R}}\chi_{AB} &= \breve{R}(\chi_{AB}) - \chi(\mathcal{L}_{\breve{R}}X_A, X_B) - \chi(X_A, \mathcal{L}_{\breve{R}}X_B) \\
&= \breve{R}(\chi_{AB}) + \chi(\mathscr{D}_A\breve{R}, X_B) + \chi(\mathscr{D}_B\breve{R}, X_A) \\
&\quad - \chi(\mathscr{D}_{\breve{R}}X_A, X_B) - \chi(\mathscr{D}_{\breve{R}}X_B, X_A).
\end{aligned} \tag{15.13}$$

We next use (3.73), the Def. 11.1 of the Riemann curvature tensor, and the torsion-free property of $\mathscr{D}$ to express the first term on the right-hand side of (15.13) as

follows:
(15.14)

$$\begin{aligned}\breve{R}(\chi_{AB}) &= \breve{R}(g(\mathscr{D}_A L, X_B)) = g(\mathscr{D}_{\breve{R}}(\mathscr{D}_A L), X_B) + g(\mathscr{D}_A L, \mathscr{D}_{\breve{R}} X_B) \\ &= g(\mathscr{D}_A(\mathscr{D}_{\breve{R}} L), X_B) + g(\mathscr{D}_{[\breve{R}, X_A]} L, X_B) + g(\mathscr{D}_A L, \mathscr{D}_{\breve{R}} X_B) - \mathscr{R}_{\breve{R}ALB}.\end{aligned}$$

Using the above torsion-free property, Lemma 4.10, and Lemma 5.1, we compute that

$$g(\mathscr{D}_A(\mathscr{D}_{\breve{R}} L), X_B) = \nabla\!\!\!\!/^2_{AB}\mu + \nabla\!\!\!\!/_A(\mu\zeta)_B - (L\mu)\chi_{AB}, \tag{15.15}$$

$$g(\mathscr{D}_{[\breve{R}, X_A]} L, X_B) = \chi([\breve{R}, X_A], X_B) = \chi(\mathscr{D}_{\breve{R}} X_A, X_B) - \chi(\mathscr{D}_A \breve{R}, X_B), \tag{15.16}$$

$$\begin{aligned}g(\mathscr{D}_A L, \mathscr{D}_{\breve{R}} X_B) &= -\zeta_A g(L, \mathscr{D}_{\breve{R}} X_B) + \chi(X_A, \mathscr{D}_{\breve{R}} X_B) \\ &= \zeta_A g(\mathscr{D}_{\breve{R}} L, X_B) + \chi(X_A, \mathscr{D}_{\breve{R}} X_B) \\ &= \mu\zeta_A\zeta_B + \zeta_A d\!\!\!/_B\mu + \chi(X_A, \mathscr{D}_{\breve{R}} X_B),\end{aligned} \tag{15.17}$$

$$\chi(\mathscr{D}_A \breve{R}, X_B) = -\mu\chi_A^{\ C}\chi_{BC} + \mu\chi_A^{\ C} k\!\!\!/_{BC}. \tag{15.18}$$

The first two terms on the right-hand side of (15.17) are the only μ^{-1}-singular terms. More precisely, using the decomposition 5.3a, we deduce that

$$\mu\zeta_A\zeta_B = \frac{1}{4}\mu^{-1} G\!\!\!/_{LA} G\!\!\!/_{LB} (\breve{R}\Psi)^2 + G^2_{(Frame)} \begin{pmatrix} \mu L\Psi \\ \breve{R}\Psi \\ \mu d\!\!\!/\Psi \end{pmatrix} \begin{pmatrix} L\Psi \\ d\!\!\!/\Psi \end{pmatrix}, \tag{15.19}$$

$$\zeta_A d\!\!\!/_B\mu = -\frac{1}{2}\mu^{-1} G\!\!\!/_{LA} d\!\!\!/_B\mu + G_{(Frame)}(d\!\!\!/\mu) \begin{pmatrix} L\Psi \\ d\!\!\!/\Psi \end{pmatrix}. \tag{15.20}$$

Substituting (15.14)-(15.18) into the right-hand side of (15.13), using (15.19)-(15.20), using Lemma 4.6 to express the factor $L\mu$ in (15.15) in terms of $G_{(Frame)}$ and Ψ, and symmetrizing over the indices AB (since $\mathcal{L}\!\!\!/_{\breve{R}}\chi$ is symmetric), we arrive at the desired identity (15.12). □

PROOF OF LEMMA 15.3. We first claim that $\mathcal{L}\!\!\!/_{\breve{R}}\chi_{AB}$ can be expressed as follows, where the capital-Latin-indexed-containing products are exact and the remaining ones are schematic:
(15.21)

$$\begin{aligned}\mathcal{L}\!\!\!/_{\breve{R}}\chi_{AB} &= \nabla\!\!\!\!/^2_{AB}\mu + \frac{1}{2}\left\{ G\!\!\!/_{AB} L\breve{R}\Psi - G\!\!\!/_{LA} d\!\!\!/_B \breve{R}\Psi - G\!\!\!/_{LB} d\!\!\!/_A \breve{R}\Psi - \mu G_{RR} \nabla\!\!\!\!/^2_{AB}\Psi \right\} \\ &\quad - \mu\chi_A^{\ C}\chi_{BC} + G_{(Frame)}\chi \begin{pmatrix} \mu L\Psi \\ \breve{R}\Psi \end{pmatrix} + G_{(Frame)} g\!\!\!/^{-1}\chi \begin{pmatrix} \mu L\Psi \\ \breve{R}\Psi \\ \mu d\!\!\!/\Psi \end{pmatrix} \\ &\quad + G_{(Frame)}(d\!\!\!/\mu) \begin{pmatrix} L\Psi \\ d\!\!\!/\Psi \end{pmatrix} + (\nabla\!\!\!\!/ G_{(Frame)}) \begin{pmatrix} \mu L\Psi \\ \breve{R}\Psi \\ \mu d\!\!\!/\Psi \end{pmatrix} \\ &\quad + \begin{pmatrix} G^2_{(Frame)} \\ G'_{(Frame)} \end{pmatrix} \begin{pmatrix} \mu L\Psi \\ \breve{R}\Psi \\ \mu d\!\!\!/\Psi \end{pmatrix} \begin{pmatrix} L\Psi \\ d\!\!\!/\Psi \end{pmatrix} + G^2_{(Frame)} g\!\!\!/^{-1} \begin{pmatrix} \mu L\Psi \\ \breve{R}\Psi \\ \mu d\!\!\!/\Psi \end{pmatrix} \begin{pmatrix} L\Psi \\ d\!\!\!/\Psi \end{pmatrix}.\end{aligned}$$

To derive (15.21), we first substitute the right-hand side of (15.8) for the curvature terms $-\frac{1}{2}\mathscr{R}_{\breve{R}ALB} - \frac{1}{2}\mathscr{R}_{\breve{R}BLA}$ on the first line on the right-hand side of (15.12). We observe that the μ^{-1}-singular terms on the second line of (15.8) exactly cancel the μ^{-1}-singular terms on the second line of (15.12).

Next, we use the decomposition (5.3a) to compute that

(15.22)

$$\nabla\!\!\!/_A(\mu\zeta)_B = \frac{1}{2}\left\{-\mathcal{G}\!\!\!/_{LB}\, d\!\!\!/_A\breve{R}\Psi + \mu\mathcal{G}\!\!\!/_{RB}\, d\!\!\!/_A L\Psi - \mu G_{LR}\nabla\!\!\!/^2_{AB}\Psi - \mu G_{RR}\nabla\!\!\!/^2_{AB}\Psi\right\}$$
$$+ \begin{pmatrix} \mu L\Psi \\ \breve{R}\Psi \\ \mu d\!\!\!/\Psi \end{pmatrix} \nabla\!\!\!/ G_{(Frame)} + G_{(Frame)}(d\!\!\!/\mu)\begin{pmatrix} L\Psi \\ d\!\!\!/\Psi \end{pmatrix}.$$

Substituting the right-hand side of (15.22) for the terms in the first braces on the first line of the right-hand side of (15.12), we arrive at the desired identity (15.21).

It remains for us to use the equation (15.21) verified by $\mathcal{L}\!\!\!/_{\breve{R}}\chi$ to derive an analogous equation for $\mathcal{L}\!\!\!/_{\breve{R}}\chi^{(Small)}$. To this end, we use the identities $\chi_{AB}^{(Small)} = \chi_{AB} - \varrho^{-1}g\!\!\!/_{AB}$, $\breve{R}\varrho = -1$, $\mathcal{L}\!\!\!/_{\breve{R}}g\!\!\!/_{AB} = {}^{(\breve{R})}\pi\!\!\!/_{AB} = -2\mu\chi_{AB} + G_{(Frame)}\begin{pmatrix} \mu L\Psi \\ \breve{R}\Psi \\ \mu d\!\!\!/\Psi \end{pmatrix}$ (see (4.1c), Lemma 7.5, (7.5f)-(7.5g), and (12.6a)) to compute that

$$\mathcal{L}\!\!\!/_{\breve{R}}\chi_{AB}^{(Small)} = \mathcal{L}\!\!\!/_{\breve{R}}\chi_{AB} - \frac{1}{\varrho^2}g\!\!\!/_{AB} + 2\frac{1}{\varrho^2}\mu g\!\!\!/_{AB} + 2\frac{1}{\varrho}\mu\chi_{AB}^{(Small)} + \frac{1}{\varrho}G_{(Frame)}\begin{pmatrix} \mu L\Psi \\ \breve{R}\Psi \\ \mu d\!\!\!/\Psi \end{pmatrix}, \tag{15.23}$$

$$\mu\chi_A^{\ C}\chi_{BC} = \mu\chi_A^{(Small)C}\chi_{BC}^{(Small)} + 2\frac{1}{\varrho}\mu\chi_{AB}^{(Small)} + \frac{1}{\varrho^2}\mu g\!\!\!/_{AB}. \tag{15.24}$$

Using equation (15.23) to substitute for $\mathcal{L}\!\!\!/_{\breve{R}}\chi_{AB}$ in equation (15.21), and substituting the right-hand side of (15.24) for the first product on the second line of (15.21), we finally arrive at the desired decomposition (15.2). □

15.2. Top-order pointwise commutator estimates corresponding to the spherical Codazzi equations

From Propositions 7.7 and 9.6 and Lemma 9.4, it follows that the terms $\mathrm{div\!\!\!/}\chi^{(Small)}$ and $\mathrm{div\!\!\!/}\hat{\chi}^{(Small)}$ are present in the commuted wave equation. If we estimated the top-order derivatives of these quantities in a naive fashion via estimates based on transport equations, then our estimates would lose a derivative. However, the next lemma shows that these quantities can be expressed in terms of $d\!\!\!/\mathrm{tr}_{g\!\!\!/}\chi^{(Small)}$ plus terms that do not lose derivatives. The main point is that later in the monograph, we will show how to use the fully modified quantities defined in Chapter 11 in conjunction with some elliptic estimates on the $S_{t,u}$ to estimate the top-order derivatives of $d\!\!\!/\mathrm{tr}_{g\!\!\!/}\chi^{(Small)}$ without losing derivatives. The lemma is essentially a version of the Codazzi equations for the Riemannian manifolds $(S_{t,u}, g\!\!\!/)$ viewed as embedded submanifolds of the Lorentzian manifold-with-boundary $(\mathcal{M}_{T_{(Bootstrap)},U_0}, g)$. In order to close our estimates, we do not need to know much about the precise structure of the terms on the right-hand side of (15.25), except for the first one.

LEMMA 15.6 (**Codazzi-type identities involving $\mathrm{di\!/\!v}\chi^{(Small)}$, $\mathrm{di\!/\!v}\hat{\chi}^{(Small)}$, and $\not{d}\mathrm{tr}_{\not{g}}\chi_{(Small)}$**). *Let $\chi_{(Small)}$ be the symmetric type $\binom{0}{2}$ tensorfield defined in (4.1c). The $S_{t,u}$ one-forms $\mathrm{di\!/\!v}\chi^{(Small)}$ and $\mathrm{di\!/\!v}\hat{\chi}^{(Small)}$ verify the following equations, where the terms on the left-hand side and the first term on the right-hand side are exact, and the remaining terms are schematic:*

(15.25)

$$\mathrm{di\!/\!v}\chi^{(Small)}, 2\mathrm{di\!/\!v}\hat{\chi}^{(Small)} = \overbrace{\not{d}\mathrm{tr}_{\not{g}}\chi^{(Small)}}^{\text{precise term}} + \sum_{i_1+i_2=1} \not{g}^{-1}(\not{\nabla}^{i_1}G_{(Frame)})\not{\nabla}^{i_2}\begin{pmatrix} L\Psi \\ \not{d}\Psi \end{pmatrix}$$
$$+ \sum_{i_1+i_2=1} (\not{\nabla}^{i_1}f(\Psi))(\not{\nabla}^{i_2}\not{d}^{\#}x)\not{d}L_{(Small)},$$

and the quantities f are smooth scalar-valued functions of Ψ.

REMARK 15.7 (**Absence of μ^{-1}**). An important structural feature of the right-hand side of (15.25) is that there are no factors of μ^{-1}.

PROOF. We apply $\not{\nabla}^A$ to both sides of (4.19a). When carrying out spherical covariant differentiation with $\not{\nabla}$, we view all lowercase Latin-indexed quantities as scalar-valued functions on $S_{t,u}$. Using the fact that $\not{\nabla}_B\not{d}^A = \not{\nabla}^A\not{d}_B$ when applied to functions and the chain rule identity $\not{d}_B g_{ab} = G_{ab}\not{d}_B\Psi$, we see that the first term on the right-hand side of the resulting identity can be written as

$$(15.26) \qquad \not{\nabla}^A\left\{g_{ab}(\not{d}_A x^a)\not{d}_B L^b_{(Small)}\right\}$$
$$= \not{d}_B\Big(\overbrace{g_{ab}(\not{d}_A x^a)\not{d}^A L^b_{(Small)}}^{\not{d}_a L^a_{(Small)}}\Big)$$
$$+ G_{ab}(\not{d}^A\Psi)(\not{d}_A x^a)\not{d}_B L^b_{(Small)} - G_{ab}(\not{d}_B\Psi)(\not{d}_A x^a)\not{d}^A L^b_{(Small)}$$
$$+ g_{ab}(\not{\Delta}x^c)\not{d}_B L^b_{(Small)} - g_{ab}(\not{\nabla}^2_{AB}x^a)\not{d}^A L^b_{(Small)}.$$

Furthermore, using (4.11c), we see that the second term on the right-hand side of the resulting identity takes the schematic form

$$(15.27) \quad -\not{\nabla}^A\Lambda^{(Tan-\Psi)}_{AB} = \not{g}^{-1}G_{(Frame)}\not{\nabla}\begin{pmatrix} L\Psi \\ \not{d}\Psi \end{pmatrix} + \not{g}^{-1}(\not{\nabla}G_{(Frame)})\begin{pmatrix} L\Psi \\ \not{d}\Psi \end{pmatrix}.$$

From (4.19b) and (4.11c), we see that the first product in parentheses on the right-hand side of (15.26) is equal to $\mathrm{tr}_{\not{g}}\chi^{(Small)} + \mathrm{tr}_{\not{g}}\Lambda^{(Tan-\Psi)} = \mathrm{tr}_{\not{g}}\chi^{(Small)} - \frac{1}{2}\not{G}_A^A L\Psi$. Combining these identities, we arrive at the desired identity (15.25) for $\mathrm{di\!/\!v}\chi^{(Small)}$. The corresponding identity for $\mathrm{di\!/\!v}\hat{\chi}^{(Small)}$ follows easily from the identity $\mathrm{di\!/\!v}\hat{\chi}^{(Small)} = \mathrm{di\!/\!v}\chi^{(Small)} - \frac{1}{2}\not{d}\mathrm{tr}_{\not{g}}\chi^{(Small)}$. □

We now provide a commuted version of Lemma 15.6.

LEMMA 15.8 (**Commutator estimates for the Codazzi equation**). *Let $1 \leq N \leq 23$ be an integer and let $\chi_{(Small)}$ be the symmetric type $\binom{0}{2}$ tensorfield defined in (4.1c). Under the small-data and bootstrap assumptions of Sects. 12.1-12.4, if ε is sufficiently small, then the following pointwise commutator estimates* *hold on* $\mathcal{M}_{T_{(Bootstrap)},U_0}$ *(see Remark* 15.1*):*

$$\left|\not{\mathcal{L}}_{\mathscr{Z}}^{N-1}\mathrm{di\!/\!v}\,\chi^{(Small)} - \not{d}\mathscr{Z}^{N-1}\mathrm{tr}_{\not{g}}\chi^{(Small)}\right|, \tag{15.28}$$

$$\left|\not{\mathcal{L}}_{\mathscr{Z}}^{N-1}\mathrm{di\!/\!v}\,\hat{\chi}^{(Small)} - \frac{1}{2}\not{d}\mathscr{Z}^{N-1}\mathrm{tr}_{\not{g}}\chi^{(Small)}\right|,$$

$$\left|\mathrm{di\!/\!v}\,\not{\mathcal{L}}_{\mathscr{Z}}^{N-1}\chi^{(Small)} - \not{d}\mathscr{Z}^{N-1}\mathrm{tr}_{\not{g}}\chi^{(Small)}\right|,$$

$$\left|\mathrm{di\!/\!v}\,\hat{\not{\mathcal{L}}}_{\mathscr{Z}}^{N-1}\hat{\chi}^{(Small)} - \frac{1}{2}\not{d}\mathscr{Z}^{N-1}\mathrm{tr}_{\not{g}}\chi^{(Small)}\right|$$

$$\lesssim \frac{1}{(1+t)^2}\left|\begin{pmatrix} \varrho L\mathscr{Z}^{\leq N}\Psi \\ \breve{R}\mathscr{Z}^{\leq N}\Psi \\ \varrho\not{d}\mathscr{Z}^{\leq N}\Psi \\ \mathscr{Z}^{\leq N}\Psi \end{pmatrix}\right| + \frac{1}{(1+t)^3}\left|\begin{pmatrix} \mathscr{Z}^{\leq N}(\upmu - 1) \\ \sum_{a=1}^{3}\varrho\left|\mathscr{Z}^{\leq N}L_{(Small)}^a\right| \end{pmatrix}\right|.$$

PROOF. We prove the estimate (15.28) for $\not{\mathcal{L}}_{\mathscr{Z}}^{N-1}\mathrm{di\!/\!v}\,\chi^{(Small)}$ in detail. The estimate for $\not{\mathcal{L}}_{\mathscr{Z}}^{N-1}\mathrm{di\!/\!v}\,\hat{\chi}^{(Small)}$ can be proved by a nearly identical argument. The estimate (15.28) for $\mathrm{di\!/\!v}\,\not{\mathcal{L}}_{\mathscr{Z}}^{N-1}\chi^{(Small)}$ then follows from the estimate for $\not{\mathcal{L}}_{\mathscr{Z}}^{N-1}\mathrm{di\!/\!v}\,\chi^{(Small)}$, the commutator estimate (12.109) with $\chi^{(Small)}$ in the role of ξ and $N-1$ in the role of N, and inequalities (12.67a) and (12.128a). The final estimate (15.28) for $\mathrm{di\!/\!v}\,\hat{\not{\mathcal{L}}}_{\mathscr{Z}}^{N-1}\hat{\chi}^{(Small)}$ follows similarly from the estimate for $\not{\mathcal{L}}_{\mathscr{Z}}^{N-1}\mathrm{di\!/\!v}\,\hat{\chi}^{(Small)}$, thanks to the commutator estimate (12.110).

To derive the estimate for $\not{\mathcal{L}}_{\mathscr{Z}}^{N-1}\mathrm{di\!/\!v}\,\chi^{(Small)}$, we first use equation (15.25) and Lemma 8.6 to deduce that

$$\begin{aligned} &\left|\not{\mathcal{L}}_{\mathscr{Z}}^{N-1}\mathrm{di\!/\!v}\,\chi^{(Small)} - \not{d}\mathscr{Z}^{N-1}\mathrm{tr}_{\not{g}}\chi^{(Small)}\right| \\ &\lesssim \left|\not{\mathcal{L}}_{\mathscr{Z}}^{N-1}\sum_{i_1+i_2=1}\not{g}^{-1}(\not{\nabla}^{i_1}G_{(Frame)})\not{\nabla}^{i_2}\begin{pmatrix} L\Psi \\ \not{d}\Psi \end{pmatrix}\right| \\ &\quad + \left|\not{\mathcal{L}}_{\mathscr{Z}}^{N-1}\sum_{i_1+i_2=1}\varrho^{-1}(\not{\nabla}^{i_1}f(\Psi))(\not{\nabla}^{i_2}\not{d}^{\#}x)\not{d}(\varrho L_{(Small)})\right|. \end{aligned} \tag{15.29}$$

We now apply the Leibniz rule to the terms on the right-hand side of (15.29). We bound the terms $\not{\mathcal{L}}_{\mathscr{Z}}^{M}\not{g}^{-1}$ by using the estimates of Lemma 12.26. We bound the terms $\not{\mathcal{L}}_{\mathscr{Z}}^{M}G_{(Frame)}$ and $\not{\mathcal{L}}_{\mathscr{Z}}^{M}\not{\nabla}G_{(Frame)}$ by using Lemma 8.6, (12.92), (12.46a), (12.47), (12.48), and the estimates of Lemma 12.25. To bound $\begin{pmatrix} \mathscr{Z}^M L\Psi \\ \not{\mathcal{L}}_{\mathscr{Z}}^{M}\not{d}\Psi \end{pmatrix}$ and $\not{\mathcal{L}}_{\mathscr{Z}}^{M}\not{\nabla}\begin{pmatrix} L\Psi \\ \not{d}\Psi \end{pmatrix}$, we use Lemma 8.6, (12.46b), (12.95), and Lemma 12.23. We bound the terms $\not{\mathcal{L}}_{\mathscr{Z}}^{M}\varrho^{-1}$ by using (12.6b). We bound the terms $\mathscr{Z}^M f(\Psi)$ by using the bootstrap assumptions (**BA**Ψ) to deduce that $\left|\mathscr{Z}^M f(\Psi)\right| \lesssim \left|\mathscr{Z}^{\leq M}\Psi\right| + 1$. We bound the terms $\not{\mathcal{L}}_{\mathscr{Z}}^{M}\not{d}f(\Psi)$ by using Lemma 8.6 to deduce that $\left|\not{\mathcal{L}}_{\mathscr{Z}}^{M}\not{d}f(\Psi)\right| \lesssim \left|\not{d}\mathscr{Z}^{\leq M}\Psi\right|$. We bound the terms $\not{\mathcal{L}}_{\mathscr{Z}}^{M}\not{d}^{\#}x = \not{\mathcal{L}}_{\mathscr{Z}}^{M}(\not{g}^{-1}\cdot\not{d}x)$ by using Lemma 8.6 and the estimates (12.57a), (12.57b), (12.60a), and (12.60b). We bound the terms $\not{\mathcal{L}}_{\mathscr{Z}}^{M}\not{\nabla}\not{d}^{\#}x = \not{\mathcal{L}}_{\mathscr{Z}}^{M}(\not{g}^{-1}\cdot\not{\nabla}\not{d}x)$ by using Lemma 8.6, (12.92) and (12.47) with $\not{d}x^i$ in the role of ξ, (12.57a), (12.57b), (12.60a), and (12.60b). We bound the terms

$\not{\mathcal{L}}_{\mathscr{Z}}^{M} \not{d}(\varrho L_{(Small)})$ by using Lemma 8.6, (12.6b), (12.46a), and (12.128a). Combining these estimates and using the bootstrap assumptions (**BA**Ψ), we arrive at the desired estimate (15.28) for the first term on the left-hand side. □

CHAPTER 16

Pointwise Estimates for the Easy Error Integrands and Identification of the Difficult Error Integrands Corresponding to the Commuted Wave Equation

Recall that we derived our energy-cone flux identities for solutions Ψ to $\upmu\square_{g(\Psi)}\Psi = 0$ in Prop. 10.13. Furthermore, similar identities hold for its higher-order analogs $\mathscr{Z}^N\Psi$, which verify $\upmu\square_{g(\Psi)}(\mathscr{Z}^N\Psi) = {}^{(\mathscr{Z}^N)}\mathfrak{F}$, where $\mathscr{Z} = \{\varrho L, \breve{R}, O_{(1)}, O_{(2)}, O_{(3)}\}$ is the set of commutation vectorfields. In this chapter, our main goal is to derive pointwise estimates for the "easy factors" in the integrands on the right-hand side of the identities of Prop. 10.13 with $\mathfrak{F} := {}^{(\mathscr{Z}^N)}\mathfrak{F}$. These pointwise estimates play an important role in Chapter 20, where we use them to bound the corresponding error integrals in terms of the fundamental L^2-controlling quantities defined in Chapter 14. Most of the integrand factors are easy to pointwise bound, but some terms found in ${}^{(\mathscr{Z}^N)}\mathfrak{F}$ are not; we postpone the analysis of the difficult terms until Chapter 17. More precisely, in Sects. 16.1-16.6, we decompose ${}^{(\mathscr{Z}^N)}\mathfrak{F}$ and identify those terms that lead to error integrals that are easy to bound. This analysis is among the most important of the entire monograph, *for our pointwise estimates for the few important terms in ${}^{(\mathscr{Z}^N)}\mathfrak{F}$, which are provided in Chapter 17, affect the blow-up rates of our high-order a priori L^2 estimates in terms of powers of $\upmu_\star^{-1}$*; the connection between our estimates for these important terms and the blow-up rates of the high-order energies will become clear in the proof of Lemma 20.10. In Sect. 16.7, we derive pointwise estimates for the remaining non-$\mathfrak{F}$ integrand factors appearing in Prop. 10.13, which are relatively easy to bound. Finally, in Sect. 16.8, we derive some simple pointwise estimates that we need to close our top-order elliptic estimates.

16.1. Preliminary analysis and the definition of harmless terms

We start by proving a preliminary lemma. Roughly speaking, when combined with Cor. 16.5, the lemma shows that difficult inhomogeneous terms in the $N-$times-commuted wave equation can arise *only from repeatedly differentiating some of the inhomogeneous terms that arise after a single commutation.* In particular, the difficult terms involve certain top-order derivatives of the eikonal function quantities $\upmu$ and $\upchi^{(Small)}$.

LEMMA 16.1 (**Basic structure of the inhomogeneous terms in the $N-$times commuted wave equation**). *Assume that Ψ verifies the wave equation*

$$\square_{g(\Psi)}\Psi = 0 \qquad (13.1)$$

and let $1 \leq N \leq 24$ be an integer. Let $\mathscr{Z}^N$ be an Nth order commutation vectorfield operator consisting of iterated vectorfields belonging to the commutation set $\mathscr{Z} =$

$\{\varrho L, \breve{R}, O_{(1)}, O_{(2)}, O_{(3)}\}$. *Assume that* $\mathscr{Z}^N$ *is of the form* $\mathscr{Z}^N = \mathscr{Z}^{N-1}Z$, *and let* ${}^{(Z)}\mathscr{J}[\Psi]$ *be the commutation current* (9.5). *Under the small-data and bootstrap assumptions of Sects.* 12.1-12.4, *if* ε *is sufficiently small, then* $\mathscr{Z}^N\Psi$ *verifies the inhomogeneous wave equation*

$$\mu\square_{g(\Psi)}(\mathscr{Z}^N\Psi) = {}^{(\mathscr{Z}^N)}\mathfrak{F}, \tag{16.2}$$

$$ {}^{(\mathscr{Z}^N)}\mathfrak{F} = \mathscr{Z}^{N-1}(\mu\mathscr{D}_\alpha{}^{(Z)}\mathscr{J}^\alpha[\Psi]) + {}^{(\mathscr{Z}^N)}\mathfrak{F}_{(Eikonal-Low)}, \tag{16.3}$$

where ${}^{(\mathscr{Z}^N)}\mathfrak{F}_{(Eikonal-Low)} = 0$ *if* $N = 1$ *and otherwise* ${}^{(\mathscr{Z}^N)}\mathfrak{F}_{(Eikonal-Low)}$ *verifies the following pointwise inequality on* $\mathcal{M}_{T_{(Bootstrap)},U_0}$:

$$\begin{aligned} &\left|{}^{(\mathscr{Z}^N)}\mathfrak{F}_{(Eikonal-Low)}\right| \\ &\lesssim \sum_{\substack{N_1+N_2+N_3\leq N-1\\ N_1,N_2\leq N-2}} \sum_{Z_1,Z_2\in\mathscr{Z}} \left(\left|\mathscr{Z}^{N_1}\mathrm{tr}_{\not g}{}^{(Z_1)}\not\pi\right| + 1\right)\left|\mathscr{Z}^{N_2}(\mu\mathscr{D}_\alpha^{(Z_2)}\mathscr{J}^\alpha[\mathscr{Z}^{N_3}\Psi])\right|. \end{aligned} \tag{16.4}$$

PROOF. We iterate the identity (9.6). Clearly the first term on the right-hand side of (16.3) arises when $\mathscr{Z}^{N-1}$ falls on the first term on the right-hand side of (9.6). Using the observation made in Remark 9.2 and the estimate $\|\mathscr{Z}^{\leq 11}\mathrm{tr}_{\not g}{}^{(Z)}\not\pi\|_{C^0(\Sigma_t^u)} \lesssim 1$ (that is, (12.60b)), we deduce that all of the remaining terms that arise are in magnitude $\lesssim$ the right-hand side of (16.4) as desired. □

The term $\mathscr{Z}^{N-1}(\mu\mathscr{D}_\alpha^{(Z)}\mathscr{J}^\alpha[\Psi])$ on the right-hand side of (16.3) is difficult to bound. The two sums on the right-hand side of (16.4) are lower-order in terms of number of derivatives of the eikonal function quantities and are relatively easy to bound.

In our analysis of $\mathscr{Z}^{N-1}(\mu\mathscr{D}_\alpha{}^{(Z)}\mathscr{J}^\alpha[\Psi])$ and ${}^{(\mathscr{Z}^N)}\mathfrak{F}_{(Eikonal-Low)}$, the vast majority of the terms that we encounter have a negligible effect on the dynamics. We call such terms "harmless." We now give a precise definition of what we mean by "harmless."

DEFINITION 16.2 (**Harmless terms**). Let $0 \leq N \leq 24$ be an integer. A "$Harmless^{\leq N}$" term is any term such that under the small-data and bootstrap assumptions of Sects. 12.1-12.4, if ε is sufficiently small, then on the spacetime domain $\mathcal{M}_{T_{(Bootstrap)},U_0}$, it verifies the following bound:

$$\begin{aligned} \left|Harmless^{\leq N}\right| &\lesssim \left|\left\{L + \frac{1}{2}\mathrm{tr}_{\not g}\chi\right\}\mathscr{Z}^{\leq N}\Psi\right| + \frac{\ln(e+t)}{(1+t)^2}\left|\begin{pmatrix} \varrho(1+\mu)L\mathscr{Z}^{\leq N}\Psi \\ \breve{R}\mathscr{Z}^{\leq N}\Psi \\ \varrho(1+\mu)\not d\mathscr{Z}^{\leq N}\Psi \\ \mathscr{Z}^{\leq N}\Psi \end{pmatrix}\right| \\ &\quad + \varepsilon\frac{\ln(e+t)}{(1+t)^3}\left|\begin{pmatrix} \mathscr{Z}^{\leq N}(\mu-1) \\ \sum_{a=1}^3 \varrho\left|\mathscr{Z}^{\leq N}L^a_{(Small)}\right| \end{pmatrix}\right|. \end{aligned} \tag{16.5}$$

REMARK 16.3 (**The main features of the terms** $Harmless^{\leq N}$). The important features of $Harmless^{\leq N}$ terms are the following.

- Various L^2 norms of the terms on the right-hand side of (16.5) can be controlled *without the need for introducing modified quantities or deriving elliptic estimates.*

- The $t-$weight factors on the right-hand side of (16.5) are strong enough that these terms have a negligible effect on the dynamics in $\mathcal{M}_{T_{(Bootstrap)},U_0}$.

In the next proposition, we identify the difficult products in the main term $\mathscr{Z}^{N-1}(\upmu\mathscr{D}_\alpha^{(Z)}\mathscr{J}^\alpha[\Psi])$ on the right-hand side of (16.3). The difficult products vary depending on the vectorfield Z from ${}^{(Z)}\mathscr{J}^\alpha[\Psi]$.

PROPOSITION 16.4 (**Identification of the key difficult error term factors**). *Let* ${}^{(Z)}\mathscr{J}[\Psi]$ *be the commutation current* (9.5) *and let* $1 \leq N \leq 24$ *be an integer. Under the small-data and bootstrap assumptions of Sects.* 12.1-12.4, *if* ε *is sufficiently small, then the following pointwise estimates hold on* $\mathcal{M}_{T_{(Bootstrap)},U_0}$:

$$\mathscr{Z}^{N-1}(\upmu\mathscr{D}_\alpha{}^{(\breve{R})}\mathscr{J}^\alpha[\Psi]) = (\breve{R}\Psi)\Delta\!\!\!/\,\mathscr{Z}^{N-1}\upmu - (\upmu d\!\!\!/^{\#}\Psi)\cdot(\upmu d\!\!\!/\mathscr{Z}^{N-1}\mathrm{tr}_{g\!\!\!/}\chi^{(Small)}) + Harmless^{\leq N}, \tag{16.6a}$$

$$\mathscr{Z}^{N-1}(\upmu\mathscr{D}_\alpha{}^{(\varrho L)}\mathscr{J}^\alpha[\Psi]) = \varrho(d\!\!\!/^{\#}\Psi)\cdot(\upmu d\!\!\!/\mathscr{Z}^{N-1}\mathrm{tr}_{g\!\!\!/}\chi^{(Small)}) + Harmless^{\leq N}, \tag{16.6b}$$

$$\mathscr{Z}^{N-1}(\upmu\mathscr{D}_\alpha{}^{(O_{(l)})}\mathscr{J}^\alpha[\Psi]) = (\breve{R}\Psi)O_{(l)}\mathscr{Z}^{N-1}\mathrm{tr}_{g\!\!\!/}\chi^{(Small)} + \mathrm{p}_{(l)}(d\!\!\!/^{\#}\Psi)\cdot(\upmu d\!\!\!/\mathscr{Z}^{N-1}\mathrm{tr}_{g\!\!\!/}\chi^{(Small)}) + Harmless^{\leq N}, \tag{16.6c}$$

where $Harmless^{\leq N}$ *terms are defined in Def.* 16.2.

We provide the proof of Prop. 16.4 in Sect. 16.5.

In the next corollary, we show that the second term on the right-hand side of (16.3) is $Harmless^{\leq N}$.

COROLLARY 16.5 (**Pointwise estimates for** ${}^{(\mathscr{Z}^N)}\mathfrak{F}_{(Eikonal-Low)}$). *Let* $1 \leq N \leq 24$ *be an integer. Under the small-data and bootstrap assumptions of Sects.* 12.1-12.4, *if* ε *is sufficiently small, then the following pointwise estimates hold for the term* ${}^{(\mathscr{Z}^N)}\mathfrak{F}_{(Eikonal-Low)}$ *from* (16.4) *on* $\mathcal{M}_{T_{(Bootstrap)},U_0}$:

$${}^{(\mathscr{Z}^N)}\mathfrak{F}_{(Eikonal-Low)} = Harmless^{\leq N}, \tag{16.7}$$

where $Harmless^{\leq N}$ *terms are defined in Def.* 16.2.

We provide the proof of Cor. 16.5 in Sect. 16.6.

The main point of the next corollary is that when a few factors of ϱL are present in the differential operator $\mathscr{Z}^N$ from (16.2), we can completely avoid the use of modified quantities and elliptic estimates in deriving our a priori L^2 estimates. Thus, in order to close our L^2 estimates, we only have to perform a detailed analysis for a handful of cases in which $\mathscr{Z}^N$ contains at most one factor of ϱL. Furthermore, corresponding to each of these handful of cases, there are only one or two terms that are difficult to estimate.

COROLLARY 16.6 (**Reduction of the L^2 analysis to a few difficult cases**). *Let* $1 \leq N \leq 24$ *be an integer and assume that the differential operator* $\mathscr{Z}^N$ *contains precisely one factor of* ϱL. *Under the small-data and bootstrap assumptions of*

Sects. 12.1-12.4, if ε is sufficiently small, then the term ${}^{(\mathscr{Z}^N)}\mathfrak{F}$ *from the right-hand side of* (16.2) *verifies one of the following two estimates on* $\mathcal{M}_{T_{(Bootstrap)},U_0}$:

$$ {}^{(\mathscr{Z}^N)}\mathfrak{F} = Harmless^{\leq N}, \tag{16.8} $$

$$ {}^{(\mathscr{Z}^N)}\mathfrak{F} = \varrho(\not{d}\Psi^{\#}) \cdot (\upmu \not{d}\mathscr{S}^{N-1}\mathrm{tr}_{\not{g}}\chi^{(Small)}) + Harmless^{\leq N}, \tag{16.9} $$

where $\mathscr{S}^{N-1}$ *is an* $(N-1)^{st}$ *order pure spatial commutation vectorfield operator (see definition* (7.2a)*) and* $Harmless^{\leq N}$ *terms are defined in Def.* 16.2.

Furthermore, if the differential operator $\mathscr{Z}^N$ *contains two or more factors of* ϱL*, then*

$$ {}^{(\mathscr{Z}^N)}\mathfrak{F} = Harmless^{\leq N}. \tag{16.10} $$

PROOF. We first assume that $\mathscr{Z}^N$ contains precisely one factor of ϱL and prove (16.8) and (16.9). We split the argument into three cases: **i)** $\mathscr{Z}^N = \mathscr{Z}^{N-1}\varrho L$, **ii)** $\mathscr{Z}^N = \mathscr{Z}^{N-1}\breve{R}$, and **iii)** $\mathscr{Z}^N = \mathscr{Z}^{N-1}O_{(l)}$. In case **i)**, the estimate (16.9) follows from (16.3), (16.6b), and Cor. 16.5. In case **ii)**, $\mathscr{Z}^{N-1}$ must contain a factor of ϱL. We will show that ${}^{(\mathscr{Z}^N)}\mathfrak{F} = Harmless^{\leq N}$ in this case. By (16.3), (16.6a), and Cor. 16.5, ${}^{(\mathscr{Z}^N)}\mathfrak{F}$ is equal to the first two terms on the right-hand side of (16.6a) $+Harmless^{\leq N}$. Using inequalities (12.46a) and (12.46c) and the bootstrap assumption $\left\|\breve{R}\Psi\right\|_{C^0(\Sigma_t^u)} \leq \varepsilon(1+t)^{-1}$ (that is, (**BA**Ψ)), we deduce that the first term on the right-hand side of (16.6a) is bounded in magnitude by

$$ \lesssim \varepsilon \frac{1}{(1+t)^3} \sum_{l,m=1}^{3} \left|O_{(l)}O_{(m)}\mathscr{Z}^{N-1}\upmu\right|. \tag{16.11} $$

Next, we repeatedly use inequalities (12.81a) and (12.81b) with $f = \upmu - 1$, (12.128a), and (12.56) to commute the factor $Z := \varrho L$ in $\mathscr{Z}^{N-1}$ all the way in front (so that it acts last), which allows us to bound the right-hand side of (16.11) by

$$ \begin{aligned} &\lesssim \varepsilon \frac{1}{(1+t)^2}\left|L\mathscr{Z}^{\leq N}\upmu\right| + \varepsilon \frac{\ln(e+t)}{(1+t)^2}\left|\left(\begin{array}{c} \varrho L \mathscr{Z}^{\leq N-1}\Psi \\ \breve{R}\mathscr{Z}^{\leq N-1}\Psi \\ \varrho\not{d}\mathscr{Z}^{\leq N-1}\Psi \\ \mathscr{Z}^{\leq N-1}\Psi \end{array}\right)\right| \\ &\quad + \varepsilon \frac{\ln(e+t)}{(1+t)^3}\left|\left(\begin{array}{c} \mathscr{Z}^{\leq N}(\upmu - 1) \\ \sum_{a=1}^{3} \varrho \left|\mathscr{Z}^{\leq N} L_{(Small)}^a\right| \end{array}\right)\right|. \end{aligned} \tag{16.12} $$

Referring to Def. 16.2 and using the estimate (12.127b) to bound $\left|L\mathscr{Z}^{\leq N}\upmu\right|$, we see that the right-hand side of (16.12) is $= Harmless^{\leq N}$ as desired. Using a similar argument, we also deduce that the second term on the right-hand side of (16.6a) is $= Harmless^{\leq N}$ as desired. This completes the proof in case **ii)**. In case **iii)**, we use arguments similar to the ones we used in the case **ii)**, except that we use (16.6c) in place of (16.6a) and we use the estimate (12.28b) to bound the factor $\mathsf{p}_{(l)}$ on the right-hand side of (16.6c). We conclude that ${}^{(\mathscr{Z}^N)}\mathfrak{F} = Harmless^{\leq N}$ in case **iii)**. We have thus proved (16.8) and (16.9).

We now prove (16.10). Under the present assumptions, in the cases **ii)** and **iii)** from above, $\mathscr{Z}^{N-1}$ must contain two factors of ϱL. In these cases, our previous argument already implies the desired estimate (16.10); in fact, our proof above relied only on the assumption that $\mathscr{Z}^{N-1}$ contains a single factor of ϱL. In case **i)**, the estimate (16.9) implies that ${}^{(\mathscr{Z}^N)}\mathfrak{F} = \varrho(\not{d}\Psi^{\#}) \cdot (\upmu \not{d} \mathscr{Z}^{N-1} \mathrm{tr}_{\not{g}} \chi^{(Small)}) + Harmless^{\leq N}$. The main point is that $\mathscr{Z}^{N-1}$ must contain at least one factor of ϱL. Hence, essentially the same arguments given for case **ii)** in the previous paragraph imply that $\varrho(\not{d}\Psi^{\#}) \cdot (\upmu \not{d} \mathscr{Z}^{N-1} \mathrm{tr}_{\not{g}} \chi^{(Small)}) = Harmless^{\leq N}$, which yields the desired estimate (16.10) in this case. □

16.2. The important terms in the top-order derivatives of the deformation tensors of the commutation vectorfields

As we noted above, the most difficult terms in the commuted wave equation (16.2) are contained in the first term $\mathscr{Z}^{N-1}(\upmu \mathscr{D}_{\alpha}^{(Z)} \mathscr{J}^{\alpha}[\Psi])$ on the right-hand side of equation (16.3). Specifically, the difficult terms arise from the top-order derivatives of the deformation tensors of the commutation vectorfields, whose first derivatives appear after a single commutation; see Prop. 9.6. The terms are difficult precisely because at top order, we need to use modified quantities and elliptic estimates to bound them in L^2. In the next three lemmas, in the three cases $Z = \varrho L$, $Z = \breve{R}$, $Z = O$, we identify these "important" difficult terms in $\mathscr{Z}^{N-1}(\upmu \mathscr{D}_{\alpha}^{(Z)} \mathscr{J}^{\alpha}[\Psi])$. *The important terms appear on the left-hand sides of the inequalities of the lemmas.*

LEMMA 16.7 (**Identification of the important top-order terms in the derivatives of ${}^{(\breve{R})}\pi$**). *Let $1 \leq N \leq 24$ be an integer. Under the small-data and bootstrap assumptions of Sects. 12.1-12.4, if ε is sufficiently small, then the following estimates hold on $\mathcal{M}_{T_{(Bootstrap)}, U_0}$:*

$$
\begin{aligned}
&\left| \not{\mathcal{L}}_{\mathscr{Z}}^{N-1} \not{\mathcal{L}}_{\breve{R}} {}^{(\breve{R})}\not{\pi}_L^{\#} + \not{d}^{\#} \mathscr{Z}^{N-1} \breve{R}\upmu \right|, \\
&\left| \mathscr{Z}^{N-1} \breve{R} \mathrm{tr}_{\not{g}} {}^{(\breve{R})}\not{\pi} + 2\upmu \not{\Delta} \mathscr{Z}^{N-1} \upmu \right| \\
&\lesssim \left| \begin{pmatrix} \varrho L \mathscr{Z}^{\leq N} \Psi \\ \breve{R} \mathscr{Z}^{\leq N} \Psi \\ \varrho \not{d} \mathscr{Z}^{\leq N} \Psi \\ \mathscr{Z}^{\leq N} \Psi \end{pmatrix} \right| + \frac{1}{1+t} \left| \begin{pmatrix} \mathscr{Z}^{\leq N}(\upmu - 1) \\ \sum_{a=1}^{3} \varrho \left| \mathscr{Z}^{\leq N} L_{(Small)}^a \right| \end{pmatrix} \right| + \frac{1}{1+t},
\end{aligned}
\tag{16.13a}
$$

$$
\begin{aligned}
&(1+t) \left| \mathscr{Z}^{N-1} \not{\mathrm{div}} {}^{(\breve{R})}\not{\pi}_L^{\#} + \not{\Delta} \mathscr{Z}^{N-1} \upmu \right|, \\
&\left| \not{\mathcal{L}}_{\mathscr{Z}}^{N-1} \not{d}^{\#} {}^{(\breve{R})}\pi_{L\breve{R}} + \not{d}^{\#} \mathscr{Z}^{N-1} \breve{R}\upmu \right|, \\
&(1+t) \left| \not{\mathcal{L}}_{\mathscr{Z}}^{N-1} \not{\mathrm{div}} {}^{(\breve{R})}\not{\pi}^{\#\#} + \upmu \not{d}^{\#} \mathscr{Z}^{N-1} \mathrm{tr}_{\not{g}} \chi^{(Small)} \right| \\
&\lesssim \left| \begin{pmatrix} \varrho L \mathscr{Z}^{\leq N} \Psi \\ \breve{R} \mathscr{Z}^{\leq N} \Psi \\ \varrho \not{d} \mathscr{Z}^{\leq N} \Psi \\ \mathscr{Z}^{\leq N} \Psi \end{pmatrix} \right| + \frac{1}{1+t} \left| \begin{pmatrix} \mathscr{Z}^{\leq N}(\upmu - 1) \\ \sum_{a=1}^{3} \varrho \left| \mathscr{Z}^{\leq N} L_{(Small)}^a \right| \end{pmatrix} \right|.
\end{aligned}
\tag{16.13b}
$$

PROOF. We first prove the bound (16.13a) for $\not{\mathcal{L}}_{\mathscr{Z}}^{N-1} \not{\mathcal{L}}_{\breve{R}} {}^{(\breve{R})}\not{\pi}_L^{\#} + \not{d}^{\#} \mathscr{Z}^{N-1} \breve{R}\upmu$. We apply $\not{\mathcal{L}}_{\mathscr{Z}}^{N-1} \not{\mathcal{L}}_{\breve{R}}$ to the $\not{g}$-dual of the right-hand side of (7.5d). By Lemma 8.6 and the Leibniz rule, the first term on right-hand side is equal to the principal term $-\not{d}^{\#} \mathscr{Z}^{N-1} \breve{R}\upmu$ plus an error term that is bounded in magnitude by

$\lesssim \sum_{\substack{N_1+N_2\leq N\\ N_2\leq N-1}} \left|\mathcal{L}_{\mathscr{Z}}^{N_1} \not{g}^{-1}\right| \left|\not{d}\mathscr{Z}^{N_2}\upmu\right|$. We bring the principal term over to the left, while by inequality (12.46a) and the estimates (12.60a), (12.60b), and (12.128a), the error term is $\lesssim$ the right-hand side of (16.13a) as desired. Hence, to conclude the desired bound, it remains only for us to bound the magnitude of $\not{\mathcal{L}}_{\mathscr{Z}}^{N-1}\not{\mathcal{L}}_{\breve{R}} = \not{\mathcal{L}}_{\mathscr{Z}}^{N}$ applied to the $\not{g}$-dual of the second product on the right-hand side of (7.5d) by the right-hand side of (16.13a). This product is of the schematic form $\xi := G_{(Frame)}\not{g}^{-1}\begin{pmatrix} \upmu L\Psi \\ \breve{R}\Psi \\ \upmu\not{d}\Psi \end{pmatrix}$. To bound $\not{\mathcal{L}}_{\mathscr{Z}}^{N}\xi$ in magnitude by the right-hand side of (16.13a), we use the Leibniz rule and the estimates (12.54) (with Ψ in the role of f), (12.58a), (12.58b), (12.60a), (12.60b), (12.128a), and the bootstrap assumptions (**BA**Ψ) to deduce that

(16.14a)

$$\begin{aligned}
\left|\not{\mathcal{L}}_{\mathscr{Z}}^{N}\xi\right| &\lesssim \sum_{N_1+N_2+N_3+N_4\leq N} \left|\not{\mathcal{L}}_{\mathscr{Z}}^{N_1} G_{(Frame)}\right| \left|\not{\mathcal{L}}_{\mathscr{Z}}^{N_2}\not{g}^{-1}\right| \left|\mathscr{Z}^{N_3}\upmu\right| \left|\not{\mathcal{L}}_{\mathscr{Z}}^{N_4}\begin{pmatrix} L\Psi \\ \not{d}\Psi \end{pmatrix}\right| \\
&\quad + \sum_{N_1+N_2+N_3\leq N} \left|\not{\mathcal{L}}_{\mathscr{Z}}^{N_1} G_{(Frame)}\right| \left|\not{\mathcal{L}}_{\mathscr{Z}}^{N_2}\not{g}^{-1}\right| \left|\mathscr{Z}^{N_3}\breve{R}\Psi\right| \\
&\lesssim \frac{\ln(e+t)}{1+t}\left|\begin{pmatrix} \varrho L\mathscr{Z}^{\leq N}\Psi \\ \varrho\not{d}\mathscr{Z}^{\leq N}\Psi \\ \mathscr{Z}^{\leq N}\Psi \end{pmatrix}\right| + \left|\breve{R}\mathscr{Z}^{\leq N}\Psi\right| \\
&\quad + \varepsilon\frac{1}{(1+t)^2}\left|\begin{pmatrix} \mathscr{Z}^{\leq N}(\upmu-1) \\ \sum_{a=1}^{3}\varrho\left|\mathscr{Z}^{\leq N}L_{(Small)}^{a}\right| \end{pmatrix}\right|,
\end{aligned}$$

(16.14b)

$$\left\|\not{\mathcal{L}}_{\mathscr{Z}}^{\leq 12}\xi\right\|_{C^0(\Sigma_t^u)} \lesssim \varepsilon\frac{1}{1+t}.$$

as desired (we use the bound (16.14b) below).

To prove the bound (16.13a) for $\mathscr{Z}^{N-1}\breve{R}\mathrm{tr}_{\not{g}}{}^{(\breve{R})}\not{\pi} + 2\upmu\not{\Delta}\mathscr{Z}^{N-1}\upmu$, we apply $\mathscr{Z}^{N-1}\breve{R}$ to the right-hand side of (7.5g). The last term is of the form

$$\xi := G_{(Frame)}\not{g}^{-1}\begin{pmatrix} \upmu L\Psi \\ \breve{R}\Psi \\ \upmu\not{d}\Psi \end{pmatrix}$$

and hence the proof of (16.14a) yields the desired bound (16.14a) for ξ. To bound the first term on the right-hand side of (7.5g), we use (12.6b) to deduce that $\left|\mathscr{Z}^{N-1}\breve{R}(\varrho^{-1}\upmu)\right| \lesssim (1+t)^{-1}\left|\mathscr{Z}^{\leq N}(\upmu-1)\right| + (1+t)^{-1}$ as desired. It remains for us to address the important term $-2\mathscr{Z}^{N-1}\breve{R}(\upmu\mathrm{tr}_{\not{g}}\chi^{(Small)})$ and to show that it is equal to $-2\upmu\mathscr{Z}^{N-1}\upmu$ plus an error term that is in magnitude $\lesssim$ the right-hand side of (16.13a). To see that this is the case, we apply the Leibniz rule. We bound all terms except the one in which all derivatives fall on $\mathrm{tr}_{\not{g}}\chi^{(Small)}$ by using (12.128a) and (12.67a). To handle the one term in which all derivatives fall on $\mathrm{tr}_{\not{g}}\chi^{(Small)}$, we use in addition the fourth inequality in (15.1). We have thus proved the desired bound.

We now prove the bound (16.13b) for $\mathscr{Z}^{N-1}\mathrm{d}\!\!/\mathrm{iv}\,{}^{(\breve{R})}\!\not{\pi}_L^{\#} + \not{\Delta}\mathscr{Z}^{N-1}\upmu$. We apply $\mathscr{Z}^{N-1}\mathrm{d}\!\!/\mathrm{iv}$ to the $\not{g}$-dual of the right-hand side of (7.5d). We begin by addressing the second term on the right-hand side, which is equal to the $\not{g}$-dual of an $S_{t,u}$ one-form ξ that is of the form $\xi := G_{(Frame)}\begin{pmatrix}\upmu L\Psi\\ \breve{R}\Psi\\ \upmu\not{d}\Psi\end{pmatrix}$. We now note that a similar proof to that of (16.14a) and (16.14b) yields that ξ verifies the bounds (16.14a) and (16.14b). Next, to derive the desired bound for $\mathscr{Z}^{N-1}\mathrm{d}\!\!/\mathrm{iv}\,\xi^{\#}$, we use the Leibniz rule to deduce that

(16.15)
$$\left|\mathscr{Z}^{N-1}\mathrm{d}\!\!/\mathrm{iv}\,\xi^{\#}\right| \lesssim \sum_{N_1+N_2\leq N-1}\left|\not{\mathcal{L}}_{\mathscr{Z}}^{N_1}\not{g}^{-1}\right|\left|\not{\nabla}\not{\mathcal{L}}_{\mathscr{Z}}^{N_2}\xi\right| + \sum_{N_1+N_2\leq N-1}\left|\not{\mathcal{L}}_{\mathscr{Z}}^{N_1}\not{g}^{-1}\right|\left|[\not{\mathcal{L}}_{\mathscr{Z}}^{N_2},\not{\nabla}]\xi\right|.$$

To deduce that the first sum on the right-hand side of (16.15) is $\lesssim$ the product of $(1+t)^{-1}$ and the right-hand side of (16.13b), we use the estimates (12.47), (12.60a), (12.60b), and the aforementioned bounds (16.14a) and (16.14b). To deduce that the second sum on the right-hand side of (16.22) is $\lesssim$ the product of $(1+t)^{-1}$ and the right-hand side of (16.13b), we combine similar reasoning with the commutator estimate (12.92), where N_2 is in the role of N. We now address the term $-\mathscr{Z}^{N-1}\not{\Delta}\upmu$ arising from the first term on the right-hand side of (7.5d). We commute $\mathscr{Z}^{N-1}$ and $\not{\Delta}$ and move the principal term $\not{\Delta}\mathscr{Z}^{N-1}\upmu$ to the left-hand side. To complete the proof of the bound for $\mathscr{Z}^{N-1}\mathrm{d}\!\!/\mathrm{iv}\,{}^{(\breve{R})}\!\not{\pi}_L^{\#} + \not{\Delta}\mathscr{Z}^{N-1}\upmu$, it remains only for us to bound the magnitude of the commutator term $[\mathscr{Z}^{N-1},\not{\Delta}]\upmu$ by the product of $(1+t)^{-1}$ and the right-hand side of (16.13b). To this end, we use the estimate (12.97) with $N-1$ in the role of N and $\upmu-1$ in the role of f and inequality (12.128a).

We now prove the bound (16.13b) for $\not{\mathcal{L}}_{\mathscr{Z}}^{N-1}\not{d}^{\#}\,{}^{(\breve{R})}\pi_{L\breve{R}} + \not{d}^{\#}\mathscr{Z}^{N-1}\breve{R}\upmu$. We apply $\not{\mathcal{L}}_{\mathscr{Z}}^{N-1}\not{d}^{\#}$ to the right-hand side of (7.5c). The quantity we must estimate is $-\not{\mathcal{L}}_{\mathscr{Z}}^{N-1}\not{d}^{\#}\breve{R}\upmu$. From Lemma 8.6 and the Leibniz rule, it follows that this quantity is equal to the principal term $-\not{d}^{\#}\mathscr{Z}^{N-1}\breve{R}\upmu$ plus an error term that is bounded in magnitude by

(16.16)
$$\lesssim \sum_{\substack{N_1+N_2\leq N-1\\ N_2\leq N-2}}\left|\not{\mathcal{L}}_{\mathscr{Z}}^{N_1}\not{g}^{-1}\right|\left|\not{d}\mathscr{Z}^{N_2}\breve{R}\upmu\right|.$$

We bring the principal term over to the left-hand side. To complete the proof, we need to show that the right-hand side of (16.16) is $\lesssim$ the right-hand side of (16.13b). To deduce the desired bound, we use the estimates (12.60a), (12.60b), (12.46a), and (12.128a).

Finally, we prove (16.13b) for the term $\not{\mathcal{L}}_{\mathscr{Z}}^{N-1}\mathrm{d}\!\!/\mathrm{iv}\,{}^{(\breve{R})}\!\not{\pi}^{\#\#} + \upmu\not{d}^{\#}\mathscr{Z}^{N-1}\mathrm{tr}_{\not{g}}\chi^{(Small)}$. We apply $\not{\mathcal{L}}_{\mathscr{Z}}^{N-1}\mathrm{d}\!\!/\mathrm{iv}$ to the double $\not{g}$-dual of the right-hand side of (7.5f). We begin by addressing the first term, which is the difficult one:

$$-2\not{\mathcal{L}}_{\mathscr{Z}}^{N-1}\not{\nabla}_B\left\{\upmu(\not{g}^{-1})^{AC}(\not{g}^{-1})^{BD}\hat{\chi}_{CD}^{(Small)}\right\}.$$

When all derivatives fall on $(\not{g}^{-1})^{BD}\hat{\chi}_{CD}^{(Small)}$, we use (12.128a) and the second inequality in (15.28) to rewrite it as the principal term $-\upmu\not{d}^{\#}\mathscr{Z}^{N-1}\mathrm{tr}_{\not{g}}\chi^{(Small)}$ plus an error term with magnitude $\lesssim$ the product of $(1+t)^{-1}$ and the right-hand side of (16.13b). We bring the principal term over to the left-hand side. Using

Lemma 8.6, we see that the remaining terms arising from the Leibniz expansion of $-2\not{\mathcal{L}}_{\mathscr{Z}}^{N-1}\not{\nabla}_B\left\{\mu(\not{g}^{-1})^{AC}(\not{g}^{-1})^{BD}\hat{\chi}_{CD}^{(Small)}\right\}$ are bounded in magnitude by

$$\lesssim \sum_{\substack{N_1+N_2+N_3+N_4\leq N-1\\ N_4\leq N-2}} \left|\not{\mathcal{L}}_{\mathscr{Z}}^{N_1}\mu\right|\left|\not{\mathcal{L}}_{\mathscr{Z}}^{N_2}\not{g}^{-1}\right|\left|\not{\mathcal{L}}_{\mathscr{Z}}^{N_3}\not{g}^{-1}\right|\left|\not{\mathcal{L}}_{\mathscr{Z}}^{N_4}\not{\nabla}\hat{\chi}^{(Small)}\right| + \sum_{N_1+N_2+N_3+N_4\leq N-1} \left|\not{d}\not{\mathcal{L}}_{\mathscr{Z}}^{N_1}\mu\right|\left|\not{\mathcal{L}}_{\mathscr{Z}}^{N_2}\not{g}^{-1}\right|\left|\not{\mathcal{L}}_{\mathscr{Z}}^{N_3}\not{g}^{-1}\right|\left|\not{\mathcal{L}}_{\mathscr{Z}}^{N_4}\hat{\chi}^{(Small)}\right|. \tag{16.17}$$

Using inequalities (12.46a) and (12.48), the estimates (12.60a), (12.60b), (12.67a), and (12.128a), the commutator estimate (12.92) with $\xi = \hat{\chi}^{(Small)}$, and (12.102) with $\xi = \chi^{(Small)}$, we deduce that the right-hand side of (16.17) is $\lesssim$ the product of $(1+t)^{-1}$ and the right-hand side of (16.13b) as desired. To complete the proof, it remains for us to show that the $\not{\mathcal{L}}_{\mathscr{Z}}^{N-1}\mathrm{di}\not{v}$ derivatives of the double $\not{g}$-dual of the remaining terms on the right-hand side of (7.5f) have magnitudes that are $\lesssim$ the product of $(1+t)^{-1}$ and the right-hand side of (16.13b). To this end, we let $\hat{\xi}^{\#\#}$ denote the remaining terms, which correspond to a symmetric type $\binom{0}{2}$ $S_{t,u}$ tensorfield ξ of the schematic form $\xi := G_{(Frame)} \otimes \begin{pmatrix} \mu L\Psi \\ \breve{R}\Psi \\ \mu\not{d}\Psi \end{pmatrix}$. We now note that a similar proof to that of (16.14a) and (16.14b) yields that ξ verifies the bounds (16.14a) and (16.14b). We now recall that our goal is to bound $\not{\mathcal{L}}_{\mathscr{Z}}^{N-1}\mathrm{di}\not{v}\hat{\xi}^{\#\#}$ in magnitude by $\lesssim$ the product of $(1+t)^{-1}$ and the right-hand side of (16.18b). To proceed, we use the Leibniz rule to deduce that inequality (16.15) holds with $\not{\mathcal{L}}_{\mathscr{Z}}^{N-1}\mathrm{di}\not{v}\hat{\xi}^{\#\#}$ in place of $\mathscr{Z}^{N-1}\mathrm{di}\not{v}\hat{\xi}^{\#\#}$ on the left-hand side and $\hat{\xi}$ in place of ξ on the right-hand side. Finally, to bound the right-hand side of (16.15) (with $\hat{\xi}$ in place of ξ) by $\lesssim$ the product of $(1+t)^{-1}$ and the right-hand side of (16.18b), we use the estimates (12.48), (12.60a), (12.60b), (12.92), (12.102), and the aforementioned bounds (16.14a) and (16.14b). □

LEMMA 16.8 (**Identification of the important top-order terms in the derivatives of** ${}^{(O_{(l)})}\pi$). *Let* $1 \leq N \leq 24$ *be an integer. Under the small-data and bootstrap assumptions of Sects.* 12.1-12.4, *if* ε *is sufficiently small, then the following estimates hold on* $\mathcal{M}_{T_{(Bootstrap)},U_0}$:

$$\left|\not{\mathcal{L}}_{\mathscr{Z}}^{N-1}\not{\mathcal{L}}_{\breve{R}}{}^{(O_{(l)})}\not{\pi}_L^{\#} + (\not{\nabla}^{2\#}\mathscr{Z}^{N-1}\mu)\cdot O_{(l)}\right|,$$
$$\left|\mathscr{Z}^{N-1}\breve{R}\mathrm{tr}_{\not{g}}{}^{(O_{(l)})}\not{\pi} - 2\rho_{(l)}\not{\Delta}\mathscr{Z}^{N-1}\mu\right|$$
$$\lesssim \left|\begin{pmatrix} \varrho L\mathscr{Z}^{\leq N}\Psi \\ \breve{R}\mathscr{Z}^{\leq N}\Psi \\ \varrho\not{d}\mathscr{Z}^{\leq N}\Psi \\ \mathscr{Z}^{\leq N}\Psi \end{pmatrix}\right| + \frac{1}{1+t}\left|\begin{pmatrix} \mathscr{Z}^{\leq N}(\mu-1) \\ \sum_{a=1}^{3}\varrho\left|\mathscr{Z}^{\leq N}L_{(Small)}^a\right| \end{pmatrix}\right|, \tag{16.18a}$$

(16.18b)

$$(1+t)\left|\mathscr{Z}^{N-1}\mathrm{d}\!\!/\mathrm{i}\!\!/\mathrm{v}\,{}^{(O_{(l)})}\not\pi_L^{\#} + O_{(l)}\mathscr{Z}^{N-1}\mathrm{tr}_{\not g}\chi^{(Small)}\right|,$$

$$\frac{(1+t)}{\ln(e+t)}\left|\mathscr{Z}^{N-1}\mathrm{d}\!\!/\mathrm{i}\!\!/\mathrm{v}\,{}^{(O_{(l)})}\not\pi_{\breve{R}}^{\#} - \left\{\mu O_{(l)}\mathscr{Z}^{N-1}\mathrm{tr}_{\not g}\chi^{(Small)} + \rho_{(l)}\not\Delta\mathscr{Z}^{N-1}\mu\right\}\right|,$$

$$\left|\not\mathcal{L}_{\mathscr{Z}}^{N-1}\not d^{\#(O_{(l)})}\pi_{L\breve{R}} + (\not\nabla^{2\#}\mathscr{Z}^{N-1}\mu)\cdot O_{(l)}\right|,$$

$$(1+t)\left|\not\mathcal{L}_{\mathscr{Z}}^{N-1}\mathrm{d}\!\!/\mathrm{i}\!\!/\mathrm{v}\,{}^{(O_{(l)})}\not\pi^{\#\#} - \rho_{(l)}\not d^{\#}\mathscr{Z}^{N-1}\mathrm{tr}_{\not g}\chi^{(Small)}\right|$$

$$\lesssim \left|\begin{pmatrix} \varrho L\mathscr{Z}^{\leq N}\underline{\Upsilon} \\ \breve{R}\mathscr{Z}^{\leq N}\Psi \\ \varrho\not d\mathscr{Z}^{\leq N}\Psi \\ \mathscr{Z}^{\leq N}\Psi \end{pmatrix}\right| + \frac{1}{1+t}\left|\begin{pmatrix} \mathscr{Z}^{\leq N}(\mu-1) \\ \sum_{a=1}^{3}\varrho\left|\mathscr{Z}^{\leq N}L^{c}_{(Small)}\right| \end{pmatrix}\right|.$$

PROOF. We first prove (16.18a) for $\not\mathcal{L}_{\mathscr{Z}}^{N-1}\not\mathcal{L}_{\breve{R}}{}^{(O_{(l)})}\not\pi_L^{\#} + (\not\nabla^{2\#}\mathscr{Z}^{N-1}\mu)\cdot O_{(l)}$. We apply $\not\mathcal{L}_{\mathscr{Z}}^{N-1}\not\mathcal{L}_{\breve{R}}$ to the $\not g$-dual of the right-hand side of (7.8d). We begin by addressing the first term, which is the difficult one: $-\not\mathcal{L}_{\mathscr{Z}}^{N-1}\not\mathcal{L}_{\breve{R}}\left\{(\not g^{-1})^{AB}\chi_{BC}^{(Small)}O_{(l)}^{C}\right\}$. When all derivatives fall on $\chi_{BC}^{(Small)}$, we use (12.34a) and the first inequality in (15.1) to rewrite this term as the principal eikonal function term

$$-(\not g^{-1})^{AB}(\not\nabla^2_{BC}\mathscr{Z}^{N-1}\mu)O_{(l)}^{C}$$

plus an error term with magnitude $\lesssim$ the right-hand side of (16.18a). We then bring the principal term over to the left-hand side. To conclude the desired inequality, it remains for us to show that all remaining terms arising from the right-hand side of (7.8d) have magnitudes that are $\lesssim$ the right-hand side of (16.18a). The remaining terms arising in the Leibniz expansion of $-\not\mathcal{L}_{\mathscr{Z}}^{N-1}\not\mathcal{L}_{\breve{R}}\left\{(\not g^{-1})^{AB}\chi_{BC}^{(Small)}O_{(l)}^{C}\right\}$ are bounded in magnitude by

$$\lesssim \sum_{\substack{N_1+N_2+N_3\leq N \\ N_2\leq N-1}} \left|\not\mathcal{L}_{\mathscr{Z}}^{N_1}\not g^{-1}\right|\left|\not\mathcal{L}_{\mathscr{Z}}^{N_2}\chi^{(Small)}\right|\left|\not\mathcal{L}_{\mathscr{Z}}^{N_3}O_{(l)}\right|. \tag{16.19}$$

From the estimates (12.60a), (12.60b), (12.67a), (12.70a), (12.70b), and (12.128a), we conclude that the right-hand side of (16.19) is $\lesssim$ the right-hand side of (16.18a) as desired. To bound the first term $\not\mathcal{L}_{\mathscr{Z}}^{N-1}\not\mathcal{L}_{\breve{R}}\left\{G_{(Frame)}\not g^{-1}\begin{pmatrix} O_{(l)} \\ \rho_{(l)}\end{pmatrix}\begin{pmatrix} L\Psi \\ \not d\Psi\end{pmatrix}\right\}$ arising from the right-hand side of (7.9a) by the right-hand side of (16.18a), we use the estimates (12.54), (12.28a), (12.28b), (12.58a), (12.58b), (12.60a), (12.60b), (12.70a), and (12.70b) and the bootstrap assumptions (**BA**Ψ). To bound the second term $\not\mathcal{L}_{\mathscr{Z}}^{N-1}\not\mathcal{L}_{\breve{R}}\left\{f(\Psi)L_{(Small)}\not d^{\#}x\right\}$ arising from the right-hand side of (7.9a) by the right-hand side of (16.18a), we use the estimates (12.54), (12.57a) (12.57b), (12.60a), (12.60b), and (12.128a), and the bootstrap assumptions (**BA**Ψ).

We now prove (16.18a) for $\mathscr{Z}^{N-1}\breve{R}\mathrm{tr}_{\not g}{}^{(O_{(l)})}\not\pi - 2\rho_{(l)}\not\Delta\mathscr{Z}^{N-1}\mu$. We apply $\mathscr{Z}^{N-1}\breve{R}$ to the right-hand side of (7.8g) and apply the Leibniz rule. We begin by addressing the first term, which is the difficult one: $2\mathscr{Z}^{N-1}\breve{R}(\rho_{(l)}\mathrm{tr}_{\not g}\chi^{(Small)})$. When all derivatives fall on $\mathrm{tr}_{\not g}\chi^{(Small)}$, we use (12.28b) and the fourth inequality in (15.1) to rewrite this term as the principal term $2\rho_{(l)}\not\Delta\mathscr{Z}^{N-1}\mu$ plus an error term with magnitude $\lesssim$ the right-hand side of (16.18a). We then bring the principal term over to the left-hand side. To complete the proof, it remains for us to show that the $\mathscr{Z}^{N-1}\breve{R}$ derivatives of the remaining terms on the right-hand side of (7.8g) have magnitudes that are $\lesssim$ the right-hand side of (16.18a). To bound the remaining terms arising from the Leibniz expansion of $2\mathscr{Z}^{N-1}\breve{R}(\rho_{(l)}\mathrm{tr}_{\not{g}}\chi^{(Small)})$, we use inequalities (12.28a), (12.28b), (12.67a), and (12.128a). To bound the magnitude of the term $\mathscr{Z}^{N-1}\breve{R}\left(\frac{\rho_{(l)}}{\varrho}\right)$ arising from the right-hand side of (7.9d) by the right-hand side of (16.18a), we use (12.6b) and (12.28a). To bound the magnitude of the term $\not{\mathcal{L}}_{\mathscr{Z}}^{N-1}\not{\mathcal{L}}_{\breve{R}}\left\{f(\Psi)\Psi(\not{d}^{\#}x)\not{d}x\right\}$ arising from the right-hand side of (7.9d) by the right-hand side of (16.18a), we use the estimates (12.57a), (12.57b), (12.60a), (12.60b), and the bootstrap assumptions (**BA**Ψ). To bound the magnitude of the remaining terms $\not{\mathcal{L}}_{\mathscr{Z}}^{N-1}\not{\mathcal{L}}_{\breve{R}}\left\{G_{(Frame)}\not{g}^{-1}\begin{pmatrix} O_{(l)} \\ \rho_{(l)} \end{pmatrix}\begin{pmatrix} L\Psi \\ \not{d}\Psi \end{pmatrix}\right\}$ arising from the right-hand side of (7.9d) by the right-hand side of (16.18a), we note that these terms have essentially the same structure as the $\not{g}$-dual of some of the terms in (7.9a). Hence, the analysis in the previous paragraph yields the desired bound.

We now prove (16.18b) for

$$\mathscr{Z}^{N-1}\mathrm{d}\not{\mathrm{i}}\mathrm{v}^{(O_{(l)})}\not{\mathcal{J}}_{\breve{R}}^{\#} - \left\{\mu O_{(l)}\mathscr{Z}^{N-1}\mathrm{tr}_{\not{g}}\chi^{(Small)} + \rho_{(l)}\not{\Delta}\mathscr{Z}^{N-1}\mu\right\}.$$

We apply $\mathscr{Z}^{N-1}\mathrm{d}\not{\mathrm{i}}\mathrm{v}$ to the $\not{g}$-dual of (7.8e) and apply the Leibniz rule to the terms on the right-hand side. We begin by addressing the first term, which is the difficult one: $\mathscr{Z}^{N-1}\not{\nabla}_A\left\{\mu(\not{g}^{-1})^{AB}\chi_{BC}^{(Small)}O_{(l)}^C\right\}$. When all derivatives fall on $(\not{g}^{-1})^{AB}\chi_{BC}^{(Small)}$, we use (12.34a), (12.128a), and the first inequality in (15.28) to deduce that this term is equal to the principal term $\mu O_{(l)}\mathscr{Z}^{N-1}\mathrm{tr}_{\not{g}}\chi^{(Small)}$ plus an error term with magnitude $\lesssim$ the product of $\ln(e+t)(1+t)^{-1}$ and the right-hand side of (16.18b). We move the principal term to the left-hand side. From the Leibniz rule and the fact that $\not{\nabla}O_{(l)} = \not{g}^{-1}\cdot\not{\nabla}O_{(l)\flat}$, where $O_{(l)\flat}$ is the $\not{g}$-dual of $O_{(l)}$, we deduce that the remaining terms arising in the Leibniz expansion of $\mathscr{Z}^{N-1}\not{\nabla}_A\left\{\mu(\not{g}^{-1})^{AB}\chi_{BC}^{(Small)}O_{(l)}^C\right\}$ are bounded in magnitude as follows:

$$\begin{aligned}
(16.20)\quad &\lesssim \sum_{\substack{N_1+N_2+N_3+N_4\leq N-1 \\ N_3\leq N-2}} \left|\not{\mathcal{L}}_{\mathscr{Z}}^{N_1}\mu\right|\left|\not{\mathcal{L}}_{\mathscr{Z}}^{N_2}\not{g}^{-1}\right|\left|\not{\mathcal{L}}_{\mathscr{Z}}^{N_3}\not{\nabla}\chi^{(Small)}\right|\left|\not{\mathcal{L}}_{\mathscr{Z}}^{N_4}O_{(l)}\right| \\
&+ \sum_{N_1+N_2+N_3+N_4\leq N-1} \left|\not{d}\not{\mathcal{L}}_{\mathscr{Z}}^{N_1}\mu\right|\left|\not{\mathcal{L}}_{\mathscr{Z}}^{N_2}\not{g}^{-1}\right|\left|\not{\mathcal{L}}_{\mathscr{Z}}^{N_3}\chi^{(Small)}\right|\left|\not{\mathcal{L}}_{\mathscr{Z}}^{N_4}O_{(l)}\right| \\
&+ \sum_{N_1+N_2+N_3+N_4+N_5\leq N-1} \left|\not{\mathcal{L}}_{\mathscr{Z}}^{N_1}\mu\right|\left|\not{\mathcal{L}}_{\mathscr{Z}}^{N_2}\not{g}^{-1}\right|\left|\not{\mathcal{L}}_{\mathscr{Z}}^{N_3}\not{g}^{-1}\right|\left|\not{\mathcal{L}}_{\mathscr{Z}}^{N_4}\chi^{(Small)}\right|\left|\not{\mathcal{L}}_{\mathscr{Z}}^{N_5}\not{\nabla}O_{(l)\flat}\right|.
\end{aligned}$$

Using inequalities (12.60a), (12.60b), (12.70a), (12.70b), (12.67a), (12.46a), (12.47), (12.48), (12.128a), and (12.92), we deduce that the right-hand side of (16.20) is $\lesssim$ the product of $\ln(e+t)(1+t)^{-1}$ and the right-hand side of (16.18b) as desired. We now address the estimates corresponding to the second term on the right-hand side of (7.8e), that is, corresponding to $\mathscr{Z}^{N-1}\left\{\rho_{(l)}\not{\Delta}\mu + (\not{d}^{\#}\rho_{(l)})\cdot\not{d}\mu\right\}$. Using Lemma 8.6, we see that these terms are equal to the principal term $\rho_{(l)}\not{\Delta}\mathscr{Z}^{N-1}\mu$, which we move to the left, plus an error term with magnitude

$$\lesssim |\rho_{(l)}| \left|[\mathscr{Z}^{N-1}, \Delta\!\!\!/]\mu\right| + \sum_{\substack{N_1+N_2 \leq N-1 \\ N_2 \leq N-2}} \left|\mathscr{Z}^{N_1}\rho_{(l)}\right| \left|\mathscr{Z}^{N_2}\Delta\!\!\!/\mu\right| + \sum_{N_1+N_2+N_3 \leq N-1} \left|\not{\mathcal{L}}_{\mathscr{Z}}^{N_1} g\!\!\!/^{-1}\right| \left|d\!\!\!/\mathscr{Z}^{N_2}\rho_{(l)}\right| \left|d\!\!\!/\mathscr{Z}^{N_3}\mu\right|. \tag{16.21}$$

Using inequalities (12.46a) and (12.46c), the estimates (12.28a), (12.28b), (12.60a), (12.60b), (12.128a), and the commutator estimate (12.97) with $\mu - 1$ in the role of f and N_2 in the role of N, we deduce that the right-hand side of (16.21) is $\lesssim$ the product of $\ln(e+t)(1+t)^{-1}$ and the right-hand side of (16.18b). To complete the proof, it remains for us to show that the $\mathscr{Z}^{N-1}\text{di}\!\!\!/\text{v}$ derivatives of the remaining terms on the right-hand side of (7.8e) have magnitudes that are $\lesssim$ the product of $\ln(e+t)(1+t)^{-1}$ and the right-hand side of (16.18b). To this end, we let $\xi^{\#}$ denote the remaining terms, which are the $g\!\!\!/$-dual of the last three (schematically written) terms on the right-hand side of (12.76). We denote these schematic terms by ξ. Our goal is to bound $\mathscr{Z}^{N-1}\text{di}\!\!\!/\text{v}\,\xi^{\#}$. To this end, we first use the Leibniz rule to deduce that inequality (16.15) holds in the present context. Next, we note that the argument given in the discussion following equation (12.76) implies that

$$\left|\not{\mathcal{L}}_{\mathscr{Z}}^{\leq N}\xi\right| \lesssim \ln(e+t)\left|\begin{pmatrix} \varrho L\mathscr{Z}^{\leq N}\Psi \\ \breve{R}\mathscr{Z}^{\leq N}\Psi \\ \varrho d\!\!\!/\mathscr{Z}^{\leq N}\Psi \\ \mathscr{Z}^{\leq N}\Psi \end{pmatrix}\right| + \frac{\ln(e+t)}{1+t}\left|\begin{pmatrix} \mathscr{Z}^{\leq N}(\mu-1) \\ \sum_{a=1}^{3}\varrho\left|\mathscr{Z}^{\leq N}L_{(Small)}^{a}\right| \end{pmatrix}\right|, \tag{16.22}$$

$$\left\|\not{\mathcal{L}}_{\mathscr{Z}}^{\leq 12}\xi\right\|_{C^0(\Sigma_t^u)} \lesssim \varepsilon\frac{\ln^2(e+t)}{1+t}. \tag{16.23}$$

From (12.47), (12.60a), (12.60b), (16.22), and (16.23), we deduce that the first sum on the right-hand side of (16.15) is $\lesssim$ the product of $\ln(e+t)(1+t)^{-1}$ and the right-hand side of (16.18b) as desired. To show that the second sum on the right-hand side of (16.15) is $\lesssim$ the product of $\ln(e+t)(1+t)^{-1}$ and the right-hand side of (16.18b), we combine similar reasoning with the commutator estimate (12.92), where N_2 is in the role of N.

The proof of the estimate (16.18b) for the term

$$\mathscr{Z}^{N-1}\text{di}\!\!\!/\text{v}\,{}^{(O_{(l)})}\pi\!\!\!/_{L}^{\#} + O_{(l)}\mathscr{Z}^{N-1}\text{tr}_{g\!\!\!/}\chi^{(Small)}$$

is similar to the one for

$$\mathscr{Z}^{N-1}\text{di}\!\!\!/\text{v}\,{}^{(O_{(l)})}\pi\!\!\!/_{\breve{R}}^{\#} - \left\{\mu O_{(l)}\mathscr{Z}^{N-1}\text{tr}_{g\!\!\!/}\chi^{(Small)} + \rho_{(l)}\Delta\!\!\!/\mathscr{Z}^{N-1}\mu\right\}$$

but easier because the terms have a similar but slightly simpler structure. We apply $\mathscr{Z}^{N-1}\text{di}\!\!\!/\text{v}$ to the $g\!\!\!/$-dual of right-hand side of (7.8d) and argue as in the previous paragraph; we omit the details.

We now prove the estimate (16.18b) for $\not{\mathcal{L}}_{\mathscr{Z}}^{N-1}d\!\!\!/^{\#}\,{}^{(O_{(l)})}\pi_{L\breve{R}} + (\nabla\!\!\!/^{2\#}\mathscr{Z}^{N-1}\mu)\cdot O_{(l)}$. We apply $\not{\mathcal{L}}_{\mathscr{Z}}^{N-1}d\!\!\!/^{\#}$ to (7.8c). The quantity we must estimate is $-\not{\mathcal{L}}_{\mathscr{Z}}^{N-1}d\!\!\!/^{\#}\left\{O_{(l)}\cdot d\!\!\!/\mu\right\}$. When all derivatives fall on μ, we obtain the principal eikonal function term $-(\nabla\!\!\!/^{2\#}\mathscr{Z}^{N-1}\mu)\cdot O_{(l)}$ plus the commutator term $([\nabla\!\!\!/^2, \not{\mathcal{L}}_{\mathscr{Z}}^{N-1}]\mu)^{\#}\cdot O_{(l)}$. We move the principal term to the left-hand side. To bound the commutator term by $\lesssim$ the right-hand side of (16.18b), we use inequality (12.95) with $\upmu - 1$ in the role of f and $N-1$ in the role of N and the estimates (12.34a) and (12.128a). From the Leibniz rule, Lemma 8.6, and the fact that $\nabla\!\!\!/ O_{(l)} = g\!\!\!/^{-1} \cdot \nabla\!\!\!/ O_{(l)\flat}$, where $O_{(l)\flat}$ is the $g\!\!\!/$-dual of $O_{(l)}$, we deduce that the remaining terms arising in the Leibniz expansion of $-\mathcal{L}\!\!\!/_{\mathscr{Z}}^{N-1} d\!\!\!/^{\#} \left\{O_{(l)} \cdot d\!\!\!/ \upmu\right\}$ are bounded in magnitude as follows:

$$\lesssim \sum_{\substack{N_1+N_2+N_3 \leq N-1 \\ N_3 \leq N-2}} \left|\mathcal{L}\!\!\!/_{\mathscr{Z}}^{N_1} g\!\!\!/^{-1}\right| \left|\mathcal{L}\!\!\!/_{\mathscr{Z}}^{N_2} O_{(l)\flat}\right| \left|\mathcal{L}\!\!\!/_{\mathscr{Z}}^{N_3} \nabla\!\!\!/^2 \upmu\right| + \sum_{\substack{N_1+N_2+N_3+N_4 \leq N-1 \\ N_4 \leq N-2}} \left|\mathcal{L}\!\!\!/_{\mathscr{Z}}^{N_1} g\!\!\!/^{-1}\right| \left|\mathcal{L}\!\!\!/_{\mathscr{Z}}^{N_2} g\!\!\!/^{-1}\right| \left|\mathcal{L}\!\!\!/_{\mathscr{Z}}^{N_3} \nabla\!\!\!/ O_{(l)\flat}\right| \left|d\!\!\!/ \mathscr{Z}^{N_4} \upmu\right|. \tag{16.24}$$

Using inequalities (12.46a), (12.46b), (12.47), (12.60a), (12.60b), (12.70a), (12.70b), (12.92), (12.95), and (12.128a), we deduce that the right-hand side of (16.24) is $\lesssim$ the right-hand side of (16.18b) as desired.

We now prove the estimate (16.18b) for

$$\mathcal{L}\!\!\!/_{\mathscr{Z}}^{N-1} \mathrm{div}\!\!\!/ \, {}^{(O_{(l)})}\pi\!\!\!/^{\#\#} - \uprho_{(l)} d\!\!\!/^{\#} \mathscr{Z}^{N-1} \mathrm{tr}_{g\!\!\!/} \upchi^{(Small)}.$$

We apply $\mathcal{L}\!\!\!/_{\mathscr{Z}}^{N-1} \mathrm{div}\!\!\!/$ to the double $g\!\!\!/$-dual of (7.8f) and apply the Leibniz rule to the terms on the right-hand side. We begin by addressing the first term, which is the difficult one: $2\mathcal{L}\!\!\!/_{\mathscr{Z}}^{N-1} \nabla\!\!\!/_B \left\{\uprho_{(l)} (g\!\!\!/^{-1})^{AC} (g\!\!\!/^{-1})^{BD} \hat{\upchi}_{CD}^{(Small)}\right\}$. When all derivatives fall on $(g\!\!\!/^{-1})^{BD} \hat{\upchi}_{CD}^{(Small)}$, we use (12.28b) and the second inequality in (15.28) to rewrite this term as the principal eikonal function term $\uprho_{(l)} d\!\!\!/^{\#} \mathscr{Z}^{N-1} \mathrm{tr}_{g\!\!\!/} \upchi^{(Small)}$ plus an error term with magnitude $\lesssim$ the product of $(1+t)^{-1}$ and the right-hand side of (16.18b). We move the principal term the left-hand side. Using Lemma 8.6, we see that the remaining terms arising from the Leibniz expansion of $2\mathcal{L}\!\!\!/_{\mathscr{Z}}^{N-1} \nabla\!\!\!/_B \left\{(g\!\!\!/^{-1})^{AC} (g\!\!\!/^{-1})^{BD} \uprho_{(l)} \hat{\upchi}_{CD}^{(Small)}\right\}$ are bounded in magnitude as follows:

$$\lesssim \sum_{\substack{N_1+N_2+N_3+N_4 \leq N-1 \\ N_4 \leq N-2}} \left|\mathcal{L}\!\!\!/_{\mathscr{Z}}^{N_1} \uprho_{(l)}\right| \left|\mathcal{L}\!\!\!/_{\mathscr{Z}}^{N_2} g\!\!\!/^{-1}\right| \left|\mathcal{L}\!\!\!/_{\mathscr{Z}}^{N_3} g\!\!\!/^{-1}\right| \left|\mathcal{L}\!\!\!/_{\mathscr{Z}}^{N_4} \nabla\!\!\!/ \hat{\upchi}^{(Small)}\right| + \sum_{N_1+N_2+N_3+N_4 \leq N-1} \left|d\!\!\!/ \mathcal{L}\!\!\!/_{\mathscr{Z}}^{N_1} \uprho_{(l)}\right| \left|\mathcal{L}\!\!\!/_{\mathscr{Z}}^{N_2} g\!\!\!/^{-1}\right| \left|\mathcal{L}\!\!\!/_{\mathscr{Z}}^{N_3} g\!\!\!/^{-1}\right| \left|\mathcal{L}\!\!\!/_{\mathscr{Z}}^{N_4} \hat{\upchi}^{(Small)}\right|. \tag{16.25}$$

Using inequalities (12.46a) and (12.48), the estimates (12.28a), (12.28b), (12.60a), (12.60b), (12.67a), and (12.128a), the commutator estimate (12.92) with $\xi = \hat{\upchi}^{(Small)}$ and N_4 in the role of N, and (12.102) with $\xi = \upchi^{(Small)}$, we deduce that the right-hand side of (16.25) is $\lesssim$ the product of $(1+t)^{-1}$ and the right-hand side of (16.18b) as desired. To complete the proof, it remains for us to show that the $\mathcal{L}\!\!\!/_{\mathscr{Z}}^{N-1} \mathrm{div}\!\!\!/$ derivatives of the double $g\!\!\!/$-dual of the remaining terms on the right-hand side of (7.8f) have magnitudes that are $\lesssim$ the product of $(1+t)^{-1}$ and the right-hand side of (16.18b). To this end, we let $\hat{\xi}^{\#\#}$ denote the remaining terms, which are the double $g\!\!\!/$-dual of the terms in equation (7.9c). Note that we have the following schematic relation for the symmetric type $\binom{0}{2}$ $S_{t,u}$ tensorfield ξ corresponding to the terms $\hat{\xi}^{\#\#}$ under consideration:

$$\xi = G_{(Frame)} \otimes \begin{pmatrix} O_{(l)} \\ \rho_{(l)} \end{pmatrix} \begin{pmatrix} L\Psi \\ \not{d}\Psi \end{pmatrix} + f(\Psi)\breve{\upsilon}\not{d}x \otimes \not{d}x. \tag{16.26}$$

We claim that the following bounds hold:

$$\left| \not{\mathcal{L}}_{\mathscr{Z}}^{\leq N} \xi \right| \lesssim \left| \begin{pmatrix} \varrho L \mathscr{Z}^{\leq N}\Psi \\ \breve{R}\mathscr{Z}^{\leq N}\Psi \\ \varrho \not{d}\mathscr{Z}^{\leq N}\Psi \\ \mathscr{Z}^{\leq N}\Psi \end{pmatrix} \right| + \frac{1}{(1+t)^2} \left| \begin{pmatrix} \mathscr{Z}^{\leq N}(\upmu - 1) \\ \sum_{a=1}^{3} \varrho \left| \mathscr{Z}^{\leq N} L_{(Small)}^a \right| \end{pmatrix} \right|, \tag{16.27}$$

(16.28)

$$\left\| \not{\mathcal{L}}_{\mathscr{Z}}^{\leq 12} \xi \right\|_{C^0(\Sigma_t^u)} \lesssim \varepsilon \frac{1}{1+t}.$$

To prove (16.27), we apply $\not{\mathcal{L}}_{\mathscr{Z}}^{\leq N}$ to ξ and apply the Leibniz rule. We bound the terms $\not{\mathcal{L}}_{\mathscr{Z}}^{M} G_{(Frame)}$ with the estimates of Lemma 12.25. We bound the terms $\begin{pmatrix} \mathscr{Z}^M L\Psi \\ \not{\mathcal{L}}_{\mathscr{Z}}^M \not{d}\Psi \\ \mathscr{Z}^M \Psi \end{pmatrix}$ with Lemma 12.23 and the bootstrap assumptions (**BA**Ψ). We bound the terms $\mathscr{Z}^M f(\Psi)$ by using the bootstrap assumptions (**BA**Ψ) to deduce that $\left|\mathscr{Z}^M f(\Psi)\right| \lesssim \left|\mathscr{Z}^{\leq M}\Psi\right| + 1$. We bound the terms $\not{\mathcal{L}}_{\mathscr{Z}}^{M} \not{d}x$ with (12.57a) and (12.57b). We bound the terms $\mathscr{Z}^M \rho_{(l)}$ with (12.28a) and (12.28b). We bound the terms $\not{\mathcal{L}}_{\mathscr{Z}}^{M} O_{(l)}$ with (12.70a) and (12.70b). In total, these estimates yield the desired bound (16.27). The bound (16.28) then follows from (16.27), (12.128a), and the bootstrap assumptions (**BA**Ψ). We now recall that our goal is to bound $\not{\mathcal{L}}_{\mathscr{Z}}^{N-1} \mathrm{d}\not{\mathrm{i}}\mathrm{v}\, \hat{\xi}^{\#\#}$ in magnitude by $\lesssim$ the product of $(1+t)^{-1}$ and the right-hand side of (16.18b). To proceed, we use the Leibniz rule to deduce that

$$\begin{aligned} \left| \not{\mathcal{L}}_{\mathscr{Z}}^{N-1} \mathrm{d}\not{\mathrm{i}}\mathrm{v}\, \hat{\xi}^{\#\#} \right| \lesssim & \sum_{N_1+N_2+N_3 \leq N-1} \left| \not{\mathcal{L}}_{\mathscr{Z}}^{N_1} \not{g}^{-1} \right| \left| \not{\mathcal{L}}_{\mathscr{Z}}^{N_2} \not{g}^{-1} \right| \left| \not{\nabla} \not{\mathcal{L}}_{\mathscr{Z}}^{N_3} \hat{\xi} \right| \\ & + \sum_{N_1+N_2+N_3 \leq N-1} \left| \not{\mathcal{L}}_{\mathscr{Z}}^{N_1} \not{g}^{-1} \right| \left| \not{\mathcal{L}}_{\mathscr{Z}}^{N_2} \not{g}^{-1} \right| \left| [\not{\mathcal{L}}_{\mathscr{Z}}^{N_3}, \not{\nabla}] \hat{\xi} \right|. \end{aligned} \tag{16.29}$$

Finally, to bound the right-hand side of (16.29) by $\lesssim$ the product of $(1+t)^{-1}$ and the right-hand side of (16.18b), we use the estimates (12.48), (12.60a), (12.60b), the commutator estimate (12.92) with $\hat{\xi}$ in the role of ξ and N_3 in the role of N, (12.102), (16.27), and (16.28). □

LEMMA 16.9 (**Identification of the important top-order terms in the derivatives of** $^{(\varrho L)}\pi$). *Let* $1 \leq N \leq 24$ *be an integer. Under the small-data and bootstrap assumptions of Sects. 12.1-12.4, if* ε *is sufficiently small, then the following estimates hold on* $\mathcal{M}_{T_{(Bootstrap)}, U_0}$:

$$\begin{aligned} & \left| \mathscr{Z}^{N-1} \breve{R} \mathrm{tr}_{\not{g}}\, {}^{(\varrho L)}\not{\pi} - 2\varrho \not{\Delta} \mathscr{Z}^{N-1} \upmu \right| \\ & \lesssim \left| \begin{pmatrix} \varrho L \mathscr{Z}^{\leq N}\Psi \\ \breve{R}\mathscr{Z}^{\leq N}\Psi \\ \varrho \not{d}\mathscr{Z}^{\leq N}\Psi \\ \mathscr{Z}^{\leq N}\Psi \end{pmatrix} \right| + \frac{1}{1+t} \left| \begin{pmatrix} \mathscr{Z}^{\leq N}(\upmu - 1) \\ \sum_{a=1}^{3} \varrho \left| \mathscr{Z}^{\leq N} L_{(Small)}^a \right| \end{pmatrix} \right|, \end{aligned} \tag{16.30a}$$

$$\left|\not{\mathcal{L}}_{\mathscr{Z}}^{N-1} \not{d}^{\#(\varrho L)}\pi_{L\breve{R}}\right|,$$

$$\left|\mathscr{Z}^{N-1}\text{di}\!\not{v}^{(\varrho L)}\not{\pi}_{\breve{R}}^{\#} - \varrho\not{\Delta}\mathscr{Z}^{N-1}\upmu\right|,$$

$$(1+t)\left|\not{\mathcal{L}}_{\mathscr{Z}}^{N-1}\text{di}\!\not{v}^{(\varrho L)}\not{\pi}^{\#\#} - \varrho\not{d}^{\#}\mathscr{Z}^{N-1}\text{tr}_{\not{g}}\chi^{(Small)}\right| \tag{16.30b}$$

$$\lesssim \left|\begin{pmatrix} \varrho L\mathscr{Z}^{\leq N}\Psi \\ \breve{R}\mathscr{Z}^{\leq N}\Psi \\ \varrho\not{d}\mathscr{Z}^{\leq N}\Psi \\ \mathscr{Z}^{\leq N}\Psi \end{pmatrix}\right| + \frac{1}{1+t}\left|\begin{pmatrix} \mathscr{Z}^{\leq N}(\upmu-1) \\ \sum_{a=1}^{3}\varrho\left|\mathscr{Z}^{\leq N}L_{(Small)}^{a}\right| \end{pmatrix}\right|.$$

PROOF. We first prove (16.30a). From (7.7g), we see that we have to estimate $2\mathscr{Z}^{N-1}\breve{R}(\varrho\text{tr}_{\not{g}}\chi^{(Small)})$. When all derivatives fall on $\text{tr}_{\not{g}}\chi^{(Small)}$, we use the third inequality in (15.1) to rewrite this term as the principal eikonal function term $2\varrho\not{\Delta}\mathscr{Z}^{N-1}\upmu$ plus an error term with magnitude $\lesssim$ the right-hand side of (16.30a). We move the principal term to the left-hand side. We claim that the remaining terms arising in the Leibniz expansion of $\mathscr{Z}^{N-1}\breve{R}(\varrho\text{tr}_{\not{g}}\chi^{(Small)})$ are error terms with magnitudes that are $\lesssim$ the right-hand side of (16.30a) as desired. To prove the claim, we use (12.6b), (12.67a), and (12.128a).

We now prove (16.30b) for $\not{\mathcal{L}}_{\mathscr{Z}}^{N-1}\not{d}^{\#(\varrho L)}\pi_{L\breve{R}}$. We apply $\not{\mathcal{L}}_{\mathscr{Z}}^{N-1}\not{d}^{\#}$ to the terms to the right of the first equality in (7.7c). From the Leibniz rule, Lemma 8.6, (12.6b), and (12.46a), we deduce that the terms of interest are bounded in magnitude by

$$\lesssim \sum_{\substack{N_1+N_2\leq N \\ N_1\leq N-1}}\left|\not{\mathcal{L}}_{\mathscr{Z}}^{N_1}\not{g}^{-1}\right|\left|\not{\mathcal{L}}_{\mathscr{Z}}^{N_2}L\upmu\right| + \frac{1}{1+t}\sum_{\substack{N_1+N_2\leq N \\ N_1\leq N-1}}\left|\not{\mathcal{L}}_{\mathscr{Z}}^{N_1}\not{g}^{-1}\right|\left|\not{\mathcal{L}}_{\mathscr{Z}}^{N_2}(\upmu-1)\right|. \tag{16.31}$$

To bound the right-hand side of (16.31) by $\lesssim$ the right-hand side of (16.30b), we use (12.60a), (12.60b), (12.127b), (12.128a), and (12.128b).

We now prove (16.30b) for $\mathscr{Z}^{N-1}\text{di}\!\not{v}^{(\varrho L)}\not{\pi}_{\breve{R}}^{\#} - \varrho\not{\Delta}\mathscr{Z}^{N-1}\upmu$. We apply $\mathscr{Z}^{N-1}\text{di}\!\not{v}$ to the $\not{g}$-dual of the right-hand side of (7.7e). We begin by addressing the first term, which is the difficult one: $\mathscr{Z}^{N-1}(\varrho\not{\Delta}\upmu)$. By the Leibniz rule and (12.6b), we see that this term is equal to the principal eikonal function term $\varrho\not{\Delta}\mathscr{Z}^{N-1}\upmu$ plus an error term that is bounded in magnitude by

$$\lesssim (1+t)\left|[\mathscr{Z}^{\leq N-1},\not{\Delta}]\upmu\right| + (1+t)\left|\not{\Delta}\mathscr{Z}^{\leq N-2}\upmu\right|. \tag{16.32}$$

From inequalities (12.46a) and (12.46c), the commutator estimate (12.97) with $N-1$ in the role of N and $\upmu$ in the role of f, and the estimate (12.128a), we deduce that the right-hand side of (16.32) is $\lesssim$ the right-hand side of (16.30b). We then bring the principal term over to the left-hand side. To complete the proof, it remains for us to show that the $\mathscr{Z}^{N-1}\text{di}\!\not{v}$ derivatives of the $\not{g}$-dual of the remaining terms on the right-hand side of (7.7e) have magnitudes that are $\lesssim$ the right-hand side of (16.30b). To this end, we let $\xi^{\#}$ denote the remaining terms, which are the $\not{g}$-dual of an $S_{t,u}$ one-form of the schematic form $\xi := \varrho G_{(Frame)}\begin{pmatrix} \upmu L\Psi \\ \breve{R}\Psi \\ \upmu\not{d}\Psi \end{pmatrix}$. We now claim that ξ verifies the following bounds:

$$\left|\not{\mathcal{L}}_{\mathscr{Z}}^{\leq N}\xi\right| \lesssim \left|\begin{pmatrix} \varrho L\mathscr{Z}^{\leq N}\Psi \\ \breve{R}\mathscr{Z}^{\leq N}\Psi \\ \varrho \not{d}\mathscr{Z}^{\leq N}\Psi \\ \mathscr{Z}^{\leq N}\Psi \end{pmatrix}\right| + \frac{1}{1+t}\left|\begin{pmatrix} \mathscr{Z}^{\leq N}(\upmu-1) \\ \sum_{a=1}^{3}\varrho\left|\mathscr{Z}^{\leq N}L_{(Small)}^{a}\right| \end{pmatrix}\right|, \tag{16.33}$$

$$\left\|\not{\mathcal{L}}_{\mathscr{Z}}^{\leq 12}\xi\right\|_{C^0(\Sigma_t^u)} \lesssim \varepsilon\frac{1}{1+t}. \tag{16.34}$$

To prove (16.33), we apply $\not{\mathcal{L}}_{\mathscr{Z}}^{\leq N}$ to ξ and apply the Leibniz rule. We bound the terms $\mathscr{Z}^{M}\varrho$ with (12.6b). We bound the terms $\not{\mathcal{L}}_{\mathscr{Z}}^{M}G_{(Frame)}$ with the estimates of Lemma 12.25. We bound the terms $\begin{pmatrix} \mathscr{Z}^{M}L\Psi \\ \mathscr{Z}^{M}\breve{R}\Psi \\ \not{\mathcal{L}}_{\mathscr{Z}}^{M}\not{d}\Psi \end{pmatrix}$ with Lemma 12.23 and the bootstrap assumptions (**BA**Ψ). We bound the terms $\mathscr{Z}^{M}\upmu$ with (12.128a). In total, these estimates yield the desired bound (16.33). The bound (16.34) then follows from (16.33), (12.128a), and the bootstrap assumptions (**BA**Ψ). We now recall that our goal is to bound $\mathscr{Z}^{N-1}\mathrm{d}\not{\mathrm{i}}\mathrm{v}\,\xi^{\#}$ in magnitude by $\lesssim$ the right-hand side of (16.18b). To proceed, we use the Leibniz rule to deduce that inequality (16.15) holds in the present context. Finally, to bound the right-hand side of (16.15) by $\lesssim$ the right-hand side of (16.30b), we use the estimates (12.47), (12.60a), (12.60b), (12.92) with N_2 in the role of N, (16.33), and (16.34).

Finally, we prove (16.30b) for the term $\not{\mathcal{L}}_{\mathscr{Z}}^{N-1}\mathrm{d}\not{\mathrm{i}}\mathrm{v}^{(\varrho L)}\not{\pi}^{\#\#} - \varrho\not{d}^{\#}\mathscr{Z}^{N-1}\mathrm{tr}_{\not{g}}\chi^{(Small)}$. We apply $\not{\mathcal{L}}_{\mathscr{Z}}^{N-1}\mathrm{d}\not{\mathrm{i}}\mathrm{v}$ to the double $\not{g}$-dual of the product on the right-hand side of (7.7f), that is, we apply $\not{\mathcal{L}}_{\mathscr{Z}}^{N-1}\not{\nabla}_{B}$ to $2\varrho(\not{g}^{-1})^{AC}(\not{g}^{-1})^{BD}\hat{\chi}_{CD}^{(Small)}$. We then apply the Leibniz rule. When all derivatives fall on $(\not{g}^{-1})^{BD}\hat{\chi}_{CD}^{(Small)}$, we use the second inequality in (15.28) to deduce that the product is equal to the principal eikonal function term $\varrho\not{d}^{A}\mathscr{Z}^{N-1}\mathrm{tr}_{\not{g}}\chi^{(Small)}$ plus an error term that is bounded in magnitude by $\lesssim (1+t)^{-1}$ times the right-hand side of (16.30b) as desired. Using (12.6b), we deduce that the remaining terms arising from the Leibniz expansion of $\not{\mathcal{L}}_{\mathscr{Z}}^{N-1}\not{\nabla}_{B}\left\{\varrho(\not{g}^{-1})^{AC}(\not{g}^{-1})^{BD}\hat{\chi}_{CD}^{(Small)}\right\}$ are bounded in magnitude as follows:

$$\lesssim (1+t)\sum_{\substack{N_1+N_2+N_3\leq N-1 \\ N_3\leq N-2}}\left|\not{\mathcal{L}}_{\mathscr{Z}}^{N_1}\not{g}^{-1}\right|\left|\not{\mathcal{L}}_{\mathscr{Z}}^{N_2}\not{g}^{-1}\right|\left|\not{\mathcal{L}}_{\mathscr{Z}}^{N_3}\not{\nabla}\hat{\chi}^{(Small)}\right|. \tag{16.35}$$

Using (12.48), (12.60a), (12.60b), (12.67a), (12.128a), and the commutator estimate (12.92) with $\xi = \hat{\chi}^{(Small)}$ and N_3 in the role of N, we deduce that the right-hand side of (16.35) is $\lesssim (1+t)^{-1}$ times the right-hand side of (16.30b) as desired. □

16.3. Crude pointwise estimates for the below-top-order derivatives of the deformation tensors of the commutation vectorfields

In Lemmas 16.10 and 16.11, we establish some rather crude pointwise estimates for the frame components of the deformation tensors that appear in the wave equation error terms (9.15) and (9.16) and their below-top-order derivatives. The lemmas collectively show that there are no difficult terms present in the below-top-order derivatives of the frame components of ${}^{(Z)}\pi$ for $Z \in \mathscr{Z}$. The quantities that we bound in Lemma 16.11 involve covariant angular differentiation. Hence, in proving the corresponding estimates, we use a few additional ingredients that are not needed in the proof of Lemma 16.10.

We now state and prove Lemma 16.10. From the point of view of proving our sharp classical lifespan theorem, the most important aspect of these estimates is that the quantities in (16.36b) can experience some logarithmic-in-time growth, while those in (16.37b) do not.

LEMMA 16.10 (**Crude pointwise estimates for the below-top-order derivatives of some frame components of** ${}^{(Z)}\pi$). *Let* $0 \leq N \leq 23$ *be an integer and let* $Z \in \mathscr{Z}$. *Under the small-data and bootstrap assumptions of Sects.* 12.1-12.4, *if* ε *is sufficiently small, then the following estimates hold on* $\mathcal{M}_{T_{(Bootstrap)},U_0}$:

$$\left| \begin{pmatrix} \mathscr{Z}^{N\,(Z)}\pi_{L\breve{R}} \\ \mathscr{Z}^{N\,(Z)}\pi_{\breve{R}R} \\ \not{\mathcal{L}}_{\mathscr{Z}}^{N\,(Z)}\not{\pi}_{\breve{R}}^{\#} \end{pmatrix} \right| \lesssim (1+t) \left| \begin{pmatrix} \varrho L \mathscr{Z}^{\leq N}\Psi \\ \breve{R}\mathscr{Z}^{\leq N}\Psi \\ \varrho \not{d}\mathscr{Z}^{\leq N}\Psi \\ \mathscr{Z}^{\leq N}\Psi \end{pmatrix} \right| + \left| \begin{pmatrix} \mathscr{Z}^{\leq N+1}(\upmu - 1) \\ \sum_{a=1}^{3} \varrho |\mathscr{Z}^{\leq N+1} L_{(Small)}^a| \end{pmatrix} \right| + 1, \tag{16.36a}$$

$$\left\| \begin{pmatrix} \mathscr{Z}^{\leq 11\,(Z)}\pi_{L\breve{R}} \\ \mathscr{Z}^{\leq 11\,(Z)}\pi_{\breve{R}R} \\ \not{\mathcal{L}}_{\mathscr{Z}}^{\leq 11\,(Z)}\not{\pi}_{\breve{R}}^{\#} \end{pmatrix} \right\|_{C^0(\Sigma_t^u)} \lesssim \ln(e+t), \tag{16.36b}$$

$$\left| \begin{pmatrix} \not{\mathcal{L}}_{\mathscr{Z}}^{N\,(Z)}\not{\pi}_L^{\#} \\ \not{\mathcal{L}}_{\mathscr{Z}}^{N\,(Z)}\not{\pi}^{\#} \\ \mathscr{Z}^{N}\mathrm{tr}_{\not{g}}{}^{(Z)}\not{\pi} \\ \not{\mathcal{L}}_{\mathscr{Z}}^{N\,(Z)}\not{\pi} \\ \not{\mathcal{L}}_{\mathscr{Z}}^{N\,(Z)}\not{\pi}^{\#\#} \end{pmatrix} \right| \lesssim \left| \begin{pmatrix} \varrho L \mathscr{Z}^{\leq N}\Psi \\ \breve{R}\mathscr{Z}^{\leq N}\Psi \\ \varrho \not{d}\mathscr{Z}^{\leq N}\Psi \\ \mathscr{Z}^{\leq N}\Psi \end{pmatrix} \right| + \frac{1}{1+t} \left| \begin{pmatrix} \mathscr{Z}^{\leq N+1}(\upmu - 1) \\ \sum_{a=1}^{3} \varrho |\mathscr{Z}^{\leq N+1} L_{(Small)}^a| \end{pmatrix} \right| + 1, \tag{16.37a}$$

(16.37b)

$$\left\| \begin{pmatrix} \not{\mathcal{L}}_{\mathscr{Z}}^{\leq 11\,(Z)}\not{\pi}_L^{\#} \\ \not{\mathcal{L}}_{\mathscr{Z}}^{\leq 11\,(Z)}\not{\pi}^{\#} \\ \mathscr{Z}^{\leq 11}\mathrm{tr}_{\not{g}}{}^{(Z)}\not{\pi} \\ \not{\mathcal{L}}_{\mathscr{Z}}^{N\,(Z)}\not{\pi} \\ \not{\mathcal{L}}_{\mathscr{Z}}^{\leq 11\,(Z)}\not{\pi}^{\#\#} \end{pmatrix} \right\|_{C^0(\Sigma_t^u)} \lesssim 1.$$

PROOF. The desired estimates for $\not{\mathcal{L}}_{\mathscr{Z}}^{N\,(Z)}\not{\pi}_L^{\#}$, $\not{\mathcal{L}}_{\mathscr{Z}}^{N\,(Z)}\not{\pi}_{\breve{R}}^{\#}$, $\not{\mathcal{L}}_{\mathscr{Z}}^{N\,(Z)}\not{\pi}$, and $\mathscr{Z}^N\mathrm{tr}_{\not{g}}{}^{(Z)}\not{\pi}$ have already been established in Lemma 12.26 and Lemma 12.32. The desired estimates for $\not{\mathcal{L}}_{\mathscr{Z}}^{N\,(Z)}\not{\pi}^{\#\#} = \not{\mathcal{L}}_{\mathscr{Z}}^{N}\left\{(\not{g}^{-1})^{2\,(Z)}\hat{\not{\pi}}\right\}$ then follow from these bounds together with Lemma 12.26, (12.102), (12.128a), and the bootstrap assumptions ($\mathbf{BA}\Psi$).

The desired estimates (16.36a) for ${}^{(Z)}\pi_{L\breve{R}}$ and ${}^{(Z)}\pi_{\breve{R}R}$ follow easily from the identities (7.5b), (7.5c), (7.7b), (7.7c), (7.8b), and (7.8c) and the estimates (12.6b), and (12.127b).

The desired estimates (16.36b) for ${}^{(Z)}\pi_{L\breve{R}}$ and ${}^{(Z)}\pi_{\breve{R}R}$ then follow from the bounds (16.36a) and (12.128a), and the bootstrap assumptions ($\mathbf{BA}\Psi$). □

LEMMA 16.11 (**Pointwise estimates for the below-top-order derivatives of $^{(Z)}\pi$ involving at least one angular derivative**). *Let $2 \leq N \leq 24$ be an integer. Under the small-data and bootstrap assumptions of Sects. 12.1-12.4, if ε is sufficiently small, then the following estimates hold on $\mathcal{M}_{T_{(Bootstrap)},U_0}$:*

(16.38a)
$$\left|\mathscr{Z}^{N-2}\text{di}\!\!/\text{v}\,^{(\breve{R})}\not\pi_L^{\#}\right|, \left|\mathscr{Z}^{N-2}\text{di}\!\!/\text{v}\,^{(O_{(l)})}\not\pi_L^{\#}\right| \lesssim \frac{1}{1+t}\left|\begin{pmatrix} \varrho L\mathscr{Z}^{\leq N-1}\Psi \\ \breve{R}\mathscr{Z}^{\leq N-1}\Psi \\ \varrho \not{d}\mathscr{Z}^{\leq N-1}\Psi \\ \mathscr{Z}^{\leq N-1}\Psi \end{pmatrix}\right| + \frac{1}{(1+t)^2}\left|\begin{pmatrix} \mathscr{Z}^{\leq N}(\upmu-1) \\ \sum_{a=1}^3 \varrho\left|\mathscr{Z}^{\leq N}L_{(Small)}^a\right| \end{pmatrix}\right|,$$

(16.38b)
$$\left\|\mathscr{Z}^{\leq 10}\text{di}\!\!/\text{v}\,^{(\breve{R})}\not\pi_L^{\#}\right\|_{C^0(\Sigma_t^u)}, \left\|\mathscr{Z}^{\leq 10}\text{di}\!\!/\text{v}\,^{(O_{(l)})}\not\pi_L^{\#}\right\|_{C^0(\Sigma_t^u)} \lesssim \varepsilon\frac{\ln(e+t)}{(1+t)^2}.$$

Furthermore, the following estimates hold on $\mathcal{M}_{T_{(Bootstrap)},U_0}$:

(16.39a)
$$\left|\not{\mathcal{L}}_{\mathscr{Z}}^{N-2}\text{di}\!\!/\text{v}\,^{(\varrho L)}\not\pi^{\#\#}\right|, \left|\not{\mathcal{L}}_{\mathscr{Z}}^{N-2}\text{di}\!\!/\text{v}\,^{(\breve{R})}\not\pi^{\#\#}\right|, \left|\not{\mathcal{L}}_{\mathscr{Z}}^{N-2}\text{di}\!\!/\text{v}\,^{(O_{(l)})}\not\pi^{\#\#}\right| \lesssim \frac{1}{1+t}\left|\begin{pmatrix} \varrho L\mathscr{Z}^{\leq N-1}\Psi \\ \breve{R}\mathscr{Z}^{\leq N-1}\Psi \\ \varrho \not{d}\mathscr{Z}^{\leq N-1}\Psi \\ \mathscr{Z}^{\leq N-1}\Psi \end{pmatrix}\right| + \frac{1}{(1+t)^2}\left|\begin{pmatrix} \mathscr{Z}^{\leq N}(\upmu-1) \\ \sum_{a=1}^3 \varrho\left|\mathscr{Z}^{\leq N}L_{(Small)}^a\right| \end{pmatrix}\right|,$$

(16.39b)
$$\left\|\not{\mathcal{L}}_{\mathscr{Z}}^{\leq 10}\text{di}\!\!/\text{v}\,^{(\breve{R})}\not\pi^{\#\#}\right\|_{C^0(\Sigma_t^u)}, \left\|\not{\mathcal{L}}_{\mathscr{Z}}^{\leq 10}\text{di}\!\!/\text{v}\,^{(\varrho L)}\not\pi^{\#\#}\right\|_{C^0(\Sigma_t^u)}, \left\|\not{\mathcal{L}}_{\mathscr{Z}}^{\leq 10}\text{di}\!\!/\text{v}\,^{(O_{(l)})}\not\pi^{\#\#}\right\|_{C^0(\Sigma_t^u)} \lesssim \varepsilon\frac{\ln(e+t)}{(1+t)^2}.$$

PROOF. We first prove (16.38a) for $\mathscr{Z}^{N-2}\text{di}\!\!/\text{v}\,^{(\breve{R})}\not\pi_L^{\#}$. From inequality (16.13b) with $N-1$ in the role of N, we deduce that $\left|\mathscr{Z}^{N-2}\text{di}\!\!/\text{v}\,^{(\breve{R})}\not\pi_L^{\#}\right| \leq \left|\not\Delta\mathscr{Z}^{N-2}\upmu\right|$ plus error terms that are $\lesssim$ the right-hand side of (16.38a) as desired. To bound $\left|\not\Delta\mathscr{Z}^{N-2}\upmu\right|$ by $\lesssim$ the right-hand side of (16.38a), we use inequalities (12.46a) and (12.46c). The inequality (16.38a) for $\mathscr{Z}^{N-2}\text{di}\!\!/\text{v}\,^{(O_{(l)})}\not\pi_L^{\#}$ follows similarly from the first inequality in (16.18b) and the fact that $\left|O_{(l)}\mathscr{Z}^{N-2}\text{tr}_{\not g}\chi^{(Small)}\right|$ is $\lesssim$ the right-hand side of (16.38a), which follows from (12.67a).

The inequality (16.39a) for $\not{\mathcal{L}}_{\mathscr{Z}}^{N-2}\text{di}\!\!/\text{v}\,^{(\varrho L)}\not\pi^{\#\#}$ follows similarly from the last inequality in (16.30b) and the fact that $\left|\varrho\not{d}^{\#}\mathscr{Z}^{N-2}\text{tr}_{\not g}\chi^{(Small)}\right|$ is $\lesssim$ the right-hand side of (16.38a), which follows from (12.46a) and (12.67a). The inequality (16.39a) for $\not{\mathcal{L}}_{\mathscr{Z}}^{N-2}\text{di}\!\!/\text{v}\,^{(\breve{R})}\not\pi^{\#\#}$ follows similarly from the last inequality in (16.13b), the bound for $\left|\not{d}^{\#}\mathscr{Z}^{N-2}\text{tr}_{\not g}\chi^{(Small)}\right|$ mentioned just above, and the inequality $\upmu \lesssim \ln(e+t)$ (that is, (12.128a)). The inequality (16.39a) for $\not{\mathcal{L}}_{\mathscr{Z}}^{N-2}\text{di}\!\!/\text{v}\,^{(O_{(l)})}\not\pi^{\#\#}$ follows similarly from the last inequality in (16.18b), the bound for $\left|\not{d}^{\#}\mathscr{Z}^{N-2}\text{tr}_{\not g}\chi^{(Small)}\right|$ mentioned just above, and the inequality $|\uprho_{(l)}| \lesssim \ln(e+t)$ (that is, (12.28b)).

The estimate (16.38b) then follows from (16.38a), (12.128a), and the bootstrap assumptions (**BA**Ψ). Similarly, the estimate (16.39b) then follows from (16.39a), (12.128a), and the bootstrap assumptions (**BA**Ψ). □

16.4. Pointwise estimates for the top-order derivatives of the outgoing null derivative of the commutation vectorfield deformation tensors

In the next lemma, we establish pointwise estimates for the top-order derivatives of some deformation tensor components where one of the derivatives is in the direction L. In contrast to the other top-order derivatives of the deformation tensors, top-order derivatives involving at least one L derivative are relatively easy to bound. In particular, to bound them in L^2, we do not need to use the modified quantities of Chapter 11.

LEMMA 16.12 (**Pointwise estimates for the top-order derivatives, involving an L derivative, of ${}^{(Z)}\pi$**). *Let $1 \leq N \leq 24$ be an integer. Under the small-data and bootstrap assumptions of Sects.* 12.1-12.4, *if ε is sufficiently small and $Z \in \mathscr{Z}$ is a commutation vectorfield, then the following estimates hold on $\mathcal{M}_{T_{(Bootstrap)},U_0}$:*

$$\left| \begin{pmatrix} \mathscr{Z}^{N-1} L {}^{(Z)}\pi_{L\breve{R}} \\ \mathscr{Z}^{N-1} L {}^{(Z)}\pi_{\breve{R}R} \\ \not{\mathcal{L}}_{\mathscr{Z}}^{N-1} \not{\mathcal{L}}_L {}^{(Z)}\not{\pi}_{\breve{R}}^{\#} \end{pmatrix} \right| \lesssim \left| \begin{pmatrix} \varrho L \mathscr{Z}^{\leq N} \Psi \\ \breve{R} \mathscr{Z}^{\leq N} \Psi \\ \varrho \not{d} \mathscr{Z}^{\leq N} \Psi \\ \mathscr{Z}^{\leq N} \Psi \end{pmatrix} \right| + \frac{1}{1+t} \left| \begin{pmatrix} \mathscr{Z}^{\leq N} (\mu - 1) \\ \sum_{a=1}^{3} \varrho \left| \mathscr{Z}^{\leq N} L_{(Small)}^a \right| \end{pmatrix} \right|, \tag{16.40a}$$

$$\begin{pmatrix} \left\| \mathscr{Z}^{\leq 11} L {}^{(Z)}\pi_{L\breve{R}} \right\|_{C^0(\Sigma_t^u)} \\ \left\| \mathscr{Z}^{\leq 11} L {}^{(Z)}\pi_{\breve{R}R} \right\|_{C^0(\Sigma_t^u)} \\ \left\| \not{\mathcal{L}}_{\mathscr{Z}}^{\leq 11} \not{\mathcal{L}}_L {}^{(Z)}\not{\pi}_{\breve{R}}^{\#} \right\|_{C^0(\Sigma_t^u)} \end{pmatrix} \lesssim \varepsilon \frac{\ln(e+t)}{1+t}. \tag{16.40b}$$

Furthermore, the following estimates hold on $\mathcal{M}_{T_{(Bootstrap)},U_0}$:

$$\left| \not{\mathcal{L}}_{\mathscr{Z}}^{N-1} \not{\mathcal{L}}_L {}^{(Z)}\not{\pi}_L^{\#} \right| \lesssim \frac{1}{1+t} \left| \begin{pmatrix} \varrho L \mathscr{Z}^{\leq N} \Psi \\ \breve{R} \mathscr{Z}^{\leq N} \Psi \\ \varrho \not{d} \mathscr{Z}^{\leq N} \Psi \\ \mathscr{Z}^{\leq N} \Psi \end{pmatrix} \right| + \frac{1}{(1+t)^2} \left| \begin{pmatrix} \mathscr{Z}^{\leq N} (\mu - 1) \\ \sum_{a=1}^{3} \varrho \left| \mathscr{Z}^{\leq N} L_{(Small)}^a \right| \end{pmatrix} \right|, \tag{16.41a}$$

$$\left\| \not{\mathcal{L}}_{\mathscr{Z}}^{\leq 11} \not{\mathcal{L}}_L {}^{(Z)}\not{\pi}_L^{\#} \right\|_{C^0(\Sigma_t^u)} \lesssim \varepsilon \frac{\ln(e+t)}{(1+t)^2}. \tag{16.41b}$$

Finally, the following estimates hold on $\mathcal{M}_{T_{(Bootstrap)},U_0}$:

(16.42a)
$$\left|\mathscr{Z}^{N-1} L \mathrm{tr}_{\not g}{}^{(Z)}\not\pi\right| \lesssim \frac{1}{1+t}\left|\begin{pmatrix} \varrho L \mathscr{Z}^{\leq N}\Psi \\ \breve{R}\mathscr{Z}^{\leq N}\Psi \\ \varrho \not d \mathscr{Z}^{\leq N}\Psi \\ \mathscr{Z}^{\leq N}\Psi \end{pmatrix}\right| + \frac{1}{(1+t)^2}\left|\begin{pmatrix} \mathscr{Z}^{\leq N}(\upmu - 1) \\ \sum_{a=1}^{3} \varrho \left|\mathscr{Z}^{\leq N} L_{(Small)}^a\right| \end{pmatrix}\right| + \frac{\ln(e+t)}{(1+t)^2},$$

(16.42b)
$$\left\|\mathscr{Z}^{\leq 11} L \mathrm{tr}_{\not g}{}^{(Z)}\not\pi\right\|_{C^0(\Sigma_t^u)} \lesssim \frac{\ln(e+t)}{(1+t)^2}.$$

PROOF. Once we have proved (16.40a), (16.41a), and (16.42a), the remaining estimates (16.40b), (16.41b), and (16.42b) then follow from (12.128a) and the bootstrap assumptions (**BA**Ψ).

It remains for us to prove (16.40a), (16.41a), and (16.42a). The proofs are very similar to the proofs of the estimates for $\mathscr{Z}^{N(Z)}\pi_{L\breve{R}}$, $\mathscr{Z}^{N(Z)}\pi_{R\breve{R}}$, $\not{\mathcal{L}}_{\mathscr{Z}}^{N}{}^{(Z)}\not\pi_L^{\#}$, $\not{\mathcal{L}}_{\mathscr{Z}}^{N}{}^{(Z)}\not\pi_{\breve{R}}^{\#}$, and $\mathscr{Z}^N \mathrm{tr}_{\not g}{}^{(Z)}\not\pi$ that we derived in Lemma 16.10, so we only highlight the three important differences. First, the relation $\varrho L \in \mathscr{Z}$ and inequality (12.6b) together imply that the estimates (16.40a), (16.41a), and (16.42a), which involve the operator $\mathscr{Z}^{N-1}L$, are better by a factor of $(1+t)^{-1}$ than their counterpart estimates from Lemma 16.10. Second, and by far most importantly, the right-hand sides of the estimates (16.40a), (16.41a), and (16.42a) involve *one fewer derivatives* of $\upmu$ and $L^i_{(Small)}$ compared to the corresponding estimates of Lemma 16.10. This crucially important fact essentially follows from inequality (12.127b), whose right-hand side involves one fewer derivatives of $L^i_{(Small)}$ compared to the number of derivatives that one would obtain by using (12.127a). Finally, unlike the estimate (16.36a), the right-hand side of the estimates (16.40a) and (16.41a) do not involve a constant "+1" type term. The reason is that the +1 term on the right-hand side of (16.36a) arises from the $2\upmu = 2 + 2(\upmu - 1)$ and $\upmu = 1 + (\upmu - 1)$ terms on the right-hand sides of (7.7b) and (7.7c); the constants are therefore annihilated by the L derivative on the left-hand side of (16.40a). □

16.5. Proof of Proposition 16.4

We now prove Prop. 16.4. Our goal is to analyze $\mathscr{Z}^{N-1}(\upmu\mathscr{D}_\alpha^{(Z)}\mathscr{J}^\alpha[\Psi])$ for each of the 5 commutation vectorfields $Z \in \mathscr{Z} = \{\varrho L, \breve{R}, O_{(1)}, O_{(2)}, O_{(3)}\}$. To this end, we first use the identity (9.13) to decompose

$$\begin{aligned}\upmu\mathscr{D}_\alpha{}^{(Z)}\mathscr{J}^\alpha[\Psi] &= \mathscr{K}_{(\pi-Danger)}^{(Z)}[\Psi] + \mathscr{K}_{(\pi-Cancel-1)}^{(Z)}[\Psi] + \mathscr{K}_{(\pi-Cancel-2)}^{(Z)}[\Psi] \\ &\quad + \mathscr{K}_{(\pi-Elliptic)}^{(Z)}[\Psi] + \mathscr{K}_{(\pi-Good)}^{(Z)}[\Psi] + \mathscr{K}_{(\Psi)}^{(Z)}[\Psi] + \mathscr{K}_{(Low)}^{(Z)}[\Psi].\end{aligned}$$

We now separately analyze $\mathscr{Z}^{N-1}$ applied to each of the above 7 terms. Throughout our analysis, we implicitly use the definition of $Harmless^{\leq N}$ terms; see Def. 16.2. We also point out that Remark 15.1 applies in some of the below analysis.

Analysis of $\mathscr{Z}^{N-1}\mathscr{K}_{(\pi-Danger)}^{(Z)}[\Psi]$: We first consider the case $Z = O_{(l)}$. We will show that

$$\mathscr{Z}^{N-1}\mathscr{K}_{(\pi-Danger)}^{(O_{(l)})}[\Psi] = (\breve{R}\Psi)O_{(l)}\mathscr{Z}^{N-1}\mathrm{tr}_{\not g}\chi^{(Small)} - Harmless^{\leq N},$$

where $\mathscr{K}_{(\pi-Danger)}^{(O_{(l)})}[\Psi]$ is defined by (9.14a). By the Leibniz rule, we have to bound the following terms:

$$(16.43)\qquad -\left(\mathscr{Z}^{N-1}\mathrm{di}\!\!/\!\!v\,{}^{(O_{(l)})}\not{\pi}_L^{\#}\right)\breve{R}\Psi - \sum_{\substack{N_1+N_2\leq N-1\\ N_1\leq N-2}}\left(\mathscr{Z}^{N_1}\mathrm{di}\!\!/\!\!v\,{}^{(O_{(l)})}\not{\pi}_L^{\#}\right)\mathscr{Z}^{N_2}\breve{R}\Psi.$$

Using the first inequality in (16.18b) and the bootstrap assumption $\left\|\breve{R}\Psi\right\|_{C^0(\Sigma_t^u)}\leq \varepsilon(1+t)^{-1}$ (that is, $(\mathbf{BA}\Psi)$), we deduce that the first term in (16.43) is equal to the top-order-eikonal function-containing product $(\breve{R}\Psi)O_{(l)}\mathscr{Z}^{N-1}\mathrm{tr}_{\not{g}}\chi^{(Small)}$ plus a quadratic error term that is bounded in magnitude by

(16.44)

$$\lesssim \varepsilon\frac{1}{(1+t)^2}\left|\begin{pmatrix}\varrho L\mathscr{Z}^{\leq N}\Psi\\ \breve{R}\mathscr{Z}^{\leq N}\Psi\\ \varrho\not{d}\mathscr{Z}^{\leq N}\Psi\\ \mathscr{Z}^{\leq N}\Psi\end{pmatrix}\right| + \varepsilon\frac{1}{(1+t)^3}\left|\begin{pmatrix}\mathscr{Z}^{\leq N}(\upmu-1)\\ \sum_{a=1}^3\varrho\left|\mathscr{Z}^{\leq N}L_{(Small)}^a\right|\end{pmatrix}\right|$$
$$= Harmless^{\leq N}$$

as desired. To bound the magnitude of the remaining sum in (16.43) by $\lesssim$ the right-hand side of (16.44), we use the second inequalities in (16.38a) and (16.38b) and the bootstrap assumptions $(\mathbf{BA}\Psi)$.

Similarly, in the case $Z=\breve{R}$, we use the first inequality in (16.13b) and the first inequalities in (16.38a) and (16.38b) to deduce that the analog of the expression (16.43) is equal to the top-order-eikonal function-containing product $(\breve{R}\Psi)\not{\Delta}\mathscr{Z}^{N-1}\upmu$ plus a quadratic error term that is bounded in magnitude by the right-hand side of (16.44).

In the case $Z=\varrho L$, we have ${}^{(\varrho L)}\not{\pi}_L^{\#}=0$ (see (7.7d)) and the desired estimate is trivial.

Analysis of $\mathscr{Z}^{N-1}\mathscr{K}_{(\pi-Cancel-1)}^{(Z)}[\Psi]$: We first consider the case $Z=O_{(l)}$. We will show that

$$\mathscr{Z}^{N-1}\mathscr{K}_{(\pi-Cancel-1)}^{(O_{(l)})}[\Psi]=Harmless^{\leq N},$$

where $\mathscr{K}_{(\pi-Cancel-1)}^{(O_{(l)})}[\Psi]$ is defined by (9.14b). By the Leibniz rule, we have to analyze the following sum:

(16.45)

$$\sum_{N_1+N_2+N_3\leq N-1}\left\{\frac{1}{2}\mathscr{Z}^{N_1}\breve{R}\mathrm{tr}_{\not{g}}{}^{(O_{(l)})}\not{\pi}-\mathscr{Z}^{N_1}\mathrm{di}\!\!/\!\!v\,{}^{(O_{(l)})}\not{\pi}_{\breve{R}}^{\#}-(\mathscr{Z}^{N_3}\upmu)\mathscr{Z}^{N_1}\mathrm{di}\!\!/\!\!v\,{}^{(O_{(l)})}\not{\pi}_L^{\#}\right\}\mathscr{Z}^{N_2}L\Psi.$$

From the second inequality in (16.18a) and the first two inequalities in (16.18b), we deduce that in the cases $N_3=0$, the terms $\rho_{(l)}\not{\Delta}\mathscr{Z}^{N_1}\upmu$ and $\upmu O_{(l)}\mathscr{Z}^{N_1}\mathrm{tr}_{\not{g}}\chi^{(Small)}$ (which are dangerous when $N_1=N-1$) cancel out of the terms in braces in (16.45), leaving only error terms. Using in addition the estimate $\upmu\lesssim\ln(e+t)$ (that is,

(12.128a)), we deduce that

$$\left|\frac{1}{2}\mathscr{Z}^{N_1}\breve{R}\mathrm{tr}_{\not g}{}^{(O_{(l)})}\not\pi - \mathscr{Z}^{N_1}\mathrm{d}\not{\mathrm{i}}\mathrm{v}^{(O_{(l)})}\not\pi^{\#}_{\breve{R}} - \upmu\mathscr{Z}^{N_1}\mathrm{d}\not{\mathrm{i}}\mathrm{v}^{(O_{(l)})}\not\pi^{\#}_{L}\right| \lesssim \left|\begin{pmatrix} \varrho L\mathscr{Z}^{\leq N_1+1}\Psi \\ \breve{R}\mathscr{Z}^{\leq N_1+1}\Psi \\ \varrho\not d\mathscr{Z}^{\leq N_1+1}\Psi \\ \mathscr{Z}^{\leq N_1+1}\Psi \end{pmatrix}\right| + \frac{1}{1+t}\left|\begin{pmatrix} \mathscr{Z}^{\leq N_1+1}(\upmu-1) \\ \sum_{a=1}^{3}\varrho|\mathscr{Z}^{\leq N_1+1}L^a_{(Small)}| \end{pmatrix}\right|. \tag{16.46}$$

Furthermore, from (16.46), (12.128a), and the bootstrap assumptions $(\mathbf{BA}\Psi)$, we deduce that for $N_1 \leq 11$, we have

$$\left\|\frac{1}{2}\mathscr{Z}^{N_1}\breve{R}\mathrm{tr}_{\not g}{}^{(O_{(l)})}\not\pi - \mathscr{Z}^{N_1}\mathrm{d}\not{\mathrm{i}}\mathrm{v}^{(O_{(l)})}\not\pi^{\#}_{\breve{R}} - \upmu\mathscr{Z}^{N_1}\mathrm{d}\not{\mathrm{i}}\mathrm{v}^{(O_{(l)})}\not\pi^{\#}_{L}\right\|_{C^0(\Sigma_t^u)} \lesssim \varepsilon\frac{\ln(e+t)}{1+t}. \tag{16.47}$$

From (16.46), (16.47), Lemma 12.23, and the bootstrap assumptions $(\mathbf{BA}\Psi)$, we deduce that the terms in the sum (16.45) with $N_3 = 0$ are $= Harmless^{\leq N}$ as desired. To bound the remaining terms in (16.45) (which have $N_3 \geq 1$), we use Lemma 12.23, (16.38a), (16.38b), (12.128a), and the bootstrap assumptions $(\mathbf{BA}\Psi)$ to bound them by

$$\begin{aligned} &\sum_{\substack{N_1+N_2+N_3\leq N-1 \\ N_1,N_2\leq N-2}} \left|(\mathscr{Z}^{N_3}\upmu)(\mathscr{Z}^{N_1}\mathrm{d}\not{\mathrm{i}}\mathrm{v}^{(O_{(l)})}\not\pi^{\#}_{L})\mathscr{Z}^{N_2}L\Psi\right| \\ &\lesssim \frac{\ln^2(e+t)}{(1+t)^3}\left|\begin{pmatrix} \varrho L\mathscr{Z}^{\leq N-1}\Psi \\ \breve{R}\mathscr{Z}^{\leq N-1}\Psi \\ \varrho\not d\mathscr{Z}^{\leq N-1}\Psi \\ \mathscr{Z}^{\leq N-1}\Psi \end{pmatrix}\right| + \frac{\ln(e+t)}{(1+t)^4}\left|\begin{pmatrix} \mathscr{Z}^{\leq N}(\upmu-1) \\ \sum_{a=1}^{3}\varrho|\mathscr{Z}^{\leq N}L^a_{(Small)}| \end{pmatrix}\right| \\ &= Harmless^{\leq N} \end{aligned} \tag{16.48}$$

as desired.

The proof is similar in the case $Z = \varrho L$. It is in fact a bit simpler because the identity (7.7d) implies that the product involving N_3 in the analog of the sum (16.45) is completely absent. The main point is that we can use (7.7d), (16.30a), and the second inequality in (16.30b) to deduce that in the analog of the sum (16.45) (where ${}^{(O_{(l)})}\not\pi$ is replaced by ${}^{(\varrho L)}\not\pi$, etc.), the terms $\varrho\not\Delta\mathscr{Z}^{N_1}\upmu$ (which are dangerous when $N_1 = N-1$) cancel out of the terms in braces, leaving only error terms. Hence, essentially the same proof as in the case $Z = O_{(l)}$ yields that

$$\mathscr{Z}^{N-1}\mathscr{K}^{(\varrho L)}_{(\pi-Cancel-1)}[\Psi] = Harmless^{\leq N} \tag{16.49}$$

as desired.

The proof is also similar in the case $Z = \breve{R}$. The main point is that we can use (7.5e), the second inequality in (16.13a), and the first inequality in (16.13b) to deduce that when $N_3 = 0$ in the analog of the sum (16.45) (where ${}^{(O_{(l)})}\not\pi$ is replaced by ${}^{(\breve{R})}\not\pi$, etc.), the terms $\upmu\not\Delta\mathscr{Z}^{N_1}\upmu$ (which are dangerous when $N_1 = N-1$) cancel from the analog of the terms in braces in (16.45), leaving only error terms. Hence, using essentially the same reasoning as in the case $Z = O_{(l)}$, we deduce that

$$\mathscr{Z}^{N-1}\mathscr{K}^{(\breve{R})}_{(\pi-Cancel-1)}[\Psi] = Harmless^{\leq N} \tag{16.50}$$

as desired.

Analysis of $\mathscr{Z}^{N-1}\mathscr{K}^{(Z)}_{(\pi-Cancel-2)}[\Psi]$ The analysis is similar to our analysis of $\mathscr{Z}^{N-1}\mathscr{K}^{(Z)}_{(\pi-Cancel-1)}[\Psi]$. We first consider the case $Z = O_{(l)}$. We will show that

$$\mathscr{Z}^{N-1}\mathscr{K}^{(O_{(l)})}_{(\pi-Cancel-2)}[\Psi] = Harmless^{\leq N},$$

where $\mathscr{K}^{(O_{(2)})}_{(\pi-Cancel-2)}[\Psi]$ is defined by (9.14c). By the Leibniz rule and Lemma 8.7, we have to analyze the following sum:

$$\sum_{N_1+N_2 \leq N-1} \left(\not{\mathcal{L}}_{\mathscr{Z}}^{N_1} \left\{ -\not{\mathcal{L}}_{\breve{R}} {}^{(O_{(l)})}\not{\pi}_L^{\#} + \not{d}^{\#} {}^{(O_{(l)})}\pi_{L\breve{R}} \right\} \right) \cdot \not{d}\mathscr{Z}^{N_2}\Psi. \tag{16.51}$$

Using the first inequality in (16.18a) and the third inequality in (16.18b), we deduce that the terms $(\not{\nabla}^{2\#}\mathscr{Z}^{N_1}\upmu)\cdot O_{(l)}$ (which are dangerous when $N_1 = N-1$) cancel out of the sum in braces in (16.51), leaving only error terms. Thanks to this cancellation, we can argue as in our analysis of the sum (16.45) to deduce that $\mathscr{Z}^{N-1}\mathscr{K}^{(O_{(l)})}_{(\pi-Cancel-2)}[\Psi] = Harmless^{\leq N}$ as desired.

Similarly, in the case $Z = \breve{R}$, we use the first inequality in (16.13a) and the second inequality in (16.13b) to deduce that the terms $\not{d}^{\#}\mathscr{Z}^{N_1}\breve{R}\upmu$ cancel from the analog of the sum in braces in (16.51), and a similar argument yields

$$\mathscr{Z}^{N-1}\mathscr{K}^{(\breve{R})}_{(\pi-Cancel-2)}[\Psi] = Harmless^{\leq N} \tag{16.52}$$

as desired. *This cancellation is critically important when* $N_1 = N-1$ *because we do not have any way to estimate the top-order terms* $\not{d}^{\#}\mathscr{Z}^{23}\breve{R}\upmu$.

In the case of the vectorfield $Z = \varrho L$, (7.7d) and the first inequality in (16.30b) together imply that there are no dangerous top-order eikonal function quantities present in the analog of the sum in braces in (16.51), and a similar argument yields the desired estimate

$$\mathscr{Z}^{N-1}\mathscr{K}^{(\varrho L)}_{(\pi-Cancel-2)}[\Psi] = Harmless^{\leq N}. \tag{16.53}$$

Analysis of $\mathscr{Z}^{N-1}\mathscr{K}^{(Z)}_{(\pi-Elliptic)}[\Psi]$ We first consider the case $Z = O_{(l)}$. We will show that

$$\mathscr{Z}^{N-1}\mathscr{K}^{(O_{(l)})}_{(\pi-Elliptic)}[\Psi] = \uprho_{(l)}(\not{d}^{\#}\Psi)\cdot(\upmu\not{d}\mathscr{Z}^{N-1}\mathrm{tr}_{\not{g}}\upchi^{(Small)}) + Harmless^{\leq N},$$

where $\mathscr{K}^{(O_{(l)})}_{(\pi-Elliptic)}$ is defined by (9.14d). By the Leibniz rule and Lemma 8.7, we have to analyze the following sum:

$$\upmu(\not{\mathcal{L}}_{\mathscr{Z}}^{N-1}\mathrm{di\!/\!v}\,{}^{(O_{(l)})}\not{\pi}^{\#\#})\cdot\not{d}\Psi + \sum_{\substack{N_1+N_2+N_3\leq N-1 \\ N_2 \leq N-2}} (\mathscr{Z}^{N_1}\upmu)(\not{\mathcal{L}}_{\mathscr{Z}}^{N_2}\mathrm{di\!/\!v}\,{}^{(O_{(l)})}\not{\pi}^{\#\#})\cdot\not{d}\mathscr{Z}^{N_3}\Psi. \tag{16.54}$$

Using the last inequality in (16.18b), the estimate $|\not{d}\Psi| \lesssim \varepsilon(1+t)^{-2}$ (that is, $(\mathbf{BA}\Psi)$), and the bound $\upmu \lesssim \ln(e+t)$ (that is, (12.128a)), we deduce that the first term in (16.54) is equal to the top-order-eikonal function-containing product $\rho_{(l)}(\not{d}\Psi)\cdot(\upmu\not{d}^{\#}\mathscr{Z}^{N-1}\mathrm{tr}_{\not{g}}\chi^{(Small)})$ plus a quadratic error term that $= Harmless^{\leq N}$ as desired.

To bound the sum in (16.54), we use the estimates (12.128a), (16.39a), and (16.39b) and the bootstrap assumptions ($\mathbf{BA}\Psi$) to deduce that it $= Harmless^{\leq N}$ as desired.

Similarly, in the case $Z = \breve{R}$, we use the third inequality in (16.13b), (12.128a), (16.39a), (16.39b), and the bootstrap assumptions ($\mathbf{BA}\Psi$) to deduce that the analog of (16.54) is equal to the top-order-eikonal function-containing product $-(\not{d}^{\#}\Psi)\cdot(\upmu\not{d}\mathscr{Z}^{N-1}\mathrm{tr}_{\not{g}}\chi^{(Small)})$ plus a quadratic error term that $= Harmless^{\leq N}$ as desired.

Similarly, in the case $Z = \varrho L$, we use the third inequality in (16.30b), (12.128a), (16.39a), (16.39b), and the bootstrap assumptions ($\mathbf{BA}\Psi$) to deduce that the analog of (16.54) is equal to the top-order-eikonal function-containing product $(\varrho\not{d}^{\#}\Psi)\cdot(\upmu\not{d}\mathscr{Z}^{\leq N-1}\mathrm{tr}_{\not{g}}\chi^{(Small)})$ plus a quadratic error term that $= Harmless^{\leq N}$ as desired.

Analysis of $\mathscr{Z}^{N-1}\mathscr{K}_{(\pi-Good)}^{(Z)}[\Psi]$: We will show that

$$\mathscr{Z}^{N-1}\mathscr{K}_{(\pi-Good)}^{(Z)}[\Psi] = Harmless^{\leq N}$$

where $\mathscr{K}_{(\pi-Good)}^{(Z)}[\Psi]$ is given by (9.14e). To begin, we use (9.14e) to deduce the following schematic identity:

(16.55)

$$\mathscr{K}_{(\pi-Good)}^{(Z)}[\Psi] = \begin{pmatrix} \upmu L\mathrm{tr}_{\not{g}}{}^{(Z)}\not{\pi} \\ L^{(Z)}\pi_{L\breve{R}} \\ L^{(Z)}\pi_{\breve{R}R} \end{pmatrix} L\Psi + (L\mathrm{tr}_{\not{g}}{}^{(Z)}\not{\pi})\breve{R}\Psi + \begin{pmatrix} \upmu\not{\mathcal{L}}_L{}^{(Z)}\not{\pi}_L^{\#} \\ \not{\mathcal{L}}_L{}^{(Z)}\not{\pi}_{\breve{R}}^{\#} \end{pmatrix} \not{d}\Psi.$$

We now apply $\not{\mathcal{L}}_{\mathscr{Z}}^{N-1}$ to (16.55) and then apply the Leibniz rule to the products on the right-hand side. Using Lemma 8.6, inequality (12.54) with $f = \Psi$, the estimates (12.128a), (16.40a), (16.40b), (16.41a), (16.41b), (16.42a), and (16.42b), and the bootstrap assumptions ($\mathbf{BA}\Psi$), we deduce that

$$(16.56)\qquad \left|\mathscr{Z}^{N-1}\mathscr{K}_{(\pi-Good)}^{(Z)}[\Psi]\right| \lesssim \frac{\ln(e+t)}{(1+t)^2}\left|\begin{pmatrix} \varrho L\mathscr{Z}^{\leq N}\Psi \\ \breve{R}\mathscr{Z}^{\leq N}\Psi \\ \varrho\not{d}\mathscr{Z}^{\leq N}\Psi \\ \mathscr{Z}^{\leq N}\Psi \end{pmatrix}\right| + \varepsilon\frac{1}{(1+t)^3}\left|\begin{pmatrix} \mathscr{Z}^{\leq N}(\upmu-1) \\ \sum_{a=1}^{3}\varrho\left|\mathscr{Z}^{\leq N}L_{(Small)}^a\right| \end{pmatrix}\right| = Harmless^{\leq N}$$

as desired.

Analysis of $\mathscr{Z}^{N-1}\mathscr{K}_{(\Psi)}^{(Z)}[\Psi]$: We will show that

$$\mathscr{Z}^{N-1}\mathscr{K}_{(\Psi)}^{(Z)}[\Psi] = Harmless^{\leq N},$$

where $\mathscr{K}_{(\Psi)}^{(Z)}[\Psi]$ is given by (9.15). To this end, we apply $\mathscr{Z}^{N-1}$ to both sides of (9.15). We first address the terms that arise from the first and second lines of the right-hand side of (9.15). Using the Leibniz rule, (12.6b), the estimates (12.128a), (16.36a), (16.36b), (16.37a), (16.37b), (12.88a), and (12.88b), and the bootstrap assumptions ($\mathbf{BA}\Psi$), we bound the terms of interest by

(16.57)

$$\begin{aligned}
&\lesssim \left| \left\{ L + \frac{1}{2}\mathrm{tr}_{\not g}\chi \right\} \mathscr{Z}^{\leq N}\Psi \right| \\
&\quad + \frac{\ln(e+t)}{(1+t)^2} \left| \begin{pmatrix} \varrho L \mathscr{Z}^{\leq N}\Psi \\ \breve{R}\mathscr{Z}^{\leq N}\Psi \\ \varrho \not d \mathscr{Z}^{\leq N}\Psi \\ \mathscr{Z}^{\leq N}\Psi \end{pmatrix} \right| + \varepsilon \frac{1}{(1+t)^3} \left| \begin{pmatrix} \mathscr{Z}^{\leq N}(\upmu - 1) \\ \sum_{a=1}^{3} \varrho \left| \mathscr{Z}^{\leq N} L_{(Small)}^a \right| \end{pmatrix} \right| \\
&= Harmless^{\leq N}
\end{aligned}$$

as desired.

We now address the terms that arise from the third and fourth lines of the right-hand side of (9.15). Using the Leibniz rule, (12.6b), the estimates (12.128a), (16.36a), (16.36b), (16.37a), (16.37b), (12.101a) (with $\leq N-1$ in the role of N), and (12.101b), and the bootstrap assumptions ($\mathbf{BA}\Psi$), we bound the terms of interest by

(16.58)

$$\begin{aligned}
&\lesssim \frac{\ln(e+t)}{(1+t)^2} \left| \begin{pmatrix} \varrho L \mathscr{Z}^{\leq N}\Psi \\ \breve{R}\mathscr{Z}^{\leq N}\Psi \\ \varrho \not d \mathscr{Z}^{\leq N}\Psi \\ \mathscr{Z}^{\leq N}\Psi \end{pmatrix} \right| + \varepsilon \frac{1}{(1+t)^3} \left| \begin{pmatrix} \mathscr{Z}^{\leq N}(\upmu - 1) \\ \sum_{a=1}^{3} \varrho \left| \mathscr{Z}^{\leq N} L_{(Small)}^a \right| \end{pmatrix} \right| \\
&= Harmless^{\leq N}
\end{aligned}$$

as desired.

Analysis of $\mathscr{Z}^{N-1}\mathscr{K}_{(Low)}^{(Z)}[\Psi]$: We will show that

$$\mathscr{Z}^{N-1}\mathscr{K}_{(Low)}^{(Z)}[\Psi] = Harmless^{\leq N},$$

where $\mathscr{K}_{(Low)}^{(Z)}[\Psi]$ is given by (9.16). To this end, we apply $\not{\mathcal{L}}_{\mathscr{Z}}^{N-1}$ to both sides of (9.16) and then apply the Leibniz rule to terms on the right-hand side. We claim that all resulting products are in magnitude $\lesssim$ the terms on the first line of (16.58) and hence are $Harmless^{\leq N}$ as desired. We bound all derivatives of all deformation tensors with the estimates of Lemma 16.10. We bound the quantities $\mathscr{Z}^M \varrho^{-1}$ with (12.6b). To bound $\begin{pmatrix} \mathscr{Z}^M L\Psi \\ \not{\mathcal{L}}_{\mathscr{Z}}^M \not d\Psi \end{pmatrix}$ and $\mathscr{Z}^M\Psi$, we use Lemma 12.23 and the bootstrap assumptions ($\mathbf{BA}\Psi$). To bound $\mathscr{Z}^M\upmu$ and $\not{\mathcal{L}}_{\mathscr{Z}}^M \not d\upmu$, we use Lemma 8.6, inequality (12.46a), and (12.128a). To bound $\not{\mathcal{L}}_{\mathscr{Z}}^M \not d^{\#}\upmu$, we use Lemma 8.6, inequality (12.46a), (12.128a), and the estimates of Lemma 12.26 (to bound the derivatives of $\not g^{-1}$). To bound $\mathscr{Z}^M \mathrm{tr}_{\not g}\chi^{(Small)}$, we use (12.67a) and (12.128a). The quantities $L\upmu$, $\mathrm{tr}_{\not g}\not k^{(Trans-\Psi)}$, $\upmu \mathrm{tr}_{\not g}\not k^{(Tan-\Psi)}$, $\zeta^{(Trans-\Psi)\#}$ and $\upmu\zeta^{(Tan-\Psi)\#}$ are all schematically of the form $\sum_{p=0}^{1} G_{(Frame)}(\not{g}^{-1})^p \begin{pmatrix} \upmu L\Psi \\ \breve{R}\Psi \\ \upmu \not{d}\Psi \end{pmatrix}$ (see (4.5), (5.4b), (5.5b), (5.4a), and (5.5a)). Hence, we can bound their derivatives by using the estimates of Lemma 12.25, the previously mentioned estimates for $\mathscr{Z}^M \upmu$, $\not{\mathcal{L}}_{\mathscr{Z}}^M \not{g}^{-1}$, and $\begin{pmatrix} \mathscr{Z}^M L\Psi \\ \not{\mathcal{L}}_{\mathscr{Z}}^M \not{d}\Psi \end{pmatrix}$, and the bootstrap assumptions (**BA**Ψ). In total, these estimates yield that the terms of interest are $\lesssim$ the right-hand side of (16.58) as desired. □

16.6. Proof of Corollary 16.5

We now prove Cor. 16.5. We first consider inequality (16.4) in the case $N_1 \leq 11$. Then from the estimate $\left\| \mathscr{Z}^{\leq 11} \mathrm{tr}_{\not{g}} {}^{(Z)}\not{\pi} \right\|_{C^0(\Sigma_t^u)} \lesssim 1$ (that is, (12.60b)), we deduce that

$$\left| {}^{(\mathscr{Z}^N)}\mathfrak{F}_{(Eikonal-Low)} \right| \lesssim \sum_{\substack{N_2+N_3 \leq N-1 \\ N_2 \leq N-2}} \sum_{Z \in \mathscr{Z}} \left| \mathscr{Z}^{N_2} (\upmu \mathscr{D}_\alpha^{(Z)} \mathscr{J}^\alpha [\mathscr{Z}^{N_3}\Psi]) \right|. \tag{16.59}$$

Our goal is to show that

$$\text{the right-hand side of (16.59)} = Harmless^{\leq N}. \tag{16.60}$$

To prove (16.60), we begin by repeating the proof of Prop. 16.4 with $N-1$ replaced by N_2 (where $N_2 \leq N-2$) and Ψ replaced by $\mathscr{Z}^{N_3}\Psi$. The same proof yields that all terms in (16.59) are $Harmless^{\leq N}$ except for the terms corresponding to the explicitly written terms in (16.6a)-(16.6c), that is, except for terms of the form $(\breve{R}\mathscr{Z}^{N_3}\Psi)\not{\Delta}\mathscr{Z}^{N_2}\upmu$, $(\breve{R}\mathscr{Z}^{N_3}\Psi)O_{(l)}\mathscr{Z}^{N_2}\mathrm{tr}_{\not{g}}\upchi^{(Small)}$, $(\upmu \not{d}^{\#} \mathscr{Z}^{N_3}\Psi) \cdot (\upmu \not{d}\mathscr{Z}^{N_2}\mathrm{tr}_{\not{g}}\upchi^{(Small)})$, $\rho_{(l)}(\not{d}^{\#} \mathscr{Z}^{N_3}\Psi) \cdot (\upmu \not{d}\mathscr{Z}^{N_2}\mathrm{tr}_{\not{g}}\upchi^{(Small)})$, and $\varrho(\not{d}^{\#} \mathscr{Z}^{N_3}\Psi) \cdot (\upmu \not{d}\mathscr{Z}^{N_2}\mathrm{tr}_{\not{g}}\upchi^{(Small)})$. To handle the first two of these terms, we use the bootstrap assumptions (**BA**Ψ), inequalities (12.46a) and (12.46c), the pointwise estimate (12.67a), inequality (12.123a), and the assumption $N_2 \leq N-2$ to deduce that

$$\sum_{\substack{N_2+N_3 \leq N-1 \\ N_2 \leq N-2}} \left| (\breve{R}\mathscr{Z}^{\leq N_3}\Psi)\not{\Delta}\mathscr{Z}^{\leq N_2}\upmu \right| \lesssim \varepsilon \frac{\ln(e+t)}{(1+t)^2} \left| \breve{R}\mathscr{Z}^{\leq N}\Psi \right| + \varepsilon \frac{1}{(1+t)^3} \left| \mathscr{Z}^{\leq N}(\upmu - 1) \right|, \tag{16.61}$$

$$\begin{aligned} &\sum_{\substack{N_2+N_3 \leq N-1 \\ N_2 \leq N-2}} \left| (\breve{R}\mathscr{Z}^{N_3}\Psi)O_{(l)}\mathscr{Z}^{N_2}\mathrm{tr}_{\not{g}}\upchi^{(Small)} \right| \\ &\lesssim \varepsilon \frac{\ln(e+t)}{(1+t)^2} \left| \begin{pmatrix} \varrho L\mathscr{Z}^{\leq N}\Psi \\ \breve{R}\mathscr{Z}^{\leq N}\Psi \\ \varrho\not{d}\mathscr{Z}^{\leq N}\Psi \\ \mathscr{Z}^{\leq N}\Psi \end{pmatrix} \right| \\ &\quad + \varepsilon \frac{1}{(1+t)^3} \left| \begin{pmatrix} \mathscr{Z}^{\leq N-1}(\upmu - 1) \\ \sum_{a=1}^{3} \varrho \left| \mathscr{Z}^{\leq N} L_{(Small)}^a \right| \end{pmatrix} \right|. \end{aligned} \tag{16.62}$$

Hence, by Def. 16.2, the terms in (16.61)-(16.62) are $= Harmless^{\leq N}$. Also using the pointwise estimates $\upmu \lesssim \ln(e+t)$ (that is, (12.128a)) and $|\rho_{(l)}| \lesssim \varepsilon \ln(e+t)$ (that is, (12.28b) and Cor. 12.47), we use a similar argument to deduce that

$$\sum_{\substack{N_2+N_3 \leq N-1 \\ N_2 \leq N-2}} \left| (\upmu \not{d}^{\#} \mathscr{Z}^{N_3} \Psi)(\upmu \not{d} \mathscr{Z}^{N_2} \mathrm{tr}_{\not{g}} \chi^{(Small)}) \right| = Harmless^{\leq N}, \tag{16.63}$$

$$\sum_{\substack{N_2+N_3 \leq N-1 \\ N_2 \leq N-2}} \left| \rho_{(l)} (\not{d}^{\#} \mathscr{Z}^{N_3} \Psi)(\upmu \not{d} \mathscr{Z}^{N_2} \mathrm{tr}_{\not{g}} \chi^{(Small)}) \right| = Harmless^{\leq N}, \tag{16.64}$$

$$\sum_{\substack{N_2+N_3 \leq N-1 \\ N_2 \leq N-2}} \left| \varrho (\not{d}^{\#} \mathscr{Z}^{N_3} \Psi)(\upmu \not{d} \mathscr{Z}^{N_2} \mathrm{tr}_{\not{g}} \chi^{(Small)}) \right| = Harmless^{\leq N}. \tag{16.65}$$

We have thus proved (16.60).

It remains for us to consider inequality (16.4) in the case that $N_1 \geq 12$ and thus $N_2 + N_3 \leq 11$. Revisiting the proof in the previous paragraph and carefully counting derivatives, we find that $\mathscr{Z}^{N_2}(\upmu \mathscr{D}_\alpha^{(Z)} \mathscr{J}^\alpha[\mathscr{Z}^{N_3}\Psi]) = Harmless^{\leq N_2+N_3+1} \leq Harmless^{\leq 12}$. Hence, by Def. 16.2, the bootstrap assumptions ($\mathbf{BA}\Psi$), the estimate $|\mathrm{tr}_{\not{g}} \chi| \lesssim (1+t)^{-1}$ (that is, (12.60b)), and inequality (12.128a), we have[1] that

$$\sum_{N_2+N_3 \leq 11} \left\| \mathscr{Z}^{N_2}(\upmu \mathscr{D}_\alpha^{(Z)} \mathscr{J}^\alpha[\mathscr{Z}^{N_3}\Psi]) \right\|_{C^0(\Sigma_t^u)} \lesssim \varepsilon \frac{1}{(1+t)^2}. \tag{16.66}$$

Then from (16.66), the estimates (12.60a) and (12.60b) for the derivatives of $\mathrm{tr}_{\not{g}} {}^{(Z)}\not{\pi}$, (16.60), and Def. 16.2, we deduce that

(16.67)

$$\begin{aligned}
&\sum_{\substack{N_1+N_2+N_3 \leq N-1 \\ N_1, N_2 \leq N-2 \\ N_2+N_3 \leq 11}} \sum_{Z_1, Z_2 \in \mathscr{Z}} \left| \mathscr{Z}^{N_1} \mathrm{tr}_{\not{g}} {}^{(Z)}\not{\pi} \right| \left| \mathscr{Z}^{N_2}(\upmu \mathscr{D}_\alpha^{(Z)} \mathscr{J}^\alpha[\mathscr{Z}^{N_3}\Psi]) \right| \\
&\lesssim \varepsilon \frac{1}{(1+t)^2} \left| \begin{pmatrix} \varrho L \mathscr{Z}^{\leq N-1}\Psi \\ \breve{R} \mathscr{Z}^{\leq N-1}\Psi \\ \varrho \not{d} \mathscr{Z}^{\leq N-1}\Psi \\ \mathscr{Z}^{\leq N-1}\Psi \end{pmatrix} \right| + \varepsilon \frac{1}{(1+t)^3} \left| \begin{pmatrix} \mathscr{Z}^{\leq N}(\upmu - 1) \\ \sum_{a=1}^3 \varrho \left| \mathscr{Z}^{\leq N} L_{(Small)}^a \right| \end{pmatrix} \right| \\
&\quad + Harmless^{\leq N} \\
&= Harmless^{\leq N}
\end{aligned}$$

as desired. □

16.7. Pointwise estimates for the error integrands involving the deformation tensors of the multiplier vectorfields

We have now derived all of the necessary pointwise estimates corresponding to the inhomogeneous terms in the commuted wave equation. However, in order to

[1] Actually, by treating the first term on the right-hand side of (16.5) with more refined arguments like those used in the proof of Lemma 12.58, we could show that the right-hand side of (16.66) can be improved to $\varepsilon \ln(e+t)(1+t)^{-3}$, but we have no need for the improvement here.

derive our desired L^2 estimates, we also have to derive pointwise estimates for the terms generated by the deformation tensors of the multiplier vectorfields, that is, for the error integrands from Def. 10.19 and Lemma 10.21. We derive the desired pointwise estimates in the next proposition.

PROPOSITION 16.13 (**Pointwise estimates for ${}^{(T)}\mathfrak{P}[\Psi]$ and ${}^{(\widetilde{K})}\mathfrak{P}[\Psi]$**). *Let ${}^{(T)}\mathfrak{P}[\Psi]$ and ${}^{(\widetilde{K})}\mathfrak{P}[\Psi]$ be the error integrands from Def. 10.19 and Lemma 10.21 and recall that $\varrho(t,u) = 1 - u + t$. There exists a constant $C > 0$ such that under the small-data and bootstrap assumptions of Sects. 12.1-12.4, if ε is sufficiently small, then we have the following upper bound for $(t',u',\vartheta) \in \mathcal{M}_{T_{(Bootstrap)},U_0}$ (without taking the absolute value on the left-hand side), where we view both the left and right-hand sides as functions of (t',u',ϑ) and t is any fixed time verifying $t' \leq t < T_{(Bootstrap)}$:*

(16.68a)

$$\begin{aligned}
{}^{(T)}\mathfrak{P}[\Psi] \leq{}& C\frac{\ln(e+t')}{(1+t')^2}\Psi^2 + C\ln(e+t')\left\{L\Psi + \frac{1}{2}\mathrm{tr}_{\not g}\chi\Psi\right\}^2 + C\frac{1}{(1+t')^2}(\breve{R}\Psi)^2 \\
&+ C\ln^2(e+t')\upmu|\not d\Psi|^2 + C\varepsilon^{1/2}\frac{\ln(e+t')}{\sqrt{\ln(e+t)-\ln(e+t')}}\upmu|\not d\Psi|^2 \\
&+ C\varepsilon\ln^2(e+t')\mathbf{1}_{\{\upmu\leq 1/4\}}|\not d\Psi|^2.
\end{aligned}$$

In addition, the following pointwise estimate holds for $(t',u',\vartheta) \in \mathcal{M}_{T_{(Bootstrap)},U_0}$:

(16.68b)

$$\begin{aligned}
&\left|{}^{(\widetilde{K})}\mathfrak{P}[\Psi] + \frac{1}{2}\varrho^2|\not d\Psi|^2[L\upmu]_-\right| \\
&\leq C\frac{\ln^3(e+t')}{1+t'}\Psi^2 + C\ln^3(e+t')(1+t')\left\{L\Psi + \frac{1}{2}\mathrm{tr}_{\not g}\chi\Psi\right\}^2 \\
&\quad + \frac{1}{2}(1+C\varepsilon)\left\{\frac{1}{\varrho(t',u')\left\{1+\ln\left(\frac{\varrho(t',u')}{\varrho(0,u')}\right)\right\}}\right\}\varrho^2(t',u')\upmu|\not d\Psi|^2 \\
&\quad + C\varepsilon\frac{(1+t')}{\ln(e+t')}\mathbf{1}_{\{\upmu\leq 1/4\}}|\not d\Psi|^2.
\end{aligned}$$

PROOF. Throughout this proof, we freely use the trivial bound $|\underline{\breve{L}}f| \lesssim \upmu|Lf| + |\breve{R}f|$ (see (3.19a)) without mentioning it each time. We also silently use the simple inequality $(Lf)^2 \leq \left\{Lf + \frac{1}{2}\mathrm{tr}_{\not g}\chi f\right\}^2 + C(1+t')^{-2}f^2$, which follows easily from the estimate $|\varrho\mathrm{tr}_{\not g}\chi| \lesssim 1$ (that is, (12.60b)). Furthermore, we silently use simple inequalities of the form $|f_1f_2| \lesssim hf_1^2 + h^{-1}f_2^2$, where $h > 0$ is allowed to depend on time.

We first prove (16.68a). We separately bound each term ${}^{(T)}\mathfrak{P}_{(i)}[\Psi]$ in the sum (10.48a). From (10.49a), (12.128a), and (12.128b), we deduce that $\left|{}^{(T)}\mathfrak{P}_{(1)}[\Psi]\right| \lesssim \varepsilon\ln(e+t')(L\Psi)^2$, which is easily seen to be $\leq$ the right-hand side of (16.68a).

To bound the term ${}^{(T)}\mathfrak{P}_{(2)}[\Psi]$ from (10.49b), we first use (12.128b), to deduce that $|\not d\Psi|^2|3\upmu L\upmu| \lesssim \varepsilon(1+t')^{-1}\upmu|\not d\Psi|^2$ as desired. We then use (13.30) to deduce that the remaining product in (10.49b) verifies

$$\frac{1}{2}(\breve{\underline{L}}\upmu + L\upmu)|{\not d}\Psi|^2 \leq C\varepsilon^{1/2}\frac{\ln(e+t')}{\sqrt{\ln(e+t)-\ln(e+t')}}\upmu|{\not d}\Psi|^2 + C\varepsilon\ln(e+t')\upmu|{\not d}\Psi|^2$$

as desired (we do *not* take the absolute value on the left-hand side because of the nature of the estimate (13.30)).

To bound the term ${}^{(T)}\mathfrak{P}_{(4)}[\Psi]$ from (10.49d), we first note that the terms in braces multiplying $(L\Psi)({\not d}^{\#}\Psi)$ are schematically of the form ${\not d}\upmu$, $\upmu{\not d}\upmu$, or $(1+\upmu)G_{(Frame)}\begin{pmatrix}\upmu L\Psi\\ \breve{R}\Psi\\ \upmu{\not d}\Psi\end{pmatrix}$ (see (5.4a) and (5.5a)). Hence, from inequality (12.46a), Lemma 12.25, the estimate (12.128a), and the bootstrap assumptions (**BA**Ψ), we deduce that the terms in braces multiplying $(L\Psi)({\not d}^{\#}\Psi)$ are in magnitude $\lesssim \varepsilon\ln^2(e+t')(1+t')^{-1}$. When $\upmu \leq 1/4$, it thus follows that $\left|{}^{(T)}\mathfrak{P}_{(4)}[\Psi]\right| \lesssim \varepsilon\ln^2(e+t')(1+t')^{-1}(L\Psi)^2 + \varepsilon\mathbf{1}_{\{\upmu\leq 1/4\}}\ln^2(e+t')(1+t')^{-1}|{\not d}\Psi|^2$ as desired. When $\upmu > 1/4$, it thus follows that $\left|{}^{(T)}\mathfrak{P}_{(4)}[\Psi]\right| \lesssim \varepsilon\ln^2(e+t')(1+t')^{-1}(L\Psi)^2 + \varepsilon\ln^2(e+t')(1+t')^{-1}\upmu|{\not d}\Psi|$ as desired.

To bound the term ${}^{(T)}\mathfrak{P}_{(3)}[\Psi]$ from (10.49c), we first note that the same reasoning we used in analyzing ${}^{(T)}\mathfrak{P}_{(4)}[\Psi]$ yields that the terms in braces multiplying $-(\breve{\underline{L}}\Psi)({\not d}^{\#}\Psi)$ are in magnitude $\lesssim \varepsilon\ln(e+t')(1+t')^{-1}$. Hence, using the bound $\upmu \lesssim \ln(e+t')$ (that is, (12.128a)), we deduce that $\left|{}^{(T)}\mathfrak{P}_{(3)}[\Psi]\right| \lesssim \varepsilon\ln^2(e+t')(1+t')^{-1}(L\Psi)^2+\varepsilon\ln(e+t')(1+t')^{-1}\upmu|{\not d}\Psi|^2+\varepsilon\ln(e+t')(1+t')^{-1}|\breve{R}\Psi||{\not d}\Psi|$. The first two terms are $\lesssim$ the right-hand side of (16.68a) as desired. Furthermore, by separately considering the cases $\upmu \leq 1/4$ and $\upmu > 1/4$ (as we did at end of the proof of the bound for ${}^{(T)}\mathfrak{P}_{(4)}[\Psi]$), we deduce that $\varepsilon\ln(e+t')(1+t')^{-1}|\breve{R}\Psi||{\not d}\Psi| \lesssim \varepsilon(1+t')^{-2}(\breve{R}\Psi)^2 + \varepsilon\mathbf{1}_{\{\upmu\leq 1/4\}}\ln^2(e+t')|{\not d}\Psi|^2 + \ln^2(e+t')\upmu|{\not d}\Psi|^2$ as desired.

To bound the term ${}^{(T)}\mathfrak{P}_{(5)}[\Psi]$ from (10.49e), we first note that the terms in braces are schematically of the form $\hat{\chi}^{(Small)}+$ the trace-free part of $G_{(Frame)}\begin{pmatrix}\upmu L\Psi\\ \breve{R}\Psi\\ \upmu{\not d}\Psi\end{pmatrix}$ (see (5.4b) and (5.5b)). Hence, using the estimate $|\hat{\xi}| \lesssim |\xi|$ for symmetric type $\binom{0}{2}$ $S_{t,u}$ tensors, the argument used to bound the terms in braces in ${}^{(T)}\mathfrak{P}_{(4)}[\Psi]$, and (12.128a), we deduce that the terms in braces are in magnitude $\lesssim \varepsilon(1+t')^{-1}$. It follows that $\left|{}^{(T)}\mathfrak{P}_{(5)}[\Psi]\right| \lesssim \varepsilon(1+t')^{-1}\upmu|{\not d}\Psi|^2$ as desired.

To bound the term ${}^{(T)}\mathfrak{P}_{(6)}[\Psi]$ from (10.49f), we first note that the terms in braces multiplying $-\frac{1}{2}(L\Psi)(\breve{\underline{L}}\Psi)$ are schematically of the form $\mathrm{tr}_{\not g}\chi$ plus the trace (with respect to ${\not g}$) of $G_{(Frame)}\begin{pmatrix}\upmu L\Psi\\ \breve{R}\Psi\\ \upmu{\not d}\Psi\end{pmatrix}$ (see (5.4b) and (5.5b)). Hence, using the fact that $|\mathrm{tr}_{\not g}\xi| \lesssim |\xi|$ for symmetric type $\binom{0}{2}$ $S_{t,u}$ tensors, the estimate $|\varrho\mathrm{tr}_{\not g}\chi| \lesssim 1$ mentioned at the beginning of the proof, and the above bound for the terms in braces in ${}^{(T)}\mathfrak{P}_{(5)}[\Psi]$, we deduce that the terms in braces are in magnitude $\lesssim (1+t')^{-1}$. Also using the estimate $\upmu \lesssim \ln(e+t')$ (that is, (12.128a)), we deduce that $\left|{}^{(T)}\mathfrak{P}_{(6)}[\Psi]\right| \lesssim (L\Psi)^2 + (1+t')^{-2}(\breve{R}\Psi)^2$, which can easily be seen to be $\leq$ the right-hand side of (16.68a) as desired.

We now prove (16.68b). We separately bound each term ${}^{(\widetilde{K})}\mathfrak{P}_{(i)}[\Psi]$ in the sum (10.45b). To bound the term ${}^{(\widetilde{K})}\mathfrak{P}_{(1)}[\Psi]$ from (10.50a), we first use the identity $\varrho\breve{\underline{L}}\varrho = \varrho(\upmu - 2)$ and the estimates (12.128a) and (12.128b) to deduce that the terms in braces multiplying $(L\Psi)^2$ are in magnitude $\lesssim \ln(e+t')(1+t')$. The desired bound now readily follows.

To bound the term ${}^{(\widetilde{K})}\mathfrak{P}_{(2)}[\Psi]$ from (10.50b), we first use (12.128a) and (13.26) to deduce that the terms in braces multiplying $\frac{1}{2}\varrho^2\upmu(|\not{d}\Psi|^2)$ are in magnitude $\leq (1+C\varepsilon)\dfrac{1}{\varrho(t',u')\left\{1+\ln\left(\frac{\varrho(t',u')}{\varrho(0,u')}\right)\right\}}$. It follows that $\left|{}^{(\widetilde{K})}\mathfrak{P}_{(2)}[\Psi]\right|$ is $\leq$ the next-to-last term on the right-hand side of (16.68b) as desired.

To bound the term ${}^{(\widetilde{K})}\mathfrak{P}_{(3)}[\Psi]$ from (10.50c), we first use essentially the same analysis that we used in analyzing ${}^{(T)}\mathfrak{P}_{(4)}[\Psi]$ to deduce that the terms in braces multiplying $\varrho^2(L\Psi)(\not{d}^{\#}\Psi)$ are in magnitude $\lesssim \varepsilon\ln(e+t')(1+t')^{-1}$. It follows that $\left|{}^{(\widetilde{K})}\mathfrak{P}_{(3)}[\Psi]\right| \lesssim \varepsilon\ln^3(e+t')(1+t')(L\Psi)^2 + \varepsilon\dfrac{1+t'}{\ln(e+t')}|\not{d}\Psi|^2$. The first term is easily seen to be $\leq$ the right-hand side of (16.68b). To bound the second term $\varepsilon\dfrac{1+t'}{\ln(e+t')}|\not{d}\Psi|^2$ by the right-hand side of (16.68b), we use the same reasoning that we used at end of the proof of the bound for ${}^{(T)}\mathfrak{P}_{(4)}[\Psi]$, where we separately considered the cases $\upmu \leq 1/4$ and $\upmu > 1/4$ (in the case $\upmu > 1/4$, $\varepsilon\dfrac{1+t'}{\ln(e+t')}|\not{d}\Psi|^2$ is $\leq$ the $C\varepsilon\cdots$ term on the second line of the right-hand side of (16.68b)).

To bound the last term ${}^{(\widetilde{K})}\mathfrak{P}_{(4)}[\Psi]$ from (10.50d), we first use (12.128a) to deduce that $|\hat{\upchi}^{(Small)}| \lesssim \varepsilon\ln(e+t')(1+t')^{-2}$. Using this bound and the fact that $|\hat{\xi}| \lesssim |\xi|$ for symmetric type $\binom{0}{2}$ $S_{t,u}$ tensors, we conclude that $\left|{}^{(\widetilde{K})}\mathfrak{P}_{(4)}[\Psi]\right| \lesssim \varepsilon\ln(e+t')\upmu|\not{d}\Psi|^2$, which is $\leq$ the $C\varepsilon\cdots$ term on the second line of the right-hand side of (16.68b) as desired. This completes the proof of the proposition. □

In order to derive our desired a priori estimates for the energy quantities $\widetilde{\mathbb{Q}}_{(\leq N)}$, we still have to derive pointwise estimates for a few more terms: the remaining error integrands present on the right-hand side of (10.29b). We derive these simple estimates in the next lemma.

LEMMA 16.14 (**Pointwise estimates for the remaining easy energy error integrands**). *Under the small-data and bootstrap assumptions of Sects.* 12.1-12.4, *if ε is sufficiently small, then the following estimates hold on $\mathcal{M}_{T_{(Bootstrap)},U_0}$:*

(16.69a)
$$\left\{\left|\breve{\underline{L}}[\varrho^2\mathrm{tr}_{\not{g}}\upchi]\right| + \left|\varrho^2\upmu(\mathrm{tr}_{\not{g}}\upchi)^2\right| + \left|\varrho^2\mathrm{tr}_{\not{g}}\upchi\,\mathrm{tr}_{\not{g}}\not{k}^{(Trans-\Psi)}\right| + \left|\varrho^2\upmu\,\mathrm{tr}_{\not{g}}\upchi\not{k}^{(Tan-\Psi)}\right|\right\}\Psi^2 \lesssim \ln(e+t)\Psi^2,$$

(16.69b)
$$\left|\upmu\square_{g(\Psi)}[\varrho^2\mathrm{tr}_{\not{g}}\upchi]\right|\Psi^2 \lesssim \frac{\ln(e+t)}{1+t}\Psi^2.$$

PROOF. We first prove (16.69a). Using (4.1c) and the identity $\underline{\breve{L}}\varrho = \upmu - 2$, we compute that $\underline{\breve{L}}[\varrho^2 \mathrm{tr}_{\not g}\chi] = 2(\upmu - 2) + \upmu L[\varrho^2 \mathrm{tr}_{\not g}\chi^{(Small)}] - 4\varrho \mathrm{tr}_{\not g}\chi^{(Small)} + 2\varrho^2 \breve{R} \mathrm{tr}_{\not g}\chi^{(Small)}$. Hence, from (12.128a) and (12.128b), we deduce that the magnitude of this term is $\lesssim \ln(e+t)$ as desired. From (12.60b) and (12.128a), we deduce that the second term in the absolute value in (16.69a) verifies $\left|\varrho^2 \upmu (\mathrm{tr}_{\not g}\chi)^2\right| \lesssim \ln(e+t)$ as desired. The third and fourth terms are schematically of the form $\varrho^2 \mathrm{tr}_{\not g}\chi G_{(Frame)} \not g^{-1} \begin{pmatrix} \upmu L\Psi \\ \breve{R}\Psi \\ \upmu \not d\Psi \end{pmatrix}$ (see (5.4b) and (5.5b)). Hence, by Lemma 12.25, the estimates (12.60b) and (12.128a), and the bootstrap assumptions (**BA**Ψ), we see that these terms are in magnitude $\lesssim \varepsilon$ as desired.

To deduce (16.69b), we use the wave operator decomposition identity (5.12a) with $\varrho^2 \mathrm{tr}_{\not g}\chi = 2\varrho + \varrho^2 \mathrm{tr}_{\not g}\chi^{(Small)}$ in the role of f and hence $Lf = 2 + L(\varrho^2 \mathrm{tr}_{\not g}\chi^{(Small)})$ and $\breve{R}f = -2 - 2\varrho \mathrm{tr}_{\not g}\chi^{(Small)} + \varrho^2 \breve{R} \mathrm{tr}_{\not g}\chi^{(Small)}$. Up to an overall minus sign, the first term on the right-hand side of (5.12a) is equal to $\upmu L L f + 2L\breve{R}f + (L\upmu)Lf$, which in the present context can be expressed as

$$
\begin{aligned}
(16.70) \qquad & \upmu LL(\varrho^2 \mathrm{tr}_{\not g}\chi^{(Small)}) - 4\varrho^{-1} L(\varrho^2 \mathrm{tr}_{\not g}\chi^{(Small)}) + 4 \mathrm{tr}_{\not g}\chi^{(Small)} \\
& + 2L(\varrho^2 \breve{R} \mathrm{tr}_{\not g}\chi^{(Small)}) + (L\upmu)\left\{2 + L(\varrho^2 \mathrm{tr}_{\not g}\chi^{(Small)})\right\}.
\end{aligned}
$$

From (12.128a) and (12.128b), we deduce that all terms in (16.70) are in magnitude $\lesssim \varepsilon(1+t)^{-1}$ as desired. The second term on the first line of the right-hand side of (5.12a) is equal to $\varrho^2 \upmu \not\Delta \mathrm{tr}_{\not g}\chi^{(Small)}$. Hence, from (12.46a), (12.46c), and (12.128a), we deduce that its magnitude is $\lesssim \varepsilon \ln^2(e+t)(1+t)^{-2}$ as desired. The term $\mathrm{tr}_{\not g}\chi \breve{R} f$ on the second line of (5.12a) is equal to $(2\varrho^{-1} + \mathrm{tr}_{\not g}\chi^{(Small)})(-2 - 2\varrho \mathrm{tr}_{\not g}\chi^{(Small)} + \varrho^2 \breve{R} \mathrm{tr}_{\not g}\chi^{(Small)})$. Hence, from (12.128a), we deduce that this term is in magnitude $\lesssim \ln(e+t)(1+t)^{-1}$ as desired. With the help of (5.4a), (5.4b), (5.5a), and (5.5b), we see that the last four terms on the right-hand side of (5.12a) are schematically of the form

$$
G_{(Frame)} \not g^{-1} \begin{pmatrix} \upmu L\Psi \\ \breve{R}\Psi \\ \upmu \not d\Psi \end{pmatrix} \begin{pmatrix} 2 + L(\varrho^2 \mathrm{tr}_{\not g}\chi^{(Small)}) \\ \varrho^2 \not d \mathrm{tr}_{\not g}\chi^{(Small)} \end{pmatrix}.
$$

Hence, by inequality (12.46a), Lemma 12.25, the estimates (12.128a) and (12.128b), and the bootstrap assumptions (**BA**Ψ), we see that these terms are in magnitude $\lesssim \varepsilon(1+t)^{-1}$ as desired. □

16.8. Pointwise estimates needed to close the elliptic estimates

In Lemma 20.20, we derive our main top-order elliptic estimates for $\upmu$ and $\chi^{(Small)}$. In the next lemma, we provide a collection of preliminary pointwise estimates that play a role in our derivation of the elliptic estimates.

LEMMA 16.15 (**Pointwise estimates needed to close the elliptic estimates**). *Let $\chi^{(Small)}$ be the symmetric type $\binom{0}{2}$ $S_{t,u}$ tensorfield defined in (4.1c) and let $O \in \mathscr{O}$ be a geometric rotation vectorfield (see Def. 7.2). Let $1 \leq N \leq 24$ be an integer and recall that $\varrho(t,u) = 1 - u + t$. Under the small-data and bootstrap assumptions of Sects. 12.1-12.4, if ε is sufficiently small, then the following* *estimates hold on* $\mathcal{M}_{T_{(Bootstrap)},U_0}$:

$$\left|\nabla\!\!\!/ \mathcal{L}\!\!\!/_{\mathscr{Z}}^{N-1}\chi^{(Small)} - \left\{\nabla\!\!\!/ \mathcal{L}\!\!\!/_{\mathscr{Z}}^{N-1}\hat{\chi}^{(Small)} + \frac{1}{2}(d\!\!\!/ \mathscr{Z}^{N-1}\mathrm{tr}_{g\!\!\!/}\chi^{(Small)})g\!\!\!/\right\}\right| \lesssim \frac{1}{(1+t)^2}\left|\begin{pmatrix} \varrho L\mathscr{Z}^{\leq N}\Psi \\ \breve{R}\mathscr{Z}^{\leq N}\Psi \\ \varrho d\!\!\!/ \mathscr{Z}^{\leq N}\Psi \\ \mathscr{Z}^{\leq N}\Psi \end{pmatrix}\right| + \frac{1}{(1+t)^3}\left|\begin{pmatrix} \mathscr{Z}^{\leq N}(\upmu-1) \\ \sum_{a=1}^{3}\varrho\left|\mathscr{Z}^{\leq N}L_{(Small)}^a\right| \end{pmatrix}\right|, \tag{16.71}$$

$$\left|\nabla\!\!\!/ \mathcal{L}\!\!\!/_{\mathscr{Z}}^{N-1}\hat{\chi}^{(Small)} - \nabla\!\!\!/ \hat{\mathcal{L}\!\!\!/}_{\mathscr{Z}}^{N-1}\hat{\chi}^{(Small)}\right| \lesssim \frac{1}{(1+t)^2}\left|\begin{pmatrix} \varrho L\mathscr{Z}^{\leq N}\Psi \\ \breve{R}\mathscr{Z}^{\leq N}\Psi \\ \varrho d\!\!\!/ \mathscr{Z}^{\leq N}\Psi \\ \mathscr{Z}^{\leq N}\Psi \end{pmatrix}\right| + \frac{1}{(1+t)^3}\left|\begin{pmatrix} \mathscr{Z}^{\leq N}(\upmu-1) \\ \sum_{a=1}^{3}\varrho\left|\mathscr{Z}^{\leq N}L_{(Small)}^a\right| \end{pmatrix}\right|, \tag{16.72}$$

$$\left|O\mathscr{Z}^{N-1}\mathrm{tr}_{g\!\!\!/}\chi^{(Small)}\right| \lesssim \varrho\left|\nabla\!\!\!/ \mathcal{L}\!\!\!/_{\mathscr{Z}}^{N-1}\chi^{(Small)}\right| + \frac{1}{1+t}\left|\begin{pmatrix} \varrho L\mathscr{Z}^{\leq N}\Psi \\ \breve{R}\mathscr{Z}^{\leq N}\Psi \\ \varrho d\!\!\!/ \mathscr{Z}^{\leq N}\Psi \\ \mathscr{Z}^{\leq N}\Psi \end{pmatrix}\right| + \frac{1}{(1+t)^2}\left|\begin{pmatrix} \mathscr{Z}^{\leq N}(\upmu-1) \\ \sum_{a=1}^{3}\varrho\left|\mathscr{Z}^{\leq N}L_{(Small)}^a\right| \end{pmatrix}\right|, \tag{16.73}$$

$$\left|\hat{\mathcal{L}\!\!\!/}_{\mathscr{Z}}^{N-1}\hat{\chi}^{(Small)}\right| \lesssim \frac{1}{1+t}\left|\begin{pmatrix} \varrho L\mathscr{Z}^{\leq N}\Psi \\ \breve{R}\mathscr{Z}^{\leq N}\Psi \\ \varrho d\!\!\!/ \mathscr{Z}^{\leq N}\Psi \\ \mathscr{Z}^{\leq N}\Psi \end{pmatrix}\right| + \frac{1}{(1+t)^2}\left|\begin{pmatrix} \mathscr{Z}^{\leq N}(\upmu-1) \\ \sum_{a=1}^{3}\varrho\left|\mathscr{Z}^{\leq N}L_{(Small)}^a\right| \end{pmatrix}\right|. \tag{16.74}$$

PROOF. The estimate (16.71) follows from (12.107) with $\chi_{(Small)}$ in the role of ξ and $N-1$ in the role of N, (12.67a), and (12.128a).

The estimate (16.72) follows from (12.108) with $\chi_{(Small)}$ in the role of ξ and $N-1$ in the role of N, (12.67a), and (12.128a).

To prove (16.73), we first use inequality (12.104) with $\chi^{(Small)}$ in the role of ξ and $O\mathscr{Z}^{N-1}$ in the role of $\mathscr{Z}^N$, together with the fact that the trace of a type $\binom{0}{2}$ $S_{t,u}$ tensor is in magnitude $\lesssim$ the magnitude of the tensor itself, in order to deduce that $\left|O\mathscr{Z}^{N-1}\mathrm{tr}_{g\!\!\!/}\chi^{(Small)}\right|$ is $\lesssim \left|\mathcal{L}\!\!\!/_O\mathcal{L}\!\!\!/_{\mathscr{Z}}^{N-1}\chi^{(Small)}\right|$ plus the right-hand side of (12.104). Using (12.128a) and the estimate (12.67a) with $N-1$ in the role of N, we deduce that the right-hand side of (12.104) is $\lesssim$ the right-hand side of (16.73) as desired. Furthermore, using (12.49), we deduce that $\left|\mathcal{L}\!\!\!/_O\mathcal{L}\!\!\!/_{\mathscr{Z}}^{N-1}\chi^{(Small)}\right| \lesssim \varrho\left|\nabla\!\!\!/ \mathcal{L}\!\!\!/_{\mathscr{Z}}^{N-1}\chi^{(Small)}\right| + \left|\mathcal{L}\!\!\!/_{\mathscr{Z}}^{N-1}\chi^{(Small)}\right|$. The first term on the right-hand side of the previous inequality is the first term on the right-hand side of (16.73), while the second is $\lesssim$ the remaining terms on the right-hand side of (16.73) by virtue of the estimate (12.67a) with $N-1$ in the role of N.

To prove (16.74), we first use inequalities (12.102) and (12.103) with $\chi^{(Small)}$ in the role of ξ and $N-1$ in the role of N to deduce that $\left|\hat{\not{\mathcal{L}}}_{\mathscr{Z}}^{N-1}\hat{\chi}^{(Small)}\right|$ is $\lesssim$ $\left|\not{\mathcal{L}}_{\mathscr{Z}}^{\leq N-1}\chi^{(Small)}\right|$ plus the terms on the right-hand sides of (12.102) and (12.103). Using inequalities (12.67a) and (12.128a), we deduce that all of these terms are $\lesssim$ the right-hand side of (16.74) as desired. □

CHAPTER 17

Pointwise Estimates for the Difficult Error Integrands Corresponding to the Commuted Wave Equation

Recall that for solutions to the commuted wave equation $\mu\square_{g(\Psi)}(\mathscr{Z}^N\Psi) = {}^{(\mathscr{Z}^N)}\mathfrak{F}$, the basic energy-cone flux identities verified by $\mathscr{Z}^N\Psi$ were established Prop. 10.13, with $\mathscr{Z}^N\Psi$ in the role of Ψ and ${}^{(\mathscr{Z}^N)}\mathfrak{F}$ in the role of $\mathfrak{F}$. In this chapter, we complete the task, initiated in Chapter 16, of deriving pointwise estimates for the error integrands on the right-hand sides of the identities of the proposition; these pointwise estimates are a preliminary ingredient in our derivation of a priori L^2 estimates, which we carry out in Chapter 20. In particular, thanks to the decomposition (16.3), Prop. 16.4, and Cor. 16.5, we have already derived pointwise estimates for all terms in ${}^{(\mathscr{Z}^N)}\mathfrak{F}$ except for a handful, and in fact, aside from the handful, the terms are $Harmless^{\leq N}$ in the sense of Def. 16.2. Our goal in this chapter is to derive pointwise estimates for the remaining terms in ${}^{(\mathscr{Z}^N)}\mathfrak{F}$ that are *not* $Harmless^{\leq N}$. These difficult terms would cause derivative loss and other problems if they were not properly handled, and to derive suitable pointwise estimates for them, we need to use the modified quantities of Chapter 11.

The pointwise estimates that we derive in Chapter 17 can be split into two classes. In Chapter 20, we use the first class of estimates to bound the difficult top-order error integrals corresponding to the timelike multiplier T. We derive the first class estimates with the help of the fully modified quantities of Chapter 11. We provide the main estimates in Prop. 17.7 of Sect. 17.4, but in order to prove the proposition, we first prove a series of auxiliary lemmas in Sects. 17.1 and 17.3. Similarly, in Chapter 20, we use the second class of estimates to bound the difficult top-order error integrals corresponding to the Morawetz multiplier $\widetilde{K}$. We derive the second class estimates with the help of the partially modified quantities of Chapter 11. These estimates are easier to derive than those of the first class. We prove some auxiliary lemmas in Sect. 17.2 and we then provide the main estimates in Sect. 17.5.

17.1. Preliminary pointwise estimates for the derivatives of the inhomogeneous terms in the transport equations for the fully modified quantities

Recall that Prop. 11.10 provides the transport equation verified by the fully modified quantity ${}^{(\mathscr{S}^N)}\mathscr{X}$ and that $\mathscr{S}^N\mathfrak{X}$ and $\mathscr{S}^{N-1(S)}\mathfrak{J}$ are two source terms in the equation (see the right-hand sides of (11.33) and (11.34)). In the next lemma, we derive pointwise estimates for these source terms.

LEMMA 17.1 (**Pointwise estimates for $\mathscr{Z}^N\mathfrak{X}$ and $\mathscr{Z}^{N-1(S)}\mathfrak{J}$**). *Let $0 \leq N \leq 24$ be an integer. Let $\mathfrak{X}$ be the quantity defined in* (11.20b). *Under the small-data and bootstrap assumptions of Sects.* 12.1-12.4, *if ε is sufficiently small, then the following estimates hold on $\mathcal{M}_{T_{(Bootstrap)},U_0}$:*

$$\left|\mathscr{Z}^N\mathfrak{X}\right| \lesssim \frac{\ln(e+t)}{1+t}\left|\begin{pmatrix} \varrho L\mathscr{Z}^{\leq N}\Psi \\ \varrho ⁄\!\!\!d\mathscr{Z}^{\leq N}\Psi \\ \mathscr{Z}^{\leq N}\Psi \end{pmatrix}\right| + \left|\breve{R}\mathscr{Z}^{\leq N}\Psi\right| + \varepsilon\frac{1}{(1+t)^2}\left|\begin{pmatrix} \mathscr{Z}^{\leq N}(\upmu-1) \\ \sum_{a=1}^3 \varrho\left|\mathscr{Z}^{\leq N}L_{(Small)}^a\right| \end{pmatrix}\right|, \tag{17.1a}$$

$$\left\|\mathscr{Z}^{\leq 12}\mathfrak{X}\right\|_{C^0(\Sigma_t^u)} \lesssim \varepsilon\frac{1}{1+t}. \tag{17.1b}$$

Furthermore, the following estimates hold on $\mathcal{M}_{T_{(Bootstrap)},U_0}$:

$$\begin{aligned}
&\left|\mathscr{Z}^N\mathfrak{X} - \left\{-G_{LL}\breve{R}\mathscr{Z}^N\Psi - \frac{1}{2}\upmu ⁄\!\!\!G_A^{\ A} L\mathscr{Z}^N\Psi - \frac{1}{2}\upmu G_{LL}L\mathscr{Z}^N\Psi + \upmu ⁄\!\!\!G_L^{\ A} ⁄\!\!\!d_A\mathscr{Z}^N\Psi\right\}\right| \\
&\lesssim \frac{\ln(e+t)}{1+t}\left|\begin{pmatrix} \varrho L\mathscr{Z}^{\leq N-1}\Psi \\ \varrho ⁄\!\!\!d\mathscr{Z}^{\leq N-1}\Psi \\ \mathscr{Z}^{\leq N-1}\Psi \end{pmatrix}\right| \\
&\quad + \left|\breve{R}\mathscr{Z}^{\leq N-1}\Psi\right| + \varepsilon\frac{1}{(1+t)^2}\left|\begin{pmatrix} \mathscr{Z}^{\leq N}(\upmu-1) \\ \sum_{a=1}^3 \varrho\left|\mathscr{Z}^{\leq N}L_{(Small)}^a\right| \end{pmatrix}\right|.
\end{aligned} \tag{17.2}$$

Furthermore, let $S \in \mathscr{S}$ be a spatial commutation vectorfield (see Def. 7.2*) and let ${}^{(S)}\mathfrak{J}$ be the inhomogeneous term defined in* (11.26). *Then the following estimates hold on $\mathcal{M}_{T_{(Bootstrap)},U_0}$:*

$$\left|\mathscr{Z}^{N-1(S)}\mathfrak{J}\right| \lesssim \frac{\ln(e+t)}{(1+t)^2}\left|\begin{pmatrix} \varrho L\mathscr{Z}^{\leq N}\Psi \\ \breve{R}\mathscr{Z}^{\leq N}\Psi \\ \varrho ⁄\!\!\!d\mathscr{Z}^{\leq N}\Psi \\ \mathscr{Z}^{\leq N}\Psi \end{pmatrix}\right| + \varepsilon\frac{\ln(e+t)}{(1+t)^3}\left|\begin{pmatrix} \mathscr{Z}^{\leq N}(\upmu-1) \\ \sum_{a=1}^3 \varrho\left|\mathscr{Z}^{\leq N}L_{(Small)}^a\right| \end{pmatrix}\right|, \tag{17.3a}$$

$$\left\|\mathscr{Z}^{\leq 11(S)}\mathfrak{J}\right\|_{C^0(\Sigma_t^u)} \lesssim \varepsilon\frac{\ln^2(e+t)}{(1+t)^3}. \tag{17.3b}$$

PROOF.
Proof of (17.1a) **and** (17.1b): We note the schematic identity

$$\mathfrak{X} = G_{(Frame)}^{\#}\begin{pmatrix} \upmu L\Psi \\ \breve{R}\Psi \\ \upmu ⁄\!\!\!d\Psi \end{pmatrix}.$$

Hence, our proofs of (16.14a) and (16.14b) also yield the desired bounds (17.1a) and (17.1b).

Proof of (17.2): We first use definition (11.20b) and Lemma 8.6 to deduce that the term in the absolute value on the left-hand side of (17.2) can be written as

(17.4)
$$
\begin{aligned}
&-[\mathscr{Z}^N, G_{LL}]\breve{R}\Psi - \frac{1}{2}[\not{\mathcal{L}}_{\mathscr{Z}}^N, \mu \not{G}_A^{\ A}]L\Psi - \frac{1}{2}[\mathscr{Z}^N, \mu G_{LL}]L\Psi + [\not{\mathcal{L}}_{\mathscr{Z}}^N, \mu \not{G}_L^{\ A}]\not{d}_A\Psi \\
&- G_{LL}[\mathscr{Z}^N, \breve{R}]\Psi - \frac{1}{2}\mu \not{G}_A^{\ A}[\not{\mathcal{L}}_{\mathscr{Z}}^N, L]\Psi - \frac{1}{2}\mu G_{LL}[\mathscr{Z}^N, L]\Psi.
\end{aligned}
$$

Hence, using the Leibniz rule and the reasoning used to prove (16.14a) and (17.1a), we deduce that the terms on the first line of (17.4) are in magnitude $\lesssim$ the right-hand side of (16.14a), *with* $N-1$ *in place of* N *in the* Ψ*-containing terms on the right-hand side*. It follows that these terms are in magnitude $\lesssim$ the right-hand side of (17.2) as desired. To deduce that the terms on the second line of (17.4) are in magnitude $\lesssim$ the right-hand side of (17.2), we use the estimates (12.58b) and (12.128a), inequalities (12.81a) and (12.81c) with $f = \Psi$, and the bootstrap assumptions (**BA**Ψ). We have thus proved the desired bound (17.2).

Proof of (17.3a) **and** (17.3b): It suffices to prove (17.3a) since (17.3b) follows from (17.3a), (12.128a), and the bootstrap assumptions (**BA**Ψ). To proceed, we apply $\not{\mathcal{L}}_{\mathscr{Z}}^{N-1}$ to the right-hand side of (11.26). We begin by bounding the term $Z^{N-1}S\mathfrak{A}$ arising from the first term on the right-hand side of (11.26), where $\mathfrak{A}$ is given by (11.13). To this end, we apply the Leibniz rule to the terms on the right-hand side of (11.13). The terms of interest can be bounded by using essentially the same argument that we used to deduce (17.1a) and (17.1b). We note that the resulting bounds are better by a factor of $(1+t)^{-2}$ compared to (17.1a) and (17.1b) because the right-hand side of (11.13) features an additional factor of $\begin{pmatrix} L\Psi \\ \not{d}\Psi \end{pmatrix}$ compared to the term $\mathfrak{X} = G_{(Frame)}^{\#}\begin{pmatrix} \mu L\Psi \\ \breve{R}\Psi \\ \mu\not{d}\Psi \end{pmatrix}$ from (17.1a) and (17.1b).

We now bound the terms arising from the last products on the first line of the right-hand side of (11.26) (that is, the product involving the terms in braces). This term can be schematically written as

$$
(17.5) \qquad \mathrm{tr}_{\not{g}}\chi \not{\mathcal{L}}_S \left\{ G_{(Frame)}^{\#} \begin{pmatrix} \mu L\Psi \\ \mu\not{d}\Psi \end{pmatrix} \right\}.
$$

Applying $\not{\mathcal{L}}_{\mathscr{Z}}^{N-1}$ to the right-hand side of (17.5) and using the Leibniz rule, we bound the resulting terms by

$$
(17.6) \qquad \lesssim \sum_{\substack{N_1+N_2+N_3+N_4 \leq N \\ N_1 \leq N-1}} \left|\mathscr{Z}^{N_1}\mathrm{tr}_{\not{g}}\chi\right| \left|\not{\mathcal{L}}_{\mathscr{Z}}^{N_2} G_{(Frame)}^{\#}\right| \left|\mathscr{Z}^{N_3}\mu\right| \left|\not{\mathcal{L}}_{\mathscr{Z}}^{N_4}\begin{pmatrix} L\Psi \\ \not{d}\Psi \end{pmatrix}\right|.
$$

Using Lemma 12.23 with $f = \Psi$, Lemma 12.26 with $N_1 \leq N-1$ in the role of N, Cor. 12.27, (12.128a), and the bootstrap assumptions (**BA**Ψ), we deduce that the right-hand side of (17.6) is $\lesssim$ the right-hand side of (17.3a) as desired.

We now bound the first product $(S\mu)L\mathrm{tr}_{\not{g}}\chi^{(Small)} = \varrho^{-2}(S\mu)L(\varrho^2\mathrm{tr}_{\not{g}}\chi^{(Small)}) - 2\varrho^{-1}(S\mu)\mathrm{tr}_{\not{g}}\chi^{(Small)}$ (recall that $L\varrho = 1$) on the second line of the right-hand side of (11.26). We apply $\mathscr{Z}^{N-1}$ to this product and use (12.6b) and the Leibniz rule, which leads to the following bound:

$$\begin{aligned}
(17.7)\qquad &\left|\mathscr{Z}^{N-1}\left\{\varrho^{-2}(S\upmu)L(\varrho^2\mathrm{tr}_{\not g}\chi^{(Small)}) - 2\varrho^{-1}(S\upmu)\mathrm{tr}_{\not g}\chi^{(Small)}\right\}\right| \\
&\lesssim \frac{1}{(1+t)^2}\sum_{\substack{N_1+N_2\leq N\\ N_1\leq N-1}}\left|\mathscr{Z}^{N_1}L(\varrho^2\mathrm{tr}_{\not g}\chi^{(Small)})\right|\left|\mathscr{Z}^{N_2}(\upmu-1)\right| \\
&\quad + \frac{1}{1+t}\sum_{\substack{N_1+N_2\leq N\\ N_1\leq N-1}}\left|\mathscr{Z}^{N_1}\mathrm{tr}_{\not g}\chi^{(Small)}\right|\left|\mathscr{Z}^{N_2}(\upmu-1)\right|.
\end{aligned}$$

We then use the estimates (12.67a), (12.127b), (12.128a), and (12.128b) to deduce that the right-hand side of (17.7) is $\lesssim$ the right-hand side of (17.3a) as desired. The remaining products on the second line of the right-hand side of (11.26) can be bounded in a similar fashion. The last two terms on the last line of the right-hand side of (11.26) can also be bounded in a similar fashion. For these estimates, it is important that $S\varrho \in \{0,-1\}$ for spatial commutation vectorfields $S \in \mathscr{S}$.

It remains for us to bound the first term on the last line of the right-hand side of (11.26). To this end, we note that the identity $[L,S] = {}^{(S)}\not\pi_L^{\#}$ (see (4.21b)) implies that the term can be written as ${}^{(S)}\not\pi_L^{\#}\cdot \not d\mathfrak{X}$. Applying $\not\mathcal{L}_{\mathscr{Z}}^{N-1}$ to this product and using the Leibniz rule, Lemma 8.6, and inequality (12.46a), we see that it suffices to bound

$$(17.8)\quad \frac{1}{1+t}\left\|\not\mathcal{L}_{\mathscr{Z}}^{\leq 11}\,{}^{(S)}\not\pi_L^{\#}\right\|_{C^0(\Sigma_t^u)}\left|\mathscr{Z}^{\leq N}\mathfrak{X}\right| + \frac{1}{1+t}\left\|\mathscr{Z}^{\leq 12}\mathfrak{X}\right\|_{C^0(\Sigma_t^u)}\left|\not\mathcal{L}_{\mathscr{Z}}^{\leq N-1}\,{}^{(S)}\not\pi_L^{\#}\right|.$$

Using (12.72a), (12.72b), (12.74a), (12.74b), (17.1a), and (17.1b), we see that (17.8) is $\lesssim$ the right-hand side of (17.3a) as desired. ☐

In the next lemma, we derive pointwise estimates for the inhomogeneous term ${}^{(\mathscr{S}^N)}\mathfrak{J}$ appearing on the right-hand side of (11.33).

REMARK 17.2 (**The need for (BA′Ψ)**). In proving Lemma 17.3, we derive the estimate (17.14), which, in the case $N = 24$, involves a term that is quadratic in $\hat{\chi}^{(Small)}$ with 12 derivatives on each factor. Bounding this term requires a bound for $\|\mathscr{Z}^{\leq 12}\hat{\chi}^{(Small)}\|_{C^0(\Sigma_t^u)}$, which we obtained with the help of the bootstrap assumption (**BA′Ψ**); see Remark 12.7.

LEMMA 17.3 (**Pointwise estimate for the inhomogeneous term** ${}^{(\mathscr{S}^N)}\mathfrak{J}$). *Let $1 \leq N \leq 24$ be an integer. Let $\mathscr{S}^N$ be an Nth order pure spatial commutation vectorfield operator (see definition* (7.2a)*) and let ${}^{(\mathscr{S}^N)}\mathfrak{J}$ be the corresponding inhomogeneous term defined in* (11.34)*. Under the small-data and bootstrap assumptions of Sects.* 12.1-12.4*, if ε is sufficiently small, then the following estimates hold on $\mathcal{M}_{T_{(Bootstrap)},U_0}$:*

(17.9)

$$\left|{}^{(\mathscr{S}^N)}\mathfrak{J}\right| \lesssim \frac{\ln(e+t)}{(1+t)^2}\left|\begin{pmatrix}\varrho L\mathscr{Z}^{\leq N}\Psi\\ \breve{R}\mathscr{Z}^{\leq N}\Psi\\ \varrho\not d\mathscr{Z}^{\leq N}\Psi\\ \mathscr{Z}^{\leq N}\Psi\end{pmatrix}\right| + \frac{\ln(e+t)}{(1+t)^3}\left|\begin{pmatrix}\mathscr{Z}^{\leq N}(\upmu-1)\\ \sum_{a=1}^3\varrho\left|\mathscr{Z}^{\leq N}L^a_{(Small)}\right|\end{pmatrix}\right|.$$

PROOF. We begin by noting that many of the commutator terms that we estimate in this proof are absent when $N = 1$. To bound the first term $\mathscr{S}^{N-1(S)}\mathfrak{J}$

on the first line of (11.34) by the right-hand side of (17.9), we simply quote the estimate (17.3a).

To bound the second term $[L, \mathscr{S}^{N-1}]S\mathfrak{X}$ on the first line of (11.34) by the right-hand side of (17.9), we first use the commutator estimate (12.83) with $N-1$ in the role of N and $S\mathfrak{X}$ in the role of f to deduce that

(17.10)

$$\begin{aligned} |[L, \mathscr{S}^{N-1}]S\mathfrak{X}| &\lesssim \varepsilon^{1/2} \frac{\ln(e+t)}{(1+t)^2} \left| \begin{pmatrix} \varrho L \mathscr{Z}^{\leq N} \mathfrak{X} \\ \breve{R} \mathscr{Z}^{\leq N} \mathfrak{X} \\ \varrho \not{d} \mathscr{Z}^{\leq N} \mathfrak{X} \\ \mathscr{Z}^{\leq N} \mathfrak{X} \end{pmatrix} \right| \\ &+ \frac{1}{1+t} \left\| \mathscr{Z}^{\leq 12} \mathfrak{X} \right\|_{C^0(\Sigma_t^u)} \left| \begin{pmatrix} \varrho L \mathscr{Z}^{\leq N-1} \Psi \\ \breve{R} \mathscr{Z}^{\leq N-1} \Psi \\ \varrho \not{d} \mathscr{Z}^{\leq N-1} \Psi \\ \mathscr{Z}^{\leq N-1} \Psi \end{pmatrix} \right| \\ &+ \frac{1}{(1+t)^2} \left\| \mathscr{Z}^{\leq 12} \mathfrak{X} \right\|_{C^0(\Sigma_t^u)} \left| \begin{pmatrix} \mathscr{Z}^{\leq N}(\upmu - 1) \\ \sum_{a=1}^3 \varrho \left| \mathscr{Z}^{\leq N} L_{(Small)}^a \right| \end{pmatrix} \right|. \end{aligned}$$

The desired bound now follows from (17.10), (17.1a), and (17.1b).

To bound the first term $-[\mathscr{S}^{N-1}, \upmu \mathrm{tr}_{\not{g}} \chi] S \mathrm{tr}_{\not{g}} \chi^{(Small)}$ on the second line of (11.34) by the right-hand side of (17.9), we first use the Leibniz rule, the decomposition $\mathrm{tr}_{\not{g}} \chi = 2\varrho^{-1} + \mathrm{tr}_{\not{g}} \chi^{(Small)}$ (see (4.1c)), and the bound $\left| \mathscr{S}^{\leq N-1} \varrho^{-1} \right| \lesssim \frac{1}{1+t}$ (which follows from (12.6a)) to deduce that

$$\begin{aligned} (17.11) \qquad & \left| [\mathscr{S}^{N-1}, \upmu \mathrm{tr}_{\not{g}} \chi] S \mathrm{tr}_{\not{g}} \chi^{(Small)} \right| \\ &\lesssim \frac{1}{1+t} \sum_{\substack{N_1+N_2 \leq N \\ N_1, N_2 \leq N-1}} \left| \mathscr{Z}^{N_1} \upmu \right| \left| \mathscr{Z}^{N_2} \mathrm{tr}_{\not{g}} \chi^{(Small)} \right| \\ &+ \sum_{\substack{N_1+N_2+N_3 \leq N \\ N_1, N_2, N_3 \leq N-1}} \left| \mathscr{Z}^{N_1} \upmu \right| \left| \mathscr{Z}^{N_2} \mathrm{tr}_{\not{g}} \chi^{(Small)} \right| \left| \mathscr{Z}^{N_3} \mathrm{tr}_{\not{g}} \chi^{(Small)} \right|. \end{aligned}$$

The desired bound now follows from (17.11), (12.67a) and (12.128a).

To bound the second term $\frac{1}{2}[\mathscr{S}^{N-1}, \mathrm{tr}_{\not{g}} \chi] S\mathfrak{X} = \frac{1}{2}[\mathscr{S}^{N-1}, 2\varrho^{-1} + \mathrm{tr}_{\not{g}} \chi^{(Small)}] S\mathfrak{X}$ (see (4.1c)) on the second line of (11.34) by the right-hand side of (17.9), we first use the Leibniz rule and the bound $\left| \mathscr{S}^M \varrho^{-1} \right| \lesssim \frac{1}{(1+t)^2}$ (which holds for $M \geq 1$ and follows from (12.6a)) to bound its magnitude by

(17.12)

$$\begin{aligned} &\lesssim \frac{1}{(1+t)^2} \left| \mathscr{Z}^{\leq N-1} \mathfrak{X} \right| \\ &+ \left\| \mathscr{Z}^{\leq 11} \mathrm{tr}_{\not{g}} \chi^{(Small)} \right\|_{C^0(\Sigma_t^u)} \left| \mathscr{Z}^{\leq N-1} \mathfrak{X} \right| + \left\| \mathscr{Z}^{\leq 12} \mathfrak{X} \right\|_{C^0(\Sigma_t^u)} \left| \mathscr{Z}^{\leq N-1} \mathrm{tr}_{\not{g}} \chi^{(Small)} \right|. \end{aligned}$$

The desired bound now follows from (17.12), (12.60a) with $N-1$ in the role of N, (12.60b), (17.1a) with $N-1$ in the role of N, and (17.1b).

We omit the proof of the desired bound for the first term on the third line of (11.34) because it can be bounded by using the same argument that we now use to bound the last term on the third line. To bound this last term $-[\mathscr{S}^{N-1},\upmu]\left(SL\mathrm{tr}_{\not{g}}\chi^{(Small)}\right)$, which can be expressed as

$$-[\mathscr{S}^{N-1},\upmu]S\left(\varrho^{-2}L(\varrho^2\mathrm{tr}_{\not{g}}\chi^{(Small)})\right)+2[\mathscr{S}^{N-1},\upmu]S\left(\varrho^{-1}\mathrm{tr}_{\not{g}}\chi^{(Small)}\right)$$

(recall that $L\varrho=1$), by the right-hand side of (17.9), we first use the Leibniz rule and (12.6a) to deduce that its magnitude is

$$\begin{aligned}(17.13)\qquad &\lesssim \frac{1}{(1+t)^2}\sum_{\substack{N_1+N_2\leq N\\ N_1,N_2\leq N-1}}\left|\mathscr{Z}^{N_1}(\upmu-1)\right|\left|\mathscr{Z}^{N_2}L(\varrho^2\mathrm{tr}_{\not{g}}\chi^{(Small)})\right|\\ &\quad+\frac{1}{1+t}\sum_{\substack{N_1+N_2\leq N\\ N_1,N_2\leq N-1}}\left|\mathscr{Z}^{N_1}(\upmu-1)\right|\left|\mathscr{Z}^{N_2}\mathrm{tr}_{\not{g}}\chi^{(Small)}\right|.\end{aligned}$$

The desired bound now follows from (17.13), (12.127a), (12.127b), (12.128a), and (12.128b).

To bound the difference on the last line of (11.34) by the right-hand side of (17.9), we use the Leibniz rule, the estimates (12.60a), (12.60b), (12.127a), (12.128a), and the bootstrap assumptions ($\mathbf{BA}\Psi$) to deduce that

(17.14)

$$\begin{aligned}&\left|2\upmu\hat{\chi}^{(Small)\#\#}\cdot\not{\mathcal{L}}_{\mathscr{S}}^N\hat{\chi}^{(Small)}-\mathscr{S}^N(\upmu\hat{\chi}^{(Small)\#\#}\cdot\hat{\chi}^{(Small)})\right|\\ &\lesssim\sum_{\substack{N_1+N_2+N_3+N_4+N_5\leq N\\ N_4,N_5\leq N-1}}\left|\mathscr{Z}^{N_1}\upmu\right|\left|\not{\mathcal{L}}_{\mathscr{Z}}^{N_2}\not{g}^{-1}\right|\left|\not{\mathcal{L}}_{\mathscr{Z}}^{N_3}\not{g}^{-1}\right|\left|\not{\mathcal{L}}_{\mathscr{Z}}^{N_4}\hat{\chi}^{(Small)}\right|\left|\not{\mathcal{L}}_{\mathscr{Z}}^{N_5}\hat{\chi}^{(Small)}\right|\\ &\lesssim\frac{\ln^2(e+t)}{(1+t)^3}\left|\begin{pmatrix}\varrho L\mathscr{Z}^{\leq N}\Psi\\ \breve{R}\mathscr{Z}^{\leq N}\Psi\\ \varrho\not{d}\mathscr{Z}^{\leq N}\Psi\\ \mathscr{Z}^{\leq N}\Psi\end{pmatrix}\right|+\frac{\ln(e+t)}{(1+t)^4}\left|\begin{pmatrix}\mathscr{Z}^{\leq N}(\upmu-1)\\ \sum_{a=1}^3\varrho\left|\mathscr{Z}^{\leq N}L_{(Small)}^a\right|\end{pmatrix}\right|.\end{aligned}$$

We have thus proved the desired estimate (17.9). □

17.2. Preliminary pointwise estimates for the derivatives of the inhomogeneous terms in the transport equations for the partially modified quantities

Recall that Lemma 11.15 provides the transport equation verified by the partially modified quantity $\varrho^{2(\mathscr{S}^{N-1})}\widetilde{\mathscr{X}}$ and that ${}^{(\mathscr{S}^{N-1})}\widetilde{\mathfrak{B}}$ is a source term in the equation. In the next lemma, we derive pointwise estimates for this source term.

LEMMA 17.4 (**Pointwise estimates for the inhomogeneous terms corresponding to the partially modified version of $\mathscr{S}^{N-1}\mathrm{tr}_{\not{g}}\chi^{(Small)}$**). *Let $1\leq N\leq 24$ be an integer. Let $\mathscr{S}^{N-1}$ be an $(N-1)^{st}$ order pure spatial commutation vectorfield operator (see definition* (7.2a)*) and let ${}^{(\mathscr{S}^{N-1})}\widetilde{\mathfrak{B}}$ be the corresponding inhomogeneous term defined in* (11.47). *Under the small-data and bootstrap assumptions of Sects.* 12.1-12.4, *if ε is sufficiently small, then the following estimates* *hold on* $\mathcal{M}_{T_{(Bootstrap)},U_0}$:

(17.15)

$$\left|{}^{(\mathscr{S}^{N-1})}\widetilde{\mathfrak{B}}\right| \lesssim \left|\begin{pmatrix} \varrho L\mathscr{Z}^{\leq N-1}\Psi \\ \breve{R}\mathscr{Z}^{\leq N-1}\Psi \\ \varrho \not{d}\mathscr{Z}^{\leq N-1}\Psi \\ \mathscr{Z}^{\leq N-1}\Psi \end{pmatrix}\right| + \frac{\ln(e+t)}{(1+t)^2}\left|\begin{pmatrix} \mathscr{Z}^{\leq N}(\upmu - 1) \\ \sum_{a=1}^3 \varrho \left|\mathscr{Z}^{\leq N} L_{(Small)}^a\right| \end{pmatrix}\right|.$$

Proof. We start by analyzing the first term $\mathscr{S}^{N-1}\widetilde{\mathfrak{B}}$ on the right-hand side of (11.47). Note that by (11.43), this term splits into two pieces: $-\mathscr{S}^{N-1}(\varrho^2\mathfrak{B})$ and $2\mathscr{S}^{N-1}(\varrho\widetilde{\mathfrak{X}})$. We now bound the first piece $-\mathscr{S}^{N-1}(\varrho^2\mathfrak{B})$; we will bound the second piece at the end of the proof. To proceed, we apply $\not{\mathcal{L}}_{\mathscr{S}}^{N-1}$ to ϱ^2 times the right-hand side of (11.15) and apply the Leibniz rule. We bound $\mathscr{S}^M\varrho$ with (12.6a). We bound the terms $\not{\mathcal{L}}_{\mathscr{S}}^M\begin{pmatrix} G_{(Frame)} \\ G'_{(Frame)} \end{pmatrix}$ with the estimates of Lemma 12.25. We bound $\not{\mathcal{L}}_{\mathscr{S}}^M\not{g}^{-1}$ and $\not{\mathcal{L}}_{\mathscr{S}}^M\mathrm{tr}_{\not{g}}\upchi$ with Lemma 12.26. We bound the terms $\not{\mathcal{L}}_{\mathscr{S}}^M\begin{pmatrix} L\Psi \\ \not{d}\Psi \end{pmatrix}$ with Lemma 12.23 and the bootstrap assumptions ($\mathbf{BA}\Psi$). In total, these estimates yield that $\left|\mathscr{S}^{N-1}(\varrho^2\mathfrak{B})\right|$ is $\lesssim$ the right-hand side of (17.15) as desired.

We next bound the second term $\mathscr{S}^{N-1}(\varrho^2|\upchi^{(Small)}|^2)$ on the right-hand side of (11.47). From (12.6b) and (12.128a), we deduce that the magnitude of this term is

$$\lesssim \varepsilon\ln(e+t)\left|\not{\mathcal{L}}_{\mathscr{S}}^{\leq N-1}\upchi^{(Small)}\right|. \tag{17.16}$$

Using inequality (12.67a) with $N-1$ in the role of N, we deduce that the right-hand side of (17.16) is $\lesssim$ the right-hand side of (17.15) as desired.

We next bound the first term $\frac{1}{2}[\mathscr{S}^{N-1},\varrho^2 G_{LL}]\not{\Delta}\Psi$ on the second line of the right-hand side of (11.47). From (12.6a) and the Leibniz rule, we deduce that the magnitude of the first term is

$$\lesssim \varrho^2\left|\mathscr{S}^{\leq 12}G_{LL}\right|\left|\mathscr{S}^{\leq N-2}\not{\Delta}\Psi\right| + \varrho^2\left|\mathscr{S}^{\leq 11}\not{\Delta}\Psi\right|\left|\mathscr{S}^{\leq N-1}G_{LL}\right|. \tag{17.17}$$

From inequalities (12.58a), (12.58b), (12.101a), and (12.101b), and the bootstrap assumptions ($\mathbf{BA}\Psi$), we deduce that the magnitude of the right-hand side of (17.17) is $\lesssim$ the right-hand side of (17.15) as desired.

We next bound the second term $\frac{1}{2}\varrho^2 G_{LL}[\mathscr{S}^{N-1},\not{\Delta}]\Psi$ on the second line of the right-hand side of (11.47). Using the commutator estimate (12.97) with $N-1$ in the role of N and $f=\Psi$, inequalities (12.55) and (12.58b), and the bootstrap assumptions ($\mathbf{BA}\Psi$), we deduce that the magnitude of this term is $\lesssim$ the right-hand side of (17.15) as desired.

We next bound the first term $[L,\mathscr{S}^{N-1}](\varrho^2\mathrm{tr}_{\not{g}}\upchi^{(Small)})$ on the third line of the right-hand side of (11.47). Using the commutator estimate (12.83) with $N-1$ in the role of N and $\varrho^2\mathrm{tr}_{\not{g}}\upchi^{(Small)}$ in the role of f, and inequalities (12.6b), (12.46a), (12.127a), and (12.128a), we deduce that the magnitude of this term is $\lesssim$ the right-hand side of (17.15) as desired.

We next bound the last term $[L, \mathscr{S}^{N-1}](\varrho^2\widetilde{\mathfrak{X}})$ on the third line of the right-hand side of (11.47). Referring to definition (11.41b), we see that we have to bound

$$[\not{\mathcal{L}}_L, \not{\mathcal{L}}_{\mathscr{S}}^{N-1}]\left\{-\frac{1}{2}\varrho^2 \not{G}_A^{\ A} L\Psi - \frac{1}{2}\varrho^2 G_{LL} L\Psi + \varrho^2 \not{G}_L^{\ A} \not{d}_A \Psi\right\}. \tag{17.18}$$

From Lemma 8.6, Lemma 12.23, inequality (12.6b), inequality (12.46a), the estimates of Cor. 12.27, the commutator estimate (12.83) with $N-1$ in the role of N and f equal to the terms in braces in (17.18), and the bootstrap assumptions (**BA**Ψ), we deduce that

$$\begin{aligned} &\left|[\not{\mathcal{L}}_L, \not{\mathcal{L}}_{\mathscr{S}}^{N-1}]\left\{-\frac{1}{2}\varrho^2 \not{G}_A^{\ A} L\Psi - \frac{1}{2}\varrho^2 G_{LL} L\Psi + \varrho^2 \not{G}_L^{\ A} \not{d}_A \Psi\right\}\right| \\ &\lesssim \frac{\ln(e+t)}{1+t}\left|\begin{pmatrix} \varrho L \mathscr{Z}^{\leq N-1}\Psi \\ \breve{R}\mathscr{Z}^{\leq N-1}\Psi \\ \varrho \not{d} \mathscr{Z}^{\leq N-1}\Psi \\ \mathscr{Z}^{\leq N-1}\Psi \end{pmatrix}\right| \\ &\quad + \frac{1}{(1+t)^2}\left|\begin{pmatrix} \mathscr{Z}^{\leq N}(\upmu - 1) \\ \sum_{a=1}^3 \varrho \left|\mathscr{Z}^{\leq N} L_{(Small)}^a\right| \end{pmatrix}\right|, \end{aligned} \tag{17.19}$$

which is $\lesssim$ the right-hand side of (17.15) as desired.

We now bound the first term $L\left\{[\varrho^2, \mathscr{S}^{N-1}]\mathrm{tr}_{\not{g}}\chi^{(Small)}\right\}$ on the last line of the right-hand side of (11.47). We first use (12.6a) and the fact that $\varrho L \in \mathscr{Z}$ to deduce that

$$\left|L\left\{[\varrho^2, \mathscr{S}^{N-1}]\mathrm{tr}_{\not{g}}\chi^{(Small)}\right\}\right| \lesssim \left|\mathscr{Z}^{\leq N-1}\mathrm{tr}_{\not{g}}\chi^{(Small)}\right|. \tag{17.20}$$

To conclude that the right-hand side of (17.20) is $\lesssim$ the right-hand side of (17.15) as desired, we use inequality (12.67a) with $N-1$ in the role of N.

We now bound the final term $L\left\{\varrho^{2(\mathscr{S}^{N-1})}\widetilde{\mathfrak{X}} - \mathscr{S}^{N-1}(\varrho^2\widetilde{\mathfrak{X}})\right\}$ on the last line of the right-hand side of (11.47). We first use definitions (11.41b) and (11.45b), the Leibniz rule, the fact that $\varrho L \in \mathscr{Z}$, Lemma 8.6, inequality (12.6b), and inequality (12.46a) to deduce that the term can be bounded as follows:

$$\begin{aligned} &\left|L\left\{\varrho^{2(\mathscr{S}^{N-1})}\widetilde{\mathfrak{X}} - \mathscr{S}^{N-1}(\varrho^2\widetilde{\mathfrak{X}})\right\}\right| \\ &\lesssim \sum_{\substack{N_1+N_2\leq N+1 \\ N_1,N_2\leq N}} \left|\not{\mathcal{L}}_{\mathscr{Z}}^{N_1} G_{(Frame)}^{\#}\right| \left|\mathscr{Z}^{N_2}\Psi\right| + \left|G_{(Frame)}^{\#}\right| \left|[\varrho L, \varrho L \mathscr{S}^{N-1}]\Psi\right|. \end{aligned} \tag{17.21}$$

To bound the first sum on the right-hand side of (17.21) by $\lesssim$ the right-hand side of (17.15), we use inequality (12.55), the estimates of Cor. 12.27 and the bootstrap assumptions (**BA**Ψ). To bound the last product on the right-hand side of (17.21), we also use the commutator estimate (12.81a) with $f = \Psi$.

To complete the proof of (17.15), it remains for us to bound the term $2\mathscr{S}^{N-1}(\varrho\widetilde{\mathfrak{X}})$ mentioned at the beginning of the proof. Using a subset of the reasoning that we used in the previous paragraph, we bound the magnitude of this term by $\lesssim$ the first sum on the right-hand side of (17.21) and then by $\lesssim$ the right-hand side of (17.15). We have therefore proved (17.15). □

Recall that Lemma 11.19 provides the transport equation verified by the partially modified quantity $\not{\mathcal{L}}\mathscr{S}^{N-1}\upmu$ and that ${}^{(\mathscr{S}^{N-1})}\mathfrak{J}$ is a source term in the equation. In the next lemma, we derive pointwise estimates for this source term.

LEMMA 17.5 (**Pointwise estimates for the inhomogeneous terms corresponding to the partially modified version of $\not{d}\mathscr{S}^{N-1}\upmu$**). *Let $1 \leq N \leq 24$ be an integer. Let $\mathscr{S}^{N-1}$ be an $(N-1)^{st}$ order pure spatial commutation vectorfield operator (see definition* (7.2a)*) and let ${}^{(\mathscr{S}^{N-1})}\mathfrak{J}$ be the corresponding $S_{t,u}$ one-form inhomogeneous term defined in* (11.53)*. Under the small-data and bootstrap assumptions of Sects.* 12.1-12.4*, if ε is sufficiently small, then the following estimates hold on $\mathcal{M}_{T_{(Bootstrap)},U_0}$:*

(17.22)
$$\varrho\left|{}^{(\mathscr{S}^{N-1})}\mathfrak{J}\right| \lesssim \left|\begin{pmatrix} \varrho L\mathscr{Z}^{\leq N-1}\Psi \\ \breve{R}\mathscr{Z}^{\leq N-1}\Psi \\ \varrho\not{d}\mathscr{Z}^{\leq N-1}\Psi \\ \mathscr{Z}^{\leq N-1}\Psi \end{pmatrix}\right| + \frac{\ln(e+t)}{(1+t)^2}\left|\begin{pmatrix} \mathscr{Z}^{\leq N}(\upmu-1) \\ \sum_{a=1}^{3}\varrho\left|\mathscr{Z}^{\leq N}L_{(Small)}^{a}\right| \end{pmatrix}\right|.$$

PROOF. We start by bounding the product $\varrho\not{\mathcal{L}}_{\mathscr{S}}^{N-1}\mathfrak{J}$ arising from the first term on the right-hand side of (11.53) where the $S_{t,u}$ one-form $\mathfrak{J}$ is defined in (11.50). We apply ϱ times $\not{\mathcal{L}}_{\mathscr{S}}^{N-1}$ to the right-hand side of (11.50) and apply the Leibniz rule. We bound the terms $\mathscr{S}^{M}G_{(Frame)}$, $\mathscr{S}^{M}LG_{(Frame)}$, and $\not{\mathcal{L}}_{\mathscr{S}}^{M}\not{d}G_{(Frame)}$ by using the fact that $\varrho L\in\mathscr{Z}$, (12.6a), Lemma 8.6, Lemma 12.23, inequality (12.46a), and the estimates of Lemma 12.25. We bound the terms $\mathscr{S}^{M}\upmu$ and $\not{\mathcal{L}}_{\mathscr{S}}^{M}\not{d}\upmu$ with Lemma 8.6 and inequality (12.128a). We bound the terms $\not{\mathcal{L}}_{\mathscr{S}}^{M}\begin{pmatrix} L\Psi \\ \breve{R}\Psi \\ \not{d}\Psi \end{pmatrix}$ with Lemma 12.23 and the bootstrap assumptions (**BA**Ψ). In total, these estimates yield that $\varrho\left|\not{\mathcal{L}}_{\mathscr{S}}^{N-1}\mathfrak{J}\right|$ is $\lesssim$ the right-hand side of (17.22) as desired.

To bound ϱ times the second term on the right-hand side of (11.53), we first use the Leibniz rule and Lemma 8.6 to deduce that

(17.23)
$$\varrho\left|[\not{\mathcal{L}}_{\mathscr{S}}^{N-1},G_{LL}]\not{d}\breve{R}\Psi\right| \lesssim \left|\not{\mathcal{L}}_{\mathscr{S}}^{\leq 12}G_{LL}\right|\left|\varrho\not{d}\mathscr{S}^{\leq N-2}\breve{R}\Psi\right| + \varrho\left|\not{d}\mathscr{S}^{\leq 12}\Psi\right|\left|\not{\mathcal{L}}_{\mathscr{S}}^{\leq N-1}G_{LL}\right|.$$

Inequalities (12.58a) and (12.58b) and the bootstrap assumptions (**BA**Ψ) then yield that the right-hand side of (17.23) is $\lesssim$ the right-hand side of (17.22) as desired.

To bound ϱ times the third term on the right-hand side of (11.53), we use inequality (12.91) with $N-1$ in the role of N and $\not{d}\upmu$ in the role of ξ, Lemma 8.6, and inequalities (12.46a) and (12.128a) to deduce the following desired bound:

(17.24)
$$\begin{aligned} &\varrho\left|[\not{\mathcal{L}}_{L},\not{\mathcal{L}}_{\mathscr{S}}^{N-1}]\not{d}\upmu\right| \\ &\lesssim \left|\begin{pmatrix} \varrho L\mathscr{Z}^{\leq N-1}\Psi \\ \varrho\not{d}\mathscr{Z}^{\leq N-1}\Psi \\ \breve{R}\mathscr{Z}^{\leq N-1}\Psi \\ \mathscr{Z}^{\leq N-1}\Psi \end{pmatrix}\right| + \frac{\ln(e+t)}{(1+t)^2}\left|\begin{pmatrix} \mathscr{Z}^{\leq N}(\upmu-1) \\ \sum_{a=1}^{3}\varrho\left|\mathscr{Z}^{\leq N}L_{(Small)}^{a}\right| \end{pmatrix}\right|. \end{aligned}$$

To bound ϱ times the fourth term on the right-hand side of (11.53), we refer to definition (11.48b) to see that we have to bound

$$\varrho[\not{\mathcal{L}}_L, \not{\mathcal{L}}_{\mathscr{S}}^{N-1}]\left\{\frac{1}{2}\upmu G_{LL}\not{d}\Psi + \upmu G_{LR}\not{d}\Psi\right\}.$$

We now derive the estimates for the term $\upmu G_{LL}\not{d}\Psi$; the estimates for the term $\upmu G_{LR}\not{d}\Psi$ can be handled using the same arguments, and we omit those details. The estimates (12.58b), (12.128a), the bootstrap assumptions ($\mathbf{BA}\Psi$), and Lemma 8.6 together imply that $\left|\not{\mathcal{L}}_{\mathscr{Z}}^{\leq 12}(\upmu G_{LL}\not{d}\Psi)\right| \lesssim \varepsilon \ln(e+t)(1+t)^{-2}$. Combining this estimate with the commutator estimate (12.91), where $\upmu G_{LL}\not{d}\Psi$ plays the role of ξ and $N-1$ plays the role of N, we deduce that

$$\begin{aligned}(17.25)\quad \varrho\left|[\not{\mathcal{L}}_L, \not{\mathcal{L}}_{\mathscr{S}}^{N-1}](\upmu G_{LL}\not{d}\Psi)\right| &\lesssim \frac{\ln(e+t)}{1+t}\left|\not{\mathcal{L}}_{\mathscr{Z}}^{\leq N-1}(\upmu G_{LL}\not{d}\Psi)\right| \\ &+ \frac{\ln(e+t)}{(1+t)^2}\left|\begin{pmatrix}\varrho L\mathscr{Z}^{\leq N-1}\Psi \\ \varrho\not{d}\mathscr{Z}^{\leq N-1}\Psi \\ \breve{R}\mathscr{Z}^{\leq N-1}\Psi \\ \mathscr{Z}^{\leq N-1}\Psi\end{pmatrix}\right| \\ &+ \frac{\ln(e+t)}{(1+t)^3}\left|\begin{pmatrix}\mathscr{Z}^{\leq N}(\upmu-1) \\ \sum_{a=1}^{3}\varrho\left|\mathscr{Z}^{\leq N}L_{(Small)}^a\right|\end{pmatrix}\right|.\end{aligned}$$

The last two terms in (17.25) are manifestly $\lesssim$ the right-hand side of (17.22) as desired. To bound the first term by $\lesssim$ the right-hand side of (17.22), we use the Leibniz rule, Lemma 8.6, the estimates (12.58a), (12.58b), (12.128a), and the bootstrap assumptions ($\mathbf{BA}\Psi$).

To bound ϱ times the last term on the right-hand side of (11.53), we use definitions (11.48b) and (11.51b) and Lemma 8.6 to deduce that the term can be rewritten as follows:

$$\begin{aligned}(17.26)\qquad \not{\mathcal{L}}_L\left\{{}^{(\mathscr{S}^{N-1})}\widetilde{\mathfrak{M}} - \not{\mathcal{L}}_{\mathscr{S}}^{N-1}\widetilde{\mathfrak{M}}\right\} &= \frac{1}{2}\not{\mathcal{L}}_L[\not{\mathcal{L}}_{\mathscr{S}}^{N-1}, \upmu G_{LL}]\not{d}\Psi \\ &+ \not{\mathcal{L}}_L[\not{\mathcal{L}}_{\mathscr{S}}^{N-1}, \upmu G_{LR}]\not{d}\Psi.\end{aligned}$$

We show how to bound the first term on the right-hand side of (17.26); the proof of the bound for the second one is identical. From the fact that $\varrho L \in \mathscr{Z}$, Lemma 8.6, the estimates (12.58b) and (12.128a), and the bootstrap assumptions ($\mathbf{BA}\Psi$), we deduce that

$$\begin{aligned}(17.27)\quad \varrho\left|\not{\mathcal{L}}_L[\not{\mathcal{L}}_{\mathscr{S}}^{N-1}, \upmu G_{LL}]\not{d}\Psi\right| &\lesssim \frac{1}{(1+t)^2}\left|\mathscr{Z}^{\leq N}(\upmu-1)\right| + \ln(e+t)\left|\not{d}\mathscr{Z}^{\leq N-1}\Psi\right| \\ &+ \ln(e+t)\left|\not{d}\mathscr{Z}^{\leq 12}\Psi\right|\left|\mathscr{Z}^{\leq N}G_{LL}\right|.\end{aligned}$$

Inequalities (12.55) and (12.58a) and the bootstrap assumptions ($\mathbf{BA}\Psi$) then yield that the right-hand side of (17.27) is $\lesssim$ the right-hand side of (17.22) as desired. □

17.3. Solving the transport equation satisfied by the fully modified version of the spatial derivatives of the trace of the re-centered null second fundamental form

In Sect. 17.4, we derive pointwise estimates for the most difficult top-order wave equation error integrands that arise from the multiplier T. Our derivation of these pointwise estimates is based on a detailed analysis of the fully modified quantities defined in Sect. 11.2. In this section, we carry out a preliminary step. Specifically, for N^{th} order pure spatial commutation vectorfield operators $\mathscr{S}^N$, we derive preliminary pointwise estimates for the corresponding fully modified quantities ${}^{(\mathscr{S}^N)}\mathscr{X}$ by inverting the transport operator in the transport equation (11.33) that they satisfy. One difficulty that we encounter is that the estimates for distinct ${}^{(\mathscr{S}^N)}\mathscr{X}$ are weakly coupled due to the presence of the important top-order term $\mu[L, \mathscr{S}^N]\mathrm{tr}_{\not{g}}\chi^{(Small)}$ on the right-hand side of (11.33). We overcome this difficulty by a deriving Gronwall-type estimate that simultaneously involves all of the Nth order quantities ${}^{(\mathscr{S}^N)}\mathscr{X}$. We provide the details in the next lemma.

LEMMA 17.6 (**Solving the transport equation satisfied by** ${}^{(\mathscr{S}^N)}\mathscr{X}$). *Let $0 \leq N \leq 24$ be an integer. Let $\mathscr{S}^N$ be an Nth order pure spatial commutation vectorfield operator (see definition (7.2a)), and let ${}^{(\mathscr{S}^N)}\mathscr{X}$, $\mathfrak{X}$, and ${}^{(\mathscr{S}^N)}\mathfrak{I}$ be the corresponding quantities appearing in Prop. 11.10. Let χ and $\chi^{(Small)}$ be the symmetric type $\binom{0}{2}$ $S_{t,u}$ tensorfields defined in (3.70) and (4.1c) and let $\varrho(t,u) = 1-u+t$ be the geometric radial variable. Under the small-data and bootstrap assumptions of Sects. 12.1-12.4, if ε is sufficiently small, then the following estimates hold on $\mathcal{M}_{T_{(Bootstrap)},U_0}$:*

(17.28)

$$
\begin{aligned}
&\varrho^2(t,u)\left|{}^{(\mathscr{S}^N)}\mathscr{X}\right|(t,u,\vartheta) \\
&\leq (1+C\varepsilon)\sup_{0\leq s\leq t}\left(\frac{\mu(s,u,\vartheta)}{\mu(0,u,\vartheta)}\right)^2 \varrho^2(0,u)\left|\mathscr{X}^{[N]}\right|(0,u,\vartheta) \\
&\quad + 2(1+C\varepsilon)\int_{s=0}^t \left(\frac{\mu(t,u,\vartheta)}{\mu(s,u,\vartheta)}\right)^2 \varrho^2(s,u)\frac{[L\mu(s,u,\vartheta)]_-}{\mu(s,u,\vartheta)}\left|\mathfrak{X}^{[N]}\right|(s,u,\vartheta)\,ds \\
&\quad + \frac{1}{2}(1+C\varepsilon)\int_{s=0}^t \left(\frac{\mu(t,u,\vartheta)}{\mu(s,u,\vartheta)}\right)^2 \varrho^2(s,u)\mathrm{tr}_{\not{g}}\chi(s,u,\vartheta)\left|\mathfrak{X}^{[N]}\right|(s,u,\vartheta)\,ds \\
&\quad + 2(1+C\varepsilon) \\
&\qquad \times \int_{s=0}^t \left(\frac{\mu(t,u,\vartheta)}{\mu(s,u,\vartheta)}\right)^2 \varrho^2(s,u)\left|\hat{\chi}^{(Small)}\right|(s,u,\vartheta)\left|\mu\not{\mathcal{L}}_{\mathscr{S}}^{\leq N}\hat{\chi}^{(Small)}\right|(s,u,\vartheta)\,ds \\
&\quad + (1+C\varepsilon)\int_{s=0}^t \left(\frac{\mu(t,u,\vartheta)}{\mu(s,u,\vartheta)}\right)^2 \varrho^2(s,u)\left\{\left|\mathfrak{I}^{[N]}\right|(s,u,\vartheta) + \left|\widetilde{\mathfrak{I}}^N\right|(s,u,\vartheta)\right\}\,ds.
\end{aligned}
$$

Above, $\left|\mathscr{X}^{[N]}\right|$ denotes a term that is $\leq \sum_{|\vec{I}|=N} c_{\vec{I}}\left|{}^{(\mathscr{S}^{\vec{I}})}\mathscr{X}\right|$, where the $c_{\vec{I}}$ are nonnegative constants verifying $\sum_{|\vec{I}|=N} c_{\vec{I}} \leq 1$. Similarly, $\left|\mathfrak{X}^{[N]}\right|$ denotes a term that is $\leq \sum_{|\vec{I}|=N} c_{\vec{I}}\left|\mathscr{S}^{\vec{I}}\mathfrak{X}\right|$, where the $c_{\vec{I}}$ are nonnegative constants verifying $\sum_{|\vec{I}|=N} c_{\vec{I}} \leq 1$, and $\left|\mathfrak{I}^{[N]}\right|$ denotes a term that is $\leq \sum_{|\vec{I}|=N} c_{\vec{I}}\left|{}^{(\mathscr{S}^{\vec{I}})}\mathfrak{I}\right|$, where the $c_{\vec{I}}$ are nonnegative constants verifying $\sum_{|\vec{I}|=N} c_{\vec{I}} \leq 1$. Furthermore, $\widetilde{\mathfrak{I}}^N$ is an inhomogeneous *term that verifies the pointwise estimate*

$$
(17.29)\qquad \left|\widetilde{\mathfrak{J}}^N\right|(t,u,\vartheta) \leq C\varepsilon\frac{\ln(e+t)}{(1+t)^2}\left|\begin{pmatrix} \varrho L\mathscr{Z}^{\leq N}\Psi \\ \breve{R}\mathscr{Z}^{\leq N}\Psi \\ \varrho{d\!\!/}\mathscr{Z}^{\leq N}\Psi \\ \mathscr{Z}^{\leq N}\Psi \end{pmatrix}\right| + C\varepsilon\frac{\ln(e+t)}{(1+t)^3}\left|\begin{pmatrix} \mathscr{Z}^{\leq N}(\upmu-1) \\ \sum_{a=1}^3 \varrho\left|\mathscr{Z}^{\leq N}L_{(Small)}^a\right| \end{pmatrix}\right|.
$$

PROOF. We view the terms in the transport equation (11.33) as functions of (s,u,ϑ). We multiply both sides of (11.33) by the integrating factor $\iota(s,u,\vartheta) := \exp\left(\int_{t'=0}^s \left\{\mathrm{tr}_{g\!\!/}\chi - 2\frac{L\upmu}{\upmu}\right\}(t',u,\vartheta)\,dt'\right)$ (corresponding to the coefficient of ${}^{(\mathscr{S}^N)}\mathscr{X}$ on the left-hand side of (11.33)) and then integrate the resulting transport equation "$L(\iota{}^{(\mathscr{S}^N)}\mathscr{X}) = \iota\times$right-hand side (11.33)" along the integral curves of $L = \frac{\partial}{\partial s}$ from $s=0$ to $s=t$. With the help of the estimate (12.163), we see that

$$
\begin{aligned}
(17.30)\qquad \iota(s,u,\vartheta) &:= \exp\left(\int_{t'=0}^s \left\{\mathrm{tr}_{g\!\!/}\chi - 2\frac{L\upmu}{\upmu}\right\}(t',u,\vartheta)\,dt'\right) \\
&= \frac{\upmu^2(0,u,\vartheta)}{\upmu^2(s,u,\vartheta)}\exp\left(\int_{t'=0}^s \mathrm{tr}_{g\!\!/}\chi(t',u,\vartheta)\,dt'\right) \\
&= (1+\mathcal{O}(\varepsilon))\frac{\varrho^2(s,u)}{\varrho^2(0,u)}\frac{\upmu^2(0,u,\vartheta)}{\upmu^2(s,u,\vartheta)}.
\end{aligned}
$$

We also use the inequality $0 < \frac{1}{2}\mathrm{tr}_{g\!\!/}\chi - 2\frac{L\upmu}{\upmu} \leq 2\frac{[L\upmu]_-}{\upmu} + \frac{1}{2}\mathrm{tr}_{g\!\!/}\chi$, which follows from (13.142), in order to bound the magnitude of the coefficient of the factor $\mathscr{S}^N\mathfrak{X}$ on the right-hand side of (11.33). These steps lead to the pointwise estimate (17.28) for $\varrho^2\left|{}^{(\mathscr{S}^N)}\mathscr{X}\right|$ except that on the right-hand side of (17.28), in place of the terms $\left|\mathscr{X}^{[N]}\right|$, $\left|\mathfrak{X}^{[N]}\right|$, and $\left|\mathfrak{J}^{[N]}\right|$, there respectively appear $\left|{}^{(\mathscr{S}^N)}\mathscr{X}\right|$, $\left|\mathscr{S}^N\mathfrak{X}\right|$, and $\left|{}^{(\mathscr{S}^N)}\mathfrak{J}\right|$, and in place of the integrand term $\left|\widetilde{\mathfrak{J}}^N\right|$, there appears the commutator term $\left|\upmu[L,\mathscr{S}^N]\mathrm{tr}_{g\!\!/}\chi^{(Small)}\right|$ from the right-hand side of (11.33).

We now claim that the following estimate holds for the commutator term mentioned in the previous sentence:

$$
\begin{aligned}
(17.31)\qquad \left|\upmu[L,\mathscr{S}^N]\mathrm{tr}_{g\!\!/}\chi^{(Small)}\right| &\leq C\varepsilon\frac{\ln(e+t)}{(1+t)^2}\sum_{|\vec{I}|=N}\left|{}^{(\mathscr{S}^{\vec{I}})}\mathscr{X}\right| \\
&\quad + C\varepsilon\frac{\ln(e+t)}{(1+t)^2}\left|\begin{pmatrix} \varrho L\mathscr{Z}^{\leq N}\Psi \\ \breve{R}\mathscr{Z}^{\leq N}\Psi \\ \varrho{d\!\!/}\mathscr{Z}^{\leq N}\Psi \\ \mathscr{Z}^{\leq N}\Psi \end{pmatrix}\right| \\
&\quad + C\varepsilon\frac{\ln(e+t)}{(1+t)^3}\left|\begin{pmatrix} \mathscr{Z}^{\leq N}(\upmu-1) \\ \sum_{a=1}^3 \varrho\left|\mathscr{Z}^{\leq N}L_{(Small)}^a\right| \end{pmatrix}\right|.
\end{aligned}
$$

Once we have shown (17.31), in order to deduce the desired estimate (17.28), we first sum the inequalities verified by each fixed $\varrho^2\left|{}^{(\mathscr{S}^N)}\mathscr{X}\right|$ (which we obtained in the previous paragraph) in order to a deduce pointwise estimate for $\sum_{|\vec{I}|=N} \varrho^2 \left| {}^{(\mathscr{S}^{\vec{I}})}\mathscr{X} \right|$. Note that on the right-hand side of this pointwise estimate, the same quantity $\sum_{|\vec{I}|=N} \varrho^2 \left| {}^{(\mathscr{S}^{\vec{I}})}\mathscr{X} \right|$ appears under a time integral (which is generated by the first term on the right-hand side of (17.31)). We then use Gronwall's inequality to derive a pointwise estimate for the quantity $\sum_{|\vec{I}|=N} \varrho^2 \left| {}^{(\mathscr{S}^{\vec{I}})}\mathscr{X} \right|$ (in which the quantity $\sum_{|\vec{I}|=N} \varrho^2 \left| {}^{(\mathscr{S}^{\vec{I}})}\mathscr{X} \right|$ no longer appears on the right-hand side). We remark that in deriving the Gronwall estimate, we use the time-integrability of the factor $\varepsilon \ln(e + t)(1+t)^{-2}$ that multiplies $\sum_{|\vec{I}|=N} \left| {}^{(\mathscr{S}^{\vec{I}})}\mathscr{X} \right|$ on the right-hand side of (17.31). The summation over $|\vec{I}| = N$ turns the constants 1, 2, etc., on the right-hand side of (17.28) into large constants C, but in the next step, the large constants will be countered by the small factor of ε multiplying the first term on the right-hand side of (17.31). In the next step, we *revisit* the original inequality verified by the fixed $\varrho^2 \left| {}^{(\mathscr{S}^{N})}\mathscr{X} \right|$ (obtained in the previous paragraph) and insert the just-obtained Gronwall estimate for $\sum_{|\vec{I}|=N} \varrho^2 \left| {}^{(\mathscr{S}^{\vec{I}})}\mathscr{X} \right|$ into the first term on the right-hand side of (17.31). Using again the time-integrability of the factor $\varepsilon \ln(e+t)(1+t)^{-2}$ that multiplies $\sum_{|\vec{I}|=N} \left| {}^{(\mathscr{S}^{\vec{I}})}\mathscr{X} \right|$ on the right-hand side of (17.31) as well as the smallness factor ε, we conclude the desired estimate (17.28) (the smallness factor ε contributes to the $C\varepsilon$ factors on the right-hand side). The remaining terms on the right-hand side of (17.31) result in the presence of the term $\left| \widetilde{\mathfrak{J}}^N \right|$ on the right-hand side of (17.28).

It remains for us to prove (17.31). As a first step, we use the pointwise estimates $\mu \lesssim \ln(e + t)$ and $\left| \mathscr{Z}^{\leq 12} \mathrm{tr}_{\not{g}} \chi^{(Small)} \right| \lesssim \varepsilon \ln(e + t)(1 + t)^{-2}$ (which follow from (12.128a)), the commutator estimate (12.83) with $\mathrm{tr}_{\not{g}} \chi^{(Small)}$ in the role of f, the bound (12.127b) for $|L(\varrho^2 \mathscr{Z}^{\leq N-1} \mathrm{tr}_{\not{g}} \chi^{(Small)})|$, the identity $L\varrho = 1$, the bound (12.67a) for $|\mathscr{Z}^{\leq N-1} \mathrm{tr}_{\not{g}} \chi^{(Small)}|$, Cor. 12.48, and inequality (12.46a) to deduce that

(17.32)

$$
\begin{aligned}
\left| \mu[L, \mathscr{S}^N] \mathrm{tr}_{\not{g}} \chi^{(Small)} \right| &\lesssim \varepsilon \frac{\ln(e+t)}{(1+t)^2} \left| \begin{pmatrix} \mu \breve{R} \mathscr{Z}^{\leq N-1} \mathrm{tr}_{\not{g}} \chi^{(Small)} \\ \sum_{l=1}^{3} \left| \mu O_{(l)} \mathscr{Z}^{\leq N-1} \mathrm{tr}_{\not{g}} \chi^{(Small)} \right| \end{pmatrix} \right| \\
&+ \varepsilon \frac{\ln^2(e+t)}{(1+t)^3} \left| \begin{pmatrix} \varrho L \mathscr{Z}^{\leq N} \Psi \\ \breve{R} \mathscr{Z}^{\leq N} \Psi \\ \varrho \not{d} \mathscr{Z}^{\leq N} \Psi \\ \mathscr{Z}^{\leq N} \Psi \end{pmatrix} \right| \\
&+ \varepsilon \frac{\ln^2(e+t)}{(1+t)^4} \left| \begin{pmatrix} \mathscr{Z}^{\leq N}(\mu - 1) \\ \sum_{a=1}^{3} \varrho \left| \mathscr{Z}^{\leq N} L^a_{(Small)} \right| \end{pmatrix} \right|.
\end{aligned}
$$

Next, we consider the first array on the right-hand side of (17.32). We repeatedly use inequalities (12.81a) and (12.81b) with $f = \mathrm{tr}_{\not{g}} \chi^{(Small)}$, (12.128a), and (12.56) to commute any factor $Z := \varrho L$ in $\breve{R} \mathscr{Z}^{\leq N-1}$ or $O \mathscr{Z}^{\leq N-1}$ all the way in front (so that it acts last). We then use the estimates described just above inequality (17.32) to bound all derivatives of $\mathrm{tr}_{\not{g}} \chi^{(Small)}$ except the pure spatial top-order ones $\mathscr{S}^N$

in terms of other variables, thus deducing that

$$\varepsilon\frac{\ln(e+t)}{(1+t)^2}\left|\begin{pmatrix} \upmu \breve{R}\mathscr{Z}^{\leq N-1}\mathrm{tr}_{\not g}\chi^{(Small)} \\ \sum_{l=1}^3 \left|\upmu O_{(l)}\mathscr{Z}^{\leq N-1}\mathrm{tr}_{\not g}\chi^{(Small)}\right| \end{pmatrix}\right| \tag{17.33}$$
$$\lesssim \varepsilon\frac{\ln(e+t)}{(1+t)^2}\sum_{|\vec{I}|=N}\left|\upmu\mathscr{S}^{\vec{I}}\mathrm{tr}_{\not g}\chi^{(Small)}\right|$$
$$+\varepsilon\frac{\ln^2(e+t)}{(1+t)^3}\left|\begin{pmatrix} \varrho L\mathscr{Z}^{\leq N}\Psi \\ \breve{R}\mathscr{Z}^{\leq N}\Psi \\ \varrho \not d\mathscr{Z}^{\leq N}\Psi \\ \mathscr{Z}^{\leq N}\Psi \end{pmatrix}\right|$$
$$+\varepsilon\frac{\ln^2(e+t)}{(1+t)^4}\left|\begin{pmatrix} \mathscr{Z}^{\leq N}(\upmu-1) \\ \sum_{a=1}^3 \varrho\left|\mathscr{Z}^{\leq N}L^a_{(Small)}\right| \end{pmatrix}\right|.$$

Using the identity $\upmu\mathscr{S}^{\vec{I}}\mathrm{tr}_{\not g}\chi^{(Small)} = {}^{(\mathscr{S}^{\vec{I}})}\mathscr{X} - \mathscr{S}^{\vec{I}}\mathfrak{X}$ (see (11.32)), we bound the first term on the right-hand side of (17.33) as follows:

$$\sum_{|\vec{I}|=N}\left|\upmu\mathscr{S}^{\vec{I}}\mathrm{tr}_{\not g}\chi^{(Small)}\right| \lesssim \sum_{|\vec{I}|=N}\left|{}^{(\mathscr{S}^{\vec{I}})}\mathscr{X}\right| + \sum_{|\vec{I}|=N}\left|\mathscr{S}^{\vec{I}}\mathfrak{X}\right|. \tag{17.34}$$

Combining (17.32), (17.33), and (17.34), and using (17.1a) to bound the terms $\left|\mathscr{S}^{\vec{I}}\mathfrak{X}\right|$ on the right-hand side of (17.34), we finally arrive at the desired bound (17.31). We have thus proved Lemma 17.6. □

17.4. Pointwise estimates for the difficult error integrands requiring full modification

In the next proposition, we derive precise pointwise estimates for the difficult error integrand products $(\breve{R}\Psi)O\mathscr{S}^{N-1}\mathrm{tr}_{\not g}\chi^{(Small)}$ and $(\breve{R}\Psi)\not\Delta\mathscr{S}^{N-1}\upmu$ from Prop. 16.4, where $\mathscr{S}^{N-1}$ is an $(N-1)^{st}$ order pure spatial commutation vectorfield operator. More precisely, the last inequality in (15.1) allows us to estimate $(\breve{R}\Psi)\breve{R}\mathscr{S}^{N-1}\mathrm{tr}_{\not g}\chi^{(Small)}$ instead of $(\breve{R}\Psi)\not\Delta\mathscr{S}^{N-1}\upmu$. We must use the fully modified quantities defined in Sect. 11.2 to derive these estimates, which play an important role in our derivation of a priori estimates for the top-order L^2 quantity $\mathbb{Q}_{(\leq 24)}$. **We stress that the large $\upmu_\star^{-1}$-degeneracy of our a priori estimates for $\mathbb{Q}_{(\leq 24)}$ (see inequality (20.8e)) stems in part from the "boxed" constants appearing on the right-hand sides of the estimates of Prop. 17.7.**

PROPOSITION 17.7 (**Pointwise estimates for $(\breve{R}\Psi)O\mathscr{S}^{N-1}\mathrm{tr}_{\not g}\chi^{(Small)}$ and $(\breve{R}\Psi)\breve{R}\mathscr{S}^{N-1}\mathrm{tr}_{\not g}\chi^{(Small)}$**). *Let $1\leq N\leq 24$ be an integer, let $\mathscr{S}^{N-1}$ be an $(N-1)^{st}$ order pure spatial commutation vectorfield operator (see definition (7.2a)), and let $O\in\mathscr{O}$ be a geometric rotation vectorfield (see definition (7.2b)). Let $\chi^{(Small)}$ be the symmetric type $\binom{0}{2}$ $S_{t,u}$ tensorfield defined in (4.1c) and let $\varrho(t,u)=1-u+t$ be the geometric radial variable. Under the small-data and bootstrap assumptions of Sects. 12.1-12.4, if ε is sufficiently small, then the following pointwise estimates hold on $\mathcal{M}_{T_{(Bootstrap)},U_0}$, where $\left|\mathscr{X}^{[N]}\right|$ denotes a term that is $\leq\sum_{|\vec{I}|=N}c_{\vec{I}}\left|{}^{(\mathscr{S}^{\vec{I}})}\mathscr{X}\right|$, where ${}^{(\mathscr{S}^{\vec{I}})}\mathscr{X}$ is defined by (11.32) and the $c_{\vec{I}}$ are nonnegative constants verifying $\sum_{|\vec{I}|=N}c_{\vec{I}}\leq 1$:*

$$
\begin{aligned}
(17.35)\quad & \left|(\breve{R}\Psi)O\mathscr{S}^{N-1}\mathrm{tr}_{\not g}\chi^{(Small)}\right|\ \left|(\breve{R}\Psi)\breve{R}\mathscr{S}^{N-1}\mathrm{tr}_{\not g}\chi^{(Small)}\right| \\
& \leq \boxed{2}\left\|\frac{L\upmu}{\upmu}\right\|_{C^0(\Sigma_t^u)}\left|\breve{R}\mathscr{S}^N\Psi\right| \\
& + \boxed{4}\,(1+C\sqrt{\varepsilon})\frac{1}{\varrho(t,u)}\left\|\frac{L\upmu}{\upmu}\right\|_{C^0(\Sigma_t^u)} \\
& \times\int_{t'=0}^t\frac{\|[L\upmu]_-\|_{C^0(\Sigma_{t'}^u)}}{\upmu_\star(t',u)}\varrho(t',u)\left|\breve{R}\mathscr{S}^N\Psi\right|(t',u,\vartheta)\,dt' \\
& + \boxed{2}\,(1+C\varepsilon)\frac{1}{\varrho^2(t,u)}\left\|\frac{L\upmu}{\upmu}\right\|_{C^0(\Sigma_t^u)} \\
& \times\int_{t'=0}^t\left\|\left(\frac{\upmu(t,\cdot)}{\upmu}\right)^2\right\|_{C^0(\Sigma_{t'}^u)}\varrho(t',u)\left|\breve{R}\mathscr{S}^N\Psi\right|(t',u,\vartheta)\,dt' \\
& + C\varepsilon\upmu_\star^{-1}(t,u)\frac{\ln^2(e+t)}{(1+t)^3}\left|\mathscr{X}^{[N]}\right|(0,u,\vartheta) \\
& + C\varepsilon\upmu_\star^{-1/2}(t,u)\frac{1}{(1+t)^2}\left|\begin{pmatrix}\varrho\sqrt{\upmu}\left\{L+\frac{1}{2}\mathrm{tr}_{\not g}\chi\right\}\mathscr{S}^{\leq N}\Psi\\ \breve{R}\mathscr{S}^{\leq N}\Psi\\ \varrho\sqrt{\upmu}\not d\mathscr{S}^{\leq N}\Psi\end{pmatrix}\right| \\
& + C\varepsilon\upmu_\star^{-1}(t,u)\frac{1}{(1+t)}\left|\begin{pmatrix}\breve{R}\mathscr{Z}^{\leq N-1}\Psi\\ \mathscr{Z}^{\leq N-1}\Psi\end{pmatrix}\right| \\
& + C\varepsilon\upmu_\star^{-3/2}(t,u)\frac{\ln(e+t)}{(1+t)^2}\left|\begin{pmatrix}\varrho\sqrt{\upmu}\left\{L+\frac{1}{2}\mathrm{tr}_{\not g}\chi\right\}\mathscr{Z}^{\leq N-1}\Psi\\ \varrho\sqrt{\upmu}\not d\mathscr{Z}^{\leq N-1}\Psi\end{pmatrix}\right| \\
& + C\varepsilon\upmu_\star^{-1}(t,u)\frac{1}{(1+t)^3}\left|\begin{pmatrix}\mathscr{Z}^{\leq N}(\upmu-1)\\ \sum_{a=1}^3\varrho\left|\mathscr{Z}^{\leq N}L_{(Small)}^a\right|\end{pmatrix}\right| \\
& + C\varepsilon\upmu_\star^{-1}(t,u)\frac{\ln^3(e+t)}{(1+t)^3} \\
& \times\int_{t'=0}^t\frac{1}{\upmu_\star(t',u)}\left|\begin{pmatrix}\varrho\sqrt{\upmu}\left\{L+\frac{1}{2}\mathrm{tr}_{\not g}\chi\right\}\mathscr{Z}^{\leq N}\Psi\\ \breve{R}\mathscr{Z}^{\leq N}\Psi\\ \varrho\sqrt{\upmu}\not d\mathscr{Z}^{\leq N}\Psi\end{pmatrix}\right|(t',u,\vartheta)\,dt' \\
& + C\varepsilon\upmu_\star^{-1}(t,u)\frac{\ln^2(e+t)}{(1+t)^3}\int_{t'=0}^t\varrho(t',u)\frac{1}{\upmu_\star(t',u)}\left|\begin{pmatrix}\breve{R}\mathscr{Z}^{\leq N-1}\Psi\\ \mathscr{Z}^{\leq N-1}\Psi\end{pmatrix}\right|(t',u,\vartheta)\,dt' \\
& + C\varepsilon\upmu_\star^{-1}(t,u)\frac{\ln^3(e+t)}{(1+t)^3} \\
& \times\int_{t'=0}^t\frac{1}{\upmu_\star^{3/2}(t',u)}\left|\begin{pmatrix}\varrho\sqrt{\upmu}\left\{L+\frac{1}{2}\mathrm{tr}_{\not g}\chi\right\}\mathscr{Z}^{\leq N-1}\Psi\\ \varrho\sqrt{\upmu}\not d\mathscr{Z}^{\leq N-1}\Psi\end{pmatrix}\right|(t',u,\vartheta)\,dt' \\
& + C\varepsilon\upmu_\star^{-1}(t,u)\frac{\ln^3(e+t)}{(1+t)^3}\int_{t'=0}^t\left|\begin{pmatrix}\upmu\hat{\not\nabla}^2\mathscr{S}^{\leq N-1}\upmu\\ \varrho\upmu\not\nabla\not\mathcal{L}_{\mathscr{S}}^{\leq N-1}\hat{\chi}^{(Small)}\end{pmatrix}\right|(t',u,\vartheta)\,dt' \\
& + C\varepsilon\upmu_\star^{-1}(t,u)\frac{\ln^2(e+t)}{(1+t)^3} \\
& \times\int_{t'=0}^t\frac{1}{(1+t')}\frac{1}{\upmu_\star(t',u)}\left|\begin{pmatrix}\mathscr{Z}^{\leq N}(\upmu-1)\\ \sum_{a=1}^3\varrho\left|\mathscr{Z}^{\leq N}L_{(Small)}^a\right|\end{pmatrix}\right|(t',u,\vartheta)\,dt'.
\end{aligned}
$$

PROOF. For the sake of efficiency, we prove inequality (17.35) only for the term $(\breve{R}\Psi)O\mathscr{S}^{N-1}\mathrm{tr}_{\not g}\chi^{(Small)}$; the proof for the term $(\breve{R}\Psi)\breve{R}\mathscr{S}^{N-1}\mathrm{tr}_{\not g}\chi^{(Small)}$ is identical. Throughout the proof, we use the notation $\mathscr{S}^N$ whenever the precise structure of an Nth order string of spatial commutation vectorfields is not important. Using Def. 11.9 (see also (11.20b)), we deduce that

(17.36)
$$\begin{aligned}
&\left|\upmu O\mathscr{S}^{N-1}\mathrm{tr}_{\not g}\chi^{(Small)} - {}^{(O\mathscr{S}^{N-1})}\mathscr{X}\right| \\
&\leq \left|G_{LL}\breve{R}\mathscr{S}^N\Psi\right| + \frac{1}{2}\left|\upmu \not G_A^{\ A}\, L\mathscr{S}^N\Psi\right| + \frac{1}{2}\left|\upmu G_{LL}L\mathscr{S}^N\Psi\right| + \left|\upmu \not G_L^{\ A}\, \not d_A\mathscr{S}^N\Psi\right| \\
&\quad + \left|\mathscr{S}^N\mathfrak{X} - \left\{-G_{LL}\breve{R}\mathscr{S}^N\Psi - \frac{1}{2}\upmu \not G_A^{\ A}\, L\mathscr{S}^N\Psi - \frac{1}{2}\upmu G_{LL}L\mathscr{S}^N\Psi + \upmu \not G_L^{\ A}\, \not d_A\mathscr{S}^N\Psi\right\}\right|.
\end{aligned}$$

From (17.36), the bootstrap assumption $\left\|\breve{R}\Psi\right\|_{C^0(\Sigma_t^u)} \leq \varepsilon(1+t)^{-1}$ (that is, $(\mathbf{BA}\Psi)$), and inequalities (12.58b) and (17.2), we deduce that

(17.37)
$$\begin{aligned}
&\left|(\breve{R}\Psi)O\mathscr{S}^{N-1}\mathrm{tr}_{\not g}\chi^{(Small)}\right| \\
&\leq \frac{1}{\varrho^2}\left|\frac{\breve{R}\Psi}{\upmu}(\varrho^2\,{}^{(O\mathscr{S}^{N-1})}\mathscr{X})\right| + \left|G_{LL}\frac{\breve{R}\Psi}{\upmu}\breve{R}\mathscr{S}^N\Psi\right| + C\varepsilon\frac{1}{(1+t)^2}\left|\begin{pmatrix}\varrho L\mathscr{S}^N\Psi \\ \varrho\not d\mathscr{S}^N\Psi\end{pmatrix}\right| \\
&\quad + C\varepsilon\frac{\ln(e+t)}{(1+t)^2}\frac{1}{\upmu}\left|\begin{pmatrix}\varrho L\mathscr{Z}^{\leq N-1}\Psi \\ \varrho\not d\mathscr{Z}^{\leq N-1}\Psi \\ \mathscr{Z}^{\leq N-1}\Psi\end{pmatrix}\right| + C\varepsilon\frac{1}{(1+t)}\frac{1}{\upmu}\left|\breve{R}\mathscr{Z}^{\leq N-1}\Psi\right| \\
&\quad + C\varepsilon\frac{1}{(1+t)^3}\frac{1}{\upmu}\left|\begin{pmatrix}\mathscr{Z}^{\leq N}(\upmu-1) \\ \sum_{a=1}^3\varrho\left|\mathscr{Z}^{\leq N}L_{(Small)}^a\right|\end{pmatrix}\right|.
\end{aligned}$$

We now bound the right-hand side of (17.37) by the right-hand side of (17.35) by using a separate argument for each term. Throughout the remainder of this proof, we use the definition of $\upmu_\star$ and in particular the estimate $\upmu_\star \leq 1$ without mentioning it each time. We also silently use the simple inequality $|Lf| \leq \left|Lf + \frac{1}{2}\mathrm{tr}_{\not g}\chi f\right| + C(1+t)^{-1}|f|$, which follows easily from the estimate $|\varrho\mathrm{tr}_{\not g}\chi| \lesssim 1$ (that is, (12.60b)). The first term on the right-hand side of (17.37) is the most difficult, and we handle it at the end of the proof. To bound the second term on the right-hand side of (17.37), we first use the transport equation (4.5) for $L\upmu$, the bootstrap assumption $\|\varrho L\Psi\|_{C^0(\Sigma_t^u)} \leq \varepsilon(1+t)^{-1}$ (that is, $(\mathbf{BA}\Psi)$), and inequality (12.58b) to deduce that

$$(17.38)\qquad \left|G_{LL}\frac{\breve{R}\Psi}{\upmu}\breve{R}\mathscr{S}^N\Psi\right| \leq 2\left\|\frac{L\upmu}{\upmu}\right\|_{C^0(\Sigma_t^u)}\left|\breve{R}\mathscr{S}^N\Psi\right| + C\varepsilon\frac{1}{(1+t)^2}\left|\breve{R}\mathscr{S}^N\Psi\right|.$$

The first term on the right-hand side of (17.38) accounts for the first term on the right-hand side of (17.35), and the second term is also easily seen to be $\leq$ the right-hand side of (17.35). The last term on the first line of the right-hand side of (17.37) and the terms on the second and third lines of the right-hand side of (17.37) are also easily seen to be $\leq$ the right-hand side of (17.35) as desired.

Bounding the difficult term piece by piece. To bound the (difficult) first term on the right-hand side of (17.37) by the right-hand side of (17.35), we use the preliminary pointwise estimates provided by Lemma 17.6, which provide a pointwise bound for $\varrho^{2}(\mathscr{S}^N)\mathscr{X}$. Specifically, we see that it suffices to bound $\frac{1}{\varrho^2}\frac{\breve{R}\Psi}{\upmu}$ times the right-hand side of (17.28); we derive the desired bounds by arguing term by term.

Bound for the term arising from the first term on the right-hand side of (17.28): To bound $\frac{1}{\varrho^2}\frac{\breve{R}\Psi}{\upmu}$ times the first term on the right-hand side of (17.28), we use the bootstrap assumption $\left\|\breve{R}\Psi\right\|_{C^0(\Sigma_t^u)} \leq \varepsilon(1+t)^{-1}$ (that is, (**BA**Ψ)) and the bound $|\upmu - 1| \lesssim \varepsilon \ln(e+t)$ (that is, (12.128a)) to deduce that

$$\begin{aligned} (17.39)\qquad & \left|\frac{1}{\varrho^2(t,u)}\frac{\breve{R}\Psi(t,u,\vartheta)}{\upmu(t,u,\vartheta)} \max_{0\leq s\leq t}\left(\frac{\upmu(s,u,\vartheta)}{\upmu(0,u,\vartheta)}\right)^2 \varrho^2(0,u)\right| |\mathscr{X}^{[N]}|(0,u,\vartheta) \\ & \leq C\varepsilon\frac{1}{\upmu_\star(t,u)}\frac{\ln^2(e+t)}{(1+t)^3}|\mathscr{X}^{[N]}|(0,u,\vartheta), \end{aligned}$$

where $|\mathscr{X}^{[N]}|$ is defined just below (17.28). Clearly the right-hand side of (17.39) is $\leq$ the right-hand side of (17.35) as desired.

Bound for the term arising from the first time integral on the right-hand side of (17.28): We now address the first time integral on the right-hand side of (17.28). To this end, we first note that by the definition of $\left|\mathfrak{X}^{[N]}\right|$, which is defined just below (17.28), the estimate (12.54) (with Ψ in the role of f) the estimate (12.58b), and inequality (17.2), we have the following bound for one of the integrand factors:

$$\begin{aligned} (17.40)\qquad \left|\mathfrak{X}^{[N]}\right| & \leq |G_{LL}\breve{R}\mathscr{S}^N\Psi| + C\frac{1}{1+t}\upmu\left|\begin{pmatrix} \varrho\left\{L+\frac{1}{2}\mathrm{tr}_{g\!\!/}\chi\right\}\mathscr{S}^{\leq N}\Psi \\ \varrho d\!\!/ \mathscr{S}^{\leq N}\Psi \end{pmatrix}\right| \\ & + C\frac{\ln(e+t)}{1+t}\left|\begin{pmatrix} \varrho\left\{L+\frac{1}{2}\mathrm{tr}_{g\!\!/}\chi\right\}\mathscr{Z}^{\leq N-1}\Psi \\ \varrho d\!\!/ \mathscr{Z}^{\leq N-1}\Psi \\ \mathscr{Z}^{\leq N-1}\Psi \end{pmatrix}\right| + C\left|\breve{R}\mathscr{Z}^{\leq N-1}\Psi\right| \\ & + C\varepsilon\frac{1}{(1+t)^2}\left|\begin{pmatrix} \mathscr{Z}^{\leq N}(\upmu-1) \\ \sum_{a=1}^3 \varrho\left|\mathscr{Z}^{\leq N}L_{(Small)}^a\right| \end{pmatrix}\right|. \end{aligned}$$

Next, we pointwise bound the product of the first and third integrand factors in the first time integral on the right-hand side of (17.28) by using Def. 13.8 and the estimates (13.38b), (13.41), and (13.42) to deduce that for $0 \leq s \leq t$, we have

(17.41)

$$\left(\frac{\upmu(t,u,\vartheta)}{\upmu(s,u,\vartheta)}\right)^2\frac{[L\upmu(s,u,\vartheta)]_-}{\upmu(s,u,\vartheta)} \leq (1-C\sqrt{\varepsilon})\frac{[L\upmu(s,u,\vartheta)]_-}{\upmu(s,u,\vartheta)} + C\varepsilon\ln^2(e+t)\frac{\ln(e+s)}{(1+s)^2}.$$

We now use (17.40), (17.41), (12.58b), the estimate $|[L\upmu]_-|(s,u,\vartheta) \leq |L\upmu|(s,u,\vartheta) \lesssim \varepsilon(1+s)^{-1}$, (which follows from (12.128b)), the bootstrap assumption $\left\|\breve{R}\Psi\right\|_{C^0(\Sigma_t^u)} \leq \varepsilon(1+t)^{-1}$ (that is, (**BA**Ψ)), and the definition of $\upmu_\star$ to deduce that $\frac{1}{\varrho^2}\frac{\breve{R}\Psi}{\upmu}$ times the first time integral on the right-hand side of (17.28) is bounded in magnitude by

(17.42)

$$
\begin{aligned}
&\leq 2(1+C\varepsilon)\frac{1}{\upmu}\frac{1}{\varrho^2}|\breve{R}\Psi|\int_{s=0}^{t}\left(\frac{\upmu(t,u,\vartheta)}{\upmu(s,u,\vartheta)}\right)^2\varrho^2(s,u)\frac{[L\upmu(s,u,\vartheta)]_-}{\upmu(s,u,\vartheta)}\left|\mathfrak{X}^{[N]}\right|(s,u,\vartheta)\,ds \\
&\leq 2(1+C\sqrt{\varepsilon})\frac{1}{\upmu}\frac{1}{\varrho^2}|\breve{R}\Psi|\int_{s=0}^{t}\varrho^2(s,u)\frac{\|[L\upmu]_-\|_{C^0(\Sigma_s^u)}}{\upmu_\star(s,u)}\left|G_{LL}\breve{R}\mathscr{S}^N\Psi\right|(s,u,\vartheta)\,ds \\
&\quad+C\varepsilon\frac{1}{\upmu_\star}\frac{\ln^3(e+t)}{(1+t)^3}\int_{s=0}^{t}\left|\breve{R}\mathscr{S}^N\Psi\right|(s,u,\vartheta)\,ds \\
&\quad+C\varepsilon\frac{1}{\upmu_\star}\frac{1}{(1+t)^3}\int_{s=0}^{t}\left|\begin{pmatrix}\varrho\left\{L+\frac{1}{2}\mathrm{tr}_{\not g}\upchi\right\}\mathscr{S}^{\leq N}\Psi\\ \varrho\not d\mathscr{S}^{\leq N}\Psi\end{pmatrix}\right|(s,u,\vartheta)\,ds \\
&\quad+C\varepsilon\frac{1}{\upmu_\star}\frac{1}{(1+t)^3}\int_{s=0}^{t}\varrho(s,u)\frac{1}{\upmu_\star(s,u)}\left|\breve{R}\mathscr{Z}^{\leq N-1}\Psi\right|(s,u,\vartheta)\,ds \\
&\quad+C\varepsilon\frac{1}{\upmu_\star}\frac{\ln(e+t)}{(1+t)^3}\int_{s=0}^{t}\frac{1}{\upmu_\star(s,u)}\left|\begin{pmatrix}\varrho\left\{L+\frac{1}{2}\mathrm{tr}_{\not g}\upchi\right\}\mathscr{Z}^{\leq N-1}\Psi\\ \varrho\not d\mathscr{Z}^{\leq N-1}\Psi\\ \mathscr{Z}^{\leq N-1}\Psi\end{pmatrix}\right|(s,u,\vartheta)\,ds \\
&\quad+C\varepsilon\frac{1}{\upmu_\star}\frac{1}{(1+t)^3}\int_{s=0}^{t}\frac{1}{(1+s)}\frac{1}{\upmu_\star(s,u)}\left|\begin{pmatrix}\mathscr{Z}^{\leq N}(\upmu-1)\\ \sum_{a=1}^3\varrho\left|\mathscr{Z}^{\leq N}L_{(Small)}^a\right|\end{pmatrix}\right|(s,u,\vartheta)\,ds.
\end{aligned}
$$

The main challenge is to bound the (difficult) first product $2(1+C\sqrt{\varepsilon})\cdots$ on the right-hand side of (17.42) by the right-hand side of (17.35); it is straightforward to see that the remaining products on the right-hand side of (17.42) are $\leq$ the right-hand side of (17.35) as desired. To bound the difficult first product on the right-hand side of (17.42), we first use the estimate (12.142), the estimate $\|[L\upmu]_-\|_{C^0(\Sigma_s^u)}\lesssim\varepsilon(1+s)^{-1}$ (which follows from (12.128b)) and the bootstrap assumption $\left\|\breve{R}\Psi\right\|_{C^0(\Sigma_t^u)}\leq\varepsilon(1+t)^{-1}$ (that is, (**BA**Ψ)) to replace the integrand factor $G_{LL}(s,u,\vartheta)$ with $G_{LL}(t,u,\vartheta)$ up to error terms, which allows us to bound the first product by

(17.43)

$$
\begin{aligned}
&\leq 2(1+C\sqrt{\varepsilon})\frac{1}{\varrho}\frac{1}{\upmu}|\breve{R}\Psi||G_{LL}|\int_{s=0}^{t}\varrho(s,u)\frac{\|[L\upmu]_-\|_{C^0(\Sigma_s^u)}}{\upmu_\star(s,u)}\left|\breve{R}\mathscr{S}^N\Psi\right|(s,u,\vartheta)\,ds \\
&\quad+C\varepsilon\frac{1}{(1+t)^3}\frac{1}{\upmu_\star}\int_{s=0}^{t}\frac{1}{\upmu_\star(s,u)}\left|\breve{R}\mathscr{S}^N\Psi\right|(s,u,\vartheta)\,ds.
\end{aligned}
$$

Clearly, the product on the second line of the right-hand side of (17.43) is $\leq$ the right-hand side of (17.35) as desired. To bound the product on the first line of the right-hand side of (17.43), we use essentially the same reasoning that we used to deduce (17.38) in order to bound it by

(17.44)

$$
\begin{aligned}
&\leq 4(1+C\sqrt{\varepsilon})\frac{1}{\varrho}\left\|\frac{L\upmu}{\upmu}\right\|_{C^0(\Sigma_t^u)}\int_{s=0}^{t}\varrho(s,u)\frac{\|[L\upmu]_-\|_{C^0(\Sigma_s^u)}}{\upmu_\star(s,u)}\left|\breve{R}\mathscr{S}^N\Psi\right|(s,u,\vartheta)\,ds \\
&\quad+C\varepsilon\frac{1}{(1+t)^3}\int_{s=0}^{t}\frac{1}{\upmu_\star(s,u)}\left|\breve{R}\mathscr{S}^N\Psi\right|(s,u,\vartheta)\,ds.
\end{aligned}
$$

Finally, we observe that both terms on the right-hand side of (17.44) are $\leq$ the right-hand side of (17.35) as desired.

Bound for the term arising from the second time integral on the right-hand side of (17.28): We now bound $\frac{1}{\varrho^2}\frac{\breve{R}\Psi}{\upmu}$ times the second time integral on the right-hand side of (17.28). We begin by using definition (4.1c) and the estimate (12.128a) to pointwise bound the integrand factor as follows: ${\mathrm{tr}}_{\not{g}}\chi(s,u,\vartheta) = 2\varrho^{-1}(s,u) + {\mathrm{tr}}_{\not{g}}\chi^{(Small)}(s,u,\vartheta) \leq 2(1+C\varepsilon)\varrho^{-1}(s,u)$. Next, we substitute this bound and the bound (17.40) for the integrand factor $\left|\mathfrak{X}^{[N]}\right|$ into the second time integral. We give a separate argument for each term on the right-hand side of (17.40). First, using the above estimate for ${\mathrm{tr}}_{\not{g}}\chi$, we bound $\frac{1}{\varrho^2}\frac{\breve{R}\Psi}{\upmu}$ times the part of the second time integral that arises from the first term on the right-hand side of (17.40) by

$$\leq (1+C\varepsilon)\frac{1}{\varrho^2}\frac{1}{\upmu}|\breve{R}\Psi| \int_{s=0}^{t} \left(\frac{\upmu(t,u,\vartheta)}{\upmu(s,u,\vartheta)}\right)^2 \varrho(s,u) \left|G_{LL}\breve{R}\mathscr{S}^N\Psi\right|(s,u,\vartheta)\,ds. \tag{17.45}$$

We now use arguments similar to the ones we used above in order to replace the integrand factor $G_{LL}(s,u,\vartheta)$ with $G_{LL}(t,u,\vartheta)$ in (17.45) up to error terms and to then bound the product $\frac{1}{\upmu}|G_{LL}||\breve{R}\Psi|$ by $2\left|\frac{L\upmu}{\upmu}\right|$ up to error terms. This line of reasoning allows us to bound the right-hand side of (17.45) by the product $\boxed{2}(1 + C\varepsilon)\frac{1}{\varrho^2(t,u)}\cdots$ on the right-hand side of (17.35) plus other nonboxed-constant-multiplied error terms on the right-hand side of (17.35).

Next, we use the estimate $|\varrho{\mathrm{tr}}_{\not{g}}\chi| \lesssim 1$ noted above and the bootstrap assumption $\left\|\breve{R}\Psi\right\|_{C^0(\Sigma_t^u)} \leq \varepsilon(1+t)^{-1}$ to bound $\frac{1}{\varrho^2}\frac{\breve{R}\Psi}{\upmu}$ times the part of the second time integral that arises from the second product on the right-hand side of (17.40) by

$$\leq C\varepsilon\frac{1}{(1+t)^3}\int_{s=0}^{t}\frac{\upmu(t,u,\vartheta)}{\upmu(s,u,\vartheta)}\left|\begin{pmatrix} \varrho\left\{L+\frac{1}{2}{\mathrm{tr}}_{\not{g}}\chi\right\}\mathscr{S}^{\leq N}\Psi \\ \varrho\not{d}\mathscr{S}^{\leq N}\Psi\end{pmatrix}\right|(s,u,\vartheta)\,ds. \tag{17.46}$$

Furthermore, from Def. 13.8 and the estimates (13.38b) and (13.42), it follows that the factor $\frac{\upmu(t,u,\vartheta)}{\upmu(s,u,\vartheta)}$ in (17.46) is $\lesssim \ln(e+t)$. In total, we see that (17.46) is $\leq$ the right-hand side of (17.35) as desired.

Using similar reasoning, we bound $\frac{1}{\varrho^2}\frac{\breve{R}\Psi}{\upmu}$ times the part of the second time integral that arises from the third, fourth, and fifth products on the right-hand side of (17.40) respectively by

$$\leq C\varepsilon\frac{1}{\upmu_\star}\frac{\ln^3(e+t)}{(1+t)^3}\int_{s=0}^{t}\left|\begin{pmatrix} \varrho\left\{L+\frac{1}{2}{\mathrm{tr}}_{\not{g}}\chi\right\}\mathscr{Z}^{\leq N-1}\Psi \\ \varrho\not{d}\mathscr{Z}^{\leq N-1}\Psi \\ \mathscr{Z}^{\leq N-1}\Psi\end{pmatrix}\right|(s,u,\vartheta)\,ds, \tag{17.47}$$

$$\leq C\varepsilon\frac{1}{\upmu_\star}\frac{\ln^2(e+t)}{(1+t)^3}\int_{s=0}^{t}\varrho(s,u)\left|\breve{R}\mathscr{Z}^{N-1}\Psi\right|(s,u,\vartheta)\,ds, \tag{17.48}$$

and

$$(17.49) \quad \leq C\varepsilon \frac{1}{\upmu_\star} \frac{\ln^2(e+t)}{(1+t)^3} \int_{s=0}^t \frac{1}{1+s} \left| \begin{pmatrix} \mathscr{Z}^{\leq N}(\upmu - 1) \\ \sum_{a=1}^3 \varrho \left| \mathscr{Z}^{\leq N} L_{(Small)}^a \right| \end{pmatrix} \right| (s,u,\vartheta)\, ds,$$

which are easily seen to be $\leq$ the right-hand side of (17.35) as desired.

Bound for the term arising from the third time integral on the right-hand side of (17.28): We now pointwise bound $\frac{1}{\varrho^2} \frac{\breve{R}\Psi}{\upmu}$ times the third time integral on the right-hand side of (17.28). We first note that (12.128a) implies the pointwise bound $\left|\hat{\upchi}^{(Small)}\right|(s,u,\vartheta) \lesssim \epsilon \ln(e+s)(1+s)^{-2}$. Also using the bootstrap assumption $\left\|\breve{R}\Psi\right\|_{C^0(\Sigma_t^u)} \leq \varepsilon(1+t)^{-1}$ (that is, $(\mathbf{BA}\Psi)$), we deduce that $\frac{1}{\varrho^2}\frac{\breve{R}\Psi}{\upmu}$ times the third time integral on the right-hand side of (17.28) is bounded in magnitude by

$$(17.50) \qquad \leq C\varepsilon \frac{\ln(e+t)}{(1+t)^3} \frac{1}{\upmu(t,u,\vartheta)} \int_{s=0}^t \frac{\upmu^2(t,u,\vartheta)}{\upmu^2(s,u,\vartheta)} \left| \upmu \not{\mathcal{L}}_{\mathscr{S}}^{\leq N} \hat{\upchi}^{(Small)} \right| (s,u,\vartheta)\, ds.$$

We now notice that the analysis of the right-hand side of (17.50) easily reduces to the case of the top-order operators $\not{\mathcal{L}}_{\mathscr{S}}^N$; in the cases of the lower-order operators $\not{\mathcal{L}}_{\mathscr{S}}^{\leq N-1}$, we use inequality (12.67a) with $\leq N-1$ in the role of N and the pointwise bound $\left|\frac{\upmu(t,u,\vartheta)}{\upmu(s,u,\vartheta)}\right| \lesssim \ln(e+t)$, (which holds for $0 \leq s \leq t$ and follows from the line of reasoning just below (17.46)) to bound the below-top-order terms on the right-hand side of (17.50) by $\leq$ the right-hand side of (17.35) as desired.

We now consider two cases depending on the structure of the top-order operators $\not{\mathcal{L}}_{\mathscr{S}}^N$ in (17.50): **i)** $\mathscr{S}^N = \breve{R}\mathscr{S}^{N-1}$, and **ii)** $\mathscr{S}^N = O\mathscr{S}^{N-1}$, where $O \in \mathscr{O}$ is a geometric rotation vectorfield. If $\mathscr{S}^N = \breve{R}\mathscr{S}^{N-1}$, then we use the third inequality in (15.1), the estimate $|\varrho \mathrm{tr}_{\not{g}}\upchi| \lesssim 1$ noted above, and the pointwise bound $\frac{\upmu(t,u,\vartheta)}{\upmu(s,u,\vartheta)} \lesssim \ln(e+t)$ (which holds for $0 \leq s \leq t$) noted above to deduce that the right-hand side of (17.50) is

$$\begin{aligned}
(17.51) \quad & \leq C\varepsilon \frac{\ln^3(e+t)}{(1+t)^3} \frac{1}{\upmu_\star(t,u)} \int_{s=0}^t \left| \upmu \hat{\not{\nabla}}^2 \mathscr{S}^{\leq N-1} \upmu \right| (s,u,\vartheta)\, ds \\
& + C\varepsilon \frac{\ln^2(e+t)}{(1+t)^3} \int_{s=0}^t \frac{1}{1+s} \left| \begin{pmatrix} \varrho\left\{L + \frac{1}{2}\mathrm{tr}_{\not{g}}\upchi\right\} \mathscr{Z}^{\leq N}\Psi \\ \breve{R}\mathscr{Z}^{\leq N}\Psi \\ \varrho \not{d}\mathscr{Z}^{\leq N}\Psi \\ \mathscr{Z}^{\leq N}\Psi \end{pmatrix} \right| (s,u,\vartheta)\, ds \\
& + C\varepsilon \frac{\ln^2(e+t)}{(1+t)^3} \int_{s=0}^t \frac{1}{(1+s)^2} \left| \begin{pmatrix} \mathscr{Z}^{\leq N}(\upmu - 1) \\ \sum_{a=1}^3 \varrho \left| \mathscr{Z}^{\leq N} L_{(Small)}^a \right| \end{pmatrix} \right| (s,u,\vartheta)\, ds.
\end{aligned}$$

It is easy to see that the right-hand side of (17.51) is $\leq$ the right-hand side of (17.35) as desired.

If $\mathscr{S}^N = O\mathscr{S}^{N-1}$, then we first use inequality (12.49) to deduce that

$$\begin{aligned} &\left|\upmu \not{\mathcal{L}}_{\mathscr{S}}^N \hat{\chi}^{(Small)}\right|(s,u,\vartheta) \\ &\lesssim \varrho \left|\upmu \not\nabla \not{\mathcal{L}}_{\mathscr{S}}^{\leq N-1} \hat{\chi}^{(Small)}\right|(s,u,\vartheta) + \left|\upmu \not{\mathcal{L}}_{\mathscr{S}}^{\leq N-1} \hat{\chi}^{(Small)}\right|(s,u,\vartheta). \end{aligned} \tag{17.52}$$

Using inequality (17.52), reasoning as in the previous case $\mathscr{S}^N = \breve{R}\mathscr{S}^{N-1}$, and using inequality (12.67a) with $N-1$ in the role of N to pointwise bound the second (lower-order) term on the right-hand side of (17.52), we conclude that in the present case $\mathscr{S}^N = O\mathscr{S}^{N-1}$, the right-hand side of (17.50) is $\leq$ the right-hand side of (17.35) as desired.

Bound for the term arising from the final time integral on the right-hand side of (17.28): We now pointwise bound $\frac{1}{\varrho^2}\frac{\breve{R}\Psi}{\upmu}$ times the final time integral on the right-hand side of (17.28). The integrand terms $\left|\mathfrak{J}^{[N]}\right|(s,u,\vartheta)$ and $\left|\widetilde{\mathfrak{J}}^N\right|(s,u,\vartheta)$ have already been suitably bounded by virtue of the definition of $\left|\mathfrak{J}^{[N]}\right|(s,u,\vartheta)$, which is defined just below (17.28), and the estimates (17.9) and (17.29) (with s in the role of t in these estimates). The bootstrap assumption $\left\|\breve{R}\Psi\right\|_{C^0(\Sigma_t^u)} \leq \varepsilon(1+t)^{-1}$ (that is, (**BAΨ**)) yields the pointwise estimate

$$\left|\frac{1}{\varrho^2}\frac{\breve{R}\Psi}{\upmu}\right| \lesssim \varepsilon \frac{1}{\upmu}(1+t)^{-3}.$$

Furthermore, earlier in the proof, we proved the integrand factor estimate $\frac{\upmu(t,u,\vartheta)}{\upmu(s,u,\vartheta)} \lesssim \ln(e+t)$. Also accounting for the integrand factor $\varrho^2(s,u)$, we see that these estimates imply that $\frac{1}{\varrho^2}\frac{\breve{R}\Psi}{\upmu}$ times the final time integral on the right-hand side of (17.28) is in magnitude $\leq$ the right-hand side of (17.35) as desired. □

In the next lemma, we provide a cruder version of Prop. 17.7 that is useful for bounding some of the top-order terms that we encounter in our analysis.

LEMMA 17.8 (**Cruder pointwise estimates for $\upmu O\mathscr{S}^{N-1}\mathrm{tr}_{\not g}\chi^{(Small)}$, $\upmu\breve{R}\mathscr{S}^{N-1}\mathrm{tr}_{\not g}\chi^{(Small)}$, and $\upmu\not\Delta\mathscr{S}^{N-1}\upmu$**). *Let $1 \leq N \leq 24$ be an integer, let $\mathscr{S}^{N-1}$ be an $(N-1)^{st}$ order pure spatial commutation vectorfield operator (see definition* (7.2a)*), and let $O \in \mathscr{O}$ be a geometric rotation vectorfield (see definition* (7.2b)*). Let $\chi^{(Small)}$ be the symmetric type $\binom{0}{2}$ $S_{t,u}$ tensorfield defined in* (3.70). *Under the small-data and bootstrap assumptions of Sects.* 12.1-12.4, *if ε is sufficiently small,* *then the following pointwise estimates hold on* $\mathcal{M}_{T_{(Bootstrap)},U_0}$:

(17.53)

$$
\begin{aligned}
& \left| \upmu O \mathscr{S}^{N-1} \mathrm{tr}_{\not g} \chi^{(Small)} \right|, \\
& \left| \upmu \breve{R} \mathscr{S}^{N-1} \mathrm{tr}_{\not g} \chi^{(Small)} \right|, \\
& \left| \upmu \not\Delta \mathscr{S}^{N-1} \upmu \right| \\
& \lesssim \left| \begin{pmatrix} \upmu L \mathscr{Z}^{\leq N} \Psi \\ \breve{R} \mathscr{Z}^{\leq N} \Psi \\ \upmu \not d \mathscr{Z}^{\leq N} \Psi \\ \mathscr{Z}^{\leq N} \Psi \end{pmatrix} \right| + \frac{\ln(e+t)}{(1+t)^2} \left| \begin{pmatrix} \mathscr{Z}^{\leq N}(\upmu - 1) \\ \sum_{a=1}^{3} \varrho \left| \mathscr{Z}^{\leq N} L_{(Small)}^a \right| \end{pmatrix} \right| \\
& + \frac{1}{\varrho} \int_{t'=0}^{t} \frac{\| [L\upmu]_- \|_{C^0(\Sigma_{t'}^u)}}{\upmu_\star(t',u)} \varrho(t',u) \left| \begin{pmatrix} \breve{R} \mathscr{Z}^{\leq N} \Psi \\ \mathscr{Z}^{\leq N} \Psi \end{pmatrix} \right| (t',u,\vartheta) \, dt' \\
& + \frac{1}{\varrho^2} \int_{t'=0}^{t} \left\| \left(\frac{\upmu(t,\cdot)}{\upmu} \right)^2 \right\|_{C^0(\Sigma_{t'}^u)} \varrho(t',u) \left| \begin{pmatrix} \breve{R} \mathscr{Z}^{\leq N} \Psi \\ \mathscr{Z}^{\leq N} \Psi \end{pmatrix} \right| (t',u,\vartheta) \, dt' \\
& + \frac{\ln^2(e+t)}{(1+t)^2} \left| \mathscr{X}^{[N]} \right| (0,u,\vartheta) \\
& + \frac{\ln^3(e+t)}{(1+t)^2} \int_{t'=0}^{t} \left| \begin{pmatrix} \varrho L \mathscr{Z}^{\leq N} \Psi \\ \breve{R} \mathscr{Z}^{\leq N} \Psi \\ \varrho \not d \mathscr{Z}^{\leq N} \Psi \\ \mathscr{Z}^{\leq N} \Psi \end{pmatrix} \right| (t',u,\vartheta) \, dt' \\
& + \varepsilon \frac{\ln^3(e+t)}{(1+t)^2} \int_{t'=0}^{t} \left| \begin{pmatrix} \upmu \hat{\not\nabla}^2 \mathscr{S}^{\leq N-1} \upmu \\ \varrho \upmu \not\nabla \not\mathcal{L}_{\mathscr{S}}^{\leq N-1} \hat{\chi}^{(Small)} \end{pmatrix} \right| (t',u,\vartheta) \, dt' \\
& + \frac{\ln^3(e+t)}{(1+t)^2} \int_{t'=0}^{t} \frac{1}{(1+t')} \frac{1}{\upmu_\star(t',u)} \left| \begin{pmatrix} \mathscr{Z}^{\leq N}(\upmu - 1) \\ \sum_{a=1}^{3} \varrho \left| \mathscr{Z}^{\leq N} L_{(Small)}^a \right| \end{pmatrix} \right| (t',u,\vartheta) \, dt'.
\end{aligned}
$$

The term $\left| \mathscr{X}^{[N]} \right|$ *appearing on the right-hand side of* (17.53) *is defined just below* (17.28).

PROOF. The estimate (17.53) for

$$\left| \upmu O \mathscr{S}^{N-1} \mathrm{tr}_{\not g} \chi^{(Small)} \right| \quad \text{and} \quad \left| \upmu \breve{R} \mathscr{S}^{N-1} \mathrm{tr}_{\not g} \chi^{(Small)} \right|$$

can be proved by using arguments similar to the ones we used in the proof of (17.35) together with inequality (12.56) with Ψ in the role of f. We omit the details (note however that the present estimates contain an extra factor of $\upmu$ compared to (17.35) and that the factor $\breve{R}\Psi$ from (17.35) is not present). The argument is in fact significantly simpler because we are no longer seeking to pair $\breve{R}\Psi$ with G_{LL} in an effort to generate the factor $L\upmu$. Hence, we do not bother to replace the integrand factor $G_{LL}(t',u,\vartheta)$ with $G_{LL}(t,u,\vartheta)$ up to error terms as we did in (17.43). Moreover, to derive (17.53), it suffices to use the following crude consequence of inequality (17.40):

$$
(17.54)\qquad \left|\mathfrak{X}^{[N]}\right| \lesssim \left|\begin{pmatrix} \breve{R}\mathscr{Z}^{\leq N}\Psi \\ \mathscr{Z}^{\leq N}\Psi \end{pmatrix}\right| + \frac{1}{1+t}\upmu\left|\begin{pmatrix} \varrho L\mathscr{Z}^{\leq N}\Psi \\ \varrho {\not d}\mathscr{Z}^{\leq N}\Psi \end{pmatrix}\right| + \varepsilon\frac{1}{(1+t)^2}\left|\begin{pmatrix} \mathscr{Z}^{\leq N}(\upmu-1) \\ \sum_{a=1}^3 \varrho\left|\mathscr{Z}^{\leq N}L_{(Small)}^a\right| \end{pmatrix}\right|.
$$

Inequality (17.53) for $\left|\upmu {\not\Delta}\mathscr{S}^{N-1}\upmu\right|$ then follows from the bound (17.53) for $\left|\upmu\breve{R}\mathscr{S}^{N-1}\mathrm{tr}_{\not g}\chi^{(Small)}\right|$, the final inequality in (15.1), and the bound $\upmu \lesssim \ln(e+t)$ (that is, (12.128a)). □

17.5. Pointwise estimates for the difficult error integrands requiring partial modification

In order to derive suitable a priori estimates for the top-order L^2 quantity $\widetilde{\mathbb{Q}}_{(\leq 24)}$, we need to derive sharpened pointwise and L^2 estimates for the partially modified versions of $\mathscr{S}^{23}\mathrm{tr}_{\not g}\chi^{(Small)}$ and ${\not d}\mathscr{S}^{23}\upmu$ defined in Sects. 11.3 and 11.4; the estimates that would follow from Prop. 12.46 are not sufficient to close the top-order estimates. We provide the sharpened pointwise estimates in the next two lemmas.

LEMMA 17.9 (**Sharp transport inequalities for partially modified versions of ${\not d}\mathscr{S}^{N-1}\upmu$ and $\mathscr{S}^{N-1}\mathrm{tr}_{\not g}\chi^{(Small)}$**). *Let $1 \leq N \leq 24$ be an integer. Let $\mathscr{S}^{N-1}$ be an $(N-1)^{st}$ order pure spatial commutation vectorfield operator (see definition (7.2a)), let ${}^{(\mathscr{S}^{N-1})}\widetilde{\mathscr{X}}$ be the corresponding partially modified quantity defined in (11.45a), and let ${}^{(\mathscr{S}^{N-1})}\widetilde{\mathscr{M}}$ be the corresponding partially modified $S_{t,u}$ one-form defined in (11.51a). Let χ be the symmetric type $\binom{0}{2}$ $S_{t,u}$ tensorfield defined in (3.70) and recall that $\varrho(t,u) = 1-u+t$. Under the small-data and bootstrap assumptions of Sects. 12.1-12.4, if ε is sufficiently small, then the following estimates hold on $\mathcal{M}_{T_{(Bootstrap)},U_0}$:*

(17.55)

$$
\begin{gathered}
\left|\left|L\left|\varrho^{(\mathscr{S}^{N-1})}\widetilde{\mathscr{M}}\right|\right| - \frac{1}{2}\left|\varrho G_{LL}{\not d}\mathscr{S}^{N-1}\breve{R}\Psi\right|\right|, \\
\left|\varrho\left\{{\not\mathcal{L}}_L + \mathrm{tr}_{\not g}\chi\right\}{}^{(\mathscr{S}^{N-1})}\widetilde{\mathscr{M}}^{\#} - \frac{1}{2}\varrho G_{LL}{\not d}^{\#}\mathscr{S}^{N-1}\breve{R}\Psi\right|, \\
\left|L\left\{\varrho^{2(\mathscr{S}^{N-1})}\widetilde{\mathscr{X}}\right\} - \frac{1}{2}\varrho^2 G_{LL}{\not\Delta}\mathscr{S}^{N-1}\Psi\right|, \\
\left|\varrho^2\left\{L + \mathrm{tr}_{\not g}\chi\right\}{}^{(\mathscr{S}^{N-1})}\widetilde{\mathscr{X}} - \frac{1}{2}\varrho^2 G_{LL}{\not\Delta}\mathscr{S}^{N-1}\Psi\right| \\
\lesssim \left|\begin{pmatrix} \varrho\left\{L + \frac{1}{2}\mathrm{tr}_{\not g}\chi\right\}\mathscr{Z}^{\leq N-1}\Psi \\ \breve{R}\mathscr{Z}^{\leq N-1}\Psi \\ \varrho{\not d}\mathscr{Z}^{\leq N-1}\Psi \\ \mathscr{Z}^{\leq N-1}\Psi \end{pmatrix}\right| + \frac{\ln(e+t)}{(1+t)^2}\left|\begin{pmatrix} \mathscr{Z}^{\leq N}(\upmu-1) \\ \sum_{a=1}^3 \varrho\left|\mathscr{Z}^{\leq N}L_{(Small)}^a\right| \end{pmatrix}\right|.
\end{gathered}
$$

PROOF.

Proof of the second inequality in (17.55): We begin by proving the inequality (17.55) for the second term on the left-hand side. From equation (11.52) and the identity $\{\not{\mathcal{L}}_L + \mathrm{tr}_{\not{g}}\chi\}\not{g}^{-1} = -2\hat{\chi}^{(Small)\#\#}$, which follows from (3.72), (4.1c), (7.4a)-(7.4b), and the fact that $L\varrho = 1$, we see that the desired inequality (17.55) will follow once we show that $\varrho\left|\hat{\chi}^{(Small)}\right|\left|{}^{(\mathscr{S}^{N-1})}\widetilde{\mathscr{M}}\right|$ and $\varrho\left|{}^{(\mathscr{S}^{N-1})}\mathfrak{J}\right|$ are each bounded in magnitude by $\lesssim$ the right-hand side of (17.55), where ${}^{(\mathscr{S}^{N-1})}\mathfrak{J}$ is the inhomogeneous term defined in (11.53). We have already proved the desired bound for $\varrho\left|{}^{(\mathscr{S}^{N-1})}\mathfrak{J}\right|$ in Lemma 17.5. To bound the remaining product, we first use (12.46a), (12.58b), and the bound $\upmu \lesssim \ln(e+t)$ (that is, (12.128a)) to deduce that $\left|{}^{(\mathscr{S}^{N-1})}\widetilde{\mathscr{M}}\right| \lesssim (1+t)^{-1}\left|\mathscr{Z}^{\leq N}(\upmu-1)\right| + \ln(e+t)\left|\not{d}\mathscr{S}^{\leq N-1}\Psi\right|$. Also using the bound $\varrho\left|\hat{\chi}^{(Small)}\right| \lesssim \varepsilon\ln(e+t)(1+t)^{-1}$, which follows from (12.128a), we conclude that $\varrho\left|\hat{\chi}^{(Small)}\right|\left|{}^{(\mathscr{S}^{N-1})}\widetilde{\mathscr{M}}\right|$ is $\lesssim$ the right-hand side of (17.55) as desired.

Proof of the first inequality in (17.55): We first note the following simple inequality, which holds for any $S_{t,u}$ one-form ξ:

$$
\begin{aligned}
(17.56)\qquad |L(|\varrho\xi|)| &\leq \left|\hat{\chi}^{(Small)}\right||\varrho\xi| + \frac{1}{2}\left|\mathrm{tr}_{\not{g}}\chi^{(Small)}\right||\varrho\xi| + \varrho\left|(\not{\mathcal{L}}_L + \mathrm{tr}_{\not{g}}\chi)\xi^{\#}\right| \\
&\leq 2\left|\chi^{(Small)}\right||\varrho\xi| + \varrho\left|(\not{\mathcal{L}}_L + \mathrm{tr}_{\not{g}}\chi)\xi^{\#}\right|.
\end{aligned}
$$

Inequality (17.56) follows in a straightforward fashion from the identities $\not{\mathcal{L}}_L\not{g} = 2\chi$, and $\not{\mathcal{L}}_L\not{g}^{-1} = -2\chi^{\#\#}$ (see (3.72) and (7.4a)-(7.4b)), the decomposition $\chi = \varrho^{-1}\not{g} + \chi^{(Small)}$ (see (4.1c)), and the fact that $L\varrho = 1$. We now set $\xi = {}^{(\mathscr{S}^{N-1})}\widetilde{\mathscr{M}}$. A suitable estimate for the term $\varrho\left|(\not{\mathcal{L}}_L + \mathrm{tr}_{\not{g}}\chi)\xi^{\#}\right|$ is provided by the already proven second inequality in (17.55). Furthermore, the same reasoning that we used in our proof of the second inequality in (17.55) also implies that $2\left|\chi^{(Small)}\right||\varrho\xi| = 2\left|\chi^{(Small)}\right||\varrho\xi^{\#}|$ is $\lesssim$ the right-hand side of (17.55). We have thus proved the desired bound for the first term on the left-hand side of (17.55).

Proof of the third inequality in (17.55): We now prove inequality (17.55) for the third term on the left-hand side. To derive the desired bound, by equation (11.46), we only have to show that $\left|{}^{(\mathscr{S}^{N-1})}\widetilde{\mathfrak{B}}\right|$ is $\lesssim$ the right-hand side of (17.55), where ${}^{(\mathscr{S}^{N-1})}\widetilde{\mathfrak{B}}$ is the inhomogeneous term defined in (11.47). The desired bound follows from Lemma 17.4 and the estimate $\varrho\left|\mathrm{tr}_{\not{g}}\chi\right| \lesssim 1$ (that is, (12.60b)).

Proof of the fourth inequality in (17.55): We begin by noting that if f is a function, then

$$
(17.57)\qquad \varrho^2\{L + \mathrm{tr}_{\not{g}}\chi\}f - L(\varrho^2 f) = \varrho^2\mathrm{tr}_{\not{g}}\chi^{(Small)}f.
$$

The identity (17.57) follows easily from the decomposition $\mathrm{tr}_{\not{g}}\chi = 2\varrho^{-1} + \mathrm{tr}_{\not{g}}\chi^{(Small)}$ (see (4.1c)) and the fact that $L\varrho = 1$. We now set $f = {}^{(\mathscr{S}^{N-1})}\widetilde{\mathscr{X}}$, that is, equal to the relevant factor in the fourth term on the left-hand side of (17.55). A suitable estimate for the term $L(\varrho^2 f)$ is provided by the already proven third inequality in (17.55). By inequality (12.128a), the magnitude of the remaining term $\varrho^2\mathrm{tr}_{\not{g}}\chi^{(Small)}f$ is $\lesssim \varepsilon\ln(e+t)|f|$. From (12.58b), Cor. 12.27, the estimate $\varrho\left|\mathrm{tr}_{\not{g}}\chi\right| \lesssim 1$ (that is, (12.60b)), and inequality (12.67a) with $N-1$ in the role of N, we deduce that $\varepsilon\ln(e+t)|f|$ is $\lesssim$ the right-hand side of (17.55) as desired. $\square$

In the next lemma, we provide time-integrated versions of the estimates of Lemma 17.9.

LEMMA 17.10 (**Sharp pointwise estimates for partially modified versions of $\not{d}\mathscr{S}^{N-1}\upmu$, $\mathscr{S}^{N-1}\mathrm{tr}_{\not{g}}\chi^{(Small)}$**). *Let $1 \leq N \leq 24$ be an integer. Let $\mathscr{S}^{N-1}$ be an $(N-1)^{st}$ order pure spatial commutation vectorfield operator (see definition (7.2a)), let ${}^{(\mathscr{S}^{N-1})}\widetilde{\mathscr{X}}$ be the corresponding partially modified quantity defined in (11.45a), and let ${}^{(\mathscr{S}^{N-1})}\widetilde{\mathscr{M}}$ be the corresponding partially modified $S_{t,u}$ one-form defined in (11.51a). Let χ be the symmetric type $\binom{0}{2}$ $S_{t,u}$ tensorfield defined in (3.70) and recall that $\varrho(t,u) = 1 - u + t$. Under the small-data and bootstrap assumptions of Sects. 12.1-12.4, if ε is sufficiently small, then the following estimates hold on $\mathcal{M}_{T_{(Bootstrap)},U_0}$:*

(17.58)

$$\begin{aligned}
\left|\varrho^{(\mathscr{S}^{N-1})}\widetilde{\mathscr{M}}\right| \leq \left|\varrho^{(\mathscr{S}^{N-1})}\widetilde{\mathscr{M}}\right|(0,u,\vartheta) & \\
+ \frac{1}{2}\left|G_{LL}\right|(t,u,\vartheta)\int_{t'=0}^{t}\left|\varrho\not{d}\mathscr{S}^{N}\Psi\right|(t',u,\vartheta)\,dt' & \\
+ C\varepsilon\int_{t'=0}^{t}\left|\not{d}\mathscr{S}^{N}\Psi\right|(t',u,\vartheta)\,dt' & \\
+ C\int_{t'=0}^{t}\left|\begin{pmatrix} \varrho\left\{L + \frac{1}{2}\mathrm{tr}_{\not{g}}\chi\right\}\mathscr{Z}^{\leq N-1}\Psi \\ \breve{R}\mathscr{Z}^{\leq N-1}\Psi \\ \varrho\not{d}\mathscr{Z}^{\leq N-1}\Psi \\ \mathscr{Z}^{\leq N-1}\Psi \end{pmatrix}\right|(t',u,\vartheta)\,dt' & \\
+ C\int_{t'=0}^{t}\frac{\ln(e+t')}{(1+t')^2}\left|\begin{pmatrix} \mathscr{Z}^{\leq N}(\upmu-1) \\ \sum_{a=1}^{3}\varrho\left|\mathscr{Z}^{\leq N}L_{(Small)}^{a}\right| \end{pmatrix}\right|(t',u,\vartheta)\,dt', &
\end{aligned}$$

(17.59)

$$\begin{aligned}
\left|\varrho^{2(\mathscr{S}^{N-1})}\widetilde{\mathscr{X}}\right| \leq \left|\varrho^{2(\mathscr{S}^{N-1})}\widetilde{\mathscr{X}}\right|(0,u,\vartheta) & \\
+ \frac{1}{2}\left|G_{LL}\right|(t,u,\vartheta)\int_{t'=0}^{t}\left|\varrho^{2}\not{\Delta}\mathscr{S}^{N-1}\Psi\right|(t',u,\vartheta)\,dt' & \\
+ C\varepsilon\int_{t'=0}^{t}\left|\not{d}\mathscr{S}^{N}\Psi\right|(t',u,\vartheta)\,dt' & \\
+ C\int_{t'=0}^{t}\left|\begin{pmatrix} \varrho\left\{L + \frac{1}{2}\mathrm{tr}_{\not{g}}\chi\right\}\mathscr{Z}^{\leq N-1}\Psi \\ \breve{R}\mathscr{Z}^{\leq N-1}\Psi \\ \varrho\not{d}\mathscr{Z}^{\leq N-1}\Psi \\ \mathscr{Z}^{\leq N-1}\Psi \end{pmatrix}\right|(t',u,\vartheta)\,dt' & \\
+ C\int_{t'=0}^{t}\frac{\ln(e+t')}{(1+t')^2}\left|\begin{pmatrix} \mathscr{Z}^{\leq N}(\upmu-1) \\ \sum_{a=1}^{3}\varrho\left|\mathscr{Z}^{\leq N}L_{(Small)}^{a}\right| \end{pmatrix}\right|(t',u,\vartheta)\,dt'. &
\end{aligned}$$

PROOF. To deduce (17.58), we integrate the first inequality of (17.55) along the integral curves of L. We use the estimate (12.142) to replace the integrand factor $G_{LL}(t',u,\vartheta)$ with $G_{LL}(t,u,\vartheta)$ up to an error term that is bounded by the second time integral on the right-hand side of (17.58). The estimate (17.58) thus follows.

The estimate (17.59) follows from the third inequality of (17.55), inequality (12.46c) with $f = \mathscr{S}^{N-1}\Psi$, and the same argument. □

In the next lemma, we derive pointwise estimates for some negligible error terms that arise when we derive a priori estimates for the top-order quantity $\widetilde{\mathbb{Q}}_{(\leq 24)}$. Specifically, these terms arise in Sect. 20.7 when, to avoid error integrals with unfavorable time-growth, we substitute the partially modified quantities $O^{(\mathscr{S}^{N-1})}\widetilde{\mathscr{X}}$ and $\text{di}\!\!/\!\!\text{v}^{(\mathscr{S}^{N-1})}\widetilde{\mathscr{M}}$ in place of $O\mathscr{S}^{N-1}\text{tr}_{\not{g}}\chi^{(Small)}$ and $\not{\Delta}\mathscr{S}^{N-1}\upmu$.

LEMMA 17.11 (**Pointwise estimates for some harmless error terms that arise from the top-order Morawetz multiplier estimates**). *Let $1 \leq N \leq 24$ be an integer. Let $\mathscr{S}^{N-1}$ be an $(N-1)^{st}$ order pure spatial commutation vectorfield operator (see definition* (7.2a)*), let ${}^{(\mathscr{S}^{N-1})}\widetilde{\mathfrak{X}}$ be the corresponding function defined in* (11.45b)*, and let ${}^{(\mathscr{S}^{N-1})}\widetilde{\mathfrak{M}}^{\#}$ be the corresponding $S_{t,u}$-tangent vectorfield that is $\not{g}$-dual to the one-form defined in* (11.51b)*. Let $O \in \mathscr{O}$ be a geometric rotation vectorfield (see definition* (7.2b)*). Under the small-data and bootstrap assumptions of Sects.* 12.1-12.4*, if ε is sufficiently small, then the following estimates hold on $\mathcal{M}_{T_{(Bootstrap)},U_0}$*

$$(\breve{R}\Psi)O^{(\mathscr{S}^{N-1})}\widetilde{\mathfrak{X}} = Harmless^{\leq N}, \tag{17.60}$$

$$(\breve{R}\Psi)\text{di}\!\!/\!\!\text{v}^{(\mathscr{S}^{N-1})}\widetilde{\mathfrak{M}}^{\#} = Harmless^{\leq N}, \tag{17.61}$$

where $Harmless^{\leq N}$ terms are defined in Def. 16.2.

PROOF. We first prove (17.60). From definition (11.45b), the Leibniz rule, the fact that $\varrho L \in \mathscr{Z}$, the estimate $|Of| \lesssim \varrho|\not{d}f|$ for functions f (see (12.34a)), Lemma 8.6, and the bootstrap assumptions ($\mathbf{BA}\Psi$), we deduce that

$$\left|(\breve{R}\Psi)O^{(\mathscr{S}^{N-1})}\widetilde{\mathfrak{X}}\right| \lesssim \varepsilon\frac{1}{(1+t)^2}\left|\not{\mathcal{L}}_{\mathscr{Z}}^{\leq 1}G_{(Frame)}^{\#}\right|\left|\begin{pmatrix} \varrho\not{d}\mathscr{Z}^{\leq N}\Psi \\ \mathscr{Z}^{\leq N}\Psi \end{pmatrix}\right|. \tag{17.62}$$

In view of Def. 16.2, we see that the desired result (17.60) follows from (17.62), Cor. 12.27, and the estimate (12.58b).

We now prove (17.61). From definition (11.51b), the Leibniz rule, inequalities (12.46a) and (12.46c), the estimate(12.58b), the estimate (12.128a), and the bootstrap assumptions ($\mathbf{BA}\Psi$), we deduce that

$$\left|(\breve{R}\Psi)\text{di}\!\!/\!\!\text{v}^{(\mathscr{S}^{N-1})}\widetilde{\mathfrak{M}}^{\#}\right| \lesssim \varepsilon\frac{\ln(e+t)}{(1+t)^2}\left|\varrho\not{d}\mathscr{Z}^{\leq N}\Psi\right|. \tag{17.63}$$

In view of Def. 16.2, we see that the desired result (17.61) follows from (17.63). □

CHAPTER 18

Elliptic Estimates and Sobolev Embedding on the Spheres

In this chapter, we provide elliptic estimates corresponding to various elliptic operators on the Riemannian manifolds $(S_{t,u}, \not{g})$, where $\not{g}$ is the metric induced by g on $S_{t,u}$ (see Def. 3.9). We use these estimates in Chapter 20 in order to close our top-order a priori L^2 estimates. Specifically, we must use the elliptic estimates when we bound the top-order derivatives of $\not{\nabla}^2 \mu$ and $\not{\nabla} \chi^{(Small)}$. We also prove a Sobolev embedding-type proposition, which we eventually use to improve the fundamental C^0 bootstrap assumptions (**BA**Ψ). To obtain these estimates, we must first derive some simple estimates for the Gaussian curvature of $\not{g}$.

18.1. Elliptic estimates

Our elliptic estimates are based on pointwise estimates for the Gaussian curvature $\mathfrak{K}$ of $\not{g}$, which we now define.

DEFINITION 18.1 (**Gaussian curvature of $\not{g}$**). Let $\{e_i\}_{i=1,2}$ be a basis for the tangent space of $S_{t,u}$ at p and let $\mathfrak{R}$ be the Riemann curvature tensor of $\not{g}$ (see definition (11.2)). At the point p, we define $\mathfrak{K}$, the Gaussian curvature $\mathfrak{K}$ of $\not{g}$, as follows:

$$\mathfrak{K} = \mathfrak{K}(\not{g}) := \frac{\mathfrak{R}_{e_1 e_2 e_1 e_2}}{\det \begin{pmatrix} \not{g}(e_1, e_1) & \not{g}(e_1, e_2) \\ \not{g}(e_2, e_1) & \not{g}(e_2, e_2) \end{pmatrix}}. \tag{18.1}$$

REMARK 18.2 (**Frame independence of the Gaussian curvature of $\not{g}$**). It is a standard result (see, for example, [**67**, Section 3.3]), based on the symmetry properties (11.3), that $\mathfrak{K}$ is independent of the basis $\{e_i\}_{i=1,2}$.

The first lemma of this section is standard and provides a connection between $\mathfrak{K}$ and the Riemann curvature tensor $\mathfrak{R}_{ABCD}$ of $\not{g}$.

LEMMA 18.3 (**Connection between the Gaussian and Riemann curvatures of $\not{g}$**). *Let $\mathfrak{K}$ and $\mathfrak{R}_{ABCD}$ respectively denote the Gaussian curvature and Riemann curvature tensor of the Riemannian metric $\not{g}$ on the two-dimensional manifold $S_{t,u}$ (see* (11.2) *for the definition of $\mathfrak{R}_{ABCD}$). Then*

$$\mathfrak{R}_{ABCD} = \mathfrak{K}\{\not{g}_{AC}\not{g}_{BD} - \not{g}_{AD}\not{g}_{BC}\}. \tag{18.2}$$

PROOF. See, for example, [**42**, Section 3.3] for a proof. □

We now state a second basic lemma that provides the well-known formula for the first variation of the scalar curvature $\mathfrak{R}$ of $\not{g}$.

LEMMA 18.4 (**Formula for the first variation of $\mathfrak{R}$**). *Let*

$$\mathfrak{R}_{AB} := (\not{g}^{-1})^{CD}\mathfrak{R}_{ACBD} \quad \textit{and} \quad \mathfrak{R} := (\not{g}^{-1})^{AB}\mathfrak{R}_{AB}$$

respectively denote the Ricci curvature tensor and scalar curvature of $\not{g}$. Let ${}^{(\lambda)}\not{g}$ be a one-parameter family of Riemannian metrics on $S_{t,u}$ verifying ${}^{(0)}\not{g} = \not{g}$ and let $\dot{\not{g}} := \frac{d}{d\lambda}|_{\lambda=0}{}^{(\lambda)}\not{g}$. Similarly, let ${}^{(\lambda)}\mathfrak{R}$ denote the scalar curvature of ${}^{(\lambda)}\not{g}$ and let $\dot{\mathfrak{R}} := \frac{d}{d\lambda}|_{\lambda=0}{}^{(\lambda)}\mathfrak{R}$. Then we have the following identity for the first variation of the scalar curvature:

$$\dot{\mathfrak{R}} = -\mathfrak{R}^{AB}\dot{\not{g}}_{AB} - \not{\Delta}\mathrm{tr}_{\not{g}}\dot{\not{g}} + (\not{g}^{-1})^{AB}(\not{g}^{-1})^{CD}\not{\nabla}^2_{AC}\dot{\not{g}}_{BD}. \tag{18.3}$$

PROOF. This is another standard result. See, for example, [**14**, Chapter 2]. □

Next, we use Lemma (18.4) to derive an expression for $\breve{R}\mathfrak{R}$.

COROLLARY 18.5 (**An expression $\breve{R}\mathfrak{R}$**). *Let $\mathfrak{R}_{AB} := (\not{g}^{-1})^{CD}\mathfrak{R}_{ACBD}$ and $\mathfrak{R} := (\not{g}^{-1})^{AB}\mathfrak{R}_{AB}$ respectively denote the Ricci curvature tensor and scalar curvature of $\not{g}$. Then $\mathfrak{R}$ verifies the following evolution equation:*

$$\breve{R}\mathfrak{R} = -\mathfrak{R}^{AB\,(\breve{R})}\not{\pi}_{AB} - \not{\Delta}\mathrm{tr}_{\not{g}}{}^{(\breve{R})}\not{\pi} + \not{\nabla}^2_{AB}{}^{(\breve{R})}\not{\pi}^{AB}. \tag{18.4}$$

PROOF. We use Lemma 18.4 in the case that ${}^{(\lambda)}\not{g}$ is the pullback (see (3.65)) of $\not{g}$ by the flow map $\varphi_{(\lambda)}$ of $\breve{R}$ (cf. the proof of Lemma 8.9 in the case $Z = O_{(l)}$). Then using Lemma 7.5 and the fact (see Remark 3.56) that the Lie derivative of a tensorfield with respect to a vectorfield is the derivative (with respect to the flow parameter) of the pullback of the tensorfield by the flow map of the vectorfield, we obtain the identities $\dot{\not{g}} = \not{\mathcal{L}}_{\breve{R}}\not{g} = {}^{(\breve{R})}\not{\pi}$ (see Lemma 7.5) and $\dot{\mathfrak{R}} = \breve{R}\mathfrak{R}$. Substituting for $\dot{\not{g}}$ and $\dot{\mathfrak{R}}$ in (18.3), we arrive at the desired identity (18.4). □

The following corollary provides an evolution equation for $\mathfrak{K}$ along the integral curves of $\breve{R}$.

COROLLARY 18.6 (**Transport equation for the Gaussian curvature $\mathfrak{K}$ of $\not{g}$**). *Let $\mathfrak{K}$ denote the Gaussian curvature of the Riemannian metric $\not{g}$ on $S_{t,u}$ and let $\varrho(t,u) = 1 - u + t$ be the geometric radial variable. Then $\varrho^2\mathfrak{K}$ verifies the following transport equation:*

$$\begin{aligned}\breve{R}(\varrho^2\mathfrak{K}) = (\varrho^2\mathfrak{K})&\left\{\frac{2(\upmu-1)}{\varrho} + \upmu\mathrm{tr}_{\not{g}}\chi^{(Small)} - \mathrm{tr}_{\not{g}}\not{k}^{(Trans-\Psi)} - \upmu\mathrm{tr}_{\not{g}}\not{k}^{(Tan-\Psi)}\right\}\\ &+ \varrho\not{\Delta}\upmu + \varrho^2\not{\Delta}(\upmu\mathrm{tr}_{\not{g}}\chi^{(Small)}) - \varrho^2\not{\Delta}\mathrm{tr}_{\not{g}}\not{k}^{(Trans-\Psi)} - \varrho^2\not{\Delta}(\upmu\mathrm{tr}_{\not{g}}\not{k}^{(Tan-\Psi)})\\ &- \varrho^2(\not{\nabla}^2)^{AB}(\upmu\chi^{(Small)}_{AB}) + \varrho^2(\not{\nabla}^2)^{AB}\not{k}^{(Trans-\Psi)}_{AB} + \varrho^2(\not{\nabla}^2)^{AB}(\upmu\not{k}^{(Tan-\Psi)}_{AB}).\end{aligned} \tag{18.5}$$

In (18.5), the $S_{t,u}$ tensors $\chi^{(Small)}$, $\not{k}^{(Trans-\Psi)}$, and $\not{k}^{(Tan-\Psi)}$ are defined by (4.1c), (5.4b), and (5.5b).

PROOF. Contracting the identity (18.2) against $(\not{g}^{-1})^{AC}$ and $(\not{g}^{-1})^{AC}(\not{g}^{-1})^{BD}$, we see that $\mathfrak{R}_{AB} = \mathfrak{K}\not{g}_{AB}$ and $\mathfrak{R} = 2\mathfrak{K}$, where $\mathfrak{R}_{AB}$ and $\mathfrak{R}$ respectively denote the Ricci curvature and the scalar curvature of $\not{g}$. The desired identity (18.5) now follows from these identities, (18.4), the identity $\breve{R}\varrho = -1$, and the identity ${}^{(\breve{R})}\not{\pi}_{AB} = -2\varrho^{-1}\upmu\not{g}_{AB} - 2\upmu\chi^{(Small)}_{AB} + 2\not{k}^{(Trans-\Psi)}_{AB} + 2\upmu\not{k}^{(Tan-\Psi)}_{AB}$ (see (7.5f)-(7.5g)). □

We proceed to the next lemma, which, based on the bootstrap assumptions and some previously proven C^0 estimates, provides pointwise estimates for the Gaussian curvature $\mathfrak{K}$ of $\not{g}$. This lemma is important because in order to derive our desired elliptic estimates, we use the following consequence of the lemma: $\mathfrak{K}$ is uniformly positive. Actually, we prove a much stronger estimate in the lemma.

LEMMA 18.7 (**Pointwise estimate for the Gaussian curvature $\mathfrak{K}$ of $\not{g}$**). *Under the small-data and bootstrap assumptions of Sects 12.1-12.4, if ε is sufficiently small, then the following pointwise estimate holds for the Gaussian curvature $\mathfrak{K}$ of $\not{g}$ on $\mathcal{M}_{T_{(Bootstrap)},U_0}$:*

$$\left|\varrho^2\mathfrak{K} - 1\right|(t,u,\vartheta) \lesssim \varepsilon\frac{\ln(e+t)}{1+t}. \tag{18.6}$$

PROOF. We fix the time t. We will apply a Gronwall argument to the quantity

$$q := \varrho^2\mathfrak{K} - 1 \tag{18.7}$$

along the integral curves of $\breve{R}$, which are contained in $\Sigma_t^{U_0}$. Recall that $\breve{R}u = 1$ and hence we can view $\breve{R} = \frac{d}{du}$ along the integral curves. We first derive suitable pointwise estimates for the non-$\mathfrak{K}$ terms on the right-hand side of (18.5) at time t. To begin, we use (5.4b) and (5.5b) to deduce the following schematic identities: $\not{k}^{(Trans-\Psi)} = G_{(Frame)}\breve{R}\Psi$, $\not{k}^{(Tan-\Psi)} = G_{(Frame)}\begin{pmatrix} L\Psi \\ \not{d}\Psi \end{pmatrix}$. Furthermore, the same identities hold for $\mathrm{tr}_{\not{g}}\not{k}^{(Trans-\Psi)}$ and $\mathrm{tr}_{\not{g}}\not{k}^{(Tan-\Psi)}$ but with $G_{(Frame)}\not{g}^{-1}$ in place of $G_{(Frame)}$. We then use these schematic identities, inequalities (12.46a), (12.46c), and (12.64), the C^0 estimates (12.58b), (12.60b), and (12.128a), and the bootstrap assumptions ($\mathbf{BA}\Psi$) to deduce that at time t, all terms on the second and third lines of the right-hand side of (18.5) have magnitudes bounded by $\lesssim \varepsilon\ln(e+t)(1+t)^{-1}$. Similarly, we deduce that the terms in braces multiplying $\varrho^2\mathfrak{K}$ in the first line on the right-hand side of (18.5) are in magnitude bounded by $\lesssim \varepsilon\ln(e+t)(1+t)^{-1}$. Inserting these estimates into (18.5), we deduce that

$$\left|\frac{d}{du}q\right| \lesssim \varepsilon\frac{\ln(e+t)}{1+t}|q| + \varepsilon\frac{\ln(e+t)}{1+t}. \tag{18.8}$$

Integrating (18.8) (along the integral curves of $\breve{R}$) from 0 to u and using the fact that $q(u=0) \equiv 0$ (that is, in view of our assumptions that the data are supported in Σ_0^1, it follows that $(S_{t,0},\not{g}(t,0,\cdot))$ is a round Euclidean sphere of radius $\varrho(t,0) = 1+t$ and Gaussian curvature $(1+t)^{-2}$), we deduce that

$$|q(u)| \leq C\varepsilon\frac{\ln(e+t)}{1+t} + C\varepsilon\frac{\ln(e+t)}{1+t}\int_{u'=0}^{u}|q(u')|\,du'. \tag{18.9}$$

Finally, applying Gronwall's inequality to (18.9) and using the fact that $0 \leq u \leq U_0 < 1$, we conclude that $|q(u)| \lesssim \varepsilon\ln(e+t)(1+t)^{-1}$, which is precisely the desired inequality (18.6). □

We now use Lemma 18.7 to derive the desired elliptic estimates. We split the estimates into two lemmas. The first lemma is for scalar-valued functions f, and we later apply it with f equal to various derivatives of $\upmu$. The second is for symmetric, trace-free type $\binom{0}{2}$ $S_{t,u}$ tensorfields ξ, and we later apply it with ξ equal to various trace-free Lie derivatives of $\hat{\chi}^{(Small)}$.

LEMMA 18.8 (**Elliptic estimates for solutions to Poisson's equation on** $S_{t,u}$). *Let f be a scalar-valued function. Under the small-data and bootstrap assumptions of Sects.* 12.1-12.4, *if ε is sufficiently small, then the following integral inequality holds for* $(t,u) \in [0, T_{(Bootstrap)}) \times [0, U_0]$:

$$\int_{S_{t,u}} \upmu^2 |\nabla\!\!\!/^2 f|^2 \, d\upsilon_{g\!\!/} \leq 4 \int_{S_{t,u}} \upmu^2 (\Delta\!\!\!/ f)^2 \, d\upsilon_{g\!\!/} + C\varepsilon^2 \frac{\ln^2(e+t)}{(1+t)^2} \int_{S_{t,u}} |d\!\!\!/ f|^2 \, d\upsilon_{g\!\!/}. \tag{18.10}$$

PROOF. First, using definition (11.2) and (18.2), we derive the commutator identity $\nabla\!\!\!/_A \Delta\!\!\!/ f = \nabla\!\!\!/^B \nabla\!\!\!/_{BA}^2 f - \mathfrak{K} d\!\!\!/_A f$. From this identity, the fact that $\nabla\!\!\!/_{AB}^2 f = \nabla\!\!\!/_{BA}^2 f$, and direct computation, we deduce that

$$\begin{aligned} \upmu^2 |\nabla\!\!\!/^2 f|^2 + \upmu^2 \mathfrak{K} |d\!\!\!/ f|^2 &= \upmu^2 (\Delta\!\!\!/ f)^2 + 2\upmu (\Delta\!\!\!/ f)(d\!\!\!/^A \upmu) d\!\!\!/_A f - 2(\nabla\!\!\!/_{AB}^2 f)(d\!\!\!/^A \upmu) d\!\!\!/^B f \\ &\quad + \nabla\!\!\!/^A \left\{ \upmu^2 (\nabla\!\!\!/_{AB}^2 f) d\!\!\!/^B f - \upmu^2 (\Delta\!\!\!/ f) d\!\!\!/_A f \right\}. \end{aligned} \tag{18.11}$$

Integrating both sides of (18.11) over the Riemannian manifold $(S_{t,u}, g\!\!/)$ and using the fact that the integral over $S_{t,u}$ of a perfect divergence vanishes, we deduce that

$$\begin{aligned} \int_{S_{t,u}} \upmu^2 \left\{ |\nabla\!\!\!/^2 f|^2 + \mathfrak{K} |d\!\!\!/ f|^2 \right\} d\upsilon_{g\!\!/} &= 2 \int_{S_{t,u}} \upmu (\Delta\!\!\!/ f)(d\!\!\!/^A \upmu) d\!\!\!/_A f \, d\upsilon_{g\!\!/} \\ &\quad - 2 \int_{S_{t,u}} \upmu (\nabla\!\!\!/_{AB}^2 f)(d\!\!\!/^A \upmu) d\!\!\!/^B f \, d\upsilon_{g\!\!/} \\ &\quad + \int_{S_{t,u}} \upmu^2 (\Delta\!\!\!/ f)^2 \, d\upsilon_{g\!\!/}. \end{aligned} \tag{18.12}$$

From (18.12) and the elementary inequalities

$$2 \left| \upmu (\Delta\!\!\!/ f)(d\!\!\!/^A \upmu) d\!\!\!/_A f \right| \leq \upmu^2 (\Delta\!\!\!/ f)^2 + |d\!\!\!/ \upmu|^2 |d\!\!\!/ f|^2, \tag{18.13}$$

$$2 \left| \upmu (\nabla\!\!\!/_{AB}^2 f)(d\!\!\!/^A \upmu) d\!\!\!/^B f \right| \leq \frac{1}{2} \upmu^2 |\nabla\!\!\!/^2 f|^2 + 2 |d\!\!\!/ \upmu|^2 |d\!\!\!/ f|^2, \tag{18.14}$$

we deduce that

$$\int_{S_{t,u}} \upmu^2 \left\{ \frac{1}{2} |\nabla\!\!\!/^2 f|^2 + \mathfrak{K} |d\!\!\!/ f|^2 \right\} d\upsilon_{g\!\!/} \leq 2 \int_{S_{t,u}} \upmu^2 (\Delta\!\!\!/ f)^2 \, d\upsilon_{g\!\!/} + 3 \int_{S_{t,u}} |d\!\!\!/ \upmu|^2 |d\!\!\!/ f|^2 \, d\upsilon_{g\!\!/}. \tag{18.15}$$

By (18.6), we have $\mathfrak{K} > 0$. Also using the inequality $|d\!\!\!/ \upmu| \lesssim \varepsilon \ln(e+t)(1+t)^{-1}$ (which follows from (12.46a) and (12.128a)) to estimate the last integral on the right-hand side of (18.15), we arrive at the desired inequality (18.10). □

LEMMA 18.9 (**Elliptic estimates for symmetric, trace-free type** $\binom{0}{2}$ **tensorfields on** $S_{t,u}$). *Let ξ be a* ***symmetric, trace-free*** *type $\binom{0}{2}$ $S_{t,u}$ tensorfield. Under the small-data and bootstrap assumptions of Sects.* 12.1-12.4, *if ε is sufficiently small, then the following integral inequality holds for* $(t,u) \in [0, T_{(Bootstrap)}) \times [0, U_0]$:

$$\int_{S_{t,u}} \upmu^2 |\nabla\!\!\!/ \xi|^2 \, d\upsilon_{g\!\!/} \leq 6 \int_{S_{t,u}} \upmu^2 |\mathrm{div}\!\!\!/ \, \xi|^2 \, d\upsilon_{g\!\!/} + C\varepsilon^2 \frac{\ln^2(e+t)}{(1+t)^2} \int_{S_{t,u}} |\xi|^2 \, d\upsilon_{g\!\!/}. \tag{18.16}$$

PROOF. Our proof relies on the following identities, which are valid for ξ verifying the hypotheses of the lemma and which we now prove:

$$\nabla\!\!\!/^2_{CA}\xi_B^{\ C} - \nabla\!\!\!/^2_{AC}\xi_B^{\ C} = 2\mathfrak{K}\xi_{AB}, \tag{18.17}$$

$$(\mathrm{div}\!\!\!/\,\xi)_B = (g\!\!\!/^{-1})^{AC}(\nabla\!\!\!/_A\xi_{BC} - \nabla\!\!\!/_B\xi_{AC}), \tag{18.18}$$

$$|\nabla\!\!\!/\xi|^2 = |\mathrm{div}\!\!\!/\,\xi|^2 + (\nabla\!\!\!/_A\xi_{BC})\nabla\!\!\!/^B\xi^{AC}. \tag{18.19}$$

The identity (18.17) is a straightforward consequence of definition (11.2), the identity (18.2), the symmetry of ξ, and the assumption that $\mathrm{tr}_{g\!\!\!/}\xi = 0$. The identity (18.18) is an immediate consequence of the assumption that $\mathrm{tr}_{g\!\!\!/}\xi = 0$. To prove (18.19), we first bring the last term on the right-hand side over to the left-hand side to deduce that (18.19) is equivalent to

$$\frac{1}{2}(\nabla\!\!\!/_A\xi_{BC} - \nabla\!\!\!/_B\xi_{AC})\left(\nabla\!\!\!/^A\xi^{BC} - \nabla\!\!\!/^B\xi^{AC}\right) = |\mathrm{div}\!\!\!/\,\xi|^2. \tag{18.20}$$

The identity (18.20) is easy to verify relative to a local $g\!\!\!/$-orthonormal frame $\{\widetilde{X}_{\widetilde{A}}\}_{\widetilde{A}=1,2}$ on $S_{t,u}$, as we now explain. Relative to such a frame, we have, with $\xi_{\widetilde{A}} := \xi\cdot\widetilde{X}_{\widetilde{A}}$, $\mathrm{tr}_{g\!\!\!/}\xi = \xi_{\widetilde{1}\widetilde{1}} + \xi_{\widetilde{2}\widetilde{2}} = 0$ and hence the left-hand side of (18.20) can be expressed as $(\nabla\!\!\!/_{\widetilde{1}}\xi_{\widetilde{2}\widetilde{1}} - \nabla\!\!\!/_{\widetilde{2}}\xi_{\widetilde{1}\widetilde{1}})^2 + (\nabla\!\!\!/_{\widetilde{1}}\xi_{\widetilde{2}\widetilde{2}} - \nabla\!\!\!/_{\widetilde{2}}\xi_{\widetilde{1}\widetilde{2}})^2 = (\nabla\!\!\!/_{\widetilde{1}}\xi_{\widetilde{2}\widetilde{1}} + \nabla\!\!\!/_{\widetilde{2}}\xi_{\widetilde{2}\widetilde{2}})^2 + (\nabla\!\!\!/_{\widetilde{1}}\xi_{\widetilde{1}\widetilde{1}} + \nabla\!\!\!/_{\widetilde{2}}\xi_{\widetilde{1}\widetilde{2}})^2$. We now observe that, in view of the frame under consideration, the right-hand side of the previous expression is equal to $|\mathrm{div}\!\!\!/\,\xi|^2$. We have therefore proved (18.19).

Next, with the help of (18.17)-(18.19), we compute that

$$\begin{aligned} \mu^2|\nabla\!\!\!/\xi|^2 + 2\mu^2\mathfrak{K}|\xi|^2 &= 2\mu^2|\mathrm{div}\!\!\!/\,\xi|^2 + 2\mu\xi^{AB}(d\!\!\!/_A\mu)(\mathrm{div}\!\!\!/\,\xi)_B - 2\mu\xi_{BC}(d\!\!\!/_A\mu)\nabla\!\!\!/^B\xi^{AC} \\ &\quad + \mathrm{div}\!\!\!/\,Y, \end{aligned} \tag{18.21}$$

$$Y^A := \mu^2\xi_{BC}\nabla\!\!\!/^B\xi^{AC} - \mu^2\xi^{AB}(\mathrm{div}\!\!\!/\,\xi)_B. \tag{18.22}$$

Integrating (18.21) over the Riemannian manifold $(S_{t,u}, g\!\!\!/)$ and using the fact that the integral over $S_{t,u}$ of a perfect divergence vanishes, we deduce that

$$\begin{aligned} &\int_{S_{t,u}} \mu^2\left\{|\nabla\!\!\!/\xi|^2 + 2\mathfrak{K}|\xi|^2\right\} dv_{g\!\!\!/} \\ &= 2\int_{S_{t,u}} \mu^2|\mathrm{div}\!\!\!/\,\xi|^2\,dv_{g\!\!\!/} + 2\int_{S_{t,u}} \mu\xi^{AB}(d\!\!\!/_A\mu)(\mathrm{div}\!\!\!/\,\xi)_B\,dv_{g\!\!\!/} \\ &\quad - 2\int_{S_{t,u}} \mu\xi_{BC}(d\!\!\!/_A\mu)\nabla\!\!\!/^B\xi^{AC}\,dv_{g\!\!\!/}. \end{aligned} \tag{18.23}$$

From (18.23) and the elementary inequalities

$$2\left|\mu\xi^{AB}(d\!\!\!/_A\mu)(\mathrm{div}\!\!\!/\,\xi)_B\right| \leq \mu^2|\mathrm{div}\!\!\!/\,\xi|^2 + |d\!\!\!/\mu|^2|\xi|^2, \tag{18.24}$$

$$2\left|\mu\xi_{BC}(d\!\!\!/_A\mu)\nabla\!\!\!/^B\xi^{AC}\right| \leq \frac{1}{2}\mu^2|\nabla\!\!\!/\xi|^2 + 2|d\!\!\!/\mu|^2|\xi|^2, \tag{18.25}$$

we deduce that

$$\int_{S_{t,u}} \mu^2\left\{\frac{1}{2}|\nabla\!\!\!/\xi|^2 + 2\mathfrak{K}|\xi|^2\right\} dv_{g\!\!\!/} \leq 3\int_{S_{t,u}} \mu^2|\mathrm{div}\!\!\!/\,\xi|^2\,dv_{g\!\!\!/} + 3\int_{S_{t,u}} |d\!\!\!/\mu|^2|\xi|^2\,dv_{g\!\!\!/}. \tag{18.26}$$

The desired estimate (18.16) now follows from (18.26), (18.6), and the estimate $|d\!\!\!/\mu| \lesssim \varepsilon\ln(e+t)(1+t)^{-1}$ noted at the end of our proof of Lemma 18.8. $\square$

18.2. Sobolev embedding

After we have derived suitable a priori L^2 estimates, we will use the following Sobolev embedding-type proposition and the subsequent corollary to improve the fundamental C^0 bootstrap assumptions (**BA**Ψ) and (**BA**$'\Psi$) for Ψ. For an alternate proof based on the isoperimetric Sobolev inequality, see [**17**, Lemma 13.1].

PROPOSITION 18.10 (C^0 **bounds for** f **in terms of** $\left\|\mathscr{O}^{\leq 2} f\right\|_{L^2(S_{t,u})}$). *Let f be a scalar-valued function and let $\mathscr{O} := \{O_{(1)}, O_{(2)}, O_{(3)}\}$ denote the set of geometric rotation vectorfields, which are $S_{t,u}$-tangent. Under the small-data and bootstrap assumptions of Sects.* 12.1-12.4, *if ε is sufficiently small, then the following Sobolev embedding estimate holds for $(t,u) \in [0, T_{(Bootstrap)}) \times [0, U_0]$:*

$$\|f\|_{C^0(S_{t,u})} \lesssim \frac{1}{1+t} \left\|\mathscr{O}^{\leq 2} f\right\|_{L^2(S_{t,u})}. \tag{18.27}$$

PROOF. We will prove the following Sobolev embedding result:

$$\varrho^2(t,u)\|f\|^2_{C^0(S_{t,u})} \leq C \int_{S_{t,u}} |f|^2 + \varrho^2 |\not{d} f|^2 + \varrho^4 |\not{\nabla}^2 f|^2 \, dv_{\not{g}}. \tag{18.28}$$

Inequality (18.27) then follows from (12.46a), (12.46b), and (18.28).

To prove (18.28), we fix t, u and define the rescaled metric $\not{h} := \varrho^{-2} \not{g}$ on $S_{t,u}$, where $\varrho = 1-u+t$. From scaling considerations, we deduce that (18.28) is equivalent to

$$\begin{aligned} \|f\|^2_{C^0(S_{t,u})} &\leq C \int_{S_{t,u}} |f|^2 + |\not{d} f|^2_{\not{h}} + |\not{\nabla}^2 f|^2_{\not{h}} \, dv_{\not{h}} \\ &= C \int_{\vartheta \in \mathbb{S}^2} |f|^2 + |\not{d} f|^2_{\not{h}} + |\not{\nabla}^2 f|^2_{\not{h}} \, dv_{\not{h}(t,u,\vartheta)}, \end{aligned} \tag{18.29}$$

where $|\cdot|_{\not{h}}$ denotes the pointwise norm of a tensor as measured by $\not{h}$, and there is no ambiguity in the meaning of $\not{\nabla}^2 f$ since this expression is invariant under rescaling $\not{g}$ by a scalar factor that is independent of ϑ. Also, $dv_{\not{h}(t,u,\vartheta)} = \sqrt{\det \not{h}(t,u,\vartheta)}$, where the determinant is taken relative to the geometric coordinates $(\vartheta^1, \vartheta^2)$ induced on $S_{t,u}$.

To prove (18.29), we let $i_{(inject)}(\not{h})(t,u)$ denote the injectivity radius[1] of $(\mathbb{S}^2, \not{h}(t,u,\cdot))$. By [**9**, Theorem 2.20], an estimate of the form (18.29) holds, and furthermore, the constant C can be bounded from above by a continuous function that depends only on a lower bound for $i_{(inject)}(\not{h})(t,u)$ and the absolute value of the Gaussian curvature $\mathfrak{K}(\not{h})(t,u,\vartheta)$ of $\not{h}(t,u,\vartheta)$ (this follows from the proof of [**9**, Theorem 2.20]). To complete the proof of the proposition, it remains only for us to derive these bounds. To proceed, we first use scaling considerations and the estimate (18.6) for $\mathfrak{K}(\not{g})$ to deduce

$$\left|\mathfrak{K}(\not{h}) - 1\right| (t,u,\vartheta) \lesssim \varepsilon \frac{\ln(e+t)}{1+t}. \tag{18.30}$$

[1] The injectivity radius at the point $\vartheta \in \mathbb{S}^2$, denoted by $i_{(inject);\vartheta}(\not{h})(t,u)$, is by definition the sup of the set of radii ρ such that the exponential map at ϑ is a diffeomorphism on the geodesic ball (corresponding to $\not{h}(t,u)$) of radius ρ. The number $i_{(inject)}(\not{h})(t,u)$ is by definition $\inf_{\vartheta \in \mathbb{S}^2} i_{(inject);\vartheta}(\not{h})(t,u)$.

The desired bounds for $|\mathfrak{K}(\not{h})|$ clearly follow from (18.30). Furthermore, since (18.30) implies in particular that $\mathfrak{K}(\not{h})$ is positive, an argument of Klingenberg (see [**12**, Theorem 5.9] or the proof given on [**19**, pg. 34]) implies that

$$i_{(inject)}(\not{h})(t,u) \geq \frac{\pi}{\sqrt{\sup_{\vartheta \in \mathbb{S}^2} \mathfrak{K}(\not{h})(t,u,\vartheta)}} \geq \pi - C\varepsilon \frac{\ln(e+t)}{1+t}. \tag{18.31}$$

The desired lower bound for $i_{(inject)}(\not{h})(t,u)$ clearly follows from (18.31). We have thus proved the proposition. □

COROLLARY 18.11 (C^0 **bounds for** $\mathscr{Z}^N\Psi$ **in terms of the energies**). *Let f be a scalar-valued function, let $\mathbb{E}[\;](t,u)$ be the energy functional defined in* (10.21a)*, and let $\mathscr{O} := \{O_{(1)}, O_{(2)}, O_{(3)}\}$ denote the set of geometric rotation vectorfields. Under the small-data and bootstrap assumptions of Sects.* 12.1-12.4*, if ε is sufficiently small, then the following estimates hold for $(t,u) \in [0, T_{(Bootstrap)}) \times [0, U_0]$:*

$$\|f\|_{C^0(S_{t,u})} \leq C\frac{1}{1+t} \sup_{(t',u') \in [0,t] \times [0,u]} \mathbb{E}^{1/2}[\mathscr{O}^{\leq 2} f](t',u'). \tag{18.32}$$

PROOF. Inequality (18.32) follows from (14.1a) and (18.27). □

CHAPTER 19

Square Integral Estimates for the Eikonal Function Quantities that Do Not Rely on Modified Quantities

In this chapter, we use the pointwise estimates of Prop. 12.46 to derive corresponding L^2 estimates for the below-top-order derivatives of $\upmu$, $L^i_{(Small)}$, and $\upchi^{(Small)}$ in terms of the fundamental L^2-controlling quantities. These estimates are easy to derive because we allow them to lose one derivative relative to Ψ and hence we do not need to invoke the modified quantities of Chapter 11. We also estimate the top-order derivatives of $\upmu$, $L^i_{(Small)}$, and $\upchi^{(Small)}$ that involve at least one L differentiation. These estimates are easy to derive without losing derivatives and without the help of the modified quantities.

19.1. Square integral estimates for the eikonal function quantities that do not rely on modified quantities

We provide the desired L^2 estimates in the next lemma.

LEMMA 19.1 (**Preliminary L^2 estimates for the below-top-order derivatives and the top-order L derivatives of $\upmu$, $L^i_{(Small)}$, and $\upchi^{(Small)}$**). *Let $0 \leq N \leq 23$ be an integer. Under the small-data and bootstrap assumptions of Sects. 12.1-12.4, if ε is sufficiently small, then the following estimates hold for $(t, u) \in [0, T_{(Bootstrap)}) \times [0, U_0]$:*

(19.1a)
$$\left\| \begin{pmatrix} \mathscr{Z}^{\leq N}(\upmu - 1) \\ \sum_{a=1}^3 \varrho \left| \mathscr{Z}^{\leq N} L^a_{(Small)} \right| \\ \varrho^2 \not{\mathcal{L}}_{\mathscr{Z}}^{\leq N-1} \upchi^{(Small)} \\ \varrho^2 \not{\mathcal{L}}_{\mathscr{Z}}^{\leq N-1} \upchi^{(Small)\#} \\ \varrho^2 \mathscr{Z}^{\leq N-1} \mathrm{tr}_{\not{g}} \upchi^{(Small)} \\ \varrho^2 \not{\mathcal{L}}_{\mathscr{Z}}^{\leq N-1} \hat{\upchi}^{(Small)} \\ \varrho^2 \not{\mathcal{L}}_{\mathscr{Z}}^{\leq N-1} \hat{\upchi}^{(Small)\#} \end{pmatrix} \right\|_{L^2(\Sigma_t^u)} \lesssim (1+t)\varepsilon + (1+t) \int_{s=0}^t \frac{\mathbb{Q}^{1/2}_{(\leq N+1)}(s,u)}{1+s} \, ds,$$

$$\left\| \begin{pmatrix} L \mathscr{Z}^{\leq N} \upmu \\ \sum_{a=1}^3 \left| L(\varrho \mathscr{Z}^{\leq N} L^a_{(Small)}) \right| \\ \varrho^2 \not{\mathcal{L}}_L \not{\mathcal{L}}_{\mathscr{Z}}^{\leq N-1} \upchi^{(Small)} \\ \not{\mathcal{L}}_L (\varrho^2 \not{\mathcal{L}}_{\mathscr{Z}}^{\leq N-1} \upchi^{(Small)\#}) \\ L(\varrho^2 \mathscr{Z}^{\leq N-1} \mathrm{tr}_{\not{g}} \upchi^{(Small)}) \\ \varrho^2 \not{\mathcal{L}}_L \not{\mathcal{L}}_{\mathscr{Z}}^{\leq N-1} \hat{\upchi}^{(Small)} \\ \not{\mathcal{L}}_L (\varrho^2 \not{\mathcal{L}}_{\mathscr{Z}}^{\leq N-1} \hat{\upchi}^{(Small)\#}) \end{pmatrix} \right\|_{L^2(\Sigma_t^u)} \lesssim \varepsilon \frac{\ln(e+t)}{1+t} + \mathbb{Q}^{1/2}_{(\leq N+1)}(t,u). \tag{19.1b}$$

Furthermore, if $0 \leq N \leq 24$ is an integer, then we have the following estimates:

(19.2a)

$$\left\| \begin{pmatrix} \mathscr{Z}^{\leq N}(\upmu - 1) \\ \sum_{a=1}^{3} \varrho \left| \mathscr{Z}^{\leq N} L^a_{(Small)} \right| \\ \varrho^2 \mathcal{L}_{\mathscr{Z}}^{\leq N-1} \chi^{(Small)} \\ \varrho^2 \mathcal{L}_{\mathscr{Z}}^{\leq N-1} \chi^{(Small)\#} \\ \varrho^2 \mathscr{Z}^{\leq N-1} \mathrm{tr}_{\not{g}} \chi^{(Small)} \\ \varrho^2 \mathcal{L}_{\mathscr{Z}}^{\leq N-1} \hat{\chi}^{(Small)} \\ \varrho^2 \mathcal{L}_{\mathscr{Z}}^{\leq N-1} \hat{\chi}^{(Small)\#} \end{pmatrix} \right\|_{L^2(\Sigma_t^u)} \lesssim (1+t)\varepsilon + (1+t) \int_{s=0}^{t} \frac{\mathbb{Q}_{(\leq N)}^{1/2}(s,u)}{1+s}\, ds + (1+t) \int_{s=0}^{t} \frac{\widetilde{\mathbb{Q}}_{(\leq N)}^{1/2}(s,u)}{(1+s)\upmu_\star^{1/2}(s,u)}\, ds,$$

(19.2b)

$$\left\| \begin{pmatrix} L \mathscr{Z}^{\leq N} \upmu \\ \sum_{a=1}^{3} \left| L(\varrho \mathscr{Z}^{\leq N} L^a_{(Small)}) \right| \\ \varrho^2 \mathcal{L}_L \mathcal{L}_{\mathscr{Z}}^{\leq N-1} \chi^{(Small)} \\ \mathcal{L}_L (\varrho^2 \mathcal{L}_{\mathscr{Z}}^{\leq N-1} \chi^{(Small)\#}) \\ L(\varrho^2 \mathscr{Z}^{\leq N-1} \mathrm{tr}_{\not{g}} \chi^{(Small)}) \\ \varrho^2 \mathcal{L}_L \mathcal{L}_{\mathscr{Z}}^{\leq N-1} \hat{\chi}^{(Small)} \\ \mathcal{L}_L (\varrho^2 \mathcal{L}_{\mathscr{Z}}^{\leq N-1} \hat{\chi}^{(Small)\#}) \end{pmatrix} \right\|_{L^2(\Sigma_t^u)} \lesssim \varepsilon \frac{\ln(e+t)}{1+t} + \mathbb{Q}_{(\leq N)}^{1/2}(t,u) + \upmu_\star^{-1/2}(t,u) \widetilde{\mathbb{Q}}_{(\leq N)}^{1/2}(t,u).$$

PROOF. To simplify the notation, we set

(19.3)

$$q_N(t,u,\vartheta) := \left| \begin{pmatrix} \mathscr{Z}^{\leq N}(\upmu - 1) \\ \sum_{a=1}^{3} \varrho \left| \mathscr{Z}^{\leq N} L^a_{(Small)} \right| \\ \varrho^2 \mathscr{Z}^{\leq N-1} \mathrm{tr}_{\not{g}} \chi^{(Small)} \end{pmatrix} \right| (t,u,\vartheta), \qquad \mathring{q}_N(u,\vartheta) := q_N(0,u,\vartheta).$$

Integrating inequality (12.127b) along the integral curves of L to derive a pointwise bound for q_N, using Lemma 12.57 to obtain an L^2 bound for q_N, and then multiplying both sides by $(1+t)^{-1}$, we deduce that

(19.4)

$$\frac{1}{1+t} \| q_N \|_{L^2(\Sigma_t^u)} \leq \frac{1}{1+t} \| \mathring{q}_N \|_{L^2(\Sigma_t^u)} + C \int_{s=0}^{t} \frac{1}{1+s} \left\| \begin{pmatrix} \varrho L \mathscr{Z}^{\leq N} \Psi \\ \varrho \not{d} \mathscr{Z}^{\leq N} \Psi \\ \breve{R} \mathscr{Z}^{\leq N} \Psi \\ \mathscr{Z}^{\leq N} \Psi \end{pmatrix} \right\|_{L^2(\Sigma_s^u)} ds + C \int_{s=0}^{t} \frac{\ln(e+s)}{(1+s)^3} \| q_N \|_{L^2(\Sigma_s^u)}\, ds.$$

To bound the first term on the right-hand side of (19.4) by $\lesssim \varepsilon$, we use Lemma 12.53 and the definition (3.85b) of the norm $\|\cdot\|_{L^2(\Sigma_t^u)}$ to deduce that $(1+t)^{-1}\|\mathring{q}_N\|_{L^2(\Sigma_t^u)} \lesssim \|\mathring{q}_N\|_{L^2(\Sigma_0^u)}$, and then Lemma 12.17, Lemma 12.30, and (12.3) to deduce that $\|\mathring{q}_N\|_{L^2(\Sigma_0^u)} \lesssim \mathring{\epsilon} \leq \varepsilon$.

To bound the second term on the right-hand side of (19.4), we first use Prop. 14.7 to deduce that the array under the time integral is bounded in the norm $\|\cdot\|_{L^2(\Sigma_s^u)}$ by $\lesssim \mathbb{Q}_{(\leq N)}^{1/2}(s,u) + \upmu_\star^{-1/2}(s,u)\widetilde{\mathbb{Q}}_{(\leq N)}^{1/2}(s,u)$. Hence, the second term is

$$\leq C\int_{s=0}^{t}\frac{1}{1+s}\mathbb{Q}_{(\leq N)}^{1/2}(s,u)\,ds + C\int_{s=0}^{t}\frac{1}{(1+s)\upmu_\star^{1/2}(s,u)}\widetilde{\mathbb{Q}}_{(\leq N)}^{1/2}(s,u)\,ds. \tag{19.5}$$

We now insert these two bounds into the right-hand side of (19.4) and apply Gronwall's inequality to the quantity $(1+t)^{-1}\|q_N\|_{L^2(\Sigma_t^u)}$, which thus leads to the desired bound (19.2a) for the first two terms and the fifth term in the array on the left-hand side.

To obtain the desired bounds (19.2a) for $\varrho^2 \not{\mathcal{L}}_{\mathscr{Z}}^{\leq N-1}\chi^{(Small)}$ and $\varrho^2 \not{\mathcal{L}}_{\mathscr{Z}}^{\leq N-1}\hat{\chi}^{(Small)}$, we first note the following inequality, which holds for any symmetric type $\binom{0}{2}$ $S_{t,u}$ tensorfield ξ:

$$\begin{aligned} \left|L(|\varrho^2\xi|)\right| &\leq \left|\varrho^2 \not{\mathcal{L}}_L\xi\right| + 2\left|\chi^{(Small)}\right|\left|\varrho^2\xi\right| \\ &\leq \left|\varrho^2 \not{\mathcal{L}}_L\xi\right| + C\varepsilon\frac{\ln(e+t)}{(1+t)^2}\left|\varrho^2\xi\right|. \end{aligned} \tag{19.6}$$

The inequality (19.6) follows in a straightforward fashion from the identity $\not{\mathcal{L}}_L \not{g}^{-1} = -2\chi^{\#\#}$ (see (3.72) and (7.4a)-(7.4b)), the decomposition $\chi = \varrho^{-1}\not{g} + \chi^{(Small)}$ (see (4.1c)), the fact that $L\varrho = 1$, and the estimate (12.128a). We now apply (19.6) with $\not{\mathcal{L}}_{\mathscr{Z}}^{\leq N-1}\chi^{(Small)}$ and $\not{\mathcal{L}}_{\mathscr{Z}}^{\leq N-1}\hat{\chi}^{(Small)}$ in the role of ξ and also use inequality (12.127b) to bound the term $\left|\varrho^2 \not{\mathcal{L}}_L\xi\right|$. A Gronwall argument similar to the one used in the previous paragraph now yields the desired estimates. We remark that the smallness of the data of $\varrho^2 \not{\mathcal{L}}_{\mathscr{Z}}^{\leq N-1}\chi^{(Small)}$ and $\varrho^2 \not{\mathcal{L}}_{\mathscr{Z}}^{\leq N-1}\hat{\chi}^{(Small)}$ follows from the same argument that we used above to deduce the smallness of $\mathring{q}_N$.

To obtain the desired bounds (19.2a) for

$$\varrho^2 \not{\mathcal{L}}_{\mathscr{Z}}^{\leq N-1}\chi^{(Small)\#} \quad \text{and} \quad \varrho^2 \not{\mathcal{L}}_{\mathscr{Z}}^{\leq N-1}\hat{\chi}^{(Small)\#},$$

we first note the following inequality, which holds for any type $\binom{1}{1}$ $S_{t,u}$ tensorfield:

$$\begin{aligned} \left|L|\xi|\right| &\leq |\not{\mathcal{L}}_L\xi| + 2\left|\chi^{(Small)}\right||\xi| \\ &\leq |\not{\mathcal{L}}_L\xi| + C\varepsilon\frac{\ln(e+t)}{(1+t)^2}|\xi|. \end{aligned} \tag{19.7}$$

The inequality (19.7) follows in a straightforward fashion from the identities $\not{\mathcal{L}}_L \not{g} = 2\chi$ and $\not{\mathcal{L}}_L \not{g}^{-1} = -2\chi^{\#\#}$ (see (3.72) and (7.4a)-(7.4b)), the decomposition $\chi = \varrho^{-1}\not{g} + \chi^{(Small)}$ mentioned above, the fact that $L\varrho = 1$, and the estimate (12.128a). We now apply (19.7) with $\varrho^2 \not{\mathcal{L}}_{\mathscr{Z}}^{\leq N-1}\chi^{(Small)\#}$ and $\varrho^2 \not{\mathcal{L}}_{\mathscr{Z}}^{\leq N-1}\hat{\chi}^{(Small)\#}$ in the role of ξ and also use inequality (12.127b) to bound the term $|\not{\mathcal{L}}_L\xi|$. A Gronwall argument similar to the one used above now yields the desired estimates. We remark that the smallness of the data of $\varrho^2 \not{\mathcal{L}}_{\mathscr{Z}}^{\leq N-1}\chi^{(Small)\#}$ and $\varrho^2 \not{\mathcal{L}}_{\mathscr{Z}}^{\leq N-1}\hat{\chi}^{(Small)\#}$ follows from the same argument that we used above to deduce the smallness of $\mathring{q}_N$. We have therefore proved (19.2a).

The proof of (19.1a) is very similar. The only difference is that we use Prop. 14.7 to bound the array under the time integral in the second term on the right-hand side of (19.4) in the norm $\|\cdot\|_{L^2(\Sigma_s^u)}$ by $\lesssim \mathbb{Q}^{1/2}_{(\leq N+1)}(s,u)$.

Inequality (19.2b) follows from the estimates (12.127b) and (19.2a), inequality (13.152), the fact that the array of Ψ-dependent terms on the right-hand side of (12.127b) is bounded in the norm $\|\cdot\|_{L^2(\Sigma_t^u)}$ by $\lesssim \mathbb{Q}^{1/2}_{(\leq N)}(t,u) + \mu_\star^{-1/2}(t,u)\widetilde{\mathbb{Q}}^{1/2}_{(\leq N)}(t,u)$ (this fact follows from Prop. 14.7), and the fact that $\mathbb{Q}_{(\leq N)}$ and $\widetilde{\mathbb{Q}}_{(\leq N)}$ are increasing in their arguments.

Inequality (19.1b) similarly follows from the estimates (12.127b) and (19.1a) and the fact that the array of Ψ-dependent terms on the right-hand side of (12.127b) is bounded in the norm $\|\cdot\|_{L^2(\Sigma_t^u)}$ by $\lesssim \mathbb{Q}^{1/2}_{(\leq N+1)}(t,u)$ (this fact follows from Prop. 14.7). □

CHAPTER 20

A Priori Estimates for the Fundamental Square-Integral-Controlling Quantities

In this chapter, we derive the most important estimates of the monograph: a priori "energy" estimates for the fundamental L^2-controlling quantities $\mathbb{Q}_{(N)}(t,u)$, $\widetilde{\mathbb{Q}}_{(N)}(t,u)$, and $\widetilde{\mathbb{K}}_{(N)}(t,u)$, $(N \leq 24)$, defined in Sect. 14.2. To this end, we first carry out the main precursor step: we use the pointwise estimates from Chapters 15 and 17 to bound the error integrals that arise in the energy-cone flux identities for the commuted wave equation. By "energy-cone flux identities," we mean the identities of Prop. 10.13 for solutions to $\mu\square_{g(\Psi)}(\mathscr{Z}^N\Psi) = {}^{(\mathscr{Z}^N)}\mathfrak{F}$. To make the analysis more tractable, we devote different sections to bounding the different kinds of error integrals. We carry out this analysis in Sects. 20.3-20.7, and in particular, we bound the most difficult error integrals in Sects. 20.6-20.7. We then combine the large number of error integral estimates into our two main propositions of this chapter, which are stated in Sect. 20.2: Propositions 20.8 and 20.9. The first proposition provides inequalities verified by the up-to-top-order $(N \leq 24)$ fundamental L^2-controlling quantities, while the second provides less degenerate inequalities for the lower-order $(N \leq 23)$ fundamental L^2-controlling quantities. The main point is that the estimates of the propositions imply that the hypotheses of the Gronwall-type Lemma 20.10 are verified by $\mathbb{Q}_{(N)}(t,u)$, $\widetilde{\mathbb{Q}}_{(N)}(t,u)$, and $\widetilde{\mathbb{K}}_{(N)}(t,u)$. Hence, the lemma yields the desired a priori estimates. We provide the proofs of Propositions 20.8 and 20.9 in Sects. 20.8 and 20.9. In Sect. 20.10, we provide the proof of the Gronwall-type lemma.

20.1. Bootstrap assumptions for the fundamental square-integral-controlling quantities

In deriving our a priori estimates, we find it convenient to make bootstrap assumptions for the L^2-controlling quantities $\mathbb{Q}_{(N)}(t,u)$, $\widetilde{\mathbb{Q}}_{(N)}(t,u)$, and $\widetilde{\mathbb{K}}_{(N)}(t,u)$ from Definitions 14.2 and 14.3. To state the bootstrap assumptions, we first recall that $\mu_\star(t,u) := \min\{1, \inf_{\Sigma_t^u} \mu\}$ and that $\varepsilon > 0$ is the small bootstrap parameter from Chapter 12. Our L^2 bootstrap assumptions are that the following inequalities hold for $(t,u) \in [0, T_{(Bootstrap)}) \times [0, U_0]$:

L^2-type Bootstrap Assumptions

(20.1a) $$\mathbb{Q}_{(N)}^{1/2}(t,u) \leq \varepsilon, \qquad (0 \leq N \leq 15),$$

(20.1b) $$\widetilde{\mathbb{Q}}_{(N)}^{1/2}(t,u) + \widetilde{\mathbb{K}}_{(N)}^{1/2}(t,u) \leq \varepsilon \ln^2(e+t), \qquad (0 \leq N \leq 15),$$

(20.1c) $$\mathbb{Q}_{(16+M)}^{1/2}(t,u) \leq \varepsilon \upmu_{\star}^{-.75-M}(t,u), \qquad (0 \leq M \leq 7),$$

(20.1d) $$\widetilde{\mathbb{Q}}_{(16+M)}^{1/2}(t,u) + \widetilde{\mathbb{K}}_{(16+M)}^{1/2}(t,u) \leq \varepsilon \ln^2(e+t) \upmu_{\star}^{-.75-M}(t,u), \qquad (0 \leq M \leq 7),$$

(20.1e) $$\mathbb{Q}_{(24)}^{1/2}(t,u) \leq \varepsilon \ln^{A_*}(e+t) \upmu_{\star}^{-8.75}(t,u),$$

(20.1f) $$\widetilde{\mathbb{Q}}_{(24)}^{1/2}(t,u) + \widetilde{\mathbb{K}}_{(24)}^{1/2}(t,u) \leq \varepsilon \ln^{A_*+2}(e+t) \upmu_{\star}^{-8.75}(t,u),$$

where $A_* > 4$ is a large constant that we choose in the proof of Lemma 20.10. Our goal in Chapter 20 is to use the results from the previous chapters to prove that for sufficiently small data, the above bootstrap assumptions in fact hold with $C\mathring{\epsilon}$ in place of ε, where $\mathring{\epsilon}$ is the size of the data as defined in Def. 12.3.

REMARK 20.1 (**We do not use the top-order bootstrap assumptions**). In our proof of the fundamental Gronwall lemma, Lemma 20.10, we do not rely on the top-order bootstrap assumptions (20.1e)-(20.1f). However, we make them anyway for the sake illustrating the expected behavior of $\mathbb{Q}_{(24)}(t,u)$, $\widetilde{\mathbb{Q}}_{(24)}(t,u)$, and $\widetilde{\mathbb{K}}_{(24)}(t,u)$.

REMARK 20.2 (**The initial smallness of $\mathbb{Q}_{(\leq 24)}^{1/2}$ and $\widetilde{\mathbb{Q}}_{(\leq 24)}^{1/2}$**). We recall that the conclusion of Lemma 14.8 is that $\mathbb{Q}_{(\leq 24)}^{1/2}(0,u), \widetilde{\mathbb{Q}}_{(\leq 24)}^{1/2}(0,u) \leq C\mathring{\epsilon}$. Moreover, Def. 14.3 implies that $\widetilde{\mathbb{K}}_{(\leq 24)}^{1/2}(0,u) = 0$. Thus, in view of inequality (12.3), we see that the bootstrap assumptions (20.1a)-(20.1f) are initially satisfied whenever $\mathring{\epsilon}$ is sufficiently small compared to ε.

20.2. Statement of the two main propositions and the fundamental Gronwall lemma

We begin by defining four classes of quantities that appear in the statements of the two main propositions. These quantities collectively provide suitable upper bounds for the various error integrals that arise in our analysis.

REMARK 20.3 (**The importance of the boxed constants**). The "boxed" constants that appear in Defs. 20.4 and 20.6 (and on the right-hand sides of the top-order estimates of Prop. 20.8) are the ones that force us to prove estimates for the top-order quantities $\mathbb{Q}_{(24)}$, $\widetilde{\mathbb{Q}}_{(24)}$, and $\widetilde{\mathbb{K}}_{(24)}$ that are allowed to blow up like a large power of $\upmu_{\star}^{-1}$. More precisely, the size of the boxed constants is directly responsible for the size of the "blow-up exponent" 8.75 on the right-hand sides of (20.1e)-(20.1f). A critically important feature of our analysis is that under our small-data assumption, **the boxed constants can be chosen to be independent of the number of times the equations are commuted with vectorfields** $Z \in \mathscr{Z}$. This prevents the blow-up exponents from growing with the number of commutations and allows us to derive nondegenerate estimates for $\mathbb{Q}_{(\leq 15)}^{1/2}(t,u)$, $\widetilde{\mathbb{Q}}_{(\leq 15)}^{1/2}(t,u)$, and $\widetilde{\mathbb{K}}_{(\leq 15)}^{1/2}(t,u)$, which are essential for deriving improvements of our bootstrap assumptions (**BA**Ψ)-(**BA**$'\Psi$).

DEFINITION 20.4 (**Quantities that bound difficult top-order error integrals generated by the multiplier** T). We define the classes of terms $\boxed{\mathbf{I})}_{(\leq N)}(t,u)$, $\mathbf{II})_{(\leq N)}(t,u)$, $\cdots$, and $\mathbf{X})_{(\leq N)}(t,u)$ to be any functions of (t,u) such that there exist a constant $0 < a < 1/2$ and a constant $C > 0$ such that under the small-data and bootstrap assumptions of Sects. 12.1-12.4, if ε is sufficiently small, then the following estimates hold for $(t,u) \in [0, T_{(Bootstrap)}) \times [0, U_0]$:

(20.2a)

$$\boxed{\mathbf{I})}_{(\leq N)} \leq \boxed{9} \int_{t'=0}^{t} \frac{\|[L\upmu]_-\|_{C^0(\Sigma_{t'}^u)}}{\upmu_\star(t',u)} \mathbb{Q}_{(\leq N)}^{1/2}(t',u) \times \int_{s=0}^{t'} \frac{\|[L\upmu]_-\|_{C^0(\Sigma_s^u)}}{\upmu_\star(s,u)} \mathbb{Q}_{(\leq N)}^{1/2}(s,u)\, ds\, dt',$$

(20.2b)

$$\mathbf{II})_{(\leq N)} \leq C\varepsilon \int_{t'=0}^{t} \frac{1}{(1+t')^{1+a}\upmu_\star(t',u)} \mathbb{Q}_{(\leq N)}^{1/2}(t',u) \times \int_{s=0}^{t'} \frac{1}{(1+s)} \frac{1}{\upmu_\star(s,u)} \mathbb{Q}_{(\leq N)}^{1/2}(s,u)\, ds\, dt',$$

(20.2c)

$$\mathbf{III})_{(\leq N)} \leq C\varepsilon \int_{t'=0}^{t} \frac{1}{(1+t')^{1+a}\upmu_\star(t',u)} \mathbb{Q}_{(\leq N)}^{1/2}(t',u) \times \int_{s=0}^{t'} \frac{1}{(1+s)} \frac{1}{\upmu_\star(s,u)} \widetilde{\mathbb{Q}}_{(\leq N)}^{1/2}(s,u)\, ds\, dt',$$

(20.2d)

$$\mathbf{IV})_{(\leq N)} \leq 9 \int_{t'=0}^{t} \frac{1}{\varrho(t',u)\left\{1+\ln\left(\frac{\varrho(t',u)}{\varrho(0,u)}\right)\right\}} \mathbb{Q}_{(\leq N)}^{1/2}(t',u) \times \int_{s=0}^{t'} \frac{\|[L\upmu]_-\|_{C^0(\Sigma_s^u)}}{\upmu_\star(s,u)} \mathbb{Q}_{(\leq N)}^{1/2}(s,u)\, ds\, dt',$$

(20.2e)

$$\boxed{\mathbf{V})}_{(\leq N)} \leq \boxed{9} \int_{t'=0}^{t} \frac{\|[L\upmu]_-\|_{C^0(\Sigma_{t'}^u)}}{\upmu_\star(t',u)} \mathbb{Q}_{(\leq N)}(t',u)\, dt',$$

(20.2f)

$$\mathbf{VI})_{(\leq N)} \leq 9 \int_{t'=0}^{t} \frac{1}{\varrho(t',u)\left\{1+\ln\left(\frac{\varrho(t',u)}{\varrho(0,u)}\right)\right\}} \mathbb{Q}_{(\leq N)}(t',u)\, dt',$$

(20.2g)

$$\mathbf{VII})_{(\leq N)} \leq C\varepsilon \int_{t'=0}^{t} \frac{1}{(1+t')^{1+a}\upmu_\star(t',u)} \mathbb{Q}_{(\leq N)}(t',u)\, dt',$$

(20.2h)

$$\mathbf{VIII})_{(\leq N)} \leq C\varepsilon \int_{t'=0}^{t} \frac{1}{(1+t')^{1+a}\upmu_\star(t',u)} \widetilde{\mathbb{Q}}_{(\leq N)}(t',u)\, dt',$$

(20.2i)

$$\mathbf{IX})_{(\leq N)} \leq C\varepsilon \frac{1}{\upmu_\star^{1/2}(t,u)} \mathbb{Q}_{(\leq N)}^{1/2}(t,u) \int_{t'=0}^{t} \frac{1}{(1+t')^{1+a}} \mathbb{Q}_{(\leq N)}^{1/2}(t',u)\, dt',$$

(20.2j)

$$\mathbf{X})_{(\leq N)} \leq C\varepsilon \frac{1}{\upmu_{\star}^{1/2}(t,u)} \mathbb{Q}_{(\leq N)}^{1/2}(t,u) \int_{t'=0}^{t} \frac{1}{(1+t')^{1+a}\upmu_{\star}^{1/2}(t',u)} \widetilde{\mathbb{Q}}_{(\leq N)}^{1/2}(t',u)\,dt'.$$

DEFINITION 20.5 (**Quantities that bound easy top-order and below–top-order error integrals generated by the multiplier** T). We define the classes of terms $\mathbf{0})_{(\leq N+1)}(t,u)$, $\mathbf{i})_{(\leq N)}(t,u)$, $\mathbf{ii})_{(\leq N)}(t,u)$, $\cdots$, $\mathbf{vi})_{(\leq N)}(t,u)$, to be any functions of (t,u) such that there exists a constant $C>0$ such that under the small-data and bootstrap assumptions of Sects. 12.1-12.4, if ε is sufficiently small, then the following estimates hold for $(t,u) \in [0, T_{(Bootstrap)}) \times [0, U_0]$:

$$\begin{aligned}
\mathbf{0})_{(\leq N+1)} \leq {} & C\varepsilon \int_{t'=0}^{t} \frac{1}{(1+t')^{3/2}\upmu_{\star}^{1/2}(t',u)} \mathbb{Q}_{(\leq N)}^{1/2}(t',u) \\
& \times \underbrace{\int_{s=0}^{t'} \frac{1}{1+s} \mathbb{Q}_{(\leq N+1)}^{1/2}(s,u)\,ds\,dt'}_{\text{depends on an order } N+1 \text{ quantity}} \\
& + C\varepsilon \int_{t'=0}^{t} \frac{1}{(1+t')^{3/2}\upmu_{\star}^{1/2}(t',u)} \mathbb{Q}_{(\leq N)}^{1/2}(t',u) \\
& \times \underbrace{\int_{s=0}^{t'} \frac{1}{(1+s)\upmu_{\star}^{1/2}(s,u)} \widetilde{\mathbb{Q}}_{(\leq N+1)}^{1/2}(s,u)\,ds\,dt',}_{\text{depends on an order } N+1 \text{ quantity}}
\end{aligned} \tag{20.3a}$$

$$\mathbf{i})_{(\leq N)} \leq C \int_{t'=0}^{t} \frac{1}{(1+t')^{3/2}} \mathbb{Q}_{(\leq N)}(t',u)\,dt', \tag{20.3b}$$

$$\mathbf{ii})_{(\leq N)} \leq C\varepsilon^{1/2} \int_{t'=0}^{t} \frac{\ln(e+t')}{(1+t')^{2}\sqrt{\ln(e+t)-\ln(e+t')}} \widetilde{\mathbb{Q}}_{(\leq N)}(t',u)\,dt', \tag{20.3c}$$

$$\mathbf{iii})_{(\leq N)} \leq C \int_{t'=0}^{t} \frac{1}{(1+t')^{3/2}} \widetilde{\mathbb{Q}}_{(\leq N)}(t',u)\,dt', \tag{20.3d}$$

$$\mathbf{iv})_{(\leq N)} \leq C \int_{u'=0}^{u} \mathbb{Q}_{(\leq N)}(t,u')\,du', \tag{20.3e}$$

$$\mathbf{v})_{(\leq N)} \leq C \frac{1}{(1+t)^{1/2}} \int_{u'=0}^{u} \widetilde{\mathbb{Q}}_{(\leq N)}(t,u')\,du', \tag{20.3f}$$

$$\mathbf{vi})_{(\leq N)} \leq C \sup_{t'\in[0,t]} \frac{\widetilde{\mathbb{K}}_{(\leq N)}(t',u)}{(1+t')^{1/2}}. \tag{20.3g}$$

DEFINITION 20.6 (**Quantities that bound difficult top-order error integrals generated by the multiplier $\widetilde{K}$**). We define the classes of terms $\boxed{\widetilde{\mathbf{I}})}_{(\leq N)}(t,u)$, $\widetilde{\mathbf{II}})_{(\leq N)}(t,u)$, $\cdots$, and $\widetilde{\mathbf{VIII}})_{(\leq N)}(t,u)$ to be any functions of (t,u) such that there exists a constant $C>0$ such that under the small-data and bootstrap assumptions of Sects. 12.1-12.4, if ε is sufficiently small, then the following estimates hold for $(t,u) \in [0, T_{(Bootstrap)}) \times [0, U_0]$:

(20.4a)

$$\boxed{\widetilde{\mathbf{I}})}_{(\leq N)} \leq \boxed{5} \int_{t'=0}^{t} \frac{\|[L\upmu]_-\|_{C^0(\Sigma_{t'}^u)}}{\upmu_\star(t',u)} \widetilde{\mathbb{Q}}_{(\leq N)}(t',u)\,dt',$$

(20.4b)

$$\widetilde{\mathbf{II}})_{(\leq N)} \leq 5 \int_{t'=0}^{t} \frac{1}{\varrho(t',u)\left\{1+\ln\left(\frac{\varrho(t',u)}{\varrho(0,u)}\right)\right\}} \widetilde{\mathbb{Q}}_{(\leq N)}(t',u)\,dt',$$

(20.4c)

$$\widetilde{\mathbf{III}})_{(\leq N)} \leq C\varepsilon \int_{t'=0}^{t} \frac{1}{(1+t')^{3/2}\upmu_\star^{1/2}(t',u)} \mathbb{Q}_{(\leq N)}(t',u)\,dt',$$

(20.4d)

$$\widetilde{\mathbf{IV}})_{(\leq N)} \leq C\varepsilon \int_{t'=0}^{t} \frac{1}{(1+t')^{3/2}\upmu_\star(t',u)} \widetilde{\mathbb{Q}}_{(\leq N)}(t',u)\,dt',$$

(20.4e)

$$\boxed{\widetilde{\mathbf{V}})}_{(\leq N)} \leq \boxed{5} \frac{\|\varrho L\upmu\|_{C^0({}^{(-)}\Sigma_{t;t}^u)}}{\upmu_\star^{1/2}(t,u)} \widetilde{\mathbb{Q}}_{(\leq N)}^{1/2}(t,u) \times \int_{t'=0}^{t} \frac{1}{\varrho(t',u)\upmu_\star^{1/2}(t',u)} \widetilde{\mathbb{Q}}_{(\leq N)}^{1/2}(t',u)\,dt',$$

(20.4f)

$$\widetilde{\mathbf{VI}})_{(\leq N)} \leq 5 \left\| \frac{\varrho L\upmu}{\upmu} \right\|_{C^0({}^{(+)}\Sigma_{t;t}^u)} \widetilde{\mathbb{Q}}_{(\leq N)}^{1/2}(t,u) \times \int_{t'=0}^{t} \left\| \sqrt{\frac{\upmu(t,\cdot)}{\upmu}} \right\|_{C^0({}^{(+)}\Sigma_{t';t}^u)} \frac{1}{\varrho(t',u)} \widetilde{\mathbb{Q}}_{(\leq N)}^{1/2}(t',u)\,dt',$$

(20.4g)

$$\widetilde{\mathbf{VII}})_{(\leq N)} \leq C\varepsilon \frac{1}{\upmu_\star^{1/2}(t,u)} \widetilde{\mathbb{Q}}_{(\leq N)}^{1/2}(t,u) \int_{t'=0}^{t} \frac{1}{(1+t')^{3/2}} \mathbb{Q}_{(\leq N)}^{1/2}(t',u)\,dt',$$

(20.4h)

$$\widetilde{\mathbf{VIII}})_{(\leq N)} \leq C\varepsilon \frac{1}{\upmu_\star^{1/2}(t,u)} \widetilde{\mathbb{Q}}_{(\leq N)}^{1/2}(t,u) \int_{t'=0}^{t} \frac{1}{(1+t')^{3/2}\upmu_\star^{1/2}(t',u)} \widetilde{\mathbb{Q}}_{(\leq N)}^{1/2}(t',u)\,dt'.$$

DEFINITION 20.7 (**Quantities that bound easy top-order and below–top-order error integrals generated by the multiplier $\widetilde{K}$**). We define the classes of terms $\widetilde{\mathbf{0}})_{(\leq N+1)}(t,u)$, $\widetilde{\mathbf{i}})_{(\leq N)}(t,u)$, $\widetilde{\mathbf{ii}})_{(\leq N)}(t,u)$, $\cdots, \widetilde{\mathbf{v}})_{(\leq N)}(t,u)$, to be any functions of (t,u) such that there exists a constant $C>0$ such that under the small-data and bootstrap assumptions of Sects. 12.1-12.4, if ε is sufficiently small, then the following estimates hold for $(t,u) \in [0, T_{(Bootstrap)}) \times [0, U_0]$:

$$\widetilde{\mathbf{0})}_{(\leq N+1)} \leq \underbrace{C\varepsilon \int_{t'=0}^{t} \frac{1}{(1+t')^2} \left(\int_{s=0}^{t'} \frac{\mathbb{Q}^{1/2}_{(\leq N+1)}(s,u)}{1+s} \, ds \right)^2 dt'}_{\text{depends on an order } N+1 \text{ quantity}} \tag{20.5a}$$

$$+ \underbrace{C\varepsilon \int_{t'=0}^{t} \frac{1}{(1+t')^2} \left(\int_{s=0}^{t'} \frac{\widetilde{\mathbb{Q}}^{1/2}_{(\leq N+1)}(s,u)}{(1+s)\upmu_\star^{1/2}(s,u)} \, ds \right)^2 dt',}_{\text{depends on an order } N+1 \text{ quantity}}$$

$$\widetilde{\mathbf{i})}_{(\leq N)} \leq C \ln^4(e+t) \mathbb{Q}_{(\leq N)}(t,u), \tag{20.5b}$$

$$\widetilde{\mathbf{ii})}_{(\leq N)} \leq 2 \int_{t'=0}^{t} \frac{1}{\varrho(t',u)\left\{1 + \ln\left(\frac{\varrho(t',u)}{\varrho(0,u)}\right)\right\}} \widetilde{\mathbb{Q}}_{(\leq N)}(t',u)\, dt', \tag{20.5c}$$

$$\widetilde{\mathbf{iii})}_{(\leq N)} \leq C \int_{t'=0}^{t} \frac{1}{(1+t')^{3/2}} \widetilde{\mathbb{Q}}_{(\leq N)}(t',u)\, dt', \tag{20.5d}$$

$$\widetilde{\mathbf{iv})}_{(\leq N)} \leq C \int_{u'=0}^{u} \widetilde{\mathbb{Q}}_{(\leq N)}(t,u')\, du', \tag{20.5e}$$

$$\widetilde{\mathbf{v})}_{(\leq N)} \leq C \widetilde{\mathbb{K}}_{(\leq N)}(t,u). \tag{20.5f}$$

We now state the two propositions. We provide their proofs in Sects. 20.8 and 20.9.

PROPOSITION 20.8 (**The main top-order energy-cone flux integral inequalities**). *Assume that* $\square_{g(\Psi)}\Psi = 0$ *and consider the quantities* $\boxed{\mathbf{I})}_{(\leq N)}$, $\mathbf{i})_{(\leq N)}$, $\cdots$ *defined in Def. 20.4-Def. 20.7. Let* $0 \leq N \leq 24$ *be an integer. There exists a constant*[1] $C > 0$ *such that under the small-data and bootstrap assumptions of Sects. 12.1-12.4, if* ε *is sufficiently small and* $\varsigma > 0$, $\widetilde{\varsigma} > 0$ *are any real numbers, then the following inequalities hold for* $(t,u) \in [0, T_{(Bootstrap)}) \times [0, U_0]$:

$$\begin{aligned} &\mathbb{Q}_{(\leq N)}(t,u) \\ &\leq \mathbb{Q}_{(\leq N)}(0,u) + C\varepsilon^3 \upmu_\star^{-1}(t,u) + C\varepsilon \mathbb{Q}_{(\leq N)}(t,u) \\ &\quad + C\varepsilon \upmu_\star^{-1}(t,u) \ln^8(e+t) \mathbb{Q}_{(\leq N-1)}(t,u) + C\varepsilon \upmu_\star^{-1}(t,u) \widetilde{\mathbb{Q}}_{(\leq N-1)}(t,u) \\ &\quad + \boxed{\mathbf{I})}_{(\leq N)}(t,u) + \mathbf{II})_{(\leq N)}(t,u) + \cdots + \mathbf{X})_{(\leq N)}(t,u) \\ &\quad + \left\{1 + \varsigma^{-1}\right\} \mathbf{i})_{(\leq N)}(t,u) + \mathbf{ii})_{(\leq N)}(t,u) + \left\{1 + \varsigma^{-1}\right\} \mathbf{iii})_{(\leq N)}(t,u) \\ &\quad + \mathbf{iv})_{(\leq N)}(t,u) + \left\{1 + \varsigma^{-1}\right\} \mathbf{v})_{(\leq N)}(t,u) + \{\varepsilon + \varsigma\} \mathbf{vi})_{(\leq N)}(t,u), \end{aligned} \tag{20.6a}$$

[1] It is understood that the same the constant C also appears on the right-hand sides of the quantities in Def. 20.4-Def. 20.7.

(20.6b)

$$\max\left\{\widetilde{\mathbb{Q}}_{(\leq N)}(t,u), \widetilde{\mathbb{K}}_{(\leq N)}(t,u)\right\}$$
$$\leq C\mathbb{Q}_{(\leq N)}(0,u) + C\varepsilon^3 \upmu_\star^{-1}(t,u) + C\varepsilon\mathbb{Q}_{(\leq N)}(t,u) + C\varepsilon\widetilde{\mathbb{Q}}_{(\leq N)}(t,u)$$
$$+ C\varepsilon\upmu_\star^{-1}(t,u)\ln^2(e+t)\mathbb{Q}_{(\leq N-1)}(t,u) + C\varepsilon\upmu_\star^{-1}(t,u)\ln^2(e+t)\widetilde{\mathbb{Q}}_{(\leq N-1)}(t,u)$$
$$+ \boxed{\widetilde{\mathbf{I}})}_{(\leq N)}(t,u) + \widetilde{\mathbf{II}})_{(\leq N)}(t,u) + \cdots + \widetilde{\mathbf{VIII}})_{(\leq N)}(t,u)$$
$$+ \left\{1+\widetilde{\varsigma}^{-1}\right\}\widetilde{\mathbf{i}})_{(\leq N)}(t,u) + \widetilde{\mathbf{ii}})_{(\leq N)}(t,u) + \left\{1+\widetilde{\varsigma}^{-1}\right\}\widetilde{\mathbf{iii}})_{(\leq N)}(t,u)$$
$$+ \left\{1+\widetilde{\varsigma}^{-1}\right\}\widetilde{\mathbf{iv}})_{(\leq N)}(t,u) + \{\varepsilon+\widetilde{\varsigma}\}\widetilde{\mathbf{v}})_{(\leq N)}(t,u).$$

PROPOSITION 20.9 (**The main below-top-order energy-cone flux integral inequalities**). *Assume that* $\square_{g(\Psi)}\Psi = 0$ *and let* $0 \leq N \leq 23$ *be an integer. Consider the quantities* $\mathbf{0})_{\leq N+1}$, $\mathbf{i})_{(\leq N)}$, $\cdots$ *defined in Defs. 20.5 and 20.7. There exists a constant* $C > 0$ *(where Footnote 1 on pg. 370 applies here as well) such that under the small-data and bootstrap assumptions of Sects. 12.1-12.4, if* ε *is sufficiently small and* $\varsigma > 0$, $\widetilde{\varsigma} > 0$ *are any real numbers, then the following inequalities hold for* $(t,u) \in [0, T_{(Bootstrap)}) \times [0, U_0]$:

(20.7a)
$$\mathbb{Q}_{(\leq N)}(t,u)$$
$$\leq \mathbb{Q}_{(\leq N)}(0,u) + C\varepsilon^3$$
$$+ \mathbf{0})_{\leq N+1}(t,u)$$
$$+ \left\{1+\varsigma^{-1}\right\}\mathbf{i})_{(\leq N)}(t,u) + \mathbf{ii})_{(\leq N)}(t,u) + \left\{1+\varsigma^{-1}\right\}\mathbf{iii})_{(\leq N)}(t,u)$$
$$+ \mathbf{iv})_{(\leq N)}(t,u) + \left\{1+\varsigma^{-1}\right\}\mathbf{v})_{(\leq N)}(t,u) + \{\varepsilon+\varsigma\}\mathbf{vi})_{(\leq N)}(t,u),$$

(20.7b)
$$\max\left\{\widetilde{\mathbb{Q}}_{(\leq N)}(t,u), \widetilde{\mathbb{K}}_{(\leq N)}(t,u)\right\}$$
$$\leq \widetilde{\mathbb{Q}}_{(\leq N)}(0,u) + C\varepsilon^3$$
$$+ \widetilde{\mathbf{0}})_{\leq N+1}(t,u)$$
$$+ \left\{1+\widetilde{\varsigma}^{-1}\right\}\widetilde{\mathbf{i}})_{(\leq N)}(t,u) + \widetilde{\mathbf{ii}})_{(\leq N)}(t,u) + \left\{1-\widetilde{\varsigma}^{-1}\right\}\widetilde{\mathbf{iii}})_{(\leq N)}(t,u)$$
$$+ \left\{1+\widetilde{\varsigma}^{-1}\right\}\widetilde{\mathbf{iv}})_{(\leq N)}(t,u) + \{\varepsilon+\widetilde{\varsigma}\}\widetilde{\mathbf{v}})_{(\leq N)}(t,u).$$

Before proving Prop. 20.8 and Prop. 20.9, we first state our main Gronwall-type lemma. The lemma allows us to derive suitable a priori estimates for $\mathbb{Q}_{(\leq N)}(t,u)$, $\widetilde{\mathbb{Q}}_{(\leq N)}(t,u)$, and $\widetilde{\mathbb{K}}_{(\leq N)}(t,u)$ based on the inequalities provided by Props. 20.8 and 20.9. We provide the proof of the lemma in Sect. 20.10. Its proof is lengthy because of the large variety of terms that are involved. However, the proof is not difficult: we already established the difficult estimates in Prop. 13.12.

LEMMA 20.10 (**The fundamental Gronwall lemma**). *Let* $\mathring{\epsilon}$ *be the size of the data as defined in Def. 12.3. Assume that on the domain* $(t,u) \in [0, T_{(Bootstrap)}) \times [0, U_0]$, *the quantities* $\mathbb{Q}_{(\leq N)}(t,u)$, $\widetilde{\mathbb{Q}}_{(\leq N)}(t,u)$, $\widetilde{\mathbb{K}}_{(\leq N)}(t,u)$ *verify the estimates of Prop. 20.8 for* $0 \leq N \leq 24$ *and the estimates of Prop. 20.9 for* $0 \leq N \leq 23$. *Assume* *in addition that the bootstrap assumptions* (20.1a)-(20.1f) *hold on the same domain. Then if $A_* > 4$ is sufficiently large, there exists a large constant $C > 0$ (depending on A_*) such that if ε is sufficiently small, then the following estimates hold for $(t, u) \in [0, T_{(Bootstrap)}) \times [0, U_0]$:*

(20.8a)
$$\mathbb{Q}^{1/2}_{(N)}(t,u) \leq C\left\{\mathring{\epsilon} + \varepsilon^{3/2}\right\}, \qquad (0 \leq N \leq 15),$$

(20.8b)
$$\widetilde{\mathbb{Q}}^{1/2}_{(N)}(t,u) + \widetilde{\mathbb{K}}^{1/2}_{(N)}(t,u) \leq C\left\{\mathring{\epsilon} + \varepsilon^{3/2}\right\}\ln^2(e+t), \qquad (0 \leq N \leq 15),$$

(20.8c)
$$\mathbb{Q}^{1/2}_{(16+M)}(t,u) \leq C\left\{\mathring{\epsilon} + \varepsilon^{3/2}\right\}\mu_\star^{-.75-M}(t,u), \qquad (0 \leq M \leq 7),$$

(20.8d)
$$\widetilde{\mathbb{Q}}^{1/2}_{(16+M)}(t,u) + \widetilde{\mathbb{K}}^{1/2}_{(16+M)}(t,u) \leq C\left\{\mathring{\epsilon} + \varepsilon^{3/2}\right\}\ln^2(e+t)\mu_\star^{-.75-M}(t,u), \qquad (0 \leq M \leq 7),$$

(20.8e)
$$\mathbb{Q}^{1/2}_{(24)}(t,u) \leq C\left\{\mathring{\epsilon} + \varepsilon^{3/2}\right\}\ln^{A_*}(e+t)\mu_\star^{-8.75}(t,u),$$

(20.8f)
$$\widetilde{\mathbb{Q}}^{1/2}_{(24)}(t,u) + \widetilde{\mathbb{K}}^{1/2}_{(24)}(t,u) \leq C\left\{\mathring{\epsilon} + \varepsilon^{3/2}\right\}\ln^{A_*+2}(e+t)\mu_\star^{-8.75}(t,u).$$

REMARK 20.11 (**Improvements of the L^2-type bootstrap assumptions**). Note that if $\mathring{\epsilon}$ and ε are sufficiently small, then the estimates (20.8a)-(20.8f) yield improvements of the bootstrap assumptions (20.1a)-(20.1f).

20.3. Estimates for all but the most difficult error integrals

In our proofs of Prop. 20.8 and Prop. 20.9, we encounter many quadratic spacetime error integrals that we must bound in terms of the L^2-type quantities $\mathbb{Q}_{(\leq N)}$, $\widetilde{\mathbb{Q}}_{(\leq N)}$, and $\widetilde{\mathbb{K}}_{(\leq N)}$. In this section, we derive estimates for all such error integrals except the most difficult top-order ones.

We begin with the next three lemmas, in which we derive estimates for most of the error integrals appearing our analysis, including the ones corresponding to the $Harmless^{\leq N}$ inhomogeneous terms (see Def. 16.2) appearing in the commuted wave equation.

LEMMA 20.12 (**Bounds for standard spacetime integrals in terms of the fundamental L^2-controlling quantities**). *Let $0 \leq N \leq 24$ be an integer. Let A be a constant and recall that $\varrho(t,u) = 1 - u + t$. Under the small-data* *and bootstrap assumptions of Sects. 12.1-12.4, if ε is sufficiently small, then the following inequalities hold for $(t,u) \in [0, T_{(Bootstrap)}) \times [0, U_0]$:*

(20.9a)

$$\int_{\mathcal{M}_{t,u}} \frac{\ln^A(e+t')}{(1+t')^2} \left| \begin{pmatrix} (1+\upmu)L\mathscr{Z}^{\leq N}\Psi \\ \breve{R}\mathscr{Z}^{\leq N}\Psi \end{pmatrix} \right| \left| \begin{pmatrix} \varrho(1+\upmu)L\mathscr{Z}^{\leq N}\Psi \\ \breve{R}\mathscr{Z}^{\leq N}\Psi \\ \varrho(\sqrt{\upmu}+\upmu)\not{d}\mathscr{Z}^{\leq N}\Psi \\ \mathscr{Z}^{\leq N}\Psi \end{pmatrix} \right| d\varpi$$

$$\overset{A}{\lesssim} \int_{t'=0}^{t} \frac{1}{(1+t')^{3/2}} \mathbb{Q}_{(\leq N)}(t',u)\, dt' + \int_{t'=0}^{t} \frac{1}{(1+t')^{3/2}} \widetilde{\mathbb{Q}}_{(\leq N)}(t',u)\, dt' + \frac{1}{(1+t)^{1/2}} \int_{u'=0}^{u} \mathbb{Q}_{(\leq N)}(t,u')\, du',$$

(20.9b)

$$\int_{\mathcal{M}_{t,u}} \frac{\ln^A(e+t)}{(1+t')^3} \left| \begin{pmatrix} (1+\upmu)L\mathscr{Z}^{\leq N}\Psi \\ \breve{R}\mathscr{Z}^{\leq N}\Psi \end{pmatrix} \right| \left| \begin{pmatrix} \mathscr{Z}^{\leq N}(\upmu-1) \\ \sum_{a=1}^{3} \varrho \left| \mathscr{Z}^{\leq N} L_{(Small)}^a \right| \end{pmatrix} \right| d\varpi$$

$$\overset{A}{\lesssim} \int_{t'=0}^{t} \frac{1}{(1+t')^{3/2}} \mathbb{Q}_{(\leq N)}(t',u)\, dt' + \int_{t'=0}^{t} \frac{1}{(1+t')^{3/2}} \widetilde{\mathbb{Q}}_{(\leq N)}(t',u)\, dt' + \frac{1}{(1+t)^{3/2}} \int_{u'=0}^{u} \mathbb{Q}_{(\leq N)}(t,u')\, du' + \varepsilon^2.$$

REMARK 20.13 (**We need the Morawetz integral to control non-$\upmu$-weighted angular derivatives of Ψ**). Note that Lemma 20.12 does not provide control of integrals involving the non $\upmu$-weighted quantity $\left|\not{d}\mathscr{Z}^{\leq N}\Psi\right|$. We obtain control of such integrals in Lemma 20.15 with the help of the Morawetz spacetime integral of Def. 14.3.

PROOF OF LEMMA 20.12. The proof is essentially a tedious exercise in Cauchy-Schwarz and integration by parts based on the estimates of Prop. 14.7. Since the estimates are all very similar, we bound only one representative product from

(20.9a). Specifically, we refer to Defs. 3.68 and 3.69 and use spacetime Cauchy-Schwarz and Prop. 14.7 to deduce that

(20.10)

$$\begin{aligned}
&\int_{\mathcal{M}_{t,u}} \frac{\ln^A(e+t')}{(1+t')^2} \left|L\mathscr{Z}^{\leq N}\Psi\right| \left|\varrho\sqrt{\upmu}\not{d}\mathscr{Z}^{\leq N}\Psi\right| \, d\varpi \\
&\lesssim \int_{t'=0}^{t} \int_{\Sigma_{t'}^{u}} \frac{\ln^A(e+t')}{(1+t')^2} \upmu \left|\varrho\not{d}\mathscr{Z}^{\leq N}\Psi\right|^2 \, d\underline{\varpi} \, dt' \\
&\quad + \int_{t'=0}^{t} \frac{\ln^A(e+t')}{(1+t')^2} \int_{u'=0}^{u} \int_{S_{t',u'}} \left|L\mathscr{Z}^{\leq N}\Psi\right|^2 \, d\upsilon_{\not{g}} \, du' \, dt' \\
&\lesssim \int_{t'=0}^{t} \frac{\ln^A(e+t')}{(1+t')^2} \widetilde{\mathbb{Q}}_{(\leq N)}(t',u) \, dt' \\
&\quad + \frac{\ln^A(e+t)}{(1+t)^2} \int_{u'=0}^{u} \int_{t'=0}^{t} \int_{S_{t',u'}} \left|L\mathscr{Z}^{\leq N}\Psi\right|^2 \, d\upsilon_{\not{g}} \, dt' \, du' \\
&\quad + \int_{t'=0}^{t} \int_{u'=0}^{u} \int_{s=0}^{t'} \int_{S_{s,u'}} \left|L\mathscr{Z}^{\leq N}\Psi\right|^2 \, d\upsilon_{\not{g}(s,u',\vartheta)} \, ds \, du' \left|\frac{d}{dt'} \frac{\ln^A(e+t')}{(1+t')^2}\right| dt' \\
&\lesssim \int_{t'=0}^{t} \frac{\ln^A(e+t')}{(1+t')^2} \widetilde{\mathbb{Q}}_{(\leq N)}(t',u) \, dt' + \frac{\ln^A(e+t)}{(1+t)^2} \int_{u'=0}^{u} \mathbb{Q}_{(\leq N)}(t,u') \, du' \\
&\quad + \int_{t'=0}^{t} \left|\frac{d}{dt'} \frac{\ln^A(e+t')}{(1+t')^2}\right| \mathbb{Q}_{(\leq N)}(t',u) \, dt',
\end{aligned}$$

where to conclude the next-to-last inequality, we integrated by parts in t', and to conclude the last inequality, we used the fact that $\mathbb{Q}_{(\leq N)}$ and $\widetilde{\mathbb{Q}}_{(\leq N)}$ are increasing in their arguments and the fact that $u \leq U_0 < 1$. Clearly, the right-hand side of (20.10) is bounded by the right-hand side of (20.9a) as desired.

The proof of (20.9b) is similar, but after using Cauchy-Schwarz and suitably distributing the t' weights, we have to estimate integrals such as

(20.11)

$$\int_{\mathcal{M}_{t,u}} \frac{1}{(1+t')^4} \left|\mathscr{Z}^{\leq N}(\upmu-1)\right|^2 \, d\varpi = \int_{t'=0}^{t} \frac{1}{(1+t')^4} \int_{\Sigma_{t'}^{u}} \left|\mathscr{Z}^{\leq N}(\upmu-1)\right|^2 \, d\underline{\varpi} \, dt'.$$

We use (19.2a) to deduce the desired bound for the right-hand side of (20.11) as follows:

(20.12)

$$\begin{aligned}
&\lesssim \int_{t'=0}^{t} \frac{1}{(1+t')^2} \varepsilon^2 \, dt' + \int_{t'=0}^{t} \frac{1}{(1+t')^2} \left(\int_{s=0}^{t'} \frac{\mathbb{Q}_{(\leq N)}^{1/2}(s,u)}{1+s} \, ds \right)^2 dt' \\
&\quad + \int_{t'=0}^{t} \frac{1}{(1+t')^2} \left(\int_{s=0}^{t'} \frac{\widetilde{\mathbb{Q}}_{(\leq N)}^{1/2}(s,u)}{(1+s)\upmu_\star^{1/2}(s,u)} \, ds \right)^2 dt' \\
&\lesssim \varepsilon^2 + \int_{t'=0}^{t} \frac{1}{(1+t')^{3/2}} \mathbb{Q}_{(\leq N)}(t',u) \, dt' + \int_{t'=0}^{t} \frac{1}{(1+t')^{3/2}} \widetilde{\mathbb{Q}}_{(\leq N)}(t',u) \, dt',
\end{aligned}$$

where in the last step, we used the estimate (13.152) to destroy the factor $\mu_\star^{1/2}(s,u)$ in the denominator. ☐

LEMMA 20.14 (**Bounds for spacetime integrals involving $\{L+\frac{1}{2}\mathrm{tr}_{\not g}\chi\}\mathscr{Z}^{\leq N}\Psi$ in terms of the fundamental L^2-controlling quantities**). *Let $0 \leq N \leq 24$ be an integer. Let A be a constant and recall that $\varrho(t,u) = 1-u+t$. Let χ be the symmetric type $\binom{0}{2}$ $S_{t,u}$ tensorfield defined in (3.70). Under the small-data and bootstrap assumptions of Sects. 12.1-12.4, if ε is sufficiently small, then the following inequalities hold for $(t,u) \in [0, T_{(Bootstrap)}) \times [0, U_0]$:*

$$\int_{\mathcal{M}_{t,u}} \ln^A(e+t') \left| \left\{ L + \frac{1}{2}\mathrm{tr}_{\not g}\chi \right\} \mathscr{Z}^{\leq N}\Psi \right| \left| \begin{pmatrix} \varrho(1+\mu)L\mathscr{Z}^{\leq N}\Psi \\ \breve{R}\mathscr{Z}^{\leq N}\Psi \\ \varrho(\sqrt{\mu}+\mu)\not d\mathscr{Z}^{\leq N}\Psi \\ \mathscr{Z}^{\leq N}\Psi \end{pmatrix} \right| d\varpi \tag{20.13a}$$
$$\overset{A}{\lesssim} \int_{t'=0}^{t} \frac{1}{(1+t')^{3/2}} \mathbb{Q}_{(\leq N)}(t',u)\,dt' + \int_{t'=0}^{t} \frac{1}{(1+t')^{3/2}} \widetilde{\mathbb{Q}}_{(\leq N)}(t',u)\,dt' + \frac{1}{(1+t)^{1/2}} \int_{u'=0}^{u} \widetilde{\mathbb{Q}}_{(\leq N)}(t,u')\,du',$$

$$\int_{\mathcal{M}_{t,u}} \varrho^2 \left| \left\{ L + \frac{1}{2}\mathrm{tr}_{\not g}\chi \right\} \mathscr{Z}^{\leq N}\Psi \right|^2 d\varpi \lesssim \int_{u'=0}^{u} \widetilde{\mathbb{Q}}_{(\leq N)}(t,u')\,du', \tag{20.13b}$$

(20.13c)
$$\int_{\mathcal{M}_{t,u}} \frac{\ln^A(e+t')}{1+t'} \left| \left\{ L + \frac{1}{2}\mathrm{tr}_{\not g}\chi \right\} \mathscr{Z}^{\leq N}\Psi \right| \left| \begin{pmatrix} \mathscr{Z}^{\leq N}(\mu-1) \\ \sum_{a=1}^{3} \varrho \left| \mathscr{Z}^{\leq N} L^a_{(Small)} \right| \end{pmatrix} \right| d\varpi$$
$$\overset{A}{\lesssim} \int_{t'=0}^{t} \frac{1}{(1+t')^{3/2}} \mathbb{Q}_{(\leq N)}(t',u)\,dt' + \int_{t'=0}^{t} \frac{1}{(1-t')^{3/2}} \widetilde{\mathbb{Q}}_{(\leq N)}(t',u)\,dt' + \int_{u'=0}^{u} \widetilde{\mathbb{Q}}_{(\leq N)}(t,u')\,du' + \varepsilon^2.$$

PROOF. The proof is very similar to that of Lemma 20.12. Since the estimates are all very similar, we estimate only one representative product from (20.13a). For example, decomposing

$$L\Psi = \left\{ L\Psi + \frac{1}{2}\mathrm{tr}_{\not{g}}\chi\Psi \right\} - \frac{1}{2}\mathrm{tr}_{\not{g}}\chi\Psi,$$

using the bound $|\mathrm{tr}_{\not{g}}\chi| \lesssim (1+t)^{-1}$ (that is, (12.60b)), referring to Defs. 3.68 and 3.69, using Prop. 14.7, and arguing as in our proof of (20.10), we deduce that

(20.14)

$$\begin{aligned}
&\int_{\mathcal{M}_{t,u}} \ln^A(e+t') \left| \left\{ L + \frac{1}{2}\mathrm{tr}_{\not{g}}\chi \right\} \mathscr{Z}^{\leq N}\Psi \right| \left| \varrho L \mathscr{Z}^{\leq N}\Psi \right| \, d\varpi \\
&\overset{A}{\lesssim} \int_{t'=0}^{t} \frac{\ln^A(e+t')}{1+t'} \int_{u'=0}^{u} \int_{S_{t',u'}} \left| \varrho \left\{ L + \frac{1}{2}\mathrm{tr}_{\not{g}}\chi \right\} \mathscr{Z}^{\leq N}\Psi \right|^2 d\upsilon_{\not{g}} \, du' \, dt' \\
&\quad + \int_{t'=0}^{t} \frac{1}{(1+t')^{3/2}} \int_{\Sigma_{t'}^u} \left| \mathscr{Z}^{\leq N}\Psi \right|^2 d\underline{\varpi} \, dt' \\
&\overset{A}{\lesssim} \frac{\ln^A(e+t)}{1+t} \int_{u'=0}^{u} \widetilde{\mathbb{Q}}_{(\leq N)}(t,u') \, du' + \int_{t'=0}^{t} \left| \frac{d}{dt'} \frac{\ln^A(e+t')}{1+t'} \right| \widetilde{\mathbb{Q}}_{(\leq N)}(t',u) \, dt' \\
&\quad + \int_{t'=0}^{t} \frac{1}{(1+t')^{3/2}} \mathbb{Q}_{(\leq N)}(t',u) \, dt',
\end{aligned}$$

where to deduce the last inequality, we integrated by parts in t' and used the facts that $\mathbb{Q}_{(\leq N)}$ and $\widetilde{\mathbb{Q}}_{(\leq N)}$ are increasing in their arguments and that $u \leq U_0 < 1$ (as in our proof of (20.10)). Clearly, the right-hand side of (20.14) is bounded by the right-hand side of (20.13a) as desired.

The bound (20.13b) follows easily from (14.16). The bounds in (20.13c) can be proved by using arguments similar to the ones we used to prove (20.9b). □

The following key lemma, which is based on the availability of the coercive Morawetz spacetime integral (14.10), provides us with control over error integrals involving non-$\upmu$-weighted angular derivatives of Ψ.

LEMMA 20.15 (**Bounds for spacetime integrals in terms of the coercive Morawetz integral $\widetilde{\mathbb{K}}$**). *Let $0 \leq N \leq 24$ be an integer. Let A be a constant and recall that $\varrho(t,u) = 1 - u + t$. Let $\mathbf{1}_{\{\upmu \leq 1/4\}}$ denote the characteristic function of* *the spacetime set* $\{(t,u,\vartheta) \mid \upmu(t,u,\vartheta) \leq 1/4\}$. *Under the small-data and bootstrap assumptions of Sects. 12.1-12.4, if* ε *is sufficiently small and* $\varsigma > 0$ *is a real number, then the following inequalities hold for* $(t,u) \in [0, T_{(Bootstrap)}) \times [0, U_0]$ *(and the constants implicit in* "$\overset{A}{\lesssim}$" *are* ***independent of*** ς*):*

$$\int_{\mathcal{M}_{t,u}} \mathbf{1}_{\{\upmu \leq 1/4\}} \frac{\ln^A(e+t')}{1+t'} \left| \not{d}\mathscr{Z}^{\leq N}\Psi \right| \left| \begin{pmatrix} (1+\upmu) L \mathscr{Z}^{\leq N}\Psi \\ \breve{R}\mathscr{Z}^{\leq N}\Psi \\ \mathscr{Z}^{\leq N}\Psi \end{pmatrix} \right| d\varpi \tag{20.15a}$$

$$\overset{A}{\lesssim} \varsigma^{-1} \int_{t'=0}^{t} \frac{1}{(1+t')^{3/2}} \mathbb{Q}_{(\leq N)}(t',u)\, dt' + \varsigma^{-1} \int_{u'=0}^{u} \mathbb{Q}_{(\leq N)}(t,u')\, du'$$

$$+ \varsigma \sup_{t' \in [0,t)} \frac{\widetilde{\mathbb{K}}_{(\leq N)}(t',u)}{(1+t')^{1/2}},$$

$$\int_{\mathcal{M}_{t,u}} \mathbf{1}_{\{\upmu \leq 1/4\}} \frac{\ln^A(e+t')}{1+t'} \left| \not{d}\mathscr{Z}^{\leq N}\Psi \right|^2 d\varpi \tag{20.15b}$$

$$\overset{A}{\lesssim} \sup_{t' \in [0,t)} \frac{\widetilde{\mathbb{K}}_{(\leq N)}(t',u)}{(1+t')^{1/2}},$$

$$\int_{\mathcal{M}_{t,u}} \mathbf{1}_{\{\upmu \leq 1/4\}} \ln^A(e+t') \left| \not{d}\mathscr{Z}^{\leq N}\Psi \right| \left| \varrho \left\{ L + \frac{1}{2}\mathrm{tr}_{\not{g}}\chi \right\} \mathscr{Z}^{\leq N}\Psi \right| d\varpi \tag{20.15c}$$

$$\overset{A}{\lesssim} \varsigma^{-1} \int_{u'=0}^{u} \widetilde{\mathbb{Q}}_{(\leq N)}(t,u')\, du' + \varsigma \widetilde{\mathbb{K}}_{(\leq N)}(t,u).$$

PROOF OF LEMMA 20.15. We prove only the estimate for the term $\ln^A(e+t')(1+t')^{-1} \left| \not{d}\mathscr{Z}^{\leq N}\Psi \right| \left| L\mathscr{Z}^{\leq N}\Psi \right|$ in (20.15a); the proofs of the remaining estimates in (20.15a) and the estimate (20.15b) are similar and we omit those details. The proof essentially involves Cauchy-Schwarz and a suitable partitioning of $[0,t]$. To proceed, we partition $[0,t) = [t_1 := 0, t_2) \cup [t_2, t_3) \cup \cdots [t_{M-2}, t_{M-1}) \cup [t_{M-1}, t_M := t)$, where $\{t_m\}_{m=1,2,\cdots,M}$ is the sequence of times defined, for $1 \leq m \leq M-1$, by $t_m := m^4 - 1$, where $M - 1 =: \max\{m \in \mathbb{N} \mid m^4 - 1 < t\}$. In particular, we have that $m = (1+t_m)^{1/4}$ (except perhaps for $m = M$) and, for $1 \leq m \leq M-1$, $\frac{(1+t_{m+1})^{1/4}}{(1+t_m)^{1/4}} \leq 2$ Referring to Defs. 3.68 and 3.69 and using the previous

inequality, Cauchy-Schwarz, Prop. 14.7, and Lemma 14.5, we deduce that
(20.16)

$$
\begin{aligned}
&\int_{\mathcal{M}_{t,u}} \frac{\ln^A(e+t')}{1+t'} \mathbf{1}_{\{\upmu \leq 1/4\}} \left| L \mathscr{Z}^{\leq N} \Psi \right| \left| {\not{d}} \mathscr{Z}^{\leq N} \Psi \right| \, d\varpi \\
&\leq \sum_{m=1}^{M-1} \left(\int_{t'=t_m}^{t_{m+1}} \int_{\Sigma_{t'}^u} \frac{(1+t')}{\ln(e+t')} \mathbf{1}_{\{\upmu \leq 1/4\}} \left| {\not{d}} \mathscr{Z}^{\leq N} \Psi \right|^2 \, d\underline{\varpi} \, dt' \right)^{1/2} \\
&\qquad \times \left(\int_{t'=t_m}^{t_{m+1}} \int_{\Sigma_{t'}^u} \frac{\ln^{2A+1}(e+t')}{(1+t')^3} \left| L \mathscr{Z}^{\leq N} \Psi \right|^2 \, d\underline{\varpi} \, dt' \right)^{1/2} \\
&\leq C \sum_{m=1}^{M-1} \frac{\widetilde{\mathbb{K}}_{(\leq N)}^{1/2}(t_{m+1},u)}{(1+t_{m+1})^{1/4}} \frac{(1+t_{m+1})^{1/4}}{(1+t_m)^{1/4}} \\
&\qquad \times \left(\int_{t'=t_m}^{t_{m+1}} \int_{\Sigma_{t'}^u} \frac{\ln^{2A+1}(e+t')}{(1+t')^{5/2}} \left| L \mathscr{Z}^{\leq N} \Psi \right|^2 \, d\underline{\varpi} \, dt' \right)^{1/2} \\
&\leq C \sup_{s \in [0,t)} \frac{\widetilde{\mathbb{K}}_{(\leq N)}^{1/2}(s,u)}{(1+s)^{1/4}} \sum_{m=1}^{M-1} \left(\int_{t'=t_m}^{t_{m+1}} \int_{\Sigma_{t'}^u} \frac{\ln^{2A+1}(e+t')}{(1+t')^{5/2}} \left| L \mathscr{Z}^{\leq N} \Psi \right|^2 \, d\underline{\varpi} \, dt' \right)^{1/2} \\
&\leq C \left(\sup_{s \in [0,t)} \frac{\widetilde{\mathbb{K}}_{(\leq N)}^{1/2}(s,u)}{(1+s)^{1/4}} \right) \left(\sum_{m=1}^{\infty} m^{-2} \right)^{1/2} \\
&\qquad \times \left(\sum_{m=1}^{M-1} m^2 \int_{t'=t_m}^{t_{m+1}} \int_{\Sigma_{t'}^u} \frac{\ln^{2A+1}(e+t')}{(1+t')^{5/2}} \left| L \mathscr{Z}^{\leq N} \Psi \right|^2 \, d\underline{\varpi} \, dt' \right)^{1/2} \\
&\leq C \left(\sup_{s \in [0,t)} \frac{\widetilde{\mathbb{K}}_{(\leq N)}^{1/2}(s,u)}{(1+s)^{1/4}} \right) \left(\int_{u'=0}^{u} \int_{\mathcal{C}_{u'}^t} \left| L \mathscr{Z}^{\leq N} \Psi \right|^2 \, d\overline{\varpi} \, du' \right)^{1/2} \\
&\leq C \left(\sup_{s \in [0,t)} \frac{\widetilde{\mathbb{K}}_{(\leq N)}^{1/2}(s,u)}{(1+s)^{1/4}} \right) \left(\int_{u'=0}^{u} \mathbb{Q}_{(\leq N)}(t,u') \, du' \right)^{1/2} \\
&\leq C\varsigma \sup_{s \in [0,t)} \frac{\widetilde{\mathbb{K}}_{(\leq N)}(s,u)}{(1+s)^{1/2}} + C\varsigma^{-1} \int_{u'=0}^{u} \mathbb{Q}_{(\leq N)}(t,u') \, du',
\end{aligned}
$$

which in turn is bounded by the right-hand side of (20.15a) as desired.

The estimate (20.15c) can be proved using similar arguments. Actually, the estimate is much simpler because there is no need to partition $[0,t]$ during the proof. □

In the next two corollaries, we extract the most important content of the previous three lemmas and present it in a manner that is relevant our proofs of Props. 20.8 and 20.9.

COROLLARY 20.16 (**Integral estimates for the Harmless spacetime error integrals corresponding to the multiplier** T). *Let* $0 \leq N \leq 24$ *be an integer and let* $\varsigma > 0$ *be a number. Under the small-data and bootstrap assumptions of Sects.* 12.1-12.4, *if* ε *is sufficiently small, then the following inequalities hold for* *the* $Harmless^{\leq N}$ *terms from Def. 16.2 for* $(t,u) \in [0, T_{(Bootstrap)}) \times [0, U_0]$ *(and the implicit constants are* ***independent of*** ς*):*

$$\begin{aligned}
(20.17)\qquad & \int_{\mathcal{M}_{t,u}} \left\{ \left|(1+2\upmu) L \mathscr{Z}^{\leq N}\Psi\right| + \left|2\breve{R}\mathscr{Z}^{\leq N}\Psi\right| \right\} \left|Harmless^{\leq N}\right| \, d\varpi \\
& \lesssim (1+\varsigma^{-1}) \int_{t'=0}^{t} \frac{1}{(1+t')^{3/2}} \mathbb{Q}_{(\leq N)}(t',u)\, dt' \\
& + (1+\varsigma^{-1}) \int_{t'=0}^{t} \frac{1}{(1+t')^{3/2}} \widetilde{\mathbb{Q}}_{(\leq N)}(t',u)\, dt' \\
& + (1+\varsigma^{-1}) \int_{u'=0}^{u} \mathbb{Q}_{(\leq N)}(t,u')\, du' \\
& + (1+\varsigma^{-1}) \frac{1}{(1+t)^{1/2}} \int_{u'=0}^{u} \widetilde{\mathbb{Q}}_{(\leq N)}(t,u')\, du' \\
& + \varsigma \sup_{t' \in [0,t]} \frac{\widetilde{\mathbb{K}}_{(\leq N)}(t',u)}{(1+t')^{1/2}} + \varepsilon^3.
\end{aligned}$$

PROOF. Referring to Def. 16.2 and using the trivial bound

$$\left|{\not d}\mathscr{Z}^N\Psi\right| \lesssim \mathbf{1}_{\{\upmu \leq 1/4\}} \left|{\not d}\mathscr{Z}^N\Psi\right| + \sqrt{\upmu}\left|{\not d}\mathscr{Z}^N\Psi\right|,$$

we see that all of the terms on the left-hand side of (20.17) can be bounded by the right-hand side of (20.17) with the help of the estimates of Lemma 20.12, Lemma 20.14, and Lemma 20.15. □

COROLLARY 20.17 (**Estimates for the Harmless error integrals corresponding to the multiplier** $\widetilde{K}$). *Let* $0 \leq N \leq 24$ *be an integer and let* $\widetilde{\varsigma} > 0$ *be a number. Recall that* $\varrho(t,u) = 1 - u + t$ *and let* χ *be the symmetric type* $\binom{0}{2}$ $S_{t,u}$ *tensorfield defined in* (3.70). *Under the small-data and bootstrap assumptions of Sects. 12.1-12.4, if* ε *is sufficiently small, then the following inequalities hold for the* $Harmless^{\leq N}$ *terms from Def. 16.2 for* $(t,u) \in [0, T_{(Bootstrap)}) \times [0, U_0]$ *(and the implicit constants are* ***independent of*** $\widetilde{\varsigma}$*):*

$$\begin{aligned}
(20.18)\qquad & \int_{\mathcal{M}_{t,u}} \varrho^2 \left| \left\{ L + \frac{1}{2}\mathrm{tr}_{\not g}\chi \right\} \mathscr{Z}^{\leq N}\Psi \right| \left|Harmless^{\leq N}\right| \, d\varpi \\
& \lesssim (1+\widetilde{\varsigma}^{-1}) \int_{t'=0}^{t} \frac{1}{(1+t')^{3/2}} \mathbb{Q}_{(\leq N)}(t',u)\, dt' \\
& + (1+\widetilde{\varsigma}^{-1}) \int_{t'=0}^{t} \frac{1}{(1+t')^{3/2}} \widetilde{\mathbb{Q}}_{(\leq N)}(t',u)\, dt' \\
& + (1+\widetilde{\varsigma}^{-1}) \int_{u'=0}^{u} \widetilde{\mathbb{Q}}_{(\leq N)}(t,u')\, du' + \widetilde{\varsigma}\widetilde{\mathbb{K}}_{(\leq N)}(t,u) + \varepsilon^3.
\end{aligned}$$

PROOF. The proof is essentially the same as that of Cor. 20.16 and is based on Lemmas 20.14 and 20.15. □

As we will see, when deriving some of our L^2 estimates for the below-top-order derivatives of Ψ, we must allow some of the error integrals to lose one derivative. This is essential, for if we instead chose to avoid losing derivatives, then we would have to use the modified quantities of Chapter 11, and the resulting L^2 estimates would be as degenerate with respect to powers of $\upmu_\star^{-1}$ as they are at the top order; such degeneracy would prevent us from deriving improvements of the bootstrap assumptions of Sect. 20.1. In the next two lemmas, we derive estimates that are sufficient for controlling the derivative-losing error integrals.

LEMMA 20.18 (**Estimates for the error integrals corresponding to the multiplier T involving a loss of one derivative**). *Let $1 \leq N \leq 23$ be an integer. Let $\chi^{(Small)}$ be the symmetric type $\binom{0}{2}$ $S_{t,u}$ tensorfield defined in (4.1c) and let $O \in \mathscr{O}$ be a geometric rotation vectorfield (see definition (7.2b)). Under the small-data and bootstrap assumptions of Sects. 12.1-12.4, if ε is sufficiently small, then the following inequalities hold for $(t,u) \in [0, T_{(Bootstrap)}) \times [0, U_0]$:*

$$
\begin{aligned}
(20.19)\quad & \int_{\mathcal{M}_{t,u}} \frac{1}{1+t'} \left|(1+2\upmu)L\mathscr{Z}^N\Psi + \breve{R}\mathscr{Z}^N\Psi\right| \left| \begin{pmatrix} \Delta\!\!\!/ \mathscr{Z}^{N-1}\upmu \\ O\mathscr{Z}^{N-1}\mathrm{tr}_{g\!\!\!/}\chi^{(Small)} \end{pmatrix} \right| d\varpi \\
& \lesssim \int_{t'=0}^{t} \frac{1}{(1+t')^{3/2}\upmu_\star^{1/2}(t',u)} \mathbb{Q}_{(\leq N)}^{1/2}(t',u) \\
& \qquad \times \int_{s=0}^{t'} \frac{1}{1+s} \mathbb{Q}_{(\leq N+1)}^{1/2}(s,u)\, ds\, dt' \\
& \quad + \int_{t'=0}^{t} \frac{1}{(1+t')^{3/2}\upmu_\star^{1/2}(t',u)} \mathbb{Q}_{(\leq N)}^{1/2}(t',u) \\
& \qquad \times \int_{s=0}^{t'} \frac{1}{(1+s)\upmu_\star^{1/2}(s,u)} \widetilde{\mathbb{Q}}_{(\leq N+1)}^{1/2}(s,u)\, ds\, dt' \\
& \quad + \int_{t'=0}^{t} \frac{1}{(1+t')^{3/2}\upmu_\star^{1/2}(t',u)} \mathbb{Q}_{(\leq N)}(t',u)\, dt' + \varepsilon^2.
\end{aligned}
$$

PROOF. Since the proofs of both of the estimates are similar, we provide only the proof of one of them. Specifically, using (12.46a) and (12.46c), we see that $\left|\Delta\!\!\!/ \mathscr{Z}^{N-1}\upmu\right| \lesssim (1+t)^{-2}\left|\mathscr{Z}^{\leq N+1}\upmu\right|$. Hence, using this bound, referring to Defs. 3.68 and 3.69, using Cauchy-Schwarz on the $\Sigma_{t'}^u$, Prop. 14.7, and (19.2a) with $N+1$ in the role of N, we deduce

$$
\begin{aligned}
(20.20)\quad & \int_{\mathcal{M}_{t,u}} \frac{1}{1+t'} \left|L\mathscr{Z}^N\Psi\right| \left|\Delta\!\!\!/ \mathscr{Z}^{N-1}\upmu\right| d\varpi \\
& \lesssim \int_{t'=0}^{t} \frac{1}{(1+t')^3} \int_{\Sigma_{t'}^u} \left|L\mathscr{Z}^N\Psi\right| \left|\mathscr{Z}^{\leq N+1}\upmu\right| d\underline{\varpi}\, dt' \\
& \lesssim \varepsilon \int_{t'=0}^{t} \frac{1}{(1+t')^2\upmu_\star^{1/2}(t',u)} \mathbb{Q}_{(\leq N)}^{1/2}(t',u)\, dt' \\
& \quad + \int_{t'=0}^{t} \frac{1}{(1+t')^2\upmu_\star^{1/2}(t',u)} \mathbb{Q}_{(\leq N)}^{1/2}(t',u) \\
& \qquad \times \int_{s=0}^{t'} \frac{1}{1+s} \mathbb{Q}_{(\leq N+1)}^{1/2}(s,u)\, ds\, dt' \\
& \quad + \int_{t'=0}^{t} \frac{1}{(1+t')^2\upmu_\star^{1/2}(t',u)} \mathbb{Q}_{(\leq N)}^{1/2}(t',u) \\
& \qquad \times \int_{s=0}^{t'} \frac{1}{(1+s)\upmu_\star^{1/2}(s,u)} \widetilde{\mathbb{Q}}_{(\leq N+1)}^{1/2}(s,u)\, ds\, dt'.
\end{aligned}
$$

To complete the proof of inequality (20.19), it remains only for us to bound the first integral on the right-hand side of (20.20) by the last two terms on the right-hand side of (20.19). To derive this bound, we use the simple inequality

$$\varepsilon \mathbb{Q}^{1/2}_{(\leq N)}(t', u) \lesssim \varepsilon^2 + \mathbb{Q}_{(\leq N)}(t', u)$$

and inequality (13.151). □

LEMMA 20.19 (**Estimates for the error integrals corresponding to the multiplier $\widetilde{K}$ involving a loss of one derivative**). *Let $1 \leq N \leq 23$ be an integer. Recall that $\varrho(t,u) = 1-u+t$, let χ be the symmetric type $\binom{0}{2}$ $S_{t,u}$ tensorfield defined in* (3.70), *and let $\chi^{(Small)}$ be the symmetric type $\binom{0}{2}$ $S_{t,u}$ tensorfield defined in* (4.1c). *Under the small-data and bootstrap assumptions of Sects.* 12.1-12.4, *if ε is sufficiently small, then the following inequalities hold for $(t,u) \in [0, T_{(Bootstrap)}) \times [0, U_0]$:*

$$(20.21) \qquad \int_{\mathcal{M}_{t,u}} \varrho \left| L \mathscr{Z}^N \Psi + \frac{1}{2} \mathrm{tr}_{\not g} \chi \mathscr{Z}^N \Psi \right| \left| \begin{pmatrix} \not\Delta \mathscr{Z}^{N-1} \upmu \\ \varrho \not d \mathscr{Z}^{N-1} \mathrm{tr}_{\not g} \chi^{(Small)} \end{pmatrix} \right| d\varpi$$
$$\lesssim \int_{t'=0}^{t} \frac{1}{(1+t')^2} \left(\int_{s=0}^{t'} \frac{\mathbb{Q}^{1/2}_{(\leq N+1)}(s,u)}{1+s} \, ds \right)^2 dt'$$
$$+ \int_{t'=0}^{t} \frac{1}{(1+t')^2} \left(\int_{s=0}^{t'} \frac{\widetilde{\mathbb{Q}}^{1/2}_{(\leq N+1)}(s,u)}{(1+s)\upmu_\star^{1/2}(s,u)} \, ds \right)^2 dt'$$
$$+ \int_{u'=0}^{u} \widetilde{\mathbb{Q}}_{(\leq N)}(t,u') \, du' + \varepsilon^2.$$

PROOF. Since the proofs of both of the estimates are similar, we provide only the proof of one of them. Specifically, referring to Defs. 3.68 and 3.69, using the bound $|\not\Delta \mathscr{Z}^{N-1}\upmu| \lesssim (1+t)^{-2} |\mathscr{Z}^{\leq N+1}\upmu|$ noted in the proof of Lemma 20.18, Cauchy-Schwarz, and Prop. 14.7, we deduce that

$$(20.22) \quad \int_{\mathcal{M}_{t,u}} \varrho \left| L \mathscr{Z}^N \Psi + \frac{1}{2} \mathrm{tr}_{\not g} \chi \mathscr{Z}^N \Psi \right| \left| \not\Delta \mathscr{Z}^{N-1} \upmu \right| d\varpi$$
$$\lesssim \int_{\mathcal{M}_{t,u}} \varrho^2 \left| L \mathscr{Z}^N \Psi + \frac{1}{2} \mathrm{tr}_{\not g} \chi \mathscr{Z}^N \Psi \right|^2 d\varpi$$
$$+ \int_{t'=0}^{t} \frac{1}{(1+t')^4} \int_{\Sigma_{t'}^u} \left| \mathscr{Z}^{\leq N+1}(\upmu - 1) \right|^2 d\underline{\varpi} \, dt'$$
$$\lesssim \int_{u'=0}^{u} \widetilde{\mathbb{Q}}_{(\leq N)}(t,u') \, du' + \int_{t'=0}^{t} \frac{1}{(1+t')^4} \int_{\Sigma_{t'}^u} \left| \mathscr{Z}^{\leq N+1}(\upmu - 1) \right|^2 d\underline{\varpi} \, dt'.$$

To bound the second spacetime integral on the right-hand side of (20.22), we use the estimate (19.2a) with $N+1$ in the role of N to deduce that

$$\begin{aligned}
(20.23) \qquad & \int_{t'=0}^{t} \frac{1}{(1+t')^4} \int_{\Sigma_{t'}^u} \left| \mathscr{Z}^{\leq N+1}(\upmu - 1) \right|^2 \, d\underline{\varpi} \, dt' \\
& \lesssim \varepsilon^2 \int_{t'=0}^{t} \frac{1}{(1+t')^2} \, dt' \\
& \quad + \int_{t'=0}^{t} \frac{1}{(1+t')^2} \left(\int_{s=0}^{t'} \frac{\mathbb{Q}_{(\leq N+1)}^{1/2}(s,u)}{1+s} \, ds \right)^2 dt' \\
& \quad + \int_{t'=0}^{t} \frac{1}{(1+t')^2} \left(\int_{s=0}^{t'} \frac{\widetilde{\mathbb{Q}}_{(\leq N+1)}^{1/2}(s,u)}{(1+s)\upmu_\star^{1/2}(s,u)} \, ds \right)^2 dt'.
\end{aligned}$$

Combining (20.22) and (20.23), we deduce the desired estimate (20.21). □

The next lemma, which is based on a combination of transport equation estimates and elliptic estimates, will allow us to derive nonsharp (but suitable for proving our sharp classical lifespan theorem) top-order Sobolev estimates for $\upmu \nabla\!\!\!/^2 \mathscr{Z}^{N-1}\upmu$ and $\upmu \nabla\!\!\!/ \mathscr{Z}^{N-1}\upchi^{(Small)}$. The main point is that the tensorfields $\upmu\hat{\nabla\!\!\!/}^2 \mathscr{Z}^{N-1}\upmu$ and $\upmu \nabla\!\!\!/ \mathscr{Z}^{N-1}\hat{\upchi}^{(Small)}$ cannot be estimated via pure energy estimates. The obstacle is that we have not been able to derive a transport equation for a modified version of them with a consistent number of derivatives on the right-hand side. To overcome this difficulty, we combine the elliptic estimates of Chapter 18 with a Gronwall argument, which allows us to bound these tensorfields in terms of other quantities that do not lose derivatives.

LEMMA 20.20. (**Nonsharp top-order L^2 estimates for $\nabla\!\!\!/^2 \mathscr{Z}^{N-1}\upmu$, $\nabla\!\!\!/^2 \mathscr{Z}^{N-1} L_{(Small)}^i$, and $\nabla\!\!\!/ \mathscr{Z}^{N-1}\upchi^{(Small)}$ in terms of $\mathbb{Q}_{(N)}$ and $\widetilde{\mathbb{Q}}_{(N)}$.**) *Let $1 \leq N \leq 24$ be an integer. Recall that $\varrho(t,u) = 1 - u + t$ and let $\upchi^{(Small)}$ be the symmetric type $\binom{0}{2}$ $S_{t,u}$ tensorfield defined in* (4.1c). *Under the small-data and bootstrap assumptions of Sects.* 12.1-12.4, *if ε is sufficiently small, then for $(t,u) \in$*

$[0, T_{(Bootstrap)}) \times [0, U_0]$, *we have the following estimates:*

(20.24a)
$$\begin{pmatrix} \left\| \mu \nabla\!\!\!/^2 \mathscr{Z}^{\leq N-1} \mu \right\|_{L^2(\Sigma_t^u)} \\ \sum_{a=1}^3 \left\| \varrho \mu \nabla\!\!\!/^2 \mathscr{Z}^{\leq N-1} L^a_{(Small)} \right\|_{L^2(\Sigma_t^u)} \\ \left\| \varrho \mu \nabla\!\!\!/ \mathcal{L}\!\!\!/_{\mathscr{Z}}^{\leq N-1} \chi^{(Small)} \right\|_{L^2(\Sigma_t^u)} \\ \left\| \varrho \mu d\!\!\!/ \mathscr{Z}^{\leq N-1} \mathrm{tr}_{g\!\!\!/} \chi^{(Small)} \right\|_{L^2(\Sigma_t^u)} \\ \left\| \varrho \mu \nabla\!\!\!/ \mathcal{L}\!\!\!/_{\mathscr{Z}}^{\leq N-1} \hat{\chi}^{(Small)} \right\|_{L^2(\Sigma_t^u)} \end{pmatrix}$$
$$\begin{aligned} &\lesssim \int_{t'=0}^t \frac{\|[L\mu]_-\|_{C^0(\Sigma_{t'}^u)}}{\mu_\star(t', u)} \mathbb{Q}^{1/2}_{(\leq N)}(t', u)\, dt' \\ &+ \int_{t'=0}^t \frac{1}{(1-t')^{3/2}\mu_\star(t', u)} \mathbb{Q}^{1/2}_{(\leq N)}(t', u)\, dt' \\ &+ \int_{t'=0}^t \frac{1}{(1+t')^{3/2}\mu_\star(t', u)} \widetilde{\mathbb{Q}}^{1/2}_{(\leq N)}(t', u)\, dt' \\ &+ \mathbb{Q}^{1/2}_{(\leq N)}(t, u) + \varepsilon \left\{ \ln \mu_\star^{-1}(t, u) + 1 \right\}, \end{aligned}$$

(20.24b)
$$\left\| \int_{t'=0}^t \left| \begin{pmatrix} \mu \nabla\!\!\!/^2 \mathscr{Z}^{\leq N-1} \mu \\ \sum_{a=1}^3 \left| \varrho \mu \nabla\!\!\!/^2 \mathscr{Z}^{\leq N-1} L^a_{(Small)} \right| \\ \varrho \mu \nabla\!\!\!/ \mathcal{L}\!\!\!/_{\mathscr{Z}}^{\leq N-1} \chi^{(Small)} \\ \varrho \mu d\!\!\!/ \mathscr{Z}^{\leq N-1} \mathrm{tr}_{g\!\!\!/} \chi^{(Small)} \\ \varrho \mu \nabla\!\!\!/ \mathcal{L}\!\!\!/_{\mathscr{Z}}^{\leq N-1} \hat{\chi}^{(Small)} \end{pmatrix} \right| (t', \cdot)\, dt' \right\|_{L^2(\Sigma_t^u)}$$
$$\begin{aligned} &\lesssim \ln(e+t)(1+t)\mathbb{Q}^{1/2}_{(\leq N)}(t, u) + \ln(e+t)(1+t)\widetilde{\mathbb{Q}}^{1/2}_{(\leq N)}(t, u) \\ &+ \varepsilon \ln(e+t)(1+t). \end{aligned}$$

We provide the proof of Lemma 20.20 in Sect. 20.5.

In the next two lemmas, we estimate some top-order error integrals. The integrals are nontrivial to bound in the sense that in order to avoid losing derivatives, we need to use the transport-elliptic estimates of Lemma 20.20. However, because of the favorable time decay factors present in the integrands, the integrals are not too difficult to estimate

LEMMA 20.21 (**Estimates for the easy top-order eikonal function quantity error integrals corresponding to the multiplier** T). *Let* $1 \leq N \leq 24$ *be an integer. Recall that* $\varrho(t, u) = 1 - u + t$ *and let* $\chi^{(Small)}$ *be the symmetric* *type* $\binom{0}{2}$ $S_{t,u}$ *tensorfield defined in* (4.1c). *Let* $\rho_{(l)}$, $(l = 1,2,3)$, *be the scalar-valued functions from Def.* 6.2. *Under the small-data and bootstrap assumptions of Sects.* 12.1-12.4, *if* ε *is sufficiently small, then the following inequalities hold for* $(t,u) \in [0, T_{(Bootstrap)}) \times [0, U_0]$:

$$\int_{\mathcal{M}_{t,u}} \left|(1+2\mu)L\mathscr{Z}^{\leq N}\Psi + 2\breve{R}\mathscr{Z}^{\leq N}\Psi\right| \times \left|\begin{pmatrix} (\mu \not{d}^{\#}\Psi)\cdot(\mu \not{d}\mathscr{Z}^{N-1}\mathrm{tr}_{\not{g}}\chi^{(Small)}) \\ \varrho(\not{d}\Psi^{\#})\cdot(\mu \not{d}\mathscr{Z}^{N-1}\mathrm{tr}_{\not{g}}\chi^{(Small)}) \\ \rho_{(l)}(\not{d}^{\#}\Psi)\cdot(\mu \not{d}\mathscr{Z}^{N-1}\mathrm{tr}_{\not{g}}\chi^{(Small)}) \end{pmatrix}\right| d\varpi \lesssim \varepsilon \mathbb{Q}_{(\leq N)}(t,u) + \varepsilon \int_{t'=0}^{t} \frac{1}{(1+t')^2}\widetilde{\mathbb{Q}}_{(\leq N)}(t',u)\,dt' + \varepsilon^3. \tag{20.25}$$

PROOF. Since the proofs of all of the estimates involving $L\mathscr{Z}^{\leq N}\Psi$, $\mu L\mathscr{Z}^{\leq N}\Psi$, and $\breve{R}\mathscr{Z}^{\leq N}\Psi$ are similar, we provide only the proof for one representative product. First, using the pointwise estimates $\mu \lesssim \ln(e+t)$ (that is, (12.128a)), $|\rho_{(l)}| \lesssim \varepsilon \ln(e+t)$ (that is, (12.28b) and Cor. 12.48), and $|\not{d}\Psi| \lesssim \varepsilon(1+t)^{-2}$ (that is, the bootstrap assumptions $(\mathbf{BA}\Psi)$), we deduce that

$$\left|\begin{pmatrix} (\mu \not{d}^{\#}\Psi)\cdot(\mu \not{d}\mathscr{Z}^{N-1}\mathrm{tr}_{\not{g}}\chi^{(Small)}) \\ \varrho(\not{d}\Psi^{\#})\cdot(\mu \not{d}\mathscr{Z}^{N-1}\mathrm{tr}_{\not{g}}\chi^{(Small)}) \\ \rho_{(l)}(\not{d}^{\#}\Psi)\cdot(\mu \not{d}\mathscr{Z}^{N-1}\mathrm{tr}_{\not{g}}\chi^{(Small)}) \end{pmatrix}\right| \lesssim \varepsilon \frac{1}{(1+t)^2}\left|\varrho\mu \not{d}\mathscr{Z}^{N-1}\mathrm{tr}_{\not{g}}\chi^{(Small)}\right|. \tag{20.26}$$

Referring to Defs. 3.68 and 3.69 and using spacetime Cauchy-Schwarz, inequalities (13.148), (13.149), (13.151), (20.24a), and (20.26), Prop. 14.7, the fact that $\mathbb{Q}_{(\leq N)}$ and $\widetilde{\mathbb{Q}}_{(\leq N)}$ are increasing in their arguments, and simple estimates of the form $ab \lesssim a^2 + b^2$, we deduce the desired estimate as follows:

$$\begin{aligned}
(20.27) \quad & \int_{\mathcal{M}_{t,u}} \left| L \mathscr{Z}^{\leq N} \Psi \right| \left| \begin{pmatrix} (\upmu \not{d}^{\#} \Psi) \cdot (\upmu \not{d} \mathscr{Z}^{N-1} \mathrm{tr}_{\not{g}} \upchi^{(Small)}) \\ \varrho (\not{d} \Psi^{\#}) \cdot (\upmu \not{d} \mathscr{Z}^{N-1} \mathrm{tr}_{\not{g}} \upchi^{(Small)}) \\ \mathsf{p}_{(l)} (\not{d}^{\#} \Psi) \cdot (\upmu \not{d} \mathscr{Z}^{N-1} \mathrm{tr}_{\not{g}} \upchi^{(Small)}) \end{pmatrix} \right| \, d\varpi \\
& \lesssim \varepsilon \int_{u'=0}^{u} \int_{\mathcal{C}_{u'}^{t}} \frac{1}{(1+t')^2} \left| L \mathscr{Z}^{\leq N} \Psi \right|^2 \, d\overline{\varpi} \, du' \\
& \quad + \varepsilon \int_{t'=0}^{t} \frac{1}{(1+t')^2} \int_{\Sigma_{t'}^{u}} \left| \varrho \upmu \not{d} \mathscr{Z}^{N-1} \mathrm{tr}_{\not{g}} \upchi \right|^2 \, d\underline{\varpi} \, dt' \\
& \lesssim \varepsilon \int_{u'=0}^{u} \mathbb{Q}_{(\leq N)}(t,u') \, du' \\
& \quad + \varepsilon \int_{t'=0}^{t} \frac{1}{(1+t')^2} \left(\int_{s=0}^{t'} \frac{\|[L\upmu]_-\|_{C^0(\Sigma_s^u)}}{\upmu_\star(s,u)} \mathbb{Q}_{(\leq N)}^{1/2}(s,u) \, ds \right)^2 dt' \\
& \quad + \varepsilon \int_{t'=0}^{t} \frac{1}{(1+t')^2} \mathbb{Q}_{(\leq N)}(t',u) \, dt' + \varepsilon \int_{t'=0}^{t} \frac{1}{(1+t')^2} \widetilde{\mathbb{Q}}_{(\leq N)}(t',u) \, dt' \\
& \quad + \varepsilon^3 \int_{t'=0}^{t} \frac{\left\{ \ln \upmu_\star^{-1}(t,u) + 1 \right\}^2}{(1+t')^2} \, dt' \\
& \lesssim \varepsilon \mathbb{Q}_{(\leq N)}(t,u) + \varepsilon \int_{t'=0}^{t} \frac{1}{(1+t')^{3/2}} \mathbb{Q}_{(\leq N)}(t',u) \left\{ \ln \upmu_\star^{-1} + 1 \right\}^2 \, dt' \\
& \quad + \varepsilon \int_{t'=0}^{t} \frac{1}{(1+t')^2} \widetilde{\mathbb{Q}}_{(\leq N)}(t',u) \, dt' + \varepsilon^3 \\
& \lesssim \varepsilon \mathbb{Q}_{(\leq N)}(t,u) + \varepsilon \int_{t'=0}^{t} \frac{1}{(1+t')^2} \widetilde{\mathbb{Q}}_{(\leq N)}(t',u) \, dt' + \varepsilon^3.
\end{aligned}$$

□

LEMMA 20.22 (**Estimates for the easy top-order eikonal function quantity error integrals corresponding to the multiplier $\widetilde{K}$**). *Let $1 \leq N \leq 24$ be an integer. Recall that $\varrho(t,u) = 1 - u + t$ and let $\upchi$ and $\upchi^{(Small)}$ be the symmetric type $\binom{0}{2}$ $S_{t,u}$ tensorfields defined in (3.70) and (4.1c). Let $\mathsf{p}_{(l)}$, $(l = 1,2,3)$, be the scalar-valued functions from Def. 3.2. Under the small-data and bootstrap assumptions of Sects. 12.1-12.4, if ε is sufficiently small, then the following inequalities hold for $(t,u) \in [0, T_{(Bootstrap)}) \times [0, U_0]$:*

(20.28)

$$\begin{aligned}
& \int_{\mathcal{M}_{t,u}} \varrho^2 \left| \left\{ L + \frac{1}{2} \mathrm{tr}_{\not{g}} \upchi \right\} \mathscr{Z}^{\leq N} \Psi \right| \left| \begin{pmatrix} (\upmu \not{d}^{\#} \Psi) \cdot (\upmu \not{d} \mathscr{Z}^{N-1} \mathrm{tr}_{\not{g}} \upchi^{(Small)}) \\ \varrho (\not{d} \Psi^{\#}) \cdot (\upmu \not{d} \mathscr{Z}^{N-1} \mathrm{tr}_{\not{g}} \upchi^{(Small)}) \\ \mathsf{p}_{(l)} (\not{d}^{\#} \Psi) \cdot (\upmu \not{d} \mathscr{Z}^{N-1} \mathrm{tr}_{\not{g}} \upchi^{(Small)}) \end{pmatrix} \right| \, d\varpi \\
& \qquad \lesssim \varepsilon \mathbb{Q}_{(\leq N)}(t,u) + \varepsilon \widetilde{\mathbb{Q}}_{(\leq N)}(t,u) + \varepsilon^3.
\end{aligned}$$

PROOF. Referring to Defs. 3.68 and 3.69 and using spacetime Cauchy-Schwarz, inequality (20.26), and Prop. 14.7, we have

(20.29)

$$\int_{\mathcal{M}_{t,u}} \varrho^2 \left| \left\{ L + \frac{1}{2}\mathrm{tr}_{\not g}\chi \right\} \mathscr{Z}^{\leq N}\Psi \right| \left| \begin{pmatrix} (\upmu \not d^{\#}\Psi) \cdot (\upmu \not d \mathscr{Z}^{N-1}\mathrm{tr}_{\not g}\chi^{(Small)}) \\ \varrho(\not d\Psi^{\#}) \cdot (\upmu \not d \mathscr{Z}^{N-1}\mathrm{tr}_{\not g}\chi^{(Small)}) \\ \uprho_{(l)}(\not d^{\#}\Psi) \cdot (\upmu \not d \mathscr{Z}^{N-1}\mathrm{tr}_{\not g}\chi^{(Small)}) \end{pmatrix} \right| d\varpi$$

$$\lesssim \varepsilon \int_{u'=0}^{u} \int_{\mathcal{C}_{u'}^t} (1+t')^2 \left| \left\{ L + \frac{1}{2}\mathrm{tr}_{\not g}\chi \right\} \mathscr{Z}^{\leq N}\Psi \right|^2 d\overline{\varpi}\, du'$$

$$+ \varepsilon \int_{t'=0}^{t} \int_{\Sigma_{t'}^u} \frac{1}{(1+t')^2} \left| \varrho\upmu \not d \mathscr{Z}^{N-1}\mathrm{tr}_{\not g}\chi^{(Small)} \right|^2 d\underline{\varpi}\, dt'$$

$$\lesssim \varepsilon \widetilde{\mathbb{Q}}_{(\leq N)}(t,u) + \varepsilon \int_{\mathcal{M}_{t,u}} \frac{1}{(1+t')^2} \left| \varrho\upmu \not d \mathscr{Z}^{N-1}\mathrm{tr}_{\not g}\chi^{(Small)} \right|^2 d\varpi.$$

The remainder of the proof now proceeds as in (20.27) but with a simple additional final step: the bound $\varepsilon \int_{t'=0}^{t} \frac{1}{(1+t')^2} \widetilde{\mathbb{Q}}_{(\leq N)}(t',u)\, dt' \lesssim \varepsilon \widetilde{\mathbb{Q}}_{(\leq N)}(t,u)$ for the integral on the right-hand side of (20.27). □

In the next lemma, we derive estimates for the error integrals corresponding to the deformation tensors of the multiplier vectorfields T and $\widetilde{K}$. Thanks to the pointwise estimates we have previously derived, the lemma is not difficult to prove.

LEMMA 20.23 (**The main estimates for the error integrals corresponding to the deformation tensors of T and $\widetilde{K}$ in terms of $\mathbb{Q}_{(N)}$, $\widetilde{\mathbb{Q}}_{(N)}$, and $\widetilde{\mathbb{K}}_{(N)}$**). *Let $0 \leq N \leq 24$ be an integer. Assume that the small-data and bootstrap assumptions of Sects. 12.1-12.4 hold for $(t,u) \in [0, T_{(Bootstrap)}) \times [0, U_0]$. There exists a constant $C > 0$ such that if ε is sufficiently small, then for $(t,u) \in [0, T_{(Bootstrap)}) \times [0, U_0]$, we have the following bounds for the non-$\mathfrak{F}$-containing integrals appearing on the right-hand sides of the inequalities of Prop. 10.13 (where the terms ${}^{(T)}\mathfrak{P}[\cdot]$ and ${}^{(\widetilde{K})}\mathfrak{P}[\cdot]$ are the energy error integrands from Def. 10.19 and Lemma 10.21, and the Morawetz spacetime integral with a good sign has been subtracted from the term ${}^{(\widetilde{K})}\mathfrak{P}[\cdot]$ on the left-hand side of (20.30b)):*

$$\int_{\mathcal{M}_{t,u}} {}^{(T)}\mathfrak{P}[\mathscr{Z}^N\Psi]\, d\varpi \tag{20.30a}$$

$$\leq C \int_{t'=0}^{t} \frac{\ln(e+t')}{(1+t')^2} \mathbb{Q}_{(N)}(t',u)\, dt'$$

$$+ C \int_{t'=0}^{t} \frac{\ln^2(e+t')}{(1+t')^2} \widetilde{\mathbb{Q}}_{(N)}(t',u)\, dt'$$

$$+ C\varepsilon^{1/2} \int_{t'=0}^{t} \frac{\ln(e+t')}{(1+t')^2\sqrt{\ln(e+t) - \ln(e+t')}} \widetilde{\mathbb{Q}}_{(N)}(t',u)\, dt'$$

$$+ C\frac{\ln(e+t)}{(1+t)^2} \int_{u'=0}^{u} \widetilde{\mathbb{Q}}_{(N)}(t,u')\, du' + C\varepsilon \sup_{t'\in[0,t]} \frac{\widetilde{\mathbb{K}}_{(N)}(t',u)}{(1+t')^{1/2}},$$

(20.30b)
$$\begin{aligned}&\int_{\mathcal{M}_{t,u}} \left|{}^{(\widetilde{K})}\mathfrak{P}[\mathscr{Z}^N\Psi] + \frac{1}{2}\varrho^2[L\upmu]_-|\not{d}\mathscr{Z}^N\Psi|^2\right| \\ &\leq C\int_{t'=0}^{t} \frac{\ln^3(e+t')}{(1+t')}\mathbb{Q}_{(N)}(t',u)\,dt' \\ &+ (1+C\varepsilon)\int_{t'=0}^{t} \frac{1}{\varrho(t',u)\ln\left(\frac{\varrho(t',u)}{\varrho(0,u)}\right)}\widetilde{\mathbb{Q}}_{(N)}(t',u)\,dt' \\ &+ \int_{u'=0}^{u} \widetilde{\mathbb{Q}}_{(N)}(t,u')\,du' \\ &+ C\varepsilon\widetilde{\mathbb{K}}_{(N)}(t,u),\end{aligned}$$

(20.30c)
$$\begin{aligned}&\int_{\Sigma_t^u} \left|\breve{\underline{L}}[\varrho^2\mathrm{tr}_{\not{g}}\chi]\right|(\mathscr{Z}^N\Psi)^2\,d\underline{\varpi} + \int_{\Sigma_t^u} \left|\varrho^2\upmu(\mathrm{tr}_{\not{g}}\chi)^2\right|(\mathscr{Z}^N\Psi)^2\,d\underline{\varpi} \\ &+ \int_{\Sigma_t^u} \left|\varrho^2\mathrm{tr}_{\not{g}}\chi\not{k}^{(Trans-\Psi)}\right|(\mathscr{Z}^N\Psi)^2\,d\underline{\varpi} + \int_{\Sigma_t^u} \left|\varrho^2\upmu\mathrm{tr}_{\not{g}}\chi\not{k}^{(Tan-\Psi)}\right|(\mathscr{Z}^N\Psi)^2\,d\underline{\varpi} \\ &\leq C\ln(e+t)\mathbb{Q}_{(N)}(t,u),\end{aligned}$$

(20.30d)
$$\begin{aligned}&\int_{\mathcal{M}_{t,u}} \upmu\left|\square_{g(\Psi)}[\varrho^2\mathrm{tr}_{\not{g}}\chi]\right|(\mathscr{Z}^N\Psi)^2\,d\varpi \\ &\leq C\ln^2(e+t)\mathbb{Q}_{(N)}(t,u).\end{aligned}$$

REMARK 20.24 (**Isolating the coercive Morawetz term $-\frac{1}{2}\varrho^2[L\upmu]_-|\not{d}\mathscr{Z}^N\Psi|^2$**). Recall that we have isolated the spacetime integral of $-\frac{1}{2}\varrho^2[L\upmu]_-|\not{d}\mathscr{Z}^N\Psi|^2$ by "removing" it from the left-hand side of (20.30b); this isolated term generates the coercive Morawetz spacetime integral (14.9a).

PROOF. To prove (20.30a) and (20.30b), we integrate inequalities (16.68a) and (16.68b) (with $\mathscr{Z}^N\Psi$ in the role of Ψ) over the spacetime region $\mathcal{M}_{t,u}$. The vast majority of the terms can be suitably bounded by using Prop. 14.7 and the fact that $\mathbb{Q}_{(\leq N)}$ and $\widetilde{\mathbb{Q}}_{(\leq N)}$ are increasing in their arguments. We also integrate by parts in t' to bound the integral of the term $C\ln(e+t')\left\{L\mathscr{Z}^N\Psi + \frac{1}{2}\mathrm{tr}_{\not{g}}\chi\mathscr{Z}^N\Psi\right\}^2$ from the right-hand side of (16.68a) by $\leq$ the sum of the second and fourth terms on the right-hand side of (20.30a), much as in the proof of (20.10). There are only two terms that cannot be bounded in this fashion. The first one is the term $\varepsilon\ln^2(e+t')\mathbf{1}_{\{\upmu\leq 1/4\}}\left|\not{d}\mathscr{Z}^N\Psi\right|^2$ from the right-hand side of (16.68a). To derive the desired bound for the corresponding spacetime integral, we simply use the estimate (20.15b). The second integral requiring special treatment is $\varepsilon\int_{\mathcal{M}_{t,u}} \frac{(1+t')}{\ln(e+t')}\mathbf{1}_{\{\upmu\leq 1/4\}}|\not{d}\mathscr{Z}^N\Psi|^2\,d\varpi$, which is generated by the last term on the right-hand side of (16.68b). To bound the integral by $\lesssim \varepsilon\widetilde{\mathbb{K}}_{(N)}(t,u)$, we simply use Lemma 14.5.

The *hypersurface integral* inequalities (20.30c) and the spacetime integral inequality (20.30d) follow easily from the pointwise estimates (16.69a) and (16.69b), Prop. 14.7, and the fact that $\mathbb{Q}_{(\leq N)}$ is increasing in its arguments. □

20.4. Difficult top-order error integral estimates

We now derive suitable estimates for the difficult top-order error integrals, which are generated when we commute the wave equation $\upmu \square_{g(\Psi)} \Psi = 0$ with a top-order pure spatial commutator vectorfield operator $\mathscr{S}^N$ (see Def. 7.2). These are the most important estimates in the monograph. We state them in the next two "primary" lemmas. In order to break up the proofs into manageable pieces, we devote the remainder of the present section to proving a series auxiliary lemmas. We then give the proof of the two lemmas of primary interest, along with Lemma 20.20 above, in Sects. 20.5-20.7.

20.4.1. Statement of the two primary lemmas. We now state the two lemmas of primary interest.

LEMMA 20.25 (**The main estimates for the difficult top-order error spacetime integrals corresponding to the multiplier** T). *Let $1 \leq N \leq 24$ be an integer, let $\mathscr{S}^{N-1}$ be an $(N-1)^{st}$ order pure spatial commutation vectorfield operator (see definition* (7.2a)*), let $O \in \mathscr{O}$ be a geometric rotation vectorfield (see definition* (7.2b)*), and let $\varsigma > 0$ be a number. Let $\upchi^{(Small)}$ be the symmetric type $\binom{0}{2}$ $S_{t,u}$ tensorfield defined in* (4.1c)*. Under the small-data and bootstrap assumptions of Sects.* 12.1-12.4*, there exist a small constant $a > 0$ and a large constant $C > 0$ such that if ε is sufficiently small, then the following estimates hold for $(t, u) \in [0, T_{(Bootstrap)}) \times [0, U_0]$ (and the constants are* ***independent of*** ς*):*

(20.31)

$$
\begin{aligned}
& \left| \int_{\mathcal{M}_{t,u}} \left\{ (1+2\upmu) L \mathscr{S}^{N-1} \breve{R} \Psi + 2 \breve{R} \mathscr{S}^{N-1} \breve{R} \Psi \right\} (\breve{R}\Psi) \not{\Delta} \mathscr{S}^{N-1} \upmu \, d\varpi \right|, \\
& \left| \int_{\mathcal{M}_{t,u}} \left\{ (1+2\upmu) L \mathscr{S}^{N-1} O \Psi + 2 \breve{R} \mathscr{S}^{N-1} O \Psi \right\} (\breve{R}\Psi) O \mathscr{S}^{N-1} \mathrm{tr}_{\not{g}} \upchi^{(Small)} \, d\varpi \right| \\
& \leq C\varepsilon \mathbb{Q}_{(\leq N)}(t,u) + C\varepsilon^3 \frac{1}{\upmu_\star(t,u)} \\
& \quad + C\varepsilon \ln^8(e+t) \frac{1}{\upmu_\star(t,u)} \mathbb{Q}_{(\leq N-1)}(t,u) + C\varepsilon \frac{1}{\upmu_\star(t,u)} \widetilde{\mathbb{Q}}_{(\leq N-1)}(t,u) \\
& \quad + \boxed{9} \int_{t'=0}^{t} \frac{\| [L\upmu]_- \|_{C^0(\Sigma_{t'}^u)}}{\upmu_\star(t',u)} \mathbb{Q}_{(\leq N)}^{1/2}(t',u) \\
& \qquad \times \int_{s=0}^{t'} \frac{\| [L\upmu]_- \|_{C^0(\Sigma_s^u)}}{\upmu_\star(s,u)} \mathbb{Q}_{(\leq N)}^{1/2}(s,u) \, ds \, dt' \\
& \quad + C\varepsilon \int_{t'=0}^{t} \frac{1}{(1+t')^{1+a} \upmu_\star(t',u)} \mathbb{Q}_{(\leq N)}^{1/2}(t',u) \\
& \qquad \times \int_{s=0}^{t'} \frac{1}{(1+s)} \frac{1}{\upmu_\star(s,u)} \mathbb{Q}_{(\leq N)}^{1/2}(s,u) \, ds \, dt' \\
& \quad + C\varepsilon \int_{t'=0}^{t} \frac{1}{(1+t')^{1+a} \upmu_\star(t',u)} \mathbb{Q}_{(\leq N)}^{1/2}(t',u)
\end{aligned}
$$

$$\times \int_{s=0}^{t'} \frac{1}{(1+s)} \frac{1}{\upmu_\star(s,u)} \widetilde{\mathbb{Q}}_{(\leq N)}^{1/2}(s,u)\, ds\, dt'$$

$$+ \boxed{9} \int_{t'=0}^{t} \frac{\| [L\upmu]_- \|_{C^0(\Sigma_{t'}^u)}}{\upmu_\star(t',u)} \mathbb{Q}_{(\leq N)}(t',u)\, dt'$$

$$+ 9 \int_{t'=0}^{t} \frac{1}{\varrho(t',u)\left\{1 + \ln\left(\frac{\varrho(t',u)}{\varrho(0,u)}\right)\right\}} \mathbb{Q}_{(\leq N)}^{1/2}(t',u)$$

$$\times \int_{s=0}^{t'} \frac{\| [L\upmu]_- \|_{C^0(\Sigma_s^u)}}{\upmu_\star(s,u)} \mathbb{Q}_{(\leq N)}^{1/2}(s,u)\, ds\, dt'$$

$$+ 9 \int_{t'=0}^{t} \frac{1}{\varrho(t',u)\left\{1 + \ln\left(\frac{\varrho(t',u)}{\varrho(0,u)}\right)\right\}} \mathbb{Q}_{(\leq N)}(t',u)\, dt'$$

$$+ C\varepsilon \int_{t'=0}^{t} \frac{1}{(1+t')^{1+a}\upmu_\star(t',u)} \mathbb{Q}_{(\leq N)}(t',u)\, dt'$$

$$+ C\varepsilon \int_{t'=0}^{t} \frac{1}{(1+t')^{1+a}\upmu_\star(t',u)} \widetilde{\mathbb{Q}}_{(\leq N)}(t',u)\, dt'$$

$$+ C\varepsilon \frac{1}{\upmu_\star^{1/2}(t,u)} \mathbb{Q}_{(\leq N)}^{1/2}(t,u) \int_{t'=0}^{t} \frac{1}{(1+t')^{1+a}} \mathbb{Q}_{(\leq N)}^{1/2}(t',u)\, dt'$$

$$+ C\varepsilon \frac{1}{\upmu_\star^{1/2}(t,u)} \mathbb{Q}_{(\leq N)}^{1/2}(t,u) \int_{t'=0}^{t} \frac{1}{(1+t')^{1+a}\upmu_\star^{1/2}(t',u)} \widetilde{\mathbb{Q}}_{(\leq N)}^{1/2}(t',u)\, dt'$$

$$+ C(1+\varsigma^{-1}) \int_{t'=0}^{t} \frac{1}{(1+t')^{3/2}} \mathbb{Q}_{(\leq N)}(t',u)\, dt'$$

$$+ C(1+\varsigma^{-1}) \int_{t'=0}^{t} \frac{1}{(1+t')^{3/2}} \widetilde{\mathbb{Q}}_{(\leq N)}(t',u)\, dt'$$

$$+ C(1+\varsigma^{-1}) \int_{u'=0}^{u} \mathbb{Q}_{(\leq N)}(t,u')\, du'$$

$$+ C(1+\varsigma^{-1}) \frac{1}{(1+t)^{1/2}} \int_{u'=0}^{u} \widetilde{\mathbb{Q}}_{(\leq N)}(t,u')\, du' + C\varsigma \sup_{t'\in[0,t]} \frac{\widetilde{\mathbb{K}}_{(\leq N)}(t',u)}{(1+t')^{1/2}}.$$

We provide the proof of Lemma 20.25 in Sect. 20.6.

LEMMA 20.26 (**The main estimates for the difficult top-order error spacetime integrals corresponding to the multiplier $\widetilde{K}$**). *Let $1 \leq N \leq 24$ be an integer, let $\mathscr{S}^{N-1}$ be an $(N-1)^{st}$ order pure spatial commutation vectorfield operator (see definition (7.2a)), let $O \in \mathscr{O}$ be a geometric rotation vectorfield (see definition (7.2b)), and let $\widetilde{\varsigma} > 0$ be a number. Recall that $\varrho(t,u) = 1 - u + t$ and let χ and $\chi^{(Small)}$ be the symmetric type $\binom{0}{2}$ $S_{t,u}$ tensorfields defined in (3.70) and (4.1c). Under the small-data and bootstrap assumptions of Sects. 12.1-12.4, there exist a small constant $a > 0$ and a large constant $C > 0$ such that if ε is sufficiently small, then the following estimates hold for $(t,u) \in [0, T_{(Bootstrap)}) \times [0, U_0]$ (and* *the constants are* ***independent of*** $\widetilde{\varsigma}$*):*

(20.32)

$$
\begin{aligned}
&\left| \int_{\mathcal{M}_{t,u}} \varrho^2 \left\{ L\mathscr{S}^{N-1}\breve{R}\Psi + \frac{1}{2}\mathrm{tr}_{\not g}\chi \mathscr{S}^{N-1}\breve{R}\Psi \right\} (\breve{R}\Psi) \not\Delta \mathscr{S}^{N-1}\upmu \, d\varpi \right|, \\
&\left| \int_{\mathcal{M}_{t,u}} \varrho^2 \left\{ L\mathscr{S}^{N-1}O\Psi + \frac{1}{2}\mathrm{tr}_{\not g}\chi \mathscr{S}^{N-1}O\Psi \right\} (\breve{R}\Psi) O \mathscr{S}^{N-1}\mathrm{tr}_{\not g}\chi^{(Small)} \, d\varpi \right| \\
&\leq C\varepsilon \mathbb{Q}_{(\leq N)}(t,u) + C\varepsilon \widetilde{\mathbb{Q}}_{(\leq N)}(t,u) + C\varepsilon^3 \frac{1}{\upmu_\star(t,u)} \\
&\quad + C\varepsilon \frac{1}{\upmu_\star(t,u)} \ln^2(e+t) \mathbb{Q}_{(\leq N-1)}(t,u) + C\varepsilon \frac{1}{\upmu_\star(t,u)} \ln^2(e+t) \widetilde{\mathbb{Q}}_{(\leq N-1)}(t,u) \\
&\quad + \boxed{5} \int_{t'=0}^{t} \frac{\|[L\upmu]_-\|_{C^0(\Sigma_{t'}^u)}}{\upmu_\star(t',u)} \widetilde{\mathbb{Q}}_{(\leq N)}(t',u) \, dt' \\
&\quad + 5 \int_{t'=0}^{t} \frac{1}{\varrho(t',u)\left\{1 + \ln\left(\frac{\varrho(t',u)}{\varrho(0,u)}\right)\right\}} \widetilde{\mathbb{Q}}_{(\leq N)}(t',u) \, dt' \\
&\quad + C\varepsilon \int_{t'=0}^{t} \frac{1}{(1+t')^{3/2}\upmu_\star^{1/2}(t',u)} \mathbb{Q}_{(\leq N)}(t',u) \, dt' \\
&\quad + C\varepsilon \int_{t'=0}^{t} \frac{1}{(1+t')^{3/2}\upmu_\star(t',u)} \widetilde{\mathbb{Q}}_{(\leq N)}(t',u) \, dt' \\
&\quad + C(1+\widetilde{\varsigma}^{-1}) \int_{t'=0}^{t} \frac{1}{(1+t')^{3/2}} \mathbb{Q}_{(\leq N)}(t',u) \, dt' \\
&\quad + C(1+\widetilde{\varsigma}^{-1}) \int_{t'=0}^{t} \frac{1}{(1+t')^{3/2}} \widetilde{\mathbb{Q}}_{(\leq N)}(t',u) \, dt' \\
&\quad + C \ln^4(e+t) \mathbb{Q}_{(N)}(t,u) \\
&\quad + \boxed{5} \frac{\|\varrho L\upmu\|_{C^0({}^{(-)}\Sigma_{t;t}^u)}}{\upmu_\star^{1/2}(t,u)} \widetilde{\mathbb{Q}}_{(\leq N)}^{1/2}(t,u) \int_{t'=0}^{t} \frac{1}{\varrho(t',u)\upmu_\star^{1/2}(t',u)} \widetilde{\mathbb{Q}}_{(\leq N)}^{1/2}(t',u) \, dt' \\
&\quad + 5 \left\| \frac{\varrho L\upmu}{\upmu} \right\|_{C^0({}^{(+)}\Sigma_{t;t}^u)} \widetilde{\mathbb{Q}}_{(\leq N)}^{1/2}(t,u) \int_{t'=0}^{t} \left\| \sqrt{\frac{\upmu(t,\cdot)}{\upmu}} \right\|_{C^0({}^{(+)}\Sigma_{t';t}^u)} \frac{1}{\varrho(t',u)} \widetilde{\mathbb{Q}}_{(\leq N)}^{1/2}(t',u) \, dt' \\
&\quad + C\varepsilon \frac{1}{\upmu_\star^{1/2}(t,u)} \widetilde{\mathbb{Q}}_{(\leq N)}^{1/2}(t,u) \int_{t'=0}^{t} \frac{1}{(1+t')^{3/2}} \mathbb{Q}_{(\leq N)}^{1/2}(t',u) \, dt' \\
&\quad + C\varepsilon \frac{1}{\upmu_\star^{1/2}(t,u)} \widetilde{\mathbb{Q}}_{(\leq N)}^{1/2}(t,u) \int_{t'=0}^{t} \frac{1}{(1+t')^{3/2}\upmu_\star^{1/2}(t',u)} \widetilde{\mathbb{Q}}_{(\leq N)}^{1/2}(t',u) \, dt' \\
&\quad + C(1+\widetilde{\varsigma}^{-1}) \int_{u'=0}^{u} \widetilde{\mathbb{Q}}_{(\leq N)}(t,u') \, du' + C\widetilde{\varsigma} \widetilde{\mathbb{K}}_{(\leq N)}(t,u).
\end{aligned}
$$

We provide the proof of Lemma 20.26 in Sect. 20.7.

20.4.2. Auxiliary lemmas. We now state and prove a series of auxiliary lemmas in order to help us prove the previous two lemmas as well as Lemma 20.20 above.

LEMMA 20.27 (**Preliminary L^2 bounds for some top-order derivatives of $\chi^{(Small)}$ and $L^i_{(Small)}$**). *Let $1 \leq N \leq 24$ be an integer. Recall that $\varrho(t,u) = 1-u+t$. Let $\chi^{(Small)}$ be the symmetric type $\binom{0}{2}$ $S_{t,u}$ tensorfield defined in* (4.1c) *and let $O \in \mathscr{O}$ be a geometric rotation vectorfield (see definition* (7.2b)*). Under the small-data and bootstrap assumptions of Sects.* 12.1-12.4*, if ε is sufficiently small, then the following estimates hold for $(t,u) \in [0, T_{(Bootstrap)}) \times [0, U_0]$:*

$$\begin{aligned}
(20.33)\quad \left\| \varrho\mu \nabla\!\!\!/ \mathcal{L}\!\!\!/_{\mathscr{Z}}^{N-1} \chi^{(Small)} \right\|_{L^2(\Sigma_t^u)} \lesssim{}& \sum_{l=1}^{3} \left\| \mu O_{(l)} \mathscr{Z}^{N-1} \mathrm{tr}_{g\!\!/} \chi^{(Small)} \right\|_{L^2(\Sigma_t^u)} \\
&+ \left\| \varrho\mu \nabla\!\!\!/ \mathcal{L}\!\!\!/_{\mathscr{Z}}^{N-1} \hat{\chi}^{(Small)} \right\|_{L^2(\Sigma_t^u)} \\
&+ \mathbb{Q}_{(\leq N)}^{1/2}(t,u) \\
&+ \int_{t'=0}^{t} \frac{1}{(1+t')^{3/2} \mu_\star^{1/2}(t',u)} \widetilde{\mathbb{Q}}_{(\leq N)}^{1/2}(t',u)\, dt' \\
&+ \varepsilon,
\end{aligned}$$

$$\begin{aligned}
(20.34)\quad \begin{pmatrix} \left\| \varrho\mu d\!\!/ \mathscr{Z}^{N-1} \mathrm{tr}_{g\!\!/} \chi^{(Small)} \right\|_{L^2(\Sigma_t^u)} \\ \left\| \mu O \mathscr{Z}^{N-1} \mathrm{tr}_{g\!\!/} \chi^{(Small)} \right\|_{L^2(\Sigma_t^u)} \end{pmatrix} \lesssim{}& \left\| \varrho\mu \nabla\!\!\!/ \mathcal{L}\!\!\!/_{\mathscr{Z}}^{N-1} \chi^{(Small)} \right\|_{L^2(\Sigma_t^u)} \\
&+ \mathbb{Q}_{(\leq N)}^{1/2}(t,u) \\
&+ \int_{t'=0}^{t} \frac{1}{(1+t')^{3/2} \mu_\star^{1/2}(t',u)} \widetilde{\mathbb{Q}}_{(\leq N)}^{1/2}(t',u)\, dt' \\
&+ \varepsilon,
\end{aligned}$$

$$\begin{aligned}
(20.35)\quad \left\| \varrho\mu \nabla\!\!\!/ \mathcal{L}\!\!\!/_{\mathscr{Z}}^{N-1} \hat{\chi}^{(Small)} \right\|_{L^2(\Sigma_t^u)} \lesssim{}& \left\| \varrho\mu \nabla\!\!\!/ \mathcal{L}\!\!\!/_{\mathscr{Z}}^{N-1} \chi^{(Small)} \right\|_{L^2(\Sigma_t^u)} \\
&+ \mathbb{Q}_{(\leq N)}^{1/2}(t,u) \\
&+ \int_{t'=0}^{t} \frac{1}{(1+t')^{3/2} \mu_\star^{1/2}(t',u)} \widetilde{\mathbb{Q}}_{(\leq N)}^{1/2}(t',u)\, dt' \\
&+ \varepsilon,
\end{aligned}$$

$$\begin{aligned}
(20.36)\quad \left\| \varrho\mu \,\mathrm{div}\!\!\!/\, \hat{\mathcal{L}\!\!\!/}_{\mathscr{Z}}^{N-1} \hat{\chi}^{(Small)} \right\|_{L^2(\Sigma_t^u)} \lesssim{}& \sum_{l=1}^{3} \left\| \mu O_{(l)} \mathscr{Z}^{N-1} \mathrm{tr}_{g\!\!/} \chi^{(Small)} \right\|_{L^2(\Sigma_t^u)} \\
&+ \mathbb{Q}_{(\leq N)}^{1/2}(t,u) \\
&+ \int_{t'=0}^{t} \frac{1}{(1+t')^{3/2} \mu_\star^{1/2}(t',u)} \widetilde{\mathbb{Q}}_{(\leq N)}^{1/2}(t',u)\, dt' \\
&+ \varepsilon,
\end{aligned}$$

$$(20.37)\qquad \left\| \varrho\, \hat{\not{\mathcal{L}}}_{\mathscr{Z}}^{N-1} \hat{\chi}^{(Small)} \right\|_{L^2(\Sigma_t^u)} \lesssim \ln(e+t)\mathbb{Q}_{(\leq N)}^{1/2}(t,u) + \int_{t'=0}^{t} \frac{1}{(1+t')\upmu_\star^{1/2}(t',u)} \widetilde{\mathbb{Q}}_{(\leq N)}^{1/2}(t',u)\,dt' + \varepsilon.$$

Furthermore, the following estimate holds for $(t,u) \in [0, T_{(Bootstrap)}) \times [0, U_0]$*:*

$$(20.38)\qquad \left\| \upmu \not{\mathcal{L}}_{\breve{R}} \not{\mathcal{L}}_{\mathscr{Z}}^{N-1} \chi^{(Small)} \right\|_{L^2(\Sigma_t^u)} \lesssim \left\| \upmu \nabla\!\!\!\!/^{\,2} \mathscr{Z}^{N-1} \upmu \right\|_{L^2(\Sigma_t^u)} + \mathbb{Q}_{(\leq N)}^{1/2}(t,u) + \int_{t'=0}^{t} \frac{1}{(1+t')^{3/2}\upmu_\star^{1/2}(t',u)} \widetilde{\mathbb{Q}}_{(\leq N)}^{1/2}(t',u)\,dt' + \varepsilon.$$

Finally, the following estimate holds for $(t,u) \in [0, T_{(Bootstrap)}) \times [0, U_0]$, $(i = 1,2,3)$*:*

$$(20.39)\qquad \left\| \varrho\upmu \Delta\!\!\!\!/\, \mathscr{Z}^{\leq N-1} L^i_{(Small)} \right\|_{L^2(\Sigma_t^u)} \lesssim \sum_{l=1}^{3} \left\| \upmu O_{(l)} \mathscr{Z}^{N-1} \mathrm{tr}_{\not{g}} \chi^{(Small)} \right\|_{L^2(\Sigma_t^u)} + \mathbb{Q}_{(\leq N)}^{1/2}(t,u) + \int_{t'=0}^{t} \frac{1}{(1+t')^{3/2}\upmu_\star^{1/2}(t',u)} \widetilde{\mathbb{Q}}_{(\leq N)}^{1/2}(t',u)\,dt' + \varepsilon.$$

PROOF. We first prove (20.33). To this end, we multiply both sides of inequalities (16.71) and (16.72) by $\varrho\upmu$ and then take the norm $\|\cdot\|_{L^2(\Sigma_t^u)}$ of both sides. Using Prop. 14.7, inequality (19.2a), the estimate $\varrho\upmu \lesssim (1+t)\ln(e+t)$ (that is, (12.128a)), and the fact that $\mathbb{Q}_{(\leq N)}$ is increasing in its arguments, we deduce that the products of $\varrho\upmu$ and the right-hand sides of (16.71) and (16.72) are bounded in the norm $\|\cdot\|_{L^2(\Sigma_t^u)}$ by

$$(20.40)\qquad \lesssim \mathbb{Q}_{(\leq N)}^{1/2}(t,u) + \int_{t'=0}^{t} \frac{1}{(1+t')^{3/2}\upmu_\star^{1/2}(t',u)} \widetilde{\mathbb{Q}}_{(\leq N)}^{1/2}(t',u)\,dt' + \varepsilon.$$

We note that the right-hand side of (20.40) is manifestly $\lesssim$ the right-hand side of (20.33). The desired estimate (20.33) now follows from inequalities (16.71) and (16.72), the triangle inequality, and the estimate (12.46a), which implies that $\left\| \upmu\varrho d\!\!\!/ \mathscr{Z}^{N-1}\mathrm{tr}_{\not{g}}\chi^{(Small)}) \right\|_{L^2(\Sigma_t^u)} \lesssim \sum_{l=1}^{3} \left\| \upmu O_{(l)} \mathscr{Z}^{N-1}\mathrm{tr}_{\not{g}}\chi^{(Small)} \right\|_{L^2(\Sigma_t^u)}$.

The estimate (20.34) for $\left\| \upmu O \mathscr{Z}^{N-1}\mathrm{tr}_{\not{g}}\chi^{(Small)} \right\|_{L^2(\Sigma_t^u)}$ can be proved in a similar fashion with the help of the pointwise inequality (16.73). The estimate (20.34) for $\left\| \varrho\upmu d\!\!\!/ \mathscr{Z}^{N-1}\mathrm{tr}_{\not{g}}\chi^{(Small)} \right\|_{L^2(\Sigma_t^u)}$ then follows from the estimate for $\left\| \upmu O \mathscr{Z}^{N-1}\mathrm{tr}_{\not{g}}\chi^{(Small)} \right\|_{L^2(\Sigma_t^u)}$ and the inequality

$$\left\| \upmu\varrho d\!\!\!/ \mathscr{Z}^{N-1}\mathrm{tr}_{\not{g}}\chi^{(Small)} \right\|_{L^2(\Sigma_t^u)} \lesssim \sum_{l=1}^{3} \left\| \upmu O_{(l)} \mathscr{Z}^{N-1}\mathrm{tr}_{\not{g}}\chi^{(Small)} \right\|_{L^2(\Sigma_t^u)}$$

noted above.

To prove the estimate (20.35), we again base our argument on the pointwise inequality (16.71), but this time we use the triangle inequality to bound $\left\| \varrho \upmu \nabla\!\!\!/ \mathcal{L}_{\mathscr{Z}}^{N-1} \hat{\upchi}^{(Small)} \right\|_{L^2(\Sigma_t^u)}$ in terms of the remaining quantities in (16.71), which we have already shown to be bounded in the norm $\| \cdot \|_{L^2(\Sigma_t^u)}$ by the right-hand side of (20.35).

To prove (20.36), we apply similar reasoning to the fourth pointwise inequality in (15.28).

To prove (20.37), we apply similar reasoning to the pointwise inequality (16.74).

To prove (20.38), we apply similar reasoning to the second pointwise inequality in (15.1).

To prove (20.39), we apply similar reasoning to the pointwise inequality (12.146). As a preliminary step, we use the third pointwise inequality in (15.28) to bound the last term on the right-hand side of (12.146) in terms of $d\!\!\!/ \mathscr{Z}^{N-1} \mathrm{tr}_{g\!\!\!/} \upchi^{(Small)}$ and the error terms on the right-hand side of (15.28). □

LEMMA 20.28 (**Preliminary nonsharp L^2 bounds for $\upmu O \mathscr{Z}^{N-1} \mathrm{tr}_{g\!\!\!/} \upchi^{(Small)}$ and $\upmu \Delta\!\!\!/ \mathscr{Z}^{N-1} \upmu$**). *Let $1 \leq N \leq 24$ be an integer. Let $\upchi^{(Small)}$ be the symmetric type $\binom{0}{2}$ $S_{t,u}$ tensorfield defined in (4.1c). Under the small-data and bootstrap assumptions of Sects. 12.1-12.4, if ε is sufficiently small, then the following estimates hold for $(t,u) \in [0, T_{(Bootstrap)}) \times [0, U_0]$:*

$$
\begin{aligned}
(20.41) \qquad & \begin{pmatrix} \left\| \upmu O \mathscr{Z}^{N-1} \mathrm{tr}_{g\!\!\!/} \upchi^{(Small)} \right\|_{L^2(\Sigma_t^u)} \\ \left\| \upmu \Delta\!\!\!/ \mathscr{Z}^{N-1} \upmu \right\|_{L^2(\Sigma_t^u)} \end{pmatrix} \\
& \lesssim \int_{t'=0}^t \frac{\| [L\upmu]_- \|_{C^0(\Sigma_{t'}^u)}}{\upmu_\star(t',u)} \mathbb{Q}_{(\leq N)}^{1/2}(t',u)\, dt' \\
& + \int_{t'=0}^t \frac{1}{(1+t')^{3/2} \upmu_\star(t',u)} \mathbb{Q}_{(\leq N)}^{1/2}(t',u)\, dt' \\
& + \mathbb{Q}_{(\leq N)}^{1/2}(t,u) + \int_{t'=0}^t \frac{1}{(1+t')^{3/2} \upmu_\star(t',u)} \widetilde{\mathbb{Q}}_{(\leq N)}^{1/2}(t',u)\, dt' \\
& + \varepsilon \int_{t'=0}^t \frac{1}{(1+t')^{3/2}} \begin{pmatrix} \left\| \upmu \hat{\nabla}\!\!\!/^2 \mathscr{Z}^{\leq N-1} \upmu \right\|_{L^2(\Sigma_{t'}^u)} \\ \left\| \varrho \upmu \nabla\!\!\!/ \mathcal{L}_{\mathscr{Z}}^{\leq N-1} \hat{\upchi}^{(Small)} \right\|_{L^2(\Sigma_{t'}^u)} \end{pmatrix} dt' \\
& + \varepsilon \left\{ \ln \upmu_\star^{-1}(t,u) + 1 \right\}.
\end{aligned}
$$

PROOF. We first assume that $\mathscr{Z}^{N-1} = \mathscr{S}^{N-1}$ is an $(N-1)^{st}$ order pure spatial commutation vectorfield operator. Then to prove (20.41), it suffices to bound the norm $\| \cdot \|_{L^2(\Sigma_t^u)}$ of the right-hand side of (17.53) by the right-hand side of (20.41). To bound the norm $\| \cdot \|_{L^2(\Sigma_t^u)}$ of the first term on the right-hand side of (17.53) by $\lesssim \mathbb{Q}_{(\leq N)}^{1/2}(t,u)$, we use Prop. 14.7.

To bound the norm $\| \cdot \|_{L^2(\Sigma_t^u)}$ of the second term on the right-hand side of (17.53) by $\lesssim \mathbb{Q}_{(\leq N)}^{1/2}(t,u) + \int_{t'=0}^t \frac{1}{(1+t')^{3/2} \upmu_\star^{1/2}(t',u)} \widetilde{\mathbb{Q}}_{(\leq N)}^{1/2}(t',u)\, dt' + \varepsilon$, we use inequality (19.2a) and the fact that $\mathbb{Q}_{(\leq N)}$ is increasing in both of its arguments.

To bound the norm $\|\cdot\|_{L^2(\Sigma_t^u)}$ of the third term on the right-hand side of (17.53) by the first term on the right-hand side of (20.41), we use Lemma 12.57 and Prop. 14.7.

To bound the norm $\|\cdot\|_{L^2(\Sigma_t^u)}$ of the fourth term on the right-hand side of (17.53), we use Lemma 12.57, the estimate (13.144) with $b = 1/2$, Prop. 14.7, and the fact that $\mathbb{Q}_{(\leq N)}$ is increasing in both of its arguments to bound it by

$$(20.42) \qquad \lesssim \mathbb{Q}^{1/2}_{(\leq N)}(t,u)\frac{1}{\varrho(t,u)}\int_{t'=0}^{t}\left\|\left(\frac{\upmu(t,\cdot)}{\upmu}\right)^2\right\|_{C^0(\Sigma_{t'}^u)} dt' \lesssim \mathbb{Q}^{1/2}_{(\leq N)}(t,u)$$

as desired.

To bound the norm $\|\cdot\|_{L^2(\Sigma_t^u)}$ of the fifth term $\frac{\ln^2(e+t)}{(1+t)^2}|\mathscr{X}^{[N]}|(0,u,\vartheta)$ on the right-hand side of (17.53), we use the definition of $\mathscr{X}^{[N]}$ (which is defined just below (17.28)), Defs. 3.68 and 3.69, inequality (12.67a), inequality (12.152), and Lemma 12.17 to deduce that

$$(20.43) \qquad \begin{aligned} \left\|\mathscr{X}^{[N]}(0,\cdot)\right\|^2_{L^2(\Sigma_t^u)} &= \int_{u'=0}^{u}\int_{S_{t,u}}\left|\mathscr{X}^{[N]}\right|^2(0,u',\vartheta)\,dv_{\not g(t,u',\vartheta)}\,du' \\ &\lesssim \varrho^2(t,u)\int_{u'=0}^{u}\int_{S_{0,u}}\left|\mathscr{X}^{[N]}\right|^2(0,u',\vartheta)\,dv_{\not g(0,u',\vartheta)}\,du' \\ &= \varrho^2(t,u)\left\|\mathscr{X}^{[N]}(0,\cdot)\right\|^2_{L^2(\Sigma_0^u)} \lesssim \varepsilon^2\varrho^2(t,u). \end{aligned}$$

From (20.43), it follows that $\frac{\ln^2(e+t)}{(1+t)^2}\left\|\mathscr{X}^{[N]}(0,\cdot)\right\|_{L^2(\Sigma_t^u)} \lesssim \varepsilon\frac{\ln^2(e+t)}{1+t}$ as desired.

To bound the norm $\|\cdot\|_{L^2(\Sigma_t^u)}$ of the sixth term on the right-hand side of (17.53), we use Lemma 12.57, Prop. 14.7, the estimate (13.152), and the fact that $\mathbb{Q}_{(\leq N)}$ and $\widetilde{\mathbb{Q}}_{(\leq N)}$ are increasing in both of their arguments to bound it by

$$(20.44) \qquad \begin{aligned} &\lesssim \frac{\ln^3(e+t)}{1+t}\int_{t'=0}^{t}\frac{1}{(1+t')\upmu_\star^{1/2}(t',u)}\mathbb{Q}^{1/2}_{(\leq N)}(t',u)\,dt' \\ &\quad + \frac{\ln^3(e+t)}{1+t}\int_{t'=0}^{t}\frac{1}{(1+t')\upmu_\star^{1/2}(t',u)}\widetilde{\mathbb{Q}}^{1/2}_{(\leq N)}(t',u)\,dt' \\ &\lesssim \mathbb{Q}^{1/2}_{(\leq N)}(t,u) + \int_{t'=0}^{t}\frac{1}{(1+t')^{3/2}\upmu_\star^{1/2}(t',u)}\widetilde{\mathbb{Q}}^{1/2}_{(\leq N)}(t',u)\,dt' \end{aligned}$$

as desired.

To bound the norm $\|\cdot\|_{L^2(\Sigma_t^u)}$ of the seventh term on the right-hand side of (17.53) by the next-to-last term on the right-hand side of (20.41), we use Lemma 12.57.

To bound the norm $\|\cdot\|_{L^2(\Sigma_t^u)}$ of the final term on the right-hand side of (17.53), we use Lemma 12.57, the estimate (19.2a), the estimates (13.150) and (13.152), and the fact that $\mathbb{Q}_{(\leq N)}$ and $\widetilde{\mathbb{Q}}_{(\leq N)}$ are increasing in both of their arguments to bound it by

(20.45)
$$\begin{aligned}
&\lesssim \varepsilon \frac{\ln^3(e+t)}{1+t} \int_{t'=0}^t \frac{1}{(1+t')} \frac{1}{\upmu_{\star}(t',u)}\, dt' \\
&+ \frac{\ln^3(e+t)}{1+t} \int_{t'=0}^t \frac{1}{(1+t')} \frac{1}{\upmu_{\star}(t',u)} \int_{s=0}^{t'} \frac{\mathbb{Q}_{(\leq N)}^{1/2}(s,u)}{1+s}\, ds\, dt' \\
&+ \frac{\ln^3(e+t)}{1+t} \int_{t'=0}^t \frac{1}{(1+t')} \frac{1}{\upmu_{\star}(t',u)} \int_{s=0}^{t'} \frac{\widetilde{\mathbb{Q}}_{(\leq N)}^{1/2}(s,u)}{(1+s)\upmu_{\star}^{1/2}(s,u)}\, ds\, dt' \\
&\lesssim \varepsilon \left\{ \ln \upmu_{\star}^{-1}(t,u) + 1 \right\} + \int_{t'=0}^t \frac{1}{(1+t')^{3/2}} \frac{1}{\upmu_{\star}(t',u)} \mathbb{Q}_{(\leq N)}^{1/2}(t',u)\, dt' \\
&+ \int_{t'=0}^t \frac{1}{(1+t')^{3/2}} \frac{1}{\upmu_{\star}(t',u)} \widetilde{\mathbb{Q}}_{(\leq N)}^{1/2}(t',u)\, dt'
\end{aligned}$$

as desired. We have thus proved the desired estimate (20.41) when $\mathscr{Z}^{N-1} = \mathscr{S}^{N-1}$.

We now assume that $\mathscr{Z}^{N-1}$ contains a factor of ϱL. We prove the desired estimate (20.41) for $\left\| \upmu \Delta\!\!\!/ \mathscr{Z}^{N-1} \upmu \right\|_{L^2(\Sigma_t^u)}$ in detail. The estimate (20.41) for $\left\| \upmu O \mathscr{Z}^{N-1} \mathrm{tr}_{g\!\!\!/} \chi^{(Small)} \right\|_{L^2(\Sigma_t^u)}$ can be proved in a similar fashion and we omit those details. To proceed, we first argue as in our proofs of (16.11)-(16.12) to deduce that

(20.46)
$$\begin{aligned}
\left| \Delta\!\!\!/ \mathscr{Z}^{N-1} \upmu \right| &\lesssim \left| \begin{pmatrix} L \mathscr{Z}^{\leq N} \Psi \\ \frac{\ln(e+t)}{1+t} \breve{R} \mathscr{Z}^{\leq N} \Psi \\ d\!\!\!/ \mathscr{Z}^{\leq N} \Psi \\ \frac{\ln(e+t)}{1+t} \mathscr{Z}^{\leq N} \Psi \end{pmatrix} \right| \\
&+ \frac{\ln(e+t)}{(1+t)^2} \left| \begin{pmatrix} \mathscr{Z}^{\leq N} (\upmu - 1) \\ \sum_{a=1}^3 \varrho \left| \mathscr{Z}^{\leq N} L_{(Small)}^a \right| \end{pmatrix} \right|.
\end{aligned}$$

We now multiply both sides of (20.46) by $\upmu$, take the norm $\| \cdot \|_{L^2(\Sigma_t^u)}$ of each side, and use the estimates $\upmu \lesssim \ln(e+t)$ (that is, (12.128a)) and (19.2a) and Prop. 14.7 to deduce that

(20.47)
$$\begin{aligned}
\left\| \upmu \Delta\!\!\!/ \mathscr{Z}^{N-1} \upmu \right\|_{L^2(\Sigma_t^u)} &\lesssim \mathbb{Q}_{(\leq N)}^{1/2}(t,u) \\
&+ \frac{\ln^2(e+t)}{1+t} \int_{s=0}^t \left\{ \frac{\mathbb{Q}_{(\leq N)}^{1/2}(s,u)}{1+s} + \frac{\widetilde{\mathbb{Q}}_{(\leq N)}^{1/2}(s,u)}{(1+s)\upmu_{\star}^{1/2}(s,u)} \right\} ds \\
&+ \varepsilon \frac{\ln^2(e+t)}{1+t}.
\end{aligned}$$

The desired estimate (20.41) for $\left\| \upmu \Delta\!\!\!/ \mathscr{Z}^{N-1} \upmu \right\|_{L^2(\Sigma_t^u)}$ now follows easily from (20.47) and the fact that $\mathbb{Q}_{(\leq N)}$ is increasing in its arguments. This completes the proof of Lemma 20.28. □

LEMMA 20.29 (**Sharp L^2 estimates for the partially modified version of $d\!\!\!/ \mathscr{S}^{N-1} \upmu$**). *Let $1 \leq N \leq 24$ and let $\mathscr{S}^{N-1}$ be an $(N-1)^{st}$ order pure spatial commutation vectorfield operator (see Def. 7.2). Let ${}^{(\mathscr{S}^{N-1})}\widetilde{\mathscr{M}}$ be the partially modified*

$S_{t,u}$ one-form defined in (11.51a)*. Recall the splitting $\Sigma_t^u = {}^{(+)}\Sigma_{t;t}^u \cup {}^{(-)}\Sigma_{t;t}^u$ from Def.* 13.8 *and that $\varrho(t,u) = 1 - u + t$. Under the small-data and bootstrap assumptions of Sects.* 12.1-12.4*, if ε is sufficiently small, then the following estimates hold for $(t,u) \in [0, T_{(Bootstrap)}) \times [0, U_0]$:*

(20.48a)

$$\begin{aligned}
&\left\| \frac{1}{\sqrt{\upmu}} \varrho (\breve{R}\Psi)^{(\mathscr{S}^{N-1})} \widetilde{\mathscr{M}} \right\|_{L^2({}^{(-)}\Sigma_{t;t}^u)} \\
&\leq \sqrt{2}(1 + C\varepsilon) \left\| \varrho L \upmu \right\|_{C^0({}^{(-)}\Sigma_{t;t}^u)} \frac{1}{\upmu_\star^{1/2}(t,u)} \int_{t'=0}^t \frac{1}{\varrho(t',u)\upmu_\star^{1/2}(t',u)} \widetilde{\mathbb{Q}}_{(\leq N)}^{1/2}(t',u)\, dt' \\
&\quad + C\varepsilon \frac{1}{\upmu_\star^{1/2}(t,u)} \int_{t'=0}^t \frac{1}{(1+t')^{3/2}} \mathbb{Q}_{(\leq N)}^{1/2}(t',u)\, dt' \\
&\quad + C\varepsilon \frac{1}{\upmu_\star^{1/2}(t,u)} \int_{t'=0}^t \frac{1}{(1+t')^{3/2}\upmu_\star^{1/2}(t',u)} \widetilde{\mathbb{Q}}_{(\leq N)}^{1/2}(t',u)\, dt' \\
&\quad + C\varepsilon \ln(e+t) \frac{1}{\upmu_\star^{1/2}(t,u)} \mathbb{Q}_{(\leq N-1)}^{1/2}(t,u) \\
&\quad + C\varepsilon \ln(e+t) \frac{1}{\upmu_\star^{1/2}(t,u)} \widetilde{\mathbb{Q}}_{(\leq N-1)}^{1/2}(t,u) \\
&\quad + C\varepsilon^2 \frac{1}{\upmu_\star^{1/2}(t,u)},
\end{aligned}$$

(20.48b)

$$\begin{aligned}
&\left\| \frac{1}{\sqrt{\upmu}} \varrho (\breve{R}\Psi)^{(\mathscr{S}^{N-1})} \widetilde{\mathscr{M}} \right\|_{L^2({}^{(+)}\Sigma_{t;t}^u)} \\
&\leq \sqrt{2}(1 + C\varepsilon) \left\| \frac{\varrho L \upmu}{\upmu} \right\|_{C^0({}^{(+)}\Sigma_{t;t}^u)} \int_{t'=0}^t \left\| \sqrt{\frac{\upmu(t,\cdot)}{\upmu}} \right\|_{C^0({}^{(+)}\Sigma_{t';t}^u)} \frac{1}{\varrho(t',u)} \widetilde{\mathbb{Q}}_{(\leq N)}^{1/2}(t',u)\, dt' \\
&\quad + C\varepsilon \frac{1}{\upmu_\star^{1/2}(t,u)} \int_{t'=0}^t \frac{1}{(1+t')^{3/2}} \mathbb{Q}_{(\leq N)}^{1/2}(t',u)\, dt' \\
&\quad + C\varepsilon \frac{1}{\upmu_\star^{1/2}(t,u)} \int_{t'=0}^t \frac{1}{(1+t')^{3/2}\upmu_\star^{1/2}(t',u)} \widetilde{\mathbb{Q}}_{(\leq N)}^{1/2}(t',u)\, dt' \\
&\quad + C\varepsilon \ln(e+t) \frac{1}{\upmu_\star^{1/2}(t,u)} \mathbb{Q}_{(\leq N-1)}^{1/2}(t,u) \\
&\quad + C\varepsilon \ln(e+t) \frac{1}{\upmu_\star^{1/2}(t,u)} \widetilde{\mathbb{Q}}_{(\leq N-1)}^{1/2}(t,u) \\
&\quad + C\varepsilon^2 \frac{1}{\upmu_\star^{1/2}(t,u)}.
\end{aligned}$$

Furthermore, the following less sharp estimate also holds:

$$
\begin{aligned}
(20.49) \qquad \left\| \varrho^{(\mathscr{S}^{N-1})} \widetilde{\mathscr{M}} \right\|_{L^2(\Sigma_t^u)} &\leq C(1+t)\ln(e+t)\mathbb{Q}_{(\leq N)}^{1/2}(t,u) \\
&+ C(1+t)\ln(e+t)\widetilde{\mathbb{Q}}_{(\leq N)}^{1/2}(t,u) \\
&+ C\varepsilon(1+t).
\end{aligned}
$$

PROOF. To prove (20.48a), we first multiply inequality (17.58) by $\frac{1}{\sqrt{\upmu}}(\breve{R}\Psi)$ and take the norm $\|\cdot\|_{L^2({}^{(-)}\Sigma_{t;t}^u)}$ of both sides. It is only for the quantity arising from the second term on the right-hand side of (17.58) that we use the subset norm $\|\cdot\|_{L^2({}^{(-)}\Sigma_{t;t}^u)}$. We bound all other terms in the (larger) norm $\|\cdot\|_{L^2(\Sigma_t^u)}$. To proceed, we use Lemma 4.6, inequalities (12.58b) and (12.128a), and the bootstrap assumptions (**BA**Ψ) to deduce that

$$
\begin{aligned}
(20.50) \qquad \frac{1}{2}G_{LL}\breve{R}\Psi &= L\upmu + \upmu\mathcal{O}\left(\varepsilon\frac{1}{(1+t)^2}\right) \\
&= L\upmu + \sqrt{\upmu}\mathcal{O}\left(\varepsilon\frac{\ln^{1/2}(e+t)}{(1+t)^2}\right) = L\upmu + \mathcal{O}\left(\varepsilon\frac{\ln(e+t)}{(1+t)^2}\right).
\end{aligned}
$$

Using (20.50), Lemma 12.57, and inequality (14.12d), we derive the desired bound for the quantity arising from the second term on the right-hand side of (17.58) as follows:

(20.51)

$$
\begin{aligned}
&\frac{1}{2}\left\| \frac{1}{\sqrt{\upmu}}|(\breve{R}\Psi)G_{LL}| \int_{t'=0}^{t} \left|\varrho \not{d}\mathscr{S}^N\Psi\right|(t',u,\vartheta)\,dt' \right\|_{L^2({}^{(-)}\Sigma_{t;t}^u)} \\
&\leq \frac{1}{2}\left\| \frac{1}{\sqrt{\upmu}}(\breve{R}\Psi)G_{LL} \right\|_{C^0({}^{(-)}\Sigma_{t;t}^u)} \left\| \int_{t'=0}^{t} \left|\varrho \not{d}\mathscr{S}^N\Psi\right|(t',u,\vartheta)\,dt' \right\|_{L^2(\Sigma_t^u)} \\
&\leq (1+C\varepsilon)\frac{1}{\upmu_\star^{1/2}(t,u)}\left\|\varrho L\upmu\right\|_{C^0({}^{(-)}\Sigma_{t;t}^u)} \int_{t'=0}^{t} \left\|\not{d}\mathscr{S}^N\Psi\right\|_{L^2(\Sigma_{t'}^u)}\,dt' \\
&\quad + C\varepsilon\frac{\ln^{1/2}(e+t)}{1+t}\int_{t'=0}^{t}\left\|\not{d}\mathscr{S}^N\Psi\right\|_{L^2(\Sigma_t^u)}\,dt' \\
&\leq \sqrt{2}(1+C\varepsilon)\left\|\varrho L\upmu\right\|_{C^0({}^{(-)}\Sigma_{t;t}^u)}\frac{1}{\upmu_\star^{1/2}(t,u)}\int_{t'=0}^{t}\frac{1}{\varrho(t',u)\upmu_\star^{1/2}(t',u)}\widetilde{\mathbb{Q}}_{(\leq N)}^{1/2}(t',u)\,dt' \\
&\quad + C\varepsilon\int_{t'=0}^{t}\frac{1}{(1+t')^{3/2}\upmu_\star^{1/2}(t',u)}\widetilde{\mathbb{Q}}_{(\leq N)}^{1/2}(t',u)\,dt'.
\end{aligned}
$$

Similarly, we use Lemma 12.57, Prop. 14.7, and the bootstrap assumption $\left\|\breve{R}\Psi\right\|_{C^0(\Sigma_t^u)} \leq \varepsilon(1+t)^{-1}$ (that is, (**BA**Ψ)) to deduce the following desired estimate for the quantity arising from the next-to-last term on the right-hand side of (17.58):

(20.52)

$$
\begin{aligned}
& C\left\| \frac{1}{\sqrt{\upmu}}(\breve{R}\Psi)\int_{t'=0}^{t}\left|\begin{pmatrix} \varrho\{L+\frac{1}{2}\mathrm{tr}_{\not g}\chi\}\mathscr{Z}^{\leq N-1}\Psi \\ \breve{R}\mathscr{Z}^{\leq N-1}\Psi \\ \varrho\not d\mathscr{Z}^{\leq N-1}\Psi \\ \mathscr{Z}^{\leq N-1}\Psi \end{pmatrix}\right|(t',u,\vartheta)\,dt'\right\|_{L^2(\Sigma_t^u)} \\
& \leq C\varepsilon\frac{1}{\upmu_\star^{1/2}(t,u)}\int_{t'=0}^{t}\frac{1}{1+t'}\mathbb{Q}_{(\leq N-1)}^{1/2}(t',u)\,dt' \\
& \quad + C\varepsilon\frac{1}{\upmu_\star^{1/2}(t,u)}\int_{t'=0}^{t}\frac{1}{(1+t')\upmu_\star^{1/2}(t',u)}\widetilde{\mathbb{Q}}_{(\leq N-1)}^{1/2}(t',u)\,dt' \\
& \leq C\varepsilon\ln(e+t)\frac{1}{\upmu_\star^{1/2}(t,u)}\mathbb{Q}_{(\leq N-1)}^{1/2}(t,u) + C\varepsilon\ln(e+t)\frac{1}{\upmu_\star^{1/2}(t,u)}\widetilde{\mathbb{Q}}_{(\leq N-1)}^{1/2}(t,u).
\end{aligned}
$$

In the last step of (20.52), we used the fact that $\mathbb{Q}_{(\leq N-1)}$ and $\widetilde{\mathbb{Q}}_{(\leq N-1)}$ are increasing in their arguments and the estimate (13.152) in order to annihilate the factor $\upmu_\star^{1/2}(t',u)$ in the denominator of the integrand.

Similarly, we bound the quantity arising from the last term on the right-hand side of (17.58) with the help of the estimate (19.2a) as follows:

(20.53)

$$
\begin{aligned}
& C\left\| \frac{1}{\sqrt{\upmu(t,u)}}(\breve{R}\Psi)\int_{t'=0}^{t}\frac{\ln(e+t')}{(1+t')^2}\left|\begin{pmatrix} \mathscr{Z}^{\leq N}(\upmu-1) \\ \sum_{a=1}^{3}\varrho\left|\mathscr{Z}^{\leq N}L_{(Small)}^a\right| \end{pmatrix}\right|(t',u,\vartheta)\,dt'\right\|_{L^2(\Sigma_t^u)} \\
& \lesssim \varepsilon^2\frac{1}{\upmu_\star^{1/2}(t,u)}\int_{t'=0}^{t}\frac{\ln(e+t')}{(1+t')^2}\,dt' \\
& \quad + \varepsilon\frac{1}{\upmu_\star^{1/2}(t,u)}\int_{t'=0}^{t}\frac{\ln(e+t')}{(1+t')^2}\int_{s=0}^{t'}\frac{1}{1+s}\mathbb{Q}_{(\leq N)}^{1/2}(s,u)\,ds\,dt' \\
& \quad + \varepsilon\frac{1}{\upmu_\star^{1/2}(t,u)}\int_{t'=0}^{t}\frac{\ln(e+t')}{(1+t')^2}\int_{s=0}^{t'}\frac{1}{(1+s)\upmu_\star^{1/2}(s,u)}\widetilde{\mathbb{Q}}_{(\leq N)}^{1/2}(s,u)\,ds\,dt' \\
& \lesssim \varepsilon^2\frac{1}{\upmu_\star^{1/2}(t,u)} + \varepsilon\frac{1}{\upmu_\star^{1/2}(t,u)}\int_{t'=0}^{t}\frac{1}{(1+t')^{3/2}}\mathbb{Q}_{(\leq N)}^{1/2}(t,u)\,dt' \\
& \quad + \varepsilon\frac{1}{\upmu_\star^{1/2}(t,u)}\int_{t'=0}^{t}\frac{1}{(1+t')^{3/2}}\widetilde{\mathbb{Q}}_{(\leq N)}^{1/2}(t,u)\,dt'.
\end{aligned}
$$

To bound the quantity arising from the first term on the right-hand side of (17.58), we first bound $\left\|[\varrho^{(\mathscr{S}^{N-1})}\widetilde{\mathscr{M}}](0,\cdot)\right\|_{L^2(\Sigma_t^u)}$ by using the definition (11.51a), Defs. 3.68 and 3.69, inequalities (12.46a) and (12.152), and Lemma 12.17 to deduce that

$$\left\| [\varrho^{(\mathscr{S}^{N-1})}\widetilde{\mathscr{M}}](0,\cdot) \right\|_{L^2(\Sigma_t^u)}^2 \tag{20.54}$$
$$= \int_{u'=0}^{u} \int_{S_{t,u}} \left| \varrho^{(\mathscr{S}^{N-1})}\widetilde{\mathscr{M}} \right|^2 (0,u',\vartheta)\, d\upsilon_{g\!\!/(t,u',\vartheta)}\, du'$$
$$\lesssim \varrho^2(t,u) \int_{u'=0}^{u} \int_{S_{0,u}} \left| \varrho^{(\mathscr{S}^{N-1})}\widetilde{\mathscr{M}} \right|^2 (0,u',\vartheta)\, d\upsilon_{g\!\!/(0,u',\vartheta)}\, du'$$
$$= \varrho^2(t,u) \left\| [\varrho^{(\mathscr{S}^{N-1})}\widetilde{\mathscr{M}}](0,\cdot) \right\|_{L^2(\Sigma_0^u)}^2 \lesssim \varepsilon^2 \varrho^2(t,u).$$

Next, using the bootstrap assumption $\left\| \breve{R}\Psi \right\|_{C^0(\Sigma_t^u)} \leq \varepsilon(1+t)^{-1}$ and the estimate (20.54), we bound the quantity arising from the first term on the right-hand side of (17.58) as follows:

$$\left\| \frac{1}{\sqrt{\upmu}}(\breve{R}\Psi)\left\{[\varrho^{(\mathscr{S}^{N-1})}\widetilde{\mathscr{M}}](0,\cdot)\right\} \right\|_{L^2(\Sigma_t^u)} \tag{20.55}$$
$$\lesssim \varepsilon \frac{1}{(1+t)} \frac{1}{\upmu_\star^{1/2}(t,u)} \left\| [\varrho^{(\mathscr{S}^{N-1})}\widetilde{\mathscr{M}}](0,\cdot) \right\|_{L^2(\Sigma_t^u)}$$
$$\lesssim \varepsilon^2 \frac{1}{\upmu_\star^{1/2}(t,u)}.$$

Finally, using the bootstrap assumption $\left\| \breve{R}\Psi \right\|_{C^0(\Sigma_t^u)} \leq \varepsilon(1+t)^{-1}$, Lemma 12.57, and inequality (14.12d), we deduce that the quantity arising from the term $C\varepsilon \int_{t'=0}^{t} \left| d\!\!/ \mathscr{S}^N \Psi \right| (t',u,\vartheta)\, dt'$ on the right-hand side of (17.58) can be bounded as follows:

$$C\varepsilon \left\| \frac{1}{\sqrt{\upmu}}(\breve{R}\Psi) \int_{t'=0}^{t} \left| d\!\!/ \mathscr{S}^N \Psi \right| (t',u,\vartheta)\, dt' \right\|_{L^2(\Sigma_t^u)} \tag{20.56}$$
$$\leq C\varepsilon^2 \frac{1}{\upmu_\star^{1/2}(t,u)} \int_{t'=0}^{t} \frac{1}{(1+t')^{3/2}\upmu_\star^{1/2}(t',u)} \widetilde{\mathbb{Q}}_{(\leq N)}^{1/2}(t',u)\, dt'.$$

We now observe that the right-hand side of (20.56) is $\lesssim$ the right-hand side of (20.48a) as desired. We have thus proved the estimate (20.48a).

The estimate (20.48b) can be proved by making minor changes to our proof of (20.48a), and we omit the details.

The proof of (20.49) is much easier. We estimate the right-hand side of (17.58) in the norm $\|\cdot\|_{L^2(\Sigma_t^u)}$ by using arguments similar to the ones we used to deduce (20.48a), but we treat the first term on the right-hand side rather crudely with the help of the estimate $\|G_{LL}\|_{C^0(\Sigma_t^u)} \lesssim 1$ (that is, (12.58b)) and without using an analog of (20.50). In doing so, we encounter the integrals

$$\int_{t'=0}^{t} \frac{1}{\varrho(t',u)\upmu_\star^{1/2}(t',u)} \widetilde{\mathbb{Q}}_{(\leq N)}^{1/2}(t',u)\, dt' \quad \text{and} \quad \int_{t'=0}^{t} \frac{1}{\varrho(t',u)} \mathbb{Q}_{(\leq N)}^{1/2}(t',u)\, dt',$$

which we respectively bound by $\lesssim \ln(e+t)\widetilde{\mathbb{Q}}_{(\leq N)}^{1/2}(t,u)$ and $\lesssim \ln(e+t)\mathbb{Q}_{(\leq N)}^{1/2}(t,u)$. In estimating the first integral, we used inequality (13.152). □

LEMMA 20.30 (**Sharp L^2 estimates for the partially modified version of $\mathscr{S}^{N-1}\mathrm{tr}_{\not g}\chi^{(Small)}$**). *Let $1 \leq N \leq 24$ and let $\mathscr{S}^{N-1}$ be an $(N-1)^{st}$ order pure spatial commutation vectorfield operator (see Def. 7.2). Let ${}^{(\mathscr{S}^{N-1})}\widetilde{\mathscr{X}}$ be the corresponding partially modified quantity defined in (11.45a). Recall the splitting $\Sigma_t^u = {}^{(+)}\Sigma_{t;t}^u \cup {}^{(-)}\Sigma_{t;t}^u$ from Def. 13.8 and that $\varrho(t,u) = 1 - u + t$. Under the small-data and bootstrap assumptions of Sects. 12.1-12.4, if ε is sufficiently small, then the following estimates hold for $(t,u) \in [0, T_{(Bootstrap)}) \times [0, U_0]$:*

(20.57a)
$$\begin{aligned}
&\left\| \frac{1}{\sqrt{\mu}}(\breve{R}\Psi)\varrho^2 {}^{(\mathscr{S}^{N-1})}\widetilde{\mathscr{X}} \right\|_{L^2({}^{(-)}\Sigma_{t;t}^u)} \\
&\leq (\sqrt{12} + C\varepsilon) \left\| \varrho L\mu \right\|_{C^0({}^{(-)}\Sigma_{t;t}^u)} \frac{1}{\mu_\star^{1/2}(t,u)} \int_{t'=0}^t \frac{1}{\varrho(t',u)\mu_\star^{1/2}(t',u)} \widetilde{\mathbb{Q}}_{(\leq N)}^{1/2}(t',u)\,dt' \\
&\quad + C\varepsilon \frac{1}{\mu_\star^{1/2}(t,u)} \int_{t'=0}^t \frac{1}{(1+t')^{3/2}} \mathbb{Q}_{(\leq N)}^{1/2}(t',u)\,dt' \\
&\quad + C\varepsilon \frac{1}{\mu_\star^{1/2}(t,u)} \int_{t'=0}^t \frac{1}{(1+t')^{3/2}\mu_\star^{1/2}(t',u)} \widetilde{\mathbb{Q}}_{(\leq N)}^{1/2}(t',u)\,dt' \\
&\quad + C\varepsilon \ln(e+t) \frac{1}{\mu_\star^{1/2}(t,u)} \mathbb{Q}_{(\leq N-1)}^{1/2}(t,u) \\
&\quad + C\varepsilon \ln(e+t) \frac{1}{\mu_\star^{1/2}(t,u)} \widetilde{\mathbb{Q}}_{(\leq N-1)}^{1/2}(t,u) \\
&\quad + C\varepsilon^2 \frac{1}{\mu_\star^{1/2}(t,u)},
\end{aligned}$$

(20.57b)
$$\begin{aligned}
&\left\| \frac{1}{\sqrt{\mu}}(\breve{R}\Psi)\varrho^2 {}^{(\mathscr{S}^{N-1})}\widetilde{\mathscr{X}} \right\|_{L^2({}^{(+)}\Sigma_{t;t}^u)} \\
&\leq (\sqrt{12} + C\varepsilon) \left\| \frac{\varrho L\mu}{\mu} \right\|_{C^0({}^{(+)}\Sigma_{t;t}^u)} \int_{t'=0}^t \left\| \sqrt{\frac{\mu(t,\cdot)}{\mu}} \right\|_{C^0({}^{(+)}\Sigma_{t';t}^u)} \frac{1}{\varrho(t',u)} \widetilde{\mathbb{Q}}_{(\leq N)}^{1/2}(t',u)\,dt' \\
&\quad + C\varepsilon \frac{1}{\mu_\star^{1/2}(t,u)} \int_{t'=0}^t \frac{1}{(1+t')^{3/2}} \mathbb{Q}_{(\leq N)}^{1/2}(t',u)\,dt' \\
&\quad + C\varepsilon \frac{1}{\mu_\star^{1/2}(t,u)} \int_{t'=0}^t \frac{1}{(1+t')^{3/2}\mu_\star^{1/2}(t',u)} \widetilde{\mathbb{Q}}_{(\leq N)}^{1/2}(t',u)\,dt' \\
&\quad + C\varepsilon \ln(e+t) \frac{1}{\mu_\star^{1/2}(t,u)} \mathbb{Q}_{(\leq N-1)}^{1/2}(t,u) \\
&\quad + C\varepsilon \ln(e+t) \frac{1}{\mu_\star^{1/2}(t,u)} \widetilde{\mathbb{Q}}_{(\leq N-1)}^{1/2}(t,u) \\
&\quad + C\varepsilon^2 \frac{1}{\mu_\star^{1/2}(t,u)}.
\end{aligned}$$

Furthermore, the following less sharp estimate also holds:

$$\left\| \varrho^{2} {}^{(\mathscr{S}^{N-1})}\widetilde{\mathscr{X}} \right\|_{L^2(\Sigma_t^u)} \leq C \ln(e+t)(1+t)\mathbb{Q}_{(\leq N)}^{1/2}(t,u) + C \ln(e+t)(1+t)\widetilde{\mathbb{Q}}_{(\leq N)}^{1/2}(t,u) + C\varepsilon(1+t). \tag{20.58}$$

PROOF. We apply essentially the same reasoning used in the proof of Lemma 20.29 to the pointwise estimate (17.59). The factors of $\sqrt{12}$ arise because in order to estimate the first term on the right-hand side of (17.59), we use inequalities (12.46c) and (14.12d), Cor. 12.48, and the elementary estimate

$$\left\| \sqrt{\sum_{l=1}^{3} \left| \not{d} O_{(l)} \mathscr{S}^{N-1}\Psi \right|^2} \right\|_{L^2(\Sigma_{t'}^u)} \leq \sqrt{3} \max_{l=1}^{3} \left\| \not{d} O_{(l)} \mathscr{S}^{N-1}\Psi \right\|_{L^2(\Sigma_{t'}^u)}. \tag{20.59}$$

□

LEMMA 20.31 (**Sharp L^2 estimates for $\not{\mathcal{L}}_L + \mathrm{tr}_{\not{g}}\chi$ applied to the partially modified version of $\not{d}\mathscr{S}^{N-1}\upmu$**). *Let $1 \leq N \leq 24$ and let $\mathscr{S}^{N-1}$ be an $(N-1)^{st}$ order pure spatial commutation vectorfield operator (see Def. 7.2). Let ${}^{(\mathscr{S}^{N-1})}\widetilde{\mathscr{M}}^{\#}$ be the corresponding $S_{t,u}$-tangent vectorfield that is $\not{g}$ dual to the partially modified $S_{t,u}$ one-form defined in (11.51a). Let χ be the symmetric type $\binom{0}{2}$ $S_{t,u}$ tensorfield defined in (3.70) and recall that $\varrho(t,u) = 1-u+t$. Under the small-data and bootstrap assumptions of Sects. 12.1-12.4, if ε is sufficiently small, then the following estimate holds for $(t,u) \in [0, T_{(Bootstrap)}) \times [0, U_0]$:*

(20.60)

$$\begin{aligned}
&\left\| \frac{1}{\sqrt{\upmu}}(\breve{R}\Psi)\varrho \left\{ \not{\mathcal{L}}_L + \mathrm{tr}_{\not{g}}\chi \right\} {}^{(\mathscr{S}^{N-1})}\widetilde{\mathscr{M}}^{\#} \right\|_{L^2(\Sigma_t^u)} \\
&\leq \sqrt{2}\frac{\|[L\upmu]_-\|_{C^0(\Sigma_t^u)}}{\upmu_\star(t,u)}\widetilde{\mathbb{Q}}_{(\leq N)}^{1/2}(t,u) + \sqrt{2}\left\| \frac{[L\upmu]_+}{\upmu} \right\|_{C^0(\Sigma_t^u)} \widetilde{\mathbb{Q}}_{(\leq N)}^{1/2}(t,u) \\
&\quad + C\varepsilon\frac{1}{(1+t)^{3/2}}\frac{1}{\upmu_\star^{1/2}(t,u)}\mathbb{Q}_{(\leq N)}^{1/2}(t,u) + C\varepsilon\frac{1}{(1+t)^{3/2}}\frac{1}{\upmu_\star^{1/2}(t,u)}\widetilde{\mathbb{Q}}_{(\leq N)}^{1/2}(t,u) \\
&\quad + C\varepsilon\frac{1}{(1+t)}\frac{1}{\upmu_\star^{1/2}(t,u)}\mathbb{Q}_{(\leq N-1)}^{1/2}(t,u) + C\varepsilon\frac{1}{(1+t)}\frac{1}{\upmu_\star(t,u)}\widetilde{\mathbb{Q}}_{(\leq N-1)}^{1/2}(t,u) \\
&\quad + C\varepsilon^2\frac{1}{(1+t)^{3/2}}\frac{1}{\upmu_\star^{1/2}(t,u)}.
\end{aligned}$$

Furthermore, the following less sharp estimate also holds:

(20.61)

$$\left\| \varrho \left\{ \not{\mathcal{L}}_L + \mathrm{tr}_{\not{g}}\chi \right\} {}^{(\mathscr{S}^{N-1})}\widetilde{\mathscr{M}}^{\#} \right\|_{L^2(\Sigma_t^u)} \leq C\mathbb{Q}_{(\leq N)}^{1/2}(t,u) + C\frac{1}{\upmu_\star^{1/2}(t,u)}\widetilde{\mathbb{Q}}_{(\leq N)}^{1/2}(t,u) + C\varepsilon.$$

PROOF. To prove (20.60), we start by multiplying the second inequality in (17.55) by $\frac{1}{\sqrt{\upmu}}(\breve{R}\Psi)$ and taking the norm $\|\cdot\|_{L^2(\Sigma_t^u)}$. To bound the norm $\|\cdot\|_{L^2(\Sigma_t^u)}$ of the quantity arising from the term $\frac{1}{2}\varrho G_{LL}\not{d}^{\#}\mathscr{S}^{N-1}\breve{R}\Psi$, we use the first equality in (20.50), the inequality

$$\left\| \frac{L\upmu}{\upmu} \right\|_{C^0(\Sigma_t^u)} \leq \frac{1}{\upmu_\star(t,u)} \|[L\upmu]_-\|_{C^0(\Sigma_t^u)} + \left\| \frac{[L\upmu]_+}{\upmu} \right\|_{C^0(\Sigma_t^u)},$$

and the inequality $\left\| \varrho\sqrt{\upmu} d\!\!/^{\#} \mathscr{S}^{N-1}\breve{R}\Psi \right\|_{L^2(\Sigma_t^u)} \leq \sqrt{2}\widetilde{\mathbb{Q}}_{(\leq N)}^{1/2}(t,u)$ (that is, (14.12d)) to deduce that

$$\begin{aligned}
&\left\| \frac{1}{2}\frac{1}{\sqrt{\upmu}}(\breve{R}\Psi)G_{LL}\varrho d\!\!/^{\#}\mathscr{S}^{N-1}\breve{R}\Psi \right\|_{L^2(\Sigma_t^u)} \\
&\leq \sqrt{2}\frac{1}{\upmu_\star(t,u)} \|[L\upmu]_-\|_{C^0(\Sigma_t^u)} \widetilde{\mathbb{Q}}_{(\leq N)}^{1/2}(t,u) + \sqrt{2}\left\| \frac{[L\upmu]_+}{\upmu} \right\|_{C^0(\Sigma_t^u)} \widetilde{\mathbb{Q}}_{(\leq N)}^{1/2}(t,u) \\
&\quad + C\varepsilon\frac{1}{(1+t)^2}\widetilde{\mathbb{Q}}_{(\leq N)}^{1/2}(t,u).
\end{aligned} \tag{20.62}$$

Clearly, the right-hand side of (20.62) is bounded by the right-hand side of (20.60) as desired.

The quantities arising from the remaining terms (which are located on the right-hand side of (17.55)) can be bounded in the norm $\|\cdot\|_{L^2(\Sigma_t^u)}$ by $\leq$ the terms on the last three lines of the right-hand side of (20.60) with the help of Prop. 14.7, the bootstrap assumption $\left\| \breve{R}\Psi \right\|_{C^0(\Sigma_t^u)} \leq \varepsilon(1+t)^{-1}$, the estimate (19.2a), inequality (13.152), and the fact that $\mathbb{Q}_{(\leq M)}$ and $\widetilde{\mathbb{Q}}_{(\leq M)}$ are increasing in both of their arguments.

The proof of (20.61) is simpler. We bound the term $\frac{1}{2}\varrho G_{LL} d\!\!/^{\#}\mathscr{S}^{N-1}\breve{R}\Psi$ from inequality (17.55) by using Prop. 14.7 and the estimate $\|G_{LL}\|_{C^0(\Sigma_t^u)} \lesssim 1$ (that is, (12.58b)) as follows:

$$\left\| \frac{1}{2}\varrho G_{LL} d\!\!/^{\#}\mathscr{S}^{N-1}\breve{R}\Psi \right\|_{L^2(\Sigma_t^u)} \lesssim \frac{1}{\upmu_\star^{1/2}(t,u)}\widetilde{\mathbb{Q}}_{(\leq N)}^{1/2}(t,u). \tag{20.63}$$

The remaining terms from inequality (17.55) can be bounded by using arguments similar to the ones we used in our proof of (20.60). □

LEMMA 20.32 (**Sharp L^2 estimates for L applied to a partially modified version of $\mathscr{S}^{N-1}\mathrm{tr}_{g\!\!/}\chi^{(Small)}$**). *Let $1 \leq N \leq 24$ and let $\mathscr{S}^{N-1}$ be an $(N-1)^{st}$ order pure spatial commutation vectorfield operator (see Def. 7.2). Let ${}^{(\mathscr{S}^{N-1})}\widetilde{\mathscr{X}}$ be the corresponding partially modified quantity defined in (11.45a). Let χ be the symmetric type $\binom{0}{2}$ $S_{t,u}$ tensorfield defined in (3.70) and recall that $\varrho(t,u) = 1-u+t$. Under the small-data and bootstrap assumptions of Sects. 12.1-12.4, if ε is* *sufficiently small, then the following estimate holds for* $(t,u) \in [0, T_{(Bootstrap)}) \times [0, U_0]$:

(20.64)

$$\begin{aligned}
&\left\| \frac{1}{\sqrt{\upmu}} (\breve{R}\Psi) L \left\{ \varrho^2 {}^{(\mathscr{S}^{N-1})}\widetilde{\mathscr{X}} \right\} \right\|_{L^2(\Sigma_t^u)} \\
&\leq \sqrt{12}(1 + C\varepsilon) \frac{\| [L\upmu]_- \|_{C^0(\Sigma_t^u)}}{\upmu(t,u)} \widetilde{\mathbb{Q}}_{(\leq N)}^{1/2}(t,u) \\
&\quad + \sqrt{12}(1 + C\varepsilon) \left\| \frac{[L\upmu]_+}{\upmu} \right\|_{C^0(\Sigma_t^u)} \widetilde{\mathbb{Q}}_{(\leq N)}^{1/2}(t,u) \\
&\quad + C\varepsilon \frac{1}{(1+t)^{3/2}} \frac{1}{\upmu_\star^{1/2}(t,u)} \mathbb{Q}_{(\leq N)}^{1/2}(t,u) + C\varepsilon \frac{1}{(1+t)^{3/2}} \frac{1}{\upmu_\star^{1/2}(t,u)} \widetilde{\mathbb{Q}}_{(\leq N)}^{1/2}(t,u) \\
&\quad + C\varepsilon \frac{1}{(1+t)} \frac{1}{\upmu_\star^{1/2}(t,u)} \mathbb{Q}_{(\leq N-1)}^{1/2}(t,u) + C\varepsilon \frac{1}{(1+t)} \frac{1}{\upmu_\star(t,u)} \widetilde{\mathbb{Q}}_{(\leq N-1)}^{1/2}(t,u) \\
&\quad + C\varepsilon^2 \frac{1}{(1+t)^{3/2}} \frac{1}{\upmu_\star^{1/2}(t,u)}.
\end{aligned}$$

Furthermore, the following less sharp estimates also hold:

(20.65)

$$\begin{aligned}
&\left\| L \left\{ \varrho^2 {}^{(\mathscr{S}^{N-1})}\widetilde{\mathscr{X}} \right\} \right\|_{L^2(\Sigma_t^u)}, \\
&\left\| \varrho^2 \left\{ L + \mathrm{tr}_{\not g}\chi \right\} {}^{(\mathscr{S}^{N-1})}\widetilde{\mathscr{X}} \right\|_{L^2(\Sigma_t^u)} \\
&\leq C \mathbb{Q}_{(\leq N)}^{1/2}(t,u) + C \frac{1}{\upmu_\star^{1/2}(t,u)} \widetilde{\mathbb{Q}}_{(\leq N)}^{1/2}(t,u) + C\varepsilon.
\end{aligned}$$

PROOF. We apply the same reasoning used in the proof of Lemma 20.31 to the third and fourth pointwise inequalities in (17.55). The factors of $\sqrt{12}$ arise because in order to bound the norm $\| \cdot \|_{L^2(\Sigma_t^u)}$ of the term $-\frac{1}{2}\varrho^2 G_{LL} \not\Delta \mathscr{Z}^{N-1}\Psi$ from the pointwise inequalities, we use inequalities (12.46c) and (14.12d), Cor. 12.48, and the elementary estimate

(20.66)

$$\left\| \sqrt{\sum_{l=1}^{3} \left| \not d O_{(l)} \mathscr{S}^{N-1}\Psi \right|^2} \right\|_{L^2(\Sigma_t^u)} \leq \sqrt{3} \max_{l=1}^{3} \left\| \not d O_{(l)} \mathscr{S}^{N-1}\Psi \right\|_{L^2(\Sigma_t^u)}.$$

□

LEMMA 20.33 (**Estimates for some easy error integrals corresponding to the partially modified version of** $O\mathscr{S}^{N-1}\mathrm{tr}_{\not g}\chi^{(Small)}$). *Let* $1 \leq N \leq 24$ *be an integer and let* $\mathscr{S}^{N-1}$ *be an* $(N-1)^{st}$ *order pure spatial commutation vectorfield operator (see definition* (7.2a)*). Let* ${}^{(\mathscr{S}^{N-1})}\widetilde{\mathscr{X}}$ *be the corresponding partially modified quantity defined in* (11.45a) *and let* $O \in \mathscr{O}$ *be a geometric rotation vectorfield (see definition* (7.2b)*). Under the small-data and bootstrap assumptions of Sects.* 12.1-12.4, *if* ε *is sufficiently small, then we have the following spacetime integral estimate for the quantity* $\mathrm{Error}_{(1)}[\mathscr{S}^{N-1}O\Psi; w; \eta]$ *defined in* (10.40a), *with*

$w := 1$ *and* $\eta := \varrho^{2(\mathscr{S}^{N-1})}\widetilde{\mathscr{X}}$, *for* $(t,u) \in [0, T_{(Bootstrap)}) \times [0, U_0]$:

(20.67a)

$$\int_{\mathcal{M}_{t,u}} \left| \mathrm{Error}_{(1)}[\mathscr{S}^{N-1}O\Psi; 1; \varrho^{2(\mathscr{S}^{N-1})}\widetilde{\mathscr{X}}] \right| \, d\varpi \lesssim \varepsilon \mathbb{Q}_{(\leq N)}(t,u) + \varepsilon \widetilde{\mathbb{Q}}_{(\leq N)}(t,u) + \varepsilon \ln^2(e+t)\mathbb{Q}_{(\leq N-1)}(t,u) + \varepsilon \ln^2(e+t)\frac{1}{\mu_\star^{1/2}(t,u)}\widetilde{\mathbb{Q}}_{(\leq N-1)}(t,u) + \varepsilon^3.$$

Furthermore, we have the following hypersurface integral estimate for the quantity $\mathrm{Error}_{(2)}[\mathscr{S}^{N-1}O\Psi; w; \eta]$ *defined in* (10.40b):

(20.67b)

$$\int_{\Sigma_t^u} \left| \mathrm{Error}_{(2)}[\mathscr{S}^{N-1}O\Psi; 1; \varrho^{2(\mathscr{S}^{N-1})}\widetilde{\mathscr{X}}] \right| \, d\underline{\varpi} \lesssim \varepsilon \mathbb{Q}_{(\leq N)}(t,u) + \varepsilon \widetilde{\mathbb{Q}}_{(\leq N)}(t,u) + \varepsilon \ln^2(e+t)\mathbb{Q}_{(\leq N-1)}(t,u) + \varepsilon \ln^2(e+t)\frac{1}{\mu_\star(t,u)}\widetilde{\mathbb{Q}}_{(\leq N-1)}(t,u) + \varepsilon^3.$$

Finally, we have the following hypersurface integral estimate for the quantity $\mathrm{Error}_{(3)}[O\mathscr{S}^{N-1}\Psi; w; \eta]$ *defined in* (10.40c):

$$\int_{\Sigma_0^u} \left| \mathrm{Error}_{(3)}[\mathscr{S}^{N-1}O\Psi; 1; \varrho^{2(\mathscr{S}^{N-1})}\widetilde{\mathscr{X}}] \right| \, d\underline{\varpi} \lesssim \varepsilon^3. \tag{20.67c}$$

PROOF. Throughout this proof, $\eta := \varrho^{2(\mathscr{S}^{N-1})}\widetilde{\mathscr{X}}$, as in the statement of the lemma. To prove (20.67a), we first express the spacetime integral as a time integral of integrals over $\Sigma_{t'}^u$: $\int_{\mathcal{M}_{t,u}} \cdots \, d\varpi = \int_{t'=0}^t \int_{\Sigma_{t'}^u} \cdots \, d\underline{\varpi}\, dt'$. We bound all spatial integrals $\int_{\Sigma_{t'}^u} \cdots \, d\underline{\varpi}$ by using Cauchy-Schwarz in the form $\int_{\Sigma_{t'}^u} |v_1 v_2 v_3| \, d\underline{\varpi} \lesssim \|v_1\|_{C^0(\Sigma_{t'}^u)} \|v_2\|_{L^2(\Sigma_{t'}^u)} \|v_3\|_{L^2(\Sigma_{t'}^u)}$. The product $v_1 v_2 v_3$ represents any of the products from the right-hand side of (10.40a) with $\mathscr{Z}^N := \mathscr{S}^{N-1}O$, where v_2 is equal to either the factor $\mathscr{S}^{N-1}O\Psi$, $O\mathscr{S}^{N-1}O\Psi = O \cdot \not{d}\mathscr{S}^{N-1}O\Psi$, or $\not{d}\mathscr{S}^{N-1}O\Psi$, v_3 is equal to either $L\eta$ or η, and v_1 is the product of the remaining factors. To bound v_2, we use the estimates $\left\|\mathscr{S}^{N-1}O\Psi\right\|_{L^2(\Sigma_{t'}^u)} \lesssim \mathbb{Q}_{(\leq N-1)}^{1/2}(t',u) + \mu_\star^{-1/2}(t',u)\widetilde{\mathbb{Q}}_{(\leq N-1)}^{1/2}(t',u)$, $\left\|O\mathscr{S}^{N-1}O\Psi\right\|_{L^2(\Sigma_{t'}^u)} \lesssim \mu_\star^{-1/2}(t',u)\widetilde{\mathbb{Q}}_{(\leq N)}^{1/2}(t',u)$, and $\left\|\not{d}\mathscr{S}^{N-1}O\Psi\right\|_{L^2(\Sigma_{t'}^u)} \lesssim (1+t')^{-1}\mu_\star^{-1/2}(t',u)\widetilde{\mathbb{Q}}_{(\leq N)}^{1/2}(t',u)$, which follow from Prop. 14.7, (12.34a), and, in the case of the estimate for $\mathscr{S}^{N-1}O\Psi$, a two-case argument that accounts for whether the left-most vectorfield in the operator $\mathscr{S}^{N-1}$ is equal to $\breve{R}$ or a rotation $O_{(l)}$. To bound v_3, we respectively bound $\|L\eta\|_{L^2(\Sigma_{t'}^u)}$ and $\|\eta\|_{L^2(\Sigma_{t'}^u)}$ via the estimates

(20.65) and (20.58). To bound v_1, we use the inequalities $\left\| \breve{R}\Psi \right\|_{C^0(\Sigma_{t'}^u)} \lesssim \varepsilon(1+t')^{-1}$, $\left\| O\breve{R}\Psi \right\|_{C^0(\Sigma_{t'}^u)} \lesssim \varepsilon(1+t')^{-1}$, $\left\| {}^{(O)}\not{\pi}_L^{\#} \right\|_{C^0(\Sigma_{t'}^u)} \lesssim \varepsilon \ln(e+t')(1+t')^{-1}$, $\left\| \mathrm{d}\not{\mathrm{iv}}\, {}^{(O)}\not{\pi}_L^{\#} \right\|_{C^0(\Sigma_{t'}^u)} \lesssim \varepsilon \ln(e+t')(1+t')^{-2}$, $\left\| \mathrm{tr}_{\not{g}}\, {}^{(O)}\not{\pi} \right\|_{C^0(\Sigma_{t'}^u)} \lesssim \varepsilon \ln(e+t')(1+t')^{-1}$, $\left\| L\breve{R}\Psi + \frac{1}{2}\mathrm{tr}_{\not{g}}\chi \breve{R}\Psi \right\|_{C^0(\Sigma_{t'}^u)} \lesssim \varepsilon \ln(e+t')(1+t')^{-3}$, and $\left\| O\left\{ L\breve{R}\Psi + \frac{1}{2}\mathrm{tr}_{\not{g}}\chi \breve{R}\Psi \right\} \right\|_{C^0(\Sigma_{t'}^u)} \lesssim \varepsilon \ln(e+t')(1+t')^{-3}$. The first two of these inequalities follow from the bootstrap assumptions (**BA**Ψ), the next two from inequality (12.47), the estimate (12.74b), and Cor. 12.48, the fifth one from (7.8g), (12.9), (12.28b), (12.34a), (12.58b), (12.128a), Cor. 12.48, and the bootstrap assumptions (**BA**Ψ), and the last two from (12.168).

Examining the products on the right-hand side of (10.40a), combining the above estimates for v_1, v_2, and v_3, using simple estimates of the form $ab \lesssim a^2 + b^2$, using the fact that $\mathbb{Q}_{(\leq N)}$, $\widetilde{\mathbb{Q}}_{(\leq N)}$, $\mathbb{Q}_{(\leq N-1)}$, and $\widetilde{\mathbb{Q}}_{(\leq N-1)}$ are increasing in their arguments, and using the estimates (13.150), (13.151), and (13.152), we deduce that all of the integrals $\int_{\mathcal{M}_{t,u}} \cdots \, d\varpi$ are $\lesssim$ one of the following:

(20.68)
$$\varepsilon \int_{t'=0}^{t} \frac{1}{(1+t')^{3/2}\upmu_\star^{1/2}(t',u)} \mathbb{Q}_{(\leq N)}(t',u)\, dt' \lesssim \varepsilon \mathbb{Q}_{(\leq N)}(t,u),$$

(20.69)
$$\varepsilon \int_{t'=0}^{t} \frac{1}{(1+t')^{3/2}\upmu_\star^{1/2}(t',u)} \widetilde{\mathbb{Q}}_{(\leq N)}(t',u)\, dt' \lesssim \varepsilon \widetilde{\mathbb{Q}}_{(\leq N)}(t,u),$$

(20.70)
$$\varepsilon \int_{t'=0}^{t} \frac{1}{(1+t')} \mathbb{Q}_{(\leq N-1)}^{1/2}(t',u)\mathbb{Q}_{(\leq N)}^{1/2}(t',u)\, dt' \lesssim \varepsilon \mathbb{Q}_{(\leq N)}(t,u) + \varepsilon \ln^2(e+t)\mathbb{Q}_{(\leq N-1)}(t,u),$$

(20.71)
$$\varepsilon \int_{t'=0}^{t} \frac{1}{(1+t')\upmu_\star^{1/2}(t',u)} \mathbb{Q}_{(\leq N-1)}^{1/2}(t',u)\widetilde{\mathbb{Q}}_{(\leq N)}^{1/2}(t',u)\, dt' \lesssim \varepsilon \widetilde{\mathbb{Q}}_{(\leq N)}(t,u) + \varepsilon \ln^2(e+t)\mathbb{Q}_{(\leq N-1)}(t,u),$$

(20.72)
$$\varepsilon \int_{t'=0}^{t} \frac{1}{(1+t')\upmu_\star^{1/2}(t',u)} \widetilde{\mathbb{Q}}_{(\leq N-1)}^{1/2}(t',u)\mathbb{Q}_{(\leq N)}^{1/2}(t',u)\, dt' \lesssim \varepsilon \mathbb{Q}_{(\leq N)}(t,u) + \varepsilon \ln^2(e+t)\widetilde{\mathbb{Q}}_{(\leq N-1)}(t,u),$$

(20.73)
$$\varepsilon \int_{t'=0}^{t} \frac{1}{(1+t')\upmu_\star(t',u)} \widetilde{\mathbb{Q}}_{(\leq N-1)}^{1/2}(t',u)\widetilde{\mathbb{Q}}_{(\leq N)}^{1/2}(t',u)\, dt' \lesssim \varepsilon \widetilde{\mathbb{Q}}_{(\leq N)}(t,u) + \varepsilon \ln^2(e+t)\frac{1}{\upmu_\star^{1/2}(t,u)}\widetilde{\mathbb{Q}}_{(\leq N-1)}(t,u),$$

$$\varepsilon^3 \int_{t'=0}^{t} \frac{1}{(1+t')^{3/2}\upmu_\star^{1/2}(t',u)}\,dt' \lesssim \varepsilon^3. \tag{20.74}$$

We have therefore proved the desired estimate (20.67a).

Applying similar reasoning to the terms on the right-hand side of (10.40b) (without having to integrate in time), we deduce (20.67b).

Finally, the estimate (20.67c) follows easily from the estimates of Sect. 12.8 and Lemma 12.30 since the integrands on the right-hand side of (10.40c) are cubic and depend on the data. □

LEMMA 20.34 (**Estimates for some easy error integrals corresponding to the partially modified version of $\Delta\mathscr{S}^{N-1}\upmu$**). *Let $1 \leq N \leq 24$ be an integer and let $\mathscr{S}^{N-1}$ be an $(N-1)^{st}$ order pure spatial commutation vectorfield operator (see Def. 7.2). Let ${}^{(\mathscr{S}^{N-1})}\widetilde{\mathscr{M}}$ be the corresponding partially modified $S_{t,u}$ one-form defined in (11.51a). Consider the $S_{t,u}$-tangent vectorfield $Y := {}^{(\mathscr{S}^{N-1})}\widetilde{\mathscr{M}}^{\#}$, and define the weight function $w = w(t,u) := \varrho^2(t,u)$, where $\varrho(t,u) = 1-u+t$. Under the small-data and bootstrap assumptions of Sects. 12.1-12.4, if ε is sufficiently small, then we have the following spacetime integral estimate for the quantity $\widetilde{\mathrm{Error}}_{(1)}[\mathscr{S}^{N-1}\breve{R}\Psi; w; Y]$ defined in (10.43a) for $(t,u) \in [0, T_{(Bootstrap)}) \times [0, U_0]$:*

$$\begin{aligned} \int_{\mathcal{M}_{t,u}} \left| \widetilde{\mathrm{Error}}_{(1)}[\mathscr{S}^{N-1}\breve{R}\Psi; \varrho^2; {}^{(\mathscr{S}^{N-1})}\widetilde{\mathscr{M}}^{\#}] \right| d\varpi & \lesssim \varepsilon \mathbb{Q}_{(\leq N)}(t,u) + \varepsilon \widetilde{\mathbb{Q}}_{(\leq N)}(t,u) \\ & \quad + \varepsilon \ln^2(e+t) \mathbb{Q}_{(\leq N-1)}(t,u) \\ & \quad + \varepsilon \ln^2(e+t) \frac{1}{\upmu_\star^{1/2}(t,u)} \widetilde{\mathbb{Q}}_{(\leq N-1)}(t,u) \\ & \quad + \varepsilon^3. \end{aligned} \tag{20.75a}$$

Furthermore, we have the following hypersurface integral estimate for the quantity $\widetilde{\mathrm{Error}}_{(2)}[\breve{R}\mathscr{S}^{N-1}\Psi; w; Y]$ defined in (10.43b):

$$\begin{aligned} \int_{\Sigma_t^u} \left| \widetilde{\mathrm{Error}}_{(2)}[\mathscr{S}^{N-1}\breve{R}\Psi; \varrho^2; {}^{(\mathscr{S}^{N-1})}\widetilde{\mathscr{M}}^{\#}] \right| d\underline{\varpi} & \lesssim \varepsilon \mathbb{Q}_{(\leq N)}(t,u) + \varepsilon \widetilde{\mathbb{Q}}_{(\leq N)}(t,u) \\ & \quad + \varepsilon \ln^2(e+t) \mathbb{Q}_{(\leq N-1)}(t,u) \\ & \quad + \varepsilon \ln^2(e+t) \frac{1}{\upmu_\star(t,u)} \widetilde{\mathbb{Q}}_{(\leq N-1)}(t,u) \\ & \quad + \varepsilon^3. \end{aligned} \tag{20.75b}$$

Finally, we have the following hypersurface integral estimate for the quantity $\widetilde{\mathrm{Error}}_{(3)}[\breve{R}\mathscr{S}^{N-1}\Psi; w; Y]$ defined in (10.43c):

$$\int_{\Sigma_0^u} \left| \widetilde{\mathrm{Error}}_{(3)}[\mathscr{S}^{N-1}\breve{R}\Psi; \varrho^2; {}^{(\mathscr{S}^{N-1})}\widetilde{\mathscr{M}}^{\#}] \right| d\underline{\varpi} \lesssim \varepsilon^3. \tag{20.75c}$$

PROOF. With the help of inequality (12.46a), we may apply reasoning similar to the reasoning that we used to prove Lemma 20.33 and we therefore omit the details.

We note that we use the estimates (20.49) and (20.61) in place of the estimates (20.58) and (20.65) used in the proof of Lemma 20.33. Moreover, we bound the factors of $Lw - \mathrm{tr}_{\not g}\chi w = -\varrho^2 \mathrm{tr}_{\not g}\chi^{(Small)}$ (this identity follows easily from (4.1c) and the fact that $L\varrho = 1$) appearing on the right-hand side of (10.43a) with the estimate (12.128a). In addition, we bound the factor $\not d\left\{L\breve{R}\Psi + \frac{1}{2}\mathrm{tr}_{\not g}\chi\breve{R}\Psi\right\}$ appearing on the right-hand side of (10.43a) with the estimates (12.46a) and (12.168). □

In Sects. 20.5-20.7, we use the auxiliary lemmas to prove the three lemmas of primary interest, namely Lemmas 20.20, 20.25, and 20.26. We begin with the proof of Lemma 20.20.

20.5. Proof of Lemma 20.20

We now prove Lemma 20.20. To deduce (20.24a), we show that

$$\begin{pmatrix} \left\| \mu \not\nabla^2 \mathscr{Z}^{\leq N-1} \mu \right\|_{L^2(\Sigma_t^u)} \\ \sum_{a=1}^3 \left\| \varrho\mu \not\nabla^2 \mathscr{Z}^{\leq N-1} L^a_{(Small)} \right\|_{L^2(\Sigma_t^u)} \\ \left\| \varrho\mu \not\nabla \not\mathcal{L}_{\mathscr{Z}}^{\leq N-1} \chi^{(Small)} \right\|_{L^2(\Sigma_t^u)} \\ \left\| \varrho\mu \not d \mathscr{Z}^{\leq N-1} \mathrm{tr}_{\not g}\chi^{(Small)} \right\|_{L^2(\Sigma_t^u)} \\ \left\| \varrho\mu \not\nabla \not\mathcal{L}_{\mathscr{Z}}^{\leq N-1} \hat{\chi}^{(Small)} \right\|_{L^2(\Sigma_t^u)} \end{pmatrix} \tag{20.76}$$
$$\lesssim \int_{t'=0}^t \frac{\|[L\mu]_-\|_{C^0(\Sigma_{t'}^u)}}{\mu_\star(t',u)} \mathbb{Q}^{1/2}_{(\leq N)}(t',u)\,dt'$$
$$+ \int_{t'=0}^t \frac{1}{(1+t')^{3/2}\mu_\star(t',u)} \mathbb{Q}^{1/2}_{(\leq N)}(t',u)\,dt' + \mathbb{Q}^{1/2}_{(\leq N)}(t,u)$$
$$+ \int_{t'=0}^t \frac{1}{(1+t')^{3/2}\mu_\star(t',u)} \widetilde{\mathbb{Q}}^{1/2}_{(\leq N)}(t',u)\,dt' + \varepsilon\left\{\ln \mu_\star^{-1}(t,u) + 1\right\}$$
$$+ \varepsilon \int_{t'=0}^t \frac{1}{(1+t')^{3/2}} \begin{pmatrix} \left\| \mu \not\nabla^2 \mathscr{Z}^{\leq N-1} \mu \right\|_{L^2(\Sigma_{t'}^u)} \\ \left\| \varrho\mu \not\nabla \not\mathcal{L}_{\mathscr{Z}}^{\leq N-1} \chi^{(Small)} \right\|_{L^2(\Sigma_{t'}^u)} \end{pmatrix} dt'.$$

The desired estimate (20.24a) then follows from applying Gronwall's inequality to the quantity on the left-hand side of (20.76).

It remains for us to prove (20.76). We first prove the desired estimate for $\begin{pmatrix} \left\| \varrho\mu \not\nabla \not\mathcal{L}_{\mathscr{Z}}^{\leq N-1} \chi^{(Small)} \right\|_{L^2(\Sigma_t^u)} \\ \left\| \varrho\mu \not\nabla \not\mathcal{L}_{\mathscr{Z}}^{\leq N-1} \hat{\chi}^{(Small)} \right\|_{L^2(\Sigma_t^u)} \end{pmatrix}$ To this end, we first use (20.33) and (20.35) to deduce that

$$\begin{pmatrix} \left\| \varrho\upmu \nabla\!\!\!\!/ \mathcal{L}\!\!\!/_{\mathscr{Z}}^{\leq N-1} \upchi^{(Small)} \right\|_{L^2(\Sigma_t^u)} \\ \left\| \varrho\upmu \nabla\!\!\!\!/ \mathcal{L}\!\!\!/_{\mathscr{Z}}^{\leq N-1} \hat{\upchi}^{(Small)} \right\|_{L^2(\Sigma_t^u)} \end{pmatrix} \tag{20.77}$$
$$\lesssim \sum_{l=1}^{3} \left\| \upmu O_{(l)} \mathscr{Z}^{\leq N-1} \mathrm{tr}_{g\!\!\!/} \upchi^{(Small)} \right\|_{L^2(\Sigma_t^u)} + \left\| \varrho\upmu \nabla\!\!\!\!/ \hat{\mathcal{L}\!\!\!/}_{\mathscr{Z}}^{N-1} \hat{\upchi}^{(Small)} \right\|_{L^2(\Sigma_t^u)}$$
$$+ \mathbb{Q}_{(\leq N)}^{1/2}(t,u) + \int_{t'=0}^{t} \frac{1}{(1+t')^{3/2} \upmu_{\star}^{1/2}(t',u)} \widetilde{\mathbb{Q}}_{(\leq N)}^{1/2}(t',u)\,dt' + \varepsilon.$$

Next, applying the elliptic estimate (18.16) to $\varrho \nabla\!\!\!\!/ \hat{\mathcal{L}\!\!\!/}_{\mathscr{Z}}^{N-1} \hat{\upchi}^{(Small)}$ and using inequalities (20.36) and (20.37), we bound the second term on the right-hand side of (20.77) as follows:

$$\left\| \varrho\upmu \nabla\!\!\!\!/ \hat{\mathcal{L}\!\!\!/}_{\mathscr{Z}}^{N-1} \hat{\upchi}^{(Small)} \right\|_{L^2(\Sigma_t^u)} \lesssim \left\| \varrho\upmu \mathrm{div\!\!\!/} \hat{\mathcal{L}\!\!\!/}_{\mathscr{Z}}^{N-1} \hat{\upchi}^{(Small)} \right\|_{L^2(\Sigma_t^u)} \tag{20.78}$$
$$+ \varepsilon \frac{\ln(e+t)}{1+t} \left\| \varrho \hat{\mathcal{L}\!\!\!/}_{\mathscr{Z}}^{N-1} \hat{\upchi}^{(Small)} \right\|_{L^2(\Sigma_t^u)}$$
$$\lesssim \sum_{l=1}^{3} \left\| \upmu O_{(l)} \mathscr{Z}^{\leq N-1} \mathrm{tr}_{g\!\!\!/} \upchi^{(Small)} \right\|_{L^2(\Sigma_t^u)}$$
$$+ \mathbb{Q}_{(\leq N)}^{1/2}(t,u)$$
$$+ \int_{t'=0}^{t} \frac{1}{(1+t')^{3/2} \upmu_{\star}^{1/2}(t',u)} \widetilde{\mathbb{Q}}_{(\leq N)}^{1/2}(t',u)\,dt'$$
$$+ \varepsilon.$$

With the exception of $\left\| \upmu O \mathscr{Z}^{\leq N-1} \mathrm{tr}_{g\!\!\!/} \upchi^{(Small)} \right\|_{L^2(\Sigma_t^u)}$, all terms on the right-hand side of (20.77) and (20.78) are clearly $\lesssim$ the right-hand side of (20.76) as desired. To bound $\left\| \upmu O \mathscr{Z}^{\leq N-1} \mathrm{tr}_{g\!\!\!/} \upchi^{(Small)} \right\|_{L^2(\Sigma_t^u)}$, we use inequality (20.41), the simple inequality $\left\| \upmu \hat{\nabla\!\!\!\!/}^2 \mathscr{Z}^{\leq N-1} \upmu \right\|_{L^2(\Sigma_t^u)} \leq \left\| \upmu \nabla\!\!\!\!/^2 \mathscr{Z}^{\leq N-1} \upmu \right\|_{L^2(\Sigma_t^u)}$, inequality (20.35), and the estimate

$$\int_{s=0}^{t'} \frac{1}{(1+s)^{3/2} \upmu_{\star}^{1/2}(s,u)} \widetilde{\mathbb{Q}}_{(\leq N)}^{1/2}(s,u)\,ds \lesssim \widetilde{\mathbb{Q}}_{(\leq N)}^{1/2}(t',u)$$

(which follows from (13.151) and the fact that $\widetilde{\mathbb{Q}}_{(\leq N)}$ is increasing in its arguments) to deduce that

$$\begin{aligned}(20.79)\quad & \left\| \upmu O \mathscr{Z}^{\leq N-1} \mathrm{tr}_{\not g}\chi^{(Small)} \right\|_{L^2(\Sigma_t^u)} \\ & \lesssim \int_{t'=0}^{t} \frac{\|[L\upmu]_-\|_{C^0(\Sigma_{t'}^u)}}{\upmu_\star(t',u)} \mathbb{Q}_{(\leq N)}^{1/2}(t',u)\,dt' \\ & + \int_{t'=0}^{t} \frac{1}{(1+t')^{3/2}\upmu_\star(t',u)} \mathbb{Q}_{(\leq N)}^{1/2}(t',u)\,dt' \\ & + \mathbb{Q}_{(\leq N)}^{1/2}(t,u) + \int_{t'=0}^{t} \frac{1}{(1+t')^{3/2}\upmu_\star(t',u)} \widetilde{\mathbb{Q}}_{(\leq N)}^{1/2}(t',u)\,dt' \\ & + \varepsilon \int_{t'=0}^{t} \frac{1}{(1+t')^{3/2}} \begin{pmatrix} \left\| \upmu \nabla^2 \mathscr{Z}^{\leq N-1}\upmu \right\|_{L^2(\Sigma_{t'}^u)} \\ \left\| \varrho\upmu \nabla \not{\mathcal{L}}_{\mathscr{Z}}^{\leq N-1} \chi^{(Small)} \right\|_{L^2(\Sigma_{t'}^u)} \end{pmatrix} dt' \\ & + \varepsilon \left\{ \ln \upmu_\star^{-1}(t,u) - 1 \right\}.\end{aligned}$$

Clearly, the right-hand side of (20.79) is $\lesssim$ the right-hand side of (20.76) as desired. We have thus bounded $\left\| \varrho\upmu \nabla \not{\mathcal{L}}_{\mathscr{Z}}^{\leq N-1} \chi^{(Small)} \right\|_{L^2(\Sigma_t^u)}$ and $\left\| \varrho\upmu \nabla \not{\mathcal{L}}_{\mathscr{Z}}^{\leq N-1} \hat{\chi}^{(Small)} \right\|_{L^2(\Sigma_t^u)}$ by the right-hand side of (20.76) as desired. Using in addition inequality (12.46a), we deduce from (20.79) that $\left\| \varrho\upmu \not{d} \mathscr{Z}^{\leq N-1} \mathrm{tr}_{\not g}\chi^{(Small)} \right\|_{L^2(\Sigma_t^u)}$ is also $\lesssim$ the right-hand side of (20.76) as desired.

The proof that $\left\| \upmu \nabla^2 \mathscr{Z}^{\leq N-1}\upmu \right\|_{L^2(\Sigma_t^u)}$ is bounded by the right-hand side of (20.76) is similar but simpler because the elliptic estimates involve a scalar quantity rather than a trace-free tensorial one. Specifically, we first use inequalities (12.46a) and (19.2a), the elliptic estimate (18.10), and the fact that $\mathbb{Q}_{(\leq N)}$ is increasing in its arguments to deduce that

(20.80)

$$\begin{aligned}\left\| \upmu \nabla^2 \mathscr{Z}^{\leq N-1}\upmu \right\|_{L^2(\Sigma_t^u)} & \lesssim \left\| \upmu \Delta \mathscr{Z}^{\leq N-1}\upmu \right\|_{L^2(\Sigma_t^u)} \\ & + \mathbb{Q}_{(\leq N)}^{1/2}(t,u) + \int_{t'=0}^{t} \frac{1}{(1+t')^{3/2}\upmu_\star^{1/2}(t',u)} \widetilde{\mathbb{Q}}_{(\leq N)}^{1/2}(t',u)\,dt' \\ & + \varepsilon.\end{aligned}$$

All terms on the right-hand side of (20.80) except for the first one are manifestly bounded by the right-hand side of (20.76) as desired. To bound the remaining term $\left\| \upmu \Delta \mathscr{Z}^{\leq N-1}\upmu \right\|_{L^2(\Sigma_t^u)}$ by the right-hand side of (20.76), we use inequality (20.41) and then argue as in the previous paragraph to bound this quantity by the right-hand side of (20.79) as desired.

To complete the proof of (20.76), it remains only for us to bound

$$\sum_{a=1}^{3} \left\| \varrho \mu \nabla\!\!\!/^2 \mathscr{Z}^{\leq N-1} L^a_{(Small)} \right\|_{L^2(\Sigma_t^u)}$$

by the right-hand side of (20.76). The proof is based on inequality (20.39) and the elliptic estimate (18.10) . We omit the details because they are very similar to the bound for $\left\| \mu \nabla\!\!\!/^2 \mathscr{Z}^{\leq N-1} \mu \right\|_{L^2(\Sigma_t^u)}$ proved in the previous paragraph.

To prove (20.24b), we first use inequalities (13.148) and (13.149) and the fact that $\mathbb{Q}_{(\leq N)}$ and $\widetilde{\mathbb{Q}}_{(\leq N)}$ are increasing in both of their arguments to deduce that the right-hand side of (20.24a) is

$$\begin{aligned}
(20.81) \qquad & \lesssim \left\{ \ln \mu_\star^{-1}(t,u) + 1 \right\} \mathbb{Q}^{1/2}_{(\leq N)}(t,u) + \left\{ \ln \mu_\star^{-1}(t,u) + 1 \right\} \widetilde{\mathbb{Q}}^{1/2}_{(\leq N)}(t,u) \\
& + \varepsilon \left\{ \ln \mu_\star^{-1}(t,u) + 1 \right\}.
\end{aligned}$$

Next, we combine inequality (20.81) with Lemma 12.57, which allows us to bound the left-hand side of (20.24b) by

$$\begin{aligned}
(20.82) \qquad & \lesssim (1+t) \int_{t'=0}^{t} \frac{1}{1+t'} \left\{ \ln \mu_\star^{-1}(t,u) + 1 \right\} \mathbb{Q}^{1/2}_{(\leq N)}(t',u)\, dt' \\
& + (1+t) \int_{t'=0}^{t} \frac{1}{1+t'} \left\{ \ln \mu_\star^{-1}(t,u) + 1 \right\} \widetilde{\mathbb{Q}}^{1/2}_{(\leq N)}(t',u)\, dt' \\
& + \varepsilon(1+t) \int_{t'=0}^{t} \frac{1}{1+t'} \left\{ \ln \mu_\star^{-1}(t,u) + 1 \right\} dt'.
\end{aligned}$$

Again using the fact that $\mathbb{Q}_{(\leq N)}$ and $\widetilde{\mathbb{Q}}_{(\leq N)}$ are increasing in both of their arguments and also using (13.152), we deduce that the right-hand side of (20.82) is

$$\lesssim \ln(e+t)(1+t)\mathbb{Q}^{1/2}_{(\leq N)}(t,u) + \ln(e+t)(1+t)\widetilde{\mathbb{Q}}^{1/2}_{(\leq N)}(t,u) + \varepsilon \ln(e+t)(1+t)$$

as desired. □

20.6. Proof of Lemma 20.25

We now use the auxiliary lemmas to prove the first lemma of primary interest, namely Lemma 20.25. We give the proof for the first integral on the left-hand side of (20.31). The proof for the second one is nearly identical and we describe the minor differences at the end of the proof. To begin, we split the integrand into two pieces as follows:

(20.83)
$$\begin{aligned}
& \left\{ (1+2\mu) L \mathscr{S}^{N-1} \breve{R}\Psi + 2\breve{R}\mathscr{S}^{N-1}\breve{R}\Psi \right\} (\breve{R}\Psi) \Delta\!\!\!/ \mathscr{S}^{N-1}\mu \\
& = (1+2\mu)(L\mathscr{S}^{N-1}\breve{R}\Psi)(\breve{R}\Psi)\Delta\!\!\!/ \mathscr{S}^{N-1}\mu + 2(\breve{R}\mathscr{S}^{N-1}\breve{R}\Psi)(\breve{R}\Psi)\Delta\!\!\!/ \mathscr{S}^{N-1}\mu.
\end{aligned}$$

The integral of the first product on the right-hand side of (20.83) is much easier to bound than that of the second; we bound the integral of the easier product at the end of the proof using a separate argument.

Bound for $\int_{\mathcal{M}_{t,u}} (\breve{R}\mathscr{S}^{N-1}\breve{R}\Psi)(2\breve{R}\Psi)\not{\Delta}\mathscr{S}^{N-1}\upmu\, d\varpi$: Our goal is to bound the spacetime integral

$$\int_{\mathcal{M}_{t,u}} (2\breve{R}\mathscr{S}^{N-1}\breve{R}\Psi)(\breve{R}\Psi)\not{\Delta}\mathscr{S}^{N-1}\upmu\, d\varpi \tag{20.84}$$

in magnitude by the right-hand side of (20.31). To this end, we first use the final inequality in (15.1), the bootstrap assumption $\left\|\breve{R}\Psi\right\|_{C^0(\Sigma_t^u)} \leq \varepsilon(1+t)^{-1}$, and Def. 16.2 to deduce that

$$(\breve{R}\Psi)\not{\Delta}\mathscr{S}^{N-1}\upmu = (\breve{R}\Psi)\breve{R}\mathscr{S}^{N-1}\mathrm{tr}_{\not{g}}\upchi^{(Small)} + Harmless^{\leq N}. \tag{20.85}$$

By Cor. 20.16, the part of the spacetime integral (20.84) involving the $Harmless^{\leq N}$ terms from (20.85) has already been bounded by the right-hand side of (20.31). It remains for us to bound the spacetime integral (20.84) with $\not{\Delta}\mathscr{S}^{N-1}\upmu$ replaced by $\breve{R}\mathscr{S}^{N-1}\mathrm{tr}_{\not{g}}\upchi^{(Small)}$. To this end, we first express it as a time integral of integrals over $\Sigma_{t'}^u$: $\int_{\mathcal{M}_{t,u}} \cdots d\varpi = \int_{t'=0}^{t}\int_{\Sigma_{t'}^u} \cdots d\underline{\varpi}\, dt'$. Next, using Cauchy-Schwarz on $\Sigma_{t'}^u$ and inequality (14.11e), we deduce that

$$\begin{aligned}
&\left|\int_{\mathcal{M}_{t,u}} (2\breve{R}\mathscr{S}^{N-1}\breve{R}\Psi)(\breve{R}\Psi)\breve{R}\mathscr{S}^{N-1}\mathrm{tr}_{\not{g}}\upchi^{(Small)}\, d\varpi\right| \\
&\leq 2\int_{t'=0}^{t} \left\|\breve{R}\mathscr{S}^{N-1}\breve{R}\Psi\right\|_{L^2(\Sigma_{t'}^u)} \left\|(\breve{R}\Psi)\breve{R}\mathscr{S}^{N-1}\mathrm{tr}_{\not{g}}\upchi^{(Small)}\right\|_{L^2(\Sigma_{t'}^u)} dt' \\
&\leq 2\int_{t'=0}^{t} \mathbb{Q}_{(\leq N)}^{1/2}(t',u) \left\|(\breve{R}\Psi)\breve{R}\mathscr{S}^{N-1}\mathrm{tr}_{\not{g}}\upchi^{(Small)}\right\|_{L^2(\Sigma_{t'}^u)} dt'.
\end{aligned} \tag{20.86}$$

The main part of the argument consists of showing that there exists a small constant $a > 0$ such that the following key inequality holds for the integrand factor on the right-hand side of (20.86):

(20.87)

$$
\begin{aligned}
&\left\| (\breve{R}\Psi)\breve{R}\mathscr{S}^{N-1}\mathrm{tr}_{\not g}\chi^{(Small)} \right\|_{L^2(\Sigma_t^u)} \\
&\leq \boxed{4.5}\frac{\|[L\mu]_-\|_{C^0(\Sigma_t^u)}}{\mu_\star(t,u)}\int_{s=0}^t \frac{\|[L\mu]_-\|_{C^0(\Sigma_s^u)}}{\mu_\star(s,u)}\mathbb{Q}_{(\leq N)}^{1/2}(s,u)\,ds \\
&\quad + C\varepsilon\frac{1}{(1+t)^{1+a}}\frac{1}{\mu_\star(t,u)}\int_{s=0}^t \frac{1}{(1+s)}\frac{1}{\mu_\star(s,u)}\mathbb{Q}_{(\leq N)}^{1/2}(s,u)\,ds \\
&\quad + C\varepsilon\frac{1}{(1+t)^{1+a}}\frac{1}{\mu_\star(t,u)}\int_{s=0}^t \frac{1}{(1+s)}\frac{1}{\mu_\star(s,u)}\widetilde{\mathbb{Q}}_{(\leq N)}^{1/2}(s,u)\,ds \\
&\quad + \boxed{4.5}\frac{\|[L\mu]_-\|_{C^0(\Sigma_t^u)}}{\mu_\star(t,u)}\mathbb{Q}_{(\leq N)}^{1/2}(t,u) \\
&\quad + 4.5\left\|\frac{[L\mu]_+}{\mu}\right\|_{C^0(\Sigma_t^u)}\int_{s=0}^t \frac{\|[L\mu]_-\|_{C^0(\Sigma_s^u)}}{\mu_\star(s,u)}\mathbb{Q}_{(\leq N)}^{1/2}(s,u)\,ds \\
&\quad + 4.5\left\|\frac{[L\mu]_+}{\mu}\right\|_{C^0(\Sigma_t^u)}\mathbb{Q}_{(\leq N)}^{1/2}(t,u) \\
&\quad + C\varepsilon\frac{1}{(1+t)^{1+a}}\frac{1}{\mu_\star(t,u)}\mathbb{Q}_{(\leq N)}^{1/2}(t,u) + C\varepsilon\frac{1}{(1+t)^{1+a}}\frac{1}{\mu_\star(t,u)}\widetilde{\mathbb{Q}}_{(\leq N)}^{1/2}(t,u) \\
&\quad + C\varepsilon\frac{\ln^3(e+t)}{1+t}\frac{1}{\mu_\star^{3/2}(t,u)}\mathbb{Q}_{(\leq N-1)}(t,u) + C\varepsilon\frac{1}{(1+t)^{3/2}}\frac{1}{\mu_\star^{3/2}(t,u)}\widetilde{\mathbb{Q}}_{(\leq N-1)}(t,u) \\
&\quad + C\varepsilon^2\frac{1}{(1+t)^{3/2}}\frac{1}{\mu_\star^{3/2}(t,u)}.
\end{aligned}
$$

Once we have shown (20.87), in order to bound the right-hand side of (20.86) by the right-hand side of (20.31), we insert inequality (20.87) (with t replaced by t') into the right-hand side of (20.86) and then integrate the inequality from $t'=0$ to $t'=t$. We use inequality (13.26) to bound the factors $\left\|\frac{[L\mu]_+}{\mu}\right\|_{C^0(\Sigma_{t'}^u)}$ arising from the right-hand side of (20.87). The factor of 2 on the right-hand side of (20.86) results in the doubling of the "boxed" constants on the right-hand side of (20.87), that is, the boxed constants in (20.31) are twice as large as the ones in (20.87). These estimates allow us to bound the integrals corresponding to all but the last five terms on the right-hand side of (20.87) by the right-hand side of (20.31). We now explain how to suitably bound these last five terms. Specifically, to bound the integral arising from the term $C\varepsilon\ln^3(e+t)(1+t)^{-1}\mu_\star^{-3/2}(t,u)\mathbb{Q}_{(\leq N-1)}(t,u)$ on the right-hand side of (20.87) by the right-hand side of (20.31), we use the fact that $\mathbb{Q}_{(\leq N-1)}$ and $\mathbb{Q}_{(\leq N)}$ are increasing in both of their arguments and inequality

(13.147) to deduce that

$$
\begin{aligned}
(20.88)\qquad & \varepsilon \int_{t'=0}^{t} \frac{\ln^3(e+t')}{(1+t')} \frac{1}{\upmu_{\star}^{3/2}(t',u)} \mathbb{Q}_{(\leq N)}^{1/2}(t',u)\mathbb{Q}_{(\leq N-1)}^{1/2}(t',u)\,dt' \\
& \lesssim \varepsilon \ln^3(e+t)\mathbb{Q}_{(\leq N)}^{1/2}(t,u)\mathbb{Q}_{(\leq N-1)}^{1/2}(t,u) \int_{t'=0}^{t} \frac{1}{(1+t')} \frac{1}{\upmu_{\star}^{3/2}(t',u)}\,dt' \\
& \lesssim \varepsilon \ln^4(e+t) \frac{1}{\upmu_{\star}^{1/2}(t',u)} \mathbb{Q}_{(\leq N)}^{1/2}(t,u)\mathbb{Q}_{(\leq N-1)}^{1/2}(t,u) \\
& \lesssim \varepsilon \ln^8(e+t) \frac{1}{\upmu_{\star}(t',u)} \mathbb{Q}_{(\leq N-1)}(t,u) + \varepsilon \mathbb{Q}_{(\leq N)}(t,u),
\end{aligned}
$$

which is clearly $\lesssim$ the right-hand side of (20.31) as desired. We can similarly bound the integrals generated by the second term on the next-to-last line of the right-hand side of (20.87) and the term on the last line of the right-hand side of (20.87), but we use inequality (13.146) in place of inequality (13.147). Finally, it is straightforward to see that we can bound the integrals generated by the two terms on the third-to-last line of the right-hand side of (20.87) by the terms $C\varepsilon \int_{t'=0}^{t} \frac{1}{(1+t')^{1+a}\upmu_{\star}(t',u)} \mathbb{Q}_{(\leq N)}(t',u)$ and $C\varepsilon \int_{t'=0}^{t} \frac{1}{(1+t')^{1+a}\upmu_{\star}(t',u)} \widetilde{\mathbb{Q}}_{(\leq N)}(t',u)$ on the right-hand side of (20.31). We have thus shown that the desired inequality (20.31) follows from (20.87).

It remains for us to prove (20.87). By Prop. 17.7, it suffices to bound the norm $\|\cdot\|_{L^2(\Sigma_t^u)}$ of the terms on the right-hand side of (17.35) by the right-hand side of (20.87). We proceed by arguing one term at a time. We begin by using the estimate $\left\|\breve{R}\mathscr{S}^N\Psi\right\|_{L^2(\Sigma_t^u)} \leq \mathbb{Q}_{(\leq N)}^{1/2}(t,u)$, which follows from inequality (14.11e), to deduce that the norm $\|\cdot\|_{L^2(\Sigma_t^u)}$ of the first product on the right-hand side of (17.35) is

$$
\begin{aligned}
(20.89)\qquad & = \boxed{2} \left\| \frac{L\upmu}{\upmu} \right\|_{C^0(\Sigma_t^u)} \left\| \breve{R}\mathscr{S}^N\Psi \right\|_{L^2(\Sigma_t^u)} \\
& \leq \boxed{2} \frac{\|[L\upmu]_-\|_{C^0(\Sigma_t^u)}}{\upmu_{\star}(t,u)} \mathbb{Q}_{(\leq N)}^{1/2}(t,u) + 2 \left\| \frac{[L\upmu]_+}{\upmu} \right\|_{C^0(\Sigma_t^u)} \mathbb{Q}_{(\leq N)}^{1/2}(t,u),
\end{aligned}
$$

which in turn is manifestly bounded by the right-hand side of (20.87).

To bound the norm $\|\cdot\|_{L^2(\Sigma_t^u)}$ of the second product on the right-hand side of (17.35), we first use Lemma 12.57 and inequality (14.11e) to bound the norm $\|\cdot\|_{L^2(\Sigma_t^u)}$ of the time integral term as follows:

$$
\begin{aligned}
(20.90)\qquad & \left\| \int_{t'=0}^{t} \frac{\|[L\upmu]_-\|_{C^0(\Sigma_{t'}^u)}}{\upmu_{\star}(t',u)} \varrho(t',u) \left| \breve{R}\mathscr{S}^N\Psi \right| (t',u,\vartheta)\,dt' \right\|_{L^2(\Sigma_t^u)} \\
& \leq (1+C\varepsilon)\varrho(t,u) \int_{t'=0}^{t} \frac{\|[L\upmu]_-\|_{C^0(\Sigma_{t'}^u)}}{\upmu_{\star}(t',u)} \mathbb{Q}_{(\leq N)}^{1/2}(t',u)\,dt'.
\end{aligned}
$$

The factor on the right-hand side of (17.35) that multiplies the first time integral is bounded in the norm $\|\cdot\|_{C^0(\Sigma_t^u)}$ by

(20.91)
$$\leq 4(1+C\sqrt{\varepsilon})\frac{1}{\varrho(t,u)}\frac{\|[L\upmu]_-\|_{C^0(\Sigma_t^u)}}{\upmu_\star(t,u)} + 4(1+C\sqrt{\varepsilon})\frac{1}{\varrho(t,u)}\left\|\frac{[L\upmu]_+}{\upmu}\right\|_{C^0(\Sigma_t^u)}.$$

We now multiply the right-hand sides of (20.90) and (20.91) and note that the resulting product is bounded by the right-hand side of (20.87) as desired. Note that this argument does not exhaust the full amount of the constant $\boxed{4.5}$ in front of the first term on the right-hand side of (20.87), but rather only $4(1+C\sqrt{\varepsilon})$.

To bound the norm $\|\cdot\|_{L^2(\Sigma_t^u)}$ of the third product on the right-hand side of (17.35), we first use Lemma 12.57, inequality (13.144) with the parameter $b>0$ chosen to be small enough so that $1+Cb\leq 1.1$ on the right-hand side of (13.144), inequality (14.11e), and the fact that $\mathbb{Q}_{(\leq N)}$ is increasing in its arguments to bound the norm $\|\cdot\|_{L^2(\Sigma_t^u)}$ of the time integral (more precisely, the second time integral on the right-hand side of (17.35)) as follows:

(20.92)
$$\begin{aligned}
&\left\|\int_{t'=0}^t \left\|\left(\frac{\upmu(t,\cdot)}{\upmu}\right)^2\right\|_{C^0(\Sigma_{t'}^u)} \varrho(t',u)\left|\breve{R}\mathscr{P}^N\Psi\right|(t',u,\vartheta)\,dt'\right\|_{L^2(\Sigma_t^u)} \\
&\leq (1+C\varepsilon)\varrho(t,u)\int_{t'=0}^t \left\|\left(\frac{\upmu(t,\cdot)}{\upmu}\right)^2\right\|_{C^0(\Sigma_{t'}^u)} \mathbb{Q}_{(\leq N)}^{1/2}(t',u)\,dt' \\
&\leq (1.1)(1+C\varepsilon)\varrho^2(t,u)\mathbb{Q}_{(\leq N)}^{1/2}(t,u) + C\varrho^2(t,u)\frac{\ln^2(e+t)}{(1+t)^b}\mathbb{Q}_{(\leq N)}^{1/2}(t,u) \\
&\leq (1.2)\varrho^2(t,u)\mathbb{Q}_{(\leq N)}^{1/2}(t,u) + C\varrho^2(t,u)\frac{1}{(1+t)^a}\mathbb{Q}_{(\leq N)}^{1/2}(t,u),
\end{aligned}$$

where $0<a<b$ is a constant. The factor on the right-hand side of (17.35) that multiplies the second time integral (which is under consideration) is bounded in the norm $\|\cdot\|_{C^0(\Sigma_t^u)}$ by the right-hand side of (20.91), except ϱ^{-1} is replaced by ϱ^{-2} and $4(1+C\sqrt{\varepsilon})$ is replaced with $2(1+C\varepsilon)$. Multiplying this factor by the first term on the right-hand side of (20.92), we see that the resulting product is bounded by the right-hand side of (20.87) as desired. Note that in bounding this term and the first one, we did not exhaust the full amount of the constant $\boxed{4.5}$ in front of the fourth term on the right-hand side of (20.87), but rather only $2+2(1.2)+C\varepsilon = 4.4+C\varepsilon$. Next, we use the estimate (12.128b) for $L\upmu$ to deduce the following easier estimate for the factor on the right-hand side of (17.35) that multiplies the second time integral: $2(1+C\varepsilon)\frac{1}{\varrho^2}\left\|\frac{L\upmu}{\upmu}\right\|_{C^0(\Sigma_t^u)} \lesssim \varepsilon(1+t)^{-3}\upmu_\star^{-1}(t,u)$. Hence, multiplying the factor by the second term on the right-hand side of (20.92), we see that the resulting product is bounded by $\lesssim \varepsilon(1+t)^{-(1+a)}\upmu_\star^{-1}(t,u)\mathbb{Q}_{(\leq N)}^{1/2}(t,u)$, which in turn is $\lesssim$ the right-hand side of (20.87) as desired.

The remaining terms on the right-hand of (17.35) are relatively easy to bound in the norm $\|\cdot\|_{L^2(\Sigma_t^u)}$ and have only a tiny effect on the dynamics. Many of the estimates we derive in this paragraph are nonoptimal. Throughout this paragraph, we silently use the coerciveness estimates provided by Prop. 14.7 as well as the fact that

$\mathbb{Q}_{(\leq N)}$ and $\widetilde{\mathbb{Q}}_{(\leq N)}$ are increasing in their arguments. To bound the norm $\|\cdot\|_{L^2(\Sigma_t^u)}$ of the product on the right-hand of (17.35) involving the factor $|\mathscr{X}^{[N]}|(0,u,\vartheta)$ by $\lesssim \varepsilon^2 \ln^2(e+t)(1+t)^{-2}\mu_\star^{-1}(t,u)$ as desired, we use inequality (20.43). Next, we see that the terms in the next line of (17.35) (which involve an array of non-time-integrated top-order terms including $\varrho\sqrt{\mu}\left\{L + \frac{1}{2}\mathrm{tr}_{\not g}\chi\right\}\mathscr{S}^{\leq N}\Psi$) are bounded in the norm $\|\cdot\|_{L^2(\Sigma_t^u)}$ by $\lesssim \varepsilon(1+t)^{-2}\mu_\star^{-1/2}(t,u)\mathbb{Q}_{(\leq N)}^{1/2}(t,u) + \varepsilon(1+t)^{-2}\mu_\star^{-1/2}(t,u)\widetilde{\mathbb{Q}}_{(\leq N)}^{1/2}(t,u)$ as desired. Similarly, we bound the terms on the next line of the right-hand side of (17.35) (which involve two arrays of non-time-integrated below-top-order terms including $\breve{R}\mathscr{Z}^{\leq N-1}\Psi$) in the norm $\|\cdot\|_{L^2(\Sigma_t^u)}$ by $\lesssim \varepsilon(1+t)^{-1}\mu_\star^{-1}(t,u)\mathbb{Q}_{(\leq N-1)}(t,u) + \varepsilon(1+t)^{-3/2}\mu_\star^{-3/2}(t,u)\widetilde{\mathbb{Q}}_{(\leq N-1)}(t,u)$. To bound the terms in the next line of the right-hand of (17.35) (which involve an array of non-time integrated terms including $\mathscr{Z}^{\leq N}(\mu-1)$) in the norm $\|\cdot\|_{L^2(\Sigma_t^u)}$ by

$$\lesssim \varepsilon\frac{1}{(1+t)^{3/2}\mu_\star(t,u)}\mathbb{Q}_{(\leq N)}^{1/2}(t,u) + \varepsilon\frac{1}{(1+t)^{3/2}\mu_\star(t,u)}\widetilde{\mathbb{Q}}_{(\leq N)}^{1/2}(t,u) + \varepsilon^2\frac{1}{(1+t)^{3/2}\mu_\star(t,u)},$$

we use (19.2a) and the estimate (13.152). To bound the terms in the next two lines of the right-hand of (17.35) (which involve an array of time-integrated top-order terms that include the term $\varrho\sqrt{\mu}\left\{L + \frac{1}{2}\mathrm{tr}_{\not g}\chi\right\}\mathscr{S}^{\leq N}\Psi$) in the norm $\|\cdot\|_{L^2(\Sigma_t^u)}$ by the sum of the second and third terms on the right-hand side of (20.87) (where $a > 0$ is a small constant), we use Lemma 12.57. To bound the terms in the next three lines of the right-hand of (17.35) (which involve two arrays of time-integrated below-top-order terms such as $\varrho\breve{R}\mathscr{Z}^{\leq N-1}\Psi$) in the norm $\|\cdot\|_{L^2(\Sigma_t^u)}$ by

$$\lesssim \varepsilon\frac{\ln^3(e+t)}{(1+t)\mu_\star^{3/2}(t,u)}\mathbb{Q}_{(\leq N-1)}(t,u) + \varepsilon\frac{1}{(1+t)^{3/2}\mu_\star^{3/2}(t,u)}\widetilde{\mathbb{Q}}_{(\leq N-1)}(t,u),$$

we use Lemma 12.57 and the estimates (13.147) and (13.150). To bound the terms on the next line of (17.35) (which involve an array of time-integrated top-order eikonal function quantities that includes the term $\hat{\nabla}^2\mathscr{S}^{\leq N-1}\mu$) in the norm $\|\cdot\|_{L^2(\Sigma_t^u)}$ by the right-hand side of (20.87), we use the simple inequality $|\hat{\nabla}^2\mathscr{S}^{\leq N-1}\mu| \lesssim |\nabla^2\mathscr{S}^{\leq N-1}\mu|$ and the estimate (20.24b). Finally, using Lemma 12.57, the estimate (19.2a), and the estimate (13.150), we see that the terms on the last two lines of (17.35) (which involve an array of time-integrated eikonal function quantities that contains $\mathscr{Z}^{\leq N}(\mu-1)$) are bounded in the norm $\|\cdot\|_{L^2(\Sigma_t^u)}$ by $\lesssim$ the sum of the second, third, and last terms on the right-hand side of (20.87). We have thus proved the desired estimate (20.87).

Bound for $\int_{\mathcal{M}_{t,u}}(1+2\mu)(L\mathscr{S}^{N-1}\breve{R}\Psi)(\breve{R}\Psi)\Delta\mathscr{S}^{N-1}\mu\,d\varpi$: We now bound the integral corresponding to the easier piece in (20.83). More precisely, our goal is to bound the spacetime integral

$$\int_{\mathcal{M}_{t,u}}(1+2\mu)(L\mathscr{S}^{N-1}\breve{R}\Psi)(\breve{R}\Psi)\Delta\mathscr{S}^{N-1}\mu\,d\varpi \tag{20.93}$$

in magnitude by the right-hand side of (20.31). To this end, we first bound the integral of the integrand piece $(L\mathscr{S}^{N-1}\breve{R}\Psi)(\breve{R}\Psi)\not{\Delta}\mathscr{S}^{N-1}\upmu$; the integral of the remaining piece $2\upmu(L\mathscr{S}^{N-1}\breve{R}\Psi)(\breve{R}\Psi)\not{\Delta}\mathscr{S}^{N-1}\upmu$ can be handled similarly and we omit most of those details. In fact, our argument below shows that the extra factor of $\upmu$ in that product makes many of the error integrals less degenerate with respect to $\upmu_\star^{-1}$, while other error integrals incur a factor of $\varepsilon\ln(e+t)$ due to the bound $\|Z\upmu\|_{C^0(\Sigma_t^u)} \lesssim \varepsilon\ln(e+t)$ for $Z \in \mathscr{Z}$ (see (12.128a)), a factor which is negligible due to the favorable time decay induced by the other factors. To proceed with the bound for (20.93), we further decompose the first factor as

$$L\mathscr{S}^{N-1}\breve{R}\Psi = \left\{L\mathscr{S}^{N-1}\breve{R}\Psi + \frac{1}{2}\mathrm{tr}_{\not{g}}\chi\mathscr{S}^{N-1}\breve{R}\Psi\right\} - \frac{1}{2}\mathrm{tr}_{\not{g}}\chi\mathscr{S}^{N-1}\breve{R}\Psi. \tag{20.94}$$

We now bound the spacetime integral corresponding to the term $\frac{1}{2}\mathrm{tr}_{\not{g}}\chi\mathscr{S}^{N-1}\breve{R}\Psi$ in (20.94). Integrating by parts on the $S_{t,u}$ in order to remove an angular derivative from $\not{\Delta}\mathscr{S}^{N-1}\upmu$ and referring to definition (4.1c), we see that it suffices to bound the spacetime integrals

$$\begin{aligned} &\int_{\mathcal{M}_{t,u}} \mathrm{tr}_{\not{g}}\chi(\breve{R}\Psi)(\not{d}\mathscr{S}^{N-1}\breve{R}\Psi)\cdot\not{d}^{\#}\mathscr{S}^{N-1}\upmu\, d\varpi \\ &+ \int_{\mathcal{M}_{t,u}} (\breve{R}\Psi)(\mathscr{S}^{N-1}\breve{R}\Psi)(\not{d}\mathrm{tr}_{\not{g}}\chi^{(Small)})\cdot\not{d}^{\#}\mathscr{S}^{N-1}\upmu\, d\varpi \\ &+ \int_{\mathcal{M}_{t,u}} \mathrm{tr}_{\not{g}}\chi(\mathscr{S}^{N-1}\breve{R}\Psi)(\not{d}\breve{R}\Psi)\cdot\not{d}^{\#}\mathscr{S}^{N-1}\upmu\, d\varpi. \end{aligned} \tag{20.95}$$

To this end, we first express all spacetime integrals as integrals over $\Sigma_{t'}^u$: $\int_{\mathcal{M}_{t,u}}\cdots d\varpi = \int_{t'=0}^t\int_{\Sigma_{t'}^u}\cdots d\underline{\varpi}\, dt'$. None of the $\int_{\Sigma_{t'}^u}$ integrals are difficult to bound because of the favorable time decay that is available. To bound the integrals $\int_{\Sigma_{t'}^u}\cdots d\underline{\varpi}$, we use Cauchy-Schwarz. We bound the quantities $\not{d}\mathscr{S}^{N-1}\breve{R}\Psi$, $\mathscr{S}^{N-1}\breve{R}\Psi$, and $\not{d}^{\#}\mathscr{S}^{N-1}\upmu$, in the norm $\|\cdot\|_{L^2(\Sigma_{t'}^u)}$ as follows with the help of Prop. 14.7, inequalities (12.46a) and (19.2a), and the estimate $\int_{s=0}^{t'}\frac{\widetilde{\mathbb{Q}}_{(\leq N)}^{1/2}(s,u)}{(1+s)\upmu_\star^{1/2}(s,u)}\,ds \lesssim \ln(e+t')\widetilde{\mathbb{Q}}_{(\leq N)}^{1/2}(t',u)$, which follows from (13.152):

$$\left\|\not{d}\mathscr{S}^{N-1}\breve{R}\Psi\right\|_{L^2(\Sigma_{t'}^u)} \lesssim \upmu_\star^{-1/2}(t',u)\mathbb{Q}_{(\leq N)}^{1/2}(t',u), \tag{20.96}$$

$$\left\|\mathscr{S}^{N-1}\breve{R}\Psi\right\|_{L^2(\Sigma_{t'}^u)} \lesssim \mathbb{Q}_{(\leq N)}^{1/2}(t',u), \tag{20.97}$$

$$\begin{aligned} \left\|\not{d}^{\#}\mathscr{S}^{N-1}\upmu\right\|_{L^2(\Sigma_{t'}^u)} &\lesssim \frac{1}{1+t'}\left\|\mathscr{Z}^N\upmu\right\|_{L^2(\Sigma_{t'}^u)} \\ &\lesssim \varepsilon + \ln(e+t')\mathbb{Q}_{(\leq N)}^{1/2}(t',u) + \ln(e+t')\widetilde{\mathbb{Q}}_{(\leq N)}^{1/2}(t',u). \end{aligned} \tag{20.98}$$

We bound the remaining factors in $\int_{\Sigma_{t'}^u}\cdots d\underline{\varpi}$ in the norm $\|\cdot\|_{C^0(\Sigma_{t'}^u)}$ as follows: $\left\|\breve{R}\Psi\right\|_{C^0(\Sigma_{t'}^u)} \leq \varepsilon(1+t')^{-1}$, $\left\|\not{d}\breve{R}\Psi\right\|_{C^0(\Sigma_{t'}^u)} \lesssim \varepsilon(1+t')^{-2}$, $\left\|\mathrm{tr}_{\not{g}}\chi\right\|_{C^0(\Sigma_{t'}^u)} \lesssim (1+t')^{-1}$, and $\left\|\not{d}\mathrm{tr}_{\not{g}}\chi^{(Small)}\right\|_{C^0(\Sigma_{t'}^u)} \lesssim \varepsilon\ln(e+t')(1+t')^{-3}$. These estimates follow from the bootstrap assumptions (**BA**Ψ), (12.46a), (12.60b), and (12.128a). Also using simple estimates of the form $ab \lesssim a^2 + b^2$, we deduce the following bound for the spacetime integrals $\int_{\mathcal{M}_{t,u}} \cdots d\varpi = \int_{t'=0}^{t} \int_{\Sigma_{t'}^u} \cdots d\underline{\varpi}\, dt'$ of interest:

$$
\begin{aligned}
(20.99) \qquad & \lesssim \varepsilon \int_{t'=0}^{t} \frac{1}{(1+t')^{3/2} \upmu_{\star}^{1/2}(t',u)} \mathbb{Q}_{(\leq N)}(t',u)\, dt' \\
& + \varepsilon \int_{t'=0}^{t} \frac{1}{(1+t')^{3/2} \upmu_{\star}^{1/2}(t',u)} \widetilde{\mathbb{Q}}_{(\leq N)}(t',u)\, dt' \\
& + \varepsilon^3 \int_{t'=0}^{t} \frac{1}{(1+t')^{3/2} \upmu_{\star}^{1/2}(t',u)}\, dt'.
\end{aligned}
$$

Using (13.151) to destroy the factor $\upmu_{\star}^{-1/2}$ in the last integral in (20.99), we conclude that (20.99) is $\lesssim$ the right-hand side of (20.31) as desired.

We now bound the spacetime integral corresponding to the first term in (20.94), that is, the integral

$$
(20.100) \qquad - \int_{\mathcal{M}_{t,u}} (\breve{R}\Psi) \left\{ L \mathscr{S}^{N-1} \breve{R}\Psi + \frac{1}{2} \mathrm{tr}_{\not{g}} \chi \mathscr{S}^{N-1} \breve{R}\Psi \right\} \not{\Delta} \mathscr{S}^{N-1} \upmu\, d\varpi.
$$

To this end, we first integrate by parts with Lemma 10.18, where the weight function w is equal to 1, $\not{d}^{\#} \mathscr{Z}^{N-1} \upmu$ plays the role of the vectorfield Y from the lemma (and hence $\mathrm{di\not{v}} Y = \not{\Delta} \mathscr{Z}^{N-1} \upmu$), and $\mathscr{Z}^N \Psi := \mathscr{S}^{N-1} \breve{R}\Psi$. We remark that if we were to treat the error integral corresponding to the piece $2\upmu (L \mathscr{S}^{N-1} \breve{R}\Psi)(\breve{R}\Psi) \not{\Delta} \mathscr{S}^{N-1} \upmu$ in detail, then at this point in the argument, the weight function w would be equal to $2\upmu$. Returning to the estimate of (20.100), we must bound the integrals on the right-hand side of (10.42) with $w := 1$ and $Y := \not{d}^{\#} \mathscr{Z}^{N-1} \upmu$. To this end, we first express all spacetime integrals as integrals over $\Sigma_{t'}^u$: $\int_{\mathcal{M}_{t,u}} \cdots d\varpi = \int_{t'=0}^{t} \int_{\Sigma_{t'}^u} \cdots d\underline{\varpi}\, dt'$. None of the $\int_{\Sigma_{t'}^u}$ integrals are difficult to bound because of the favorable time decay that is available. Specifically, when bounding integrals over $\int_{\Sigma_{t'}^u}$, we use Cauchy-Schwarz in the form $\int_{\Sigma_{t'}^u} |v_1 v_2 v_3|\, d\underline{\varpi} \lesssim \|v_1\|_{C^0(\Sigma_{t'}^u)} \|v_2\|_{L^2(\Sigma_{t'}^u)} \|v_3\|_{L^2(\Sigma_{t'}^u)}$. The product $v_1 v_2 v_3$ represents any of the spacetime integrands on the right-hand side of (10.42) with $\mathscr{Z}^N := \mathscr{S}^{N-1} \breve{R}$, where v_2 is equal to either the factor $\not{d} \mathscr{S}^{N-1} \breve{R}\Psi$ or $\mathscr{S}^{N-1} \breve{R}\Psi$, v_3 is equal to either $\not{d}^{\#} \mathscr{S}^{N-1} \upmu$ or $\left\{ \not{\mathcal{L}}_L + \mathrm{tr}_{\not{g}} \chi \right\} \not{d}^{\#} \mathscr{S}^{N-1} \upmu$, and v_1 is the product of the remaining factors. To bound v_2 and the first type of v_3 term, namely $\not{d}^{\#} \mathscr{Z}^{N-1} \upmu$, in the norm $\| \ \|_{L^2(\Sigma_{t'}^u)}$, we simply quote the estimates (20.96)-(20.98). To bound the second type of v_3 term, namely, $\left\{ \not{\mathcal{L}}_L + \mathrm{tr}_{\not{g}} \chi \right\} \not{d}^{\#} \mathscr{S}^{N-1} \upmu$, in the norm $\| \cdot \|_{L^2(\Sigma_{t'}^u)}$, we rely on the identity $\left\{ \not{\mathcal{L}}_L + \mathrm{tr}_{\not{g}} \chi \right\} \not{g}^{-1} = -2 \hat{\chi}^{(Small)}$, which follows in a straightforward fashion from the identity $\not{\mathcal{L}}_L \not{g}^{-1} = -2\chi^{\#\#}$ (see (3.72)) and (7.4a)-(7.4b)), the decomposition $\chi = \varrho^{-1} \not{g} + \chi^{(Small)}$ (see (4.1c)), and the fact that $L\varrho = 1$. Using this identity, inequality (12.46a), and the bounds $\left\| \hat{\chi}^{(Small)} \right\|_{C^0(\Sigma_{t'}^u)} \lesssim \varepsilon \frac{\ln(e+t')}{(1+t')^2}$ and $\|[L, O]\|_{C^0(\Sigma_{t'}^u)} \lesssim \varepsilon \ln(e+t')(1+t')^{-1}$, which follow from (4.21b), (12.74b), (12.128a), and Cor. 12.48, we deduce the following pointwise estimate:

(20.101)

$$\begin{aligned}\left|\{\not{\mathcal{L}}_L + \mathrm{tr}_{\not{g}}\chi\} \not{d}^{\#} \mathscr{S}^{N-1}\upmu\right| &\lesssim \left|\not{d}L\mathscr{S}^{N-1}\upmu\right| + \left|\hat{\chi}^{(Small)}\right| \left|\not{d}\mathscr{S}^{N-1}\upmu\right| \\ &\lesssim \frac{1}{1+t'}\sum_{l=1}^{3}\left|LO_{(l)}\mathscr{Z}^{N-1}\upmu\right| \\ &\quad + \frac{1}{1+t'}\sum_{l=1}^{3}\left|[L, O_{(l)}]\right| \left|\not{d}\mathscr{Z}^{N-1}\upmu\right| \\ &\quad + \frac{1}{1+t'}\left|\hat{\chi}^{(Small)}\right|\sum_{l=1}^{3}\left|O_{(l)}\mathscr{Z}^{N-1}\upmu\right| \\ &\lesssim \frac{1}{1+t'}\left|L\mathscr{Z}^{\leq N}\upmu\right| + \varepsilon\frac{\ln(e+t')}{(1+t')^3}\left|\mathscr{Z}^{\leq N}(\upmu - 1)\right|,\end{aligned}$$

where all terms above are evaluated at time t'. We then use inequalities (13.152), (19.2a), and (19.2b) and the fact that $\mathbb{Q}_{(\leq N)}$ and $\widetilde{\mathbb{Q}}_{(\leq N)}$ are increasing in their arguments to bound the norm $\|\cdot\|_{L^2(\Sigma_{t'}^u)}$ of the right-hand side of (20.101), which leads to the bound

$$\begin{aligned}\left\|\{\not{\mathcal{L}}_L + \mathrm{tr}_{\not{g}}\chi\} \not{d}^{\#} \mathscr{S}^{N-1}\upmu\right\|_{L^2(\Sigma_{t'}^u)} &\lesssim \varepsilon\frac{\ln(e+t')}{(1+t')^2} + \frac{1}{1+t'}\mathbb{Q}_{(\leq N)}^{1/2}(t', u) \\ &\quad + \frac{1}{1+t'}\upmu_\star^{-1/2}(t', u)\widetilde{\mathbb{Q}}_{(\leq N)}^{1/2}(t', u).\end{aligned} \tag{20.102}$$

To bound v_1, that is, the remaining integrand factors on the right-hand side of (10.42), in the norm $\|\cdot\|_{C^0(\Sigma_{t'}^u)}$, we use the estimates

$$\begin{aligned}\left\|\breve{R}\Psi\right\|_{C^0(\Sigma_{t'}^u)} &\lesssim \varepsilon(1+t')^{-1}, \\ \left\|\not{d}\breve{R}\Psi\right\|_{C^0(\Sigma_{t'}^u)} &\lesssim \varepsilon(1+t')^{-2}, \\ \left\|\mathrm{tr}_{\not{g}}\chi\right\|_{C^0(\Sigma_{t'}^u)} &\lesssim (1+t')^{-1}, \\ \left\|\not{d}\mathrm{tr}_{\not{g}}\chi^{(Small)}\right\|_{C^0(\Sigma_{t'}^u)} &\lesssim \varepsilon\ln(e+t')(1+t')^{-3}, \\ \left\|\left\{L + \frac{1}{2}\mathrm{tr}_{\not{g}}\chi\right\}\breve{R}\Psi\right\|_{C^0(\Sigma_{t'}^u)} &\lesssim \varepsilon\ln(e+t')(1+t')^{-3}, \\ \left\|\not{d}\left\{L + \frac{1}{2}\mathrm{tr}_{\not{g}}\chi\right\}\breve{R}\Psi\right\|_{C^0(\Sigma_{t'}^u)} &\lesssim \varepsilon\ln(e+t')(1+t')^{-4}.\end{aligned}$$

All of these C^0 estimates except for the last two were justified just above. The last two follow from (12.46a) and (12.168).

Using the above estimates and simple estimates of the form $ab \lesssim a^2 + b^2$, we conclude that the following bound holds for the magnitude of the *spacetime integrals* on the right-hand side of (10.42):

$$
\begin{aligned}
(20.103) \qquad & \lesssim \varepsilon \int_{t'=0}^{t} \frac{1}{(1+t')^{3/2}\upmu_{\star}} \mathbb{Q}_{(\leq N)}(t',u)\,dt' \\
& + \varepsilon \int_{t'=0}^{t} \frac{1}{(1+t')^{3/2}\upmu_{\star}(t',u)} \widetilde{\mathbb{Q}}_{(\leq N)}(t',u)\,dt' \\
& + \varepsilon^3 \int_{t'=0}^{t} \frac{1}{(1+t')^{3/2}}\,dt' \\
& \lesssim \varepsilon \int_{t'=0}^{t} \frac{1}{(1+t')^{3/2}\upmu_{\star}} \mathbb{Q}_{(\leq N)}(t',u) \\
& + \varepsilon \int_{t'=0}^{t} \frac{1}{(1-t')^{3/2}\upmu_{\star}(t',u)} \widetilde{\mathbb{Q}}_{(\leq N)}(t',u)\,dt' \\
& + \varepsilon^3.
\end{aligned}
$$

We now note that the right-hand side of (20.103) is manifestly $\lesssim$ the right-hand side of (20.31) as desired.

To bound the two Σ_t^u integrals on the right-hand side of (10.42) (that is, the explicitly written one and the one corresponding to the integrand (10.43b)), we use the same Cauchy-Schwarz estimate $\int_{\Sigma_t^u} |v_1 v_2 v_3|\,d\underline{\varpi} \lesssim \|v_1\|_{C^0(\Sigma_t^u)} \|v_2\|_{L^2(\Sigma_t^u)} \|v_3\|_{L^2(\Sigma_t^u)}$ that we used above, where in the present context, v_1 is equal to either $\breve{R}\Psi$ or $\not{d}\breve{R}\Psi$, v_2 is as above (see the discussion following equation (20.100)), and $v_3 = \not{d}^{\#} \mathscr{Z}^{N-1}\upmu$. To obtain the desired bounds, it suffices to use the estimates for v_1 and v_2 proved above and the following estimate for v_3, which follows from (12.46a) and (19.2a):

(20.104)

$$
\left\| \not{d}^{\#} \mathscr{Z}^{N-1}\upmu \right\|_{L^2(\Sigma_t^u)} \lesssim \int_{t'=0}^{t} \frac{\mathbb{Q}_{(\leq N)}^{1/2}(t',u)}{1+t'}\,dt' + \int_{t'=0}^{t} \frac{\widetilde{\mathbb{Q}}_{(\leq N)}^{1/2}(t',u)}{(1-t')\upmu_{\star}^{1/2}(s,u)}\,dt' + \varepsilon.
$$

Using these estimates for v_1, v_2, and v_3, the estimate (13.152), the fact that $\mathbb{Q}_{(\leq N)}$ and $\widetilde{\mathbb{Q}}_{(\leq N)}$ are increasing in both of their arguments, and simple estimates of the form $ab \lesssim a^2 + b^2$, we bound the magnitude of the integrals $\int_{\Sigma_t^u} \cdots$ under consideration by

$$
\begin{aligned}
(20.105) \qquad & \lesssim \varepsilon \frac{1}{\upmu_{\star}^{1/2}(t,u)} \mathbb{Q}_{(\leq N)}^{1/2}(t,u) \int_{t'=0}^{t} \frac{1}{(1+t')^{3/2}} \mathbb{Q}_{(\leq N)}^{1/2}(t',u)\,dt' \\
& + \varepsilon \frac{1}{\upmu_{\star}^{1/2}(t,u)} \mathbb{Q}_{(\leq N)}^{1/2}(t,u) \int_{t'=0}^{t} \frac{1}{(1+t')^{3/2}\upmu_{\star}^{1/2}(t',u)} \widetilde{\mathbb{Q}}_{(\leq N)}^{1/2}(t',u)\,dt' \\
& + \varepsilon \frac{1}{1+t} \mathbb{Q}_{(\leq N)}(t,u) + \varepsilon^3 \frac{1}{(1+t)} \frac{1}{\upmu_{\star}(t,u)},
\end{aligned}
$$

where the time integrals in (20.105) are generated by the right-hand side of (20.104). Clearly, the right-hand side of (20.105) is $\lesssim$ the right-hand side of (20.31) as desired.

The initial data hypersurface integral $\int_{\Sigma_0^u} \cdots$ on the right-hand side of (10.42) (corresponding to the integrand (10.43c)) can be treated in the same way as the $\int_{\Sigma_t^u} \cdots$ integrals. Specifically, its magnitude is bounded by the right-hand side of (20.105) with $t := 0$. Hence, in view of the fact that $\upmu|_{\Sigma_0} = 1 + \mathcal{O}(\mathring{\epsilon})$ (see (12.22b)),

Lemma 14.8, and (12.3), we see that the magnitude of the $\int_{\Sigma_0^u} \cdots$ integral is $\lesssim \varepsilon^3$ as desired.

Minor changes needed to bound the second integral on the left-hand side of (20.31). The overall strategy for bounding the integral

$$\int_{\mathcal{M}_{t,u}} \left\{(1+2\upmu)L\mathscr{S}^{N-1}O\Psi + \breve{R}\mathscr{S}^{N-1}O\Psi\right\}(\breve{R}\Psi)O\mathscr{S}^{N-1}\mathrm{tr}_{\not{g}}\chi^{(Small)}\, d\varpi$$

is the same as the one we used for bounding the first integral. We split the integrand as in (20.83). The difficult part of the proof is showing the analog of (20.87), namely that $\left\|(\breve{R}\Psi)O\mathscr{S}^{N-1}\mathrm{tr}_{\not{g}}\chi^{(Small)}\right\|_{L^2(\Sigma_t^u)}$ is bounded by the right-hand side of (20.87). However, this proof is exactly the same as our proof of (20.87). More precisely, we again use Prop. 17.7 to reduce the proof of the bound for $\left\|(\breve{R}\Psi)O\mathscr{S}^{N-1}\mathrm{tr}_{\not{g}}\chi^{(Small)}\right\|_{L^2(\Sigma_t^u)}$ to bounding the norm $\|\cdot\|_{L^2(\Sigma_t^u)}$ of the right-hand side of inequality (17.35), which we already accomplished in the argument following (20.87). The rest of the proof of the bound for $\left\|(\breve{R}\Psi)O\mathscr{S}^{N-1}\mathrm{tr}_{\not{g}}\chi^{(Small)}\right\|_{L^2(\Sigma_t^u)}$ is also identical to the proof of the bound for $\left\|(\breve{R}\Psi)\not{\Delta}\mathscr{S}^{N-1}\upmu\right\|_{L^2(\Sigma_t^u)}$.

Some minor changes are needed to bound the spacetime integral corresponding to the easier piece in (20.83), namely the integral

$$\int_{\mathcal{M}_{t,u}} (1+2\upmu)(L\mathscr{S}^{N-1}O\Psi)(\breve{R}\Psi)O\mathscr{S}^{N-1}\mathrm{tr}_{\not{g}}\chi^{(Small)}\, d\varpi. \tag{20.106}$$

We once again use the splitting $(1+2\upmu)(L\mathscr{S}^{N-1}O\Psi) = L\mathscr{S}^{N-1}O\Psi + 2\upmu L\mathscr{S}^{N-1}O\Psi$. As we described just below (20.93), it suffices to treat the integral corresponding to the piece $L\mathscr{S}^{N-1}O\Psi$ since the integral corresponding to $2\upmu L\mathscr{S}^{N-1}O\Psi$ can be treated using similar arguments. To proceed, we further decompose $L\mathscr{S}^{N-1}O\Psi$ as in (20.94).

To bound the spacetime integral corresponding to the term $\frac{1}{2}\mathrm{tr}_{\not{g}}\chi\mathscr{S}^{N-1}O\Psi$ from the analog of (20.94), we use an analog of the integration by parts identity (20.95) in order to remove the vectorfield O from the factor $O\mathscr{S}^{N-1}\mathrm{tr}_{\not{g}}\chi^{(Small)}$. This integration by parts also leads to the presence of some additional lower-order integrals containing the factor $\mathrm{tr}_{\not{g}}{}^{(O)}\not{\pi}$ (see Lemma 10.15). This factor in fact enhances the decay of the corresponding integrals over $\Sigma_{t'}^u$ due to the bound $\left\|\mathrm{tr}_{\not{g}}{}^{(O)}\not{\pi}\right\|_{C^0(\Sigma_{t'}^u)} \lesssim \varepsilon \ln(e+t')(1+t')^{-1}$ noted in the proof of Lemma 20.33.

In order to bound the spacetime integral corresponding to the term

$$\left\{L\mathscr{S}^{N-1}O\Psi + \frac{1}{2}\mathrm{tr}_{\not{g}}\chi\mathscr{S}^{N-1}O\Psi\right\}$$

from the analog of (20.94), we integrate by parts to remove O from the factor $O\mathscr{S}^{N-1}\mathrm{tr}_{\not{g}}\chi^{(Small)}$. This is the analog of the integration by parts performed just after equation (20.100) to remove an angular derivative off of $\not{\Delta}\mathscr{S}^{N-1}\upmu$. To carry out this integration by parts, in place of Lemma 10.18 used above, we now use Lemma 10.17, with the functions w and η from the lemma defined by

$w := 1$ and $\eta := \mathscr{S}^{N-1}\mathrm{tr}_{\not g}\chi^{(Small)}$ and $\mathscr{S}^{N-1}O\Psi$ in the role of $\mathscr{Z}^N\Psi$. We remark that if we were to treat the error integral corresponding to the piece $2\upmu(L\mathscr{S}^{N-1}O\Psi)(\breve{R}\Psi)O\mathscr{S}^{N-1}\mathrm{tr}_{\not g}\chi^{(Small)}$ in detail, then at this point in the argument, the function w would be equal to $2\upmu$. Returning to the present estimate with $w := 1$, we bound the integrals on the right-hand side of the integration by parts identity (10.39) by using arguments very similar to the ones we used in the discussion following equation (20.100). As before, these estimates require us to bound various factors in the integrals on the right-hand side of (10.39) in the norm $\|\cdot\|_{C^0(\Sigma_t^u)}$, which we accomplish using estimates mentioned above as well as a few additional ones noted in the proof of Lemma 20.33, such as $\left\|{}^{(O)}\not\pi_L^{\#}\right\|_{C^0(\Sigma_{t'}^u)} \lesssim \varepsilon\ln(e+t')(1+t')^{-1}$, $\left\|\mathrm{d}\not{\mathrm{i}}\mathrm{v}\,{}^{(O)}\not\pi_L^{\#}\right\|_{C^0(\Sigma_{t'}^u)} \lesssim \varepsilon\ln(e+t')(1+t')^{-2}$, and $\left\|\mathrm{tr}_{\not g}{}^{(O)}\not\pi\right\|_{C^0(\Sigma_{t'}^u)} \lesssim \varepsilon\ln(e+t')(1+t')^{-1}$. In total, this line of reasoning allows us to conclude that the error integral (20.106) verifies the same bounds as the error integral (20.100) treated above. This completes the proof of Lemma 20.25.

□

20.7. Proof of Lemma 20.26

We now use the auxiliary lemmas to prove the second lemma of primary interest, namely Lemma 20.26. We provide complete details for the estimate of

(20.107)
$$-\int_{\mathcal{M}_{t,u}}\left\{L\mathscr{S}^{N-1}O\Psi+\frac{1}{2}\mathrm{tr}_{\not g}\chi\mathscr{S}^{N-1}O\Psi\right\}(\breve{R}\Psi)(\varrho^2 O\mathscr{S}^{N-1}\mathrm{tr}_{\not g}\chi^{(Small)})\,d\varpi.$$

At the end of the proof, we sketch the minor changes needed to estimate the other integral on the left-hand side of (20.32). In order to avoid error integrals that lead to damaging top-order estimates, we need to replace the factor $\varrho^2\mathscr{S}^{N-1}\mathrm{tr}_{\not g}\chi^{(Small)}$ in (20.107) with the partially modified quantity $\varrho^{2(\mathscr{S}^{N-1})}\widetilde{\mathscr{X}}$. We recall that ${}^{(\mathscr{S}^{N-1})}\widetilde{\mathscr{X}}$ is defined in (11.45a); for convenience, we state:

(20.108)
$$\begin{aligned}\varrho^{2(\mathscr{S}^{N-1})}\widetilde{\mathscr{X}} := {}& \varrho^2\mathscr{S}^{N-1}\mathrm{tr}_{\not g}\chi^{(Small)}\\ &-\frac{1}{2}\varrho^2\not G_A^{\ A}\,L\mathscr{S}^{N-1}\Psi-\frac{1}{2}\varrho^2 G_{LL}L\mathscr{S}^{N-1}\Psi+\varrho^2\not G_L^{\ A}\,\not d_A\mathscr{S}^{N-1}\Psi.\end{aligned}$$

The reason that we must estimate the quantity (20.108) rather than $\varrho^2\mathscr{S}^{N-1}\mathrm{tr}_{\not g}\chi^{(Small)}$ is that we need to use the sharp L^2 estimates (20.57a), (20.57b), and (20.64) for $\varrho^{2(\mathscr{S}^{N-1})}\widetilde{\mathscr{X}}$; if we did not exploit these sharp L^2 estimates, then we would encounter error integrals that could grow in time at a rate that is just damaging enough to spoil our top-order a priori L^2 estimates.

Upon making this replacement, we generate an additional error integral equal to the vectorfield O applied to the difference between $\varrho^2\mathscr{S}^{N-1}\mathrm{tr}_{\not g}\chi^{(Small)}$ and $\varrho^{2(\mathscr{S}^{N-1})}\widetilde{\mathscr{X}}$. Equivalently, in view of definitions (11.45a) and (11.45b), we generate the following additional error integrals:

$$\int_{\mathcal{M}_{t,u}}\varrho^2\left\{L\mathscr{S}^{N-1}O\Psi+\frac{1}{2}\mathrm{tr}_{\not g}\chi\mathscr{S}^{N-1}O\Psi\right\}(\breve{R}\Psi)O^{(\mathscr{S}^{N-1})}\widetilde{\mathfrak{X}}\,d\varpi. \tag{20.109}$$

From the estimate (17.60) and Cor. 20.17, it follows that the magnitude of the integral (20.109) can be bounded by the non-boxed-constant-multiplied integrals on the right-hand side of (20.32).

Thus, to complete the proof, we have to estimate the main error integral

$$-\int_{\mathcal{M}_{t,u}} \left\{ L\mathscr{S}^{N-1}O\Psi + \frac{1}{2}\mathrm{tr}_{\not g}\chi\mathscr{S}^{N-1}O\Psi \right\} (\breve{R}\Psi)O(\varrho^{2(\mathscr{S}^{N-1})}\widetilde{\mathscr{X}})\, d\varpi. \tag{20.110}$$

We first integrate by parts with Lemma 10.17 (with $w := 1$ and $\eta := \varrho^{2(\mathscr{S}^{N-1})}\widetilde{\mathscr{X}}$ in the lemma) and therefore have to estimate both the spacetime and the hypersurface integrals on the right-hand side of (10.39). All of these integrals except the first two (the most difficult ones) were bounded by the non-boxed-constant-multiplied integrals on the right-hand side of (20.32) in Lemma 20.33.

We now bound the first difficult integral, which is the first spatial integral $\int_{\Sigma_t^u} \cdots$ on the right-hand side of (10.39).

Estimate of $\int_{\Sigma_t^u} (O\mathscr{S}^{N-1}O\Psi)(\breve{R}\Psi)\left(\varrho^{2(\mathscr{S}^{N-1})}\widetilde{\mathscr{X}}\right) d\underline{\varpi}$: To bound this integral, we first use the estimate (12.34a), inequality (14.12d), and Cor. 12.48 to deduce that

$$\begin{aligned} \left\|\sqrt{\upmu}O\mathscr{S}^{N-1}O\Psi\right\|_{L^2(\Sigma_t^u)} &\leq (1+C\varepsilon)\left\|\varrho\sqrt{\upmu}\not{d}\mathscr{S}^{N-1}O\Psi\right\|_{L^2(\Sigma_t^u)} \\ &\leq \sqrt{2}(1+C\varepsilon)\widetilde{\mathbb{Q}}_{(\leq N)}^{1/2}(t,u). \end{aligned} \tag{20.111}$$

Then from Cauchy-Schwarz and (20.111), we deduce that

$$\begin{aligned} &\left|\int_{\Sigma_t^u} (O\mathscr{S}^{N-1}O\Psi)(\breve{R}\Psi)\left(\varrho^{2(\mathscr{S}^{N-1})}\widetilde{\mathscr{X}}\right) d\underline{\varpi}\right| \\ &\leq \sqrt{2}(1+C\varepsilon)\left\|\frac{1}{\sqrt{\upmu}}(\breve{R}\Psi)\left(\varrho^{2(\mathscr{S}^{N-1})}\widetilde{\mathscr{X}}\right)\right\|_{L^2(\Sigma_t^u)} \widetilde{\mathbb{Q}}_{(\leq N)}^{1/2}(t,u). \end{aligned} \tag{20.112}$$

Hence, splitting $\Sigma_t^u = {}^{(-)}\Sigma_{t;t}^u \cup {}^{(+)}\Sigma_{t;t}^u$ (see Def. 13.8), applying Lemma 20.30, and using simple inequalities of the form $ab \lesssim a^2 + b^2$, we bound the right-hand of

(20.112) side by

$$
\begin{aligned}
(20.113)\quad & \leq (\sqrt{24} + C\varepsilon)\widetilde{\mathbb{Q}}_{(\leq N)}^{1/2}(t, u)\, \|\varrho L\upmu\|_{C^0({}^{(-)}\Sigma_{t;t}^u)} \frac{1}{\upmu_\star^{1/2}(t, u)} \\
& \quad \times \int_{t'=0}^t \frac{1}{\varrho(t', u)\upmu_\star^{1/2}(t', u)} \widetilde{\mathbb{Q}}_{(\leq N)}^{1/2}(t', u)\, dt' \\
& \quad + (\sqrt{24} + C\varepsilon)\widetilde{\mathbb{Q}}_{(\leq N)}^{1/2}(t, u) \left\| \frac{\varrho L\upmu}{\upmu} \right\|_{C^0({}^{(+)}\Sigma_{t;t}^u)} \\
& \quad \times \int_{t'=0}^t \left\| \sqrt{\frac{\upmu(t, \cdot)}{\upmu}} \right\|_{C^0({}^{(+)}\Sigma_{t';t}^u)} \frac{1}{\varrho(t', u)} \widetilde{\mathbb{Q}}_{(\leq N)}^{1/2}(t', u)\, dt' \\
& \quad + C\varepsilon \frac{1}{\upmu_\star^{1/2}(t, u)} \widetilde{\mathbb{Q}}_{(\leq N)}^{1/2}(t, u) \int_{t'=0}^t \frac{1}{(1 + t')^{3/2}} \mathbb{Q}_{(\leq N)}^{1/2}(t', u)\, dt' \\
& \quad + C\varepsilon \frac{1}{\upmu_\star^{1/2}(t, u)} \widetilde{\mathbb{Q}}_{(\leq N)}^{1/2}(t, u) \int_{t'=0}^t \frac{1}{(1 + t')^{3/2}\upmu_\star^{1/2}(t', u)} \widetilde{\mathbb{Q}}_{(\leq N)}^{1/2}(t', u)\, dt' \\
& \quad + C\varepsilon \widetilde{\mathbb{Q}}_{(\leq N)}(t, u) \\
& \quad + C\varepsilon \frac{1}{\upmu_\star(t, u)} \ln^2(e + t)\mathbb{Q}_{(\leq N-1)}(t, u) \\
& \quad + C\varepsilon \frac{1}{\upmu_\star(t, u)} \ln^2(e + t)\widetilde{\mathbb{Q}}_{(\leq N-1)}(t, u) \\
& \quad + C\varepsilon^3 \frac{1}{\upmu_\star(t, u)}.
\end{aligned}
$$

Since $\sqrt{24} < \boxed{5}$, we observe that all terms on the right-hand side of (20.113) are manifestly bounded by the right-hand side of (20.32) as desired.

To complete the proof, it remains only for us to estimate the second difficult integral, which is the first spacetime integral $\int_{\mathcal{M}_{t,u}} \cdots$ on the right-hand side of (10.39) (with $w := 1$ and $\eta := \varrho^{2}(\mathscr{S}^{N-1})\widetilde{\mathscr{X}}$ in (10.39)).

Estimate of $-\int_{\mathcal{M}_{t,u}} (O\mathscr{S}^{N-1}O\Psi)(\breve{R}\Psi)L\left(\varrho^{2}(\mathscr{S}^{N-1})\widetilde{\mathscr{X}}\right) d\varpi$: To bound this spacetime integral, we first express it as a time integral of integrals over $\Sigma_{t'}^u$: $\int_{\mathcal{M}_{t,u}} \cdots d\varpi = \int_{t'=0}^t \int_{\Sigma_{t'}^u} \cdots d\underline{\varpi}\, dt'$. We then use Cauchy-Schwarz and the estimate (20.111) to deduce that

$$
\begin{aligned}
(20.114)\quad & \left| \int_{\Sigma_{t'}^u} (O\mathscr{S}^{N-1}O\Psi)(\breve{R}\Psi)L\left(\varrho^{2}(\mathscr{S}^{N-1})\widetilde{\mathscr{X}}\right) d\varpi \right| \\
& \leq \sqrt{2}(1 + C\varepsilon)\widetilde{\mathbb{Q}}_{(\leq N)}^{1/2}(t', u) \left\| \frac{1}{\sqrt{\upmu}}(\breve{R}\Psi)L\left(\varrho^{2}(\mathscr{S}^{N-1})\widetilde{\mathscr{X}}\right) \right\|_{L^2(\Sigma_{t'}^u)}.
\end{aligned}
$$

We now use the key L^2 estimate (20.64), simple estimates of the form $ab \lesssim a^2 + b^2$, and inequality (13.26) to bound the last factor on the right-hand side of (20.114), thereby concluding that the right-hand side of (20.114) is bounded by

$$\begin{aligned}(20.115)\qquad &\leq \sqrt{24}(1+C\varepsilon)\frac{\|[L\upmu]_-\|_{C^0(\Sigma_{t'}^u)}}{\upmu_\star(t',u)}\widetilde{\mathbb{Q}}_{(\leq N)}(t',u)\\ &+\sqrt{24}(1+C\varepsilon)\frac{1}{\varrho(t',u)\left\{1+\ln\left(\frac{\varrho(t',u)}{\varrho(0,u)}\right)\right\}}\widetilde{\mathbb{Q}}_{(\leq N)}(t',u)\\ &+C\varepsilon\frac{1}{(1+t')^{3/2}}\mathbb{Q}_{(\leq N)}(t',u)+C\varepsilon\frac{1}{(1+t')^{3/2}}\frac{1}{\upmu_\star(t',u)}\widetilde{\mathbb{Q}}_{(\leq N)}(t',u)\\ &+C\varepsilon\frac{1}{(1+t')}\frac{1}{\upmu_\star^{1/2}(t',u)}\mathbb{Q}_{(\leq N-1)}^{1/2}(t',u)\widetilde{\mathbb{Q}}_{(\leq N)}^{1/2}(t',u)\\ &+C\varepsilon\frac{1}{(1+t')}\frac{1}{\upmu_\star(t',u)}\widetilde{\mathbb{Q}}_{(\leq N-1)}^{1/2}(t',u)\widetilde{\mathbb{Q}}_{(\leq N)}^{1/2}(t',u)\\ &+C\varepsilon^2\frac{1}{(1+t')^{3/2}}\frac{1}{\upmu_\star^{1/2}(t',u)}\widetilde{\mathbb{Q}}_{(\leq N)}^{1/2}(t',u).\end{aligned}$$

We now integrate inequality (20.115) dt'. Noting that $\sqrt{24}<\boxed{5}$, we observe that the time integrals of the first four products on the right-hand side of (20.115) are bounded by the right-hand side of (20.32) as desired. To bound the time integrals of the last three products on the right-hand side of (20.115), we first note that we can bound all factors $\mathbb{Q}_{(\leq N)}^{1/2}(t',u)$, $\widetilde{\mathbb{Q}}_{(\leq N)}^{1/2}(t',u)$, $\mathbb{Q}_{(\leq N-1)}^{1/2}(t',u)$, and $\widetilde{\mathbb{Q}}_{(\leq N-1)}^{1/2}(t',u)$ by their values at time t and then pull these factors out of the time integrals (since these factors are increasing in t'). We then use the estimates (13.150), (13.151), and (13.152) to deduce that the remaining time integrals are bounded as follows: $\int_{t'=0}^t \frac{1}{(1+t')}\upmu_\star^{-1/2}(t',u)\,dt' \leq C\ln(e+t)$, $\int_{t'=0}^t \frac{1}{(1+t')^{3/2}}\upmu_\star^{-1/2}(t',u)\,dt' \leq C$, and $\int_{t'=0}^t \frac{1}{1+t'}\upmu_\star^{-1}(t',u)\,dt' \leq C\ln(e+t)\left\{\ln\upmu_\star^{-1}(t,u)+1\right\}$. Also using simple inequalities of the form $ab \lesssim a^2+b^2$, we see that in total, the time integrals of the last three products on the right-hand side of (20.115) are bounded by the terms on the first two lines on right-hand side of (20.32). We have thus bounded the integral $\int_{\mathcal{M}_{t,u}}\cdots\,d\varpi$ by the right-hand side of (20.32) as desired. This completes our proof of the desired bound for the second integral on the left-hand side of (20.32).

Minor changes needed to bound

$$-\int_{\mathcal{M}_{t,u}}\varrho^2\left\{L\mathscr{S}^{N-1}\breve{R}\Psi+\frac{1}{2}\mathrm{tr}_{\not g}\chi\mathscr{S}^{N-1}\breve{R}\Psi\right\}(\breve{R}\Psi)\not\Delta\mathscr{S}^{N-1}\upmu\,d\varpi.$$

To bound the first spacetime integral on the left-hand side of (20.32), we use the same overall strategy that we used in bounding the second one. We again need to use a modified quantity in analogy with the quantity from equation (20.108). More precisely, we view $\not\Delta\mathscr{S}^{N-1}\upmu = \mathrm{d}\not{\mathrm{i}}\mathrm{v}\not d^{\#}\mathscr{S}^{N-1}\upmu$ and replace $\not d^{\#}\mathscr{S}^{N-1}\upmu$ with the partially modified $S_{t,u}$-tangent vectorfield ${}^{(\mathscr{S}^{N-1})}\widetilde{\mathscr{M}}^{\#}$, which is $\not g$-dual to the one-form ${}^{(\mathscr{S}^{N-1})}\widetilde{\mathscr{M}}$ defined in (11.51a). For convenience, we state the following identity:

$$(20.116)\qquad {}^{(\mathscr{S}^{N-1})}\widetilde{\mathscr{M}}^{\#} = \not{d}^{\#}\mathscr{S}^{N-1}\upmu + \frac{1}{2}\upmu G_{LL}\not{d}^{\#}\mathscr{S}^{N-1}\Psi + \upmu G_{LR}\not{d}^{\#}\mathscr{S}^{N-1}\Psi.$$

The reason that we must estimate the quantity (20.116) rather than $\not{d}^{\#}\mathscr{S}^{N-1}\upmu$ is that we need to use the sharp L^2 estimates of Lemma 20.29 and Lemma 20.31. This replacement leads to the generation of an additional error integral that is of the form (20.109) but with $\text{di}\not{v}\,{}^{(\mathscr{S}^{N-1})}\widetilde{\mathfrak{M}}^{\#}$ (see definition (11.51b)) in place of the factor $O^{(\mathscr{S}^{N-1})}\widetilde{\mathfrak{X}}$. This additional error integral can be suitably bounded with the help of the estimate (17.61) and Cor. 20.17, in analogy with the way we bounded the integral (20.109).

The difficult part of the analysis is bounding the remaining spacetime integral involving ${}^{(\mathscr{S}^{N-1})}\widetilde{\mathscr{M}}^{\#}$, which is the analog of the integral (20.110). To estimate this integral, we integrate by parts to remove the $\text{di}\not{v}$ operator from ${}^{(\mathscr{S}^{N-1})}\widetilde{\mathscr{M}}^{\#}$. Rather than integrating by parts with Lemma 10.17 as above, this time we use Lemma 10.18, where the role of Y from the lemma is played by ${}^{(\mathscr{S}^{N-1})}\widetilde{\mathscr{M}}^{\#}$ and the weight function w is equal to ϱ^2 (and hence $Ow = OLw = \not{d}w = 0$) because of the presence of this weight in the integrand. After integrating by parts, we have to estimate both the spacetime and the hypersurface integrals on the right-hand side of (10.42). All of these integrals except the first two (the most difficult ones) were bounded by the non-boxed-constant-multiplied integrals on the right-hand side of (20.32) in Lemma 20.34. This leaves the two difficult error integrals to estimate: **a)** the first hypersurface error integral on the right-hand side of (10.42), which we first bound via Cauchy-Schwarz and (14.12d) as follows:

$$(20.117)\qquad \begin{aligned} &\left| \int_{\Sigma_t^u} \varrho^2(\breve{R}\Psi)(\not{d}\mathscr{S}^{N-1}\breve{R}\Psi)^{(\mathscr{S}^{N-1})}\widetilde{\mathscr{M}}^{\#}\, d\underline{\varpi} \right| \\ &\quad \leq \sqrt{2}(1+C\varepsilon)\widetilde{\mathbb{Q}}_{(\leq N)}^{1/2}(t,u) \left\| \frac{1}{\sqrt{\upmu}}\varrho(\breve{R}\Psi)^{(\mathscr{S}^{N-1})}\widetilde{\mathscr{M}} \right\|_{L^2(\Sigma_t^u)}, \end{aligned}$$

and **b)** the first spacetime integral on the right-hand side of (10.42), which we first bound via Cauchy-Schwarz and (14.12d) as follows:

$$(20.118)\qquad \begin{aligned} &\left| \int_{\mathcal{M}_{t,u}} \varrho^2(\breve{R}\Psi)(\not{d}\mathscr{S}^{N-1}\breve{R}\Psi)\left\{\not{\mathcal{L}}_L + \text{tr}_{\not{g}}\chi\right\}{}^{(\mathscr{S}^{N-1})}\widetilde{\mathscr{M}}^{\#}\, d\varpi \right| \\ &\leq \sqrt{2}(1+C\varepsilon)\int_{t'=0}^{t} \widetilde{\mathbb{Q}}_{(\leq N)}^{1/2}(t',u) \left\| \frac{1}{\sqrt{\upmu}}\varrho(\breve{R}\Psi)\left\{\not{\mathcal{L}}_L + \text{tr}_{\not{g}}\chi\right\}{}^{(\mathscr{S}^{N-1})}\widetilde{\mathscr{M}}^{\#} \right\|_{L^2(\Sigma_{t'}^u)} dt'. \end{aligned}$$

To derive a suitable bound for the right-hand side of (20.117), we argue as in our proof of (20.113), using the sharp estimates provided by Lemma 20.29 in place of Lemma 20.30 to estimate the last factor $\|\cdots\|_{L^2(\Sigma_t^u)}$ on the right-hand side of (20.117). This line of reasoning allows us to bound the right-hand side of (20.117) by $\leq$ the right-hand side of (20.32) as desired.

To derive a suitable bound for the last integrand factor on the right-hand side of (20.118), we argue as in our proof of (20.115), using the sharp estimate provided by Lemma 20.31 in place of Lemma 20.32 to bound the the last factor $\|\cdots\|_{L^2(\Sigma_{t'}^u)}$.

This line of reasoning allows us to bound the right-hand side of (20.118) by $\lesssim$ the right-hand side of (20.32) as desired. This completes the proof of Lemma 20.26. □

20.8. Proof of Proposition 20.8

We now use the previously derived estimates estimates to prove Prop. 20.8. Let $0 \leq N \leq 24$ be an integer. Let $Z \in \mathscr{Z} = \{\varrho L, \breve{R}, O_{(1)}, O_{(2)}, O_{(3)}\}$ and for $N \geq 1$, let $\mathscr{Z}^N$ be an Nth order commutation vectorfield differential operator of the form $\mathscr{Z}^N = \mathscr{Z}^{N-1} Z$. We recall that by (16.2)-(16.4) and Cor. 16.5, we have

$$\mu \square_{g(\Psi)}(\mathscr{Z}^N \Psi) = {}^{(\mathscr{Z}^N)}\mathfrak{F}, \tag{20.119}$$

$${}^{(\mathscr{Z}^N)}\mathfrak{F} = \mathscr{Z}^{N-1}(\mu \mathscr{D}_\alpha^{(Z)} \mathscr{J}^\alpha) + Harmless^{\leq N}, \tag{20.120}$$

where $Harmless^{\leq N}$ terms are defined in Def. 16.2. We now state and separately consider four cases depending on the structure of $\mathscr{Z}^N$. It is easy to see that the four cases exhaust all possibilities.

<u>The four cases for the structure of $\mathscr{Z}^N$</u>

Using Lemma 16.1, Prop. 16.4, Cor. 16.5, and Cor. 16.6, we deduce the following estimates in the four cases.

(1) $\mathscr{Z}^N$ contains more than one factor of ϱL. Then

$${}^{(\mathscr{Z}^N)}\mathfrak{F} = Harmless^{\leq N}.$$

(2) $\mathscr{Z}^N$ contains precisely one factor of ϱL. Then one of the following two possibilities must occur:

$${}^{(\mathscr{Z}^N)}\mathfrak{F} = Harmless^{\leq N},$$

$${}^{(\mathscr{Z}^N)}\mathfrak{F} = \varrho(\not{d}\Psi^{\#}) \cdot (\mu \not{d} \mathscr{S}^{N-1} \mathrm{tr}_{\not{g}} \chi^{(Small)}) + Harmless^{\leq N},$$

where $\mathscr{S}^{N-1}$ is an $(N-1)^{st}$ order pure spatial commutation vectorfield operator (see definition (7.2a)).

(3) $\mathscr{Z}^N = \mathscr{S}^{N-1}\breve{R}$, where $\mathscr{S}^{N-1}$ is an $(N-1)^{st}$ order pure spatial commutation vectorfield operator. Then

$${}^{(\mathscr{Z}^N)}\mathfrak{F} = (\breve{R}\Psi) \not{\Delta} \mathscr{S}^{N-1}\mu - (\mu \not{d}^{\#}\Psi) \cdot (\mu \not{d} \mathscr{S}^{N-1} \mathrm{tr}_{\not{g}} \chi^{(Small)}) + Harmless^{\leq N}.$$

(4) $\mathscr{Z}^N = \mathscr{S}^{N-1} O_{(l)}$, where $O_{(l)} \in \{O_{(1)}, O_{(2)}, O_{(3)}\}$ is a geometric rotation vectorfield and $\mathscr{S}^{N-1}$ is an $(N-1)^{st}$ order pure spatial commutation vectorfield operator. Then

$${}^{(\mathscr{Z}^N)}\mathfrak{F} = (\breve{R}\Psi) O_{(l)} \mathscr{S}^{N-1} \mathrm{tr}_{\not{g}} \chi^{(Small)} + \rho_{(l)}(\not{d}^{\#}\Psi) \cdot (\mu \not{d} \mathscr{S}^{N-1} \mathrm{tr}_{\not{g}} \chi) + Harmless^{\leq N}.$$

Before proving the proposition in detail, we first give an overview. We must derive suitable estimates for the error integrals on the right-hand sides of the energy-cone flux identities (10.29a) and (10.29b), where $\mathscr{Z}^N\Psi$ is in the role of Ψ, and the inhomogeneous term $\mathfrak{F}$ appearing in (10.29a) and (10.29b) is equal to the term ${}^{(\mathscr{Z}^N)}\mathfrak{F}$ given in equation (20.120). This difficult analysis has already been carried out in the previous lemmas. That analysis allows us to bound $\mathbb{E}[\mathscr{Z}^N\Psi]$ and $\mathbb{F}[\mathscr{Z}^N\Psi]$ by the right-hand side of (20.6a), and $\widetilde{\mathbb{E}}[\mathscr{Z}^N\Psi]$, $\widetilde{\mathbb{F}}[\mathscr{Z}^N\Psi]$, and $\widetilde{\mathbb{K}}[\mathscr{Z}^N\Psi]$ by the right-hand side of (20.6b). We will then take the max of these estimates over all such operators of the form $\mathscr{Z}^N$, $0 \leq N \leq 24$, and then the sup over t and u. This will immediately imply the desired top-order inequalities (20.6a) and (20.6b) for $\mathbb{Q}_{(\leq N)}$, $\widetilde{\mathbb{Q}}_{(\leq N)}$, and $\mathbb{K}_{(\leq N)}$.

Estimate for $\mathbb{Q}_{(\leq N)}$: We now derive the desired inequality (20.6a). In our analysis, we implicitly use the quantities defined in Defs. 20.4 and 20.5. We now carry out the main step, which is deriving suitable bounds for the quantities $\mathbb{E}[\mathscr{Z}^N\Psi](t,u) + \mathbb{F}[\mathscr{Z}^N\Psi](t,u)$ on the left-hand side of (10.29a) (where $\mathscr{Z}^N\Psi$ is in the role of Ψ). To obtain the desired bounds, we bound the right-hand side of (10.29a) by the right-hand side of (20.6a). To this end, we first analyze the difficult error integral

$$-\int_{\mathcal{M}_{t,u}} \left\{(1+2\upmu)(L\mathscr{Z}^N\Psi) + 2\breve{R}\mathscr{Z}^N\Psi\right\} {}^{(\mathscr{Z}^N)}\mathfrak{F}\, d\varpi$$

on the right-hand side of (10.29a), which is absent when $N=0$. We separately consider the exhaustive cases **1** – **4** (depending on the structure of $\mathscr{Z}^N$) stated above. We bound all error integrals of the form

$$\int_{\mathcal{M}_{t,u}} \left|(1+2\upmu)(L\mathscr{Z}^N\Psi) + 2\breve{R}\mathscr{Z}^N\Psi\right| Harmless^{\leq N}\, d\varpi$$

in magnitude by the right-hand side of (20.6a) with Cor. 20.16, which in particular yields the desired bound in case **1**.

In case **2**, we bound the error integral

$$\int_{\mathcal{M}_{t,u}} \left\{(1+2\upmu)(L\mathscr{Z}^N\Psi) + 2\breve{R}\mathscr{Z}^N\Psi\right\} \varrho(\not{d}\Psi^{\#})\cdot(\upmu\not{d}\mathscr{Z}^{N-1}\mathrm{tr}_{\not{g}}\chi^{(Small)})\, d\varpi$$

in magnitude by the right-hand side of (20.6a) with Lemma 20.21.

In case **3**, we bound the dangerous error integral

$$\int_{\mathcal{M}_{t,u}} \left\{(1+2\upmu)(L\mathscr{S}^{N-1}\breve{R}\Psi) + 2\breve{R}\mathscr{S}^{N-1}\breve{R}\Psi\right\} (\breve{R}\Psi)\not{\Delta}\mathscr{S}^{N-1}\upmu\, d\varpi$$

in magnitude by the right-hand side of (20.6a) with Lemma 20.25. We stress that the error terms on the right-hand side of Lemma 20.25 are the only ones that result in the difficult capital Roman numeral error terms $\boxed{\mathbf{I})}_{(\leq N)}$, $\mathbf{II})_{(\leq N)}$, $\cdots$ on the right-hand side of (20.6a). Furthermore, we bound the error integral

$$\int_{\mathcal{M}_{t,u}} \left\{(1+2\upmu)(L\mathscr{S}^{N-1}\breve{R}\Psi) + 2\breve{R}\mathscr{S}^{N-1}\breve{R}\Psi\right\} (\upmu\not{d}^{\#}\Psi)\cdot(\upmu\not{d}\mathscr{S}^{N-1}\mathrm{tr}_{\not{g}}\chi^{(Small)})\, d\varpi$$

in magnitude by the right-hand side of (20.6a) with Lemma 20.21.

In case **4**, we bound the dangerous error integral

$$\int_{\mathcal{M}_{t,u}} \left\{(1+2\upmu)(L\mathscr{S}^{N-1}O_{(l)}\Psi) + 2\breve{R}\mathscr{S}^{N-1}O_{(l)}\Psi\right\} (\breve{R}\Psi)O_{(l)}\mathscr{S}^{N-1}\mathrm{tr}_{\not{g}}\chi^{(Small)}\, d\varpi$$

in magnitude by the right-hand side of (20.6a) with Lemma 20.25. Furthermore, we bound the error integral

$$\int_{\mathcal{M}_{t,u}} \left\{(1+2\upmu)(L\mathscr{S}^{N-1}O_{(l)}\Psi) + 2\breve{R}\mathscr{S}^{N-1}O_{(l)}\Psi\right\} \rho_{(l)}(\not{d}^{\#}\Psi)\cdot(\upmu\not{d}\mathscr{S}^{N-1}\mathrm{tr}_{\not{g}}\chi)\, d\varpi$$

in magnitude by the right-hand side of (20.6a) with Lemma 20.21.

We now bound the error integral on the right-hand side of (10.29a) that does not contain the inhomogeneous term factor $\mathfrak{F} = {}^{(\mathscr{Z}^N)}\mathfrak{F}$. That is, we bound the integral

$$-\frac{1}{2}\int_{\mathcal{M}_{t,u}} \mu Q^{\alpha\beta}[\mathscr{Z}^N\Psi]^{(T)}\pi_{\alpha\beta}\,d\varpi.$$

This integral is bounded by the right-hand side of (20.6a) by (20.30a) (see definition (10.45a)).

Finally, we note that the terms $\mathbb{E}[\mathscr{Z}^N\Psi](0,u)$ corresponding to the first term on the right-hand side of (10.29a) are trivially bounded by $\leq \mathbb{Q}_{(\leq N)}(0,u)$.

We have thus bounded the sum $\mathbb{E}[\mathscr{Z}^N\Psi](t,u)+\mathbb{F}[\mathscr{Z}^N\Psi](t,u)$ by the right-hand side of (20.6a) as desired. Taking the max of these estimates over all such operators of the form $\mathscr{Z}^N$, $0\leq N\leq 24$, taking the sup over t and u, and recalling Def. 14.2, we conclude the desired inequality (20.6a).

Estimate for $\widetilde{\mathbb{Q}}_{(\leq N)}$: We now derive the desired inequality (20.6b). In our analysis, we implicitly use the quantities defined in Defs. 20.6 and 20.7. We now carry out the main step, which is deriving suitable bounds for the quantities $\widetilde{\mathbb{E}}[\mathscr{Z}^N\Psi](t,u)+\widetilde{\mathbb{F}}[\mathscr{Z}^N\Psi](t,u)$ on the left-hand side of (10.29b) (where $\mathscr{Z}^N\Psi$ is in the role of Ψ). To obtain the desired bounds, we **a)** show that the nonpositive integral $-\widetilde{\mathbb{K}}[\mathscr{Z}^N\Psi](t,u)$ (see definition (14.9a)) is present on the right-hand side of (10.29b) and hence we can bring it over to the left-hand side to generate a coercive spacetime integral and **b)** bound the remaining integrals on the right-hand side of (10.29b) in magnitude by the right-hand side of (20.6b). To accomplish **a)**, we simply appeal to definition (14.9a) (with $\mathscr{Z}^N\Psi$ in the role of Ψ), which shows that $-\widetilde{\mathbb{K}}[\mathscr{Z}^N\Psi]$ is equal to the spacetime integral of the first term on the right-hand side of (10.48b). We emphasize that by (10.45b), $\widetilde{\mathbb{K}}[\mathscr{Z}^N\Psi]$ is in fact part of the integral $-\frac{1}{2}\int_{\mathcal{M}_{t,u}} \mu Q^{\alpha\beta}[\mathscr{Z}^N\Psi]\left\{{}^{(\widetilde{K})}\pi_{\alpha\beta}-\varrho^2\mathrm{tr}_{\not{g}}\chi g_{\alpha\beta}\right\}d\varpi$ and hence it would have appeared on the right-hand side of (20.6b) as $-\widetilde{\mathbb{K}}[\mathscr{Z}^N\Psi]$ if we did not bring it to the left.

We now accomplish **b)**; the proof is similar to the bound we derived for $\mathbb{Q}_{(\leq N)}$ above. Our goal is to bound the right-hand side of (10.29b) (except for the coercive term $\widetilde{\mathbb{K}}[\mathscr{Z}^N\Psi]$ that was addressed in **a)**) by the right-hand side of (20.6b). To this end, we first analyze the difficult error integral

$$-\int_{\mathcal{M}_{t,u}} \varrho^2\left\{L\mathscr{Z}^N\Psi+\frac{1}{2}\mathrm{tr}_{\not{g}}\chi\mathscr{Z}^N\Psi\right\}{}^{(\mathscr{Z}^N)}\mathfrak{F}\,d\varpi$$

on the right-hand side of (10.29b), which is absent when $N=0$. We consider the same four cases (based on the structure of $\mathscr{Z}^N$) that we did in our estimates for $\mathbb{Q}_{(\leq N)}$. We bound all error integrals of the form

$$\int_{\mathcal{M}_{t,u}} \varrho^2\left|L\mathscr{Z}^N\Psi+\frac{1}{2}\mathrm{tr}_{\not{g}}\chi\mathscr{Z}^N\Psi\right| Harmless^{\leq N}\,d\varpi$$

in magnitude by the right-hand side of (20.6b) with the help of Cor. 20.17, which in particular yields the desired bound in case **1**.

In case **2**, we bound the error integral

$$\int_{\mathcal{M}_{t,u}} \varrho^2\left\{L\mathscr{Z}^N\Psi+\frac{1}{2}\mathrm{tr}_{\not{g}}\chi\mathscr{Z}^N\Psi\right\}\varrho(\not{d}\Psi^{\#})\cdot(\mu\not{d}\mathscr{Z}^{N-1}\mathrm{tr}_{\not{g}}\chi^{(Small)})\,d\varpi$$

in magnitude by the right-hand side of (20.6b) with Lemma 20.22.

In case **3**, we bound the dangerous error integral

$$\int_{\mathcal{M}_{t,u}} \varrho^2 \left\{ L\mathscr{S}^{N-1}\breve{R}\Psi + \frac{1}{2}\mathrm{tr}_{\not g}\chi\mathscr{S}^{N-1}\breve{R}\Psi \right\} (\breve{R}\Psi)\not\Delta\mathscr{S}^{N-1}\upmu\, d\varpi$$

in magnitude by the right-hand side of (20.6b) with Lemma 20.26. We stress that the error terms on the right-hand side of Lemma 20.26 are the only ones that result in the difficult capital Roman numeral error terms $\boxed{\widetilde{\mathbf{I}})}_{(\leq N)}$, $\widetilde{\mathbf{II}})_{\leq N}$, $\cdots$ on the right-hand side of (20.6b). Furthermore, we bound the error integral

$$\int_{\mathcal{M}_{t,u}} \varrho^2 \left\{ L\mathscr{S}^{N-1}\breve{R}\Psi + \frac{1}{2}\mathrm{tr}_{\not g}\chi\mathscr{S}^{N-1}\breve{R}\Psi \right\} (\upmu\not d^{\#}\Psi)\cdot(\upmu\not d\mathscr{S}^{N-1}\mathrm{tr}_{\not g}\chi^{(Small)})\, d\varpi$$

in magnitude by the right-hand side of (20.6b) with Lemma 20.22.

In case **4**, we bound the dangerous error integral

$$\int_{\mathcal{M}_{t,u}} \varrho^2 \left\{ L\mathscr{S}^{N-1}O_{(l)}\Psi + \frac{1}{2}\mathrm{tr}_{\not g}\chi\mathscr{S}^{N-1}O_{(l)}\Psi \right\} (\breve{R}\Psi)O_{(l)}\mathscr{S}^{N-1}\mathrm{tr}_{\not g}\chi^{(Small)}\, d\varpi$$

in magnitude by the right-hand side of (20.6b) with Lemma 20.26. Furthermore we bound the error integral

$$\int_{\mathcal{M}_{t,u}} \varrho^2 \left\{ L\mathscr{S}^{N-1}O_{(l)}\Psi + \frac{1}{2}\mathrm{tr}_{\not g}\chi\mathscr{S}^{N-1}O_{(l)}\Psi \right\} \uprho_{(l)}(\not d^{\#}\Psi)\cdot(\upmu\not d\mathscr{S}^{N-1}\mathrm{tr}_{\not g}\chi)\, d\varpi$$

in magnitude by the right-hand side of (20.6b) with Lemma 20.22.

We now bound the error integrals on the right-hand side of (10.29b) that do not involve the inhomogeneous term factor $\mathfrak{F} = {}^{(\mathscr{Z}^N)}\mathfrak{F}$. That is, we bound the first, second, third fourth, sixth and seventh (final) integrals on the right-hand side of (10.29b) (with $\mathscr{Z}^N$ in the role of Ψ). The coercive "Morawetz part" of the sixth integral (that is, the next-to-last integral) was handled in **a)**, while the remaining part (see definitions (10.45b) and (10.48b)) is bounded by the right-hand side of (20.6b) by (20.30b). The remaining error integrals are bounded by (20.30c) and (20.30d), where in bounding the hypersurface integrals $-\frac{1}{4}\int_{\Sigma_0^u}\cdots$, we use (20.30c) and (20.30d) with $t := 0$ and the fact that $\widetilde{\mathbb{Q}}_{(\leq N)}(0,u) \leq C\mathbb{Q}_{(\leq N)}(0,u)$ (see Lemma 14.8).

Finally, we note that by Lemma 14.8, the terms $\widetilde{\mathbb{E}}[\mathscr{Z}^N\Psi](0,u)$ corresponding to the first term on the right-hand side of (10.29b) are bounded by $\leq \widetilde{\mathbb{Q}}_{(\leq N)}(0,u) \leq C\mathbb{Q}_{(\leq N)}(0,u)$.

We have thus bounded the sum $\widetilde{\mathbb{E}}[\mathscr{Z}^N\Psi](t,u) + \widetilde{\mathbb{F}}[\mathscr{Z}^N\Psi](t,u) + \widetilde{\mathbb{K}}[\mathscr{Z}^N\Psi](t,u)$ by the right-hand side of (20.6b) as desired. Taking the max of these estimates over all such operators of the form $\mathscr{Z}^N$, $0 \leq N \leq 24$, taking the sup over t and u, and recalling Def. 14.2, we conclude the desired inequality (20.6b). This completes the proof of Prop. 20.8. □

20.9. Proof of Proposition 20.9

We now use the previously derived estimates to prove Prop. 20.9. The proof is very similar to the proof of Prop. 20.8 with only one key change: we estimate all of the error integrals that cause top-order L^2 degeneracy with respect to $\upmu_\star^{-1}$

in a *different way*. Specifically, we estimate these integrals in terms of higher-order L^2 quantities (that is, these estimates lose one derivative), which will result in the presence of the terms $\mathbf{0})_{(\leq N+1)}$ and $\widetilde{\mathbf{0}})_{(\leq N+1)}$ (see definitions (20.3a) and (20.5a)) on the right-hand sides of (20.7a) and (20.7b). The major gain is the following: because of this alternate strategy, the top-order error terms $\boxed{\mathbf{I})}_{(\leq N)} - \mathbf{X})_{(\leq N)}$ and $\boxed{\widetilde{\mathbf{I}})}_{(\leq N)} - \widetilde{\mathbf{VIII}})_{(\leq N)}$ which are present on the right-hand sides of the top-order estimates (20.6a)-(20.6b), are *not present on the right-hand sides of the below-top-order estimates* (20.7a) *and* (20.7b). For the same reason, many terms on the first two lines of the right-hand sides of (20.6a)-(20.6b) are absent from the below-top-order estimates (20.7a) and (20.7b).

We now provide the details for the proof of (20.7a). We repeat the proof of (20.6a) and note that the only reason that the worst terms $\boxed{\mathbf{I})}_{(\leq N)} - \mathbf{X})_{(\leq N)}$ are present on the right-hand side of (20.6a) is because we had to use them as an upper bound for the dangerous error integrals

(20.121)
$$-\int_{\mathcal{M}_{t,u}} \left\{(1+2\upmu)(L\mathscr{Z}^N\Psi) + (\breve{R}\mathscr{Z}^N\Psi)\right\}(\breve{R}\Psi)\begin{pmatrix} \not{\Delta}\mathscr{S}^{N-1}\upmu \\ O_{(l)}\cdot\mathscr{S}^{N-1}\mathrm{tr}_{\not{g}}\chi^{(Small)} \end{pmatrix} d\varpi$$

from cases **3** and **4** above. Similarly, the error integrals (20.121) are the only reason that the terms on the first two lines of the right-hand side of (20.6a) containing the factor $\upmu_\star^{-1}$ are present (we note, however, that the harmless term $C\varepsilon^3$ on the right-hand side of (20.7a) is generated by many error integrals). Since we are no longer bounding a top-order quantity, to bound the integral (20.121), rather than using the argument given in the proof of (20.6a), we instead allow a permissible loss of one derivative by using the bootstrap assumption $\left\|\breve{R}\Psi\right\|_{C^0(\Sigma_t^u)} \leq \varepsilon(1+t)^{-1}$ (that is, $(\mathbf{BA}\Psi)$) and the derivative-losing Lemma 20.18 to deduce that (20.121) is bounded in magnitude by

(20.122)
$$\begin{aligned}
&\lesssim \varepsilon\int_{t'=0}^{t}\frac{1}{(1+t')^{3/2}\upmu_\star^{1/2}(t',u)}\mathbb{Q}_{(\leq N)}^{1/2}(t',u) \\
&\qquad\times\int_{s=0}^{t'}\frac{1}{1+s}\mathbb{Q}_{(\leq N+1)}^{1/2}(s,u)\,ds\,dt' \\
&\quad+\varepsilon\int_{t'=0}^{t}\frac{1}{(1+t')^{3/2}\upmu_\star^{1/2}(t',u)}\mathbb{Q}_{(\leq N)}^{1/2}(t',u) \\
&\qquad\times\int_{s=0}^{t'}\frac{1}{(1+s)\upmu_\star^{1/2}(s,u)}\widetilde{\mathbb{Q}}_{(\leq N+1)}^{1/2}(s,u)\,ds\,dt' \\
&\quad+\varepsilon\int_{t'=0}^{t}\frac{1}{(1+t')^{3/2}\upmu_\star^{1/2}(t',u)}\mathbb{Q}_{(\leq N)}(t',u)\,dt' + \varepsilon^3
\end{aligned}$$

as desired (see definition (20.3a)). We have thus proved inequality (20.7a).

The proof of (20.7b) is similar. More precisely, we repeat the proof of (20.6b) and make the following key change, which completely eliminates the dangerous terms $\boxed{\widetilde{\mathbf{I}})}_{(\leq N)} - \widetilde{\mathbf{VIII}})_{(\leq N)}$ that are present on the right-hand side of the top-order estimate (20.6b) as well as the terms on the first two lines of the right-hand side of (20.6b) (aside from the initial data term $C\mathbb{Q}_{(\leq N)}(0,u)$ and the term $C\varepsilon^3\upmu_\star^{-1}(t,u)$, which appears in the ameliorated form $C\varepsilon^3$): we bound the integrals

(20.123)
$$-\int_{\mathcal{M}_{t,u}} \varrho^2 \left\{ L\mathscr{Z}^N\Psi + \frac{1}{2}\mathrm{tr}_{\not g}\chi\mathscr{Z}^N\Psi \right\} (\breve{R}\Psi) \begin{pmatrix} \not\Delta \mathscr{S}^{N-1}\upsilon \\ O_{(l)}\mathscr{S}^{N-1}\mathrm{tr}_{\not g}\chi^{(Small)} \end{pmatrix} d\varpi$$

by using the bootstrap assumption $\left\| \breve{R}\Psi \right\|_{C^0(\Sigma_t^u)} \lesssim \varepsilon(1+t)^{-1}$ and the derivative-losing Lemma 20.19 to deduce that (20.123) is bounded in magnitude by

$$\begin{aligned}
(20.124) \qquad &\lesssim \varepsilon \int_{t'=0}^{t} \frac{1}{(1+t')^2} \left(\int_{s=0}^{t'} \frac{\mathbb{Q}_{(\leq N+1)}^{1/2}(s,u)}{1+s} \, ds \right)^2 dt' \\
&+ \varepsilon \int_{t'=0}^{t} \frac{1}{(1+t')^2} \left(\int_{s=0}^{t'} \frac{\widetilde{\mathbb{Q}}_{(\leq N+1)}^{1/2}(s,u)}{(1+s)\upmu_\star^{1/2}(s,u)} \, ds \right)^2 dt' \\
&+ \varepsilon \int_{u'=0}^{u} \widetilde{\mathbb{Q}}_{(\leq N)}(t,u') \, du' + \varepsilon^3
\end{aligned}$$

as desired (see definition (20.5a)). We have thus proved inequality (20.7b). This completes the proof of Prop. 20.9. □

20.10. Proof of Lemma 20.10

Before proving Lemma 20.10, we first provide the following simple lemma, which will be used to estimate the right-hand side of (20.3c).

LEMMA 20.35 (**Integral estimate for an unusual Gronwall factor**). *Let $t \geq 0$. There exists a constant $C > 0$* ***independent of*** *t such that*

$$(20.125) \qquad \int_{t'=0}^{t} \frac{\ln^5(e+t')}{(1+t')^2\sqrt{\ln(e+t)-\ln(e+t')}} \, dt' \leq C.$$

PROOF OF LEMMA 20.35. We make the change of variables $\uptau' := \ln(e+t')$, $\uptau := \ln(e+t)$, $d\uptau' = \frac{dt'}{e+t'}$. We first consider the case $\uptau \geq 2$. In this case, we split the integration domain into the intervals $[1, \uptau/2]$ and $[\uptau/2, \uptau]$. Noting that the function $\exp(-\uptau')$ is decreasing on the domain $\uptau' \in [1,\infty)$, that the function $\sqrt{\uptau - \uptau'}$ is decreasing on the domain $\uptau' \in [1,\uptau]$, that the function $(\uptau')^5$ is increasing on the domain $\uptau' \in [1,\infty)$, and that $\int_{\uptau'=1}^{\infty} \exp(-\uptau')\uptau'^5 \, d\uptau' < \infty$, we bound the two pieces as follows:

$$
\begin{aligned}
(20.126)\qquad & \int_{t'=0}^{t} \frac{\ln^5(e+t')}{(1+t')^2\sqrt{\ln(e+t)-\ln(e+t')}}\,dt' \\
& \leq C \int_{t'=0}^{t} \frac{\ln^5(e+t')}{(e+t')^2\sqrt{\ln(e+t)-\ln(e+t')}}\,dt' \\
& \leq C \int_{\tau'=1}^{\tau/2} \frac{\exp(-\tau')(\tau')^5}{\sqrt{\tau-\tau'}}\,d\tau' + \int_{\tau'=\tau/2}^{\tau} \frac{\exp(-\tau')(\tau')^5}{\sqrt{\tau-\tau'}}\,d\tau' \\
& \leq \frac{C}{\sqrt{\tau}} \int_{\tau'=1}^{\infty} \exp(-\tau')\tau'^5\,d\tau' + C\exp(-\tau/2)\tau^5 \int_{\tau'=\tau/2}^{\tau} \frac{1}{\sqrt{\tau-\tau'}}\,d\tau' \\
& \leq \frac{C}{\sqrt{\tau}} + C\exp(-\tau/2)\tau^{11/2} \leq C.
\end{aligned}
$$

To obtain the last inequality in (20.126), we used the simple fact that the functions $\frac{1}{\sqrt{\tau}}$ and $\exp(-\tau)\tau^{11/2}$ are uniformly bounded from above on the domain $\tau \in [1,\infty)$. This completes the proof of lemma in the case $\tau \geq 2$.

In the remaining case, we have $1 \leq \tau \leq 2$. We do not split the integration domain. Instead, we simply use the fact that the function $\exp(-\tau')(\tau')^5$ is uniformly bounded from above on the domain $\tau' \in [1,\infty)$ in order to deduce

$$
\begin{aligned}
(20.127)\qquad & \int_{t'=0}^{t} \frac{\ln^5(e+t')}{(1+t')^2\sqrt{\ln(e+t)-\ln(e+t')}}\,dt' \\
& \leq C \int_{\tau'=1}^{\tau} \frac{\exp(-\tau')(\tau')^5}{\sqrt{\tau-\tau'}}\,d\tau' \\
& \leq C \int_{\tau'=1}^{\tau} \frac{1}{\sqrt{\tau-\tau'}}\,d\tau' \\
& \leq C\sqrt{\tau} \leq C.
\end{aligned}
$$

This completes the proof of the lemma. □

We now prove Lemma 20.10.

Proof of the estimates for the top-order quantities. Throughout the proof, we often use the estimate $\varrho(t,u) \approx 1+t$ (on $\mathcal{M}_{T_{(Bootstrap)},U_0}$) and the fact that $\mathbb{Q}_{(\leq N)}(t,u)$ and $\widetilde{\mathbb{Q}}_{(\leq N)}(t,u)$ are increasing in their arguments without explicitly mentioning them each time. We first prove the desired (difficult) estimates for the top-order quantities $\mathbb{Q}_{(\leq 24)}(t,u)$, $\widetilde{\mathbb{Q}}_{(\leq 24)}(t,u)$, and $\widetilde{\mathbb{K}}_{(\leq 24)}(t,u)$. Our argument involves the following functions:

$$\iota_1(t,u) := \exp(u), \tag{20.128}$$

$$\iota_2(t,u) := \exp\left(\int_{s=0}^{t} \frac{1}{(1+s)^{1+a}}\,ds\right), \tag{20.129}$$

$$\iota_3(t,u) := 1 - \ln\left(\frac{\varrho(t,u)}{\varrho(0,u)}\right), \tag{20.130}$$

$$\iota(t,u) := \iota_1^{2P}(t,u)\iota_2^{2P}(t,u)\iota_3^{2A_*}(t,u)\upmu_\star^{-17.5}(t,u), \tag{20.131}$$

$$\widetilde{\iota}(t,u) := \iota_1^{2P}(t,u)\iota_2^{2P}(t,u)\iota_3^{2A_*+4}(t,u)\upmu_\star^{-17.5}(t,u), \tag{20.132}$$

where $P > 0$ and $A_* > 4$ are constants to be determined below, and $0 < a < 1/2$ is the small positive constant on the right-hand sides of (20.2b), (20.2c), (20.2g), and (20.2h). Note the important discrepancy factor $\iota_3^4(t,u)$ between $\iota(t,u)$ and $\widetilde{\iota}(t,u)$. The functions $\iota^{-1}(t,u)$ and $\widetilde{\iota}^{-1}(t,u)$ can be thought of as products of approximate integrating factors for the inequalities of Prop. 20.8 and hence $\iota(t,u)$ and $\widetilde{\iota}(t,u)$ are connected to the expected behavior of $\mathbb{Q}_{(\leq 24)}(t,u)$, $\widetilde{\mathbb{Q}}_{(\leq 24)}(t,u)$, and $\widetilde{\mathbb{K}}_{(\leq 24)}(t,u)$. Note that $\iota_1(t,u), \iota_2(t,u), \iota_3(t,u), \upmu_\star^{-1}(t,u), \iota(t,u), \widetilde{\iota}(t,u)$ are *increasing* in u, that $\iota_1(t,u), \iota_2(t,u), \iota_3(t,u)$ are also increasing in t, and that $\upmu_\star^{-1}(t,u), \iota(t,u)$ and $\widetilde{\iota}(t,u)$ are *approximately increasing in t* in the sense that

$$\upmu_\star^{-1}(t_1,u) \leq (1+C\sqrt{\varepsilon})\upmu_\star^{-1}(t_2,u), \qquad \text{if } t_1 \leq t_2, \tag{20.133}$$

$$\iota(t_1,u) \leq (1+C\sqrt{\varepsilon})\iota_1(t_2,u), \qquad \text{if } t_1 \leq t_2, \tag{20.134}$$

$$\widetilde{\iota}(t_1,u) \leq (1+C\sqrt{\varepsilon})\widetilde{\iota}(t_2,u), \qquad \text{if } t_1 \leq t_2. \tag{20.135}$$

To deduce (20.133)-(20.135), we have used the approximate monotonicity inequality (13.45). Note also that for a fixed P, $\iota_1^P(t,u)$ and $\iota_2^P(t,u)$ are uniformly bounded from above by a positive constant for $(t,u) \in [0, T_{(Bootstrap)}) \times [0, U_0]$. Throughout this proof, we often use these increasing/approximately increasing properties without explicitly mentioning them every time.

In order to generate sufficient smallness that will allow us to absorb error integrals during this Gronwall-type argument, we use the crucially important estimates of Prop. 13.12 together with the following simple inequalities:

$$\int_{u'=0}^{u} \iota_1^P(t,u')\,du' \leq \frac{1}{P}\iota_1^P(t,u), \tag{20.136}$$

$$\int_{t'=0}^{t} \frac{1}{(1+t')^{1+a}}\iota_2^P(t',u)\,dt' \leq \frac{1}{P}\iota_2^P(t,u), \tag{20.137}$$

$$\int_{t'=0}^{t} \frac{1}{\varrho(t',u)\left\{1+\ln\left(\frac{\varrho(t',u)}{\varrho(0,u)}\right)\right\}}\iota_3^{A_*}(t',u)\,dt' \leq \frac{1}{A_*}\iota_3^{A_*}(t,u). \tag{20.138}$$

Inequalities (20.136)-(20.138) are straightforward to verify by direct computation. The "smallness factors" that we exploit are the factors $\frac{1}{P}$ and $\frac{1}{A_*}$ as well as the bootstrap parameter ε.

To proceed, we define the following rescaled functions $q(t,u)$ and $\widetilde{q}(t,u)$, which we would like to show are small on the domain of interest:

$$q(t,u) := \sup_{(\hat{t},\hat{u})\in[0,t]\times[0,u]} \iota^{-1}(\hat{t},\hat{u})\mathbb{Q}_{(\leq 24)}(\hat{t},\hat{u}), \tag{20.139}$$

$$\widetilde{q}(t,u) := \sup_{(\hat{t},\hat{u})\in[0,t]\times[0,u]} \widetilde{\iota}^{-1}(\hat{t},\hat{u}) \max\left\{\widetilde{\mathbb{Q}}_{(\leq 24)}(\hat{t},\hat{u}), \widetilde{\mathbb{K}}(\hat{t},\hat{u})\right\}. \tag{20.140}$$

Specifically, in order to prove (20.8e) and (20.8f), it suffices to prove that there exist constants $P > 0$, $A_* > 4$, and $C > 1$ such that the following estimates hold for $(t,u) \in [0, T_{(Bootstrap)}) \times [0, U_0]$:

$$q(t,u) \leq C\left\{\mathring{\epsilon}^2 + \varepsilon^3\right\}, \tag{20.141}$$

$$\widetilde{q}(t,u) \leq C\left\{\mathring{\epsilon}^2 + \varepsilon^3\right\}. \tag{20.142}$$

The reason that the bounds (20.141) and (20.142) are sufficient for proving (20.8e) and (20.8f) is that $\iota_1(t,u)$ and $\iota_2(t,u)$ are each bounded from above by a uniform positive constant on the domain $(t,u) \in [0,\infty) \times [0,U_0]$, while $\iota_3(t,u) \approx \ln(e+t)$.

Our first goal is to derive a suitable bound for $\widetilde{q}(t,u)$ in terms of $q(t,u)$. More precisely, we will show that if $P > 1$ and $A_* > 4$ are large enough, then the following bound holds for $(t,u) \in [0, T_{(Bootstrap)}) \times [0, U_0]$:

$$\widetilde{q}(t,u) \leq C\left\{\mathring{\epsilon}^2 + \varepsilon^3 + q(t,u)\right\}. \tag{20.143}$$

To this end, we will prove that there exist constants $0 < \alpha < 1$ and $C > 0$ such that

$$\widetilde{q}(t,u) \leq C\left\{\mathring{\epsilon}^2 + \varepsilon^3 + q(t,u)\right\} + \alpha\widetilde{q}(t,u). \tag{20.144}$$

The desired estimate (20.143) easily follows from (20.144) by absorbing the product $\alpha\widetilde{q}(t,u)$ on the right-hand side of (20.144) back into the left.

In order to prove (20.144), we evaluate both sides of (20.6b) at $(\hat{t},\hat{u})$, multiply both sides of (20.6b) by $\widetilde{\iota}^{-1}(\hat{t},\hat{u})$, and then take $\sup_{(\hat{t},\hat{u})\in[0,t]\times[0,u]}$. From definition (20.140), we see that the left-hand side of the resulting inequality is precisely the term $\widetilde{q}(t,u)$ on the left-hand side of (20.144). The most difficult terms on the right-hand side of the resulting inequality are the ones involving "boxed constants terms," that is, terms arising from $\boxed{\widetilde{\mathbf{I}})}_{(\leq 24)}$ and $\boxed{\widetilde{\mathbf{V}})}_{(\leq 24)}$ (see Def. 20.6). We now derive a suitable bound for these difficult terms. We give complete details for bounding the term corresponding to $\boxed{\widetilde{\mathbf{I}})}_{(\leq 24)}$. We then provide abbreviated proofs for the remaining terms. To handle the term corresponding to $\boxed{\widetilde{\mathbf{I}})}_{(\leq 24)}$ (see definition (20.4a)), we have to bound

$$\sup_{(\hat{t},\hat{u})\in[0,t]\times[0,u]} \left\{\widetilde{\iota}^{-1}(\hat{t},\hat{u})\boxed{\widetilde{\mathbf{I}})}_{(\leq 24)}(\hat{t},\hat{u})\right\}. \tag{20.145}$$

To this end, we multiply and divide by $\mu_\star^{17.5}(t',\hat{u})$ in the integral on the right-hand side of the bound (20.4a) for $\boxed{\widetilde{\mathbf{I}})}_{(\leq 24)}(\hat{t},\hat{u})$ and use the fact that ι_1, ι_2, ι_3 are increasing in both of their arguments, thereby bounding the terms in braces in (20.145) by

$$
\begin{aligned}
(20.146)\qquad & \leq \boxed{5}\, \widetilde{\iota}^{-1}(\hat{t}, \hat{u}) \int_{t'=0}^{\hat{t}} \frac{\|[L\upmu]_-\|_{C^0(\Sigma_{t'}^{\hat{u}})}}{\upmu_\star(t', \hat{u})} \widetilde{\mathbb{Q}}_{(\leq 24)}(t', \hat{u})\, dt' \\
& \leq \boxed{5}\, \widetilde{\iota}^{-1}(\hat{t}, \hat{u}) \left\{ \sup_{(t',u')\in[0,\hat{t}]\times[0,\hat{u}]} \upmu_\star^{17.5}(t', u') \widetilde{\mathbb{Q}}_{(\leq 24)}(t', u') \right\} \\
& \quad \times \int_{t'=0}^{\hat{t}} \frac{\|[L\upmu]_-\|_{C^0(\Sigma_{t'}^{\hat{u}})}}{\upmu_\star(t', \hat{u})} \upmu_\star^{-17.5}(t', \hat{u})\, dt' \\
& \leq \boxed{5}\, \widetilde{\iota}^{-1}(\hat{t}, \hat{u}) \iota_1^{2P}(\hat{t}, \hat{u}) \iota_2^{2P}(\hat{t}, \hat{u}) \iota_3^{2A_*+4}(\hat{t}, \hat{u}) \widetilde{q}(\hat{t}, \hat{u}) \\
& \quad \times \int_{t'=0}^{\hat{t}} \frac{\|[L\upmu]_-\|_{C^0(\Sigma_{t'}^{\hat{u}})}}{\upmu_\star(t', \hat{u})} \upmu_\star^{-17.5}(t', \hat{u})\, dt' \\
& = \boxed{5}\, \widetilde{q}(\hat{t}, \hat{u}) \upmu_\star^{17.5}(\hat{t}, \hat{u}) \int_{t'=0}^{\hat{t}} \frac{\|[L\upmu]_-\|_{C^0(\Sigma_{t'}^{\hat{u}})}}{\upmu_\star(t', \hat{u})} \upmu_\star^{-17.5}(t', \hat{u})\, dt' \\
& \leq (1 + C\sqrt{\varepsilon}) \boxed{5} \times \frac{1}{17.5} \widetilde{q}(t, u).
\end{aligned}
$$

In the last step of (20.146), we used the crucially important integral estimate (13.143) and the fact that $\widetilde{q}$ is increasing in both of its arguments. We have therefore bounded (20.145) by the right-hand side of (20.146).

A similar argument based on the crucially important estimate (13.145a) leads to the following bound (see definition (20.4e)):

$$
(20.147)\qquad \sup_{(\hat{t},\hat{u})\in[0,t]\times[0,u]} \left\{ \widetilde{\iota}^{-1}(\hat{t}, \hat{u}) \boxed{\widetilde{\mathbf{V}}}_{(\leq 24)}(\hat{t}, \hat{u}) \right\} \leq (1 + C\sqrt{\varepsilon}) \boxed{5} \times \frac{1}{8.25} \widetilde{q}(t, u).
$$

The important point is that the constants $\frac{5}{17.5}$ and $\frac{5}{8.25}$ from the estimates (20.146) and (20.147) verify $\frac{5}{17.5} + \frac{5}{8.25} < .9 < 1$. This sum makes up the bulk of the constant α on the right-hand side of (20.144). The estimates (20.146) and (20.147) are the only two that force us to prove bounds for $\widetilde{\mathbb{Q}}_{(\leq 24)}(t, u)$ that involve large degeneracy with respect to powers of $\upmu_\star^{-1}$. As we will see, the remaining terms on the right-hand side of (20.6b) can be suitably bounded by choosing P and A_* to be sufficiently large.

To bound the term corresponding to $\widetilde{\mathbf{II}})_{(\leq 24)}$ (see definition (20.4b)), we use a similar argument based on the estimate (20.138) and the increasing/approximately increasing properties of the integrating factors to deduce that

$$
(20.148)\qquad \sup_{(\hat{t},\hat{u})\in[0,t]\times[0,u]} \left\{ \widetilde{\iota}^{-1}(\hat{t}, \hat{u}) \widetilde{\mathbf{II}})_{(\leq 24)}(\hat{t}, \hat{u}) \right\} \leq \frac{C}{2A_* + 4} \widetilde{q}(t, u).
$$

By choosing A_* to be large in (20.148), we can ensure that $\frac{C}{2A_* + 4}$ is as small as we need it to be.

Next, using a similar argument based on the estimate (13.146) (with $A = 0$ and $a = 1/2$), we deduce that (see definitions (20.4c), (20.4d), (20.4g), and (20.4h))

$$\sup_{(\hat{t},\hat{u})\in[0,t]\times[0,u]} \left\{ \widetilde{\iota}^{-1}(\hat{t},\hat{u}) \widetilde{\mathbf{III}})_{(\leq 24)}(\hat{t},\hat{u}) \right\} \leq C\varepsilon q, \tag{20.149}$$

$$\sup_{(\hat{t},\hat{u})\in[0,t]\times[0,u]} \left\{ \widetilde{\iota}^{-1}(\hat{t},\hat{u}) \widetilde{\mathbf{IV}})_{(\leq 24)}(\hat{t},\hat{u}) \right\} \leq C\varepsilon \widetilde{q}, \tag{20.150}$$

$$\sup_{(\hat{t},\hat{u})\in[0,t]\times[0,u]} \left\{ \widetilde{\iota}^{-1}(\hat{t},\hat{u}) \widetilde{\mathbf{VII}})_{(\leq 24)}(\hat{t},\hat{u}) \right\} \leq C\varepsilon q^{1/2}\widetilde{q}^{1/2} \leq C\varepsilon q + C\varepsilon \widetilde{q}, \tag{20.151}$$

$$\sup_{(\hat{t},\hat{u})\in[0,t]\times[0,u]} \left\{ \widetilde{\iota}^{-1}(\hat{t},\hat{u}) \widetilde{\mathbf{VIII}})_{(\leq 24)}(\hat{t},\hat{u}) \right\} \leq C\varepsilon \widetilde{q}. \tag{20.152}$$

To bound the term corresponding to $\widetilde{\mathbf{VI}})_{(\leq 24)}$ (see definition (20.4f)), we use a similar argument based on the estimate (13.145b) to deduce that

$$\sup_{(\hat{t},\hat{u})\in[0,t]\times[0,u]} \left\{ \widetilde{\iota}^{-1}(\hat{t},\hat{u}) \widetilde{\mathbf{VI}})_{(\leq 24)}(\hat{t},\hat{u}) \right\} \leq \frac{C}{A_* + 1/2}\widetilde{q}(t,u) + C\varepsilon \widetilde{q}(t,u). \tag{20.153}$$

By choosing A_* to be large and ε to be small in the previous estimate, we can ensure that $\frac{C}{A_* + 1/2}$ and $C\varepsilon$ are as small as we need them to be. We have thus bounded all of the terms generated by the capital Roman numeral terms $\boxed{\widetilde{\mathbf{I}})}_{(\leq N)} - \widetilde{\mathbf{VIII}})_{(\leq N)}$ on the right-hand side of (20.6b) (with $N = 24$).

To bound the terms corresponding to the easy terms $\widetilde{\mathbf{i}})_{(\leq 24)} - \widetilde{\mathbf{v}})_{(\leq 24)}$ (see Def. 20.7) on the right-hand side of (20.6b), we use similar arguments to deduce that

$$(1 + \widetilde{\varsigma}^{-1}) \sup_{(\hat{t},\hat{u})\in[0,t]\times[0,u]} \left\{ \widetilde{\iota}^{-1}(\hat{t},\hat{u}) \widetilde{\mathbf{i}})_{(\leq 24)}(\hat{t},\hat{u}) \right\} \leq C(1 + \widetilde{\varsigma}^{-1}) q(t,u), \tag{20.154}$$

$$\sup_{(\hat{t},\hat{u})\in[0,t]\times[0,u]} \left\{ \widetilde{\iota}^{-1}(\hat{t},\hat{u}) \widetilde{\mathbf{ii}})_{(\leq 24)}(\hat{t},\hat{u}) \right\} \leq \frac{C}{2A_* + 4}\widetilde{q}(t,u), \tag{20.155}$$

$$(1 + \widetilde{\varsigma}^{-1}) \sup_{(\hat{t},\hat{u})\in[0,t]\times[0,u]} \left\{ \widetilde{\iota}^{-1}(\hat{t},\hat{u}) \widetilde{\mathbf{iii}})_{(\leq 24)}(\hat{t},\hat{u}) \right\} \leq \frac{C}{P}(1 + \widetilde{\varsigma}^{-1}) \widetilde{q}(t,u), \tag{20.156}$$

$$(1 + \widetilde{\varsigma}^{-1}) \sup_{(\hat{t},\hat{u})\in[0,t]\times[0,u]} \left\{ \widetilde{\iota}^{-1}(\hat{t},\hat{u}) \widetilde{\mathbf{iv}})_{(\leq 24)}(\hat{t},\hat{u}) \right\} \leq \frac{C}{P}(1 + \widetilde{\varsigma}^{-1}) \widetilde{q}(t,u), \tag{20.157}$$

$$(\varepsilon + \widetilde{\varsigma}) \sup_{(\hat{t},\hat{u})\in[0,t]\times[0,u]} \left\{ \widetilde{\iota}^{-1}(\hat{t},\hat{u}) \widetilde{\mathbf{v}})_{(\leq 24)}(\hat{t},\hat{u}) \right\} \leq C(\varepsilon + \widetilde{\varsigma}) \widetilde{q}(t,u). \tag{20.158}$$

We make the following clarifying remarks concerning the above estimates. The estimate (20.154) relies on the inequality $\widetilde{\iota}^{-1}(t,u)\ln^4(e+t) \leq C\iota^{-1}(t,u)$. The estimate (20.155) was already proved in (20.148). The estimate (20.156) relies on inequality (20.137) in analogy with the way that we used (13.143) to deduce (20.146). Similarly, the estimate (20.157) relies on inequality (20.136) in analogy with the way that we used (13.143) to deduce (20.146). The estimate involving (20.158) follows easily from the definitions.

In order to complete the proof of (20.144), it remains for us to bound the terms arising from the terms on the first two lines of the right-hand side of (20.6b). We begin by bounding the terms generated by the terms on the first line. By

Lemma 14.8, the term corresponding to $C\mathbb{Q}_{(\leq 24)}(0,u)$ is $\leq C\mathring{\epsilon}^2$. The term corresponding to $C\varepsilon^3\upmu_\star^{-1}(t,u)$ can easily be bounded by $\leq C\varepsilon^3$. The term corresponding to $C\varepsilon\mathbb{Q}_{(\leq N)}(t,u)$ can easily be bounded by $\leq C\varepsilon q(t,u)$, while the term corresponding to $C\varepsilon\widetilde{\mathbb{Q}}_{(\leq N)}(t,u)$ can easily bounded by $\leq C\varepsilon\widetilde{q}(t,u)$.

We now bound the terms generated by the terms on the second line of the right-hand side of (20.6b). To bound the term generated by the first term on the second line, we use the bootstrap assumption (20.1c) for $\mathbb{Q}_{(\leq 23)}(t,u)$ to deduce that

$$\sup_{(\hat{t},\hat{u})\in[0,t]\times[0,u]}\left\{C\varepsilon\widetilde{\iota}^{-1}(\hat{t},\hat{u})\upmu_\star^{-1}(\hat{t},\hat{u})\ln^2(e+\hat{t})\mathbb{Q}_{(\leq 23)}(\hat{t},\hat{u})\right\} \\ \leq C\varepsilon^3\sup_{(\hat{t},\hat{u})\in[0,t]\times[0,u]}\upmu_\star(\hat{t},\hat{u})\leq C\varepsilon^3. \tag{20.159}$$

Using a similar argument based on the bootstrap assumption (20.1d) for $\widetilde{\mathbb{Q}}_{(\leq 23)}(t,u)$, we also bound the term generated by the second term on the second line of the right-hand side of (20.6b) by $\leq C\varepsilon^3$.

Combining all of the above estimates, we deduce that

$$\widetilde{q}(t,u)\leq C\left\{\mathring{\epsilon}^2+\varepsilon^3\right\}+C\left\{1+\widetilde{\varsigma}^{-1}+\varepsilon\right\}q(t,u) \\ +\left\{\frac{5}{17.5}+\frac{5}{8.25}+\frac{C}{A_*}+\frac{C}{P}(1+\widetilde{\varsigma}^{-1})+C\widetilde{\varsigma}+C\sqrt{\varepsilon}\right\}\widetilde{q}(t,u). \tag{20.160}$$

The bound (20.144) now follows from (20.160) if we first choose and *fix* $\widetilde{\varsigma}$ to be sufficiently small, then choose A_* and P to be sufficiently large, and finally choose ε to be sufficiently small. We remark that later in the proof, just below inequality (20.189), we may need to further enlarge A_* and P.

We now use the estimate (20.143) to help us derive the desired estimate (20.141) for $q(t,u)$. The desired estimate (20.142) for $\widetilde{q}$ then follows easily from (20.143) and (20.141). To prove (20.141), we will argue as in our proof of (20.144) to show that there exist constants $0<\upbeta<1$ and $C>0$ such that

$$q(t,u)\leq C\left\{\mathring{\epsilon}^2+\varepsilon^3\right\}+\upbeta q(t,u). \tag{20.161}$$

Clearly, the desired bound (20.141) follows once we have shown (20.161). Our proof of (20.161) is very similar to our proof of (20.144). More precisely, in order to prove (20.161), for each term on the right-hand side of (20.6a) involving an integral, we find an effective approximate integrating factor and then multiply both sides of (20.6a) by the product of all the integrating factors. Specifically, we evaluate both sides of (20.6a) at $(\hat{t},\hat{u})$, multiply both sides of (20.6a) by $\iota^{-1}(\hat{t},\hat{u})$ (see (20.131)), and then take $\sup_{(\hat{t},\hat{u})\in[0,t]\times[0,u]}$. From definition (20.139), we see that the left-hand side of the resulting inequality is precisely the term $q(t,u)$ on the left-hand side of (20.161). The most difficult terms on the right-hand side of the resulting inequality are the ones involving "boxed constants terms" that is, terms arising from $\boxed{\mathbf{I})}_{(\leq 24)}$ and $\boxed{\mathbf{V})}_{(\leq 24)}$ (see Def 20.4). We now derive a suitable bound for these difficult terms. To handle the term corresponding to $\boxed{\mathbf{I})}_{(\leq 24)}$ (see definition (20.2a)), we have to bound

$$\sup_{(\hat{t},\hat{u})\in[0,t]\times[0,u]}\left\{\iota(\hat{t},\hat{u})\boxed{\mathbf{I})}_{(\leq 24)}(\hat{t},\hat{u})\right\}. \tag{20.162}$$

Using an argument similar to the one we used to prove inequality (20.146), and in particular using the crucially important integral estimate (13.143) *twice* since the right-hand side of (20.2a) involves two time integrations, we deduce that (20.162) is

$$\leq (1 + C\sqrt{\varepsilon})\boxed{9} \times \frac{1}{17.5} \times \frac{1}{8.75} q(t,u). \tag{20.163}$$

A similar argument based on the same crucially important estimate (13.143) leads to the following bound for the term involving $\boxed{\mathbf{V})}_{(\leq 24)}$ (see definition (20.2e)):

$$\sup_{(\hat{t},\hat{u})\in[0,t]\times[0,u]} \left\{ \iota(\hat{t},\hat{u}) \boxed{\mathbf{V})}_{(\leq 24)}(\hat{t},\hat{u}) \right\} \leq \boxed{9} \times \frac{1}{17.5}. \tag{20.164}$$

As before, the important point is that the constants $\boxed{9} \times \frac{1}{17.5} \times \frac{1}{8.75}$ and $\boxed{9} \times \frac{1}{17.5}$ from the estimates (20.163) and (20.164) verify $9 \times \frac{1}{17.5} \times \frac{1}{8.75} + \boxed{9} \times \frac{1}{17.5} < .58 < 1$. As we will see, the remaining terms on the right-hand side of (20.6a) can be suitably bounded by choosing P and A_* to be sufficiently large.

In our analysis of the remaining terms, in order to connect the quantity $\widetilde{\mathbb{Q}}_{(\leq 24)}(t,u)$ to the quantity $q(t,u)$ and to account for the $\iota_3^4(t,u)$ discrepancy between $\iota(t,u)$ and $\widetilde{\iota}(t,u)$, we use the following estimate, which follows from the definitions of the quantities involved, the approximate monotonicity of the integrating factors, and (20.143):

(20.165)
$$\begin{aligned}
\iota^{-1}(t,u) & \sup_{(t',u')\in[0,t]\times[0,u]} \left\{ \iota_1^{-2P}(t',u')\iota_3^{-4}(t',u')\widetilde{\mathbb{Q}}_{(\leq 24)}(t',u') \right\} \\
& \leq C\iota_1^{-2P}(t,u) \sup_{(t',u')\in[0,t]\times[0,u]} \\
& \quad \left\{ \iota_1^{-2P}(t',u')\iota_2^{-2P}(t',u')\iota_3^{-(2A_*+4)}(t',u')\upmu_\star^{17.5}(t',u')\widetilde{\mathbb{Q}}_{(\leq 24)}(t',u') \right\} \\
& \leq C\iota_1^{-2P}(t,u) \left\{ \mathring{\epsilon}^2 + \varepsilon^3 + q(t,u) \right\}.
\end{aligned}$$

Furthermore,

(20.166) inequality (20.165) also holds with the factors ι_1^{-2P} on the left and right replaced with any of $1, \iota_2^{-2P}, \iota_3^{-2A_*}$, or $\upmu_\star^{17.5}$,

(20.167) and the same estimates hold with $\widetilde{\mathbb{K}}_{(\leq 24)}$ in place of $\widetilde{\mathbb{Q}}_{(\leq 24)}$.

We now claim that the following estimates hold for the remaining terms on the last three lines of the right-hand side of (20.6a) (see Def. 20.4 and Def. 20.5):

$$\sup_{(\hat{t},\hat{u})\in[0,t]\times[0,u]} \left\{\iota^{-1}(\hat{t},\hat{u})\mathbf{II})_{(\leq 24)}(\hat{t},\hat{u})\right\} \leq C\varepsilon q(t,u), \tag{20.168}$$

$$\sup_{(\hat{t},\hat{u})\in[0,t]\times[0,u]} \left\{\iota^{-1}(\hat{t},\hat{u})\mathbf{III})_{(\leq 24)}(\hat{t},\hat{u})\right\} \leq C\left\{\mathring{\epsilon}^2+\varepsilon^3+\varepsilon q(t,u)\right\}, \tag{20.169}$$

$$\sup_{(\hat{t},\hat{u})\in[0,t]\times[0,u]} \left\{\iota^{-1}(\hat{t},\hat{u})\mathbf{IV})_{(\leq 24)}(\hat{t},\hat{u})\right\} \leq \frac{C}{A_*}q(t,u), \tag{20.170}$$

$$\sup_{(\hat{t},\hat{u})\in[0,t]\times[0,u]} \left\{\iota^{-1}(\hat{t},\hat{u})\mathbf{VI})_{(\leq 24)}(\hat{t},\hat{u})\right\} \leq \frac{C}{A_*}q(t,u), \tag{20.171}$$

$$\sup_{(\hat{t},\hat{u})\in[0,t]\times[0,u]} \left\{\iota^{-1}(\hat{t},\hat{u})\mathbf{VII})_{(\leq 24)}(\hat{t},\hat{u})\right\} \leq C\varepsilon q(t,u), \tag{20.172}$$

$$\sup_{(\hat{t},\hat{u})\in[0,t]\times[0,u]} \left\{\iota^{-1}(\hat{t},\hat{u})\mathbf{VIII})_{(\leq 24)}(\hat{t},\hat{u})\right\} \leq C\left\{\mathring{\epsilon}^2+\varepsilon^3+\varepsilon q(t,u)\right\}, \tag{20.173}$$

$$\sup_{(\hat{t},\hat{u})\in[0,t]\times[0,u]} \left\{\iota^{-1}(\hat{t},\hat{u})\mathbf{IX})_{(\leq 24)}(\hat{t},\hat{u})\right\} \leq C\varepsilon q(t,u), \tag{20.174}$$

$$\sup_{(\hat{t},\hat{u})\in[0,t]\times[0,u]} \left\{\iota^{-1}(\hat{t},\hat{u})\mathbf{X})_{(\leq 24)}(\hat{t},\hat{u})\right\} \leq C\left\{\mathring{\epsilon}^2+\varepsilon^3+\varepsilon q(t,u)\right\}, \tag{20.175}$$

and

(20.176)
$$(1+\varsigma^{-1})\sup_{(\hat{t},\hat{u})\in[0,t]\times[0,u]} \left\{\iota^{-1}(\hat{t},\hat{u})\mathbf{i})_{(\leq 24)}(\hat{t},\hat{u})\right\} \leq \frac{C}{P}(1+\varsigma^{-1})q(t,u),$$

$$\sup_{(\hat{t},\hat{u})\in[0,t]\times[0,u]} \left\{\iota^{-1}(\hat{t},\hat{u})\mathbf{ii})_{(\leq 24)}(\hat{t},\hat{u})\right\} \leq C\left\{\mathring{\epsilon}^2+\varepsilon^3\right\}+C\varepsilon^{1/2}q(t,u), \tag{20.177}$$

(20.178)
$$(1+\varsigma^{-1})\sup_{(\hat{t},\hat{u})\in[0,t]\times[0,u]} \left\{\iota^{-1}(\hat{t},\hat{u})\mathbf{iii})_{(\leq 24)}(\hat{t},\hat{u})\right\} \leq \frac{C}{P}(1+\varsigma^{-1})\left\{\mathring{\epsilon}^2+\varepsilon^3+q(t,u)\right\},$$

(20.179)
$$\sup_{(\hat{t},\hat{u})\in[0,t]\times[0,u]} \left\{\iota^{-1}(\hat{t},\hat{u})\mathbf{iv})_{(\leq 24)}(\hat{t},\hat{u})\right\} \leq \frac{C}{P}q(t,u),$$

(20.180)
$$(1+\varsigma^{-1})\sup_{(\hat{t},\hat{u})\in[0,t]\times[0,u]} \left\{\iota^{-1}(\hat{t},\hat{u})\mathbf{v})_{(\leq 24)}(\hat{t},\hat{u})\right\} \leq \frac{C}{P}(1+\varsigma^{-1})\left\{\mathring{\epsilon}^2+\varepsilon^3+q(t,u)\right\},$$

(20.181)
$$(\varepsilon+\varsigma)\sup_{(\hat{t},\hat{u})\in[0,t]\times[0,u]} \left\{\iota^{-1}(\hat{t},\hat{u})\mathbf{vi})_{(\leq 24)}(\hat{t},\hat{u})\right\} \leq C(\varepsilon+\varsigma)\left\{\mathring{\epsilon}^2+\varepsilon^3+q(t,u)\right\}.$$

We now explain how to derive the bounds (20.168)-(20.175). The bound (20.170) (see definition (20.2d)) can be proved by using an argument similar to the one used to prove (20.146). More precisely, we handle the inner time integral on the right-hand side of (20.2d) with the key integral estimate (13.143), and then we handle the outer time integral on the right-hand side of (20.2d) with the help of inequality (20.138), which provides the smallness factor $\frac{1}{A_*}$ on the right-hand side of (20.170).

To obtain the bound (20.169) (see definition (20.2c)), we first use the approximate monotonicity of the integrating factors to bound the term in braces on the left-hand side of (20.169) as follows:

$$\begin{aligned}(20.182)\quad &\leq C\varepsilon\iota^{-1}(\hat{t},\hat{u})\int_{t'=0}^{\hat{t}}\frac{1}{(1+t')^{1+a}\mu_\star(t',\hat{u})}\mathbb{Q}_{(\leq N)}^{1/2}(t',u)\\ &\qquad\times\int_{s=0}^{t'}\frac{1}{(1+s)}\frac{1}{\mu_\star(s,\hat{u})}\widetilde{\mathbb{Q}}_{(\leq N)}^{1/2}(s,\hat{u})\,ds\,dt'\\ &\leq C\varepsilon\iota^{-1}(\hat{t},\hat{u})\left\{\sup_{(t',u')\in[0,\hat{t}]\times[0,\hat{u}]}\mu_\star^{8.75}(t',u')\mathbb{Q}_{(\leq 24)}^{1/2}(t',u')\right\}\\ &\qquad\times\left\{\sup_{(t',u')\in[0,\hat{t}]\times[0,\hat{u}]}\iota_3^{-2}(t',u')\mu_\star^{8.75}(t',u')\widetilde{\mathbb{Q}}_{(\leq 24)}^{1/2}(t',u')\right\}\\ &\qquad\times\int_{t'=0}^{\hat{t}}\frac{1}{(1+t')^{1+a}\mu_\star^{9.75}(t',\hat{u})}\iota_3^2(t',\hat{u})\int_{s=0}^{t'}\frac{1}{(1+s)}\frac{1}{\mu_\star^{9.75}(s,\hat{u})}\,ds\,dt'.\end{aligned}$$

We bound the inner time integral on the right-hand side of (20.182) with the estimate (13.147) and then the outer time integral with the estimate (13.146), which yields that the right-hand side of (20.182) is

$$\begin{aligned}(20.183)\quad &\leq C\varepsilon\mu_\star^{-17.5}(\hat{t},\hat{u})\iota^{-1}(\hat{t},\hat{u})\left\{\sup_{(t',u')\in[0,\hat{t}]\times[0,\hat{u}]}\mu_\star^{8.75}(t',u')\mathbb{Q}_{(\leq 24)}^{1/2}(t',u')\right\}\\ &\qquad\times\left\{\sup_{(t',u')\in[0,\hat{t}]\times[0,\hat{u}]}\iota_3^{-2}(t',u')\mu_\star^{8.75}(t',u')\widetilde{\mathbb{Q}}_{(\leq 24)}^{1/2}(t',u')\right\}.\end{aligned}$$

Using the monotonicity of ι_1, ι_2, and ι_3 in both of their arguments and the bound (20.143), we deduce that the right-hand side of (20.183) is

$$(20.184)\qquad \leq C\varepsilon q^{1/2}(\hat{t},\hat{u})\widetilde{q}^{1/2}(\hat{t},\hat{u})\leq C\left\{\mathring{\epsilon}^2+\varepsilon^3+\varepsilon q(t,u)\right\}.$$

The desired estimate (20.169) now follows from taking $\sup_{(\hat{t},\hat{u})\in[0,t]\times[0,u]}$ in inequality (20.184).

The bound (20.168) (see definition (20.2b)) can be proved by using arguments similar to the ones we used to prove (20.169). The same remarks apply to the bounds (20.174) and (20.175) (see, respectively, definitions (20.2i) and (20.2j)).

The bound (20.171) (see definition (20.2f)) can be proved by using the same argument we used to prove (20.148).

The bound (20.172) (see definition (20.2g)) can similarly be proved by using inequality (13.146) with $A=0$ and $B=17.5+1=18.5$. The bound (20.173) (see definition (20.2h)) can similarly be proved with the help of the estimate (20.143).

We now explain how to derive the easier estimates (20.176)-(20.181). The bound (20.176) (see definition (20.3b)) can be proved by using an argument similar to the one we used to prove (20.146), but with ι_2^{-2P} in place of $\mu_\star^{-17.5}$ and the estimate (20.137) in place of the estimate (13.143).

To obtain the bound (20.177) (see definition (20.3c)), we use inequality (20.166) in the case of the constant function 1 to bound the term in braces on the left-hand side of (20.177) as follows:

$$
\begin{aligned}
(20.185)\qquad & C\varepsilon^{1/2}\iota^{-1}(\hat{t},\hat{u}) \int_{t'=0}^{\hat{t}} \frac{\ln(e+t')}{(1+t')^2\sqrt{\ln(e+\hat{t})-\ln(e+t')}}\widetilde{\mathbb{Q}}_{(\leq 24)}(t',\hat{u})\,dt' \\
& \leq C\varepsilon^{1/2}\left\{\mathring{\epsilon}^2+\varepsilon^3+q(t,u)\right\}\int_{t'=0}^{\hat{t}} \frac{\ln^5(e+t')}{(1+t')^2\sqrt{\ln(e+t)-\ln(e+t')}}\,dt'.
\end{aligned}
$$

The desired estimate (20.177) now follows from (20.185) and Lemma 20.35.

To obtain the bound (20.178) (see definition (20.3d)), we use inequality (20.166) in the case of the function ι_2^{-2P} to bound the term in braces on the left-hand side of (20.178) as follows:

(20.186)

$$
\begin{aligned}
& C(1+\varsigma^{-1})\iota^{-1}(\hat{t},\hat{u}) \int_{t'=0}^{\hat{t}} \frac{1}{(1+t')^{3/2}}\widetilde{\mathbb{Q}}_{(\leq 24)}(t',\hat{u})\,dt' \\
& \leq C(1+\varsigma^{-1})\left\{\mathring{\epsilon}^2+\varepsilon^3+q(t,u)\right\}\iota_2^{-2P}(\hat{t},\hat{u})\int_{t'=0}^{\hat{t}} \frac{\iota_3^4(t',u)}{(1+t')^{3/2}}\iota_2^{2P}(t',u)\,dt'.
\end{aligned}
$$

The desired bound (20.178) now follows from (20.186) and inequality (20.137).

The bound (20.179) (see definition (20.3e)) can be proved by using an argument similar to the one we used to prove (20.176), but with ι_1^{-2P} in place of ι_2^{-2P} and the estimate (20.136) in place of the estimate (20.137).

To obtain the bound (20.181) (see definition (20.3g)), we first use inequality (20.167) in the case of the constant function 1 to deduce that

$$
\begin{aligned}
(20.187)\qquad \iota^{-1}(t',\hat{u})\frac{\widetilde{\mathbb{K}}_{(\leq 24)}(t',\hat{u})}{(1+t')^{1/2}} &= \iota_3^4(t',\hat{u})\iota^{-1}(t',\hat{u})\frac{\left\{\iota_3^{-4}(t',\hat{u})\widetilde{\mathbb{K}}_{(\leq 24)}(t',\hat{u})\right\}}{(1+t')^{1/2}} \\
&\leq C\frac{\iota_3^4(t',\hat{u})}{(1+t')^{1/2}}\left(\mathring{\epsilon}^2+\varepsilon^3+q(t',\hat{u})\right).
\end{aligned}
$$

Hence, the term in braces on the left-hand side of (20.181) can be bounded as follows:

(20.188)

$$
\begin{aligned}
\leq C\iota^{-1}(\hat{t},\hat{u})\sup_{t'\in[0,\hat{t}]}\frac{\widetilde{\mathbb{K}}_{(\leq 24)}(t',\hat{u})}{(1+t')^{1/2}} &\leq C\sup_{t'\in[0,\hat{t}]}\left\{\iota^{-1}(t',\hat{u})\frac{\widetilde{\mathbb{K}}_{(\leq 24)}(t',\hat{u})}{(1+t')^{1/2}}\right\} \\
&\leq C\sup_{t'\in[0,\hat{t}]}\left\{\frac{\iota_3^4(t',u)}{(1+t')^{1/2}}\left(\mathring{\epsilon}^2+\varepsilon^3+q(t',u)\right)\right\} \\
&\leq C\left\{\mathring{\epsilon}^2+\varepsilon^3+q(t,u)\right\}.
\end{aligned}
$$

The desired bound (20.181) now easily follows from inequality (20.188).

To obtain the bound (20.180) (see definition (20.3f)), we use an argument similar to the one used to prove (20.181), but we use inequality (20.166) in the case of the constant function 1 in place of inequality (20.167).

In order to complete the proof of (20.161), it remains for us to bound the terms arising from the terms on the first two lines of the right-hand side of (20.6a). The terms generated by the three terms on the first line are respectively bounded from above by $\leq C\mathring{\epsilon}^2$, $C\varepsilon^3$, and $C\varepsilon q(t,u)$. The first of these bounds follows from Lemma 14.8, while the next two are easy to derive. The terms generated by the two terms on the second line of the right-hand side of (20.6a) are bounded by $\leq C\varepsilon^3$. To see that this is the case, we argue as in our proof of (20.159).

Combining all of the above bounds, we arrive at the following analog of (20.160):

(20.189)
$$q(t,u) \leq C(1+\varsigma^{-1}+\varsigma)\left\{\mathring{\epsilon}^2+\varepsilon^3\right\} + \left\{\frac{9}{17.5\times 8.75}+\frac{9}{17.5}+\frac{C}{A_*}+\frac{C}{P}(1+\varsigma^{-1})+C\varsigma+C\varepsilon^{1/2}\right\}q(t,u).$$

The desired bound (20.161) thus follows from (20.189) if we first choose and *fix* ς to be sufficiently small, then choose A_* and P to be sufficiently large (at least as large as they were chosen to be in the part of the proof following inequality (20.160)), and finally choose ε to be sufficiently small.

Proof of the estimates for the just-below-top-order quantities. We now prove the desired estimates for the below-top-order quantities $\mathbb{Q}_{(\leq N)}(t,u)$, $\widetilde{\mathbb{Q}}_{(\leq N)}(t,u)$, and $\widetilde{\mathbb{K}}_{(\leq N)}(t,u)$ for $0 \leq N \leq 23$. These estimates are much easier to prove than the top-order estimates. We give complete details for the estimates of $\mathbb{Q}_{(\leq 23)}(t,u)$, $\widetilde{\mathbb{Q}}_{(\leq 23)}(t,u)$, and $\widetilde{\mathbb{K}}_{(\leq 23)}(t,u)$. We then indicate the minor changes in the proof needed to derive the desired estimates for $\mathbb{Q}_{(\leq N)}(t,u)$, $\widetilde{\mathbb{Q}}_{(\leq N)}(t,u)$, and $\widetilde{\mathbb{K}}_{(\leq N)}(t,u)$ when $0 \leq N \leq 22$.

To begin, in place of (20.131) and (20.132), we define

$$\iota(t,u) := \iota_1^{2P}(t,u)\iota_2^{2P}(t,u)\mu_\star^{-15.5}(t,u), \tag{20.190}$$

$$\widetilde{\iota}(t,u) := \iota_1^{2P}(t,u)\iota_2^{2P}(t,u)\iota_3^4(t,u)\mu_\star^{-15.5}(t,u), \tag{20.191}$$

where ι_1, ι_2, and ι_3 are defined in (20.128)-(20.130). Notice in particular that the power of $\mu_\star^{-1}$ in (20.190) and (20.191) has been reduced by 2 and that there is no factor $\iota_3(t,u)$ in (20.190). Next, in place of (20.139) and (20.140), we define (with ι and $\widetilde{\iota}$ defined in (20.190)-(20.191))

$$q(t,u) := \sup_{(\hat{t},\hat{u})\in[0,t]\times[0,u]} \iota^{-1}(\hat{t},\hat{u})\mathbb{Q}_{(\leq 23)}(\hat{t},\hat{u}), \tag{20.192}$$

$$\widetilde{q}(t,u) := \sup_{(\hat{t},\hat{u})\in[0,t]\times[0,u]} \widetilde{\iota}^{-1}(\hat{t},\hat{u}) \max\left\{\widetilde{\mathbb{Q}}_{(\leq 23)}(\hat{t},\hat{u}), \widetilde{\mathbb{K}}_{(\leq 23)}(\hat{t},\hat{u})\right\}. \tag{20.193}$$

Our goal is to prove the following analogs of (20.141) and (20.142):

$$q(t,u) \leq C\left\{\mathring{\epsilon}^2+\varepsilon^3\right\}, \tag{20.194}$$

$$\widetilde{q}(t,u) \leq C\left\{\mathring{\epsilon}^2+\varepsilon^3\right\}. \tag{20.195}$$

The desired estimates (20.8c) and (20.8d) for $\mathbb{Q}_{(\leq 23)}$, $\widetilde{\mathbb{Q}}_{(\leq 23)}$, and $\widetilde{\mathbb{K}}_{(\leq 23)}$ easily follow from (20.194)-(20.193) and the definitions of the quantities involved.

In order to prove (20.194) and (20.193), we first prove the following analog of (20.143):

$$\widetilde{q}(t,u) \leq C\left\{\mathring{\epsilon}^2+\varepsilon^3+q(t,u)\right\}. \tag{20.196}$$

In order to prove (20.196), we will prove the following analog of (20.144):

$$\widetilde{q}(t,u) \leq C\left\{\mathring{\epsilon}^2+\varepsilon^3+q(t,u)\right\}+\alpha\widetilde{q}(t,u), \tag{20.197}$$

where $0<\alpha<1$ and $C>0$ are constants.

We now claim that the following estimates hold for the terms on the second through fourth lines of the right-hand side of (20.7b) in the case $N = 23$ (see Def. 20.7):

$$\sup_{(\hat{t},\hat{u})\in[0,t]\times[0,u]} \left\{\widetilde{\iota}^{-1}(\hat{t},\hat{u})\widetilde{\mathbf{0}}_{(\leq 24)}(\hat{t},\hat{u})\right\} \leq C\varepsilon^3, \tag{20.198}$$

$$(1+\widetilde{\varsigma}^{-1}) \sup_{(\hat{t},\hat{u})\in[0,t]\times[0,u]} \left\{\widetilde{\iota}^{-1}(\hat{t},\hat{u})\widetilde{\mathbf{i}}_{(\leq 23)}(\hat{t},\hat{u})\right\} \leq C(1+\widetilde{\varsigma}^{-1})q(t,u), \tag{20.199}$$

$$\sup_{(\hat{t},\hat{u})\in[0,t]\times[0,u]} \left\{\widetilde{\iota}^{-1}(\hat{t},\hat{u})\widetilde{\mathbf{ii}}_{(\leq 23)}(\hat{t},\hat{u})\right\} \leq \frac{1}{2}\widetilde{q}(t,u), \tag{20.200}$$

$$(1+\widetilde{\varsigma}^{-1}) \sup_{(\hat{t},\hat{u})\in[0,t]\times[0,u]} \left\{\widetilde{\iota}^{-1}(\hat{t},\hat{u})\widetilde{\mathbf{iii}}_{(\leq 23)}(\hat{t},\hat{u})\right\} \leq \frac{C}{P}(1+\widetilde{\varsigma}^{-1})\widetilde{q}(t,u), \tag{20.201}$$

$$(1+\widetilde{\varsigma}^{-1}) \sup_{(\hat{t},\hat{u})\in[0,t]\times[0,u]} \left\{\widetilde{\iota}^{-1}(\hat{t},\hat{u})\widetilde{\mathbf{iv}}_{(\leq 23)}(\hat{t},\hat{u})\right\} \leq \frac{C}{P}(1+\widetilde{\varsigma}^{-1})\widetilde{q}(t,u), \tag{20.202}$$

$$(\varepsilon+\widetilde{\varsigma}) \sup_{(\hat{t},\hat{u})\in[0,t]\times[0,u]} \left\{\widetilde{\iota}^{-1}(\hat{t},\hat{u})\widetilde{\mathbf{v}}_{(\leq 23)}(\hat{t},\hat{u})\right\} \leq C(\varepsilon+\widetilde{\varsigma})\widetilde{q}(t,u). \tag{20.203}$$

The bounds (20.199)-(20.202) can be proved by using arguments similar to the ones we used to prove (20.154)-(20.158). We note that the factor 1/2 on the right-hand side of (20.200) arises from the fact that the constant on the right-hand side of (20.5c) is precisely 2.

To obtain (20.198), we set $N = 23$ in (20.5a), insert the already proven estimates $\mathbb{Q}^{1/2}_{(\leq 24)}(t,u) \leq C\left\{\mathring{\epsilon} + \varepsilon^{3/2}\right\}\ln^{A_*}(e+t)\upmu_\star^{-8.75}(t,u)$ and $\widetilde{\mathbb{Q}}^{1/2}_{(\leq 24)}(t,u) \leq C\left\{\mathring{\epsilon} + \varepsilon^{3/2}\right\}\ln^{A_*+2}(e+t)\upmu_\star^{-8.75}(t,u)$, which follow from (20.141)-(20.142) (and the constant A_* has already been chosen above), and use the integral estimates (13.146) and (13.147) to deduce that

$$\begin{aligned}
\widetilde{\mathbf{0}}_{(\leq 24)}(\hat{t},\hat{u}) &\leq C\varepsilon^3 \int_{t'=0}^{\hat{t}} \frac{\ln^{2A_*-4}(e+t')}{(1+t')^2}\left(\int_{s=0}^{t'} \frac{1}{(1+s)\upmu_\star^{8.75+.5}(s,u)}\,ds\right)^2 dt' \\
&\leq C\varepsilon^3 \int_{t'=0}^{\hat{t}} \frac{1}{(1+t')^{3/2}}\left(\frac{1}{\upmu_\star^{8.25}(t',u)}\right)^2 dt' \leq C\varepsilon^3\upmu_\star^{-15.5}(\hat{t},\hat{u}).
\end{aligned} \tag{20.204}$$

The desired estimate (20.198) now follows from (20.204) and the definitions of the quantities involved.

In order to complete the proof of (20.197), it remains for us to bound the terms arising from the terms on the first line of the right-hand side of (20.7b). Arguing as in our proof of the bounds for the terms on the first line of the right-hand side of (20.6b), we deduce that the two corresponding terms are respectively bounded by $\leq C\mathring{\epsilon}^2$ and $C\varepsilon^3$.

Combining all of the above bounds, we arrive at the following analog of (20.160):

$$\widetilde{q}(t,u) \leq C\left\{\mathring{\epsilon}^2+\varepsilon^3\right\} + C\left\{1+\widetilde{\varsigma}^{-1}\right\}q(t,u) + \left\{\frac{1}{2}+\frac{C}{P}(1+\widetilde{\varsigma}^{-1})+C\widetilde{\varsigma}+C\varepsilon\right\}\widetilde{q}(t,u), \tag{20.205}$$

and the desired bound (20.197) thus follows if first $\widetilde{\varsigma}$ is chosen to be sufficiently small and *fixed*, then P is chosen to be sufficiently large, and finally ε is chosen to be sufficiently small.

We now use inequality (20.196) to help us derive the estimate (20.194). To prove (20.194), we will argue as in our proof of (20.196) to show that there exist constants $0 < \beta < 1$ and $C > 0$ such that

$$q(t,u) \leq C\left\{\mathring{\epsilon}^2 + \varepsilon^3\right\} + \beta q(t,u). \tag{20.206}$$

The desired bound (20.194) follows easily once we have shown (20.206).

We claim that the following estimates hold for the terms on the second through fourth lines of the right-hand side of (20.7a) in the case $N = 23$ (see Def. 20.5):

$$\sup_{(\hat{t},\hat{u})\in[0,t]\times[0,u]} \left\{\iota^{-1}(\hat{t},\hat{u})\mathbf{0})_{(\leq 24)}(\hat{t},\hat{u})\right\} \leq C\varepsilon^3, \tag{20.207}$$

$$(1+\varsigma^{-1}) \sup_{(\hat{t},\hat{u})\in[0,t]\times[0,u]} \left\{\iota^{-1}(\hat{t},\hat{u})\mathbf{i})_{(\leq 23)}(\hat{t},\hat{u})\right\} \leq \frac{C}{P}(1+\varsigma^{-1})q(t,u), \tag{20.208}$$

$$\sup_{(\hat{t},\hat{u})\in[0,t]\times[0,u]} \left\{\iota^{-1}(\hat{t},\hat{u})\mathbf{ii})_{(\leq 23)}(\hat{t},\hat{u})\right\} \leq C\left\{\mathring{\epsilon}^2 + \varepsilon^3 + \varepsilon^{1/2}q(t,u)\right\}, \tag{20.209}$$

$$(1+\varsigma^{-1}) \sup_{(\hat{t},\hat{u})\in[0,t]\times[0,u]} \left\{\iota^{-1}(\hat{t},\hat{u})\mathbf{iii})_{(\leq 23)}(\hat{t},\hat{u})\right\} \leq \frac{C}{P}(1+\varsigma^{-1})\left\{\mathring{\epsilon}^2 + \varepsilon^3 + q(t,u)\right\}, \tag{20.210}$$

$$\sup_{(\hat{t},\hat{u})\in[0,t]\times[0,u]} \left\{\iota^{-1}(\hat{t},\hat{u})\mathbf{iv})_{(\leq 23)}(\hat{t},\hat{u})\right\} \leq \frac{C}{P}q(t,u), \tag{20.211}$$

$$(1+\varsigma^{-1}) \sup_{(\hat{t},\hat{u})\in[0,t]\times[0,u]} \left\{\iota^{-1}(\hat{t},\hat{u})\mathbf{v})_{(\leq 23)}(\hat{t},\hat{u})\right\} \leq \frac{C}{P}(1+\varsigma^{-1})\left\{\mathring{\epsilon}^2 + \varepsilon^3 + q(t,u)\right\}, \tag{20.212}$$

$$(\varepsilon+\varsigma) \sup_{(\hat{t},\hat{u})\in[0,t]\times[0,u]} \left\{\iota^{-1}(\hat{t},\hat{u})\mathbf{vi})_{(\leq 23)}(\hat{t},\hat{u})\right\} \leq C(\varepsilon+\varsigma)\left\{\mathring{\epsilon}^2 + \varepsilon^3 + q(t,u)\right\}. \tag{20.213}$$

The estimate (20.207) can be proved by using an argument similar to the one we used to prove (20.198) with one small change: because of the factor $\mathbb{Q}_{(\leq N)}^{1/2}$ on the right-hand side of (20.3a), we also need to use the bootstrap assumption (20.1c) for $\mathbb{Q}_{(\leq 23)}^{1/2}(t,u)$ in addition to the already proven estimates

$$\mathbb{Q}_{(\leq 24)}^{1/2}(t,u) \leq C\left\{\mathring{\epsilon} + \varepsilon^{3/2}\right\}\ln^{A_*}(e+t)\mu_\star^{-8.75}(t,u)$$

and

$$\widetilde{\mathbb{Q}}_{(\leq 24)}^{1/2}(t,u) \leq C\left\{\mathring{\epsilon} + \varepsilon^{3/2}\right\}\ln^{A_*+2}(e+t)\mu_\star^{-8.75}(t,u)$$

mentioned above. The estimates (20.208)-(20.213) can be proved by using arguments similar to the ones we used to prove (20.176)-(20.181).

In order to complete the proof of (20.206), it remains for us to bound the terms arising from the terms on the first line of the right-hand side of (20.7a). Arguing as in our proof of the bounds for the terms on the first line of the right-hand side of (20.6a), we deduce that the two corresponding terms are respectively bounded by $\leq C\mathring{\epsilon}^2$ and $C\varepsilon^3$.

Combining all of the above bounds, we arrive at the following analog of (20.189):

$$q(t,u) \leq C(1+\varsigma^{-1}+\varsigma)\left\{\mathring{\epsilon}^2+\varepsilon^3\right\} + \left\{\frac{C}{P}(1+\varsigma^{-1}) + C\varsigma + C\varepsilon\right\} q(t,u), \tag{20.214}$$

and the desired bound (20.206) thus follows if first ς is chosen to be sufficiently small and *fixed*, then P is chosen to be sufficiently large, and finally ε is chosen to be sufficiently small.

Further descent. We now explain how to inductively derive the desired estimates for the remaining lower-order quantities $\mathbb{Q}_{(\leq N)}(t,u)$, $\widetilde{\mathbb{Q}}_{(\leq N)}(t,u)$, $\widetilde{\mathbb{K}}_{(\leq N)}(t,u)$, $0 \leq N \leq 22$ by descending. At each step in the descent, we define the approximate integrating factors (20.190)-(20.191) with one key change: at each step, we reduce the power of $\mu_\star^{-1}$ by two on the right-hand sides of (20.190)-(20.191). Moreover, starting at the case $N = 15$, the factor involving $\mu_\star^{-1}$ is completely absent from the right-hand side of (20.190)-(20.191). At each step, we prove the estimates (20.198)-(20.203) and (20.207)-(20.213) (with 24 replaced by $N+1$ and 23 replaced by N) by using essentially the same arguments that we used in the case $N = 23$. One small change is needed starting at $N = 15$; $N = 15$ is the first instance in which the $\mu_\star^{-1}$ degeneracy is completely absent. Specifically, when proving the desired estimates for $N \leq 15$, we use inequalities (13.151) and (13.152) to help prove the corresponding analogs of the estimates (20.198) and (20.207); inequalities (13.151) and (13.152) are the ones that break the $\mu_\star^{-1}$ degeneracy. This completes the proof of Lemma 20.10.

□

CHAPTER 21

Local Well-Posedness and Continuation Criteria

In this chapter, we sketch the proof of a proposition that provides local well-posedness and related continuation criteria for the covariant wave equation $\square_{g(\Psi)}\Psi = 0$. We state some results in terms of the rectangular coordinates (t, x^1, x^2, x^3) and others in terms of the geometric coordinates $(t, u, \vartheta^1, \vartheta^2)$. The rectangular coordinates are a natural coordinate system for showing that the solution exists for short times and thus for initiating the bootstrap argument that we use in the proof of the sharp classical lifespan theorem (Theorem 22.1). On the other hand, the geometric coordinates are the ones we have used throughout the monograph to derive sharp estimates.

21.1. Local well-posedness and continuation criteria

We now provide the proposition. For convenience, we assume the amount of regularity on the data that we use in proving our sharp classical lifespan theorem; in view of Prop. 1.6, we see that this assumption is highly nonoptimal.

PROPOSITION 21.1 (**Local well-posedness and continuation criteria**). *Let $N = 24$, let $0 < U_0 < 1$ be a constant, and let $\Sigma_0^{U_0} = \cup_{u\in[0,U_0]} S_{0,u} \subset \mathbb{R}^3$ be the annular region foliated by the level sets $S_{0,u}$ of the function $u = 1 - r$ defined on $\Sigma_0^{U_0}$. Let $(\mathring{\Psi} := \Psi|_{\Sigma_0}, \mathring{\Psi}_0 := \partial_t\Psi|_{\Sigma_0})$ be initial data for the covariant wave equation $\square_{g(\Psi)}\Psi = 0$ (that is, equation* (2.1a)*) that are compactly supported in the unit Euclidean ball Σ_0^1 (and hence have vanishing trace on the outer sphere $S_{0,0}$). Assume that the metric $g(\Psi)$ verifies* (2.4) *and $(g^{-1})^{00} \equiv -1$. Let $\mathring{\epsilon} := \|\mathring{\Psi}\|_{H_e^{N+1}(\Sigma_0^1)} + \|\mathring{\Psi}_0\|_{H_e^{N}(\Sigma_0^1)}$ denote the size of the data*[1] *as defined in Def.* 12.3*, and let $\mathcal{H}$ be the set of real numbers*[2] *b such that the following conditions hold:*

- *The rectangular components $g_{\mu\nu}(\cdot)$, $(\mu, \nu = 0, 1, 2, 3)$, are smooth on a neighborhood of b.*
- *$g_{00}(b) < 0$.*
- *The eigenvalues of the 3×3 matrix $\underline{g}_{ij}(b)$ (see Def.* 3.9*), $(i, j = 1, 2, 3)$, are positive.*

Part I): Local well-posedness.

<u>***Local well-posedness.***</u> *Assume that there is a compact subset $\mathfrak{K} \subset$ interior$(\mathcal{H})$ such that $\mathring{\Psi}(\Sigma_0^{U_0}) \subset \mathfrak{K}$. Then these data launch a unique classical solution Ψ to the*

[1]Because we are studying only the influence of the nontrivial portion of the data belonging to the subset $\Sigma_0^{U_0}$ of Σ_0^1, we could replace the data norms $\|\cdot\|_{H_e^{\cdot}(\Sigma_0^1)}$ with $\|\cdot\|_{H_e^{\cdot}(\Sigma_0^{U_0})}$ without altering any of the conclusions of the proposition.

[2]$\mathcal{H}$ can be viewed as a set of Ψ for which the metric $g(\Psi)$ is Lorentzian and for which the hypersurfaces Σ_t are spacelike.

equation $\square_{g(\Psi)}\Psi = 0$ *and a unique outgoing eikonal function* u *that is a classical solution to* $(g^{-1})^{\alpha\beta}(\Psi)\partial_\alpha u \partial_\beta u = 0$, *that takes on the initial condition* $1-r$ *along* $\Sigma_0^{U_0}$, *and that verifies* $0 < \upmu < \infty$ *along* $\Sigma_0^{U_0}$, *where* $\upmu := -\frac{1}{(g^{-1})^{\alpha\beta}(\Psi)\partial_\alpha t \partial_\beta u} > 0$ *is the inverse foliation density from Def.* 3.15. *The solution exists on a nontrivial spacetime region of the form* $\mathcal{M}_{T_{(Local)},U_0}$ *(see definition* (3.6e)*) for some* $T_{(Local)} > 0$. *Moreover, there exists a compact subset* $\mathfrak{K}'$ *such that* $\mathfrak{K} \subset \mathfrak{K}' \subset$ *interior*$(\mathcal{H})$ *and such that* $\Psi(\mathcal{M}_{T_{(Local)},U_0}) \subset \mathfrak{K}'$. *Furthermore, on* $\mathcal{M}_{T_{(Local)},U_0}$, *each* $S_{t,u}$ *is an embedded*[3] *two-dimensional sphere,* $\sum_{a=1}^3 |\partial_a u| > 0$, *the one-form with rectangular components* $(\partial_1 u, \partial_2 u, \partial_3 u)$ *on* $\Sigma_0^{U_0}$ *is inward-pointing relative to* $S_{t,u}$, $(g^{-1})^{\alpha\beta}(\Psi)\partial_\alpha t \partial_\beta u < 0$, *and* $0 < \upmu < \infty$. *In addition, on* $\mathcal{M}_{T_{(Local)},U_0}$, $\upmu$, *the scalar-valued functions* $L^i_{(Small)}$, $(i = 1, 2, 3)$, *which are defined by* (3.5), (3.12), *and* (4.1a), *and the geometric angular coordinates* $(\vartheta^1, \vartheta^2)$ *constructed in Chapter* 3, *are* C^{N-2} *functions of the rectangular coordinates. A similar statement holds (with, in some cases, a different degree of differentiability) for the rectangular components* Ξ^μ, $\chi_{\mu\nu}^{(Small)}$, L^μ, $R^i_{(Small)}$, R^μ, $\breve{R}^\mu$, $\underline{L}^\mu$, *and* O^μ, $(\mu, \nu = 0, 1, 2, 3)$, *and for the rectangular components of all of the other geometric quantities defined throughout the monograph.*

The (open-at-the-top) region $\mathcal{M}_{T_{(Local)},U_0} = \{\cup_{u\in[0,U_0]}\mathcal{C}_u^{T_{(Local)}}\}\backslash\Sigma_{T_{(Local)}}^{U_0}$ *is foliated by level sets* $\mathcal{C}_u^{T_{(Local)}}$ *of the eikonal function* u, *where each* $\mathcal{C}_u^{T_{(Local)}}$ *is a truncated null hypersurface of the metric* $g(\Psi)$ *that is foliated by spheres* $S_{t,u} = \mathcal{C}_u^{T_{(Local)}} \cap \Sigma_t^{U_0}$. *That is, we have* $\mathcal{C}_u^{T_{(Local)}} = \cup_{t\in[0,T_{(Local)}]}S_{t,u}$. *Moreover, we have* $\Sigma_t^{U_0} = \cup_{u\in[0,U_0]}S_{t,u}$.

The solution has the following regularity properties relative to the rectangular coordinates:

$$\Psi,\ u \in C^{N-1}(\mathcal{M}_{T_{(Local)},U_0}), \tag{21.1a}$$

$$\partial^{\vec{I}}\Psi,\ \partial^{\vec{I}}u \in C([0,T_{(Local)}), H_e^{N+1-|\vec{I}|}(\Sigma_t^{U_0})), \qquad |\vec{I}| \leq N+1, \tag{21.1b}$$

where $\vec{I}$ *denotes a multi-index corresponding to repeated differentiation with respect to the rectangular spacetime coordinate vectorfields* ∂_ν, $(\nu = 0, 1, 2, 3)$, *and* $H_e^N(\Sigma_t^{U_0})$ *is the standard Euclidean Sobolev space involving order* $\leq N$ *rectangular spatial coordinate partial derivatives along* $\Sigma_t^{U_0}$.

In addition, the solution depends continuously on the data. Furthermore, if $\mathring{\epsilon}$ *is sufficiently small, then the existence time* $T_{(Local)}$ *from above can be bounded from below by* $f(\mathring{\epsilon}^{-1})$, *where* f *is a continuous increasing function such that* $f(\mathring{\epsilon}^{-1}) \uparrow \infty$ *as* $\mathring{\epsilon} \downarrow 0$.

Regularity of the change of variables map and other geometric quantities. *The change of variables map* $\Upsilon : [0, T_{(Local)}) \times [0, U_0] \times \mathbb{S}^2 \to \mathcal{M}_{T_{(Local)},U_0}$ *from geometric to rectangular coordinates (see Def.* 3.66 *and Lemma* 3.67*) is a* C^{N-2} *diffeomorphism with an everywhere positive Jacobian determinant. Hence, on* $\mathcal{M}_{T_{(Local)},U_0}$, *we have that the scalar-valued functions* Ψ, u, $\upmu$, $L^i_{(Small)}$, *etc. are many times continuously differentiable with respect to the geometric coordinates* (t, u, ϑ). *The same statement holds for the rectangular vectorfield components* L^μ,

[3] In particular, for short times, the $S_{t,u}$ do not "self-intersect."

R^μ, $\breve{R}^\mu$, $\breve{\underline{L}}^\mu$, O^μ, and the rectangular components of all of the other geometric quantities defined throughout the monograph.

Regularity of the geometric norms and energies. *The geometric norm and energy quantities appearing on the left-hand sides of (22.6)-(22.14p) are well-defined, continuous functions of (t,u) on $[0, T_{(Local)}) \times [0, U_0]$. Furthermore, if $\mathring{\epsilon}$ is sufficiently small, then at $(t,u) = (0, U_0)$, these quantities are all bounded by $\lesssim \mathring{\epsilon}$.*

Part II): Continuation criteria. *Let $T > 0$ be a time such that on $\mathcal{M}_{T,U_0}$, the solution exists and has all of the properties stated in **Part I)**. Assume that none of the following 4 breakdown scenarios occur:*

(**1**) $\inf_{\mathcal{M}_{T,U_0}} \upmu = 0$.
(**2**) $\sup_{\mathcal{M}_{T,U_0}} \upmu = \infty$.
(**3**) *There exists a sequence $p_n \in \mathcal{M}_{T,U_0}$ such that $\Psi(p_n)$ escapes every compact subset of $\mathcal{H}$ as $n \to \infty$.*
(**4**) $\sup_{\mathcal{M}_{T,U_0}} \max_{\kappa=0,1,2,3} |\partial_\kappa \Psi| = \infty$.

In addition, assume that the following condition is verified:

(**5**) *The change of variables map Υ extends to the compact set $[0,T] \times [0,U_0] \times \mathbb{S}^2$ as a (global) C^1 diffeomorphism onto its image.*

*Then there exists a $\Delta > 0$ such that Ψ, u, $\upmu$, Υ, ϑ^1, ϑ^2, L^μ, $L^i_{(Small)}$, Ξ^μ, and all of the other quantities can be uniquely extended (where Ψ and u are solutions) to a strictly larger region of the form $\mathcal{M}_{T+\Delta,U_0}$ on which they have all of the properties stated in **Part I)**.*

REMARK 21.2 (**Controlling top-order derivatives of** u). The proof of the finiteness of the quantities (22.14j)-(22.14p), which play an essential role in our proof of the sharp classical lifespan theorem, is highly nontrivial. These quantities involve some 26th geometric derivatives of u, whereas (21.1b) yields only the finiteness of the L^2 norms of the 25th rectangular derivatives of u. The gain of one derivative is made possible by the following key ingredients:

- The special structure of the right-hand side of the divergence identity (9.13) and of the deformation tensors of the commutation vectorfields $\mathscr{Z}$ (see Prop. 7.7).
- The special structure of the equations verified by the modified quantities from Chapter 11.
- The availability of the elliptic estimates of Sect. 18.1.

SKETCH OF A PROOF. **Discussion of aspects of the proof of Part I) involving rectangular coordinates.** These aspects of the proposition can be proved with rather standard techniques based on deriving a priori energy estimates on regions of the form $\mathcal{M}_{t,U_0}$. For the main ideas behind the proof of local well-posedness for the wave equation (2.1a) relative to the rectangular coordinates, readers may consult, for example, [**30**, Chapter VI]. In this step of the proof, it is convenient to construct a development (that is, a local solution region) that is determined by the data in a slightly larger region[4] $\Sigma_0^{U_0+\delta}$, where $\delta > 0$ is a small constant. After Ψ has been solved for relative to rectangular coordinates, we can then solve for the eikonal

[4] If the data are specified only in $\Sigma_0^{U_0}$, then we can use a Sobolev extension operator to extend them to data on Σ_0.

function u and deduce its regularity properties relative to the rectangular coordinates. To achieve this, we first use the eikonal equation $(g^{-1})^{\alpha\beta}(\Psi)\partial_\alpha u \partial_\beta u = 0$ to solve for $\partial_t u = f(\Psi, \partial_1 u, \partial_2 u, \partial_3 u) > 0$, where f is smooth. We then apply standard L^2-type energy methods to this scalar equation, which leads to the existence of u and its properties. In particular, the fact that the $S_{t,u}$ are embedded spheres for short times and $u \in [0, U_0]$ follows from the fact that for short times, $\sum_{a=1}^3 |\partial_a u|$ is uniformly bounded from above and uniformly from below strictly away from 0. This step is one instance in which we rely on the fact that Ψ and u have been solved for in a development of $\Sigma_0^{U_0+\delta}$; this allows us to avoid[5] showing that the inner sphere $S_{t,U_0+\delta}$ does not self-intersect for short times. The remaining quantities $\upmu$, $L^\mu_{(Small)}$, etc. are constructed out of Ψ and u. Hence, their existence and regularity properties are straightforward consequences of the properties of Ψ and u.

Discussion of aspects of the proof of Part I) involving geometric coordinates. We first note that we have already sketched a proof that Ψ, u, $\upmu$, $L^\mu_{(Small)}$, etc. are many times differentiable with respect to the rectangular coordinates on $\mathcal{M}_{T_{(Local)},U_0}$.

We now show that (shrinking $T_{(Local)}$ if necessary) the change of variables map Υ from geometric to rectangular coordinates (see Def. 3.66 and Lemma 3.67) is a bijection from $[0, T_{(Local)}) \times [0, U_0] \times \mathbb{S}^2$ to $\mathcal{M}_{T_{(Local)},U_0}$. Since t is the same function in both coordinate systems, we have to show only that for any $t \in [0, T_{(Local)})$, $\Upsilon(t, \cdot) : [0, U_0] \times \mathbb{S}^2 \to \Sigma_t^{U_0}$ is bijective. To this end, let $\varphi_t|_{\Sigma_0^{U_0}}$ denote the restriction of the flow map φ_t of L (see Remark 3.56) to $\Sigma_0^{U_0}$. The main point is that the estimates proved relative to rectangular coordinates guarantee that the rectangular components $L^\nu = Lx^\nu$ are C^{N-2} functions of the rectangular coordinates on $\mathcal{M}_{t,U_0}$. Hence, since $Lt = 1$, it follows from the standard theory of ODEs that $\varphi_t|_{\Sigma_0^{U_0}}$ is, relative to the rectangular spatial coordinates, a bijection from $\Sigma_0^{U_0}$ to $\Sigma_t^{U_0}$. In view of the manner in which we constructed the geometric coordinates in Chapter 3 (see especially Sect. 3.5), we see that $\Upsilon(t, u, \vartheta)$ is equal to $\varphi_t|_{\Sigma_0^{U_0}}$ pre-composed with the C^∞ diffeomorphism $[0, U_0] \times \mathbb{S}^2 \to \Sigma_0^{U_0}$ that maps the geometric coordinates $(u, \vartheta) \in [0, U_0] \times \mathbb{S}^2$ along $\Sigma_0^{U_0}$ to the rectangular spatial coordinates (x^1, x^2, x^3) along $\Sigma_0^{U_0}$. We have thus shown the bijectivity of Υ.

Furthermore, examining the proof of Lemma 3.67, we see that the Jacobian of Υ^{-1} is the C^{N-3} matrix $\dfrac{\partial(t, u, \vartheta^1, \vartheta^2)}{\partial(x^0, x^1, x^2, x^3)}$, which has the form $\begin{pmatrix} 1 & \mathbf{0}_{1\times 3} \\ *_{3\times 1} & M \end{pmatrix}^{-1}$, where at $t = 0$, M is an invertible 3×3 matrix. Hence, at least for short times, the Jacobian of Υ^{-1} remains invertible, and by the inverse function theorem, Υ has the same regularity as Υ^{-1}. We remark that the Ξ^i, $(i = 1, 2, 3)$, are the terms in the Jacobian (see equation (3.83)) with the least regularity (specifically, they are elements of C^{N-3}); their regularity can be derived with the help of the transport equation (3.47), expressed relative to the rectangular coordinates (see Footnote 1 on pg. 189).

[5]In our proof of Theorem 22.1, we do not consider an enlarged region $\Sigma_0^{U_0+\delta}$. Thus, under the small-data assumptions of the theorem, we provide an argument grounded in algebraic topology showing that the spheres $S_{t,u}$ (including the inner sphere S_{t,U_0}) do not self-intersect during the solution's classical lifespan.

The statements concerning the finiteness and the (t, u)-continuity of the geometric norms and energies (22.6)-(22.14p) are difficult to derive. However, the only difficult step is obtaining a priori estimates for these quantities and, in particular, avoiding derivative loss in some of the top-order eikonal function quantities (see Remark 21.2). In our proof of Theorem 22.1, we show how to derive such a priori estimates and hence we do not repeat the lengthy argument here. To bound the geometric norms and energies at $(t, u) = (0, U_0)$ by $\lesssim \mathring{\epsilon}$, we use the small-data estimates derived in Sect 12.8.

Discussion of the proof of Part II). The continuation criteria of **Part II)** can also be proved by using mostly standard arguments, with one exception that we describe below; see, for example, [**77**] for the main ideas behind a proof. The main idea is that once we rule out the possible breakdown in the hyperbolic character of the equations, the possible blow-up of the first rectangular derivatives of various quantities, and the possible degeneracy of the region $\mathcal{M}_{T,U_0}$, we can then ensure that the solutions Ψ and u can be continued relative to rectangular coordinates. Furthermore, assuming that the condition 5 holds and using the extendibility of the solution relative to the rectangular coordinates, it is straightforward to show that we can extend the change of variables map Υ to a strictly larger set $[0, T + \Delta) \times [0, U_0] \times \mathbb{S}^2$ (for some $\Delta > 0$) as a diffeomorphism onto its image. We can then derive the desired properties, on the domain $[0, T + \Delta) \times [0, U_0] \times \mathbb{S}^2$, for all of the quantities of interest relative to the geometric coordinates by using the same arguments given in the proof of local well-posedness.

We now provide the proof of the one somewhat subtle aspect, namely showing that if none of the 4 breakdown scenarios occur, then u remains regular in the following sense:

$$0 < \inf_{\mathcal{M}_{T,U_0}} \sum_{a=1}^{3} |\partial_a u| \leq \sup_{\mathcal{M}_{T,U_0}} \sum_{a=1}^{3} |\partial_a u| < \infty. \tag{21.2}$$

Assuming that none of the 4 breakdown scenarios occur, we have in particular that the 3×3 matrices $\underline{g}_{ij}(\Psi)$ and $(\underline{g}^{-1})^{ij}(\Psi)$ are uniformly positive definite on $\mathcal{M}_{T,U_0}$ (because their eigenvalues are bounded from above and uniformly from below strictly away from 0). Hence, the inequalities in (21.2) follow from the identity $(\underline{g}^{-1})^{ab}\partial_a u \partial_b u = \mu^{-2}$ (see (3.24)). $\square$

CHAPTER 22

The Sharp Classical Lifespan Theorem

In this chapter, we state and prove our sharp classical lifespan theorem, which is the main theorem of the monograph. It guarantees that under a small-data assumption, the solution persists unless the quantity $\upmu_{\star}$ defined in (13.6) becomes 0 in finite time, in which case some rectangular derivatives of Ψ blow-up and a shock singularity has formed. We also prove Cor. 22.4, which shows that if the data have "very small" angular derivatives, then this property is propagated by the solution. We use the theorem and the corollary in Chapter 23 in our proof of finite-time shock formation for an open set of nearly spherically symmetric small data.

22.1. The sharp classical lifespan theorem

We now state and prove the main theorem of the monograph.

THEOREM 22.1 (**The sharp classical lifespan theorem together with estimates**). *Let* ($\mathring{\Psi} := \Psi|_{\Sigma_0}, \mathring{\Psi}_0 := \partial_t \Psi|_{\Sigma_0}$) *be initial data for the covariant wave equation* (2.1a) *under the assumption*[1] (2.3). *Assume that the data are supported in the Euclidean unit ball* Σ_0^1. *Let* $\mathring{\epsilon} = \|\mathring{\Psi}\|_{H_e^{25}(\Sigma_0^1)} + \|\mathring{\Psi}_0\|_{H_e^{24}(\Sigma_0^1)}$ *be the size of the data as defined in Def.* 12.3. *Assume that the data verify the hypotheses of Prop.* 21.1 *(the local well-posedness proposition) and let* $0 < U_0 < 1$ *be a constant. Let* Ψ *denote the solution corresponding to the data existing on a nontrivial region of the form* $\mathcal{M}_{T_{(Local)},U_0}$ *(see definition* (3.6e)*) Recall that* $\upmu_{\star}(t,u) := \min\{1, \min_{\Sigma_t^u} \upmu\}$, *t denotes the Minkowski time coordinate, u is the eikonal function (with initial data* $u|_{\Sigma_0} = 1 - r$, *where* $r = \sqrt{\sum_{a=1}^3 (x^a)^2}$*), and* $\varrho(t,u) := 1 - u + t$ *is the geometric radial variable. There exist large constants* $C > 0$, $C_{(Lower-Bound)} > 0$, *and* $A_* > 4$ *and a small constant* $\epsilon_0 > 0$ *such that if* $\mathring{\epsilon} \leq \epsilon_0$, *then the following statements hold true. In the statements, the constants can depend on* U_0 *and the nonlinearities in* (2.1a) *and in particular,* C *can blow-up as* $U_0 \uparrow 1$.

Existence as long as $\upmu_{\star} > 0$ ***and a classical lifespan lower bound***. *Let*

$$T_{(Lifespan);U_0} := \sup\{t \in [0,\infty) \mid \inf\{\upmu_{\star}(s,U_0) \mid s \in [0,t)\} > 0\}. \tag{22.1}$$

Then $T_{(Lifespan);U_0}$ *is the classical lifespan of the solution in the region determined by the portion of the data in* Σ_0 *belonging to the exterior of the Euclidean sphere* S_{0,U_0} *of radius* $1 - U_0$ *centered at the origin (see* (3.4b) *and* (3.6d)*). That is,* Ψ *can be extended as a classical solution (relative to both the geometric and the rectangular coordinates) to the region* $\mathcal{M}_{T_{(Lifespan);U_0},U_0}$ *on which it has all of the*

[1] As we described in Chapter 2, this assumption is easy to eliminate.

properties stated in Prop. 21.1. Furthermore, if $T_{(Lifespan);U_0} < \infty$, *then*

$$\sup_{\mathcal{M}_{T_{(Lifespan);U_0},U_0}} \max_{\nu=0,1,2,3} |\partial_\nu \Psi| = \infty. \tag{22.2}$$

In addition, with $\mathring{\upmu}(u,\vartheta) = \upmu(0,u,\vartheta)$, $G_{LL} = \frac{d}{d\Psi} g_{\alpha\beta}(\Psi) L^\alpha L^\beta$, *and* ${}^{(+)}\mathring{\aleph} = {}^{(+)}\mathring{\aleph}(\vartheta)$ *as in Def. 3.32, we have the following estimates for* $0 \leq s \leq t < T_{(Lifespan);U_0}$:

$$\left| \upmu(s,u,\vartheta) - \left\{ \mathring{\upmu}(u,\vartheta) + \frac{1}{2} \ln \left(\frac{\varrho(s,u)}{\varrho(0,u)} \right) [\varrho G_{LL} \breve{R}\Psi](t,u,\vartheta) \right\} \right| \leq C\mathring{\epsilon}, \tag{22.3a}$$

$$\left| G_{LL}(t,u,\vartheta) - {}^{(+)}\mathring{\aleph}(\vartheta) \right| \leq C\mathring{\epsilon}, \tag{22.3b}$$

$$|\mathring{\upmu}(u,\vartheta) - 1| \leq C\mathring{\epsilon}, \tag{22.3c}$$

$$\left| \varrho \breve{R}\Psi \right| (t,u,\vartheta) \leq C\mathring{\epsilon}. \tag{22.3d}$$

Furthermore, $\upmu_\star(t,U_0) > 0$ *whenever* $t < \exp\left(\left\{ C_{(Lower-Bound)} \mathring{\epsilon} \right\}^{-1} \right)$, *and hence*

$$T_{(Lifespan);U_0} > \exp\left(\frac{1}{C_{(Lower-Bound)} \mathring{\epsilon}} \right). \tag{22.4}$$

<u>$G_{LL}(t,u,\vartheta) = 0 \implies \upmu(s,u,\vartheta)$ ***is large for*** $0 \leq s \leq t$.</u> *For*

$$(t,u) \in [0, T_{(Lifespan);U_0}) \times [0,U_0] \quad \text{and } 0 \leq s \leq t,$$

let $\Sigma_{s;t}^{U_0;*} := \{(s,u,\vartheta) \in \Sigma_s^{U_0} \mid G_{LL}(t,u,\vartheta) = 0\}$. *Then we have the following estimate:*

$$\inf_{\Sigma_{s;t}^{U_0;*}} \upmu \geq 1 - C\mathring{\epsilon}. \tag{22.5}$$

<u>Ψ ***remains bounded at low levels***</u>. *The following* C^0 *estimates*[2] *for the lower-order derivatives of* Ψ *hold on the domain* $(t,u) \in [0, T_{(Lifespan);U_0}) \times [0,U_0]$:

$$\left\| \mathscr{Z}^N \Psi \right\|_{C^0(\Sigma_t^u)} \leq C\mathring{\epsilon} \frac{1}{1+t}, \qquad (N \leq 13). \tag{22.6}$$

In (22.6), $\mathscr{Z}^N$ *denotes an arbitrary* N^{th} *order differential operator corresponding to repeated differentiation with respect to commutation vectorfields belonging to the set* $\mathscr{Z}$ *defined in (7.1). Moreover, if* $T_{(Lifespan);U_0} < \infty$, *then the quantities* $\mathscr{Z}^{\leq 12}\Psi$ *extend as continuous functions of* (t,u,ϑ) *to* $\Sigma_{T_{(Lifespan);U_0}}^{U_0}$.

<u>***The rectangular metric components remain bounded at low levels***</u>. *The following* C^0 *estimates for the lower-order derivatives of the rectangular components* $g_{\mu\nu}$, $(\mu,\nu = 0,1,2,3)$, *of the spacetime metric and the rectangular spatial components* $\not{g}_{ij}$, $(i,j = 1,2,3)$, *of the metric induced by* g *on* $S_{t,u}$ *(see Def. 3.9) hold on the domain* $(t,u) \in [0, T_{(Lifespan);U_0}) \times [0,U_0]$:

$$\left\| \mathscr{Z}^N \{g_{\mu\nu} - m_{\mu\nu}\} \right\|_{C^0(\Sigma_t^u)} \leq C\mathring{\epsilon} \frac{1}{1+t}, \qquad (N \leq 13), \tag{22.7a}$$

$$\left\| \mathscr{Z}^N \left\{ \not{g}_{ij} - \left(\delta_{ij} - \frac{x^i x^j}{\varrho^2} \right) \right\} \right\|_{C^0(\Sigma_t^u)} \leq C\mathring{\epsilon} \frac{\ln(e+t)}{1+t}, \qquad (N \leq 12). \tag{22.7b}$$

[2]See Remark 3.71 concerning our use of the norm $\| \cdot \|_{C^0(\Sigma_t^u)}$.

In (22.7a), $m_{\mu\nu} = \mathrm{diag}(-1,1,1,1)$ *denotes the Minkowski metric. Moreover, if* $T_{(Lifespan);U_0} < \infty$, *then the quantities* $\mathscr{Z}^{\leq 12}\{g_{\mu\nu} - m_{\mu\nu}\}$ *and*

$$\mathscr{Z}^{\leq 11}\left\{ \not{g}_{ij} - \left(\delta_{ij} - \frac{x^i x^j}{\varrho^2}\right)\right\}$$

extend as continuous functions of (t,u,ϑ) *to* $\Sigma^{U_0}_{T_{(Lifespan);U_0}}$.

The eikonal function quantities remain bounded at low levels. *For* $(t,u) \in [0, T_{(Lifespan);U_0}) \times [0,U_0]$, *the following* C^0 *estimates hold for the inverse foliation density* $\upmu$, *the rectangular components* $L^i_{(Small)} = L^i - \frac{x^i}{\varrho}$ *and* $R^i_{(Small)} = R^i + \frac{x^i}{\varrho}$, *and the* $S_{t,u}$ *tensorfield* $\chi^{(Small)} = \chi - \frac{\not{g}}{\varrho}$:

$$\left\|\mathscr{Z}^N(\upmu - 1)\right\|_{C^0(\Sigma_t^u)} \leq C\mathring{\epsilon}\ln(e+t), \quad (N \leq 13), \tag{22.8a}$$

$$\left\|\mathscr{Z}^N L^i_{(Small)}\right\|_{C^0(\Sigma_t^u)}, \left\|\mathscr{Z}^N R^i_{(Small)}\right\|_{C^0(\Sigma_t^u)} \leq C\mathring{\epsilon}\frac{\ln(e+t)}{1+t}, \quad (N \leq 13), \tag{22.8b}$$

$$\left\|\not{\mathcal{L}}^N_{\mathscr{Z}}\chi^{(Small)}\right\|_{C^0(\Sigma_t^u)} \leq C\mathring{\epsilon}\frac{\ln(e+t)}{(1+t)^2}, \quad (N \leq 12). \tag{22.8c}$$

Moreover, if $T_{(Lifespan);U_0} < \infty$, *then the quantities* $\mathscr{Z}^{\leq 12}\upmu$, $\mathscr{Z}^{\leq 12}L^i_{(Small)}$, $\mathscr{Z}^{\leq 12}R^i_{(Small)}$, *and* $\not{\mathcal{L}}^{\leq 11}_{\mathscr{Z}}\chi^{(Small)}$ *extend as continuous functions of* (t,u,ϑ) *to* $\Sigma^{U_0}_{T_{(Lifespan);U_0}}$.

Behavior of the change of variables map Υ. *If* $T_{(Lifespan);U_0} < \infty$, *then the change of variables map (from geometric to rectangular coordinates)* $\Upsilon : [0, T_{(Lifespan);U_0}) \times [0,U_0] \times \mathbb{S}^2 \to \mathcal{M}_{T_{(Lifespan);U_0},U_0}$, $\Upsilon(t,u,\vartheta) = t, x^1, x^2, x^3)$, *extends as a* C^{11} *function to* $[0, T_{(Lifespan);U_0}] \times [0,U_0] \times \mathbb{S}^2$. *In addition,* Υ *is a bijection from*[3] $[0, T_{(Local)}) \times [0,U_0] \times \mathbb{S}^2$ *to* $\mathcal{M}_{T_{(Local)},U_0}$ *with a positive Jacobian determinant. Furthermore, if* $T_{(Lifespan);U_0} < \infty$, *then on* $[0, T_{(Lifespan);U_0}] \times [0,U_0] \times \mathbb{S}^2$, *the Jacobian determinant of* Υ *vanishes precisely on the subset*

(22.9)
$$\{(T_{(Lifespan);U_0}, u, \vartheta) \in \{T_{(Lifespan);U_0}\} \times [0,U_0] \times \mathbb{S}^2 \mid \upmu(T_{(Lifespan);U_0}, u, \vartheta) = 0\}.$$

What happens when $\upmu \to 0$. *In the spacetime subset*

$$\big\{(t,u,\vartheta) \in [0, T_{(Lifespan);U_0}] \times [0,U_0] \times \mathbb{S}^2 \mid \upmu(t,u,\vartheta) \leq 1/4 \text{ and } G_{LL}(t,u,\vartheta) \neq 0\big\},$$

we have

$$L\upmu(t,u,\vartheta) \leq -\frac{2}{\varrho(t,u)\left\{1 + \ln\left(\frac{\varrho(t,u)}{\varrho(0,u)}\right)\right\}}, \tag{22.10}$$

$$|R\Psi|(t,u,\vartheta) \geq \frac{1}{\upmu(t,u,\vartheta)} \frac{1}{\varrho(t,u)\left\{1 + \ln\left(\frac{\varrho(t,u)}{\varrho(0,u)}\right)\right\}} \frac{1}{|G_{LL}(t,u,\vartheta)|}, \tag{22.11}$$

[3] Recall that $\mathcal{M}_{T_{(Local)},U_0}$ is, by definition, "open at the top."

where the vectorfield R *(see definition* (3.19b)*) verifies the Euclidean length estimate* $|R - (-\partial_r)|_e \lesssim \mathring{\epsilon} \ln(e+t)(1+t)^{-1}$. *Here,* $|V|_e^2 := \delta_{ab}V^aV^b$ *and* $\partial_r = \frac{x^a}{r}\partial_a$ *is the standard Euclidean radial vectorfield.*

If $T_{(Lifespan);U_0} < \infty$, *then let*

(22.12)
$$\Sigma^{U_0}_{T_{(Lifespan);U_0};(Blow-up)} := \Big\{(T_{(Lifespan);U_0}, u, \vartheta) \in \{T_{(Lifespan);U_0}\} \times [0, U_0] \times \mathbb{S}^2 \mid \upmu(T_{(Lifespan);U_0}, u, \vartheta) = 0\Big\}.$$

In particular, from (22.5) *and the fact that* G_{LL} *extends to* $\Sigma^{U_0}_{T_{(Lifespan);U_0}}$ *as a continuous function of* (t, u, ϑ)*, we see that no point* $p = (T_{(Lifespan);U_0}, u, \vartheta)$ *with* $G_{LL}(p) = 0$ *can belong to* $\Sigma^{U_0}_{T_{(Lifespan);U_0};(Blow-up)}$. *Hence, it follows from* (22.11) *that at any point in* $\Sigma^{U_0}_{T_{(Lifespan);U_0};(Blow-up)}$, ***the near-Euclidean-unit-length derivative*** $R\Psi$ ***blows up.***

A hierarchy of L^2 ***estimates for*** Ψ. *The following estimates hold for the* L^2*-based quantities* $\mathbb{Q}_{(N)}$ *and* $\widetilde{\mathbb{Q}}_{(N)}$ *defined in Def.* 14.2 *and the spacetime Morawetz integral* $\widetilde{\mathbb{K}}_{(N)}$ *defined in Def.* 14.3 *on the domain* $(t, u) \in [0, T_{(Lifespan);U_0}) \times [0, U_0]$:

$$\mathbb{Q}^{1/2}_{(N)}(t, u) \leq C\mathring{\epsilon}, \qquad (0 \leq N \leq 15), \tag{22.13a}$$

$$\widetilde{\mathbb{Q}}^{1/2}_{(N)}(t, u) + \widetilde{\mathbb{K}}^{1/2}_{(N)}(t, u) \leq C\mathring{\epsilon} \ln^2(e+t), \qquad (0 \leq N \leq 15), \tag{22.13b}$$

$$\mathbb{Q}^{1/2}_{(16+M)}(t, u) \leq C\mathring{\epsilon}\upmu_\star^{-.75-M}(t, u), \qquad (0 \leq M \leq 7), \tag{22.13c}$$

$$\widetilde{\mathbb{Q}}^{1/2}_{(16+M)}(t, u) + \widetilde{\mathbb{K}}^{1/2}_{(16+M)}(t, u) \leq C\mathring{\epsilon} \ln^2(e+t)\upmu_\star^{-.75-M}(t, u), \qquad (0 \leq M \leq 7), \tag{22.13d}$$

$$\mathbb{Q}^{1/2}_{(24)}(t, u) \leq C\mathring{\epsilon} \ln^{A_*}(e+t)\upmu_\star^{-8.75}(t, u), \tag{22.13e}$$

$$\widetilde{\mathbb{Q}}^{1/2}_{(24)}(t, u) + \widetilde{\mathbb{K}}^{1/2}_{(24)}(t, u) \leq C\mathring{\epsilon} \ln^{A_*+2}(e+t)\upmu_\star^{-8.75}(t, u). \tag{22.13f}$$

A hierarchy of L^2 ***estimates for the eikonal function quantities.*** *The following* L^2 *estimates hold for the eikonal function quantities on the domain* $(t, u) \in [0, T_{(Lifespan);U_0}) \times [0, U_0]$:

$$\left\|\mathscr{Z}^N(\upmu - 1)\right\|_{L^2(\Sigma_t^u)} \leq C\mathring{\epsilon}(1+t)\ln(e+t), \qquad (0 \leq N \leq 15), \tag{22.14a}$$

$$\left\|\mathscr{Z}^N L^i_{(Small)}\right\|_{L^2(\Sigma_t^u)} \leq C\mathring{\epsilon}\ln(e+t), \qquad (0 \leq N \leq 15), \tag{22.14b}$$

$$\left\|\not{\mathcal{L}}^N_{\mathscr{Z}}\chi^{(Small)}\right\|_{L^2(\Sigma_t^u)} \leq C\mathring{\epsilon}\frac{\ln(e+t)}{1+t}, \qquad (0 \leq N \leq 14), \tag{22.14c}$$

(22.14d) $$\left\|\mathscr{Z}^{16+M}\upmu\right\|_{L^2(\Sigma_t^u)} \leq C\mathring{\epsilon}(1+t)\ln^3(e+t)\upmu_\star^{-.25-M}(t,u), \quad (0 \leq M \leq 7),$$

(22.14e) $$\left\|\mathscr{Z}^{16+M}L_{(Small)}^i\right\|_{L^2(\Sigma_t^u)} \leq C\mathring{\epsilon}\ln^3(e+t)\upmu_\star^{-.25-M}(t,u), \quad (0 \leq M \leq 7),$$

(22.14f) $$\left\|\mathcal{L}_{\mathscr{Z}}^{15+M}\chi^{(Small)}\right\|_{L^2(\Sigma_t^u)} \leq C\mathring{\epsilon}\frac{\ln^3(e+t)}{1+t}\upmu_\star^{-.25-M}(t,u), \quad (0 \leq M \leq 7),$$

(22.14g) $$\left\|\mathscr{Z}^{24}\upmu\right\|_{L^2(\Sigma_t^u)} \leq C\mathring{\epsilon}(1+t)\ln^{A_*+3}(e+t)\upmu_\star^{-8.25}(t,u),$$

(22.14h) $$\left\|\mathscr{Z}^{24}L_{(Small)}^i\right\|_{L^2(\Sigma_t^u)} \leq C\mathring{\epsilon}\ln^{A_*+3}(e+t)\upmu_\star^{-8.25}(t,u),$$

(22.14i) $$\left\|\mathcal{L}_{\mathscr{Z}}^{23}\chi^{(Small)}\right\|_{L^2(\Sigma_t^u)} \leq C\mathring{\epsilon}\frac{\ln^{A_*+3}(e+t)}{1+t}\upmu_\star^{-8.25}(t,u),$$

(22.14j) $$\left\|L\mathscr{Z}^{24}\upmu\right\|_{L^2(\Sigma_t^u)} \leq C\mathring{\epsilon}\ln^{A_*+2}(e+t)\upmu_\star^{-9.25}(t,u),$$

(22.14k) $$\left\|L\left[\varrho\mathscr{Z}^{24}L_{(Small)}^i\right]\right\|_{L^2(\Sigma_t^u)} \leq C\mathring{\epsilon}\ln^{A_*+2}(e+t)\upmu_\star^{-9.25}(t,u),$$

(22.14l) $$\left\|\mathcal{L}_L\left[\varrho^2\mathcal{L}_{\mathscr{Z}}^{23}\chi^{(Small)}\right]\right\|_{L^2(\Sigma_t^u)} \leq C\mathring{\epsilon}\ln^{A_*+2}(e+t)\upmu_\star^{-9.25}(t,u),$$

(22.14m) $$\left\|\upmu\mathcal{L}_{\breve{R}}\mathcal{L}_{\mathscr{Z}}^{23}\chi^{(Small)}\right\|_{L^2(\Sigma_t^u)} \leq C\mathring{\epsilon}\ln^{A_*}(e+t)\upmu_\star^{-8.75}(t,u),$$

(22.14n) $$\left\|\upmu\nabla\!\!\!/^2\mathscr{Z}^{23}\upmu\right\|_{L^2(\Sigma_t^u)} \leq C\mathring{\epsilon}\ln^{A_*}(e+t)\upmu_\star^{-8.75}(t,u),$$

(22.14o) $$\left\|\upmu\nabla\!\!\!/^2\mathscr{Z}^{23}L_{(Small)}^i\right\|_{L^2(\Sigma_t^u)} \leq C\mathring{\epsilon}\frac{\ln^{A_*}(e+t)}{1+t}\upmu_\star^{-8.75}(t,u),$$

(22.14p) $$\left\|\upmu\nabla\!\!\!/\mathcal{L}_{\mathscr{Z}}^{23}\chi^{(Small)}\right\|_{L^2(\Sigma_t^u)} \leq C\mathring{\epsilon}\frac{\ln^{A_*}(e+t)}{1+t}\upmu_\star^{-8.75}(t,u).$$

REMARK 22.2 (**Global existence when the classic null condition holds**). When Klainerman's classic null condition [**45**] is verified, that is, when ${}^{(+)}\aleph \equiv 0$ (see Def. 3.32), one can show, under a revised set of bootstrap assumptions, that the estimate (22.3b) can be improved to $|G_{LL}|(t,u,\vartheta) \leq C\mathring{\epsilon}(1+t)^{-1}$. Consequently, Theorem 22.1 implies, in particular via the estimates (22.3a) and (22.3d), that $\upmu_\star$ never vanishes and hence the solution exists globally in the spacetime region bounded by the inner outgoing null cone $\mathcal{C}_{U_0}$ and the outer flat outgoing null cone $\mathcal{C}_0$.

REMARK 22.3 (**Extensions to data with additional regularity**). Theorem 22.1 can of course be extended to apply to data of higher Sobolev regularity, that is, with the smallness of $\|\mathring{\Psi}\|_{H_e^{25}(\Sigma_0^1)} + \|\mathring{\Psi}_0\|_{H_e^{24}(\Sigma_0^1)}$ replaced by the smallness of $\|\mathring{\Psi}\|_{H^{N_{(Top)}+1}(\Sigma_0^1)} + \|\mathring{\Psi}_0\|_{H^{N_{(Top)}}(\Sigma_0^1)}$, where $N_{(Top)} \geq 25$ is an integer. In this case, the estimates (22.13e)-(22.13f) hold with 24 replaced by $N_{(Top)}$, the estimates (22.13c)-(22.13d) hold with $16+M$ replaced by $N_{(Top)} - 8 + M$, the estimates (22.13a)-(22.13a) hold with $0 \leq N \leq 15$ replaced by $0 \leq N \leq N_{(Top)} - 9$, and similarly for the other estimates in the theorem.

Alternatively, we could assume the smallness of $\|\mathring{\Psi}\|_{H_e^{25}(\Sigma_0^1)} + \|\mathring{\Psi}_0\|_{H_e^{24}(\Sigma_0^1)}$ and that $\|\mathring{\Psi}\|_{\dot{H}_e^{M+1}(\Sigma_0^1)} + \|\mathring{\Psi}_0\|_{\dot{H}_e^{M}(\Sigma_0^1)} < \infty$ (without smallness) for $M = 25, 26, \ldots, N_{(Top)}$.

Here, $\|\cdot\|_{\dot{H}_e^M(\Sigma_0^1)}$ denotes the standard homogeneous Sobolev norm corresponding to M^{th} order rectangular derivatives along Σ_0^1. The results of Theorem 22.1 would of course hold verbatim, and we could also prove estimates of the form $\mathbb{Q}_{(25+M)}^{1/2}(t,u) \leq f(\|\mathring{\Psi}\|_{H_e^{25+M}(\Sigma_0^1)} + \|\mathring{\Psi}_0\|_{H_e^{24+M}(\Sigma_0^1)}) \ln^A(e+t)\upmu_\star^{-B}(t,u)$, where f is a smooth increasing function and A, B are positive constants depending on M. We could also prove similar estimates for $\widetilde{\mathbb{Q}}_{(25+M)}^{1/2}(t,u)$. We could prove these higher-order estimates by using arguments similar to the ones we used to prove (20.8e)-(20.8f). In particular, in order to avoid derivative loss, our proofs of these estimates would rely on the modified quantities of Chapter 11, at least at the top order.

PROOF OF THEOREM 22.1. Let $C_* > 1$ be a constant (we will adjust C_* throughout the proof). We define

(22.15)

$T_{(Max);U_0} :=$ The supremum of the set of times $T_{(Bootstrap)} \geq 0$ such that:

- Ψ, u, $\upmu$, $L^i_{(Small)}$, Υ, and all of the other quantities defined throughout the monograph exist classically on $\mathcal{M}_{T_{(Bootstrap)},U_0}$ and all of the solution properties from **Part I)** of Prop. 21.1 hold on $\mathcal{M}_{T_{(Bootstrap)},U_0}$.
- $\inf\left\{\upmu_\star(t,U_0) \mid t \in [0,T_{(Bootstrap)})\right\} > 0$.
- The fundamental C^0 bootstrap assumptions $(\mathbf{BA}\Psi)$ – $(\mathbf{BA}'\Psi)$ hold with $\varepsilon := C_*\mathring{\epsilon}$ for $(t,u) \in \times[0,T_{(Bootstrap)}) \times [0,U_0]$.
- The L^2-type bootstrap assumptions (20.1a) – (20.1f) hold with $\varepsilon := C_*\mathring{\epsilon}$ for $(t,u) \in \times[0,T_{(Bootstrap)}) \times [0,U_0]$.

From Proposition 21.1, we deduce that if $\mathring{\epsilon}$ is sufficiently small and C_* is sufficiently large, then $T_{(Max);U_0} > 0$.

In the first part of the proof, we focus mostly on deriving the quantitative estimates stated in the theorem, but with the time $T_{(Lifespan);U_0}$ defined by (22.1) replaced by the time $T_{(Max);U_0}$ defined by (22.15). In the second part of the proof, we focus mostly on proving the qualitative statements, on showing that $T_{(Max);U_0} = T_{(Lifespan);U_0}$, and on showing that $T_{(Lifespan);U_0}$ is the classical lifespan of the solution. Note that the definitions imply that $T_{(Max);U_0} \leq T_{(Lifespan);U_0}$ and hence to show that $T_{(Max);U_0} = T_{(Lifespan);U_0}$, we have to show only that $T_{(Max);U_0} \geq T_{(Lifespan);U_0}$.

We now derive quantitative estimates. By enlarging C_* if necessary, we may assume that the estimates (20.8a)-(20.8f) of Lemma 20.10 hold with the constant C on the right-hand sides replaced by $\frac{C_*}{4}$. We then see that if $C_*\varepsilon^{1/2} = C_*^{3/2}\mathring{\epsilon}^{1/2} < 1$, then the L^2 bootstrap assumptions (20.1a)-(20.1f) hold for $(t,u) \in \times[0,T_{(Max);U_0})\times [0,U_0]$ with the factor $\varepsilon = C_*\mathring{\epsilon}$ on the right-hand side replaced by the *strictly smaller* quantity $\frac{1}{2}C_*\mathring{\epsilon}$. This fact follows from the following simple calculations:

$$\frac{1}{4}C_*\left\{\mathring{\epsilon} + \varepsilon^{3/2}\right\} = \frac{1}{4}\left\{1 + C_*\varepsilon^{1/2}\right\}\varepsilon \leq \frac{1}{2}\varepsilon = \frac{1}{2}C_*\mathring{\epsilon}. \tag{22.16}$$

In particular, the estimates (22.13a)-(22.13f) follow from (22.16). Furthermore, using Cor. 18.11, Def. 14.2, inequality (12.46a), and the fact that $\varrho L \in \mathscr{Z}$, and enlarging the constant C_* from (22.16) if necessary, we deduce that the C^0 bootstrap assumptions (**BA**Ψ) for Ψ hold for $(t,u) \in \times[0,T_{(Max);U_0}) \times [0,U_0]$ with ε on the right-hand side of (**BA**Ψ) replaced by the right-hand side of (22.16) and the norm $C^0(\Sigma_t^u)$ on the left-hand side replaced by $C^0(S_{t,u})$. Similarly, in view of the estimate (22.13c) with $M = 0$ and the estimate (13.152), we obtain (again enlarging C_* if necessary) $\int_{t'=0}^{t} \left\| \begin{pmatrix} \mathscr{Z}^{\leq 14}\Psi \\ \varrho L \mathscr{Z}^{\leq 13}\Psi \\ \varrho \not{d} \mathscr{Z}^{\leq 13}\Psi \end{pmatrix} \right\|_{C^0(S_{t',u})} dt' \leq \frac{1}{2} C_* \mathring{\epsilon} \ln(e+t)$.
Hence, we have shown that the C^0 bootstrap assumptions (**BA**$'\Psi$) for Ψ hold for $(t,u) \in \times[0,T_{(Max);U_0}) \times [0,U_0]$ with ε on the right-hand side of (**BA**$'\Psi$) replaced by the right-hand side of (22.16) and the norm $C^0(\Sigma_{t'}^u)$ on the left-hand side replaced by $C^0(S_{t',u})$. To recover the fact that for $N \leq 14$, $\mathscr{Z}^N\Psi$ is an element of $C^0(\Sigma_t^u)$ and not just $C^0(S_{t,u})$, we simply need nonquantitative estimates showing that $\mathscr{Z}^N\Psi$ is jointly continuous in u and ϑ. To this end, we note that at each point in $\mathcal{M}_{T_{(Max);U_0},U_0}$, the vectorfields belonging to the commutation set $\mathscr{Z}$ have span equal to that of the geometric coordinate derivative frame $\left\{ \frac{\partial}{\partial t}, \frac{\partial}{\partial u}, \frac{\partial}{\partial \vartheta^1}, \frac{\partial}{\partial \vartheta^2} \right\}$ and have components relative to this frame that are many times differentiable with respect to $(t,u,\vartheta^1,\vartheta^2)$. Hence, the desired nonquantitative fact follows from the L^2 bootstrap assumptions (20.1a) and (20.1c), the assumption that $\mu_\star > 0$, and standard Sobolev embedding $H^2(\Sigma_t^u) \hookrightarrow C^0(\Sigma_t^u)$.

Throughout the remainder of the proof, we silently use the estimate $\varrho(t,u) \approx 1+t$ (on $\mathcal{M}_{T_{(Max);U_0},U_0}$) and the fact that $\mathbb{Q}_{(\leq N)}$ and $\widetilde{\mathbb{Q}}_{(\leq N)}$ are increasing in their arguments.

The estimate (22.6) now follows from the above reasoning and (22.16). Furthermore, since $\varrho L \in \mathscr{Z}$, the estimate (22.6) implies that $\left\| L\mathscr{Z}^{\leq 12}\Psi \right\|_{C^0(\Sigma_t^{U_0})}$ is uniformly bounded for $0 \leq t < T_{(Max);U_0}$. Hence, recalling that $L = \frac{\partial}{\partial t}$, we conclude that if $T_{(Max);U_0} < \infty$, then the functions $\mathscr{Z}^{\leq 12}\Psi$ extend as continuous functions of (t,u,ϑ) to $\Sigma_{T_{(Max);U_0}}^{U_0}$.

The estimates (22.7a) and (22.7b) now follow from inequalities (12.14a) and (12.14c), Cor. 12.48, and (22.16). As in the previous paragraph, the estimates (22.7a) and (22.7b) imply that if $T_{(Max);U_0} < \infty$, then the quantities $\mathscr{Z}^{\leq 12}\{g_{\mu\nu} - m_{\mu\nu}\}$, etc., extend as continuous functions of (t,u,ϑ) to $\Sigma_{T_{(Max);U_0}}^{U_0}$.

The estimates (22.8a)-(22.8c) now follow from (12.128a), (22.6), and the fact that $R_{(Small)}^i = -L_{(Small)}^i$ plus a smooth function of Ψ that vanishes at $\Psi = 0$ (see (2.4)-(2.6) and (4.3)). As in the previous two paragraphs, the estimates (22.8a)-(22.8c) imply that if $T_{(Max);U_0} < \infty$, then the quantities $\mathscr{Z}^{\leq 11}\mu$, etc., extend as continuous functions of (t,u,ϑ) to $\Sigma_{T_{(Max);U_0}}^{U_0}$.

The estimates (22.14n)-(22.14p) now follow from inserting the already proven estimates for $\mathbb{Q}_{(\leq 24)}$ and $\widetilde{\mathbb{Q}}_{(\leq 24)}$ into inequality (20.24a) and using the inequalities (13.143) and (13.146).

The estimate (22.14m) now follows from inserting the already proven estimates for $\mathbb{Q}_{(\leq 24)}$ and $\widetilde{\mathbb{Q}}_{(\leq 24)}$ and (22.14n) into inequality (20.38) and using inequality (13.146).

The estimates (22.14j)-(22.14l) now follow from inserting the already proven estimates for $\mathbb{Q}_{(\leq 24)}$ and $\widetilde{\mathbb{Q}}_{(\leq 24)}$ into inequality (19.2b).

The estimates (22.14g)-(22.14i) now follow from inserting the already proven estimates for $\mathbb{Q}_{(\leq 24)}$ and $\widetilde{\mathbb{Q}}_{(\leq 24)}$ into inequality (19.2a) and using inequality (13.147).

The estimates (22.14d)-(22.14f) now follow from inserting the already proven estimates for $\mathbb{Q}_{(\leq 16+M)}$ and $\widetilde{\mathbb{Q}}_{(\leq 16+M)}$ into inequality (19.2a) and using inequality (13.147).

The estimates (22.14a)-(22.14c) now follow from inserting the already proven estimates for $\mathbb{Q}_{(\leq 16)}$ and $\widetilde{\mathbb{Q}}_{(\leq 16)}$ into inequality (19.1a) and, in the case of $\left\|\mathscr{Z}^{15}(\upmu - 1)\right\|_{L^2(\Sigma_t^u)}$, $\left\|\mathscr{Z}^{15} L_{(Small)}^i\right\|_{L^2(\Sigma_t^u)}$, and $\left\|\not{\mathcal{L}}_{\mathscr{Z}}^{14}\chi^{(Small)}\right\|_{L^2(\Sigma_t^u)}$, using inequality (13.152).

The estimate (22.3a) now follows from (13.18) and (22.16). The estimate (22.3b) now follows from (12.141), (12.142), and (22.16). The estimate (22.3c) follows from (12.21a). The estimate (22.3d) is a special case of the already proven estimate (22.6).

The estimate (22.10) was proved as (13.28). Inequality (22.11) then follows from the identity $R = \upmu^{-1}\breve{R}$, definition (13.3a), and the estimates (13.11) and (22.10). To estimate the Euclidean length of the difference of R and $-\partial_r$, we use the identity $\partial_r = \frac{x^a}{r}\partial_a$, (4.1b), and the estimates (12.8b) (by Cor. 12.48 and (22.16), this estimate is valid with $\varepsilon^{1/2}$ replaced by $C\mathring{\epsilon}$) and (22.8b) to deduce the desired bound as follows:

$$|R - (-\partial_r)|_e \lesssim \sum_{a=1}^{3}\left|\frac{x^a}{r}\right|\left|1 - \frac{r}{\varrho}\right| + \sum_{a=1}^{3}\left|R_{(Small)}^a\right| \lesssim \mathring{\epsilon}\frac{\ln(e+t)}{1+t}. \tag{22.17}$$

We now prove the statements concerning the behavior of the change of variables map Υ in the case $T_{(Max);U_0} < \infty$. Using equation (3.82) and the argument that we used in the proof of Cor. 12.61, it is straightforward to see that the extendability of Υ to $\Sigma_{T_{(Max);U_0}}^{U_0}$ as a C^{11} function of $(t, u, \vartheta^1, \vartheta^2)$ follows after we prove that **i)** $\mathscr{Z}^{\leq 12} L^i$, $\mathscr{Z}^{\leq 12}\breve{R}^i$, $\mathscr{Z}^{\leq 11}\Xi^i$, and $\mathscr{Z}^{\leq 11} X_A^i$ extend as continuous functions of $(t, u, \vartheta^1, \vartheta^2)$ to $\Sigma_{T_{(Max);U_0}}^{U_0}$ and **ii)** for each $Z \in \mathscr{Z}$, the components of Z relative to the geometric coordinates extend as C^{11} functions of $(t, u, \vartheta^1, \vartheta^2)$ to $\Sigma_{T_{(Max);U_0}}^{U_0}$. The result **ii)** follows from the estimates of Lemma 12.60, the argument that we used in the proof of Cor. 12.61, and the argument that we used to prove the continuous extendability of $\mathscr{Z}^{\leq 12}\Psi$. We note for later use that **ii)** and Lemma 4.10 imply that the vectorfields $\not{\mathcal{L}}_{\mathscr{Z}}^{\leq 11} X_A = \mathcal{L}_{\mathscr{Z}}^{\leq 11} X_A$ also continuously extend as functions of $(t, u, \vartheta^1, \vartheta^2)$ to $\Sigma_{T_{(Max);U_0}}^{U_0}$. To prove **i)**, we first note that the previous arguments have shown that $\mathscr{Z}^{\leq 12} L^i$ and $\mathscr{Z}^{\leq 12}\breve{R}^i = \mathscr{Z}^{\leq 12}(\upmu R^i)$ extend as continuous functions of $(t, u, \vartheta^1, \vartheta^2)$ to $\Sigma_{T_{(Max);U_0}}^{U_0}$. Using similar reasoning and the estimate (12.174c), we conclude that $\mathscr{Z}^{\leq 11}\Xi^i$ extends as a continuous function of $(t, u, \vartheta^1, \vartheta^2)$ to $\Sigma_{T_{(Max);U_0}}^{U_0}$. To handle the rectangular components X_A^i, we first use similar reasoning and the estimate (12.57b) to deduce that the $S_{t,u}$ one-form

$\not d\mathscr{Z}^{\leq 11}x^i$ extends as a continuous function of $(t, u, \vartheta^1, \vartheta^2)$ to $\Sigma_{T_{(Max);U_0}}^{U_0}$. Using the aforementioned continuous extendability of the vectorfields $\not{\mathcal{L}}_{\mathscr{Z}}^{\leq 11} X_A$ and the identity $X_A^i = X_A \cdot \not d x^i$, we conclude that $\mathscr{Z}^{\leq 11} X_A^i$ extends as a continuous function of $(t, u, \vartheta^1, \vartheta^2)$ to $\Sigma_{T_{(Max);U_0}}^{U_0}$. We have thus proved the extendability of Υ as a C^{11} function of $(t, u, \vartheta^1, \vartheta^2)$.

With the help of the above results, we now show that either **a)** $T_{(Max);U_0} = \infty$ or **b)** $T_{(Max);U_0} < \infty$ and $\inf\left\{\mu_\star(t, U_0) \mid t \in [0, T_{(Max);U_0})\right\} = 0$. We note that after we have shown that either **a)** or **b)** must hold, it follows from the comments made in the second paragraph of the proof that $T_{(Max);U_0} = T_{(Lifespan);U_0}$ (where $T_{(Lifespan);U_0}$ is defined in (22.1)). Furthermore, in the case $T_{(Lifespan);U_0} < \infty$, it follows from the estimates (22.11) and (22.17) that the vanishing of $\mu_\star(t, U_0)$ at time $T_{(Lifespan);U_0}$ leads to the blow-up result (22.2).

It remains for us show that **a)** or **b)** must hold. It suffices to show that it is impossible to have $T_{(Max);U_0} < \infty$ unless $\inf\left\{\mu_\star(t, U_0) \mid t \in [0, T_{(Max);U_0})\right\} = 0$. To proceed, we assume for the sake of deriving a contradiction that: $T_{(Max);U_0} < \infty$ and $\inf\left\{\mu_\star(t, U_0) \mid t \in [0, T_{(Max);U_0})\right\} > 0$. We now rule out, on the spacetime domain $\mathcal{M}_{T_{(Max);U_0}, U_0}$, the 4 breakdown scenarios from **Part II)** of Prop. 21.1, and we also show that the condition 5 for Υ is verified (when ε is sufficiently small). Scenario 1 is ruled out by assumption. Scenario 2 is ruled out by the estimate (12.128a). Scenario 3 is ruled out by the bootstrap assumptions (**BA**Ψ) and the fact that $g_{\mu\nu}^{(Small)}(\cdot)$ is a smooth function of Ψ that vanishes at $\Psi = 0$ (see (2.4)-(2.6)).

In the next paragraph, we show that Υ is a C^{11} diffeomorphism from $[0, T_{(Max);U_0}] \times [0, U_0] \times \mathbb{S}^2$ onto its image and hence Υ verifies the condition 5 from **Part II)** of Prop. 21.1. Given this fact, we can easily rule out the breakdown scenario 4 with the help of the following additional fact: Ψ extends to $\Sigma_{T_{(Max);U_0}}^{U_0}$ as a C^{12} function of (t, u, ϑ). This additional fact follows from the previously shown fact that the functions $\mathscr{Z}^{\leq 12}\Psi$ extend as continuous functions of (t, u, ϑ) to $\Sigma_{T_{(Max);U_0}}^{U_0}$ and the argument that we used in our proof that Υ extends to $\Sigma_{T_{(Max);U_0}}^{U_0}$ as a C^{11} function of (t, u, ϑ). Combining this fact with the regular nature of Υ^{-1} (established in the next paragraph), we conclude that $\partial_\kappa \Psi$, $(\kappa = 0, 1, 2, 3)$, extends to $\mathcal{M}_{T_{(Max);U_0}, U_0} \cup \Sigma_{T_{(Max);U_0}}^{U_0}$ as a continuous function of the *rectangular coordinates* and that $\sup_{\mathcal{M}_{T_{(Max);U_0}, U_0} \cup \Sigma_{T_{(Max);U_0}}^{U_0}} \max_{\kappa=0,1,2,3} |\partial_\kappa \Psi| < \infty$. Therefore, we have ruled out the breakdown scenario 4.

It remains for us to show that condition 5 from **Part II)** of Prop. 21.1 holds. We first recall that by assumption, Υ is a bijection from the domain $[0, T_{(Max);U_0}) \times [0, U_0] \times \mathbb{S}^2$ onto its image. Moreover, we have already shown that Υ extends as a C^{11} function of (t, u, ϑ) to the compact set $[0, T_{(Max);U_0}] \times [0, U_0] \times \mathbb{S}^2$. Furthermore, from equation (3.81), the estimates (12.128a), (12.160), and (22.6), and the assumption $\inf\left\{\mu_\star(t, U_0) \mid t \in [0, T_{(Max);U_0})\right\} > 0$, we conclude that the Jacobian determinant of Υ is strictly positive and finite on $[0, T_{(Max);U_0}] \times [0, U_0] \times \mathbb{S}^2$. From the inverse function theorem, we conclude that Υ extends to $[0, T_{(Max);U_0}] \times [0, U_0] \times \mathbb{S}^2$ as a locally invertible C^{11} function. We now show that Υ is a C^{11} (global) diffeomorphism from $[0, T_{(Max);U_0}] \times [0, U_0] \times \mathbb{S}^2$ onto its image. To this end, we first note that inequality (12.187), the assumption $\inf\left\{\mu_\star(t, U_0) \mid t \in [0, T_{(Max);U_0})\right\} > 0$, and the estimate $r \geq (1 - \mathcal{O}(\varepsilon))\varrho \geq (1 - \mathcal{O}(\varepsilon))(1 - U_0)$ (which follows from (12.8b) and Cor. 12.48) imply that $\sum_{a=1}^{3} |\partial_a u|$ is uniformly bounded from above and below strictly away from 0. It follows that the closed outgoing null cone portions $\mathcal{C}_u^{T_{(Max)};U_0}$ corresponding to two distinct values $u \in [0, U_0]$ cannot intersect. Hence, to prove the global invertibility of Υ, it remains only for us to show that[4] for each $u \in [0, U_0]$, distinct integral curves of L, which rule $\mathcal{C}_u^{T_{(Max)};U_0}$, do not intersect[5] at time $T_{(Max);U_0}$. Equivalently, we show that for each fixed $(t, u) \in [0, T_{(Max);U_0}] \times [0, U_0]$, the map $\Upsilon(t, u, \cdot)$, defined on the domain $\mathbb{S}^2$, is injective. To this end, we define, relative to the rectangular coordinates, the map $F(t, x^1, x^2, x^3) : \Sigma_t \backslash \{(t, 0, 0, 0)\} \to \mathbb{S}^2 \subset \mathbb{R}^3$ by $F(t, x^1, x^2, x^3) = (x^1/r, x^2/r, x^3/r)$. Consider the composition $F \circ \Upsilon(t, u, \cdot) : \mathbb{S}^2 \to \mathbb{S}^2$, which is C^{11} in view of the estimate $r \geq (1 - \mathcal{O}(\varepsilon))(1 - U_0)$ mentioned above. Just below, we show that $F \circ \Upsilon(t, u, \cdot)$ is a C^{11} immersion.[6] Given this fact, it follows from standard topological arguments, based on the compactness and connectedness of $\mathbb{S}^2$, that $F \circ \Upsilon(t, u, \cdot)$ is a covering map.[7] Then, since $\mathbb{S}^2$ is simply connected, it is a basic result of algebraic topology (see [**65**, Theorem 54.4] and note that $\mathbb{S}^2$ has a trivial fundamental group since it is simply connected) that $F \circ \Upsilon(t, u, \cdot)$ must be a C^{11} diffeomorphism from $\mathbb{S}^2$ to $\mathbb{S}^2$, which yields the desired injectivity of $\Upsilon(t, u, \cdot)$. It remains for us to show that for each fixed $(t, u) \in [0, T_{(Max);U_0}] \times [0, U_0]$, $F \circ \Upsilon(t, u, \cdot)$ is an immersion. That is, we must show that the 3×2 matrix $\dfrac{\partial(F \circ \Upsilon)}{\partial(\vartheta^1, \vartheta^2)}$ is rank 2. To proceed, we first compute that the 3×4 matrix $\dfrac{\partial F}{\partial(t, x^1, x^2, x^3)}$ is rank 2 with kernel K given by $K = \text{span}\{(1, 0, 0, 0)^\intercal, (0, x^1/r, x^2/r, x^3/r)^\intercal\}$, where $\intercal$ denotes the transpose. Next, we note that inequality (12.187) implies that the Euclidean outward normal to $S_{t,u}$ in Σ_t at the point $(t, x^1, x^2, x^3) = \Upsilon(t, u, \vartheta)$ has rectangular components equal to $(0, x^1/r, x^2/r, x^3/r)^\intercal$ up to a Σ_t-tangent error vector of Euclidean length $\mathcal{O}\left(\varepsilon \ln(e + t)(1 + t)^{-1}\right)$. Hence, since the two columns of the 4×2 matrix $\dfrac{\partial \Upsilon}{\partial(\vartheta^1, \vartheta^2)}$ are tangent to $S_{t,u}$, they do not belong to K whenever ε is sufficiently small. Since the two columns are also linearly dependent, it follows that their union with K forms a basis of $\mathbb{R}^4$. It thus follows from the chain rule that the 3×2 matrix $\dfrac{\partial(F \circ \Upsilon)}{\partial(\vartheta^1, \vartheta^2)}$ is rank 2 as desired.

In total, we conclude from **Part II)** of Prop. 21.1 that there exists a $\Delta > 0$ such that Ψ, u, $\upmu$, $L^i_{(Small)}$, etc. can be extended as solutions, for some $\Delta > 0$, to a strictly larger region of the form $\mathcal{M}_{T_{(Max);U_0}+\Delta, U_0}$ on which $\upmu_\star$ is uniformly positive. Furthermore, the proposition also yields that the quantities appearing on the left-hand sides of (20.1a)-(20.1f) are continuous in t and u. In view of this

[4]Recall that the angular coordinates ϑ^A are, by construction, constant along the integral curves of L.

[5]Equivalently, we must show that if $q \in \mathcal{C}_u^{T_{(Max)};U_0}$, then no point in $\mathcal{C}_u^{T_{(Max)};U_0}$ can belong to the null cut locus of q.

[6]Recall that $F \circ \Upsilon(t, u, \cdot)$ is said to be an immersion if the Jacobian $\dfrac{\partial(F \circ \Upsilon)}{\partial(\vartheta^1, \vartheta^2)}$ is rank 2 at each point in its domain $\mathbb{S}^2$.

[7]A continuous surjective function $f : X \to \mathbb{S}^2$ is said to be a covering map of $\mathbb{S}^2$ if for every $p \in \mathbb{S}^2$, there exists a neighborhood Ω of p such that $f^{-1}(\Omega)$ is a disjoint union of open sets in X, each of which is mapped homeomorphically onto Ω by f.

fact and the bound (22.16), we deduce that if Δ is small enough, then the energy bootstrap assumptions (20.1a)-(20.1f) hold for $(t, u) \in \times[0, T_{(Max);U_0} + \Delta) \times [0, U_0]$ with ε replaced by $\frac{3}{4}\varepsilon$. We have thus arrived at the desired contradiction of the definition of $T_{(Max);U_0}$.

To prove (22.4), that is, that there exists a constant $C_{(Lower-Bound)} > 0$ such that $\mu_\star(t, U_0) > 0$ whenever $t < \exp\left(\{C_{(Lower-Bound)}\mathring{\epsilon}\}^{-1}\right)$, we only have to combine (22.3a), (22.3c), and (22.3d). Similar reasoning also yields (22.5).

Finally, to conclude, in the case $T_{(Lifespan);U_0} < \infty$, that the Jacobian determinant of Υ vanishes precisely on the subset (22.9), we use equation (3.81), the estimate (22.6) for Ψ, and inequality (12.160) (which, by (22.16), is valid with ε replaced by $C\mathring{\epsilon}$). □

22.2. More precise control over angular derivatives

We now state and prove a corollary of Theorem 22.1 that provides additional information about some of the angular derivatives of Ψ. Roughly, the corollary allows us to infer that nearly spherically symmetric data launch nearly spherically solutions, even though the wave equation (2.1a) is not necessarily invariant under Euclidean rotations. The estimates from the corollary play an important role in our proof of small data shock formation, which is based on arguments that are valid for nearly spherically symmetric solutions.

COROLLARY 22.4 (**More precise control over the lower-order angular derivatives**). *Assume the hypotheses and conclusions of Theorem 22.1. In particular, let $\mathring{\epsilon}$ be the size of the data, and let $T_{(Lifespan);U_0}$ and U_0 be the numbers appearing in the statement of the theorem. Let $\mathbb{E}[\cdot]$, $\mathbb{F}[\cdot]$, $\widetilde{\mathbb{E}}[\cdot]$, $\widetilde{\mathbb{F}}[\cdot]$, and $\widetilde{\mathbb{K}}[\cdot]$ be the energies, cone fluxes, and Morawetz spacetime integral from Defs. 10.11 and 14.3. Let $\mathscr{O}$ be the set of geometric rotation vectorfields (see definition (7.2b)) and let*

$$\mathring{\delta} := \max_{2 \leq |\vec{I}| \leq 4} \mathbb{E}^{1/2}[\mathscr{O}^{\vec{I}}\Psi](0, U_0). \tag{22.18}$$

There exists a constant $C > 0$ such that the following estimates hold for $(t, u) \in [0, T_{(Lifespan);U_0}) \times [0, U_0]$:

$$\begin{aligned} &\max_{2 \leq |\vec{I}| \leq 4} \left\{\mathbb{E}^{1/2}[\mathscr{O}^{\vec{I}}\Psi](t, u) + \mathbb{F}^{1/2}[\mathscr{O}^{\vec{I}}\Psi](t, u)\right\} \\ &\leq C(\mathring{\delta} + \mathring{\epsilon}^{3/2}), \end{aligned} \tag{22.19a}$$

$$\begin{aligned} &\max_{2 \leq |\vec{I}| \leq 4} \left\{\widetilde{\mathbb{E}}^{1/2}[\mathscr{O}^{\vec{I}}\Psi](t, u) + \widetilde{\mathbb{F}}^{1/2}[\mathscr{O}^{\vec{I}}\Psi](t, u) + \widetilde{\mathbb{K}}^{1/2}[\mathscr{O}^{\vec{I}}\Psi](t, u)\right\} \\ &\leq C(\mathring{\delta} + \mathring{\epsilon}^{3/2}) \ln^2(e + t), \end{aligned} \tag{22.19b}$$

$$\begin{aligned} &\|\not{\Delta}\Psi\|_{C^0(\Sigma_t^u)} \\ &\leq C(\mathring{\delta} + \mathring{\epsilon}^{3/2}) \frac{1}{(1 + t)^3}. \end{aligned} \tag{22.19c}$$

PROOF. The estimate (22.19c) follows from (22.19a) and inequalities (12.46a), (12.46c), and (18.32). Hence, we need only to prove inequalities (22.19a) and (22.19b).

To this end, we commute the wave equation (2.1a) with iterated rotational differential operators $\mathscr{O}^{\vec{I}}$, where $2 \leq |\vec{I}| \leq 4$, to derive an equation of the form

$$\mu \square_{g(\Psi)} \mathscr{O}^{\vec{I}} \Psi = {}^{(\mathscr{O}^{\vec{I}})}\mathfrak{F}. \tag{22.20}$$

Using (16.3), (16.4), (7.8a)-(7.9d), (9.13), and the C^0 estimates of Theorem 22.1, we see that ${}^{(\mathscr{O}^{\vec{I}})}\mathfrak{F}$ is *quadratically small* in the sense that its magnitude is bounded by a decaying function of time multiplied by $\mathring{\epsilon}^2$. This quadratic smallness is in particular based on the fact that the deformation tensors ${}^{(O)}\pi$ of the rotations completely vanish for the background solution $\Psi \equiv 0$ (see (7.8a)-(7.9d)). The reason is that the rotations O are equal to the Euclidean rotations when $\Psi \equiv 0$, and the Euclidean rotations are Killing fields[8] of the background Minkowski metric. We remark that this stands in contrast to ${}^{(\breve{R})}\pi$ and ${}^{(\varrho L)}\pi$, which, even in the flat case $\Psi \equiv 0$, do not vanish.

We now revisit the energy-flux identities (10.29a)-(10.29b), where $\mathscr{O}^{\vec{I}}\Psi$ plays the role of Ψ and ${}^{(\mathscr{O}^{\vec{I}})}\mathfrak{F}$ plays the role of $\mathfrak{F}$. To simplify the notation, we set

$$\mathbb{Q}_{(\mathscr{O})}(t,u) := \max_{2 \leq |\vec{I}| \leq 4} \left\{ \mathbb{E}[\mathscr{O}^{\vec{I}}\Psi](t,u) + \mathbb{F}[\mathscr{O}^{\vec{I}}\Psi](t,u) \right\}, \tag{22.21}$$

$$\widetilde{\mathbb{Q}}_{(\mathscr{O})}(t,u) := \max_{2 \leq |\vec{I}| \leq 4} \left\{ \widetilde{\mathbb{E}}[\mathscr{O}^{\vec{I}}\Psi](t,u) + \widetilde{\mathbb{F}}[\mathscr{O}^{\vec{I}}\Psi](t,u) \right\}, \tag{22.22}$$

$$\widetilde{\mathbb{K}}_{(\mathscr{O})}(t,u) := \max_{2 \leq |\vec{I}| \leq 4} \widetilde{\mathbb{K}}[\mathscr{O}^{\vec{I}}\Psi](t,u), \tag{22.23}$$

where $\widetilde{\mathbb{K}}[\cdot](t,u)$ is the Morawetz integral from (14.9a). Our goal is to use the already proven non-$\mu_\star^{-1}$-degenerate estimates for the lower-order derivatives of various quantities implied by Theorem 22.1 in order to prove that the following estimates hold for $(t,u) \in [0, T_{(Lifespan);U_0}) \times [0, U_0]$:

$$\begin{aligned} &\mathbb{Q}_{(\mathscr{O})}(t,u) \\ &\leq C\mathring{\delta}^2 + C\mathring{\epsilon}^3 + C\int_{t'=0}^t \frac{\ln(e+t')}{(1+t')^2}\mathbb{Q}_{(\mathscr{O})}(t',u)\,dt' \\ &\quad + C\int_{t'=0}^t \frac{\ln^2(e+t')}{(1+t')^2}\widetilde{\mathbb{Q}}_{(\mathscr{O})}(t',u)\,dt' \\ &\quad + C\mathring{\epsilon}^{1/2}\int_{t'=0}^t \frac{\ln(e+t')}{(e+t')^2\sqrt{\ln(e+t)-\ln(e+t')}}\widetilde{\mathbb{Q}}_{(\mathscr{O})}(t',u)\,dt' \\ &\quad + C\frac{\ln(e+t)}{(1+t)^2}\int_{u'=0}^u \widetilde{\mathbb{Q}}_{(\mathscr{O})}(t,u')\,du', \end{aligned} \tag{22.24}$$

[8] That is, for $O_{(Flat)} \in \mathscr{O}_{(Flat)}$ (see definitions (6.2) and (7.3)), we have $\mathcal{L}_{O_{(Flat)}} m = 0$, where m is the Minkowski metric.

$$
\begin{aligned}
(22.25) \qquad & \max\left\{\widetilde{\mathbb{Q}}_{(\mathscr{O})}(t,u), \widetilde{\mathbb{K}}_{(\mathscr{O})}(t,u)\right\} \\
& \leq C\mathring{\delta}^2 + C\mathring{\epsilon}^3 \ln^4(e+t) \\
& + C\int_{t'=0}^{t} \frac{\ln^3(e+t')}{(1+t')}\mathbb{Q}_{(\mathscr{O})}(t',u)\,dt' + C\ln^2(e+t)\mathbb{Q}_{(\mathscr{O})}(t,u) \\
& + (1+C\varepsilon)\int_{t'=0}^{t} \frac{1}{\varrho(t',u)\ln\left(\frac{\varrho(t',u)}{\varrho(0,u)}\right)}\widetilde{\mathbb{Q}}_{(\mathscr{O})}(t',u)\,dt' \\
& + C\int_{u'=0}^{u} \widetilde{\mathbb{Q}}_{(\mathscr{O})}(t,u')\,du'.
\end{aligned}
$$

Once we have shown (22.24)-(22.25), the desired estimates (22.19a)-(22.19b) follow from applying a Gronwall-type argument to the system of inequalities (22.24)-(22.25) in the variables $\mathbb{Q}_{(\mathscr{O})}$, $\widetilde{\mathbb{Q}}_{(\mathscr{O})}$, and $\widetilde{\mathbb{K}}_{(\mathscr{O})}$. Because the argument is much simpler than the Gronwall estimates that we proved above for the below-top-order quantities $\mathbb{Q}_{(\leq 23)}$, $\widetilde{\mathbb{Q}}_{(\leq 23)}$, and $\widetilde{\mathbb{K}}_{(\leq 23)}(t,u)$ (the proof of those estimates starts just above equation (20.190)), we omit the details. However, we remark that we again use Lemma 20.35 to handle the next-to-last term on the right-hand side of (22.24).

In order to prove (22.24)-(22.25), we have to estimate the integrals on the right-hand sides of (10.29a)-(10.29b), where $\mathscr{O}^{\vec{I}}\Psi$ plays the role of Ψ and ${}^{(\mathscr{O}^{\vec{I}})}\mathfrak{F}$ plays the role of $\mathfrak{F}$. For each fixed $\vec{I}$, we show how to deduce an inequality of the form (22.24)-(22.25), but with $\mathbb{E}[\mathscr{O}^{\vec{I}}\Psi](t,u) + \mathbb{F}[\mathscr{O}^{\vec{I}}\Psi](t,u)$ on the left-hand side of (22.24) in place of $\mathbb{Q}_{(\mathscr{O})}(t,u)$ and $\widetilde{\mathbb{E}}[\mathscr{O}^{\vec{I}}\Psi](t,u) + \widetilde{\mathbb{F}}[\mathscr{O}^{\vec{I}}\Psi](t,u) + \widetilde{\mathbb{K}}[\mathscr{O}^{\vec{I}}\Psi](t,u)$ on the left-hand side of (22.25) in place of $\widetilde{\mathbb{Q}}_{(\mathscr{O})}(t,u) + \widetilde{\mathbb{K}}_{(\mathscr{O})}(t,u)$. We then take the max of both sides of the inequalities over all relevant quantities with $2 \leq |\vec{I}| \leq 4$ in order to deduce (22.24)-(22.25).

To begin the detailed analysis, we remark that we will handle the integrals involving ${}^{(\mathscr{O}^{\vec{I}})}\mathfrak{F}$ at the end of the proof. From the assumption (22.18), we see that the terms $\mathbb{E}[\mathscr{O}^{\vec{I}}\Psi](0,u)$ arising from the first term on the right-hand side of (10.29a) are $\leq \mathring{\delta}^2$ and hence are $\leq$ the right-hand side of (22.24) as desired. We now explain how we bound the integral $\int_{\mathcal{M}_{t,u}} \mu Q^{\alpha\beta}[\mathscr{O}^{\vec{I}}\Psi]^{(T)}\pi_{\alpha\beta}\,d\varpi$ from the right-hand side of (10.29a) by $\leq$ the right-hand side of (22.24). To derive the desired bound, we examine the proof of inequality (20.30a) and conclude that the same inequality holds true with $\mathscr{O}^{\vec{I}}\Psi$ in place of $\mathscr{Z}^N\Psi$ on the left-hand side and $\mathbb{Q}_{(\mathscr{O})}$, $\widetilde{\mathbb{Q}}_{(\mathscr{O})}$, and $\widetilde{\mathbb{K}}_{(\mathscr{O})}$ respectively in place of $\mathbb{Q}_{(N)}$, $\widetilde{\mathbb{Q}}_{(N)}$, and $\widetilde{\mathbb{K}}_{(N)}$ on the right-hand side. Theorem 22.1 implies that Lemma 20.23 holds with ε replaced by $C\mathring{\epsilon}$ and furthermore, by (22.13b), that the last term on the right-hand side of (20.30a) (with $\widetilde{\mathbb{K}}_{(\mathscr{O})}$ in the role of $\widetilde{\mathbb{K}}_{(N)}$) is $\lesssim \mathring{\epsilon}^3$ and hence is bounded by the right-hand side of (22.24) as desired.

We now address the terms on the right-hand side of (10.29b) that do not involve the integrand factor ${}^{(\mathscr{O}^{\vec{I}})}\mathfrak{F}$, where $\mathscr{O}^{\vec{I}}\Psi$ plays the role of Ψ. The integrals over the hypersurface Σ_0^u are quadratic in $\mathscr{O}^{\vec{I}}\Psi$ and are controlled by the data. Hence, by (14.1a) and the assumption (22.18), the magnitudes of these integrals are $\lesssim \mathring{\delta}^2$ and thus $\leq$ the right-hand side of (22.25) as desired. Clearly, the same bound holds for the first term $\widetilde{\mathbb{E}}[\cdot](0,u)$ on the right-hand side of (10.29b). To bound the integrals over the hypersurface Σ_t^u and the last spacetime integral on the right-hand side of (10.29b) by $\leq$ the right-hand side of (22.25), we examine the proofs of inequalities (20.30c) and (20.30d) and conclude that the same inequalities hold true with $\mathscr{O}^{\vec{I}}\Psi$ in place of $\mathscr{Z}^N\Psi$ on the left-hand side and $\mathbb{Q}_{(\mathscr{O})}$ in place of $\mathbb{Q}_{(N)}$ on the right-hand side. Hence, the integrals under consideration are in magnitude $\lesssim \ln^2(e+t)\mathbb{Q}_{(\mathscr{O})}(t,u)$ as desired. To handle the spacetime integral $-\frac{1}{2}\int_{\mathcal{M}_{t,u}} \upmu Q^{\alpha\beta}[\mathscr{O}^{\vec{I}}\Psi]\left\{{}^{(\widetilde{K})}\pi_{\alpha\beta} - \varrho^2 \mathrm{tr}_{\not g}\chi g_{\alpha\beta}\right\} d\varpi$ from the right-hand side of (10.29b), we first appeal to definitions (14.9a) (with $\mathscr{O}^{\vec{I}}\Psi$ in the role of Ψ) and (10.48b), which show that $-\widetilde{\mathbb{K}}[\mathscr{O}^{\vec{I}}\Psi]$ is equal to one of the terms that make up this integral. We bring the term $\widetilde{\mathbb{K}}[\mathscr{O}^{\vec{I}}\Psi]$ over to the left-hand side of (10.29b) as a positive integral. We then bound the remaining part of this spacetime integral in magnitude by $\leq$ the right-hand side of (22.25) with the help of (20.30b). More precisely, examining the proof of (20.30b), we see that the inequality holds true with with $\mathscr{O}^{\vec{I}}\Psi$ in place of $\mathscr{Z}^N\Psi$ on the left-hand side and $\mathbb{Q}_{(\mathscr{O})}$, $\widetilde{\mathbb{Q}}_{(\mathscr{O})}$, and $\widetilde{\mathbb{K}}_{(\mathscr{O})}$ respectively in place of $\mathbb{Q}_{(N)}$, $\widetilde{\mathbb{Q}}_{(N)}$, and $\widetilde{\mathbb{K}}_{(N)}$ on the right-hand side, which yields the desired bounds. We remark that the argument given at the end of the previous paragraph yields that the last term on the right-hand side of (20.30b) (with $\widetilde{\mathbb{K}}_{(\mathscr{O})}$ in the role of $\widetilde{\mathbb{K}}_{(N)}$) is $\lesssim \mathring{\epsilon}^3\ln^4(e+t)$ as desired.

To complete the proof of (22.24)-(22.25), it remains for us to show that the second term on the right-hand side of (10.29a) and the third-from-last term on the right-hand side of (10.29b) are bounded as follows:

$$\left|\int_{\mathcal{M}_{t,u}} \left\{(1+2\upmu)(L\mathscr{O}^{\vec{I}}\Psi)^{(\mathscr{O}^{\vec{I}})}\mathfrak{F} + 2(\breve{R}\mathscr{O}^{\vec{I}}\Psi)\right\}{}^{(\mathscr{O}^{\vec{I}})}\mathfrak{F}\, d\varpi\right| \lesssim \mathring{\epsilon}^3, \tag{22.26}$$

$$\left|\int_{\mathcal{M}_{t,u}} \varrho^2\left\{L\mathscr{O}^{\vec{I}}\Psi + \frac{1}{2}\mathrm{tr}_{\not g}\chi\mathscr{O}^{\vec{I}}\Psi\right\}{}^{(\mathscr{O}^{\vec{I}})}\mathfrak{F}\, d\varpi\right| \lesssim \mathring{\epsilon}^3\ln^4(e+t). \tag{22.27}$$

The most important ingredient in the proofs of (22.26) and (22.27) is the fact that all integrand terms on the left-hand sides are in fact *cubically* small. We stress again that we have cubic smallness only because all commutation vectorfields under consideration are rotations. More precisely, to prove (22.26) and (22.27), we first note that the estimates (16.3) and (16.4) imply that for $|\vec{I}| \leq 4$, the terms ${}^{(\mathscr{O}^{\vec{I}})}\mathfrak{F}$ are $\lesssim$ the right-hand side of (16.4) with $N=5$. Hence, by (16.7), we deduce that for $|\vec{I}| \leq 4$, the terms ${}^{(\mathscr{O}^{\vec{I}})}\mathfrak{F}$ are $Harmless^{\leq 5}$ (see definition (16.5)) and note that by Theorem 22.1, in the estimate ${}^{(\mathscr{O}^{\vec{I}})}\mathfrak{F} = Harmless^{\leq 5}$, we can replace the ε factor on the right-hand side of (16.5) with with $C\mathring{\epsilon}$. Moreover, in the present context, there is additional structure present in the terms beyond the structure indicated on the right-hand side of (16.5). The additional structure is that $|{}^{(\mathscr{O}^{\vec{I}})}\mathfrak{F}|$ is $\lesssim$ a modified version of the right-hand side of (16.5) (with $N=5$) in which the terms on the first line of the right-hand side are also multiplied by a $\mathring{\epsilon}$ factor. This factor is simply a reflection of the quadratic smallness of ${}^{(\mathscr{O}^{\vec{I}})}\mathfrak{F}$. Hence, revisiting the proofs of (20.17) and (20.18), we find that, thanks to this $\mathring{\epsilon}$ factor, the left-hand sides of (22.26) and (22.27) are $\lesssim \mathring{\epsilon}$ times the terms on the right-hand sides of (20.17) and (20.18) with $N=5$ and with $C\mathring{\epsilon}$ in place of ε. Also using the estimates (22.13a) and (22.13b) to respectively bound the terms on the right-hand sides of (20.17) and (20.18) by $\lesssim \mathring{\epsilon}^2$ and $\lesssim \mathring{\epsilon}^2\ln^4(e+t)$, we conclude the desired estimates (22.26) and (22.27). □

CHAPTER 23

Proof of Shock Formation for Nearly Spherically Symmetric Data

In this chapter, we use the estimates of Theorem 22.1 and Cor. 22.4 to show that there exists an open set $\mathfrak{S}$ of nearly spherically symmetric small data that launch shock forming solutions. For technical reasons to be explained, the set $\mathfrak{S}$ consists of data on $\Sigma_{-1/2}$ that are compactly supported in the Euclidean ball of radius $1/2$ centered at the origin. We state the main result as Theorem 23.9. As we described in Sect. 2.11.3, the theorem can be extended to show shock formation for a significantly larger set of compactly supported small data.

By Theorem 22.1, in order to show that a shock forms, we must show that μ vanishes in finite time. The main idea of the proof is to use the estimates (22.3a)-(22.3d) of Theorem 22.1 to effectively reduce the problem to proving that the term $\varrho G_{LL} \breve{R}\Psi$ from inequality (22.3a) becomes negative along some integral curve of $L = \frac{\partial}{\partial t}|_{u,\vartheta}$; see Sects. 2.8 and 2.9 for an overview of the argument. The main term that we have yet to suitably estimate is the product $\varrho \breve{R}\Psi$. To estimate it, we consider the wave equation $\mu \square_{g(\Psi)} \Psi = 0$ in the form (5.13a) (with $f = \Psi$), which, thanks to the estimates of Theorem 22.1 and Cor. 22.4, can be viewed as a transport equation along the integral curves of L with small error sources. The role of Cor. 22.4 is that it allows us to prove that for nearly spherically symmetric data, the linear term $\varrho \not\Delta \Psi$ on the right-hand side of (5.13a) has a sufficiently small amplitude and can be treated as a negligible error term, even near $t = 0$; all of the remaining terms on the right-hand side are quadratic and therefore are much more easily seen to be negligible error terms. In total, for data in $\mathfrak{S}$, this line of reasoning allows us to prove that for sufficiently large times, the following estimate holds along some integral curve of L (that is, a curve with fixed geometric coordinates (u_*, ϑ_*)):

$$\mu(t, u_*, \vartheta_*) \leq 1 + C\mathring{\epsilon} - \alpha_* \ln \left(\frac{\varrho(t, u_*)}{\varrho(0, u_*)} \right), \tag{23.1}$$

where $0 < \alpha_* = \mathcal{O}(\mathring{\epsilon})$ is a small factor[1] depending on the data. It easily follows from (23.1) that $\mu_\star(t, U_0) := \min\{\inf_{\Sigma_t^{U_0}} \mu, 1\}$ vanishes by the time $t \sim \exp\left(\frac{1 + C\mathring{\epsilon}}{\alpha_*}\right)$.

[1]For the initial data $(\mathring{\Psi}, \mathring{\Psi}_0)$ treated by Theorem 23.9, our analysis shows that α_* can be chosen uniformly in some neighborhood of $(\mathring{\Psi}, \mathring{\Psi}_0)$ in $H_e^{25}(\Sigma_0^1) \times H_e^{24}(\Sigma_0^1)$.

23.1. Preliminary pointwise estimates based on approximate transport equations

We begin our detailed analysis with the following simple lemma, which shows that for the solutions Ψ under consideration, the equation $\upmu\square_{g(\Psi)}\Psi = 0$ can be treated as a transport equation with small error sources.

LEMMA 23.1 (**Wave equation approximately reduces to a transport equation**). *Assume the hypotheses and conclusions of Theorem 22.1 and Cor. 22.4. In particular, let* $\mathring{\epsilon}$ *be the size of the data* ($\mathring{\Psi} := \Psi|_{\Sigma_0^1}, \mathring{\Psi}_0 := \partial_t\Psi|_{\Sigma_0^1}$) *as defined in Def. 12.3, let* $\mathring{\delta}$ *be the initial size of the geometric rotational derivatives of* Ψ *(of order two through four) defined in (22.18), and let* $T_{(Lifespan);U_0}$ *and* U_0 *be the numbers appearing in the statement of the theorem and corollary. Recall that* $\varrho(t,u) = 1 - u + t$. *Then the following estimates hold on the spacetime domain* $\mathcal{M}_{T_{(Lifespan);U_0},U_0}$:

$$\left|L\breve{\underline{L}}[\varrho\Psi](t,u,\vartheta)\right| \leq C(\mathring{\delta} + \mathring{\epsilon}^{3/2})\frac{\ln(e+t)}{(1+t)^2}, \tag{23.2a}$$

$$\left|[\varrho\breve{R}\Psi](t,u,\vartheta) - \frac{1}{2}\breve{\underline{L}}[\varrho\Psi](0,u,\vartheta)\right| \leq C(\mathring{\delta} + \mathring{\epsilon}^{3/2}) + C\mathring{\epsilon}\frac{\ln(e+t)}{1+t}, \tag{23.2b}$$

$$\left|[\varrho\breve{R}\Psi](t,u,\vartheta) - \frac{1}{2}\left\{r\mathring{\Psi}_0 - r\partial_r\mathring{\Psi} - \mathring{\Psi}\right\}(u,\vartheta)\right| \leq C(\mathring{\delta} + \mathring{\epsilon}^{3/2}) + C\mathring{\epsilon}\frac{\ln(e+t)}{1+t}, \tag{23.2c}$$

where $\partial_r = \frac{x^a}{r}\partial_a$ *is the standard Euclidean radial vectorfield.*

PROOF. To deduce (23.2a), we use the wave equation $\upmu\square_{g(\Psi)}\Psi = 0$ in the form (5.13a) with $f = \Psi$. By Cor. 22.4, the linear term $\varrho\Delta\!\!\!/\Psi$ on the right-hand side of (5.13a) verifies the pointwise bound

$$|\varrho\Delta\!\!\!/\Psi| \lesssim (\mathring{\delta} + \mathring{\epsilon}^{3/2})\frac{1}{(1+t)^2}, \tag{23.3}$$

which is $\lesssim$ the right-hand side of (23.2a). From (4.6a), (4.6b), (5.4a), (5.4b), (5.5a), and (5.5b), we see that the remaining terms on the right-hand side of (5.13a) (aside from $\upmu\square_{g(\Psi)}\Psi$, which vanishes by assumption) are of the schematic form

$$\begin{aligned}\varrho(\upmu-1)\Delta\!\!\!/\Psi + (\upmu-1)L\Psi + G_{(Frame)}\begin{pmatrix}\upmu L\Psi\\ \breve{R}\Psi\end{pmatrix}\Psi \\ + \varrho G_{(Frame)}\not{g}^{-1}\begin{pmatrix}\upmu L\Psi\\ \breve{R}\Psi\\ \upmu\not{d}\Psi\end{pmatrix}\begin{pmatrix}L\Psi\\ \not{d}\Psi\end{pmatrix} + \varrho\mathrm{tr}_{\not{g}}\chi^{(Small)}\breve{R}\Psi.\end{aligned} \tag{23.4}$$

In particular, the terms in (23.4) are quadratically small, and from inequality (12.46c), the estimates (12.58b) and (12.128a), the bootstrap assumptions (**BA**Ψ), and the fact that Theorem 22.1 implies that these estimates hold with ε replaced by $C\mathring{\epsilon}$, we deduce that these terms are bounded in magnitude by $\lesssim \mathring{\epsilon}^2\ln(e+t)(1+t)^{-2}$ as desired. We have thus proved (23.2a).

To prove (23.2b), we first integrate (23.2a) along the integral curves of L to deduce that

$$\left|\breve{\underline{L}}[\varrho\Psi](t,u,\vartheta) - \breve{\underline{L}}[\varrho\Psi](0,u,\vartheta)\right| \lesssim \mathring{\delta} + \mathring{\epsilon}^{3/2}. \tag{23.5}$$

Next, noting that the identities $\breve{\underline{L}}\Psi = 2\breve{R}\Psi + \upmu L\Psi$ (see (3.19a)) and $\breve{\underline{L}}\varrho = \upmu - 2$ (see Lemma 3.18) imply that $\breve{\underline{L}}[\varrho\Psi] = 2\varrho\breve{R}\Psi + (\upmu - 2)\Psi + \varrho\upmu L\Psi$, and using the estimates $|(\upmu - 2)\Psi|$, $|\varrho\upmu L\Psi| \lesssim \mathring{\epsilon}\ln(e+t)(1+t)^{-1}$, which follow from the estimates stated just after (23.4), we deduce that

$$\left|\varrho\breve{R}\Psi(t,u,\vartheta) - \frac{1}{2}\breve{\underline{L}}[\varrho\Psi](t,u,\vartheta)\right| \lesssim \mathring{\epsilon}\frac{\ln(e+t)}{1+t}. \tag{23.6}$$

The desired estimate (23.2b) now follows from (23.5) and (23.6).

Next, once we have shown that

$$\breve{\underline{L}}[\varrho\Psi](0,u,\vartheta) = r\mathring{\Psi}_0(u,\vartheta) - r\partial_r\mathring{\Psi}(u,\vartheta) - \mathring{\Psi}(u,\vartheta) + \mathcal{O}(\mathring{\epsilon}^2), \tag{23.7}$$

the estimate (23.2c) follows from inequalities (23.2b) and (23.7). To prove (23.7), we first further expand the term $\breve{\underline{L}}[\varrho\Psi]$ as follows:

$$\breve{\underline{L}}[\varrho\Psi](0,u,\vartheta) = \left\{2\varrho\breve{R}\Psi + \varrho L\Psi - \Psi + (\upmu - 1)\Psi + \varrho(\upmu - 1)L\Psi\right\}(0,u,\vartheta). \tag{23.8}$$

The last two products on the right-hand side of (23.8) are quadratically small and hence from the estimates stated just after (23.4), they are $\lesssim \mathring{\epsilon}^2$. Next, using the estimates

$$\varrho(0,u,\vartheta) = r(0,u,\vartheta) = 1 - u, \tag{23.9}$$

$$\breve{R}^0(0,u,\vartheta) = 0, \tag{23.10}$$

$$\breve{R}^i(0,u,\vartheta) = -\frac{x^i}{r} + \mathcal{O}(\mathring{\epsilon}), \tag{23.11}$$

$$L^0(0,u,\vartheta) = 1, \tag{23.12}$$

$$L^i(0,u,\vartheta) = \frac{x^i}{r} + \mathcal{O}(\mathring{\epsilon}), \tag{23.13}$$

$$\|\mathring{\Psi}\|_{C^0(\Sigma_0^1)}, \sum_{a=1}^{3}\|\partial_a\mathring{\Psi}\|_{C^0(\Sigma_0^1)}, \|\mathring{\Psi}_0\| = \mathcal{O}(\mathring{\epsilon}) \tag{23.14}$$

(where (23.11) and (23.13) follow from (3.19b), Def. 4.1, Lemma 12.16, (12.17), and standard Euclidean Sobolev embedding), we can further expand the first two terms on the right-hand side of (23.8) as

$$[2\varrho\breve{R}\Psi](0,u,\vartheta) = -2r\partial_r\Psi(0,u,\vartheta) + \mathcal{O}(\mathring{\epsilon}^2), \tag{23.15}$$

$$[\varrho L\Psi](0,u,\vartheta) = r\partial_t\Psi(0,u,\vartheta) + r\partial_r\Psi(0,u,\vartheta) + \mathcal{O}(\mathring{\epsilon}^2). \tag{23.16}$$

The estimate (23.7) now follows from (23.8), the sentence after (23.8), (23.15), and (23.16). □

As we will see in Lemma 23.4, the following data-dependent function appearing in Lemma 23.1 plays a central role in our proof of shock formation.

DEFINITION 23.2 (**The data-dependent function that drives shock formation**). Let $\underline{L}_{(Flat)} := \partial_t - \partial_r$ denote the standard Minkowskian ingoing null vectorfield. Let $(\mathring{\Psi} := \Psi|_{\Sigma_0^{U_0}}, \mathring{\Psi}_0 := \partial_t\Psi|_{\Sigma_0^{U_0}})$ be initial data for the wave equation (2.1a). We define the following data-dependent function:

$$\mathbb{S}[(\mathring{\Psi},\mathring{\Psi}_0)] := \frac{1}{4}\underline{L}_{(Flat)}(r\Psi)|_{(\Psi,\partial_t\Psi)=(\mathring{\Psi},\mathring{\Psi}_0)} = \frac{1}{4}\left\{r\mathring{\Psi}_0 - r\partial_r\mathring{\Psi} - \mathring{\Psi}\right\}. \tag{23.17}$$

REMARK 23.3 (**The origin of $\mathbb{S}[(\mathring{\Psi}, \mathring{\Psi}_0)]$**). Note that $\mathbb{S}[(\mathring{\Psi}, \mathring{\Psi}_0)]$ is equal to half of the term in braces on the left-hand side of (23.2c).

In the next lemma, we insert the estimates of Lemma 23.1 into some estimates proved in Theorem 22.1. We arrive at the key pointwise estimate (23.18) verified by $\upmu$, which is the main ingredient in our proof of finite-time shock formation.

LEMMA 23.4 (**The key estimate that drives the shock formation**). *Let* $(\mathring{\Psi} := \Psi|_{\Sigma_0^{U_0}}, \mathring{\Psi}_0 := \partial_t \Psi|_{\Sigma_0^{U_0}})$ *be initial data for the wave equation* (2.1a). *Assume the hypotheses and conclusions of Lemma* 23.1 *and Lemma* 23.4. *In particular, let* $\mathring{\epsilon}$ *be the size of the data as defined in Def.* 12.3 , *let* $\mathring{\delta}$ *be the initial size of the geometric rotational derivatives of* Ψ *(of order two through four) defined by* (22.18), *and recall that* $\varrho(t,u) = 1-u+t$. *Let* ${}^{(+)}\mathring{\aleph}$ *be as in Def.* 3.32 *and let* $\mathbb{S}[(\mathring{\Psi}, \mathring{\Psi}_0)](\cdots)$, *viewed as a function of the geometric coordinates* (u, ϑ), *be as in Def.* 23.2. *Then the following pointwise estimate holds on the spacetime domain* $\mathcal{M}_{T_{(Lifespan)};U_0},U_0$:

$$\left|\upmu(t,u,\vartheta) - 1 - {}^{(+)}\mathring{\aleph}(\vartheta)\mathbb{S}[(\mathring{\Psi}, \mathring{\Psi}_0)](u,\vartheta) \ln\left(\frac{\varrho(t,u)}{\varrho(0,u)}\right)\right| \le C\left(\mathring{\delta} + \mathring{\epsilon}^{3/2}\right)\ln\left(\frac{\varrho(t,u)}{\varrho(0,u)}\right) + C\mathring{\epsilon}. \tag{23.18}$$

PROOF. The estimate (23.18) follows from the estimates (22.3a)-(22.3d) and (23.2c). □

23.2. Existence of small, stable, shock-generating data

In Lemma 23.8, we construct an open set of small initial data with properties that lead to finite-time shock formation in their solutions. We provide the proof of shock formation in the subsequent Theorem 23.9, which is located in Sect. 23.3.

As a preliminary step, we prove the following lemma, which is based on an argument of John given in [**38**]. The lemma sets the stage for proving our shock formation result, roughly by showing that compactly supported spherically symmetric data always induce the "shock driving" sign in the quantity that drives shock formation. It is a spherically symmetric analog of Prop. 2.17 for wave equations of the form (2.1a).

LEMMA 23.5 (**Nontrivial spherically symmetric data force $\breve{\mathbb{S}}$ to take on both signs**). *Let* $(\breve{\Psi}, \breve{\Psi}_0) \in C^1(\mathbb{R}^3) \times C^0(\mathbb{R}^3)$ *be data for the wave equation* (2.1a) ***given along*** $\Sigma_{-1/2}$ *with support contained in the Euclidean ball of radius* $1/2$ *centered at the origin. Assume that the data are spherically symmetric in the sense that they depend on only*[2] *the Euclidean radial coordinate* $r = \sqrt{\sum_{a=1}^{3}(x^a)^2}$ *and not on the Euclidean angular coordinates* (θ^1, θ^2), *that is,* $\breve{\Psi} = \breve{\Psi}(r)$ *and* $\breve{\Psi}_0 = \breve{\Psi}_0(r)$. *For* $p \in [-1/2, 1/2]$, *we define the data-dependent function* $\breve{\mathbb{S}}[(\breve{\Psi}, \breve{\Psi}_0)]$ *by*

$$\breve{\mathbb{S}}[(\breve{\Psi}, \breve{\Psi}_0)](p) := \frac{1}{4}\left\{p\breve{\Psi}_0(|p|) - |p|\breve{\Psi}'(|p|) - \breve{\Psi}(|p|)\right\}. \tag{23.19}$$

[2]Since equation (2.1a) is not generally invariant under the Euclidean rotations (see definition (6.2)), spherically symmetric data do not generally launch spherically symmetric solutions.

If the data are nontrivial, then

$$\min_{p\in[-1/2,1/2]} \breve{\mathbb{S}}[(\breve{\Psi},\breve{\Psi}_0)](p) < 0 < \max_{p\in[-1/2,1/2]} \breve{\mathbb{S}}[(\breve{\Psi},\breve{\Psi}_0)](p). \tag{23.20}$$

REMARK 23.6 (**Relationship between $\breve{\mathbb{S}}$ and $\mathbb{S}$**). Note that for $p > 0$, $\breve{\mathbb{S}}[(\breve{\Psi},\breve{\Psi}_0)](p) = \mathbb{S}[(\breve{\Psi},\breve{\Psi}_0)](p)$, where the latter function is defined by (23.17). For $p < 0$, $\breve{\mathbb{S}}[(\breve{\Psi},\breve{\Psi}_0)](p)$ can be viewed as an extension of $\mathbb{S}[(\breve{\Psi},\breve{\Psi}_0)](\cdot)$ constructed by extending the data $(\breve{\Psi},\breve{\Psi}_0)$ to be even functions of their argument.

PROOF OF LEMMA 23.5. Throughout, we suppress the dependence of $\breve{\mathbb{S}}$ on $(\breve{\Psi},\breve{\Psi}_0)$. We view the data as *even* functions of the real variable $p \in [-1/2,1/2]$ that verify $(\breve{\Psi},\breve{\Psi}_0) \in C^1([-1/2,1/2]) \times C^0([-1/2,1/2])$ and $\breve{\Psi}(-1/2) = \breve{\Psi}(1/2) = \breve{\Psi}_0(-1/2) = \breve{\Psi}_0(1/2) = 0$. Note that $4\breve{\mathbb{S}}(p) = p\breve{\Psi}_0(p) - \frac{d}{dp}(p\breve{\Psi}(p))$. It easily follows from this identity and the previous discussion that

$$\int_{p=-1/2}^{1/2} \breve{\mathbb{S}}(p)\,dp = 0. \tag{23.21}$$

We now show that if $\min_{p\in[-1/2,1/2]} \breve{\mathbb{S}}(p) > 0$, then the data must completely vanish. We therefore assume that $\min_{p\in[-1/2,1/2]} \breve{\mathbb{S}}(p) > 0$. It then follows from (23.21) that $\breve{\mathbb{S}}(p) \equiv 0$ for $p \in [-1/2,1/2]$. But since $\breve{\Psi}(p)$ and $\breve{\Psi}_0(p)$ are even functions, we have the following identities:

$$0 = 4\left\{\breve{\mathbb{S}}(p) + \breve{\mathbb{S}}(-p)\right\} = -2\frac{d}{dp}(p\breve{\Psi}(p)), \tag{23.22}$$

$$0 = 4\left\{\breve{\mathbb{S}}(p) - \breve{\mathbb{S}}(-p)\right\} = 2p\breve{\Psi}_0(p). \tag{23.23}$$

Since $\breve{\Psi}$ and $\breve{\Psi}_0$ are compactly supported, we easily conclude from (23.22) and (23.23) that they are trivial. We have thus proved that the first inequality in (23.20) must hold when the data are nontrivial. The second inequality can be proved in a similar fashion. ☐

In the next lemma, we show that for small spherically symmetric data given on $\Sigma_{-1/2}$, the function $\breve{\mathbb{S}}[(\breve{\Psi},\breve{\Psi}_0)]$ effectively determines the function $\mathbb{S}[(\mathring{\Psi},\mathring{\Psi}_0)]$ corresponding to the "data" induced by the solution on Σ_0^1.

LEMMA 23.7 (**Quantitative relationship between $\mathbb{S}[(\mathring{\Psi},\mathring{\Psi}_0)]$ and $\breve{\mathbb{S}}[(\breve{\Psi},\breve{\Psi}_0)]$**). *Let $(\breve{\Psi},\breve{\Psi}_0)$ be spherically symmetric data for the wave equation* (2.1a) ***given along*** $\Sigma_{-1/2}$ *with support contained in the Euclidean ball of radius* 1/2 *centered at the origin, as in Lemma* 23.5. *Let*

$$\epsilon = \epsilon[(\breve{\Psi},\breve{\Psi}_0)] := \|\breve{\Psi}\|_{H_e^{25}(\Sigma_{-1/2})} + \|\breve{\Psi}_0\|_{H_e^{24}(\Sigma_{-1/2})} < \infty \tag{23.24}$$

denote their size,[3] *and let Ψ denote the corresponding solution. Let $(\mathring{\Psi} := \Psi|_{\Sigma_0^1}$, $\mathring{\Psi}_0 := \partial_t\Psi|_{\Sigma_0^1})$ be the "data" along*[4] *Σ_0 induced by Ψ. Then if ϵ is sufficiently small,*

---

[3]As always in this monograph H_e^N is the standard Euclidean Sobolev space involving rectangular spatial derivatives.

[4]Note that standard domain of dependence considerations imply that $\mathring{\Psi}$ and $\mathring{\Psi}_0$ have support contained in Σ_0^1.

Ψ persists beyond time 0 and along Σ_0^1, we have

$$\mathring{\epsilon}[(\mathring{\Psi}, \mathring{\Psi}_0)] := \|\mathring{\Psi}\|_{H_e^{25}(\Sigma_0^1)} + \|\mathring{\Psi}_0\|_{H_e^{24}(\Sigma_0^1)} \lesssim \epsilon. \tag{23.25}$$

Similarly, if $\mathring{\epsilon}$ is sufficiently small, then we have

$$\epsilon[(\check{\Psi}, \check{\Psi}_0)] \lesssim \mathring{\epsilon}[(\mathring{\Psi}, \mathring{\Psi}_0)]. \tag{23.26}$$

Furthermore, let $\mathbb{S}[(\mathring{\Psi}, \mathring{\Psi}_0)]$ be the data-dependent function from Def. 23.2 viewed as a function of the Euclidean spherical coordinates[5] *(r, θ^1, θ^2) on Σ_0^1, and let $\check{\mathbb{S}}[(\check{\Psi}, \check{\Psi}_0)]$ be the function of the real variable $p \in [-1/2, 1/2]$ as defined by (23.19). If ϵ is sufficiently small and $r_0 \in [0,1]$, then then the following estimate holds for $(r, \theta) \in [r_0, 1] \times \mathbb{S}^2$:*

$$\mathbb{S}[(\mathring{\Psi}, \mathring{\Psi}_0)](r, \theta) = \check{\mathbb{S}}[(\check{\Psi}, \check{\Psi}_0)](r - 1/2) + \mathcal{O}\left(\frac{\epsilon^2}{r_0}\right), \tag{23.27}$$

where the implicit constant in $\mathcal{O}$ can be chosen to be independent of r_0, r, and θ.

PROOF. The persistence result and inequality (23.25) are standard results that can be proved using the same techniques that we described in sketch of the proof of the local well-posedness proposition (that is, Prop. 21.1). More generally, for $t \in [-1/2, 0]$, we have the following standard estimate:

$$\sum_{M=0}^{25} \|\partial_t^M \Psi\|_{H_e^{25-M}(\Sigma_t)} \lesssim \epsilon. \tag{23.28}$$

Inequality (23.26) can be proved in the same way, by solving the wave equation with data $(\mathring{\Psi}, \mathring{\Psi}_0)$ given on Σ_0 backwards in time, down to the hypersurface $\Sigma_{-1/2}$.

Next, we reduce the analysis to that of an approximately spherically symmetric problem over the time interval $t \in [-1/2, 0]$. To this end, we first note that (2.4)-(2.6) and (2.12) allow us to express the nonlinear wave equation (2.1a) as a perturbation of the spherically symmetric linear wave equation as follows:

$$\partial_t^2(r\Psi) - \partial_r^2(r\Psi) = r\not{\Delta}_{\not{m}}\Psi + r\mathcal{N}(\Psi, \partial^2\Psi) + r\mathcal{N}(\Psi)(\partial\Psi, \partial\Psi), \tag{23.29}$$

where $\not{\Delta}_{\not{m}}$ is the angular Laplacian corresponding to the metric $\not{m}$ induced by the Minkowski metric m on the Euclidean spheres of constant t and r and $\mathcal{N}(\cdot, \cdot)$, $\mathcal{N}(\Psi)(\cdot, \cdot)$ are nonlinear terms that are quadratically small in their arguments. We now show that for $t \in [-1/2, 0]$, we have

$$L_{(Flat)}(\underline{L}_{(Flat)}(r\Psi)) = \mathcal{O}\left(\frac{\epsilon^2}{r_0}\right), \tag{23.30}$$

where $L_{(Flat)} = \partial_t + \partial_r$, $\underline{L}_{(Flat)} = \partial_t - \partial_r$, and $L_{(Flat)}\underline{L}_{(Flat)} = \partial_t^2 - \partial_r^2$. Moreover, in (23.30) and throughout the rest of this proof, $\mathcal{O}\left(\frac{\epsilon^2}{r_0}\right)$ denotes a term $\mathfrak{T}$ such that for $r_0 \in [0,1]$, we have $\max_{(t,r,\theta) \in [-1/2,1] \times [r_0,1] \times \mathbb{S}^2} |\mathfrak{T}(t, r, \theta)| \leq C\frac{\epsilon^2}{r_0}$, with C independent of r_0 (when ϵ is sufficiently small). To prove (23.30), we first deduce from (23.28) that for $t \in [-1/2, 0]$, the two quadratic terms $r\mathcal{N}(\cdots)$ on the right-hand side of (23.29) are $\mathcal{O}(\epsilon^2)$, which in particular implies that they are $\mathcal{O}\left(\frac{\epsilon^2}{r_0}\right)$. To show that

[5]Note that $\mathbb{S}[(\mathring{\Psi}, \mathring{\Psi}_0)]$ is not generally spherically symmetric, even when $\mathring{\Psi}$ and $\mathring{\Psi}_0$ are; see Footnote 2 on pg. 470.

$r\not{\Delta}_{\not{h}}\Psi = \mathcal{O}\left(\frac{\epsilon^2}{r_0}\right)$, we can argue as in the proof of Cor. 22.4, but using the Euclidean rotations $O_{(Flat)}$ (see Def. 6.1) as commutators instead of the geometric ones. More precisely, since the data (given along $\Sigma_{-1/2}$) are spherically symmetric, we have the following relation, valid for $1 \leq M \leq 25$: $\left\|\mathscr{O}_{(Flat)}^M \Psi\right\|_{L_e^2(\Sigma_{-1/2})} = 0$. In the previous equation, $\mathscr{O}_{(Flat)}^M$ denotes a Mth order differential operator corresponding to repeated differentiation with respect to the elements of $\mathscr{O}_{(Flat)}$, that is, with respect to the Euclidean rotations. Furthermore, for $t \in [-1/2, 0]$, we have the following estimate, valid for $1 \leq M \leq 24$:

$$\left\|\mathscr{O}_{(Flat)}^M \Psi\right\|_{L_e^2(\Sigma_t)}, \sum_{\alpha=0}^{3} \left\|\partial_\alpha \mathscr{O}_{(Flat)}^M \Psi\right\|_{L_e^2(\Sigma_t)} \lesssim \epsilon^2. \tag{23.31}$$

To prove (23.31), we commute equation (2.1a) with in between 1 and 24 Euclidean rotations, derive standard Minkowskian energy estimates, use the fact that all error terms in the commuted equations are quadratically small (and thus by (23.28), they are $\mathcal{O}(\epsilon^2)$), and use the assumption that the left-hand side of (23.31) vanishes when $t = -1/2$. The desired estimate $r\not{\Delta}_{\not{h}}\Psi = \mathcal{O}\left(\frac{\epsilon^2}{r_0}\right)$ then follows from the bound $r^2\not{\Delta}_{\not{h}}\Psi = \mathcal{O}(\epsilon^2)$, which in turn follows from (23.31), the Minkowskian analog of Lemma 12.22, and Sobolev embedding along the Euclidean spheres of constant t and r (that is, the Euclidean analogs of (12.46a), (12.46c), and (18.28), with r in the role of ϱ). We have thus shown (23.30).

We now recall that (see Def. 23.2) $\mathbb{S}[(\mathring{\Psi}, \mathring{\Psi}_0)] = \frac{1}{4}\underline{L}_{(Flat)}(r\Psi)|_{(\Psi,\partial_t\Psi)=(\mathring{\Psi},\mathring{\Psi}_0)}$, where, as above, $\underline{L}_{(Flat)} = \partial_t - \partial_r$. In view of this identity and (23.19), we see that the desired estimate (23.27) will follow once we establish the following estimate for $t \in [-1/2, 0]$:

$$\begin{aligned}
[\underline{L}_{(Flat)}(r\Psi)](t, r, \theta) &= (r - t - 1/2)\,\check{\Psi}_0\left(|1/2 + t - r|\right) \\
&\quad - |1/2 + t - r|\,\check{\Psi}'\left(|1/2 + t - r|\right) \\
&\quad - \check{\Psi}\left(|1/2 + t - r|\right) + \mathcal{O}\left(\frac{\epsilon^2}{r_0}\right),
\end{aligned} \tag{23.32}$$

where we set $t = 0$ in (23.32) to conclude (23.27).

To prove (23.32), we decompose the solution Ψ of (23.30) as $\Psi(t, r, \theta) = \Psi_{(Hom)}(t, r) + \Psi_{(Inhom)}(t, r, \theta)$ where **i)** $\Psi_{(Hom)}(t, r)$ is the truly spherically symmetric solution to the homogeneous linear wave equation

$$L_{(Flat)}(\underline{L}_{(Flat)}(r\Psi_{(Hom)})) = 0$$

with data $(\check{\Psi}, \check{\Psi}_0)$ (given at $t = -1/2$), for which we have the explicit formula

(23.33)

$$\begin{aligned}
r\Psi_{(Hom)}(t, r) &= \frac{1}{2}(r + 1/2 + t)\check{\Psi}(r + 1/2 + t) + \frac{1}{2}(r - 1/2 - t)\check{\Psi}(|r - 1/2 - t|) \\
&\quad + \frac{1}{2}\int_{s=|1/2+t-r|}^{s=1/2+t-r} s\check{\Psi}_0(s)\,ds,
\end{aligned}$$

and **ii)** $\Psi_{(Inhom)}(t,r,\theta)$ is the solution to the inhomogeneous equation (23.30) with trivial data (given at $t=-1/2$), which, by Duhamel's principle, verifies

$$r\Psi_{(Inhom)}(t,r,\theta)=\frac{1}{2}\int_{s=-1/2}^{t}\int_{r'=|r-(t-s)|}^{r'=r+t-s}\mathcal{O}\left(\frac{\epsilon^2}{r_0}\right)|_{(s,r',\theta)}\,dr'\,ds, \tag{23.34}$$

where the term $\mathcal{O}\left(\frac{\epsilon^2}{r_0}\right)|_{(s,r',\theta)}$ is precisely the right-hand side of (23.30), viewed as a function of the spherical coordinates (s,r',θ). The desired expression (23.32) follows in a straightforward fashion from these considerations and from simple calculations based on the formulas (23.33) and (23.34). We note that $\Psi_{(Inhom)}$ is completely responsible for the term $\mathcal{O}\left(\frac{\epsilon^2}{r_0}\right)$ in (23.32). □

We now construct an open set $\mathfrak{S}$ of small data along $\Sigma_{-1/2}$ that lead to finite-time shock formation.

LEMMA 23.8 (**Small, nontrivial spherically symmetric data lead to stable shock formation**). *We assume that the null condition fails in the covariant wave equation* (2.1a) *and we define*

$$^{(+)}\aleph_* := \max_{\vartheta\in\mathbb{S}^2}|^{(+)}\mathring{\aleph}(\vartheta)|>0, \tag{23.35}$$

where $^{(+)}\mathring{\aleph}$ *is as in Def.* 3.32. *Let* $\vartheta_*\in\mathbb{S}^2$ *be any angular direction such that* $|^{(+)}\mathring{\aleph}(\vartheta_*)|={}^{(+)}\aleph_*$. *Then there exists a nonempty* ***open***[6] *set* $\mathfrak{S}\subset H_e^{25}(\Sigma_{-1/2})\times H_e^{24}(\Sigma_{-1/2})$ *of* ϑ_**-dependent data* ***given along*** $\Sigma_{-1/2}$ *with support contained in the Euclidean ball of radius* $1/2$ *and featuring the following properties. In the statements below,* $(\check{\Psi},\check{\Psi}_0)$ *denotes data on* $\Sigma_{-1/2}$, $(\mathring{\Psi}:=\Psi|_{\Sigma_0^1},\mathring{\Psi}_0:=\partial_t\Psi|_{\Sigma_0^1})$ *denotes the "data" induced on* Σ_0^1 *(see Footnote 4 on pg. 471) by the solution* Ψ, *and* $\mathring{\epsilon}=\mathring{\epsilon}[(\mathring{\Psi},\mathring{\Psi}_0)]:=\|\mathring{\Psi}\|_{H_e^{25}(\Sigma_0^1)}+\|\mathring{\Psi}_0\|_{H_e^{24}(\Sigma_0^1)}$ *denotes the size of the latter, as in* (12.2).

(**1**) $\mathfrak{S}$ *is by definition the union of open neighborhoods* $\Omega[(\check{\Psi}^{(Radial)},\check{\Psi}_0^{(Radial)})]$ *of spherically symmetric data* $(\check{\Psi}^{(Radial)},\check{\Psi}_0^{(Radial)})$, *as described in item* **3**.

(**2**) *If* $(\check{\Psi}^{(Radial)},\check{\Psi}_0^{(Radial)})\in H_e^{25}(\Sigma_{-1/2})\times H_e^{24}(\Sigma_{-1/2})$ *are nontrivial and spherically symmetric in the sense that they depend on only the Euclidean radial variable* r, *and* $\lambda>0$ *is sufficiently small (where the required smallness depends on* $(\check{\Psi}^{(Radial)},\check{\Psi}_0^{(Radial)})$*), then* $(\lambda\check{\Psi}^{(Radial)},\lambda\check{\Psi}_0^{(Radial)})\in\mathfrak{S}$.

(**3**) *For any spherically symmetric data* $(\check{\Psi}^{(Radial)},\check{\Psi}_0^{(Radial)})\in\mathfrak{S}$, *there exist a number* $U_0\in(0,1)$, *an open (in* $H_e^{25}(\Sigma_{-1/2})\times H_e^{24}(\Sigma_{-1/2})$*) subset* $\Omega[(\check{\Psi}^{(Radial)},\check{\Psi}_0^{(Radial)})]\subset\mathfrak{S}$ *containing it, a data-dependent constant* $\alpha_*:=\alpha_*[(\check{\Psi}^{(Radial)},\check{\Psi}_0^{(Radial)})]>0$, *and a number* $u_*\in(0,U_0)$ *such that if* $(\check{\Psi},\check{\Psi}_0)\in\Omega[(\check{\Psi}^{(Radial)},\check{\Psi}_0^{(Radial)})]$, *then the following inequalities hold for the data* $(\mathring{\Psi},\mathring{\Psi}_0)$ *induced on* $\Sigma_0^{U_0}$ *by the solution* Ψ:

$$^{(+)}\mathring{\aleph}(\vartheta_*)\mathbb{S}[(\mathring{\Psi},\mathring{\Psi}_0)](u_*,\vartheta_*)+C\left(\mathring{\delta}+\mathring{\epsilon}^{3/2}\right)<-\alpha_*<0. \tag{23.36}$$

---

[6]By "open," we mean open relative to the topology corresponding to the function space $H_e^{25}(\Sigma_{-1/2})\times H_e^{24}(\Sigma_{-1/2})$.

In (23.36), $\mathbb{S}[(\mathring{\Psi}, \mathring{\Psi}_0)]$ *is the function defined in* (23.17) *and* $C\left(\mathring{\delta} + \mathring{\epsilon}^{3/2}\right)$ *is the factor in the first product on the right-hand side of* (23.18). *Furthermore,* $\alpha_* > 0$ *(see Lemma 23.5) is defined by*

(23.37)
$$\alpha_* := \begin{cases} \frac{1}{2}{}^{(+)}\aleph_* \max_{p\in[-1/2,1/2]} \check{\mathbb{S}}[(\check{\Psi}^{(Radial)}, \check{\Psi}_0^{(Radial)})](p) & \text{if } {}^{(+)}\mathring{\aleph}(\vartheta_*) < 0, \\ -\frac{1}{2}{}^{(+)}\aleph_* \min_{p\in[-1/2,1/2]} \check{\mathbb{S}}[(\check{\Psi}^{(Radial)}, \check{\Psi}_0^{(Radial)})](p) & \text{if } {}^{(+)}\mathring{\aleph}(\vartheta_*) > 0. \end{cases}$$

Moreover, there exists a constant $C > 0$ *such that*

(23.38)
$$\alpha_* \leq C\mathring{\epsilon}.$$

(4) *For all data in* $\mathfrak{S}$, *the "data"* $(\mathring{\Psi} := \Psi|_{\Sigma_0^1}, \mathring{\Psi}_0 := \partial_t \Psi|_{\Sigma_0^1})$ *induced by the solution* Ψ *are small and nearly spherically symmetric in the sense that the hypotheses and conclusions of Theorem* 22.1 *and Cor.* 22.4 *are verified.*
(5) *All constants denoted by* C *can depend on* U_0.

PROOF. The main idea of the proof is to first use an approximate scaling argument to show that given any nontrivial spherically symmetric data $(\check{\Psi}^{(Radial)}, \check{\Psi}_0^{(Radial)})$ on $\Sigma_{-1/2}$, we can rescale it by an overall factor $\lambda > 0$ so that the desired inequalities are verified by the rescaled data $(\lambda\check{\Psi}^{(Radial)}, \lambda\check{\Psi}_0^{(Radial)})$. We then show that all of the relevant quantities that we need to control are stable under small general (not necessarily spherically symmetric) perturbations. The set $\mathfrak{S}$ will therefore be the union of open neighborhoods, each of which contains one of the spherically symmetric elements $(\lambda\check{\Psi}^{(Radial)}, \lambda\check{\Psi}_0^{(Radial)})$ for some sufficiently small $\lambda > 0$.

We assume throughout the proof that ${}^{(+)}\mathring{\aleph}(\vartheta_*) < 0$; the case ${}^{(+)}\mathring{\aleph}(\vartheta_*) > 0$ can be treated using similar arguments. To begin, we let $(\check{\Psi}^{(Radial)}, \check{\Psi}_0^{(Radial)}) \in H_e^{25}(\Sigma_{-1/2}) \times H_e^{24}(\Sigma_{-1/2})$ be nontrivial spherically symmetric data on $\Sigma_{-1/2}$ supported in the Euclidean ball of radius 1/2 centered at the origin. For $\lambda > 0$, let ${}^{(\lambda)}\Psi$ denote the solution launched by the rescaled data $(\lambda\check{\Psi}^{(Radial)}, \lambda\check{\Psi}_0^{(Radial)})$, and let $({}^{(\lambda)}\mathring{\Psi} := {}^{(\lambda)}\Psi|_{\Sigma_0^1}, {}^{(\lambda)}\mathring{\Psi}_0 := \partial_t {}^{(\lambda)}\Psi|_{\Sigma_0^1})$ denote the "data" induced on Σ_0^1 by ${}^{(\lambda)}\Psi$.

In the remainder of the proof, we will assume that λ is sufficiently small, where the required smallness is allowed to depend on $(\check{\Psi}^{(Radial)}, \check{\Psi}_0^{(Radial)})$. Moreover, we allow the implicit constants in "$\mathcal{O}(\cdots)$," "$\lesssim$," and "$\approx$" to depend on $(\check{\Psi}^{(Radial)}, \check{\Psi}_0^{(Radial)})$ (and, as always, on U_0). In particular, from (23.25) and (23.26), we see that for the nontrivial spherically symmetric data under consideration, we have

(23.39)
$$\begin{aligned} \lambda \approx \mathring{\epsilon} &:= \mathring{\epsilon}[({}^{(\lambda)}\mathring{\Psi}, {}^{(\lambda)}\mathring{\Psi}_0)] \\ &\approx \epsilon := \epsilon[(\lambda\check{\Psi}^{(Radial)}, \lambda\check{\Psi}_0^{(Radial)})] \\ &= \lambda\epsilon[(\check{\Psi}^{(Radial)}, \check{\Psi}_0^{(Radial)})] \end{aligned}$$

where $\epsilon[\cdots]$ is defined in (23.24).

From (23.20), (23.27), (23.39), and the fact that $\check{\mathbb{S}}[\cdot]$ scales linearly in its data argument $\cdot$, we deduce that there exist a parameter $U_0 \in (0, 1)$ and a number $u_* \in (0, U_0)$ such that relative to the geometric coordinates (u, ϑ) along Σ_0^1, we have

$$(23.40)\qquad \begin{aligned} &{}^{(+)}\mathring{\aleph}(\vartheta_*)\mathbb{S}[({}^{(\lambda)}\mathring{\Psi}, {}^{(\lambda)}\mathring{\Psi}_0)](u_*,\vartheta_*) \\ &= -\lambda^{(+)}\aleph_* \sup_{p\in[-1/2,1/2]} \breve{\mathbb{S}}[(\breve{\Psi}^{(Radial)}, \breve{\Psi}_0^{(Radial)})](p) + \mathcal{O}(\lambda^2) \\ &< 0. \end{aligned}$$

We now claim that for the data $({}^{(\lambda)}\mathring{\Psi}, {}^{(\lambda)}\mathring{\Psi}_0)$, the factor $C\left(\mathring{\delta} + \mathring{\epsilon}^{3/2}\right)$ in the first product on the right-hand side of (23.18) is $\mathcal{O}(\lambda^{3/2})$. It then follows from this estimate and (23.40) that whenever λ is sufficiently small, inequality (23.36) holds with the role of $(\mathring{\Psi}, \mathring{\Psi}_0)$ on the left-hand side played by $({}^{(\lambda)}\mathring{\Psi}, {}^{(\lambda)}\mathring{\Psi}_0)$, and $\alpha_* = \alpha_*[(\lambda\breve{\Psi}^{(Radial)}, \lambda\breve{\Psi}_0^{(Radial)})]$ is defined in (23.37) (note the factor of 1/2 in the definition), where $(\lambda\breve{\Psi}^{(Radial)}, \lambda\breve{\Psi}_0^{(Radial)})$ is in the role of $(\breve{\Psi}^{(Radial)}, \breve{\Psi}_0^{(Radial)})$ on the right-hand side of (23.37). Moreover, using (23.39), we find that

$$\alpha_*[(\lambda\breve{\Psi}^{(Radial)}, \lambda\breve{\Psi}_0^{(Radial)})] \lesssim \mathring{\epsilon}[({}^{(\lambda)}\mathring{\Psi}, {}^{(\lambda)}\mathring{\Psi}_0)],$$

which yields inequality (23.38) (where we again stress that $(\lambda\breve{\Psi}^{(Radial)}, \lambda\breve{\Psi}_0^{(Radial)})$ is in the role of $(\breve{\Psi}^{(Radial)}, \breve{\Psi}_0^{(Radial)})$ on the right-hand side of (23.37) and $\mathring{\epsilon} = \mathring{\epsilon}[({}^{(\lambda)}\mathring{\Psi}, {}^{(\lambda)}\mathring{\Psi}_0)]$ in this part of the proof).

It remains for us to establish that for the data $({}^{(\lambda)}\mathring{\Psi}, {}^{(\lambda)}\mathring{\Psi}_0)$, we have the following bound for the factor on the right-hand side of (23.18): $C\left(\mathring{\delta} + \mathring{\epsilon}^{3/2}\right) \lesssim \lambda^{3/2}$. To obtain $C\mathring{\epsilon}^{3/2} \lesssim \lambda^{3/2}$, we simply use (23.39). We now show that the quantity $\mathring{\delta}$, which is defined by (22.18) and appears in the estimate (23.18), verifies the desired bound

$$(23.41)\qquad \mathring{\delta} \lesssim \lambda^2.$$

To prove (23.41), we first use (23.31) and (23.39) to deduce that

$$(23.42)\qquad \sum_{M=1}^{24} \left\| \mathscr{O}_{(Flat)}^{M}\,{}^{(\lambda)}\Psi \right\|_{L_e^2(\Sigma_0^1)} + \sum_{\alpha=0}^{3}\sum_{M=1}^{24} \left\| \partial_\alpha \mathscr{O}_{(Flat)}^{M}\,{}^{(\lambda)}\Psi \right\|_{L_e^2(\Sigma_0^1)} \lesssim \lambda^2.$$

Furthermore, from (23.25) and (23.39), we deduce that

$$(23.43)\qquad \left\| {}^{(\lambda)}\Psi \right\|_{H_e^{25}(\Sigma_0^1)} + \left\| \partial_t {}^{(\lambda)}\Psi \right\|_{H_e^{24}(\Sigma_0^1)} \lesssim \lambda.$$

From (23.42) and (23.43), we see that the solution is small along Σ_0^1 and spherically symmetric up to $\mathcal{O}(\lambda^2)$. In particular, from the smallness bound (23.43), it follows from the analysis of Sect. 12.8 that along Σ_0^1, all of the tensorfields defined throughout the monograph have rectangular components equal to their Minkowskian counterparts plus an error bounded by $\mathcal{O}(\lambda)$, that is, $\mu = 1 + \mathcal{O}(\lambda)$, $L^i = L^i_{(Flat)} + \mathcal{O}(\lambda)$, $\breve{R}^i = R^i_{(Flat)} + \mathcal{O}(\lambda)$, $O^i = O^i_{(Flat)} + \mathcal{O}(\lambda)$, etc. Furthermore, the same $\mathcal{O}(\lambda)$ bounds hold for the rectangular coordinate partial derivatives of these components. In view of the expression (6.5) for $O_{(Flat;l)}$, the expression (6.6), and the definition (22.18) of $\mathring{\delta}$, the desired bound (23.41) follows easily from (23.42) and the above observations.

To finish the proof of the lemma, it remains only for us to show that the inequalities proved above for the small spherically symmetric data $(\lambda\breve{\Psi}^{(Radial)}, \lambda\breve{\Psi}_0^{(Radial)})$ on $\Sigma_{-1/2}$ remain valid under small nonsymmetric perturbations that belong to

$H_e^{25}(\Sigma_{-1/2}) \times H_e^{24}(\Sigma_{-1/2})$ and that are supported in the Euclidean ball of radius $1/2$ centered at the origin. This follows easily from standard Euclidean Sobolev embedding without changing the parameter U_0 associated to the spherically symmetric data because inequality (23.36) is strict and because Prop. 21.1 implies that on the interval $t \in [-1/2, 0]$, the solution varies continuously with respect to the data given along $\Sigma_{-1/2}$. □

23.3. Proof of shock formation for small, nearly spherically symmetric data

In this section, we prove our main shock formation theorem. Specifically, we prove that the open set $\mathfrak{S}$ of data from Lemma 23.8 launches solutions that develop a shock singularity in finite time.

THEOREM 23.9 (**Finite-time shock formation for data in $\mathfrak{S}$**). *Assume that the null condition fails in the covariant wave equation* (2.1a) *under the assumption*[7] (2.3). *In particular, the quantity* $^{(+)}\mathring{\aleph}$ *from Def.* 3.32 *verifies the positivity condition* (23.35). *Let* $(\check{\Psi} := \Psi|_{\Sigma_{-1/2}}, \check{\Psi}_0 := \partial_t \Psi|_{\Sigma_{-1/2}})$ *be data given along* $\Sigma_{-1/2}$. *Assume that* $(\check{\Psi}, \check{\Psi}_0)$ *belong to the open set* $\mathfrak{S}$ *(of small, nearly spherically symmetric data) from Lemma* 23.8. *In particular, the data are supported in the Euclidean ball of radius* $1/2$ *centered at the origin in* $\Sigma_{-1/2}$. *Let* Ψ *denote the corresponding solution. Let* $U_0 \in (0, 1)$ *and* $\alpha_* > 0$ *be the data-neighborhood-dependent parameters appearing in the statement of Lemma* 23.8 *and let* $T_{(Lifespan);U_0}$ *be the classical lifespan of* Ψ *appearing in the statement of Theorem* 22.1 *and Cor.* 22.4. *Let* $\mathring{\epsilon} = \mathring{\epsilon}[(\mathring{\Psi}, \mathring{\Psi}_0)] := \|\mathring{\Psi}\|_{H_e^{25}(\Sigma_0^1)} + \|\mathring{\Psi}_0\|_{H_e^{24}(\Sigma_0^1)}$ *denote the size of the "data"* $(\mathring{\Psi} := \Psi|_{\Sigma_0^1}, \mathring{\Psi}_0 := \partial_t \Psi|_{\Sigma_0^1})$ *(which are supported in the Euclidean unit ball* Σ_0^1*) induced by the solution on* Σ_0^1 *and let* $\mathring{\delta}$ *be the size of the geometric rotational derivatives of* Ψ *(of order two through four) on* $\Sigma_0^{U_0}$, *as defined in* (22.18). *Note that by the definition of* $\mathfrak{S}$, *the hypotheses and conclusions of Theorem* 22.1 *and Cor.* 22.4 *hold for data in* $\mathfrak{S}$. *In particular, we have* $\mathring{\epsilon} \leq \epsilon_0$, *where* ϵ_0 *is the small positive constant from Theorem* 22.1. *Then* $T_{(Lifespan);U_0} < \infty$ *precisely because* μ *vanishes for the first time at one or more points in* $\Sigma_{T_{(Lifespan);U_0}}$. *Thus, according to Theorem* 22.1, *this vanishing signifies the onset of shock formation. Furthermore, there exists a constant* $C > 0$ *depending on* U_0 *such that*

$$T_{(Lifespan);U_0} \leq \exp\left(\frac{1 + C\mathring{\epsilon}}{\alpha_*}\right). \tag{23.44}$$

REMARK 23.10 (**Beyond nearly spherically symmetric data**). Although the data in $\mathfrak{S}$ are nearly spherically symmetric, equation (2.1a) is not generally invariant under Euclidean rotations. Thus, even when the data are exactly spherically symmetric, the solution does not generally inherit this property. It might therefore seem a bit unnatural to prove shock formation for the data in $\mathfrak{S}$. It is only for technical convenience that we have treated these data in detail; in Sect. 2.11.3, we outlined how to extend our results to prove shock formation for a much larger set of compactly supported small data.

PROOF. Recall that (see equation (22.1)) $T_{(Lifespan);U_0}$ is characterized by

$$T_{(Lifespan);U_0} = \sup\{t \in [0, \infty) \mid \inf\{\mu_\star(s, U_0) \mid s \in [0, t)\} > 0\},$$

[7]As we described in Chapter 2, this assumption is easy to eliminate.

where $\upmu_{\star}$ is defined in (13.6). Let u_* and ϑ_* be as in Lemma 23.8. From inequality (23.18), the properties of the data in $\mathfrak{S}$ that are stated in Lemma 23.8 (see especially (23.36)), and the definition (13.6) of $\upmu_{\star}$, it follows that

$$\upmu_{\star}(t, U_0) \leq \upmu(t, u_*, \vartheta_*) \leq 1 + C\mathring{\epsilon} - \alpha_* \ln\left(\frac{\varrho(t, u_*)}{\varrho(0, u_*)}\right). \tag{23.45}$$

It now easily follows from (23.38), (23.45), and the definition $\varrho(t,u) = 1 - u + t$ that $\upmu_{\star}(t, U_0)$ must vanish before the time on the right-hand side of (23.44). $\square$

APPENDIX A

Extension of the Results to a Class of Non-Covariant Wave Equations

In this appendix, we sketch how to extend the results of Theorems 22.1 and 23.9 to non-covariant equations of the form

$$(h^{-1})^{\alpha\beta}(\partial\Phi)\partial_\alpha\partial_\beta\Phi = 0, \tag{A.1}$$

where

$$h_{\mu\nu} = m_{\mu\nu} + h_{\mu\nu}^{(Small)}(\partial\Phi) \tag{A.2}$$

and $h_{\mu\nu}^{(Small)}(0) = 0$. As in (2.3), in order to simplify the calculations, we make the non-essential assumption

$$(h^{-1})^{00}(\partial\Phi) = -1. \tag{A.3}$$

We note that Remark 23.10 also applies to equation (A.1). That is, our results can be extended to show finite-time shock formation in solutions to equation (A.1) for a significantly larger set of initial data than the one afforded by Theorem 23.9.

REMARK A.1 (**A big difference between equation** (2.1a) **and equation** (A.1)). Even though equation (A.1) does not contain any semilinear terms, it nonetheless exhibits small-data future shock formation[1] whenever ${}^{(+)}\aleph \not\equiv 0$, where ${}^{(+)}\aleph$ is defined in (A.8). We recall that in contrast, small-data global existence holds for equation (2.1a) if we delete the semilinear terms that are present when the equation is expanded relative to the rectangular coordinates; see Remark 2.9.

A.1. From the scalar quasilinear wave equation to the equivalent system of covariant wave equations

The main idea behind extending the results is that the overwhelming majority of the analysis of solutions to (A.1) is essentially the same as the analysis we used in treating the covariant scalar equation $\square_{g(\Psi)}\Psi = 0$. As we will see, the reason is that we can differentiate equation (A.1) to transform it into a coupled system of equations that are closely related to equation (2.1a). Specifically, we will commute the wave equation (A.1) with the rectangular derivatives $\frac{\partial}{\partial x^\nu} = \partial_\nu$. This motivates the following definition.

[1]That is, an analog of Theorem 23.9, showing that nearly spherically symmetric small data lead to finite-future-time shock formation, holds for equation (A.1).

Definition A.2 (**The quantities Ψ_ν**). We define the scalar functions Ψ_ν, $(\nu = 0, 1, 2, 3)$, and the array $\vec{\Psi}$ as follows:

$$\Psi_\nu := \partial_\nu \Phi, \tag{A.4}$$

$$\vec{\Psi} := (\Psi_0, \Psi_1, \Psi_2, \Psi_3). \tag{A.5}$$

We view $\{\Psi_\nu\}_{\nu=0,1,2,3}$ as the new unknowns.

Remark A.3 (**Avoiding redundancy**). Our analysis of solutions to equation (A.1) requires analogs of all of the geometric quantities that we used in analyzing the scalar equation (2.1a), including the eikonal function u (see Def. 3.5) the frames $\{L, \breve{R}, X_1, X_2\}$ and $\{L, R, X_1, X_2\}$ (see Def. 3.30), etc. To keep the discussion short in this appendix, we only provide definitions of a few of these quantities; we hope that the reader can infer the appropriate definitions of the remaining ones by making straightforward modifications to the definitions given in the case of the scalar equation (2.1a).

The following functions of $\vec{\Psi}$, which are analogs of the functions $G_{\mu\nu}$ from (2.14), play a fundamental role in our analysis.

Definition A.4 (**Metric component derivative functions**). We define

$$H_{\mu\nu}^\lambda = H_{\mu\nu}^\lambda(\vec{\Psi}) := \frac{\partial h_{\mu\nu}^{(Small)}(\vec{\Psi})}{\partial \Psi_\lambda}. \tag{A.6}$$

It is straightforward to calculate that the analog of equation (4.4b) is the following identity for the Christoffel symbols of h relative to the rectangular coordinates:

(A.7)

$$\Gamma_{\alpha\kappa\beta} = \frac{1}{2}\left\{\partial_\alpha h_{\kappa\beta} + \partial_\beta h_{\alpha\kappa} - \partial_\kappa h_{\alpha\beta}\right\} = \frac{1}{2}\left\{H_{\kappa\beta}^\lambda \partial_\alpha \Psi_\lambda + H_{\alpha\kappa}^\lambda \partial_\beta \Psi_\lambda - H_{\alpha\beta}^\lambda \partial_\kappa \Psi_\lambda\right\}.$$

Recall that for equation (2.1a), the future null condition failure factor is provided in equation (3.33). We now provide, for equation (A.1), the appropriate analog in the region $\{t \geq 0\}$.

Definition A.5 (**Future null condition failure factor for the equation $(h^{-1})^{\alpha\beta}(\partial\Phi)\partial_\alpha\partial_\beta\Phi = 0$**). We define ${}^{(+)}\aleph$, the future null condition failure factor corresponding to equation (A.1), by

$${}^{(+)}\aleph := m_{\kappa\lambda} H_{\mu\nu}^\kappa(\vec{\Psi} = 0) L_{(Flat)}^\lambda L_{(Flat)}^\mu L_{(Flat)}^\nu, \tag{A.8}$$

where $m_{\mu\nu} = \text{diag}(-1, 1, 1, 1)$ is the Minkowski metric, $L_{(Flat)} = \partial_t + \partial_r$, and $\partial_r = \frac{x^a}{r}\partial_a$ is the standard Euclidean radial vectorfield.

We now explain the relevance of Def. A.5. Our discussion here closely parallels the discussion in Sect. 2.4. We first note that the nonlinear terms in equation (A.1) verify Klainerman's classic null condition [**45**] (see the discussion in Sect. 2.4) if and only if $m_{\kappa\lambda} H_{\mu\nu}^\kappa(\vec{\Psi} = 0)\ell^\lambda \ell^\mu \ell^\nu = 0$ for all Minkowski-null vectors ℓ. Furthermore, the previous equation holds if and only if ${}^{(+)}\aleph \equiv 0$; see Footnote 16 on pg. 40. In particular, when ${}^{(+)}\aleph \equiv 0$, the methods of [**47**] and [**15**] yield small-data global existence. Moreover, upon Taylor expanding equation (A.1) to quadratic order around the 0 solution and then decomposing the quadratic terms relative to the "flat frame" (2.19), we find that ${}^{(+)}\aleph$ is the coefficient of the term

$(R_{(Flat)}\Phi)R_{(Flat)}(R_{(Flat)}\Phi)$. As we mentioned in Remark A.1, when ${}^{(+)}\aleph$ is non-trivial, equation (A.1) exhibits small-data future shock formation caused *precisely* by the term $(R_{(Flat)}\Phi)R_{(Flat)}(R_{(Flat)}\Phi)$ (that is, the methods of [**47**] and [**15**] imply that in the region $\{t \geq 0\}$ the other quadratic terms are compatible with small-data future-global existence).

REMARK A.6 (**Past null condition failure factor for equation** (A.1)). We could also study shock formation in the region $\{t \leq 0\}$. As we described in Remark 3.35, in this region, the relevant past null condition failure factor ${}^{(-)}\aleph$ is defined by replacing $L_{(Flat)}$ with $-\partial_t + \partial_r$ in equation (A.8).

In the next lemma, we provide the wave equations verified by the $\Psi_\nu = \partial_\nu \Phi$, $(\nu = 0, 1, 2, 3)$. For reasons to be explained, we also commute equation (A.1) with the *Minkowski scaling vectorfield* $S := x^\alpha \partial_\alpha$, which is a conformal Killing field[2] of the Minkowski metric.

LEMMA A.7 (**The structure of the commuted equations**). *Assume that* Φ *verifies equation* (A.1). *Let* $\Psi \in \{\partial_\nu \Phi\}_{\nu=0,1,2,3} = \{\Psi_\nu\}_{\nu=0,1,2,3}$. *Then the scalar function* Ψ *verifies the following equation relative to the* ***rectangular Minkowski*** *coordinates:*

$$\partial_\alpha \left((h^{-1})^{\alpha\beta}\partial_\beta \Psi\right) = H^{\mu\alpha\beta}(\vec{\Psi})\left\{\partial_\beta \Psi_\alpha \partial_\mu \Psi - \partial_\mu \Psi_\alpha \partial_\beta \Psi\right\}, \tag{A.9}$$

$$H^{\mu\alpha\beta}(\vec{\Psi}) := (h^{-1})^{\alpha\alpha'}(h^{-1})^{\beta\beta'} H^{\mu}_{\alpha'\beta'}, \tag{A.10}$$

where $H_{\mu\nu}^{\lambda}$ *is defined in* (A.6).

Similarly, let S *be the Minkowski scaling vectorfield defined by*

$$S := x^\alpha \partial_\alpha, \tag{A.11}$$

and let

$$\Psi := S\Phi - \Phi = x^\alpha \Psi_\alpha - \Phi. \tag{A.12}$$

Then the scalar function Ψ *verifies the following equation relative to the* ***rectangular Minkowski*** *coordinates:*

$$\partial_\alpha \left((h^{-1})^{\alpha\beta}\partial_\beta \Psi\right) = H^{\mu\alpha\beta}(\vec{\Psi})\left\{\partial_\beta \Psi_\alpha \partial_\mu \Psi - \partial_\mu \Psi_\alpha \partial_\beta \Psi\right\}. \tag{A.13}$$

PROOF. Equation (A.9) follows from commuting (A.1) with ∂_ν and performing straightforward calculations that rely on the symmetry property $\partial_\alpha \Psi_\beta = \partial_\beta \Psi_\alpha$. Similarly, (A.13) follows from commuting (A.1) with S and performing straightforward calculations. □

Just below, we verify that (A.9) and (A.13) can be rewritten as covariant wave equations whose *right-hand* sides have a good null structure. The following functions play a role in our identification of this geometric structure.

DEFINITION A.8 (**The quantities** Ω^ν). We define the rectangular components Ω^ν, $(\nu = 0, 1, 2, 3)$, to be the following scalar-valued functions of $\vec{\Psi}$:

$$\Omega^\nu = \Omega^\nu(\vec{\Psi}) := \frac{1}{\sqrt{|\det h|(\vec{\Psi})}} \frac{\partial \sqrt{|\det h|(\vec{\Psi})}}{\partial \Psi_\nu} = \frac{1}{2} \frac{\partial \ln\left(|\det h|(\vec{\Psi})\right)}{\partial \Psi_\nu}, \tag{A.14}$$

where the determinant in (A.14) is taken relative to rectangular coordinates.

[2]That is, $\mathcal{L}_S m$ is equal to a scalar function multiple of m.

In the next lemma, we rewrite the equations from Lemma A.7 as covariant wave equations with inhomogeneous terms.

LEMMA A.9 (**Rewriting the equations as covariant wave equations**). *The system of wave equations* (A.9) *can be equivalently expressed as the following system of covariant wave equations in the unknowns* $\vec{\Psi} := (\Psi_0, \Psi_1, \Psi_2, \Psi_3)$, *where* $\Psi_\nu := \partial_\nu \Phi$ *is viewed to be a scalar function under covariant differentiation:*

$$\square_{h(\vec{\Psi})}\Psi_\nu = \mathscr{Q}(\partial\vec{\Psi}, \partial\Psi_\nu), \qquad (\nu = 0,1,2,3). \tag{A.15}$$

In (A.15), $\square_{h(\vec{\Psi})}\Psi$ *is the covariant wave operator of* h, *and it can be expressed as follows relative to the rectangular coordinates:*

$$\square_{h(\vec{\Psi})}\Psi := \frac{1}{\sqrt{|\mathrm{det} h|}}\partial_\alpha\left(\sqrt{|\mathrm{det} h|}(h^{-1})^{\alpha\beta}\partial_\beta\Psi\right). \tag{A.16}$$

Moreover, relative to the rectangular coordinates, the term $\mathscr{Q}(\partial\vec{\Psi}, \partial\Psi_\nu)$ *on the right-hand side of* (A.15) *can be expressed as*

$$\mathscr{Q}(\partial\vec{\Psi}, \partial\Psi) := H^{\mu\alpha\beta}\left\{\partial_\beta\Psi_\alpha\partial_\mu\Psi - \partial_\mu\Psi_\alpha\partial_\beta\Psi\right\} + (h^{-1})^{\alpha\beta}\Omega^\lambda\partial_\alpha\Psi_\lambda\partial_\beta\Psi. \tag{A.17}$$

The quantities $H^{\alpha\beta\mu}$ *and* Ω^λ *are defined in* (A.10) *and* (A.14).

Similarly, the wave equation (A.13) *for the scalar-valued function* $\Psi := S\Phi - \Phi$ *can be expressed as*

$$\square_{h(\vec{\Psi})}\Psi = \mathscr{Q}(\partial\vec{\Psi}, \partial\Psi), \tag{A.18}$$

where $\mathscr{Q}(\partial\vec{\Psi}, \partial\Psi)$ *is defined the same way as in* (A.17).

PROOF. Lemma A.9 follows from Lemma A.7, Def. A.8, and straightforward computations. □

The next lemma is at the heart of the connection between equations (A.1) and (2.1a). It shows shows that the quadratic terms $\mathscr{Q}(\cdot,\cdot)$ from Lemma A.9 have a special null structure and hence have a negligible effect on the dynamics of small-data solutions. Before we state the lemma, we first state the definitions of the standard null forms relative to h.

DEFINITION A.10 (**Standard null forms relative to** h). We define the standard null forms $\mathscr{Q}_{(0)}$ and $\mathscr{Q}_{(\alpha\beta)}$, which act on pairs of functions ϕ and $\widetilde{\phi}$, as follows:

$$\mathscr{Q}_{(0)}(\partial\phi, \partial\widetilde{\phi}) := (h^{-1})^{\alpha\beta}\partial_\alpha\phi\partial_\beta\widetilde{\phi}, \tag{A.19a}$$

$$\mathscr{Q}_{(\alpha\beta)}(\partial\phi, \partial\widetilde{\phi}) := \partial_\alpha\phi\partial_\beta\widetilde{\phi} - \partial_\alpha\widetilde{\phi}\partial_\beta\phi. \tag{A.19b}$$

We now recall an important property of the standard null forms. By expanding $h^{-1}(\vec{\Psi})$ relative to the frame $\{L, R, X_1, X_2\}$ as $(h^{-1})^{\mu\nu} = -L^\mu L^\nu - (L^\mu R^\nu + R^\mu L^\nu) + (\not{h}^{-1})^{\mu\nu}$ (as in (3.52c)) and noting that $(\not{h}^{-1})^{\mu\nu}$ is a smooth function of $\vec{\Psi}$ and the rectangular components L^μ, R^μ, we see that

$$\mathscr{Q}_{(0)}(\partial\phi, \partial\widetilde{\phi}) = f_1 L\phi L\widetilde{\phi} + f_2 L\phi R\widetilde{\phi} + f_3 R\phi L\widetilde{\phi} + f_4^{AB}\not{d}_A\phi\not{d}_B\widetilde{\phi}, \tag{A.20}$$

where f_1, f_2, and f_3 are smooth scalar-valued functions of $\vec{\Psi}$ and the rectangular components of the vectorfields L and R, and the same statement holds for the rectangular components $f_4^{\alpha\beta}$ of the symmetric type $\binom{2}{0}$ $S_{t,u}$ tensor f_4. The important point is that there are no quadratic terms on the right-hand side of (A.20) involving two $\mathcal{C}_u^t$-transversal derivatives, that is, there are no terms proportional to $R\phi R\widetilde{\phi}$. Similarly, upon expanding $\mathscr{Q}_{(\alpha\beta)}(\partial\phi, \partial\widetilde{\phi})$ relative to the non-rescaled frame $\{L, R, X_1, X_2\}$, we find that the coefficients depend smoothly on $\vec{\Psi}$ and the rectangular components of the vectorfields L and R and, in view of the antisymmetry of $\mathscr{Q}_{(\alpha\beta)}(\partial\phi, \partial\widetilde{\phi})$ under interchanging ϕ and $\widetilde{\phi}$, that there are no terms proportional to $R\phi R\widetilde{\phi}$.

We capture the absence of the component $R\phi R\widetilde{\phi}$ with the following definition, which is tailored to the system of wave equations (A.9) and which is relevant in the region $\{t \geq 0\}$.

DEFINITION A.11 (**Future strong null condition**). Let $\mathcal{N}(\vec{\Psi}, \partial\vec{\Psi})$ be a nonlinear term depending smoothly on $\vec{\Psi}$ and $\partial\vec{\Psi}$, where $\partial\vec{\Psi}$ denotes the first partial derivatives of $\vec{\Psi}$ with respect to the rectangular spacetime coordinates. We say that $\mathcal{N}(\vec{\Psi}, \partial\vec{\Psi})$ verifies the future strong null condition if, when expressed relative to the non-rescaled frame $\{L, R, X_1, X_2\}$, it depends linearly or not at all on the elements of $\{R\Psi_\alpha\}_{\alpha=0,1,2,3}$.

REMARK A.12 (**Future strong null condition for quadratic derivative nonlinearities**). We now clarify the notion of the future strong null condition in the special case that $\mathcal{N}(\vec{\Psi}, \partial\vec{\Psi})$ depends precisely quadratically on $\partial\vec{\Psi}$ with coefficients depending on $\vec{\Psi}$. In this case, relative to the frame $\{L, R, X_1, X_2\}$, we can expand

(A.21)

$$
\begin{aligned}
\mathcal{N}(\vec{\Psi}, \partial\vec{\Psi}) = & \sum_{\alpha,\beta=0}^{3} f_{1;\alpha\beta} L\Psi_\alpha L\Psi_\beta + \sum_{\alpha,\beta=0}^{3} f_{2;\alpha\beta} L\Psi_\alpha R\Psi_\beta + \sum_{\alpha,\beta=0}^{3} f_{3;\alpha\beta} R\Psi_\alpha R\Psi_\beta \\
& + \sum_{\alpha,\beta=0}^{3} f_{4;\alpha\beta}^{A} \not{d}_A \Psi_\alpha L\Psi_\beta + \sum_{\alpha,\beta=0}^{3} f_{5;\alpha\beta}^{A} \not{d}_A \Psi_\alpha R\Psi_\beta \\
& + \sum_{\alpha,\beta=0}^{3} f_{6;\alpha\beta}^{AB} \not{d}_A \Psi_\alpha \not{d}_B \Psi_\beta,
\end{aligned}
$$

where $f_{1;\alpha\beta}$, $f_{2;\alpha\beta}$ $f_{3\alpha\beta}$, and the rectangular components $f_{4;\alpha\beta}^{\gamma}$, $f_{5;\alpha\beta}^{\gamma}$, and $f_{6;\alpha\beta}^{\gamma\delta}$ are smooth scalar-valued functions of $\vec{\Psi}$ and the rectangular components of the vectorfields L and R. In this case, the future strong null condition is equivalent to the following condition: $f_{3;\alpha\beta} \equiv 0$ for $\alpha, \beta = 0, 1, 2, 3$.

Note that the null form terms $\mathscr{Q}_{(0)}(\partial\Psi_\mu, \partial\Psi_\nu)$ and $\mathscr{Q}_{(\alpha\beta)}(\partial\Psi_\mu, \partial\Psi_\nu)$, which are the relevant nonlinearities in the system (A.15), verify the future strong null condition.

REMARK A.13 (**Dependence of the definition on the frame**). Note that our definition of the future strong null condition is tied to our choice of vectorfields L and R in the decomposition $(h^{-1})^{\mu\nu} = -L^\mu L^\nu - (L^\mu R^\nu + R^\mu L^\nu) + (\not{h}^{-1})^{\mu\nu}$ (see (3.52d)). This may seem a bit unnatural. However, the null form terms $\mathscr{Q}_{(0)}(\partial\Psi_\mu, \partial\Psi_\nu)$ and $\mathscr{Q}_{(\alpha\beta)}(\partial\Psi_\mu, \partial\Psi_\nu)$ verify the future strong null condition independent of the particular choice of L and R. More precisely, if we decompose $(h^{-1})^{\mu\nu} := -\widetilde{L}^\mu \widetilde{L}^\nu - (\widetilde{L}^\mu \widetilde{R}^\nu + \widetilde{R}^\mu \widetilde{L}^\nu) + (\widetilde{\not{h}^{-1}})^{\mu\nu}$, where $\widetilde{L}$ and $\widetilde{R}$ are vectorfields satisfying $(\widetilde{\not{h}^{-1}})^{\alpha\nu} L_\alpha = 0$, $(\widetilde{\not{h}^{-1}})^{\alpha\nu} R_\alpha = 0$, $h(\widetilde{L}, \widetilde{L}) = 0$, $h(\widetilde{L}, \widetilde{R}) = -1$, and $h(\widetilde{R}, \widetilde{R}) = 1$, then $\mathscr{Q}_{(0)}$ still enjoys the property (A.20), but with L and R on the right-hand side respectively replaced by $\widetilde{L}$ and $\widetilde{R}$ and the differentiations $\nabla\!\!\!/_A$ replaced by differentiations in a direction that is h-orthogonal to span$\{\widetilde{L}, \widetilde{R}\}$. Similar remarks apply for $\mathscr{Q}_{(\alpha\beta)}$.

REMARK A.14 (**The future strong null condition is nonlinear in nature**). Note that the future strong null condition is a nonlinear condition tied to the dynamic metric h. In particular, it is not necessarily verified by cubic terms $(\partial\vec{\Psi})^3$, which could contain dangerous factors involving two or more derivatives in the R direction.

REMARK A.15 (**Past strong null condition**). We could also formulate a past strong null condition, which would be relevant in the region $\{t \leq 0\}$. To define such a condition, one would first need to replace the frame $\{L, R, X_1, X_2\}$ with an analogous frame in which the analog of L is outgoing as we head towards the past. We would then define the past strong null condition by replacing, in definition (A.21), L with that new outgoing null vectorfield.

We now provide the aforementioned lemma.

LEMMA A.16 (**Special null structure of the inhomogeneous terms**). *The quadratic term $\mathscr{Q}(\partial\vec{\Psi}, \partial\Psi)$ on the right-hand sides of* (A.15) *and* (A.18) *verifies the future strong null condition.*

PROOF. Lemma A.16 follows from comparing the right-hand sides of (A.15) and (A.18) to the null forms $\mathscr{Q}_{(0)}(\cdot,\cdot)$ and $\mathscr{Q}_{(\alpha\beta)}(\cdot,\cdot)$ from Def. A.10 and from using the properties of the standard null forms revealed in equation (A.20) and the discussion following it. □

The important consequence of Lemma A.16 is that the quadratic terms $\mathscr{Q}(\partial\vec{\Psi}, \partial\Psi)$ on the right-hand sides of (A.15) and (A.18) have a negligible influence on the future dynamics of small-data solutions, even those that form shocks. Specifically, because of their special null structure, the $\mathscr{Z}^N$ derivatives of $\upmu$ times[3] $\mathscr{Q}(\cdot,\cdot)$ are $Harmless^{\leq N}$ in the sense of Def. 16.2 (with $\vec{\Psi}$ in the role of Ψ) and are therefore easy to treat.

A.2. The main new estimate needed at the top order

Lemma A.16 implies that $\vec{\Psi}$ verifies the coupled system (A.15) of covariant wave equations with quadratic inhomogeneous terms that have a special null structure. We now explain how to prove an analog of Theorem 22.1 for the system. The main point is that we can derive L^2 estimates for each component Ψ_ν, which verifies a scalar wave equation, by using essentially the same arguments that we used to treat the scalar equation (2.1a). To control $\vec{\Psi}$, we replace the L^2-type controlling quantity $\mathbb{Q}_{(N)}$ from Def. 14.2 with

(A.22)
$$\mathbb{Q}_{(N)}(t,u) := \max_{|\vec{I}|=N} \max_{\nu=0,1,2,3} \sup_{(t',u')\in[0,t]\times[0,u]} \left\{ \mathbb{E}[\mathscr{Z}^{\vec{I}}\Psi_\nu](t',u') + \mathbb{F}[\mathscr{Z}^{\vec{I}}\Psi_\nu](t',u') \right\},$$

and similarly for the other controlling quantities of Chapter 14.

---

[3]Recall that in our analysis, we always work with the $\upmu$-weighted wave operator $\upmu\Box_{h(\vec{\Psi})}\Psi_\nu$.

REMARK A.17 (**Potentially non-optimal definition of the L^2-type controlling quantities**). The definition (A.22) involving the max over $\nu = 0, 1, 2, 3$ is somewhat crude in the sense that it does not account for any tensorial structure in the index ν. This crude treatment might lead to "boxed constants" (see Remark 20.3) that are somewhat larger than the ones we encountered in our study of the scalar equation (2.1a). In turn, compared to the case of equation (2.1a), larger boxed constants would increase the possible blow-up rate of our top-order L^2 estimates with respect to powers of $\mu_\star^{-1}$ and therefore would increase the number of derivatives that we need to close the estimates. Hence, it may be advantageous to instead work with more geometric L^2-controlling quantities involving frame contractions in the index ν such as $\mathbb{E}[R^a \mathscr{Z}^{\vec{I}} \Psi_a] + \mathbb{F}[R^a \mathscr{Z}^{\vec{I}} \Psi_a]$, $\mathbb{E}[L^\alpha \mathscr{Z}^{\vec{I}} \Psi_\alpha] + \mathbb{F}[L^\alpha \mathscr{Z}^{\vec{I}} \Psi_\alpha]$, etc. The ensuing discussion suggests that such quantities might be better adapted to the most degenerate terms that we encounter in the analysis in the sense that they might lead to estimates featuring smaller boxed constants; we do not pursue this issue further here.

As we explained in the previous paragraph, the quadratic terms on the right-hand side of (A.15) are easy to treat without invoking any new ideas. However, we do need one new ingredient to close the estimates, one that is needed for the top-order L^2 estimates for the Ψ_ν. To clarify matters, we now recall the analogous situation that arose in our analysis of the scalar equation (2.1a). At various points in our argument for deriving top-order L^2 estimates, it was essential that we could use equation (4.5) to make the replacement

$$G_{LL} \breve{R} \Psi = 2L\mu + \text{Error}, \tag{A.23}$$

where the error terms are small and decay at an integrable-in-time rate. One example of the importance of this identity was explained in detail just before equation (2.64). The main point is that we can treat the coupled system $\vec{\Psi}$ by separately analyzing each Ψ_ν, but we need an analog of (A.23) *for each of the four* Ψ_ν. An analog of the factor $G_{LL}\breve{R}\Psi$ from (A.23) arises when we derive top-order L^2 estimates for the scalar component Ψ_ν in equation (A.15). Specifically, the analog is $-H_{LL}^L \breve{R} \Psi_\nu$, where $H_{LL}^L := H_{\alpha\beta}^\kappa L^\alpha L^\beta L_\kappa$. On the other hand, as we will see in equation (A.26a) below, μ verifies a transport equation of the form $L\mu = -\frac{1}{2} H_{LL}^L R^a \breve{R} \Psi_a + \text{Error}$. Hence, it is not immediately apparent that for solutions to the system (A.15), the following analog of (A.23), which plays a key role in closing the top-order L^2 estimates, should hold for $\nu = 0, 1, 2, 3$:

$$\left| H_{LL}^L \breve{R} \Psi_\nu \right| \leq 2 \left| L\mu \right| + \text{Error}. \tag{A.24}$$

We stress that inequality (A.24) is exactly the new ingredient mentioned above that we need to close the top-order L^2 estimates. We dedicate the remainder of this appendix to deriving the estimate (A.24). We prove the final estimate of interest in Prop. A.20.

In the next lemma, we provide the analog of equation (4.5), that is, the transport equation equation verified by μ. Our main goal is to identify the dangerous slowly decaying component $\omega^{(Trans-\vec{\Psi})}$, which arises from the failure of the classic null condition and which is strong enough to drive μ to 0 in finite time.

LEMMA A.18 (**The transport equation verified by $\upmu$**). *In the case of equation* (A.1), *the inverse foliation density $\upmu$ defined in* (3.9) *verifies the following transport equation:*

$$L\upmu = \omega^{(Trans-\vec{\Psi})} + \upmu\omega^{(Tan-\vec{\Psi})} := \omega, \tag{A.25}$$

where

$$\omega^{(Trans-\vec{\Psi})} := -\frac{1}{2}H_{LL}^{L}R^{a}\breve{R}\Psi_{a}, \tag{A.26a}$$

$$\begin{aligned}\omega^{(Tan-\vec{\Psi})} := & -\frac{1}{2}H_{LL}^{L}R^{a}L\Psi_{a} - \frac{1}{2}H_{LL}^{R}R^{a}L\Psi_{a} + \frac{1}{2}\not{H}_{LL}^{A}R^{a}\not{d}_{A}\Psi_{a} \\ & - \frac{1}{2}H_{LL}^{\lambda}L\Psi_{\lambda} - H_{LR}^{\lambda}L\Psi_{\lambda}.\end{aligned} \tag{A.26b}$$

In equations (A.26a)-(A.26b), $H_{LL}^{L} := H_{\alpha\beta}^{\kappa}L^{\alpha}L^{\beta}L_{\kappa}$, $\not{H}_{LL}^{\lambda} := H_{\alpha\beta}^{\kappa}L^{\alpha}L^{\beta}\not{\Pi}_{\kappa}^{\ \lambda}$, $\not{H}_{LL}^{A} := H_{\alpha\beta}^{\kappa}L^{\alpha}L^{\beta}\not{\Pi}_{\kappa}^{\ \lambda}\not{h}_{\lambda\mu}X_{A}^{\mu}$, *etc., where $\not{h}_{\mu\nu}$ is the Riemannian metric induced on $S_{t,u}$ by $h_{\mu\nu}$ (defined as in Def.* 3.9 *with h in place of g), $\not{\Pi}_{\nu}^{\ \mu}$ is the $S_{t,u}$ projection tensorfield (defined as in* (3.39b) *but with the indices on L and R lowered with h rather than g), and the rectangular component function $H_{\mu\nu}^{\lambda}$ is defined in* (A.6).

PROOF. The proof is very similar to the proof of Lemma 4.6. The main difference is that we use the symmetry property $\partial_{\alpha}\Psi_{\beta} = \partial_{\beta}\Psi_{\alpha}$ to rewrite

$$\begin{aligned}H_{LL}^{\lambda}\breve{R}\Psi_{\lambda} &= \upmu H_{LL}^{\lambda}R^{a}\partial_{\lambda}\Psi_{a} \\ &= -H_{LL}^{L}R^{a}\breve{R}\Psi_{a} - \upmu H_{LL}^{L}R^{a}L\Psi_{a} - \upmu H_{LL}^{R}R^{a}L\Psi_{a} + \upmu\not{H}_{LL}^{A}R^{a}\not{d}_{A}\Psi_{a}.\end{aligned} \tag{A.27}$$

□

The next lemma provides the key technical estimates that allow us to conclude (A.24).

LEMMA A.19 (**Approximate identities verified by $\breve{R}\Psi_{\nu}$**). *Under the small-data and bootstrap assumptions of Sects.* 12.1-12.4 *(with $\vec{\Psi}$ in place of Ψ), if ε is sufficiently small, then for $i = 1, 2, 3$, the following estimates hold for $(t, u) \in [0, T_{(Bootstrap)}) \times [0, U_0]$:*

$$\left\| \breve{R}\Psi_{0} - R^{a}\breve{R}\Psi_{a} \right\|_{C^{0}(\Sigma_{t}^{u})} \leq C\varepsilon\frac{\ln(e+t)}{(1+t)^{2}}, \tag{A.28a}$$

$$\left\| \breve{R}\Psi_{i} + \frac{x^{i}}{r}R^{a}\breve{R}\Psi_{a} \right\|_{C^{0}(\Sigma_{t}^{u})} \leq C\varepsilon\frac{\ln(e+t)}{(1+t)^{2}}. \tag{A.28b}$$

PROOF. We first sketch the proof of (A.28a). From (4.1b), (12.8a), and ($\mathbf{AUX}LR_{(Small)}$) (which together imply that $|R^{a}| \lesssim 1$) and ($\mathbf{BA}\Psi$), we see that (A.28a) easily follows for $t \leq 1$. Hence, we have to address only the case $t > 1$. To this end, we let $S = x^{\alpha}\partial_{\alpha}$ be the Minkowskian scaling vectorfield. Using direct computation and (4.1b), we compute that

$$\begin{aligned}\breve{R}(S\Phi - \Phi) &= t\breve{R}\Psi_{0} + x^{a}\breve{R}\Psi_{a} \\ &= t\breve{R}\Psi_{0} - \varrho R^{a}\breve{R}\Psi_{a} + \varrho R_{(Small)}^{a}\breve{R}\Psi_{a},\end{aligned} \tag{A.29}$$

where $\varrho(t, u) = 1 - u + t$. From (4.3) (with h in place of g), ($\mathbf{BA}\Psi$), (12.5), and (12.128a), we deduce that the last term on the right-hand side of (A.29) is bounded in magnitude by $\lesssim \varepsilon \ln(e+t)(1-t)^{-1}$. From this estimate, equation (A.29), and (**BA**Ψ), we find that inequality (A.28a) for $t \geq 1$ will follow once we show that

$$\left\| \breve{R}(S\Phi - \Phi) \right\|_{C^0(\Sigma_t^u)} \leq C\varepsilon \frac{1}{1+t}. \tag{A.30}$$

To prove (A.30), we will derive energy estimates for the wave equation (A.18) verified by $\Psi := S\Phi - \Phi$. To this end, we first use equation (A.18) and Lemma A.16 to deduce that

$$\upmu \square_{g(\vec{\Psi})}(S\Phi - \Phi) = \mathfrak{F} \tag{A.31}$$

and that there exists a smooth function f such that $\mathfrak{F}$ can be expressed as $f(\vec{\psi}, \upmu, L^1, L^2, L^3)$ times terms that are quadratic in the derivatives of $S\Phi - \Phi$ relative to the rescaled frame $L, \breve{R}, X_1, X_2$, with at least one factor involving a derivative that is tangent to $\mathcal{C}_u^t$. Hence, we can apply essentially the same reasoning that we used to derive the estimates (22.13a)-(22.13b) in order to deduce that

$$\begin{aligned} &\mathbb{E}^{1/2}[\mathscr{Z}^{\leq 2}\breve{R}(S\Phi - \Phi)] + \mathbb{F}^{1/2}[\mathscr{Z}^{\leq 2}\breve{R}(S\Phi - \Phi)] \\ &\leq C\mathring{\epsilon}, \end{aligned} \tag{A.32}$$

$$\begin{aligned} &\widetilde{\mathbb{E}}^{1/2}[\mathscr{Z}^{\leq 2}\breve{R}(S\Phi - \Phi)] - \widetilde{\mathbb{F}}^{1/2}[\mathscr{Z}^{\leq 2}\breve{R}(S\Phi - \Phi)] + \widetilde{\mathbb{K}}^{1/2}[\mathscr{Z}^{\leq 2}\breve{R}(S\Phi - \Phi)] \\ &\leq C\mathring{\epsilon} \ln^2(e+t). \end{aligned} \tag{A.33}$$

Then from the Sobolev embedding estimate (18.32) and (A.32), we deduce the desired bound (A.30).

To prove (A.28b), we first use the identity $\partial_j \Psi_i = \partial_i \Psi_j$ to compute that

$$\breve{R}\Psi_i = \frac{x^i}{r}\frac{x^a}{r}\breve{R}\Psi_a + \upmu R^a V \Psi_a, \tag{A.34}$$

where V is the $\Sigma_t^{U_0}$-tangent vectorfield (depending on i) with rectangular components

$$V^j = \delta_i^j - \frac{x^i}{r}\frac{x^j}{r}. \tag{A.35}$$

From the bound $|R^a| \lesssim 1$ noted above, (4.1b), (4.3) (with h in place of g), and (A.34), the estimates (12.8b) and (12.128a), and the bootstrap assumptions (**BA**Ψ), we see that we may replace the factor $\frac{x^a}{r}$ in the first product on the right-hand side of (A.34) with R^a up to an error term that is in magnitude $\lesssim \varepsilon \frac{\ln(e+t)}{(1+t)^2}$. Hence, the desired bound (A.28b) will follow once we show that for $a = 1, 2, 3$, we have

$$\upmu |V\Psi_a| \lesssim \varepsilon \frac{\ln(e+t)}{(1+t)^2}. \tag{A.36}$$

To prove (A.36), we first note that since the Euclidean length of V is $\lesssim 1$ and since V is tangent to the spheres in $\Sigma_t^{U_0}$ of constant Euclidean radius, it is a standard fact (that can be proved by arguing as in our proof of (12.46a)) that

$$|V\Psi_a| \lesssim \frac{1}{r}\sum_{l=1}^{3} |O_{(Flat;l)}\Psi_a|, \tag{A.37}$$

where $O_{(Flat;l)}$ are the Euclidean rotations from Def. 6.1. The desired bound (A.36) now follows from (6.5), (**BA**Ψ), (12.8b), (12.28b), (12.128a), Cor. 12.48, and (A.37). □

We now use the above estimates to derive the main estimate of interest, namely (A.24).

PROPOSITION A.20 (**Sharp pointwise estimates for $H_{LL}^{L}\breve{R}\Psi_{\nu}$ in terms of $L\upmu$**). *Under the small-data and bootstrap assumptions of Sects.* 12.1-12.4 *(with $\vec{\Psi}$ in place of Ψ), if ε is sufficiently small, then the following estimates hold on the spacetime region $\mathcal{M}_{T_{(Bootstrap)},U_0}$, $(\nu = 0, 1, 2, 3)$:*

$$\left|H_{LL}^{L}\breve{R}\Psi_{\nu}\right| \leq 2\left|L\upmu\right| + C\varepsilon\frac{\ln(e+t)}{(1+t)^2}. \tag{A.38}$$

PROOF. We consider equation (A.25). Using an argument similar to the one we used to prove (17.38), we deduce that $|\omega^{(Tan-\vec{\Psi})}| \lesssim \varepsilon(1+t)^{-2}$ and $|H_{LL}^{L}| \lesssim 1$. Using these two inequalities, the simple bound $\upmu \lesssim \ln(e+t)$ (see (12.128a)), and inequality (A.28a), and examining the expression (A.26a) for $\omega^{(Trans-\vec{\Psi})}$, we conclude the desired estimate (A.38) in the case $\nu = 0$.

To prove the desired estimate (A.38) in the cases $\nu = 1, 2, 3$, we first note that (A.28b) and the trivial bound $\left|\frac{x^i}{r}\right| \leq 1$ together imply that for $i = 1, 2, 3$, we have $\left|\breve{R}\Psi_i\right| \leq \left|R^a\breve{R}\Psi_a\right| + C\varepsilon\frac{\ln(e+t)}{(1+t)^2}$. The desired estimate now follows from this inequality and the argument given in the previous paragraph. □

APPENDIX B

Summary of Notation and Conventions

In this appendix, we collect some important notation and conventions that we use throughout the monograph so that the reader can refer to it as needed. For many items, we also provide page numbers indicating the most fundamental point of reference. We remark that in some parts of Chapter 1, we deviate from the notation presented here.

B.1. Coordinates

(x^0, x^1, x^2, x^3) are a fixed set of rectangular spacetime coordinates on $\mathbb{R}^{1+3}$, relative to which the Minkowski metric m has components $m_{\mu\nu} = \text{diag}(-1, 1, 1, 1)$. We often use the alternate notation $t = x^0$.	35
$r = \sqrt{\sum_{a=1}^{3}(x^a)^2}$ is the Euclidean radial coordinate.	36
$\theta = (\theta^1, \theta^2)$ are local Euclidean angular coordinates corresponding to spherical coordinates on Minkowski spacetime, that is, local angular coordinates on the Euclidean spheres of constant t and r.	88
u denotes the eikonal function.	82
$\vartheta = (\vartheta^1, \vartheta^2)$ are local geometric angular coordinates on the spheres $S_{t,u}$.	87
$(t, u, \vartheta^1, \vartheta^2)$ are the geometric coordinates.	87
$\varrho = 1 - u + t$ is the geometric radial variable.	82

B.2. Indices

Lowercase Greek indices μ, ν, etc. belong to $\{0,1,2,3\}$ and correspond to components with respect to the *rectangular spacetime* coordinates x^0, x^1, x^2, x^3.
Lowercase Latin indices i, j, etc. belong to $\{1,2,3\}$ and correspond to components with respect to the *rectangular spatial* coordinates x^1, x^2, x^3.
Lowercase Greek indices are lowered and raised with the spacetime metric g and its inverse g^{-1}, and *not with the Minkowski metric.*
Capital Latin letters A, B etc. belong to $\{1,2\}$ and correspond to components with respect to the geometric coordinate partial derivative frame $\left\{X_1 = \frac{\partial}{\partial \vartheta^1}, X_2 = \frac{\partial}{\partial \vartheta^2}\right\}$. These indices are lowered and raised with $\not{g}$ and $\not{g}^{-1}$.
We use Einstein's summation convention in that repeated indices are summed over their respective ranges.
Indices decorated with a tilde, such as $\widetilde{A}$, are treated in the same way as their non-decorated counterparts.

B.3. Constants

C denotes a uniform constant that is free to vary from line to line.
If we want to emphasize that the constant C depends on a quantity Q, then we use notation such as "C_Q."
We use the notation $f_1 \lesssim f_2$ to indicate that there exists a uniform constant $C > 0$ such that $f_1 \leq C f_2$.
We sometimes use the alternate notation $\mathcal{O}(f_2)$ to denote a quantity f_1 that verifies $\|f_1\| \lesssim f_2$.
If we want to emphasize that the implicit constant C depends on an a quantity Q, then we use the alternate notation $f_1 \overset{Q}{\lesssim} f_2$.

B.4. Spacetime subsets

Σ_t = the hypersurface of constant time t.	83
$\mathcal{C}_u$ = the outgoing null hypersurface equal to the corresponding level set of the eikonal function.	83
Σ_t^u = the portion of Σ_t in between $\mathcal{C}_0$ and $\mathcal{C}_u$.	83
$\mathcal{C}_u^t$ = the portion of $\mathcal{C}_u$ in between Σ_0 and Σ_t.	83
$\mathcal{M}_{t,u}$ = the open-at-the-top spacetime region trapped in between Σ_0^u, Σ_t^u, $\mathcal{C}_u^t$, and $\mathcal{C}_0^t$.	83

B.5. Metrics, musical notation, and inner products

$g = g(\Psi)$ denotes the spacetime metric. Relative to rectangular coordinates, $g_{\mu\nu}(\Psi) = m_{\mu\nu} + g_{\mu\nu}^{(Small)}(\Psi)$, where $m_{\mu\nu} = \text{diag}(-1,1,1,1)$ and $g_{\mu\nu}^{(Small)}(\Psi = 0) = 0$.	35
Relative to rectangular coordinates, $G_{\mu\nu}(\Psi) = \frac{d}{d\Psi} g_{\mu\nu}(\Psi)$ and $G'_{\mu\nu}(\Psi) = \frac{d}{d\Psi} G_{\mu\nu}(\Psi)$.	38
$\underline{g}$ denotes the first fundamental form of Σ_t, that is, $\underline{g}_{ij} = g_{ij}$.	83
$\underline{g}^{-1}$ denotes the inverse first fundamental form of Σ_t.	83
$\not{g}$ denotes the first fundamental form of $S_{t,u}$.	83
$\not{g}^{-1}$ denotes the inverse first fundamental form of $S_{t,u}$.	83
We denote the $\not{g}$-dual of an $S_{t,u}$ one-form ξ by $\xi^{\#}$. In particular, $(\xi^{\#})^A = (\not{g}^{-1})^{AB}\xi_B$. We often use the abbreviated notation $\xi^A = (\not{g}^{-1})^{AB}\xi_B$.	97
Similarly, if Y is an $S_{t,u}$-tangent vector, then $Y_\flat$ denotes the $\not{g}$-dual of Y, which is an $S_{t,u}$ covector. That is, $(Y_\flat)_A = (\not{g}^{-1})_{AB}Y^B$. We often abbreviate $Y_A = (Y_\flat)_A$.	97
Similarly, we denote the $\not{g}$-dual of a symmetric type $\binom{0}{2}$ $S_{t,u}$ tensor ξ by $\xi^{\#}$ and we denote its double $\not{g}$-dual by $\xi^{\#\#}$. In particular, $(\xi^{\#})^A_{\ B} = (\not{g}^{-1})^{AC}\xi_{CB}$ and $(\xi^{\#\#})^{AB} = (\not{g}^{-1})^{AC}(\not{g}^{-1})^{BD}\xi_{CD}$.	97
We use similar notation to denote the $\not{g}$-duals of general type $\binom{m}{0}$ and type $\binom{0}{n}$ $S_{t,u}$ tensors, and we use abbreviations similar to the ones mentioned above for vectors and covectors.	97
$g(X,Y) = g_{\alpha\beta}X^\alpha Y^\beta$ denotes the inner product of the vectors X and Y with respect to the metric g. We use similar notation for inner products with respect to the metrics $\not{g}$ or $\underline{g}$.	83
$\langle X, Y\rangle = g(X,Y)$.	83

B.6. Eikonal function quantities

The eikonal function u verifies the eikonal equation $(g^{-1})^{\alpha\beta}\partial_\alpha u \partial_\beta u = 0$, has $\partial_t u > 0$, and has the initial condition $u\|_{\Sigma_0} = 1 - r$.	82
$\upmu = -\left\{(g^{-1})^{\alpha\beta}\partial_\alpha u \partial_\beta t\right\}^{-1}$ denotes the inverse foliation density.	84
$\upmu_\star(t,u) = \min\left\{1, \min_{\Sigma_t^u} \upmu\right\}$.	247
$L_{(Geo)}^\nu = -(g^{-1})^{\nu\alpha}\partial_\alpha u$ denotes the $\mathcal{C}_u$-tangent outgoing null geodesic vectorfield.	84
$L^\nu = \upmu L_{(Geo)}^\nu$ denotes a rescaled outgoing null vectorfield.	85
$L_{(Small)}^i = L^i - \varrho^{-1}x^i$ is a re-centered version of L^i.	107
$R^\mu = -L^\mu - (g^{-1})^{0\mu}$ is a decomposition of R^μ (with $R^0 = 0$).	108
$\breve{R}^i = \upmu R^i$.	86
$R_{(Small)}^i = R^i + \varrho^{-1}x^i$ is a re-centered version of R^i.	107
$\chi_{AB} = \frac{1}{2}\not{\mathcal{L}}_L \not{g}_{AB}$ is the null second fundamental form of $\mathcal{C}_u$ relative to g.	99
$\chi_{AB}^{(Small)} = \chi_{AB} - \varrho^{-1}\not{g}_{AB}$ is a re-centered version of χ_{AB}.	107
We informally refer to any of $\upmu$ L^i, $L_{(Small)}^i$, R^i, $R_{(Small)}^i$, χ, or $\chi^{(Small)}$ as the "eikonal function quantities."	107

B.7. Additional tensorfields related to the connection coefficients

$k = \frac{1}{2}\mathcal{L}_N \underline{g}$ is the second fundamental form of Σ_t relative to g.	99
$\zeta_A = g(\mathscr{D}_A L, R) = \upmu^{-1} g(\mathscr{D}_A L, \breve{R})$.	117
$\not{k}_{AB} = \upmu^{-1}\not{k}_{AB}^{(Trans-\Psi)} + \not{k}_{AB}^{(Tan-\Psi)}$ is a splitting of $\not{k}_{AB}$.	117
$\zeta_A = \upmu^{-1}\zeta_A^{(Trans-\Psi)} + \zeta_A^{(Tan-\Psi)}$ is a splitting of ζ_A.	117
$\eta_A = \zeta_A + \upmu^{-1}\not{d}_A \upmu$.	119
$\underline{\chi}_{AB} = 2\upmu\not{k}_{AB} - \upmu\chi_{AB}$.	119

B.8. Frame vectorfields and the timelike unit normal to the constant-time hypersurfaces

L is $\mathcal{C}_u^t$-tangent, outward pointing, and verifies $g(L,L)=0$ and $Lt=1$.	
$L=\frac{\partial}{\partial t}$ relative to the geometric coordinates.	88
$\breve{R}$ is Σ_t^u-tangent, $S_{t,u}$-orthogonal, inward pointing, and verifies $g(\breve{R},\breve{R})=\mu^2$, $\breve{R}u=1$.	86
$\breve{R}=\frac{\partial}{\partial u}-\Xi$ relative to the geometric coordinates, where Ξ is $S_{t,u}$-tangent.	93
$R=\mu^{-1}\breve{R}$.	86
$\breve{\underline{L}}=\mu L+2\breve{R}$ is an inward pointing null vectorfield that verifies $g(L,\breve{\underline{L}})=-2\mu$.	85
$\underline{L}=\mu^{-1}\breve{\underline{L}}$ is an inward pointing null vectorfield that verifies $g(L,\underline{L})=-2$.	85
For $A=1,2$, $X_A=\frac{\partial}{\partial\vartheta^A}$ is the geometric angular coordinate partial derivative vectorfield.	89
$N=L+R$ is the future-directed unit normal to Σ_t.	86
$L_{(Flat)}=\partial_t+\partial_r$ is a Minkowskian analog of L.	84
$\underline{L}_{(Flat)}=\partial_t-\partial_r$ is a Minkowskian analog of $\underline{L}$.	84
$\{L,R,X_1,X_2\}$ denotes the non-rescaled frame.	89
$\{L,\breve{R},X_1,X_2\}$ denotes the rescaled frame.	89
$\{L,\underline{L},X_1,X_2\}$ denotes the non-rescaled null frame.	89
$\{L,\breve{\underline{L}},X_1,X_2\}$ denotes the rescaled null frame.	89

B.9. Contraction and component notation

If ξ is a symmetric type $\binom{0}{2}$ spacetime tensor and V, W are vectors, then $\xi_{VW} = \xi_{\alpha\beta}V^\alpha W^\beta$ and ξ_V is the one-form with rectangular components $(\xi_V)_\nu = \xi_{\alpha\nu}V^\alpha$.	90
If ξ is a symmetric type $\binom{2}{0}$ spacetime tensor, then $\xi_{VW} = \xi^{\alpha\beta}V_\alpha W_\beta$. We use similar contraction notation for tensors ξ of any type.	90
We use abbreviations such as $\xi_A = \xi_{X_A}$ when contracting against the $S_{t,u}$ frame vectors X_1 and X_2.	89
We sometimes use a "·" to emphasize contraction between two tensors. For example, if T is a symmetric type $\binom{2}{0}$ $S_{t,u}$ tensorfield and ξ is a $S_{t,u}$ one-form, then $(\mathrm{di}\not{v}T) \cdot \xi = (\nabla\!\!\!/_A T^{AB})\xi_B$.	
If V is a spacetime vector, then $V = V^L L + V^{\breve{R}}\breve{R} + V^A X_A$ denotes its decomposition relative to the rescaled frame $\{L, \breve{R}, X_1, X_2\}$. We have $V^L = -V_L - \mu^{-1}V_{\breve{R}}$, $V^{\breve{R}} = -\mu^{-1}V_L$, and $V^A = (\not{g}^{-1})^{AB}V_B$.	91

B.10. Projection operators and frame components

$\underline{\Pi}$ denotes the type $\binom{1}{1}$ tensorfield that projects onto Σ_t^u.	91
$\Pi\!\!\!/$ denotes the type $\binom{1}{1}$ tensorfield that projects onto $S_{t,u}$.	91
If ξ is a spacetime tensor, then $\not{\xi} = \Pi\!\!\!/\xi$ is the projection of ξ onto $S_{t,u}$.	92
If ξ is a type $\binom{0}{2}$ spacetime tensor, then $\not{\xi}_V = \Pi\!\!\!/(\xi_V)$.	92
$G_{(Frame)} = (G_{LL}, G_{LR}, G_{RR}, \not{G}_L, \not{G}_R, \not{G})$ is the array of components of $G_{\mu\nu}$ relative to the non-rescaled frame $\{L, R, X_1, X_2\}$.	100
$G'_{(Frame)} = (G'_{LL}, G'_{LR}, G'_{RR}, \not{G}'_L, \not{G}'_R, \not{G}')$ is the array of components of $G'_{\mu\nu}$ relative to the non-rescaled frame $\{L, R, X_1, X_2\}$.	100

B.11. Tensor products, traces, and contractions

$\mathrm{tr}_g \pi = (g^{-1})^{\alpha\beta} \pi_{\alpha\beta}$ denotes the g-trace of the type $\binom{0}{2}$ spacetime tensor $\pi_{\mu\nu}$.	95
$\mathrm{tr}_{\not g} \xi = (\not g^{-1})^{AB} \xi_{AB}$ denotes the $\not g$-trace of the type $\binom{0}{2}$ $S_{t,u}$ tensor ξ_{AB}.	95
$\hat{\xi}_{AB} = \xi_{AB} - \frac{1}{2} \mathrm{tr}_{\not g} \xi \not g_{AB}$ denotes the trace-free part of the type $\binom{0}{2}$ $S_{t,u}$ tensor ξ_{AB}.	96
$(\xi \otimes \omega)_{AB} = \xi_A \omega_B$ denotes the tensor product of the $S_{t,u}$ one-forms ξ_A and ω_A.	
$(\xi \hat{\otimes} \omega)_{AB} = \frac{1}{2}(\xi_A \omega_B + \xi_B \omega_A) - \frac{1}{2}(\not g^{-1})^{CD} \xi_C \omega_D \not g_{AB}$ denotes the symmetrized trace-free part of the type $\binom{0}{2}$ $S_{t,u}$ tensor $(\xi \otimes \omega)_{AB}$.	96

B.12. The size of the data and the bootstrap parameter

$\mathring{\epsilon} = \|\mathring{\Psi}\|_{H_e^{25}(\Sigma_0^1)} + \|\mathring{\Psi}_0\|_{H_e^{24}(\Sigma_0^1)}$ is the size of the initial data on Σ_0^1.	182
ε is the small bootstrap parameter featured in the fundamental bootstrap assumption (**BA**Ψ) and (**BA**$'\Psi$) and the auxiliary bootstrap assumptions (**AUX**μ), (**AUX**$LR_{(Small)}$), and (**AUX**$\chi^{(Small)}$).	182
$\mathring{\epsilon} \leq \varepsilon$ is a running assumption used throughout the analysis.	182
In Chapter 23, $\epsilon = \|\check{\Psi}\|_{H_e^{25}(\Sigma_{-1/2})} + \|\check{\Psi}_0\|_{H_e^{24}(\Sigma_{-1/2})}$ denotes the size of the data $(\check{\Psi}, \check{\Psi}_0)$ on $\Sigma_{-1/2}$.	471

B.13. Commutation vectorfields

$O_{(Flat;l)}$ denotes the Euclidean rotation vectorfield with rectangular spatial components $O^j_{(Flat;l)} = \epsilon_{laj} x^a$, where ϵ_{ijk} is the fully antisymmetric symbol normalized by $\epsilon_{123} = 1$.	123
$O_{(l)} = \not\Pi O_{(Flat;l)}$ denotes an $S_{t,u}$-tangent "geometric" rotation vectorfield.	123
$\mathscr{Z} = \{\varrho L, \breve{R}, O_{(1)}, O_{(2)}, O_{(3)}\}$ denotes the full set of commutation vectorfields.	127
$\mathscr{S} = \{\breve{R}, O_{(1)}, O_{(2)}, O_{(3)}\}$ are the spatial commutation vectorfields.	127
$\mathscr{O} = \{O_{(1)}, O_{(2)}, O_{(3)}\}$ are the $S_{t,u}$-tangent rotation vectorfields.	127
$\mathscr{O}_{(Flat)} = \{O_{(Flat;1)}, O_{(Flat;2)}, O_{(Flat;3)}\}$ are the Euclidean rotation vectorfields.	128

B.14. Differential operators and commutator notation

$\partial_\mu = \frac{\partial}{\partial x^\mu}$ denotes a rectangular coordinate partial derivative vectorfield.	35
$\frac{\partial}{\partial t}, \frac{\partial}{\partial u}, \frac{\partial}{\vartheta^1}, \frac{\partial}{\vartheta^2}$ denote the geometric coordinate partial derivative vectorfields.	88
$Vf = V^\alpha \partial_\alpha f$ denotes the V-directional derivative of a function f.	8
We sometimes use the alternate notation $\partial_t = \partial_0$.	
$\partial_r = \frac{x^a}{r}\partial_a$ denotes the Euclidean radial vectorfield.	39
df = standard differential of a function f on spacetime.	96
$\not{d}f = \not{\Pi} df$, where f is a function defined on spacetime and $\not{\Pi}$ denotes projection onto $S_{t,u}$. Alternatively, $\not{d}f$ can be viewed as the inherent angular differential of a function f defined on $S_{t,u}$.	96
$\mathscr{D}$ = Levi-Civita connection of g.	83
∇ = Levi-Civita connection of the Minkowski metric m.	83
$\not{\nabla}$ = Levi-Civita connection of $\not{g}$.	83
$\not{d}_A = X_A \cdot \not{d}$, $\mathscr{D}_A = \mathscr{D}_{X_A}$, $\not{\nabla}_A = \not{\nabla}_{X_A}$, and similarly for other differential operators.	
$\mathscr{D}^2_{XY} = X^\alpha Y^\beta \mathscr{D}_\alpha \mathscr{D}_\beta$ and similarly for other connections (contractions against X and Y are taken *after* the two covariant differentiations).	84
$\not{\nabla}^2 \xi$ denotes the second $S_{t,u}$ covariant derivative of ξ.	
$\not{\Delta} f = \mathrm{tr}_{\not{g}} \not{\nabla}^2 f$ denotes the angular Laplacian of f.	83
$\hat{\not{\nabla}}^2 f$ denotes the trace-free part of the second $S_{t,u}$ covariant derivative of the function f.	136
If Y is an $S_{t,u}$-tangent vectorfield, then $\mathrm{di\not{v}}\, Y = \not{\nabla}_A Y^A$ denotes its angular divergence.	135
If ξ is a type $\binom{1}{1}$ $S_{t,u}$ tensorfield, then $\mathrm{di\not{v}}\,\xi$ is the $S_{t,u}$ one-form with components $(\mathrm{di\not{v}}\,\xi)_A = \not{\nabla}_B \xi^B{}_A$.	135
We similarly define the angular divergence of symmetric type $\binom{2}{0}$ or symmetric type $\binom{0}{2}$ $S_{t,u}$ tensorfields.	135
If ξ is a symmetric type $\binom{0}{2}$ $S_{t,u}$ tensorfield, then $\check{\not{\nabla}}_A \xi_{BC} = \frac{1}{2}\{\not{\nabla}_A \xi_{BC} + \not{\nabla}_B \xi_{AC} - \not{\nabla}_C \xi_{AB}\}$.	136
$\mathcal{L}_V \xi$ denotes the Lie derivative of ξ with respect to V.	97
$[V, W] = \mathcal{L}_V W$ when V and W are vectorfields.	
More generally, if P and Q are two operators, then $[P, Q] = PQ - QP$ denotes their commutator.	
$\underline{\mathcal{L}}_V \xi = \underline{\Pi} \mathcal{L}_V \xi$ is the Σ_t-projected Lie derivative of ξ with respect to V.	98
$\not{\mathcal{L}}_V \xi = \not{\Pi} \mathcal{L}_V \xi$ is the $S_{t,u}$-projected Lie derivative of ξ with respect to V.	98
If ξ is a type $\binom{0}{2}$ $S_{t,u}$ tensorfield, then $\hat{\not{\mathcal{L}}}_V \xi$ denotes the trace-free $S_{t,u}$-projected Lie derivative of ξ with respect to V.	135

B.15. Floor and ceiling functions and repeated differentiation

If M is a non-negative integer, then $\lfloor M/2 \rfloor = M/2$ for M even and $\lfloor M/2 \rfloor = (M-1)/2$ for M odd, while $\lceil M/2 \rceil = M/2$ for M even and $\lceil M/2 \rceil = (M+1)/2$ for M odd.	210
$\mathscr{Z} = \{Z_{(i)}\}_{i=1}^{5}$ is a labeling of the commutation vectorfields. If $\vec{I} = (\iota_1, \iota_2, \cdots, \iota_N)$ is a multi-index of order $\lvert\vec{I}\rvert := N$ with $\iota_1, \iota_2, \cdots, \iota_N \in \{1,2,3,4,5\}$, then $\mathscr{Z}^{\vec{I}} := Z_{(\iota_1)} Z_{(\iota_2)} \cdots Z_{(\iota_N)}$ denotes the corresponding N^{th} order differential operator.	140
When we are not concerned with the precise structure of the multi-index $\vec{I}$, we abbreviate $\mathscr{Z}^N = Z_{(\iota_1)} Z_{(\iota_2)} \cdots Z_{(\iota_N)}$. Similarly, $\not{\mathcal{L}}_{\mathscr{Z}}^N = \not{\mathcal{L}}_{Z_{(\iota_1)}} \not{\mathcal{L}}_{(Z_{\iota_2})} \cdots \not{\mathcal{L}}_{Z_{(\iota_N)}}$ denotes an Nth order $S_{t,u}$-projected Lie derivative operator. Similarly, $\hat{\not{\mathcal{L}}}_{\mathscr{Z}}^N = \hat{\not{\mathcal{L}}}_{Z_{(\iota_1)}} \hat{\not{\mathcal{L}}}_{Z_{(\iota_2)}} \cdots \hat{\not{\mathcal{L}}}_{Z_{(\iota_N)}}$ denotes an Nth order trace-free $S_{t,u}$-projected Lie derivative operator.	140
If f is a function, then $\lvert \mathscr{Z}^{\leq N} f \rvert$ denotes any term that is $\leq \sum_{\lvert\vec{I}\rvert \leq N} c_{\vec{I}} \left\lvert \mathscr{Z}^{\vec{I}} f \right\rvert$, where the $c_{\vec{I}}$ are non-negative constants that verify $\sum_{\lvert\vec{I}\rvert \leq N} c_{\vec{I}} \leq 1$. Similarly, $\lvert \mathscr{Z}^N f \rvert$ denotes any term that is $\leq \sum_{\lvert\vec{I}\rvert = N} c_{\vec{I}} \left\lvert \mathscr{Z}^{\vec{I}} f \right\rvert$, where the $c_{\vec{I}}$ are non-negative constants that verify $\sum_{\lvert\vec{I}\rvert = N} c_{\vec{I}} \leq 1$. Similarly, if V is a vectorfield, then $\lvert V \mathscr{Z}^{\leq N} f \rvert$ denotes any term that is $\leq \sum_{\lvert\vec{I}\rvert \leq N} c_{\vec{I}} \left\lvert V \mathscr{Z}^{\vec{I}} f \right\rvert$, where the $c_{\vec{I}}$ are non-negative constants that verify $\sum_{\lvert\vec{I}\rvert \leq N} c_{\vec{I}} \leq 1$. Similarly, $\lvert V \mathscr{Z}^N f \rvert$ denotes any term that is $\leq \sum_{\lvert\vec{I}\rvert = N} c_{\vec{I}} \left\lvert V \mathscr{Z}^{\vec{I}} f \right\rvert$, where the $c_{\vec{I}}$ are non-negative constants that verify $\sum_{\lvert\vec{I}\rvert = N} c_{\vec{I}} \leq 1$. Similarly, $\lvert \not{d} \mathscr{Z}^{\leq N} f \rvert$ denotes any term that is $\leq \sum_{\lvert\vec{I}\rvert \leq N} c_{\vec{I}} \left\lvert \not{d} \mathscr{Z}^{\vec{I}} f \right\rvert$, where the $c_{\vec{I}}$ are non-negative constants that verify $\sum_{\lvert\vec{I}\rvert \leq N} c_{\vec{I}} \leq 1$. We use similar notation for other expressions.	140
If ξ is an $S_{t,u}$ tensor, then $\left\lvert \not{\mathcal{L}}_{\mathscr{Z}}^{\leq N} \xi \right\rvert$ denotes any term that is $\leq \sum_{\lvert\vec{I}\rvert \leq N} c_{\vec{I}} \left\lvert \not{\mathcal{L}}_{\mathscr{Z}}^{\vec{I}} \xi \right\rvert$, where the $c_{\vec{I}}$ are non-negative constants that verify $\sum_{\lvert\vec{I}\rvert \leq N} c_{\vec{I}} \leq 1$. Similarly, $\left\lvert \not{\mathcal{L}}_{\mathscr{Z}}^{N} \xi \right\rvert$ denotes any term that is $\leq \sum_{\lvert\vec{I}\rvert = N} c_{\vec{I}} \left\lvert \not{\mathcal{L}}_{\mathscr{Z}}^{\vec{I}} \xi \right\rvert$, where the $c_{\vec{I}}$ are non-negative constants that verify $\sum_{\lvert\vec{I}\rvert = N} c_{\vec{I}} \leq 1$. Similarly, if V is a vectorfield, then $\left\lvert \not{\mathcal{L}}_V \not{\mathcal{L}}_{\mathscr{Z}}^{\leq N} \xi \right\rvert$ denotes any term that is $\leq \sum_{\lvert\vec{I}\rvert \leq N} c_{\vec{I}} \left\lvert \not{\mathcal{L}}_V \not{\mathcal{L}}_{\mathscr{Z}}^{\vec{I}} \xi \right\rvert$, where the $c_{\vec{I}}$ are non-negative constants that verify $\sum_{\lvert\vec{I}\rvert \leq N} c_{\vec{I}} \leq 1$. Similarly, $\left\lvert \not{\mathcal{L}}_V \not{\mathcal{L}}_{\mathscr{Z}}^{N} \xi \right\rvert$ denotes any term that is $\leq \sum_{\lvert\vec{I}\rvert = N} c_{\vec{I}} \left\lvert \not{\mathcal{L}}_V \not{\mathcal{L}}_{\mathscr{Z}}^{\vec{I}} \xi \right\rvert$, where the $c_{\vec{I}}$ are non-negative constants that verify $\sum_{\lvert\vec{I}\rvert = N} c_{\vec{I}} \leq 1$. We use similar notation for other expressions, including the case in which trace-free $S_{t,u}$-projected Lie derivatives $\hat{\not{\mathcal{L}}}$ are present instead of ordinary $S_{t,u}$-projected Lie derivatives $\not{\mathcal{L}}$. We use similar notation with $\mathscr{S}$, $\mathscr{O}$, or $\mathscr{O}_{(Flat)}$ in place of $\mathscr{Z}$ when the derivatives are all, respectively, spatial derivatives, geometric rotation derivatives, or Euclidean rotation derivatives.	140

B.16. Area and volume forms

$dv_{\slashed{g}(t,u,\vartheta)} = \sqrt{\det \slashed{g}(t,u,\vartheta)} d\vartheta$ denotes the area form on $S_{t,u}$ induced by g, where the determinant is taken relative to the geometric angular coordinates $\vartheta = (\vartheta^1, \vartheta^2)$.	102
$d\underline{\varpi} = d\underline{\varpi}(t,u',\vartheta) = dv_{\slashed{g}(t,u',\vartheta)} \, du'$ denotes a volume form on $\Sigma_t^{U_0}$. $\mu \, d\underline{\varpi}$ is the volume form on Σ_t^u induced by g.	102
$d\overline{\varpi} = d\overline{\varpi}(t',u,\vartheta) = dv_{\slashed{g}(t',u,\vartheta)} \, dt'$ denotes a volume form on $\mathcal{C}_u^t$.	102
$d\varpi = d\varpi(t',u',\vartheta) = dv_{\slashed{g}(t',u',\vartheta)} \, du' \, dt'$ denotes a volume form on $\mathcal{M}_{t,u}$. $\mu \, d\varpi$ is the volume form on $\mathcal{M}_{t,u}$ induced by g.	102

B.17. Norms

$\|\xi\|^2 = (\slashed{g}^{-1})^{B_1 \widetilde{B}_1} \cdots (\slashed{g}^{-1})^{B_n \widetilde{B}_n} \slashed{g}_{A_1 \widetilde{A}_1} \cdots \slashed{g}_{A_m \widetilde{A}_m} \xi^{A_1 \cdots A_m}_{B_1 \cdots B_n} \xi^{\widetilde{A}_1 \cdots \widetilde{A}_m}_{\widetilde{B}_1 \cdots \widetilde{B}_n}$ denotes the square of the norm of the type $\binom{m}{n}$ $S_{t,u}$ tensor ξ.	97
For $k \geq 0$, $C^k(\Omega)$ denotes the set of $S_{t,u}$ tensorfields that, relative to the coordinates induced on Ω by the geometric coordinates $(t,u,\vartheta) \in [0,\infty) \times [0,U_0] \times \mathbb{S}^2$, are k-times continuously differentiable.	103
$\|f\|_{C^0(\Omega)} = \sup_{p \in \Omega} |f(p)|$.	103
Throughout the monograph, whenever we state an estimate implying that $\|f\|_{C^0(\Omega)} < \infty$, we are also implicitly indicating that $f \in C^0(\Omega)$.	
$\|f\|^2_{H^N_e(\Sigma_t)} = \sum_{|\vec{I}| \leq N} \int_{\Sigma_t} (\partial^{\vec{I}} f)^2 \, d^3 x$, where $\partial^{\vec{I}}$ is a multi-indexed differential operator involving repeated differentiation with respect to the spatial coordinate partial derivative vectorfields and $d^3 x$ is the volume form corresponding to the standard Euclidean metric on Σ_t.	15
$\|\xi\|^2_{L^2(S_{t,u})} = \int_{\vartheta \in \mathbb{S}^2} |\xi|^2(t,u,\vartheta) dv_{\slashed{g}} = \int_{S_{t,u}} |\xi|^2 dv_{\slashed{g}}$.	103
$\|\xi\|^2_{L^2(\Sigma_t^u)} = \int_{u'=0}^u \int_{\vartheta \in \mathbb{S}^2} |\xi|^2(t,u',\vartheta) dv_{\slashed{g}} \, du' = \int_{\Sigma_t^u} |\xi|^2 \, d\underline{\varpi}$.	103
$\|\xi\|^2_{L^2(\mathcal{C}_u^t)} = \int_{t'=0}^t \int_{\vartheta \in \mathbb{S}^2} |\xi|^2(t',u,\vartheta) dv_{\slashed{g}} \, dt' = \int_{\mathcal{C}_u^t} |\xi|^2 \, d\overline{\varpi}$.	103
$\|\xi\|^2_{L^2(\mathcal{M}_{t,u})} = \int_{t'=0}^t \int_{u'=0}^u \int_{\vartheta \in \mathbb{S}^2} |\xi|^2(t',u',\vartheta) dv_{\slashed{g}} \, du' \, dt' = \int_{\mathcal{M}_{t,u}} |\xi|^2 \, d\varpi$.	103
We use similar notation for the norms $\|\cdot\|_{L^2(\Omega)}$ of tensorfields defined on subsets Ω of $S_{t,u}$, Σ_t^u, $\mathcal{C}_u^t$, or $\mathcal{M}_{t,u}$.	

B.18. Energy-momentum tensorfield, multiplier vectorfields, and compatible currents

$Q_{\mu\nu}[\Psi] = \mathscr{D}_\mu\Psi\mathscr{D}_\nu\Psi - \frac{1}{2}g_{\mu\nu}(g^{-1})^{\alpha\beta}\mathscr{D}_\alpha\Psi\mathscr{D}_\beta\Psi$ denotes the energy-momentum tensorfield associated to Ψ.	152
$T = (1+2\upmu)L + 2\breve{R}$ denotes the timelike multiplier vectorfield.	153
$\widetilde{K} = \varrho^2 L$ denotes the Morawetz multiplier vectorfield.	153
${}^{(T)}J^\nu[\Psi] = Q^\nu{}_\alpha[\Psi]T^\alpha$ denotes the compatible current associated to T.	155
${}^{(\widetilde{K})}J^\nu[\Psi] = Q^\nu{}_\alpha[\Psi]\widetilde{K}^\alpha$ denotes the compatible current associated to $\widetilde{K}$.	155
${}^{(Correction)}J^\nu[\Psi] = \frac{1}{2}\left\{\varrho^2 \mathrm{tr}_{\not g}\chi\Psi\mathscr{D}^\nu\Psi - \frac{1}{2}\Psi^2\mathscr{D}^\nu[\varrho^2\mathrm{tr}_{\not g}\chi]\right\}$ denotes a correction current.	155
${}^{(\widetilde{K}+Correction)}J^\nu[\Psi] = {}^{(\widetilde{K})}J^\nu[\Psi] + {}^{(Correction)}J^\nu[\Psi]$.	155

B.19. Square-integral-controlling quantities

$\mathbb{E}[\Psi](t,u)$ denotes the energy of Ψ along Σ_t^u corresponding the multiplier T.	153
$\mathbb{F}[\Psi](t,u)$ denotes the cone flux of Ψ along $\mathcal{C}_u^t$ corresponding the multiplier T.	157
$\widetilde{\mathbb{E}}[\Psi](t,u)$ denotes the energy of Ψ along Σ_t^u corresponding the multiplier $\widetilde{K}$, including a correction term.	156
$\widetilde{\mathbb{F}}[\Psi](t,u)$ denotes the cone flux of Ψ along $\mathcal{C}_u^t$ corresponding the multiplier $\widetilde{K}$, including a correction term.	157
$\mathbb{Q}_{(N)}(t,u) = \max_{\lvert\vec{I}\rvert=N}\sup_{(t',u')\in[0,t]\times[0,u]}\left\{\mathbb{E}[\mathscr{Z}^{\vec{I}}\Psi](t',u') + \mathbb{F}[\mathscr{Z}^{\vec{I}}\Psi](t',u')\right\}$.	279
$\widetilde{\mathbb{Q}}_{(N)}(t,u) = \max_{\lvert\vec{I}\rvert=N}\sup_{(t',u')\in[0,t]\times[0,u]}\left\{\widetilde{\mathbb{E}}[\mathscr{Z}^{\vec{I}}\Psi](t',u') + \widetilde{\mathbb{F}}[\mathscr{Z}^{\vec{I}}\Psi](t',u')\right\}$.	279
$\mathbb{Q}_{(\leq N)}(t,u) = \max_{0\leq M\leq N}\mathbb{Q}_{(M)}(t,u)$.	279
$\widetilde{\mathbb{Q}}_{(\leq N)}(t,u) = \max_{0\leq M\leq N}\widetilde{\mathbb{Q}}_{(M)}(t,u)$.	279
$\widetilde{\mathbb{K}}[\Psi](t,u) = \frac{1}{2}\int_{\mathcal{M}_{t,u}}\varrho^2[L\upmu]_-\lvert\not d\Psi\rvert^2\,d\varpi$ denotes the coercive Morawetz spacetime integral associated to Ψ.	280
$\widetilde{\mathbb{K}}_{(N)}(t,u) = \max_{\lvert\vec{I}\rvert=N}\widetilde{\mathbb{K}}[\mathscr{Z}^{\vec{I}}\Psi](t,u)$.	279
$\widetilde{\mathbb{K}}_{(\leq N)}(t,u) = \max_{0\leq M\leq N}\widetilde{\mathbb{K}}_{(M)}(t,u)$.	279

B.20. Modified quantities

$\mathscr{X} = \upmu \mathrm{tr}_{\not g}\upchi^{(Small)} + \mathfrak{X}$ is the lowest-order fully modified version of $\mathrm{tr}_{\not g}\upchi^{(Small)}$.	173
${}^{(\mathscr{S}^N)}\mathscr{X} = \upmu \mathscr{S}^N \mathrm{tr}_{\not g}\upchi^{(Small)} + \mathscr{S}^N \mathfrak{X}$ is a higher-order fully modified version of $\mathrm{tr}_{\not g}\upchi^{(Small)}$.	175
$\widetilde{\mathscr{X}} = \mathrm{tr}_{\not g}\upchi^{(Small)} + \widetilde{\mathfrak{X}}$ is the lowest-order partially modified version of $\mathrm{tr}_{\not g}\upchi^{(Small)}$.	177
${}^{(\mathscr{Z}^N)}\widetilde{\mathscr{X}} = \mathscr{Z}^N \mathrm{tr}_{\not g}\upchi^{(Small)} + {}^{(\mathscr{Z}^N)}\widetilde{\mathfrak{X}}$ is a higher-order partially modified version of $\mathrm{tr}_{\not g}\upchi^{(Small)}$.	178
$\widetilde{\mathscr{M}} = \not d\upmu + \widetilde{\mathfrak{M}}$ is the lowest-order partially modified version of $\not d\upmu$.	178
${}^{(\mathscr{Z}^N)}\widetilde{\mathscr{M}} = \not d \mathscr{Z}^N \upmu + {}^{(\mathscr{Z}^N)}\widetilde{\mathfrak{M}}$ is a higher-order partially modified version of $\not d\upmu$.	179

B.21. Curvature tensors

$\mathscr{R}_{\mu\nu\alpha\beta}$ is the Riemann curvature tensor of g.	169
$\mathscr{R}(W, X, Y, Z) = g(-\mathscr{D}^2_{WX}Y + \mathscr{D}^2_{XW}Y, Z)$ indicates our curvature sign convention. We use the same sign convention for the Riemann curvature tensor of $\not g$.	169
$\mathscr{R}_{\alpha\beta}$ is the Ricci curvature tensor of g.	144
$\mathfrak{R}_{ABCD}$ is the Riemann curvature tensor of $\not g$.	169
$\mathfrak{R}_{AB}$ is the Ricci curvature tensor of $\not g$.	354
$\mathfrak{R}$ is the scalar curvature tensor of $\not g$.	354
$\mathfrak{K}$ is the Gaussian curvature of $\not g$.	353

B.22. Omission of the independent variables in some expressions

Many of our pointwise estimates are stated in the form $\lvert f_1\rvert \lesssim h(t,u)\lvert f_2\rvert$ for some function h. Unless we otherwise indicate, it is understood that both f_1 and f_2 are evaluated at the point with geometric coordinates (t,u,ϑ).
Unless we otherwise indicate, in integrals $\int_{S_{t,u}} f\, d\upsilon_{g\!\!\!/}$, the integrand f and the area form $d\upsilon_{g\!\!\!/}$ are viewed as functions of (t,u,ϑ) and ϑ is the integration variable.
Unless we otherwise indicate, in integrals $\int_{\Sigma_t^u} f\, d\underline{\varpi}$, the integrand f and the volume form $d\underline{\varpi}$ are viewed as functions of (t,u',ϑ) and (u',ϑ) are the integration variables.
Unless we otherwise indicate, in integrals $\int_{\mathcal{C}_u^t} f\, d\overline{\varpi}$, the integrand f and the volume form $d\overline{\varpi}$ are viewed as functions of (t',u,ϑ) and (t',ϑ) are the integration variables.
Unless we otherwise indicate, in integrals $\int_{\mathcal{M}_{t,u}} f\, d\varpi$, the integrand f and the volume form $d\varpi$ are viewed as functions of (t',u',ϑ) and (t',u',ϑ) are the integration variables.

B.23. Data and functions relevant for the proof of shock formation

$(\mathring{\Psi},\mathring{\Psi}_0)$ are data for the wave equation on Σ_0 that are compactly supported in the Euclidean unit ball centered at the origin.	81
$(\check{\Psi},\check{\Psi}_0)$ are data for the wave equation on $\Sigma_{-1/2}$ that are compactly supported in the Euclidean ball of radius $1/2$ centered at the origin.	470
$\mathbb{S}[(\mathring{\Psi},\mathring{\Psi}_0)]$ is the data-dependent function that drives nearly spherically symmetric small-data shock formation.	469
$\check{\mathbb{S}}[(\check{\Psi},\check{\Psi}_0)]$ is the data-dependent function that effectively determines the behavior of $\mathbb{S}[(\mathring{\Psi},\mathring{\Psi}_0)]$	470

Bibliography

[1] S. Alinhac, *The null condition for quasilinear wave equations in two space dimensions I*, Invent. Math. **145** (2001), no. 3, 597–618. MR1856402 (2002i:35127)

[2] ———, *The null condition for quasilinear wave equations in two space dimensions. II*, Amer. J. Math. **123** (2001), no. 6, 1071–1101. MR1867312 (2003e:35193)

[3] Serge Alinhac, *Blowup for nonlinear hyperbolic equations*, Progress in Nonlinear Differential Equations and their Applications, 17, Birkhäuser Boston, Inc., Boston, MA, 1995. MR1339762 (96h:35109)

[4] ———, *Blowup of small data solutions for a class of quasilinear wave equations in two space dimensions. II*, Acta Math. **182** (1999), no. 1, 1–23. MR1687180 (2000d:35148)

[5] ———, *Blowup of small data solutions for a quasilinear wave equation in two space dimensions*, Ann. of Math. (2) **149** (1999), no. 1, 97–127. MR1680539 (2000d:35147)

[6] ———, *A minicourse on global existence and blowup of classical solutions to multidimensional quasilinear wave equations*, Journées "Équations aux Dérivées Partielles" (Forges-les-Eaux, 2002), 2002, pp. Exp. No. I, 33. MR1968197 (2004b:35227)

[7] ———, *An example of blowup at infinity for a quasilinear wave equation*, Astérisque **284** (2003), 1–91. Autour de l'analyse microlocale. MR2003417 (2005a:35197)

[8] Vladimir I. Arnold, *Ordinary differential equations*, Universitext, Springer-Verlag, Berlin, 2006. Translated from the Russian by Roger Cooke, Second printing of the 1992 edition. MR2242407

[9] Thierry Aubin, *Nonlinear analysis on manifolds. Monge-Ampère equations*, Grundlehren der Mathematischen Wissenschaften [Fundamental Principles of Mathematical Sciences], vol. 252, Springer-Verlag, New York, 1982. MR681859 (85j:58002)

[10] Frederick Bloom, *Mathematical problems of classical nonlinear electromagnetic theory*, Pitman Monographs and Surveys in Pure and Applied Mathematics, vol. 63, Longman Scientific & Technical, Harlow; copublished in the United States with John Wiley & Sons, Inc., New York, 1993. MR1369083 (98c:78001)

[11] James Challis, *On the velocity of sound*, Philos. Magazine **32** (1848), 494–499.

[12] Jeff Cheeger and David G. Ebin, *Comparison theorems in Riemannian geometry*, AMS Chelsea Publishing, Providence, RI, 2008. Revised reprint of the 1975 original. MR2394158 (2009c:53043)

[13] Geng Chen, Robin Young, and Qingtian Zhang, *Shock formation in the compressible Euler equations and related systems*, J. Hyperbolic Differ. Equ. **10** (2013), no. 1, 149–172. MR3043493

[14] Bennett Chow, Peng Lu, and Lei Ni, *Hamilton's Ricci flow*, Graduate Studies in Mathematics, vol. 77, American Mathematical Society, Providence, RI, 2006. MR2274812 (2008a:53068)

[15] Demetrios Christodoulou, *Global solutions of nonlinear hyperbolic equations for small initial data*, Comm. Pure Appl. Math. **39** (1986), no. 2, 267–282. MR820070 (87c:35111)

[16] ———, *The action principle and partial differential equations*, Annals of Mathematics Studies, vol. 146, Princeton University Press, Princeton, NJ, 2000. MR1739321 (2003a:58001)

[17] ———, *The formation of shocks in 3-dimensional fluids*, EMS Monographs in Mathematics, European Mathematical Society (EMS), Zürich, 2007. MR2284927 (2008e:76104)

[18] Demetrios Christodoulou and Sergiu Klainerman, *Asymptotic properties of linear field equations in Minkowski space*, Comm. Pure Appl. Math. **43** (1990), no. 2, 137–199. MR1038141 (91a:58202)

[19] ———, *The global nonlinear stability of the Minkowski space*, Princeton Mathematical Series, vol. 41, Princeton University Press, Princeton, NJ, 1993. MR1316662 (95k:83006)

[20] Demetrios Christodoulou and André Lisibach, *Shock development in spherical symmetry*, Annals of PDE **2** (2016), no. 1, 1–246.

[21] Demetrios Christodoulou and Shuang Miao, *Compressible flow and Euler's equations*, Surveys of Modern Mathematics, vol. 9, International Press, Somerville, MA; Higher Education Press, Beijing, 2014. MR3288725

[22] Constantine M. Dafermos, *Hyperbolic conservation laws in continuum physics*, Third, Grundlehren der Mathematischen Wissenschaften [Fundamental Principles of Mathematical Sciences], vol. 325, Springer-Verlag, Berlin, 2010. MR2574377

[23] K. O. Friedrichs, *Symmetric hyperbolic linear differential equations*, Comm. Pure Appl. Math. **7** (1954), 345–392. MR0062932 (16,44c)

[24] Lars Gårding, *Solution directe du problème de Cauchy pour les équations hyperboliques*, La théorie des équations aux dérivées partielles. Nancy, 9-15 avril 1956, 1956, pp. 71–90. MR0116142 (22 #6937)

[25] ———, *Cauchy's problem for hyperbolic equations*, Treizième congrès des mathématiciens scandinaves, tenu à Helsinki 18-23 août 1957, 1958, pp. 104–109. MR0116143 (22 #6938)

[26] Yan Guo and A. Shadi Tahvildar-Zadeh, *Formation of singularities in relativistic fluid dynamics and in spherically symmetric plasma dynamics*, Nonlinear partial differential equations (Evanston, IL, 1998), 1999, pp. 151–161. MR1724661

[27] Sigurdur Helgason, *The Radon transform*, Second, Progress in Mathematics, vol. 5, Birkhäuser Boston, Inc., Boston, MA, 1999. MR1723736

[28] Gustav Holzegel, Sergiu Klainerman, Jared Speck, and Willie Wai-Yeung Wong, *Small-data shock formation in solutions to 3d quasilinear wave equations: An overview*, Journal of Hyperbolic Differential Equations **13** (2016), no. 01, 1–105, available at `http://www.worldscientific.com/doi/pdf/10.1142/S0219891616500016`.

[29] Lars Hörmander, *The lifespan of classical solutions of nonlinear hyperbolic equations*, Pseudodifferential operators (Oberwolfach, 1986), 1987, pp. 214–280. MR897781 (88j:35024)

[30] ———, *Lectures on nonlinear hyperbolic differential equations*, Mathématiques & Applications (Berlin) [Mathematics & Applications], vol. 26, Springer-Verlag, Berlin, 1997. MR1466700 (98e:35103)

[31] A. Jeffrey, *The formation of magnetoacoustic shocks*, J. Math. Anal. Appl. **11** (1965), 139–150. MR0197036 (33 #5220)

[32] A. Jeffrey and M. Teymur, *Formation of shock waves in hyperelastic solids*, Acta Mech. **20** (1974), 133–149. MR0356643 (50 #9113)

[33] Alan Jeffrey and Viktor P. Korobeinikov, *Formation and decay of electromagnetic shock waves*, Zeitschrift für angewandte Mathematik und Physik ZAMP **20** (1969), no. 4, 440–447. MR0197036 (33 #5220)

[34] F. John and S. Klainerman, *Almost global existence to nonlinear wave equations in three space dimensions*, Comm. Pure Appl. Math. **37** (1984), no. 4, 443–455. MR745325 (85k:35147)

[35] Fritz John, *Formation of singularities in one-dimensional nonlinear wave propagation*, Comm. Pure Appl. Math. **27** (1974), 377–405. MR0369934 (51 #6163)

[36] ———, *Blow-up for quasilinear wave equations in three space dimensions*, Comm. Pure Appl. Math. **34** (1981), no. 1, 29–51. MR600571 (83d:35096)

[37] ———, *Formation of singularities in elastic waves*, Trends and applications of pure mathematics to mechanics (Palaiseau, 1983), 1984, pp. 194–210. MR755727 (85h:73018)

[38] ———, *Blow-up of radial solutions of $u_{tt} = c^2(u_t)\Delta u$ in three space dimensions*, Mat. Apl. Comput. **4** (1985), no. 1, 3–18. MR808321 (87c:35114)

[39] ———, *Existence for large times of strict solutions of nonlinear wave equations in three space dimensions for small initial data*, Comm. Pure Appl. Math. **40** (1987), no. 1, 79–109. MR865358 (87m:35128)

[40] ———, *Solutions of quasilinear wave equations with small initial data. The third phase*, Nonlinear hyperbolic problems (Bordeaux, 1988), 1989, pp. 155–184. MR1033282 (91e:35132)

[41] ———, *Nonlinear wave equations, formation of singularities*, University Lecture Series, vol. 2, American Mathematical Society, Providence, RI, 1990. Seventh Annual Pitcher Lectures delivered at Lehigh University, Bethlehem, Pennsylvania, April 1989. MR1066694 (91g:35001)

[42] Jürgen Jost, *Riemannian geometry and geometric analysis*, Fifth, Universitext, Springer-Verlag, Berlin, 2008. MR2431897 (2009g:53036)

[43] Joseph B. Keller and Lu Ting, *Periodic vibrations of systems governed by nonlinear partial differential equations*, Comm. Pure Appl. Math. **19** (1966), 371–420. MR0205520 (34 #5347)
[44] S. Klainerman, *On "almost global" solutions to quasilinear wave equations in three space dimensions*, Comm. Pure Appl. Math. **36** (1983), no. 3, 325–344. MR697468 (84e:35102)
[45] Sergiu Klainerman, *Long time behaviour of solutions to nonlinear wave equations*, Proceedings of the International Congress of Mathematicians, Vol. 1, 2 (Warsaw, 1983), 1984, pp. 1209–1215. MR804771
[46] ———, *Uniform decay estimates and the Lorentz invariance of the classical wave equation*, Comm. Pure Appl. Math. **38** (1985), no. 3, 321–332. MR784477 (86i:35091)
[47] ———, *The null condition and global existence to nonlinear wave equations*, Nonlinear systems of partial differential equations in applied mathematics, Part 1 (Santa Fe, N.M., 1984), 1986, pp. 293–326. MR837683 (87h:35217)
[48] ———, *A commuting vectorfields approach to Strichartz-type inequalities and applications to quasi-linear wave equations*, Internat. Math. Res. Notices **5** (2001), 221–274. MR1820023 (2001k:35210)
[49] Sergiu Klainerman and Andrew Majda, *Formation of singularities for wave equations including the nonlinear vibrating string*, Comm. Pure Appl. Math. **33** (1980), no. 3, 241–263. MR562736 (81f:35080)
[50] Sergiu Klainerman and Igor Rodnianski, *Improved local well-posedness for quasilinear wave equations in dimension three*, Duke Math. J. **117** (2003), no. 1, 1–124. MR1962783 (2004b:35233)
[51] ———, *Rough solutions of the Einstein-vacuum equations*, Ann. of Math. (2) **161** (2005), no. 3, 1143–1193. MR2180400 (2007d:58051)
[52] Sergiu Klainerman, Igor Rodnianski, and Jeremie Szeftel, *The bounded L^2 curvature conjecture*, Invent. Math. **202** (2015), no. 1, 91–216. MR3402797
[53] Steven G. Krantz and Harold R. Parks, *Geometric integration theory*, Cornerstones, Birkhäuser Boston, Inc., Boston, MA, 2008. MR2427002 (2009m:49075)
[54] Peter D. Lax, *Development of singularities of solutions of nonlinear hyperbolic partial differential equations*, J. Mathematical Phys. **5** (1964), 611–613. MR0165243 (29 #2532)
[55] ———, *Hyperbolic systems of conservation laws and the mathematical theory of shock waves*, Society for Industrial and Applied Mathematics, Philadelphia, Pa., 1973. Conference Board of the Mathematical Sciences Regional Conference Series in Applied Mathematics, No. 11. MR0350216 (50 #2709)
[56] Jean Leray, *Hyperbolic differential equations*, The Institute for Advanced Study, Princeton, N. J., 1953. MR0063548 (16,139a)
[57] Hans Lindblad, *Counterexamples to local existence for quasilinear wave equations*, Math. Res. Lett. **5** (1998), no. 5, 605–622. MR1666844 (2000a:35171)
[58] ———, *Global solutions of quasilinear wave equations*, Amer. J. Math. **130** (2008), no. 1, 115–157. MR2382144 (2009b:58062)
[59] Hans Lindblad and Igor Rodnianski, *The global stability of Minkowski space-time in harmonic gauge*, Annals of Mathematics **171** (2010), no. 3, 1401–1477.
[60] Tai Ping Liu, *Development of singularities in the nonlinear waves for quasilinear hyperbolic partial differential equations*, J. Differential Equations **33** (1979), no. 1, 92–111. MR540819 (80g:35075)
[61] A. Majda, *Compressible fluid flow and systems of conservation laws in several space variables*, Applied Mathematical Sciences, vol. 53, Springer-Verlag, New York, 1984. MR748308 (85e:35077)
[62] Andrew Majda, *The existence and stability of multidimensional shock fronts*, Bull. Amer. Math. Soc. (N.S.) **4** (1981), no. 3, 342–344. MR609047 (82g:35074)
[63] ———, *The existence of multidimensional shock fronts*, Mem. Amer. Math. Soc. **43** (1983), no. 281, v+93. MR699241 (85f:35139)
[64] ———, *The stability of multidimensional shock fronts*, Mem. Amer. Math. Soc. **41** (1983), no. 275, iv+95. MR683422 (84e:35100)
[65] J.R. Munkres, *Topology*, Second, Prentice-Hall, 2000.
[66] Barrett O'Neill, *Semi-Riemannian geometry*, Pure and Applied Mathematics, vol. 103, Academic Press Inc. [Harcourt Brace Jovanovich Publishers], New York, 1983. With applications to relativity. MR719023 (85f:53002)

[67] Peter Petersen, *Riemannian geometry*, Second, Graduate Texts in Mathematics, vol. 171, Springer, New York, 2006. MR2243772 (2007a:53001)

[68] Ivan Petrovskii, *Über das Cauchysche Problem für Systeme von partiellen Differentialgleichungen*, Rec. Math. [Mat. Sbornik] N.S. **44** (1937), no. 2, 815–870.

[69] Siméon D. Poisson, *Mémoire sur la théorie du son*, J. Ecole Polytechnique **7** (1808), 319–392.

[70] Amalkumar Raychaudhuri, *Relativistic cosmology. i*, Phys. Rev. **98** (1955May), 1123–1126.

[71] Bernhard Riemann, *Über die Fortpflanzung ebener Luftwellen von endlicher Schwingungsweite*, Abhandlungen der Kniglichen Gesellschaft der Wissenschaften in Göttingen **8** (1860), 43-66.

[72] Manuel Salas, *The curious events leading to the theory of shock waves*, Shock Waves **16** (July 2007), no. 6, 477–487.

[73] Julius Schauder, *Das Anfangswertproblem einer quasilinearen hyperbolischen Differentialgleichung zweiter Ordnung in beliebiger Anzahl von unabhängigen veränderlichen* (ger), Fundamenta Mathematicae **24** (1935), no. 1, 213–246.

[74] Thomas C. Sideris, *Formation of singularities in three-dimensional compressible fluids*, Comm. Math. Phys. **101** (1985), no. 4, 475–485. MR815196 (87d:35127)

[75] Hart F. Smith and Daniel Tataru, *Sharp local well-posedness results for the nonlinear wave equation*, Ann. of Math. (2) **162** (2005), no. 1, 291–366. MR2178963 (2006k:35193)

[76] Christopher D. Sogge, *Lectures on non-linear wave equations*, Second, International Press, Boston, MA, 2008. MR2455195 (2009i:35213)

[77] Jared Speck, *The non-relativistic limit of the Euler-Nordström system with cosmological constant*, Rev. Math. Phys. **21** (2009), no. 7, 821–876. MR2553428

[78] Robert M. Wald, *General relativity*, University of Chicago Press, Chicago, IL, 1984. MR757180 (86a:83001)

[79] Q. Wang, *A geometric approach for sharp local well-posedness of quasilinear wave equations*, ArXiv e-prints (August 2014), available at https://arxiv.org/abs/1408.3780.

Index

Selected Published Titles in This Series

214 **Jared Speck,** Shock Formation in Small-Data Solutions to 3D Quasilinear Wave Equations, 2016

213 **Harold G. Diamond and Wen-Bin Zhang (Cheung Man Ping),** Beurling Generalized Numbers, 2016

212 **Pandelis Dodos and Vassilis Kanellopoulos,** Ramsey Theory for Product Spaces, 2016

211 **Charlotte Hardouin, Jacques Sauloy, and Michael F. Singer,** Galois Theories of Linear Difference Equations: An Introduction, 2016

210 **Jason P. Bell, Dragos Ghioca, and Thomas J. Tucker,** The Dynamical Mordell–Lang Conjecture, 2016

209 **Steve Y. Oudot,** Persistence Theory: From Quiver Representations to Data Analysis, 2015

208 **Peter S. Ozsváth, András I. Stipsicz, and Zoltán Szabó,** Grid Homology for Knots and Links, 2015

207 **Vladimir I. Bogachev, Nicolai V. Krylov, Michael Röckner, and Stanislav V. Shaposhnikov,** Fokker–Planck–Kolmogorov Equations, 2015

206 **Bennett Chow, Sun-Chin Chu, David Glickenstein, Christine Guenther, James Isenberg, Tom Ivey, Dan Knopf, Peng Lu, Feng Luo, and Lei Ni,** The Ricci Flow: Techniques and Applications: Part IV: Long-Time Solutions and Related Topics, 2015

205 **Pavel Etingof, Shlomo Gelaki, Dmitri Nikshych, and Victor Ostrik,** Tensor Categories, 2015

204 **Victor M. Buchstaber and Taras E. Panov,** Toric Topology, 2015

203 **Donald Yau and Mark W. Johnson,** A Foundation for PROPs, Algebras, and Modules, 2015

202 **Shiri Artstein-Avidan, Apostolos Giannopoulos, and Vitali D. Milman,** Asymptotic Geometric Analysis. Part I, 2015

201 **Christopher L. Douglas, John Francis, André G. Henriques, and Michael A. Hill, Editors,** Topological Modular Forms, 2014

200 **Nikolai Nadirashvili, Vladimir Tkachev, and Serge Vlăduţ,** Nonlinear Elliptic Equations and Nonassociative Algebras, 2014

199 **Dmitry S. Kaliuzhnyi-Verbovetskyi and Victor Vinnikov,** Foundations of Free Noncommutative Function Theory, 2014

198 **Jörg Jahnel,** Brauer Groups, Tamagawa Measures, and Rational Points on Algebraic Varieties, 2014

197 **Richard Evan Schwartz,** The Octagonal PETs, 2014

196 **Silouanos Brazitikos, Apostolos Giannopoulos, Petros Valettas, and Beatrice-Helen Vritsiou,** Geometry of Isotropic Convex Bodies, 2014

195 **Ching-Li Chai, Brian Conrad, and Frans Oort,** Complex Multiplication and Lifting Problems, 2014

194 **Samuel Herrmann, Peter Imkeller, Ilya Pavlyukevich, and Dierk Peithmann,** Stochastic Resonance, 2014

193 **Robert Rumely,** Capacity Theory with Local Rationality, 2013

192 **Messoud Efendiev,** Attractors for Degenerate Parabolic Type Equations, 2013

191 **Grégory Berhuy and Frédérique Oggier,** An Introduction to Central Simple Algebras and Their Applications to Wireless Communication, 2013

190 **Aleksandr Pukhlikov,** Birationally Rigid Varieties, 2013

189 **Alberto Elduque and Mikhail Kochetov,** Gradings on Simple Lie Algebras, 2013

188 **David Lannes,** The Water Waves Problem, 2013

For a complete list of titles in this series, visit the AMS Bookstore at **www.ams.org/bookstore/survseries/**.